Pigments, Extenders, and Particles in Surface Coatings and Plastics

## Springer Nature More Media App

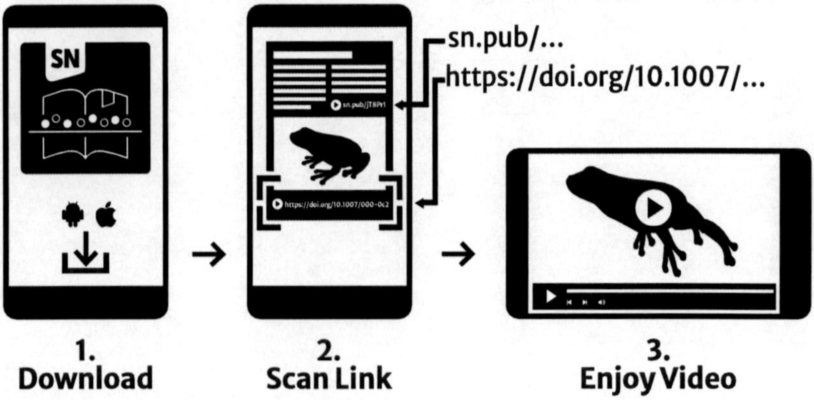

Michael Diebold · Steven De Backer ·
Philipp M. Niedenzu · Brett R. Hester ·
Frank A. C. Vanhecke

# Pigments, Extenders, and Particles in Surface Coatings and Plastics

## Fundamentals and Applications to Coatings, Plastics and Paper Laminate Formulation

Springer

Michael Diebold
Titanium Technologies
Chemours
Wilmington, DE, USA

Steven De Backer
Titanium Technologies
Chemours Belgium BVBA
Kallo, Belgium

Philipp M. Niedenzu
The Chemours Discovery Hub, Newark
The Chemours Company
Newark, DE, USA

Brett R. Hester
The Chemours Discovery Hub
The Chemours Company
Newark, DE, USA

Frank A. C. Vanhecke
Titanium Technologies
Chemours Belgium BVBA
Kallo, Belgium

This work contains media enhancements, which are displayed with a "play" icon. Material in the print book can be viewed on a mobile device by downloading the Springer Nature "More Media" app available in the major app stores. The media enhancements in the online version of the work can be accessed directly by authorized users.

ISBN 978-3-030-99082-4 ISBN 978-3-030-99083-1 (eBook)
https://doi.org/10.1007/978-3-030-99083-1



This Springer imprint is published by the registered company Springer Nature Switzerland AG
The registered company address is: Gewerbestrasse 11, 6330 Cham, Switzerland

# Preface

The goal of this book is to provide the reader with the knowledge necessary to effectively formulate paints, plastics, and décor paper for laminates. Our intention is that this book would be useful both as an introduction of concepts to the beginning formulator and as well as a review or reference for the experienced formulator.

Although books already exist that discuss many aspects of formulation, we find that these books primarily focus on the chemical nature of the raw materials rather than on their physical behavior as particles. For example, polymers are the topic of numerous books, but these books focus on the vast and rich chemistry of polymers, rather than their particulate properties. In a similar manner, books exist that discuss the extenders found in paint, plastics, and paper laminate applications, but these mainly focus on their mineral and chemical properties. Similarly, additives are also addressed in other books, but again the focus is on their chemistries.

By contrast, the interest of this book is the particulate aspects of the small particles found in paints, plastics, and paper laminates. In many ways, the contributions of these particles to the important properties of these materials are of equal or even of greater importance than that of the polymer or additives. As such, many of these important material properties, especially appearance properties, are best understood on a particle basis. For this reason, we believe that a book devoted entirely to the ways that the small particles encountered in these applications can affect and control the end-use properties of paints, plastics, and paper laminates is fully justified.

Our premise is that information alone does not provide this knowledge, but that understanding does. A book providing information only could be as short as a single sentence—"Systematically alter the concentrations and identities of the pigment, extender and resin particles until optimized"—perhaps with supplemental tables listing the various materials commonly available for these purposes. Clearly, such a book would be of little more value than a simple internet search of suppliers' offerings.

Practical knowledge of formulation comes from understanding the effects of material selection, concentration, and manufacturing steps on the performance of the final product. Which materials are likely to give the desired properties? What undesired side effects might they cause? What gives these materials these properties, and how might alternative materials perform differently? What other considerations must be taken into account? Such an understanding requires not only a significant investment of time and attention but also ready access to explanatory materials such as books, literature papers, patents, and manufacturers' brochures.

Our intention in writing this book is that it be a single source of these explanatory materials. Different aspects of much of this information are already available in a piecemeal way, but stitching this information together requires much labor, not only in finding it but also in weeding out incorrect or conflicting information and weaving the remainder into a consistent, complete, and coherent narrative. Our objective was to unify this material in just such a way.

In addition, we wanted this book to be as comprehensive as possible, which meant the inclusion of topics that do not fall directly within the category of small particle properties. For example, we do not limit our discussion of color pigments to simply their types and compositions but include aspects of how color is perceived, measured, and controlled. While color perception may seem far afield from the properties of small particles, it must be understood if the formulator is to truly master the use of these particles to their best advantage.

As for the subject matter itself: it may strike the reader as odd that the technologies of three very different industries—coatings, plastics, and paper laminates—would be discussed in one book. However, at a very basic level, these materials are essentially the same, differing only in details. They all consist of a continuous layer of polymeric resin into which are dispersed and embedded small particles—a sort of plum pudding, to borrow a phrase used over a century ago to describe a model of the atom.[1] Because of the fundamental similarity of these three materials, descriptions of certain important aspects of them overlap. If properly designed, expositions of these overlapping aspects can be of equal use to formulators in all three industries.

But what of the differing details? These fall into four categories. First, the concentration of particles within the polymer matrix varies by application. For some coatings, this can be significantly greater than 50% by volume, whereas for some plastics, this can be less than 1% by volume. Particle concentrations in paper laminates are intermediate between these, but closer to concentrated paints. Second, the typical thickness of these materials also varies widely according to application. Dry paint films are typically a few tens of microns thick, paper laminates on the order of 500 microns, and plastics can be as thin as coatings (plastics films) or as thick as several centimeters.

The third difference is the manner in which the particles are dispersed in the final product. In paints, particles are dispersed as a liquid slurry, and the final paint film is formed by evaporating the solvent (typically water) from the applied wet paint. In thermosetting plastics, the particles are dispersed in the molten polymer, which is then allowed to solidify, freezing the particles in place. Finally, in paper laminates, large agglomerates of particles are trapped in a paper fiber matrix, and then hot resin is pressed, under considerable pressure, into and throughout the paper, displacing

[1] Although this atomic model was later determined to be incorrect in the case of atomic structure, we believe that it is an appropriate and useful description of most coatings, plastics, and laminates.

the air voids within while maintaining the structure of the particles. Here again, the particles are frozen in place when the molten matrix solidifies.

Finally, the appearance of these materials can differ widely—not only between material types (coatings, plastics, or paper laminates) but also within a given type. The appearance differences are due to the two different ways that light can interact with a particle—it can be absorbed by the particle, or it can be deflected, or scattered, by it. The same particles are used to do this in all three material types. However, the details regarding how these particles are used, and their effect on the appearance properties of the final material, vary according to the material type.

Bearing in mind the similarities and differences between coatings, plastics, and paper laminates, we arranged our chapters according to topic, and arranged the topics such that they build on one another. This arrangement is shown in the table below.

| Topic | Chapter Number and Description |
|---|---|
| Introduction to Small Particles | 1. The Behavior of Small Particles<br>2. The Physical Properties of Small Particles |
| Optical Properties | 3. Light Scattering 1 - The Physics of Light Scattering<br>4. Light Scattering 2 - Light scattering in Crowded Systems<br>5. Color 1 - Seeing Color<br>6. Color 2 - Measuring Color |
| Particle Types | 7. White Pigments<br>8. Color Pigments<br>9. Extender Particles<br>10. Resin Particles |
| Aspects of Formulation | 11. Dispersion of Small Particles in Liquid Paints<br>12. Dispersion Small Particles in Plastics<br>13. Measuring the Optical Properties of Paints and Plastics<br>14. Paint and Plastics Durability |
| Formulation | 15. Formulating with Color<br>16. Formulating Paints with Particles<br>17. Formulating Plastics with Particles<br>18. Application to Décor Paper for Use in Laminates |
| Future Looking | 19. Issues and Trends |

We believe such an ordering of chapters is the most logical way to develop the understanding necessary for successful formulation. With this arrangement we lay out the fundamentals that apply universally to the three material types in the first three sections, then have chapters specific to the formulation considerations of each material type. Finally, we describe specific formulation principles for each end use application. Our ambition is not that the entirety of these chapters be of interest to all readers, but that the entirety of the interests of any one reader be contained within a subgroup of the chapters.

An endeavor as broad as this one is beyond the abilities of a small group of people, and the authors are indebted to their colleagues and other experts in the industries discussed who have contributed to the creation of this book. These include, in chapter order:

Dr. Ebrahim Najafi, Ms. Heidi Huezo, and Mr. Roger Senigo, for the electron micrograph images found throughout the book.

Dr. Grant Bleier and Ms. Christine Greentaner, for some of the pictures used throughout the book, and Mr. Robert Shiffer for his careful lab work.

Dr. Massih Pasha, Dr. Ebrahim Najafi, Dr. Scott Brown, and Dr. Mitchell Chinn, for valuable discussions regarding many aspects of Chap. 2.

Mr. Mark Ryan, for reviewing the color pigment chapter.

Special thanks to Mr. Richard Bogner—*Ahlstrom-Munksjö*, and Dr. Andreas Overberg—*Felix Schoeller Technocell*, for reviewing the décor paper for laminate chapter and providing in-depth comments regarding laminate applications.

Dr. Mitchell Chinn, for reviewing the décor paper for laminate chapter.

Ms. Annemie Verhaege, for making significant contributions to the regulation and legislation discussion in the final chapter.

Dr. Hanne Blau, for reviewing the entirety of the book and providing us with the time and other resources necessary to produce it.

Our greatest debt of gratitude is owed to Dr. Austin H. Reid, Jr., who will be familiar to those in the plastics industry for his many contributions to it. The quality of this book was greatly improved by Dr. Reid's sharp editorial eye and keen technical observations, and it is no exaggeration to say that this book would not exist without his valuable input to the book as well as his kind and generous mentoring of the lead author throughout his career. For these, we are truly grateful.

Kallo, Belgium
Newark, Delaware, USA
January 2022

Dr. Michael Diebold
Dr. Steven De Backer
Dr. Philipp M. Niedenzu
Dr. Frank A. C. Vanhecke
Dr. Brett R. Hester

# Contents

## Part IV Aspects of Formulation

# Part I
# Introduction to Small Particles

# Chapter 1
# The Behavior of Small Particles

## Contents



## Introduction

We encounter objects of various sizes every day, some of which we can see with our naked eyes, while others are too small to resolve. For example, fish eggs and larvae can span sizes between 100 microns and 1 mm, placing them on the cusp of what we are able to observe visually. Even though they are too small for us to see, we know that materials such as soot, vehicle exhaust or pollen are also composed of particles, based on their characteristics and behaviors.

This book is concerned with the nature and effects that particles commonly found in paints, plastics, and paper laminates have on end-use properties. As a prerequisite to

**Supplementary Information** The online version contains supplementary material available at https://doi.org/10.1007/978-3-030-99083-1_1. The videos can be accessed individually by clicking the DOI link in the accompanying figure caption or by scanning this link with the SN More Media App.

 
M. Diebold et al., *Pigments, Extenders, and Particles in Surface Coatings and Plastics*, https://doi.org/10.1007/978-3-030-99083-1_1

this, we must define precisely what, in our context, a particle is. The word "particle", as it is commonly used, is quite general and, because of this, somewhat vague. In broad terms, we can define a particle as an object having a specific size and exhibiting both physical and chemical properties. For our purposes, we will define a particle as an object that is smaller than approximately 100 microns and will refer to materials larger than this as "macroscopic objects". We will further divide particles into three size categories, with "large" particles being roughly 20 microns or larger, "small" particles spanning the range from 0.1 micron (100 nm) to 20 microns, and "nano" particles being smaller than this.

Our reason for dividing particles into these three groups is that many physical properties of particles are determined by their size, and the particles grouped according to these different size categories behave differently from one another in many very important ways. In our everyday life, we mainly deal with macroscopic objects—from grains of salt to carry-on luggage. In coatings, plastics, and paper laminates we deal predominantly with objects that are much smaller, with a range of characteristic sizes between roughly 0.2 and 20 microns—that is, objects that we defined as small particles in the paragraph above. Because most of our experiences are with large particles and macroscopic objects, and because the behaviors of large and small particles are very different from one another, as we will see, it is worthwhile to discuss the properties of small particles as a prelude to understanding their effects on many paints, plastics, and paper laminate properties.

## Physical Laws Governing Small Particles

We are familiar with the behaviors of objects that are large enough to hold in our hands. Objects of this size, such as grains of salt or ball bearings, flow easily and independently of one another, separate easily, are significantly acted on by gravity, and pack efficiently. However, the behaviors of objects that are smaller than this—objects that we defined as "small particles"—are much different. We rarely knowingly encounter small particles in everyday life and so are not familiar with many of these differences.

We will begin our analysis of small particle behavior with a review of the important forces that act on and between these particles.

### *The Forces of Nature*

Physicists have identified four fundamental forces from which all other forces are derived. These are the weak force, the strong force, electromagnetism[1], and gravity.

[1] As the name implies, the electromagnetic force can be split into two inter-related forces, the electrostatic force (which causes attraction and repulsion of charged species at rest) and the magnetic

Two of these forces are familiar to us in our everyday lives, and two, the weak force and the strong force, are quite foreign. We are unfamiliar with these two forces because their reach is subatomic—the strong force is important only at distances similar to those of the atomic nucleus, while the weak force operates at distances that are a small fraction (0.1%) of the size of an individual proton. The weak force is roughly a million times less intense than the strong force.

Electromagnetic forces and gravity operate over a much larger length scale than strong and weak forces. As an extreme, gravity can cause galaxies separated by distances of many millions of light-years to aggregate into galactic clusters. Both gravitational and electromagnetic forces decrease in strength as the square of distance, and the equations describing these forces are very similar:

$$F_{gravity} = G\frac{m_1 m_2}{r^2}, \text{ where } G = 6.67 \times 10^{-11}\frac{\text{Nm}^2}{\text{kg}^2} \tag{1.1}$$

$$F_{electrostatic} = k\frac{q_1 q_2}{r^2}, \text{ where } k = 8.99 \times 10^{9}\frac{\text{Nm}^2}{\text{C}^2} \tag{1.2}$$

Note the large difference in the values of the constants in these equations ($G$ and $k$). Although we cannot equate coulombs to kilograms, we can say that these are convenient units to describe the number of electrons and quantities of mass with which we are accustomed to interacting. The difference in magnitude of these constants—20 orders of magnitude—suggests a very large difference in the relative strengths of these two forces. This is, in fact, the case. As a point of comparison, consider two electrons separated from one another by some distance d. These electrons will be attracted to one another by gravity and repelled from one another by charge repulsion. The electrostatic force of repulsion between these electrons is a remarkable $4.17 \times 10^{42}$ times stronger than the attractive force of gravity [1].

The fact that gravity is by far the weakest of the four forces may come as a surprise since gravity plays such a dominant role in our everyday lives. However, if we look carefully, we can see indications that this force is, in fact, quite weak. For example, a relatively weak 10-gram magnet can easily pull a paperclip off of a table—in the process of completely overcoming the gravitational attraction of the $5.972 \times 10^{24}$ kg mass of the Earth. Similarly, when rubbed on fur, a balloon can be attached to a ceiling, again overcoming the gravitational attraction of the Earth, in this case through electrostatic, rather than magnetic, interactions.

We may wonder why, if gravity is over forty orders of magnitude weaker than the electric force, it dominates our lives. There are two reasons that, together, explain this. The first is that there are two types of electric charges, positive and negative, and the interactions among them are both attractive and repulsive, unlike gravity, which operates solely on an attractive basis. The second reason is, paradoxically, that the electrical forces are so very strong.

force (which arises from charged species in motion). These forces were united by Maxwell in the mid-nineteenth century.

The first of these reasons suggests a way that electrical forces can be minimized—charge cancellation. Because there are two opposing types of electrical charge, differently charged subatomic particles can balance one another. This is what occurs in free atoms and in molecules, where there are equal numbers of protons and electrons. Ions, on the other hand, are charged, but oppositely charged counterions are nearly always found in close proximity, again balancing the total charge of the system.

Not only can most electrical charges be balanced on a very small scale, they almost always are. This is explained by the second reason that the effect of gravitational force on Earth dominates that of the electric force. Both the attractive and repulsive forces are very strong between charged materials. The strength of the repulsive forces makes it difficult for unbalanced charges to build up in a given region of space, and the strength of the attractive forces assures that any buildup of charge will quickly be neutralized by charges of the opposing sign. The net effect is that, due to the high strength of attraction between oppositely charged materials, Nature will find a way to bring oppositely charged materials together, canceling their interactions with other materials or objects.

The net result of these factors is that the electrical charges are almost always balanced on a very small scale (on the order of a few nanometers), and so their influences do not typically extend beyond the molecular scale. Gravitational forces, on the other hand, are always attractive and so cannot be shielded or canceled in the same way as electrical forces. Gravitational forces instead always add onto one another, combining in a way that extends their influence to the galactic scale.

## *The Nature of Forces*

We are familiar with the most obvious ways that electrical energy reveals itself—the shock we get crossing a room on a low humidity day, the benefit we get from plugging appliances into electric outlets, the light we get by turning on a switch, and so forth. However, there are also manifestations of electrical energy that are not so obvious. One of these is the attraction of small particles to one another and to surfaces. These attractions are generally referred to as van der Waals forces, a term that includes three related types of interactions:

- those between permanent dipoles—the Keesom force,
- those between a permanent dipole and an induced dipole—the Debye force, and
- those between non-polar atoms or molecules arising from instantaneous dipoles—the London dispersion force.

Note that unlike electrical forces, which decrease as the square of distance, the magnitude of van der Waals forces decreases as the sixth power of surface separation distance and so recedes quite rapidly as touching particles move away from one another. They are therefore only important at distances of no more than a few nanometers.

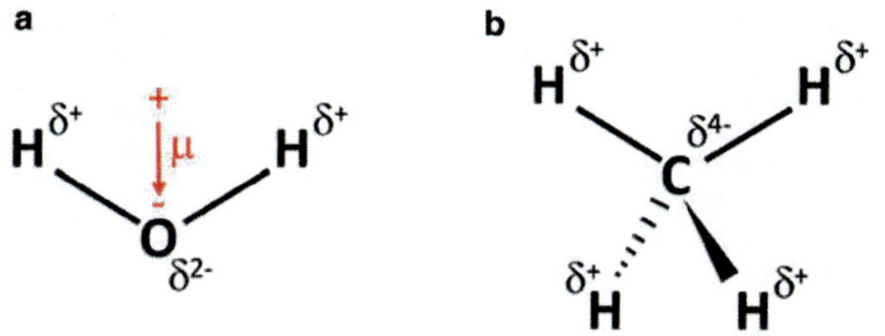


**Fig. 1.1** Charge separation in molecules. δ indicates a fractional charge. **a** Water. **b** Methane. Dipole moment in (**a**) indicated by arrow

The three components of van der Waals forces arise from electric dipoles. Dipoles are atoms or molecules for which there is a slight separation of electric charge. Permanent dipoles are found in molecules when atoms of different electronegativities are bound to one another and the symmetry of the molecule is not high enough to balance these charges. For example, the hydrogen atoms in a water molecule are less electronegative than the oxygen atom. This generates a slight positive charge on the hydrogen atoms (designated as $\delta^+$) and a negative charge double that on the oxygen atom ($\delta^{2-}$). This is shown in Fig. 1.1a. Because the hydrogen atoms are on one side of the water molecule, and the oxygen atom on the other, there is a slight separation of electrical charges in the molecule. This separation is referred to as the dipole moment (designated as $\mu$) and is shown in Fig. 1.1a as an arrow pointing to the negative part of the molecule. Molecules with dipoles are termed polar, and those without them are termed non-polar. As an example of the latter, there are slight charge separations in methane ($CH_4$), just as in water, yet methane has no net dipole moment (Fig. 1.1b). This is because of the high (tetrahedral) symmetry of this molecule, which causes all vectors of charge separation to cancel.

When polar molecules are arranged in a liquid or solid, they align themselves to maximize attractive electrostatic forces and minimize repulsive attractive forces. This means that the positive end of one molecule aligns next to the negative end of another. This is shown pictorially in Fig. 1.2. These interactions result in a net attraction between the molecules and a net stabilization when molecules with dipoles assemble.

There is also a net attraction between molecules with a dipole moment (polar molecules) and those without one (non-polar molecules). In this case, as the molecules approach one another, the positive part of the polar molecule pulls towards it the electrons in the non-polar molecule, while the negative part of the polar molecule is aligned with the remainder of the non-polar molecule. Note that this remainder is partially positively charged because some electron density flowed from it towards the positive end of the polar molecule.

The separation of charge seen in a non-polar molecule that comes in contact with a polar one is called an induced dipole. While the interactions between polar and non-polar molecules are attractive, they are not as strong as the interactions between permanent dipoles.

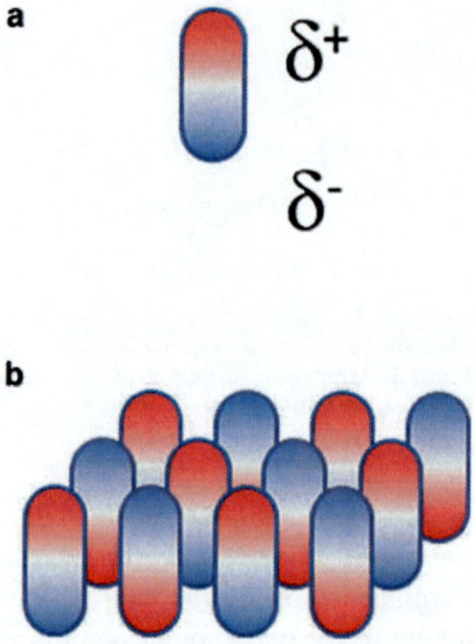


**Fig. 1.2** Alignment of molecular dipoles **a** individual molecule. **b** Assembly of molecules. Red shading indicates areas of excess positive charge; blue shading indicates areas of excess negative charge

Finally, we have the situation where two non-polar molecules come close together. This, too, results in an attraction between them. The electrons within any molecule are in constant motion, and at any given instant they could be imbalanced—that is, there could be a net positive charge in one area of the molecule and a net negative charge in another. This charge separation occurs spontaneously and is temporary. This is referred to as an instantaneous dipole in the molecule.

When two non-polar molecules are in contact with one another, and one molecule develops an instantaneous dipole, an induced dipole is created in the second, much as occurs when a non-polar molecule is brought in contact with a polar molecule. As was the case with polar/polar and polar/non-polar interactions, instantaneous dipole moments cause an attraction between molecules. The attractions between non-polar molecules are, however, significantly weaker than the other two types of dipolar interactions.

The three ways that dipoles can align are shown in Fig. 1.3. Taken together, these forces are called the van der Waals forces, and they account for the assembly of atoms or molecules that are not chemically bound to one another (for example, they account for the liquification of methane at low temperatures).

In our discussion of Fig. 1.2, we noted that molecules in solids typically orient themselves in a way that maximizes the attractive electric forces and minimizes the repulsive forces. Because of this, the surfaces of these typically have alternating positive and negative patches. Other molecules coming into contact with these surfaces will generally align themselves according to these interactions. This explains the well-known phenomenon of the adsorption of a gas or a dissolved molecular species onto a particle surface.

This also explains the attraction particles have towards other particles. This attraction is a surface phenomenon—surface atoms act as shields and prevent the dipole

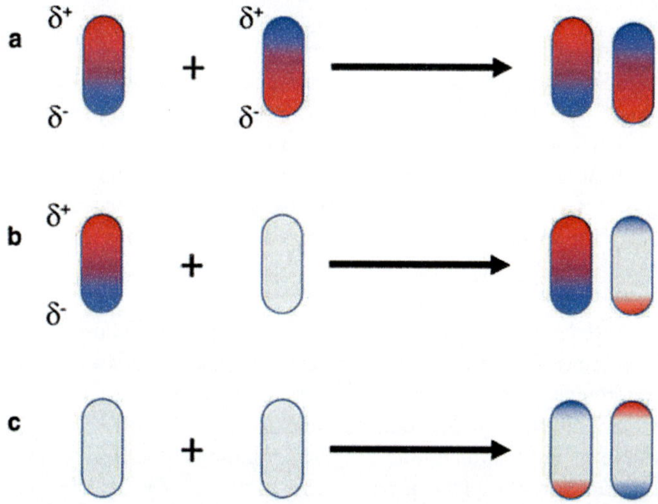


**Fig. 1.3** Dipolar interactions between molecules. **a** Permanent dipole–permanent dipole (Keesom force); **b** permanent dipole–induced dipole (Debye force); **c** Instantaneous dipole–induced dipole (London force)

charges from penetrating deeply into a particle. Because of this, we find the attraction strongest for particles with high specific surface areas.[2] Particles that have high specific surface areas are either porous or, more commonly, quite small. A sizeable fraction of the atoms and molecules within particles that are smaller than about half a micron are located at or near the particle surface, where they can participate in particle–particle interactions. The high proportion of surface atoms and molecules, and the sharp drop in interaction strength with distance, cause small particles to be more strongly attracted to one another (and to surfaces in general) than large particles.

## *The Balance of Forces in Large and Small Particles*

The differences in behaviors of large and small particles, on the one hand, and nanoparticles, on the other, are due to differences in the physical laws that they follow. Large and small particles follow what are considered the classical laws of physics—the laws discovered by scientists several centuries ago. As a group, we will call these Newtonian laws. These include, for example, the force law (force = mass times acceleration) and the law of gravity (see Eq. 1.1 above).

Nanoparticles, however, follow a different set of laws. These are the laws of quantum mechanics, where uncertainties in location and motion (momentum) are

[2] Specific surface area is the surface area of a particle divided by its weight. A collection of smaller particles has a higher specific surface area than a collection of the same weight of larger particles.

linked to one another. Quantum behavior is quite foreign to us, and the implications of quantum mechanics are difficult to perceive and understand at an intuitive level.

While nanoparticles follow quantum mechanical laws, both large and small particles follow the same laws (the Newtonian laws described above). As such we might expect their behaviors to be similar. This is not the case, however. Although large and small particles follow the same laws, the relative importance of the different laws is quite different between the two particle size regimes. As discussed above, large particles are governed primarily by gravity. Small particles, on the other hand, are governed primarily by surface forces.[3] These include friction and the natural attraction of one surface to another.

Gravitational forces scale as the volume of a particle ($d^3$, where d is the particle diameter), whereas surface forces depend on the surface area of a particle ($d^2$). The relative importance of the two for a specific particle can be described by the ratio of the forces. This ratio ($d^3/d^2$) changes linearly with particle size (d). For large particles (large d), gravitational forces ($d^3$) are dominant, whereas for small particles (small d), surface forces ($d^2$) are dominant (Fig. 1.4). This can also be understood by comparing the fraction of molecules (or atoms) in a particle are that near the particle surface to the fraction of those in the bulk. Those molecules (or atoms) near the surface can be influenced by other surfaces, as discussed above, while those away from the surface are affected only by gravity.

The remainder of this chapter will focus on comparing and contrasting the behaviors of large particles—behaviors that we are familiar with through our everyday lives—to those of small particles—which we rarely encounter. For many, the most common opportunity for interacting with small particles is in the kitchen, where many food staples are in powder form. These powders often exhibit small particle behavior, and an excellent point of reference for this chapter is the different behaviors of large and small particles of the same material—sugar. One can use granular sugar (a large particle) and powdered sugar[4] (a small particle) to show the similarities and differences between these two size regimes, and readers are encouraged to confirm the observations made below for themselves by using these forms of sugar as examples. Some of these differences are also demonstrated in the video associated with Fig. 1.5.

---

[3] There is a second distinction between large and small particles that affects certain small particle behavior. This distinction applies when particles are so small as to approach the atomic or molecular dimensions. For macroscopic objects, we can ignore the fact that material is made up of indivisible units (atoms or molecules), but we sometimes cannot do so for small particles. Brownian motion is an example of this. Random imbalances of gas or liquid molecules striking a macroscopic object are much too slight for detection, while for small particles, such as grains of pollen, the imbalance is great enough to cause the particle to move. This is manifested as the random (Brownian) motion of these materials in air or water. An explanation of Brownian motion was one of Einstein's first contributions to physics, and the fact that it arises from a discontinuity in structure (i.e., the indivisibility of atoms) was determined by Perrin and was the basis for Perrin being awarded the Nobel prize in physics in 1908.

[4] Sometimes called confectioners' or 10X sugar. It is called icing sugar in the UK.

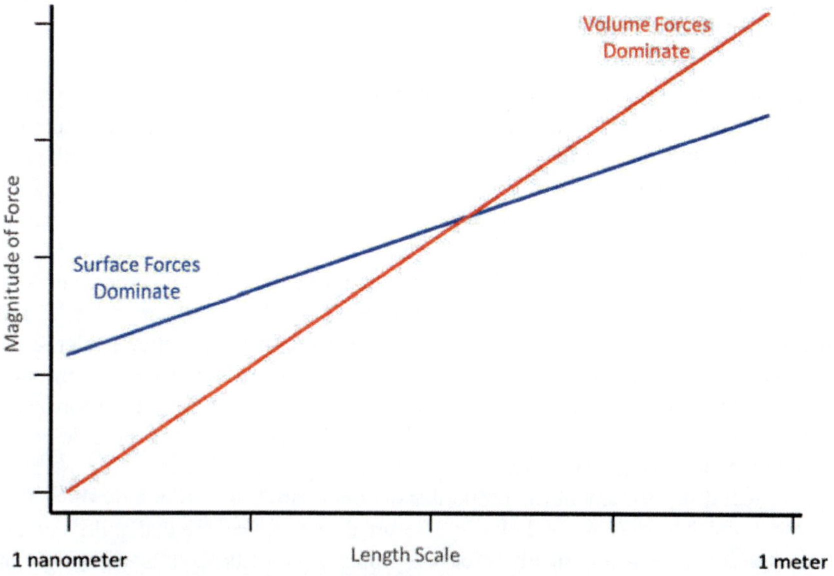


**Fig. 1.4** Schematic of the relative importance of volume and surface forces as a function of particle size (both axes are logarithmic)

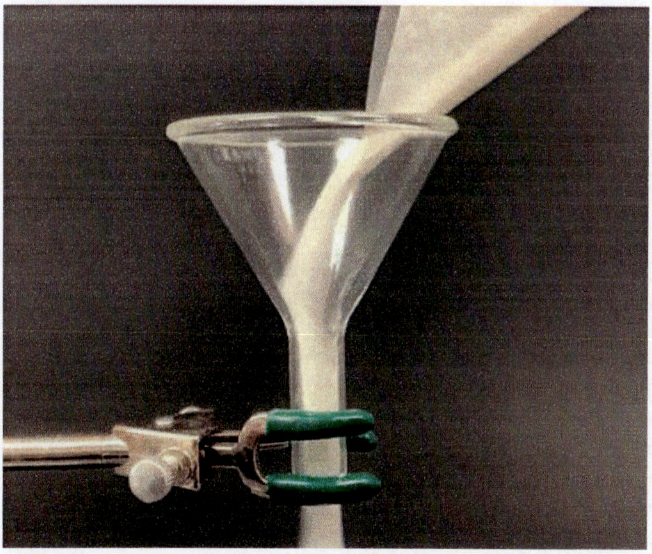

**Fig. 1.5** Video demonstration of the behavior differences between granular and powdered sugars

## Properties of Small Particles

There are many properties common to all small particles that determine, in part or in whole, their physical behaviors. These properties can be broadly divided into two groups—physical properties and chemical properties. For the most part, chemical behaviors are determined by the chemical nature of the particle surface, since atoms below the surface are not available for chemical reaction. These behaviors can be altered by modification of the particle surface, usually through the adsorption of a small molecule or ion, or by the deposition of a new solid onto the particle surface.

By contrast, physical behaviors are determined by both the surface properties and the dimensional characteristics of the particle. Here, the chemical nature or elemental composition of the particle is not as important. Instead, aspects of the particle such as size, shape, weight, density, coefficient of friction, etc., control these properties. In this sense, we are concerned with the "object" characteristics of the particle rather than the reactivity characteristics.

We will discuss many of the properties of small particles, particularly those that have relevance to coatings and plastics. Before doing so, we note that, unless otherwise specified, we will typically refer to a property on the basis of a unit weight of particles, rather than on the basis of a single particle or a set number of particles. For example, a single large particle clearly has a higher surface area than a single small particle of the same shape. However, a gram of large particles will have a smaller surface area than a gram of small particles. Our reason for defining properties on a mass basis is that particles are formulated into paints, plastics, and paper laminates on a weight or volume basis, rather than on a number basis, and so it is on this basis that we will discuss their attributes.

### *Shape*

All particles, regardless of their size, have specific shapes. This can include roughly uniform shapes, such as a sphere, cube or block, irregular shapes, such as plates or rods, and all shapes in between. It is most convenient to classify shapes based on the number of dominant dimensions a particle has. We live in a three-dimensional world, and therefore there are three shape categories (Fig. 1.6). Three-dimensional, blocky shapes, such as a cube or a sphere, are roughly the same size in all directions. Plates can be considered two-dimensional—they extend in breadth and width but have very little thickness. Finally, needle-like or rod-shaped (acicular) particles extend in a single dimension (length), with negligible breadth and width.

Particles acquire a given shape in one of two ways. First, they can be formed with a preferred shape. This occurs when a small seed particle begins to grow. Growth typically occurs by the addition of "building blocks" to an existing particle. When building blocks are added to a growing crystal, they often do not attach to all faces equally. This is because some faces of the crystal are more energetic than others

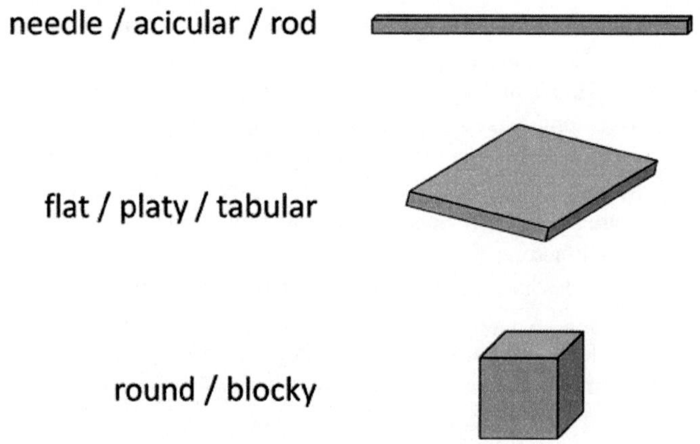


**Fig. 1.6** Representative particle shapes

and so are more likely to be attached to an incoming building block. This causes crystal growth on certain faces to be more rapid than on other faces, with the result that particle growth may occur preferentially in certain directions. This leads to the formation of plates or needles.

Once a particle is formed, it can also be shaped through attrition. When particles are ground, they are broken into smaller particles. This can occur through impact or shear. The former crushes the particles, while the latter pulls them into pieces. The origin of impact forces is obvious—a material collides with the particle. Shear forces, on the other hand, are forces that vary in magnitude across a distance perpendicular to the direction of the force. When shear is applied to a particle, one part of the particle experiences a greater force than another part of the particle (Fig. 1.7).

In general, regardless of the shapes of the original particles, attrition results in the formation of particles that are blocky or spherical. We can see this in the spherical and ellipsoidal shapes of small stones polished in a river or on an ocean shore. The

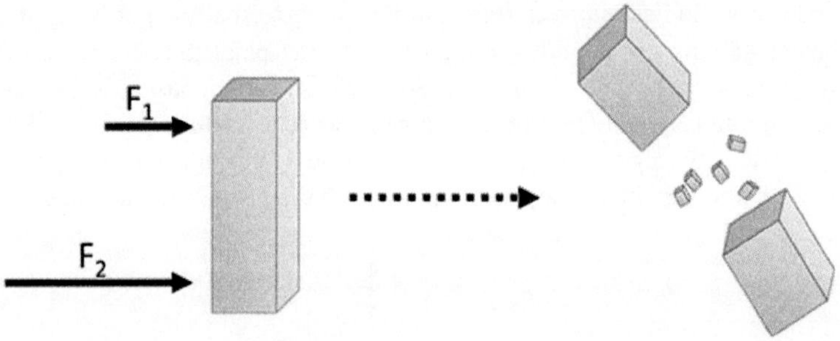


**Fig. 1.7** Unbalanced (shear) forces applied to a large particle, resulting in that particle breaking into smaller particles

reason for this is that when one or two dimensions of the crystal are longer than the other(s), the shear forces are greatest between the two ends of the crystal that are the furthest apart. This is analogous to the greater ease of snapping a pencil in half perpendicular to its long axis as opposed to cleaving it in half lengthwise.[5]

Particles of all three shape types (block, plate, or needle) can be found in paints and plastics. In some cases, the shapes bring a desired property, such as those discussed below, while in other cases, the shapes are simply those that are the most convenient for the powder manufacturer to produce.

Rod-shaped particles have a positive impact on two paint film properties. First, the rods near the film surface cause surface protrusions. These protrusions enhance the adhesion between layers of paint (the so-called paint "tooth"). Second, rods within the film will reinforce it, enhancing its physical strength and resistance to abrasion. This is analogous to the incorporation of rebar in concrete.[6]

Plate-shaped particles (typically extenders) are often used to enhance the protection from water penetration that the paint provides to the substrate. Because of surface tension, platy particles will orient parallel to the film surface as the paint dries. Since water cannot penetrate through an extender, it must follow a long and tortuous path if it is to travel from the surface of the paint to the substrate. This increase in the distance that the water must travel to reach the substrate will enhance the protection of the substrate by the paint. In addition, special effects pigment particles are often platy particles that orient themselves in the paint or plastic (see Chap. 8). This causes the interactions between the pigment particles and light to have an angle dependence.

## *True Density and Bulking Value*

There are two types of density that are important to particles: true density and bulk density. True density is the density of a solid object made of a specific material, while bulk density is the density of a collection of particles, such as a bag or container of particles. Bulk density is inevitably lower than true density because any collection of particles, no matter how tightly packed, will contain interparticle voids.

True density and its reciprocal (the bulking value, normally reported as liters per kilogram or gallons per pound) are important to the paint industry for economic reasons. Paint makers purchase their ingredients on a weight basis (e.g., dollars per kilogram) and mostly sell them on a volume basis (e.g., dollars per liter). Because of this, low-density (high bulking value) fillers have an advantage over high-density fillers since a unit weight of low-density particles has a greater volume than an equal weight of their high-density counterparts.

---

[5] This can also be understood in terms of the number of bonds that must be disrupted to create the break. This number will be a function of the area of the cross section of the break. The smallest cross-sectional area will be perpendicular to the longest dimension of the particle.

[6] The name "rebar" comes from the shortened conjunction of the words "reinforcing bar".

The true density of a particle also is important when the particle is suspended in a liquid (i.e., in slurries and paints) because this, in part, determines the settling rate of that particle. The force by which gravity affects a particle in such situations is moderated by the difference in density between the particle and the liquid. Dense particles are therefore more prone to settling than less dense particles and this must be taken into account whenever a slurry or paint is stored.

We will discuss the bulk density of particles in detail in a later section on bulk properties.

## *Surface Properties*

As discussed above, the sizes of particles found in paints and plastics can be small enough that a significant fraction of the individual molecules or atoms is close to the particle surface. For this reason, many of the behaviors associated with small particles are controlled by the surface of the particle. The extent to which the surface of a particle affects its behavior depends on two factors: the amount of surface (i.e., the surface area) and the chemical and physical state of the surface (i.e., the energetics of the surface atoms or the roughness of the particle surface). We will begin our discussion on surface properties with surface area.

### Surface Area

Surface area is a physical property of all particles. Surface area increases when large particles are broken down into smaller particles. It is easy to see why this is the case—each time a particle is broken, two new surfaces are created. As shown in Fig. 1.8, surface area increases linearly as particle size decreases—that is, cutting a cube apart in half in all three dimensions increases the total surface area by a factor

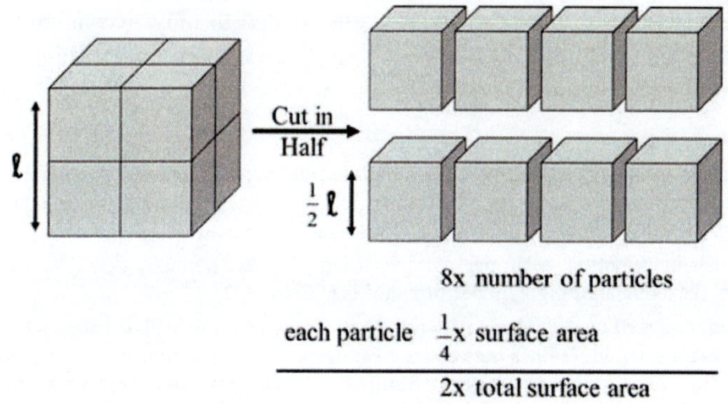


**Fig. 1.8** Increase in total surface area as particle size is reduced

**Table 1.1** Geometric surface area as a function of edge length of a cube

| Length | Area ($m^2/g$) |
| --- | --- |
| 1 mm | 0.003 |
| 10 μm | 0.3 |
| 1 μm | 3 |
| 10 nm | 300 |

of two (the surface area of each small particle is one quarter that of the larger particle, but there are eight small particles).[7]

For most particles found in paints and plastics, surface areas vary between 1 $m^2/g$ and 80 $m^2/g$, although in a few exceptional cases (such as diatomaceous earth), surface areas can be much higher. Table 1.1 shows surface area values for cubes of a density of 2 g/cc (common for extender particles) as a function of the length of the cube edge. To give an appreciation for the magnitude of surface area for small particles, a tonne of pigmentary $TiO_2$ has a surface area of roughly 15 square kilometers.[8]

## Surface Energy

In addition to area, surfaces have an energy characteristic that will vary from one type of particle to another. Surface energy is an important determinant of the bulk and handling characteristics of a powder.

To understand the origin of surface energy, consider what happens when a piece of quartz is broken. Quartz (crystalline silicon dioxide) is a network solid, meaning that there is a network of chemical bonds that link all of the atoms together (see also Chap. 14).[9] Before breaking, each interior silicon atom is chemically bonded to four oxygen atoms, and each interior oxygen atom is bonded to two silicon atoms (Fig. 1.9a). This arrangement of the atoms maximizes chemical bonding interactions and leads to an energetically favored state.

When the quartz crystal breaks a series of Si–O bonds are disrupted (Fig. 1.9b). Some silicon atoms on the new surfaces are bonded to only two or three oxygen atoms, rather than the optimal number of four. This is a high-energy, highly reactive state for the silicon atoms. The silicon atomic orbitals that had been directed to

[7] Recall that surface area is reported on a weight basis rather than a particle basis. On a per particle basis, surface area decreases as the square of particle diameter. On a per gram basis, it increases linearly with decrease in particle diameter, as shown here.

[8] This includes contributions from nano-sized particles that are normally deposited on the $TiO_2$ particle surfaces to enhance pigment performance (see Chapter 7).

[9] This can be compared to a telephone network. Here, a network of wires joins all of the telephones together. There are no direct lines between a given telephone and every other telephone; instead, the lines go from one telephone to the next. Similarly, in a network solid, there are no direct bonds between a single atom and every other atom. Rather, an atom is bonded to a few of its nearest neighbors, which, in turn, are bonded to other nearby atoms, and so on.

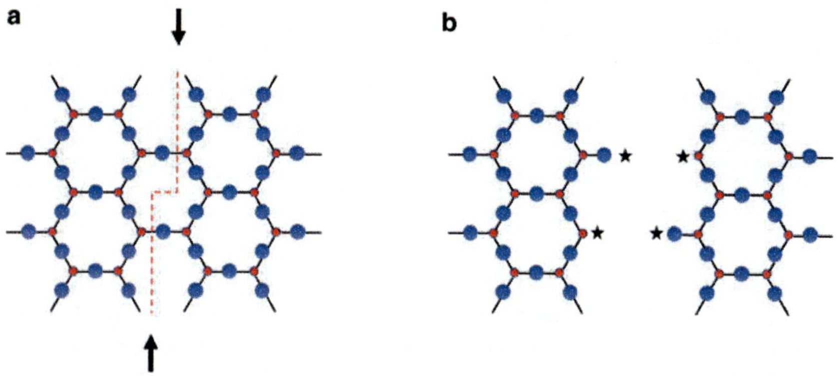


**Fig. 1.9** Creation of surface energy in quartz through bond breakage. Red circles are silicon atoms and blue circles are oxygen atoms. Note that each silicon atom is bonded to one oxygen atom that is above or below the plane of the page. **a** Before breaking. The break occurs at the arrows and the dotted line shows where the new surfaces will be. **b** After breaking. Stars indicate high-energy unsaturated atoms

oxygen atoms and were a part of an energetically favorable interaction now point to empty space. In the same way, the oxygen atoms on the new surface are also in a high-energy state.[10] The excess energy that the surface atoms have relative to those atoms in the bulk is called surface energy.

The amount of surface energy for any particle will depend on two factors. The first is the strength of the bonds that must be broken to create a surface—the greater the energy required to break these bonds, the greater the energy per unit area of surface (energy must be conserved and the chemical energy of the bonds is transformed into surface energy). The second parameter is simply the amount of surface—obviously the greater the surface area, the greater the surface energy. The total surface energy is the product of the energy per unit area (which depends on the chemical composition of the particle) and the total surface area of the particle (which depends on particle size and shape).

In general, nature is driven towards low energy states, which in this case means that the high-energy surface atoms are driven to react in ways that lower their energies. To some extent, the excess energy of these atoms can be lowered by relaxation of the surface—some atoms physically move, migrating to lower energy positions where they can interact with a higher number of atoms than they interacted with in their original configuration. In addition, water, oxygen, and even nitrogen from the air adsorb onto the highest energy (most reactive) sites on the surface. Water, in particular, is very effective at decreasing surface energy. The oxygen atom in a water molecule has two lone pairs of electrons that are available for bonding to silicon atoms. The hydrogen atoms in the water can hydrogen bond to the surface oxygen atoms.

[10] Breaking these bonds can also lead to charge separation (e. g., $SiO^-$ or $Si^+$ surface groups).

**Table 1.2** Example strengths for different bonding interactions

| Bond type | Example | Bond energy (kJ/mol) | Melting point (°C) | Surface energy (mJ/m$^2$) |
|---|---|---|---|---|
| Van der Waals | $N_2$ | 3 | −196 | 12 |
| Hydrogen | $H_2O$ | 21 | 0 | 65 |
| Covalent | Diamond | 347 | 4027 | 600 |

The surface energy of a particle depends on the weakest set of bonds that hold it together. We can divide most chemical bonds into one of three groups. In order from weakest to strongest, they are van der Waals forces, hydrogen bonding, and covalent bonding. Examples of the bond energies of materials experiencing these different interactions are given in Table 1.2, along with their melting points, which are rough measures of the strength of the bonds holding these materials together, and their surface energies.

An additional means by which surface energies can be decreased is through contact with other surfaces. In the extreme case, this could lead to chemical bonds forming across the surfaces, which would be the reverse of breaking the particle in half. However, even in the absence of chemical bond formation, the close contact of two surfaces will decrease surface energy due to van der Waals interactions between atoms on the two surfaces. Said differently, the two particles will be held together by van der Waals forces acting between the two surfaces.

A consequence of this is that surfaces are “sticky”. While this is true for all surfaces, it is most apparent for small particles since these particles have high surface areas (so high amounts of surface energy) and low weights. We can see this by comparing the stickiness of particles against a wall. A relatively large particle, such as a grain of sand, will, if pressed against a wall and released, readily slide down the wall because the gravitational force on the particle is greater than the van der Waals forces between the particle and the wall. However, for a small particle, such as a speck of smoke, the balance between surface area (and therefore surface energy) and weight (gravitational energy) favors the adhesion of the smoke particle to the wall—the wall blackens. The small smoke particle will not slide down the wall because the gravitational force is now less than the friction generated by van der Waals forces holding the particle against the wall.

## Friction

Friction is a force that acts at the junction of surfaces. It is defined as any force that hinders (or entirely prevents) touching bodies from sliding across or over one another. It is a vector quantity (i.e., it has both a magnitude and a direction), and is always in a direction parallel to the surface of contact and opposite the direction of motion or applied force if any.

The magnitude of the frictional force between two bodies is determined by two factors. The first is the strength of the force holding the bodies together. In our everyday lives, this is normally gravity. For example, the strength of friction between a book and a table on which it is placed depends on the weight of the book. However, friction can result from any applied force. For example, consider a book pressed against a wall. The book does not slide to the floor because the friction between the book and the wall,[11] which is created by applying a force to the book, prevents it from moving.

The second factor is the identity of the two bodies. For example, friction is higher between rubber and asphalt than between rubber and ice, and for this reason, a car can be stopped more quickly when it is on dry pavement than on icy pavement.

Because small particles have high surface areas, the importance of friction is much greater for these particles than for larger ones. This is manifested in the bulk behaviors of the particles.

## Bulk Density, Bulk Flow, and Powder Compressibility

Density is simply the mass of a material that fills a unit volume. The density of a solid mass with no internal voids or porosity is termed the "true" density of that material, as discussed earlier. The true density of a material depends primarily on its composition,[12] and for any given material it is a constant that can be measured accurately and reproducibly.

The density of a collection of particles is invariably lower than that of a solid mass because of the void spaces between the particles [2]. This density is referred to as the loose bulk density of the material if the material is not under any stress. For obvious reasons, the loose bulk density is determined by both the true density of the particles themselves and the efficiency with which they pack together. For uniform macroscopic spheres, such as ball bearings, packing is generally very efficient and can approach the limit of random closest packing (approximately 54–58% solid and 42–46% void space). Particle shape can also play a role in packing efficiency—when randomly arranged, blocky particles pack the most efficiently, and acicular (rod-like) particles the least.

The packing efficiencies of small particles, in comparison to large particles or small objects (e.g., sand or salt grains), are low and are dependent on particle size and surface characteristics such as surface chemistry and surface morphology [3]. The dependence of bulk density on particle size is shown in Table 1.3 for silica particles.

In this table, we specify that the bulk density values are "loose" bulk densities. This is important because the bulk density of a sample can be changed quite easily. The

---

[11] The friction between the book and the hand applying the force is also critical.

[12] Although density can be affected by temperature (thermal expansion), this effect is small and can be easily taken into account. In addition, different crystalline phases of the same material may have different densities.

**Table 1.3** Bulk density and void fraction as a function of particle size

| Material | Particle size (μ) | True density (g/cc) | Loose bulk density (g/cc) | Void fraction (%) |
|---|---|---|---|---|
| $SiO_2$ sand | 1,000 | 2.6 | 1.6 | 38 |
| $SiO_2$ microbeads | 1.5 | 2.6 | 0.88 | 66 |
| Fumed $SiO_2$ | 0.015 | 2.2 | 0.05 | 98 |

sample can be compressed, increasing the bulk density, or "loosened", for example by shaking the particles in air. This can be understood on the basis of friction. A sample of particles will naturally settle under gravity until it reaches a certain density. During this settling, the number of particle–particle contacts increases. Settling ceases and bulk density is set, when the frictional forces between the particles are greater than the gravitational forces that act on them. Applying an outside force, such as by using a piston in a pellet press or even rapping a small sample on a tabletop, augments the gravitational force. The particles then move to a denser state, where the increased friction from additional particle–particle contacts again equals the other forces acting on the particles.

Bulk density can be measured either when the particles have been exposed only to gravitational forces (loose bulk density) or when the particles have been subjected to an additional force, in which case the bulk density is described by the nature of that force (e.g., "tapped" density).

The degree to which a given amount of additional force causes the particles to densify is referred to as the compressibility of the particles. Needless to say, compressibility is a function of the particle size, the particle surface, the strength of the additional force, and the coefficient of friction. Compressibility is greater for particles that are initially loosely packed since the degree to which the particle packing can increase is greater for loose particles than for particles that are already packed tightly.

The effect of particle size on bulk density and compressibility is shown in Fig. 1.10. The same weight of two samples of sugar has been added to two test tubes. The sugars differ in particle size (540 microns vs. 60 microns). The smaller particles, due to stronger interparticle interactions, occupy a larger volume and so have a lower bulk density. The heights of the powder beds were marked, and then the test tubes rapped sharply on a tabletop ten times. After tapping, we see that the small particles have compressed to a greater extent than the large particles (the volume of the small particles decreased by 38%, while that of the large particles by only 2%). The small particles are more compressible because these particles initially have more void volume. Despite their greater compressibility, the small particles still occupy a greater volume than the same weight of larger particles after compressing. Note that the total frictional forces resisting compression (and, roughly, the total area of contact between the particles) are the same for each bed of particles since each has been acted on by the same forces.

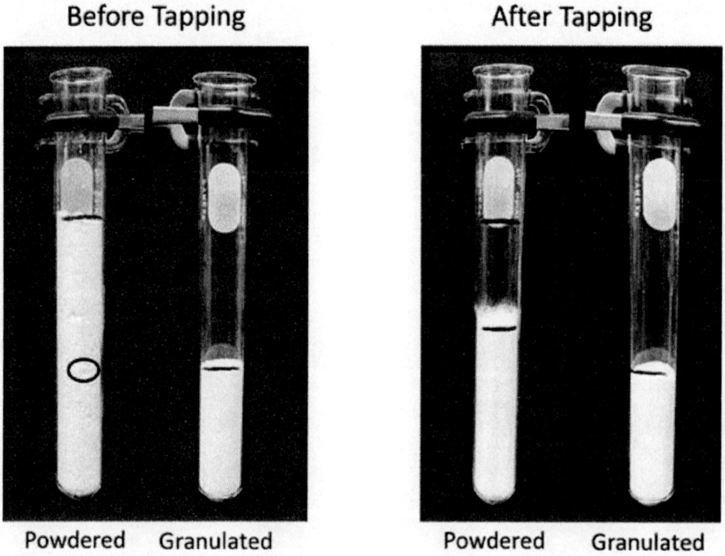


**Fig. 1.10** Test tubes showing compaction of equal weights of granulated and powdered sugar. Note large void in untapped powdered sample (circled)

A second important bulk property of small particles is their flow characteristics. If put in motion, large particles will interact with one another when they collide, but otherwise move independently. On the other hand, small particles tend to move as a coherent mass—all of the particles move with respect to their surroundings, but they do not move with respect to one another. The smaller particles remain locked in place due to friction, in a manner analogous to their resistance to compression. Here, we see that the balance between the volume force (gravity) and the surface force (friction) results in very different behaviors. For large particles gravity dominates, and both the initiation and continuation of flow are more facile than with their smaller, higher surface area counterparts.

At extreme conditions, when particles are very small and the surface chemistry of the particles leads to strong interparticle forces, conveying and handling the material in pneumatic lines and storing it in silos become challenging. For dilute-phase pneumatic conveying, the particles need to be dispersed in the gas phase, and significant energy and high turbulences are required to initiate and maintain this degree of dispersion. The challenge for storage is initiating the flow of the particles from the bottom of the storage silo when removing the particles for use. Such a silo may contain several tens of tonnes of titanium dioxide, yet when the outlet at the bottom of the silo is opened, flow might not occur, even, in extreme cases, when the chute diameter approaches one meter. Instead, the particles remain "hung up" in the silo.

This observation leads to an obvious question: what is holding the weight of the titanium dioxide particles, if not the bottom of the silo where the outlet is located? It is the walls of the silo that support the load through a combination of wall friction,

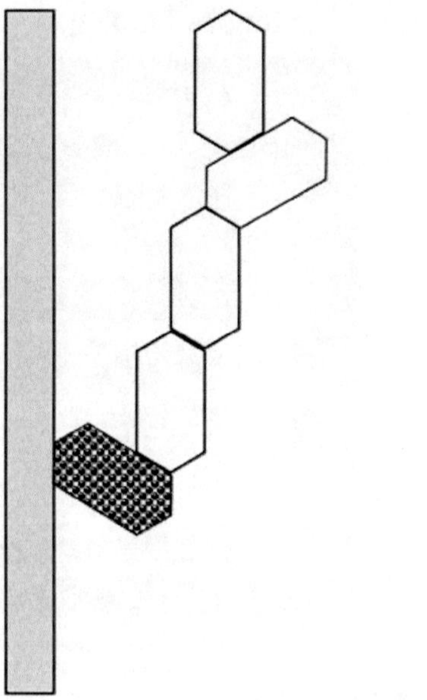


**Fig. 1.11** Frictional resistance to flow of a particle in contact with the wall of a bin

interparticle cohesion, and interparticle friction. Particles that touch the walls are held in place by wall friction, and these particles, in turn, hold particles that are touching them in place by internal friction (Fig. 1.11). A simple calculation shows that a single particle in contact with the wall can support the weight of as many as one million other particles. Such is the importance of friction for particles in this size regime.

Finally, we may wonder about the packing structure of small particles. For example, a sample of pigmentary $TiO_2$ is roughly 80% voids. How are these particles arranged? We first note that the particles cannot be distributed evenly at the single-particle scale (Fig. 1.12a). Here, there are far too few contacts to support the weight of the particle bed. This is also inconsistent with electron micrographs that show a tight packing of the particles at this length scale (Fig. 1.12b).

Instead, the particles pack in a fractal arrangement.[13] Here particles are densely packed into small micro-clumps (Fig. 1.13, top), consistent with what is seen in the electron micrograph in Fig. 1.12b. These micro-clumps, in turn, pack together densely into meso-clumps (Fig. 1.13, middle), which in turn pack together into macro-clumps (Fig. 1.13, bottom). If we assume a packing density of 50%, then the micro-clumps are 50% void, the meso-clumps are 75% void, and the macro-clumps are 87.5% void.

[13] A fractal is an object that appears the same at different length scales. An example is the appearance of a coastline—the general character of a coastline looks the same from orbit as it does from an airplane. This is known as the coastline paradox. This situation can be contrasted with the appearance of a tree—we can easily tell the scale at which we observe the tree by the features that we see.

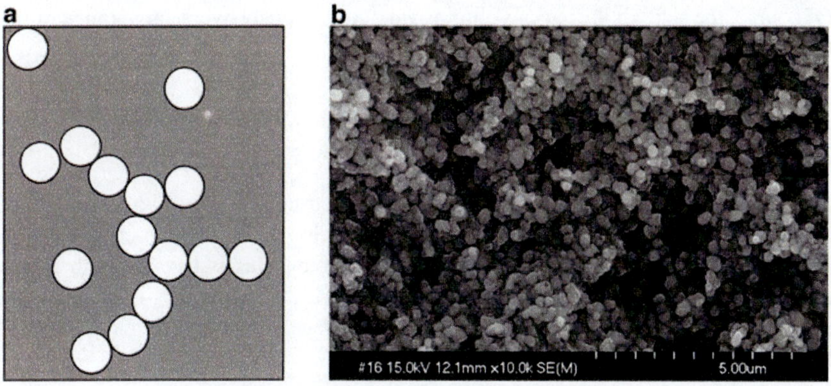


**Fig. 1.12** Arrangements of small particles. **a** Even arrangement at 80% void volume. **b** Actual packing of pigmentary $TiO_2$ with a bulk void volume of 80%

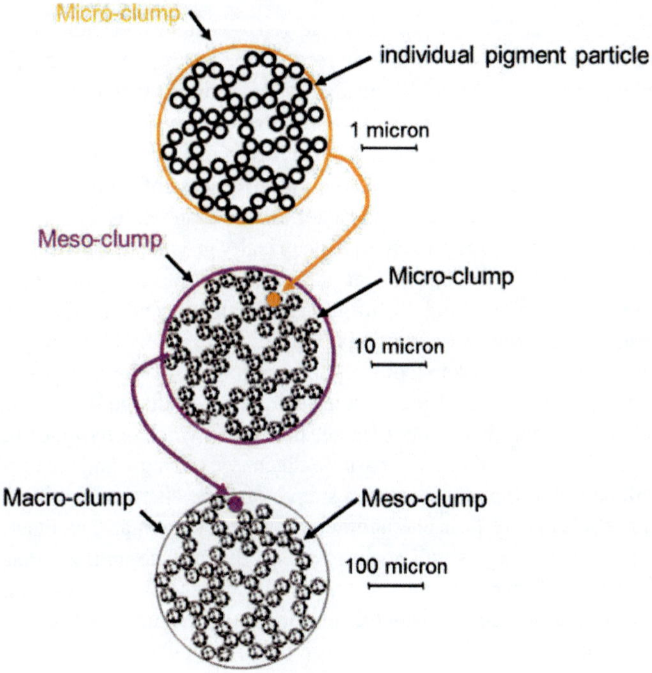


**Fig. 1.13** Fractal packing arrangement of pigmentary $TiO_2$

Of course, the sizes of the various clumps vary continuously over a large range rather than falling into just these three categories.

## Surface Charge—General

The surfaces of many materials can hold an electrical charge—either a negative charge due to an excess of electrons or surface anions or a positive charge due to a deficit of electrons or an excess of surface cations. This is true for both macro and microparticles, and in both air and in slurry. However, the origin, control, and stability of accumulated electrical charge on a surface are quite different in the wet and dry states.

## Surface Charges in Dry Powders

The charges on dry particles are typically referred to as static electricity and are found when there is an excess or deficit of electrons on the particle. For energetic reasons, the extra electrons will travel to the surface of the particles where they are arranged in a way that minimizes electron–electron repulsive energy. For spherical particles, this will be an even distribution on the particle surface. For other particles, charge tends to accumulate based on surface curvature. Regions of high curvature (edges, corners, or protrusions) accumulate a greater charge than relatively flat regions.

When a charged surface approaches an electrically grounded (earthed) surface, the excess charge will travel through the air and discharge into the grounded surface as a spark. The distance over which this discharge occurs depends on the magnitude of the charge and environmental conditions (temperature, relative humidity, etc.). The surface density of the electrons controls the magnitude of the charge—as density increases, the voltage of the surface electrons increases. Charge will accumulate on the surface until the number of charges per unit area, which determines the surface voltage, exceed a threshold and the particle discharges to a grounded surface.

For small particles, the most common source of static charge is the friction generated when particles move past one another, a process known as triboelectrification. As discussed above, charges will accumulate on the surface until a certain surface density is exceeded. Obviously, the greater the surface area, the greater the number of charge units that can accumulate. For this reason, the accumulation of static electricity is greater for smaller particles than for large ones. As a consequence, considerable charge (sometimes thousands of volts relative to ground) can build up in a stream of moving small particles, and for safety reasons, it is essential that equipment through which particles are conveyed is electrically grounded.

### Surface Charges in Liquid Slurries

The surfaces of slurried or suspended particles can also be charged but by a much different mechanism. As outlined in the discussion on surface energy, the surfaces of particles tend to be high energy, and therefore chemically reactive. This leads to the adsorption of molecules or ions onto suspended particle surfaces. In some cases, these absorptions occur at reactive sites on the surface, and the reactivities of these sites are dependent on the chemical state of the site, the nature of the adsorbing molecule or ion, and the identity of the liquid medium (water or organic solvent). Certain combinations of surface site and adsorbing species can be quite energetically favored, leading to strong adsorption.

Molecular salts, including acids and bases, tend to be soluble in water because the water molecule is polar. When dissolved in water, these materials dissociate into charged species, with the charges being either negative (anions such as hydroxide or chloride) or positive (cations such as protons or sodium ions). Even in the absence of foreign dissolved species, water has both anionic and cationic species within it. These are the hydroxide ($OH^-$) and proton ($H^+$) ions that spontaneously form when water molecules dissociate. The concentrations of these species change when an acid or base is dissolved in the water.

The surfaces of most inorganic oxides are covered with oxide groups, hydroxyl groups, and/or adsorbed water molecules. These groups can react with dissolved hydroxyl groups or with dissolved protons. These reactions include the removal of a proton from a surface hydroxyl or adsorbed water molecule, the addition of a hydroxide ion onto the surface, the removal of a hydroxyl group from the surface, or the adsorption of a proton onto the surface. The first two of these give the surface a net negative charge, and the last two of these give the surface a net positive charge. Note that both positively and negatively charged surface species can be present on a given surface at the same time because a surface can have more than one type of reactive site. Important aspects of surface charge formation are given in greater detail in Chap. 2.

## *Interactions with Light*

To some degree, all particles interact with light. These interactions can take two forms: light absorption and light scattering. In principle, both types of interactions occur for all particles. However, in most cases, one or the other of these interactions dominates.

Light absorption by atoms, molecules, and particles occurs when the energy of a photon matches the energy required to affect some change in the material. This change can be an increase in the rotational rate of a molecule, an increase in the amount of a vibration between atoms in molecule or particle, or an increase in the electronic energy. These different transitions require different amounts of energy. Typically, rotational changes occur when microwave radiation is absorbed, vibrational changes when infrared radiation is absorbed, and electronic energy changes when visible or ultraviolet radiation is absorbed. The exact frequency of light absorbed depends on the chemical identity of the absorbing species. This leads to, as an example, different visible light wavelengths being absorbed by different materials, which explains why different materials have different colors. This is discussed in greater detail in Chap. 8.

Light scattering occurs when light changes direction as the result of encountering a particle. This change in direction occurs without a change in frequency (color) or intensity (energy) of the light. The degree to which the direction of the light is changed for a particle isolated in a medium depends on four factors: the wavelength of light, the size of the scattering particle, the refractive index of the particle, and the refractive index of the surrounding medium. In general, scattering is greatest when the size of the scattering species is close to the wavelength of light being scattered and when the refractive index of the particle is very different from that of the surrounding medium. This is discussed in greater detail in Chaps. 3 and 4.

Even weak scattering, which is defined as a small change in direction for a given scattering event, can be important if there are many such events. An example of this is the blue color of the sky during the day, and the red color seen when the sun is on the horizon. The separation of blue light from red light is caused by differences in the light scattering strength of nitrogen and oxygen molecules in the air as a function of the wavelength of light. More specifically, blue light is scattered by these molecules more strongly than red light. Although each scattering event results in only a small change in direction of the light, the thickness of the atmosphere (almost 100 km) ensures that there is a multitude of such events.

## Summary

Microscopic particles behave in many ways quite differently than their larger counterparts. Because our everyday experiences are mostly with macroscopic particles or objects, microscopic particle behavior can be both unexpected and foreign to us. The factor that differentiates the behavior of large and small particles is the balance between particle volume and particle surface area. On an equal weight basis, microscopic particles have a much higher surface area than macroscopic particles, and as such their behaviors are dominated by surface forces whereas the behaviors of large particles are dominated by gravitational forces.

Regardless of particle composition, there are many important characteristics that are common to all the small particles that are routinely used in the coatings and plastics industries. These include particle shape, particle size and size distribution, surface charge, light interactions, and various bulk powder properties such as bulk density, bulk flow, and compressibility.[14] By understanding the source and nature of surface forces, we can understand these properties of powder samples.

[14] The measurement of these properties is detailed in Chap. 2.

## References

1. Feynman, R.P., Leighton, R.B., Sands, M.: The Feynman Lectures on Physics, Vol. I: The New Millennium Edition: Mainly Mechanics, Radiation, and Heat. Basic Books (2011)
2. Hassanpour, A., Hare, C., Pasha, M. (eds.): Powder Flow: Theory. Royal Society of Chemistry, Characterisation and Application (2019)
3. Castellanos, A.: The relationship between attractive interparticle forces and bulk behavior in dry and uncharged fine powders. Adv. Phys. **54**(4), 263 (2005)

# Chapter 2
# The Physical Properties of Small Particles

## Contents





M. Diebold et al., *Pigments, Extenders, and Particles in Surface Coatings and Plastics*, https://doi.org/10.1007/978-3-030-99083-1_2

## Introduction

There is a wide range of physical and chemical analyses by which we can characterize small particles. Each technique reveals a different aspect of the particle composition, behavior, or both. By understanding the analytical options available, the paint or plastics formulator can design experiments to elucidate particle properties that affect the end-use properties of importance in a specific product or application. In some cases, the results of an analysis can be directly related to a specific end-use property, while in other cases the results from different techniques must be combined to gain a more complete understanding of particle performance.

Most of the techniques that we will consider in this chapter are discussed in books devoted entirely to each individual method, an indicator of the high level of detail and complexity that has been developed on them. Due to space limitations, as well as to keep within the intended scope of this book, we will only briefly summarize the details of each technique, with the hope that this information is enough to give formulators an understanding of the scope and limitations of the different characterization methods available to them.

## Particle Size and Size Distribution

The defining attribute of the particles discussed in this book is their size. As detailed in Chap. 1, our interest is in particles with sizes ranging from a few tens of nanometers to roughly 20 microns. Most particle properties are sensitive to size within this range, and for this reason, it is important to characterize particle sizes as completely as possible. There are several techniques for measuring size and they do not always agree. In some cases, this is because of limitations in the assumptions that are made about the measurement technique or its interpretation, while in other cases, it is due to the very definition of what, exactly, a particle is.

### *Defining a Particle*

We have been using the term "small particle" to indicate a single object of a certain size. While this definition may appear explicit, there is in fact some ambiguity in it. This ambiguity originates from the tendency of particles to stick to one another with varying strengths of attachment. Individual particles are referred to as primary particles. Two or more primary particles that are attached to one another but can be easily separated by typical dispersion processes are referred to as agglomerates. This is typical of particles that are held together by physical attractions. By contrast, aggregates are groups of particles that are held so strongly together that they cannot be separated by particle dispersion processes. Aggregates are typically treated as if they are single, larger particles. Primary particles in aggregates are held together by chemical bonds. Aggregates are often created when particles are heated to high temperatures during their production.

## *Defining a Particle Size*

The meaning of particle size is quite simple when all particles have a regular shape (e.g., a sphere or cube). In such instances a number of distances can be used to characterize size, for example, diameter for a sphere, side or diagonal length for a cube, etc. However, except for resin particles, it is nearly always the case that particle shapes are not homogenous across the entire population of particles. Such distributions of particles are known as "polydisperse". A distribution in which all particles are of the same is known as "monodisperse".

In such cases, how do we characterize the size of the particle with a single number? There are several options available. We could, for example, measure the longest dimension (of importance if we wish to filter the material), or, if the particles are of uniform shape, a specific dimension of the particle (e.g., the length of rod-like particles). One of the more common methods is to characterize size based on the size of a sphere that would have the same properties. For example, a sphere of the same volume as the particle, or of the same area as the particle, or that would settle at the same rate as the particle, or that has the same diameter as the maximum or minimum length of the particle. It is critical that if we are measuring particle size for a specific purpose—such as determining the amount of time it would remain suspended in air, or how well it scatters light—we use the measurement parameter that is of the most relevance.

When interpreting the results of a given type of particle size measurement, it is important to keep in mind the basis by which size is defined. This is particularly true when reconciling differences in the results of different types of measurement.

## *Particle Size Distributions*

It is rare that all the particles of a given material have the same size. Instead, a distribution of sizes is common. Such a distribution typically follows a log-normal distribution. This distribution is skewed to the right when plotted on a linear X-axis but takes the form of a symmetric bell curve when plotted with a logarithmic X-axis (Fig. 2.1) [1]. This curve type is predicted theoretically for small particles that are attritted from larger particles (i.e., made from the top down) and also for particles that are grown by adding atoms or molecules to smaller (seed) crystals (i.e., made from the bottom up) [2–5].

Particle size distribution curves can be presented in two ways: as a probability density function and as a cumulative distribution function. Figure 2.2a is an example

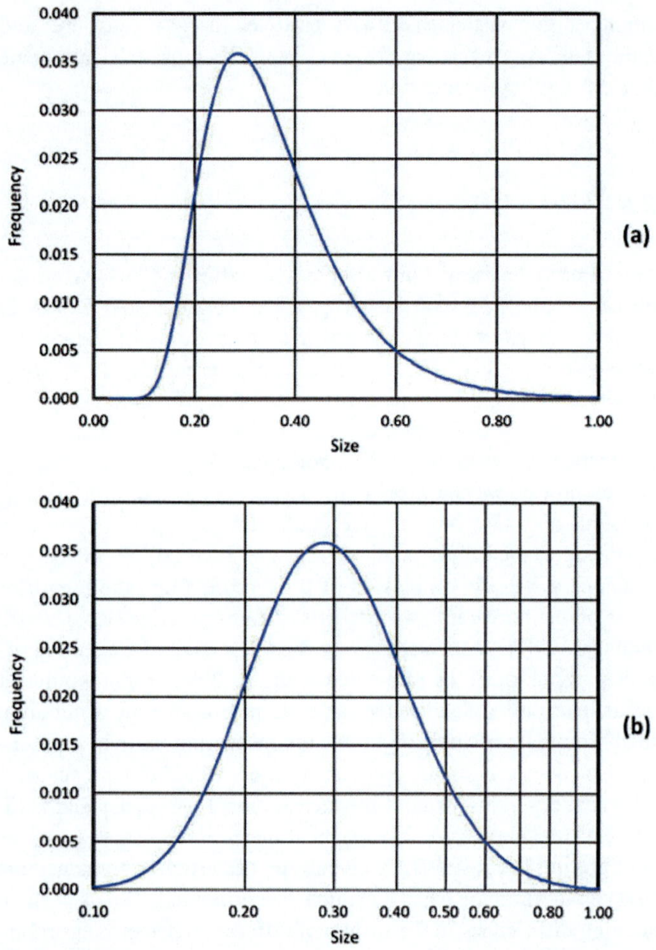


**Fig. 2.1** Log-normal distribution. **a** Plotted on a linear X-axis. **b** Plotted on a logarithmic X-axis

of the probability density function for a normal (Gaussian) distribution. Here, we find the probability of a particle with a size between $x_1$ and $x_2$ as the area under the curve between these two sizes. When presented as a cumulative distribution function (Fig. 2.2b), the y-value corresponding to an x-value is the fraction of particles with size x or lower. In this way of presenting the distribution, the probability of a particle with a size between $x_1$ and $x_2$ is the difference in y-values between these two x-values. Both ways of expressing the particle size distribution are equally valid, and the information contained within these graphs is the same. They are simply presented on a different basis.

Because larger particles can be broken up into millions of smaller particles, it is important when we describe the size characteristics of a sample of powder that we

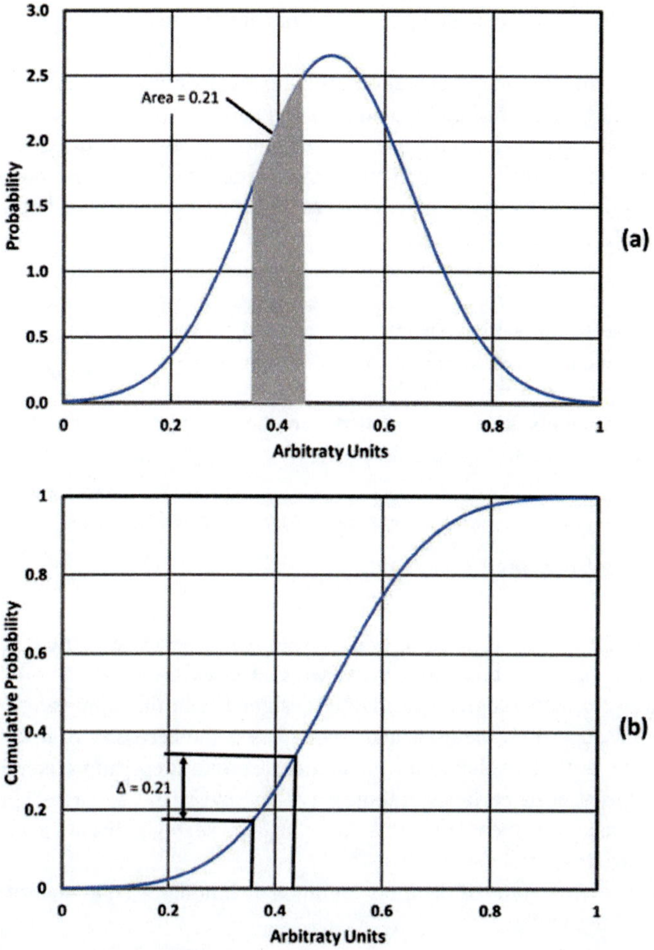


**Fig. 2.2** Different types of particle size distribution curves. **a** Probability density function. **b** Cumulative probability density function

specify whether we are making this characterization on a number basis or a weight basis. For example, if we begin with two 15 micron particles, and crush one of the particles into 3.375 million particles that are 0.1 micron in diameter, then we can describe the average particle size based on weight as being 7.55 microns (there is an equal weight of 15 micron particles and 0.1 micron particles), while the average size per particle would be a very small amount greater than 0.1 micron based on the number (there is less than 1 ppm of large particles). In most cases in the coatings and plastics industries, we choose to characterize particle size on a weight basis rather than a number basis. This is because we develop a given formulation based on the weight of the constituent particles, not on their number.

The attractive forces between particles of the dimensions typically seen in paints and plastics (0.1 micron–20 microns) are strong enough that these particles are rarely present as individuals, but rather are found bound together in agglomerates. The sizes of agglomerates are controlled by a number of factors, but as typically encountered in the coatings and plastics industries, these agglomerates are composed of tens to millions of primary particles. When measuring particle size, it is important to bear in mind that agglomerates would be counted as a single large particle if left intact. For this reason, the sample is typically sheared prior to measurement, with the intention that this will disrupt, at least temporarily, the agglomerates. Ideally, the shear intensity will match that of the process used to manufacture the paint or plastic so that the particle size distribution as measured will match the distribution in use.

Most particle size measurements are done in a slurry, and dispersion is accomplished through one of the methods outlined in Chap. 11. In the case of in-process slurries or final paints, no additional dispersion beyond that used to make the slurry or paint is used for particle size testing.

## Analytic Techniques

Several distinct technologies have been developed to measure particle size [6–8]. Some are technologies that already existed and could be used for this purpose, while others are technologies specifically designed with this application in mind. A common aspect of all methods and techniques is the necessity of testing a truly representative sample of the larger population. Certain preparation techniques can inadvertently segregate particles in some way—by size, shape, electrical charge, etc., and it is incumbent on the analyst to assure that such segregations are avoided (or at least minimized).

Below we review some of the more common techniques used to quantify particle size and size distribution.

### Microscopy

Perhaps the most obvious means of characterizing particle size is to image the particles of interest and physically determine the dimensions of a large number of particles from these images. Computer algorithms are available to measure these dimensions both rapidly and accurately. Optical microscopes can be used on particles at the large end of the range of our interest, but in most cases, electron microscopes are needed for this type of analysis (these instruments are covered in greater detail in a later section of this chapter).

While there is much to be said of the old adages that a picture is worth a thousand words and that seeing is believing, there are several potential sources of error in these measurements. As discussed above, sampling inhomogeneity, which is always a concern, is especially relevant to microscopy, for three reasons. First, the deposition of the particles onto the microscope grid can result in the preferential placement of particles based on their size. Second, we are looking at an infinitesimally small number of particles compared to the number of particles present in even 1 gram of material, and even sampling thousands of particles may not give a true representation of particle variability. Finally, there is the concern that irregularly shaped particles show a preferential orientation when placed on the microscope grid, potentially accentuating certain particle directions while de-emphasizing others.

## X-Ray Line Broadening

Solids can be classified as crystalline or amorphous depending on both the short-range and long-range order of their constituent atoms. In a crystalline solid, the atoms are arranged in a regular, repeating pattern. Light of the appropriate wavelength (roughly the atomic scale, meaning light in the X-ray range), when directed at a crystalline material, will be scattered by the atoms within it. Since these atoms are in a repeating arrangement, under certain conditions, which depend on the wavelength of light, the direction of the light and the distances between repeating atoms in the crystal, the light scattered by different crystalline elements will constructively interfere, resulting in a strong intensity of the light in specific directions. This scattering is referred to as X-ray diffraction.

When crystal sizes are small the number of repeating units is limited, and the implicit assumption in this process—that the atoms repeat infinitely in all directions—breaks down. This leads to a broadening of the X-ray beams that scatter from the crystal. By measuring the breadth of the beams, one can calculate the size of the crystal [9]. This technique can be applied to crystals of sizes up to about 0.20 microns.

There are several limitations to this technique. The first is that it can be applied only to crystalline materials. It also assumes that the crystals are perfect—any imperfections or deviations, for instance, due to the partial replacement of one component (such as sodium) by another (such as potassium), or due to any stress on the crystal itself, will also broaden the beam. Finally, what we may consider a single or primary particle may in fact be made of several individual crystalline domains that are not aligned with one another. For example, $TiO_2$ particles typically consist of twins or higher-order intergrowths (Fig. 2.3).

## Light Scattering

Perhaps the most common method of determining the size of particles of interest to us is the use of light scattering techniques. This is because particles in this size domain typically scatter light strongly, which is often the very reason that these size domains are used. There are two broad classes of light scattering technologies used for this purpose: dynamic light scattering and static light scattering.

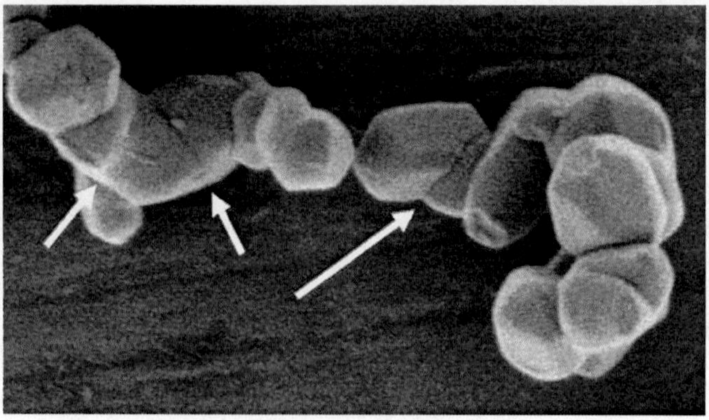

**Fig. 2.3** Representative twinning of $TiO_2$ particles (curtesy of E. Najafi and H. Huezo, Chemours)

Dynamic light scattering (also known as photon correlation spectroscopy) relies on the principles of Brownian motion [10]. In this measurement, laser light is directed at a suspension of the particles of interest. Light scattered by different particles will combine constructively or destructively with light scattered by other particles, modifying the total scattering intensity. By monitoring intensity over short periods of time (microseconds), the velocities of the particles can be determined. Particle sizes can then be calculated based on the measured velocities.

This technique assumes that the light is scattered equally in all directions (Rayleigh scattering), which is only true for particle sizes that are significantly shorter than the wavelength of light. This creates an upper particle size limit for this technique of roughly one-half microns, restricting its application primarily to nanoparticles. The size reported is the hydrodynamic diameter of the particle, which is the size of a sphere that will diffuse through the solvent in the same way as the particle.

In the static light scattering measurement, particle sizes from a few tenths of a micron to as large as a few millimeters are characterized by measuring the degree to which a focused laser beam is deflected when passed through a dilute suspension of particles. In this particle size regime, light scattering occurs by a different mechanism (Mie scattering—see Chap. 3) than in the regime probed by dynamic light scattering. In Mie scattering, light is deflected by a certain angle, rather than being completely randomized as in Rayleigh scattering. The intensity of light as a function of deflection angle is used to calculate the particle size distribution. The resolution of static light scattering measurements is dependent in part on the distance between the sample and detector, since diverging light beams spread over distance, and so these instruments tend to be larger than those based on other principles.

An important aspect of both of these types of light scattering-based techniques is that the suspensions probed must be very dilute (e.g., 0.1% by weight or less). This is because these techniques rely on the light being scattered only once. At higher particle concentrations, the degree of multiple light scattering, which interferes with the interpretation of the scattering results, is unacceptable.

### Acoustic Spectroscopy

A major drawback of the light scattering technologies discussed above is that they can be done only on dilute slurries—much more dilute than a typical paint. This restriction does not apply to acoustic spectroscopy, which is fully capable of measuring particle sizes in concentrated systems (50% or more by volume) [11–13]. An additional benefit of this technique is that dynamic processes such as flocculation or dispersion can be followed over time.

The basis of this method is that sound waves are attenuated when they pass through a dispersion of particles in a liquid. The exact form of this attenuation, and its frequency dependence, is determined by a number of factors that include the particle size distribution. This technique can measure particle sizes from roughly tens of nanometers to millimeters.

### Sedimentation

Sedimentation techniques rely on the separation of particles based on the rates at which they settle in a liquid. Several factors control this rate, such as particle and solvent densities, viscosity, etc., as shown in Stokes' law (Eq. 2.1), but when these are controlled, sedimentation rates are determined by particle size.

$$v = \frac{2}{9\eta}(\rho_p - \rho_l)gr^2 \tag{2.1}$$

where:

$v$ is the sedimentation velocity,

$\eta$ is the viscosity of the liquid,

$\rho_p$ is the density of the particle,

$\rho_l$ is the density of the liquid,

g is the external force (either gravitational or centrifugal), and

r is the particle diameter.

Sedimentation occurs when particles are set in motion by an external force. In theory, gravity can provide this force, but, for particles in the size ranges of interest to us, the time needed for separation by gravity can be too long to be practical. Instead, slurries or suspensions are typically rotated at high speeds and so separated by centrifugal forces [14]. In some instruments, the particles are collected based on size, and then the amount of each size is determined [15]. In others, the separated particles are passed through an X-ray source and detector [16]. The degree of X-ray intensity attenuation is determined by the concentration of the particles.

**Table 2.1** Size Comparisons for Hegman Gauge

| Hegman units | Microns | Mils |
|---|---|---|
| 0 | 101.6 | 4.0 |
| 1 | 88.9 | 3.5 |
| 2 | 76.2 | 3.0 |
| 3 | 63.5 | 2.5 |
| 4 | 50.8 | 2.0 |
| 5 | 38.1 | 1.5 |
| 6 | 25.4 | 1.0 |
| 7 | 12.7 | 0.5 |
| 8 | 0 | 0.0 |

## Large Aggregates

Our interest is often in the size of particle aggregates rather than the size of the individual particles themselves. This is particularly true when evaluating the degree to which particles are separated after a dispersion process (see Chaps. 11 and 12), where aggregate size can be as high as several hundred microns. The method used to quantify these large aggregates is dependent on the end-use application (coatings or plastics).

To determine the aggregate content of a liquid paint (or of a particle slurry used to make a liquid paint), the material of interest is placed on a metal block into which a tapered channel has been machined. A metal bar ("doctor blade") is then used to draw the paint down through the channel, from the deep end of the groove to the narrow end (which is flush with the block surface). The channel is characterized by its width, its length, and its maximum depth, and can be customized to a specific paint. The most common block is the "Hegman gauge", which has a maximum channel depth of 100 microns (4 mils[1]).

The drawdown is examined by eye under a bright light or automatically with a camera. The surface should be smooth and glossy at the deep end of the channel, but then will appear rough at a certain depth. This is known as the fineness line for the dispersion and represents the size of the majority of the aggregates remaining in the slurry or paint grind. When large ("oversize") particles or aggregates are present, they cause a line in the drawdown due to being dragged by the doctor's blade. The size of the aggregate can be estimated based on the initial position of this line in the channel. The side of the channel typically has markings indicating depth, in units of microns, mils, or "Hegman Units", as indicated in Table 2.1.

There are two situations in which a drawdown is used. The first is to test the quality of a finished paint or slurry during production. It is often the case that the degree of dispersion of a paint batch is monitored over time, to determine when the dispersion process is complete and the paint is ready to proceed to the next stage of production. The second is to intentionally use less energy than required for complete

[1] A mil is one-thousandth of an inch.

grind ("low shear dispersion"). Here, the intent is to determine the ease or speed of dispersion. This analysis is typically done to characterize the raw materials of the paint, rather than the paint itself, and is often a product quality release test for, for example, titanium dioxide pigments.

Undispersed aggregates present in plastics can be quantified using a variety of techniques. A common technique for plastic film application is to illuminate a semi-translucent film from behind [17]. Undispersed matter is highlighted as a dark spot on the film using either the human eye or, more commonly, optical cameras. The dark spots are counted for a defined weight or area of film. This highly relied upon technique has the drawback of being incapable of identifying the composition of the dark spot. The spot could be gels of polymer or undispersed solids.

A second method is to extrude a molten plastic that contains particles through a series of screens, referred to as a "screen pack". The layer of screens is often made with a gradation of mesh opening sizes. After a predetermined amount of polymer melt has been passed through the screen pack, the screens are removed, the plastic residue is removed and the amount of trapped solid material is determined with a variety of analytical techniques. Typically, X-ray fluorescence techniques are used to identify the elemental components of the retained solid. The intensity of the fluorescence is calibrated with standards to provide a quantitative value of retained matter from the sample.

An alternative is to extrude the melt through the screen pack while monitoring pressure [18]. This extrusion technique provides a "filter pressure value" which is the pressure differential when extruding resin through the screen pack before and after the sample has been processed. If desired, the trapped material on the screen pack can be examined under a microscope to determine the chemical composition of the aggregated particles.

## Undertone of $TiO_2$ Pigment

While it is not possible to see individual $TiO_2$ particles by eye, there are perceptible differences in the appearance of $TiO_2$ pigment based on particle size. This is because the efficiency of visible light scattering by a $TiO_2$ particle is partly determined by the size of the particle. Slightly smaller particles preferentially scatter shorter wavelength light (blue), while slightly larger particles scatter all wavelengths relatively evenly (neutral).

To test for $TiO_2$ undertone, the $TiO_2$ pigment of interest and a black pigment (typically carbon black or black iron oxide) are mulled together in mineral oil, and the color is determined. Alternatively, for plastics applications, films containing the two types of pigments can be cast. Although subtle, the differences in appearance can be readily measured and can be seen by eye in side-by-side comparisons. An

**Fig. 2.4** Colors of carbon black/$TiO_2$ mixtures in plastic films. **a** 0.23 micron mean particle size. **b** 0.26 micron mean particle size

example of the colors for plastics films of $TiO_2$ grades at the two ends of the size range used in pigments (means sizes of 0.23 and 0.26 microns) is shown in Fig. 2.4.[2]

## Gas Absorption Measurements

Much can be learned about particle surfaces by measuring their interactions with other molecules. Of particular interest are their interactions with inert gas molecules [19],[3] which are discussed in this section, and their interactions in water suspensions with dissolved ions, which are discussed later in the section on surface charge.

There is generally an energetically favorable interaction between surfaces and the gas molecules to which they are exposed, as this leads to a reduction in surface energy. As a result, gases generally adsorb onto surfaces. The amount of gas absorbed and the strength of this adsorption are functions of the concentration of the gas (gas

[2] It is remarkable that we can visually distinguish particle size differences of 30 nm. By contrast, most people are unable to visually detect the size difference of a colored billiard ball (57 mm) and the white cue ball (63.5 mm)—a difference that is over two million times greater. One can even more easily detect different colloidal gold particle sizes of the similar magnitude as the difference in $TiO_2$ particle size, since the wavelength of light most strongly scattered by a colloidal gold particle is quite sensitive to particle size. This remarkable achievement is possible because these particle sizes are matched to visible light, and because we are capable of discerning subtle changes in color (as discussed in Chapter 5, we can distinguish roughly 10 million colors).

[3] For simplicity we will refer in this section to the adsorbing species as molecules, but atomic gases, such as helium, argon or krypton, absorb onto surfaces in the same way.

pressure is usually used as a proxy for this), the temperature of this exposure, the area occupied by a single gas molecule, the surface area, and the surface energy (gas molecules attach more strongly to higher energy surfaces). We can therefore calculate surface area and surface energy if we measure the pressure of a set amount of gas that is exposed to the particles, the temperature, the weight of the sample, and use standard values for the area occupied by a gas molecule (i.e., the adsorption footprint). In addition, when 50 nm or smaller pores are present, their size and volume can be calculated.

In the typical gas adsorption experiment, a known quantity of the powder of interest is heated under a vacuum to remove any initially adsorbed species. The sample is then cooled (typically to liquid nitrogen temperatures) and known quantities of a gas (typically nitrogen) are introduced step-wise into the sample vessel. Between each dose, the gas is allowed to equilibrate with the surface and the gas pressure is measured. The number of non-adsorbed gas molecules can be calculated from the measured gas pressure using the ideal gas law and, by difference with the amount of gas added to the vessel, the number of adsorbed molecules can be determined.

Ideally, a single layer of molecules will adsorb onto the surface (Langmuir adsorption). However, there are two related factors that complicate this process. The first is that the gas molecules can adsorb onto already adsorbed molecules (i.e., the adsorbed molecules are themselves adsorption sites). This is very similar to the formation of the liquid phase of the gas, in the sense that the interactions between the already adsorbed molecules and the gaseous molecules are comparable to the interactions between condensed molecules and their counterparts in the gaseous state.

The second complication is that surfaces are not completely flat. Surface features such as edges and corners tend to have higher surface energies than the flat areas of the particle and so the gas molecules will preferentially adsorb onto them. However, of greater importance is the presence of pores within some particles, and the pores between particles. The adsorbing gas molecules can condense in these pores, increasing the uptake of the particles. While this complicates our interpretation of surface area, it is in some ways beneficial because it provides us with a means to calculate the volumes of pores as a function of their size over a certain size range.

As a result of these complications, the isotherms for adsorbed molecules can deviate greatly from an ideal curve. Six different adsorption types have been classified by the IUPAC. These are shown in Fig. 2.5. Note that, in some instances, the curves show a hysteresis—that is, surface coverage as a function of gas pressure is different when we are adding molecules to the gas phase (i.e., increasing pressure) than when we are removing molecules from the gas phase.

## *Surface Area*

For most particles, the results of the gas adsorption measurement can be modeled using the "BET" equations, so named using the initials of its three developers (Brunauer, Emmett and Teller) [20]. These equations apply when multiple layers

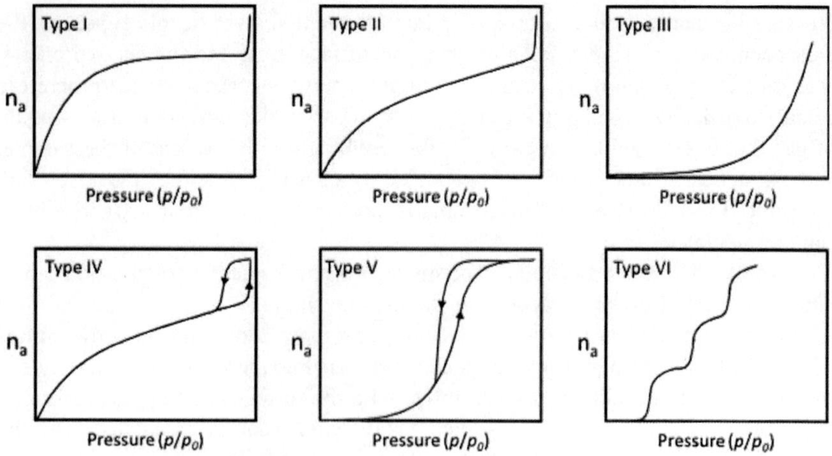


**Fig. 2.5** Shapes of the six types of adsorption isotherms. $n_a$ is the number of adsorbed molecules

of gas molecules adsorb onto the particle surface (Fig. 2.6). In this situation, there are two energies of importance—the energy released when a gas molecule attaches to the particle surface, and the energy released when a gas molecule attaches to an already present molecule. The energy associated with the latter interactions is approximately the heat of condensation for that molecule. The energy of interaction between the gas and the surface is a function of the chemical identities of the gas

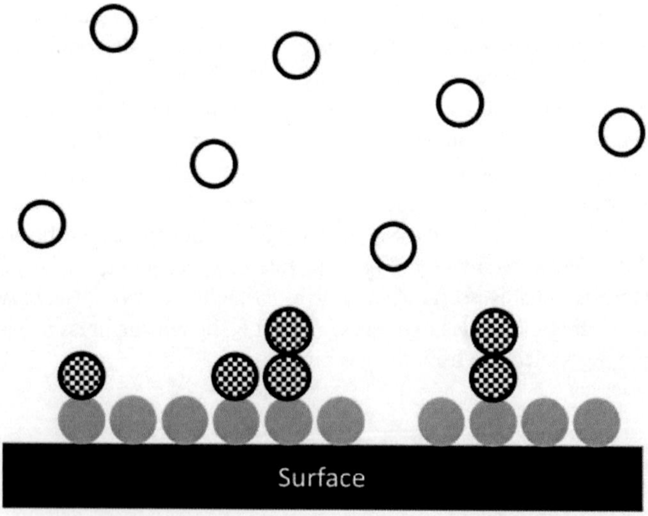


**Fig. 2.6** Interactions between gas molecules and particle surfaces. Open circles are gaseous molecules; gray circles are molecules adsorbed onto the particle surface; patterned circles are molecules condensed onto one another and the adsorbed molecules

and particle as well as a function of surface energy—the higher the surface energy, the greater the energy of this interaction.

The fractional surface coverage θ can be found based on the pressure measurements and a constant denoted as "c" by using Eq. (2.2):

$$\theta = \frac{cp}{(p_0 - p)(1 + (c - 1)(p - 1p_0))} \tag{2.2}$$

where:

θ is the fraction of surface covered by one or more molecules (θ = 1 for monolayer coverage),

p is the measured (equilibrium) pressure,

$p_0$ is the saturated pressure, which is the vapor pressure of the condensed (liquid) molecules at the adsorption temperature, and

c is a unitless constant.

In this equation and elsewhere, $P/P_0$ is referred to as the relative pressure of the gas.

This relationship is often expressed using Eq. (2.3):

$$\frac{1}{\nu_a[(p_0 - p) - 1]} = \frac{(c - 1)}{\nu_m}(p_0 - p) + \frac{1}{\nu_m c} \tag{2.3}$$

where:

$\nu_a$ is the amount of gas adsorbed,

$\nu_m$ is the amount of gas in one monolayer, and

p, $p_0$, and c are as in Eq. (2.2).

Note that $\nu_a/\nu_m$ from Eq. (2.3) equals the θ term in Eq. (2.2).

The advantage of Eq. (2.3) is that plotting $\frac{1}{\nu_a[(p_0-p)-1]} = \frac{(c-1)}{\nu_m}(p_0 - p) + \frac{1}{\nu_m c}$ against $\nu_m = \frac{1}{Sl+Int}$ gives a straight line from which the monolayer gas amount ($\nu_m$) and the constant c can be calculated by using the slope of the line (Sl) and intercept (Int) using Eqs. (2.4) and (2.5):

$$\nu_m = \frac{1}{Sl + Int} \tag{2.4}$$

$$c = 1 + \frac{Sl}{Int} \tag{2.5}$$

Caution must be exercised when interpreting the measured surface area values for different grades of pigmentary $TiO_2$. All $TiO_2$ pigment particles used in paints, and many used in plastics, are surface treated with a few percent alumina and/or silica, in order to improve end-use performance properties (this is discussed in detail in Chap. 7). Importantly, the alumina is present as small (nano) particles affixed to the pigment surface. Although the size of the $TiO_2$ particles themselves are essentially the same in all pigment grades, the surface areas of these materials can vary from roughly 10 $m^2/g$ to 50 $m^2/g$, depending on the surface treatment. For comparison, the surface area of pure $TiO_2$ particles before surface treatment is roughly 6–7 $m^2/g$.

The BET technique overestimates the surface areas of particles that contain micropores (smaller than 2 nm). This is because these pores fill with condensed gas at the same pressures as the gas molecules that coat the flat surfaces of the particles, and so the information about surface adsorption is convoluted with the information about pore filling. It is difficult to separate the volume of gas that is condensed in these pores from the volume adsorbed on the particle surfaces, although some methods to do this have been proposed [21].

## *Pore Characterization*

There are two fundamentally different types of pores that can exist within a powder. The first, intra-particle pores, are found within the individual particles. An excellent example of this is the pores found in particles of diatomaceous earth (Fig. 2.7). Diatomaceous earth is composed of fossilized skeletons of diatoms (microscopic sea organisms). These particles have a nearly endless variety of shapes and sizes (up

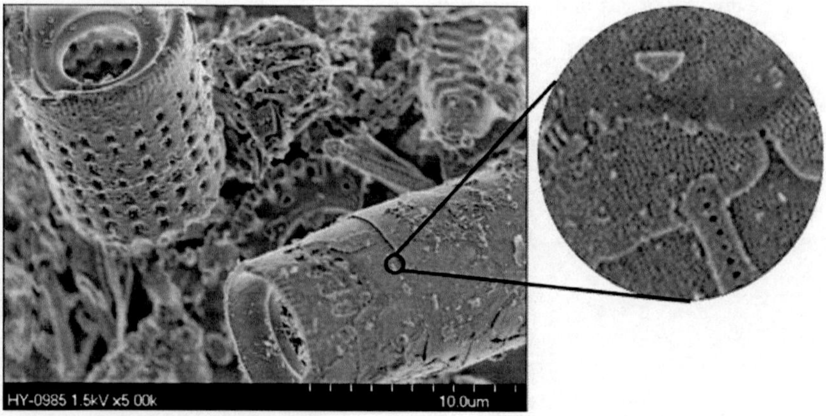


**Fig. 2.7** Electron micrograph of diatomaceous earth sample showing pores with a wide range of diameters

**Table 2.2** IUPAC pore size definitions [22]

| Designation | Pore size | Approximate pressure range ($p/p_o$) |
|---|---|---|
| Micropore | pore < 2 nm | <0.15 |
| Mesopore | 2 nm < pore < 50 nm | 0.15–1.0 |
| Macropore | pore > 50 nm | >0.35 |

to approximately 1 mm), and nearly all have an extensive pore content with sizes varying from a few nanometers to a few microns or more.

The second type of pores is those that exist between particles. These pores are significantly larger than the intra-particle pores and are strongly affected both by the same factors that affect both bulk density and by the measurement technique itself. That is, the inter-particle pore structure is affected by the arrangement of the particles in the sample.

Intra-particle pore sizes have been categorized by the IUPAC into three ranges, as shown in Table 2.2.

This classification is based in part on the practical consideration of pore size detection limits for different pore size techniques. Pore size distributions are presented in the same manner as particle size distributions (we can, for this purpose, consider pores to be air "particles" within a solid matrix). These presentation methods are discussed above in the section on particle size and size distribution.

A number of different techniques have been developed to measure the pore size distribution of materials [23, 24]. These methods fall into two categories. The first is high-pressure infiltration of a non-wetting liquid (typically mercury) into the pores. A non-wetting liquid is used because it will only penetrate a pore when forced to do so using pressure. For idealized cylindrical pores, the pressure required to fill a pore is determined by Washburn's equation (Eq. 2.6) and is inversely proportional to the pore diameter [25]. By monitoring the amount of liquid filling the pores as a function of pressure, one can construct a pore size distribution of the material.

$$p = \frac{-4\gamma\cos\theta}{D_p} \tag{2.6}$$

where:

$\gamma$ = surface tension,

$\theta$ = contact angle (typically 180° for a non-wetting liquid),

p = pressure, and

$D_p$ = pore diameter.

The minimum pore size detectable by this method is dependent on the maximum pressure that can be applied, but in most commercial instruments is in the range from 3 to 6 nm (i.e., near the lower limit of the definition of a mesopore).

Micropores and mesopores can be measured using the second category of techniques—inert gas adsorption and desorption. These techniques are an extension of those used to determine surface area, and commercial instruments are typically capable of both types of measurement. In this measurement, the sample is evacuated and heated, then placed under a high vacuum and the adsorbing gas is added, similar to the BET measurement. However, for pore size analysis, the pressure is first increased until all the pores are filled. Once the sample has equilibrated, the pressure is decreased stepwise. It is the desorption data that is then used in the pore distribution calculation (as shown in Fig. 2.5, the adsorption and desorption curves deviate when pores are present).

The distinguishing factor between micropores and mesopores is their size in relation to the adsorbing molecules (Fig. 2.8). In this figure, all molecules are chemically identical, but those within the pore that are not in direct contact with a surface and are close enough to be perturbed by it, are shown in white. In mesopores, the pore diameter is large compared to the adsorbing molecules, and so the majority of molecules

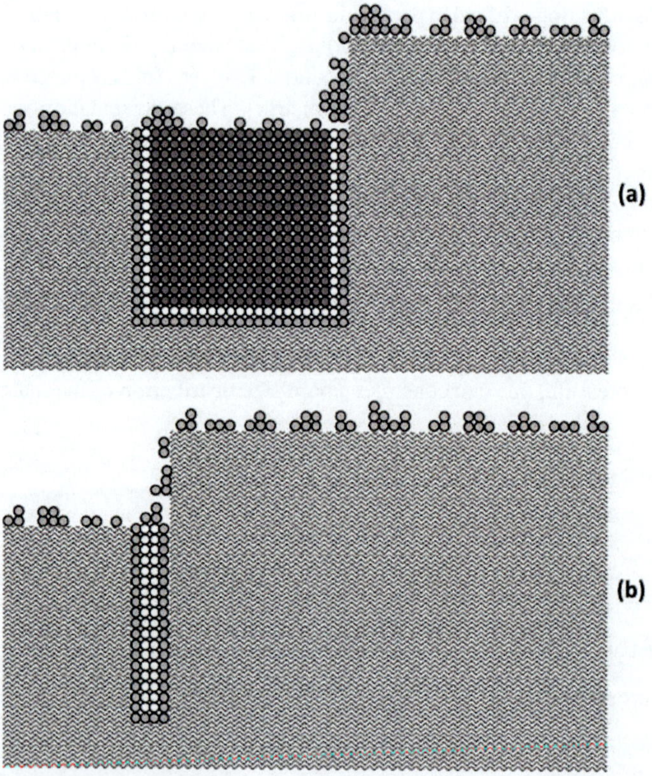

**Fig. 2.8** Filling of pores by an adsorbent. **a** Mesopores. **b** Micropores. Circles are adsorbing molecules. White circles are molecules perturbed by a surface, but not in contact with it. Dark circles are molecules in an environment similar to its liquid state

found within the pores are in an environment similar to that of the molecules in the liquid state (i.e, the proportion of molecules attached to a surface or strongly perturbed by it is low). In Fig. 2.8a, these molecules are indicated by the darkest circles. We will refer to the collection of these molecules as the condensed phase. By contrast, although not all the molecules in a micropore are attached to a surface, those that are not are still close enough that their environment is significantly perturbed by the surface in comparison to the environment in the liquid state, so there are no condensed phase molecules in these pores (Fig. 2.8b).

The most common mathematical model for mesopore size determination is the BJH method, so named using the initials of its three developers (Barret, Joyner and Halenda) [26]. This model assumes (a) that gas molecules in the condensed phase within a filled mesopore volatilize sharply at a certain pressure that is dependent on the pore size diameter and (b) that the molecules adsorbed onto the walls of the pore or near enough to the wall to be perturbed by it are desorbed over a range of pressures rather than all at one pressure.

A different model is needed for micropores, since most of the molecules occupying these pores are perturbed by the pore surface, and only a small fraction, if any, can be considered to be in the condensed phase. In this case, density functional theory (DFT) is used to calculate the pore size distribution [27–29]. This theory, which is based on statistical mechanics, bridges the gap between the macroscopic and molecular scales. Here, the equilibrium particle density is determined based on the direct interactions between the surfaces and the gas molecules.

## Surface Energy, Contact Angle, and Wettability

Particles interact with their environments through their surfaces, and so surface properties are very important with regard to both the ease with which the particles will be wetted by a liquid (solvent or water for paints, molten polymer for plastic) and with which they can be subsequently dispersed in a milling step. This aspect of particles is determined by their surface energy.

As described in Chap. 1, surface energy is the energy required to disrupt the interactions between atoms or molecules within the bulk of a solid when the solid is broken and a new surface is created. We can group surface energies into two classes based on both their magnitude and the type of solid. It is relatively easy to disrupt solids that are held together through weak forces, such as van der Waals forces, and as such, they are considered low surface energy materials. These include most paints and plastics, which are composed of organic polymer molecules that are held together by relatively weak intermolecular forces.

On the other hand, solids that are held together by networks of bonds, which includes most of the particles found in paints and plastics,[4] require high energies to

[4] The exceptions are resin particles in paints, which are organic and so are low surface energy materials.

disrupt, and as such these materials are high surface energy materials. An organic surface treatment is often applied to some types of particles (e.g., pigmentary $TiO_2$ or calcium carbonate) in part to decrease the surface energy and favorably change certain particle properties.

Broadly speaking, we can say that atoms and molecules can lower their energy by associating with other atoms or molecules. In the bulk of a solid or liquid, this interaction is with other atoms or molecules of the same type. However, the surface atoms and molecules can interact with any molecules[5] with which they come into contact, particularly gas molecules. Therefore, we see that the energy required to break a particle into two (i.e., the surface energy of a solid or liquid[6]) is a special case of a more general type of interaction—that between molecules of any sort.

We can characterize the energetics of a surface based on how that surface interacts with different molecules or atoms—not just other molecules or atoms of the same type. We would expect to see different degrees of interaction, which can be quantified based on the energy released on adsorption when different atoms or molecules are the interacting species.

For example, one method to quantify the surface energy of a particle is by determining the interactions between its surface and a gas molecule. Returning to the previous section, we noted that the c-value in Eq. (2.2) is a function of surface energy. More precisely, it is proportional to the exponential of the difference in heat of adsorption for the first layer on the surface to the heat of condensation of the adsorbing molecule (i.e., the heat released when the gas condenses to a liquid), as shown in Eq. (2.7).

$$c \propto e^{\frac{q_a - q_L}{RT}} \tag{2.7}$$

where:

c is the constant in Eq. (2.2),

$q_a$ is the heat of absorption for the first layer of gas molecules (gray circles in Fig. 2.6),

$q_L$ is the heat of condensation of the gas molecules,

R is the gas law constant, and

T is the absolute temperature.

The quantity $q_L$ is determined in part by the surface energy of the particle, and so c is related directly to particle surface energy. In this case, c tells us about the energetics of the interaction between the surface and the adsorbing atom or molecule.

Of greater interest to the paint or plastics producer is the degree to which particle surfaces interact with the liquid molecules used in that specific paint or plastics

[5] Following the practice of the last section, we will refer here to the interacting species as molecules, but such species can be atoms as well.

[6] The surface energy of a liquid is sometimes referred to as its surface tension.

manufacture. In this case, the situation is significantly more complex than when considering only the interactions between a surface and a gas. Here, we must consider the strength of the interactions between the surface and the liquid molecules as well as the strength of interactions between the liquid molecules with one another (this is a function of the surface tension of the liquid, which is surface energy per unit area).

This latter consideration is an important one. When we immerse a particle into a liquid we are, in effect, creating a liquid surface in a manner similar to when we create a solid surface by breaking a solid particle apart. In this case, however, the new surface is not in contact with the atmosphere, but rather is in contact with the particle surface. The criterion that determines the ease with which a particle will immerse in a liquid is not the absolute strength of the interaction between the liquid and surface particles, but rather it is the difference in interaction strengths between this interaction and the interactions between the liquid molecules with one another.

A convenient way to measure the interactions between a surface and a liquid is through contact angle [30]. The most common technique for measuring this is to place a drop of the liquid on the surface and optically measure the angle at the air/surface/liquid interface (Fig. 2.9). This is normally done using large surfaces (on the order of several $mm^2$; for example, a paint film), rather than particles, although this measurement is sometimes done on pressed pellets. In this case, care must be taken to ensure that the pellet surface is completely smooth because surface roughness can alter the apparent angle at the interface. Powders can also be measured by penetration of the test liquid into a sample of uncompressed powder, as discussed below.

Contact angle $\theta$[7] is determined by the relative interaction energies between the liquid and solid ($\gamma_{ls}$) compared to the interactions between the liquid molecules with one another (i.e., the surface tension or surface energy of the liquid, $\gamma_l$), and the surface energy of the solid ($\gamma_s$).[8] This relationship is given by Young's equation:

$$\gamma_s = \gamma_{ls} + \gamma_l \cdot cos\theta \tag{2.8}$$

When the liquid is water, we refer to $\theta$ as the water contact angle. If $\theta > 90°$, the surface is hydrophobic; if $\theta < 90°$, the surface is hydrophilic, and if $\theta = 0°$ the surface is completely wetted.

There is a complication to this analysis that arises when measuring contact angle on a powder pellet, even a smooth one, which is that there can be a difference in advancing and receding contact angles (this complication is termed contact angle hysteresis). This difference can be quantified in a dynamic contact angle measurement. In this test, a droplet is applied to the surface of interest. Additional liquid is then injected into this droplet, causing it to grow laterally—that is, the intersection between air, droplet, and solid surface is pushed outward. The angle measured at this

[7] This should not be confused with the surface coverage in Eq. 2. Equation 2, on the one hand, and Eqs. 8, 9, and 11, on the other, use the symbol $\theta$ in a very different way.

[8] Properly speaking, we must specify the materials on both sides of a surface when discussing surface area. In situations of interest to us, the surface interface with a liquid or solid is air (or gas), and the subscripts for the symbols for liquid and solid surface energies will include the letter "g"—e.g., $\gamma_{lg}$ and $\gamma_{sg}$.

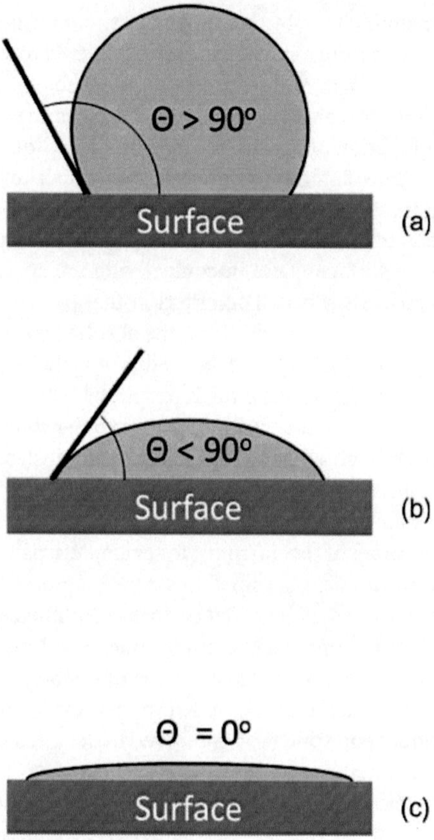


**Fig. 2.9** Contact angle measurements. **a** Poor wetting. **b** Good wetting. **c** Complete wetting

moving interface is the advancing contact angle. Similarly, liquid can be pulled out of the droplet and the receding contact angle measured.

The degree of contact angle hysteresis is dependent on a number of factors, including the liquid used, the chemical homogeneity of the particle surfaces, and, importantly, any roughness at the pellet surface, if present. This situation is very complex and the subject of much active research.

An alternative means of measuring the contact angle of a particle is to measure the penetration of a liquid into a particle bed using the Washburn capillary rise technique [31]. In this test, a powder is loaded into a tube with a porous bottom and the tube is then immersed in the wetting liquid. The height of the rising liquid front is then measured as a function of time (Fig. 2.10). In practice, the amount of infiltrating liquid is most often determined by weight gain of the column, which is attached to a tensiometer for mass change measurements. The contact angle can then be calculated by Eqs. (2.9) or (2.10). Equation (2.9) is often rearranged into Eq. (2.11), with contact angle being determined from the slope of the line relating the square of height to immersion time.

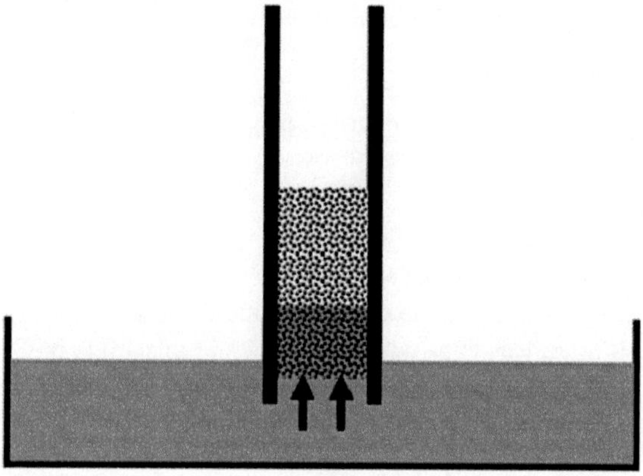

**Fig. 2.10** Infiltration of a liquid into a powder column

$$cos\theta = \frac{\eta}{C\rho^2\gamma_l} \cdot \frac{m^2}{t} \tag{2.9}$$

where:

C is a material constant that is calculated as shown in Eq. (2.10),

$\rho$ is the liquid density,

$\gamma_l$ is the surface energy (tension) of the liquid,

m is the mass of the liquid that has penetrated the powder, and

t is time.

$$C = \frac{r_{eff} \cdot A^2 \cdot \varepsilon^2}{2} \tag{2.10}$$

where:

$r_{eff}$ is the effective radius of the pores,

A is the cross-sectional area of the tube containing the powder, and

$\varepsilon$ is the porosity of the powder within the tube.

Alternatively:

$$h^2 = \frac{r_{eff} \cdot \gamma_l \cdot cos\theta}{2\eta} t \tag{2.11}$$

where:

h is the height of the penetrating liquid,

$r_{eff}$ is the effective radius of the pores,

$\gamma_l$ is the surface energy (tension) of the liquid,

$\eta$ is the viscosity of the liquid, and

t is time.

The constant "C" in Eq. (2.9)[9] can be measured, rather than calculated, by performing this test using a low surface energy liquid such as hexane, which can be assumed to have a contact angle of zero. Equation (2.9) is then rearranged to calculate C, since all other variables will be known, and this value of C is then used for other liquids.

The importance of contact angle for powders is that it is related to the wetting dynamics of the powder. As discussed in Chap. 11, an important aspect of the dispersion of particles into the liquid carrier of a paint is the ease with which the surface wets. During wetting the solid/air interfaces of the particles are replaced by solid/liquid interfaces. Wetting can be enhanced by adding a wetting agent, or surfactant, to the liquid. This reduces the surface energy, or tension, of the liquid. Referring back to Eq. (2.8), we see that a decrease in surface tension ($\gamma_l$) is offset by an increase in the cosine of the contact angle, which means a decrease in the contact angle.

## Bulk Density, Bulk Flow, and Powder Compressibility

The bulk behavior of particles is determined by the degree of attraction that the particles have towards one another [32]. In the extreme, these attractions can cause particles to irreversibly stick to one another, but even at lower levels, the degree of attraction has a significant impact on bulk properties because this attraction causes resistance to particle motion. When particles that are near one another, but not touching, move away from one another, their separation distance increases, which requires energy. For particles that are touching one another, or touching the walls of a bin or container, movement is impeded by friction, which, in turn, is controlled by the strength of attraction between particles and one another or with the bin walls.

The degree to which particles can flow past one another affects three properties of the powder: the density at which it settles, the degree to which the entire powder can flow, and the extent to which a powder compresses when placed under pressure. We will consider the measurement of each of these bulk properties, but will first discuss aspects of particle packing.

[9] This should not be confused with the constant c in the gas adsorption equations.

## *Particle Packing*

Unlike true density, the loose bulk density of powder samples is highly dependent on the measurement technique, the sample preparation technique, and the stress state of the sample. Some sample preparation procedures are designed to minimize bulk density (e.g., gentle sifting of the powder into a measurement container) and therefore give a lower limit for bulk density. Other procedures densify the pigment in a known and (hopefully) reproducible manner. There are two common methods to do this. The first is tapping a filled container on the countertop until the occupied volume no longer changes. This is known as the "tapped" or "tamped" bulk density of the material. The second is to press the powder into a pellet, which gives the ultimate packed or compressed bulk density of the material.

The prevalence of voids in collections of small particles, such as those introduced into the chamber of a pellet press, leads to the concept of powder compressibility. If we place the piston in the press and apply pressure with a hydraulic ram, the powder bed compresses. In doing so, we add work energy to the system (in the form of the force applied by the ram multiplied by the distance the piston moves). This energy is used to overcome inter-particle forces (i.e., friction and cohesion) and wall friction, as well as to rearrange the particles (which mainly takes place at the beginning of the compaction). Because the particles become more concentrated, the number of particle-particle contacts (known as the coordination number) increases. At any given time in this process, the piston compresses the powder bed to the point at which the force applied by the ram equals the combined forces of (1) inter-particle friction, (2) inter-particle cohesion, and (3) particle–wall friction or adhesion. If the force applied to the piston is increased, the powder will densify further until, again, the combined frictional forces equal the force applied by the piston.

The compressibility response of powder beds can be compared to the placement of a weight onto a compression spring. The spring compresses until its restoring force equals the gravitational force of the applied weight. There is a crucial difference between the spring and the bed of particles, however: when the weight is removed from the spring, it returns to its original state. When the force from the hydraulic ram is removed from the pellet press, the particles do not return to their original packing state, since doing so would require restoring motion, which in turn would require the particles to slide past and over one another. Additional energy would also be required to overcome the higher gravitational potential of the original particle arrangement.

For this reason, the bulk density of small particles is very sensitive to the history of the sample. If a sample is compressed, for example by stacking bags of particles on top of each other, or by being driven over a bumpy road, then it will retain its higher bulk density even after the compression force is removed. On the other hand, if a sample is handled carefully, without jarring or compression, then the bulk density will remain low. Because of this, it is not possible to state the bulk density of a sample with a single value. Instead, bulk density values should be reported in a way that indicates the sample history (e.g., tapped or loose).

### *Measurement*

Bulk density is simply measured by weighing a known volume of the powder. As indicated above, it is essential that powders be handled in a controlled and repeatable way.

There is no universally agreed-on definition of powder flowability, nor is there a means of measuring it in a way that is applicable to every situation. However, for bulk storage purposes, critical "rathole" diameters are often used as a means to characterize the expected flow from the silo outlet [33, 34]. The term "ratholing" refers to the tendency of powders to exhibit "funnel flow" out of a silo (Fig. 2.11). Here, a core channel immediately above the outlet forms, with the powder within the channel flowing while the rest of the powder remains in place. The width of the channel is referred to as the critical rathole index and is measured in units of distance (typically meters or feet for silos used industrially).

The critical rathole diameter for a powder depends on the compressibility of the powder as well as the size, shape, and materials of construction of the silo.

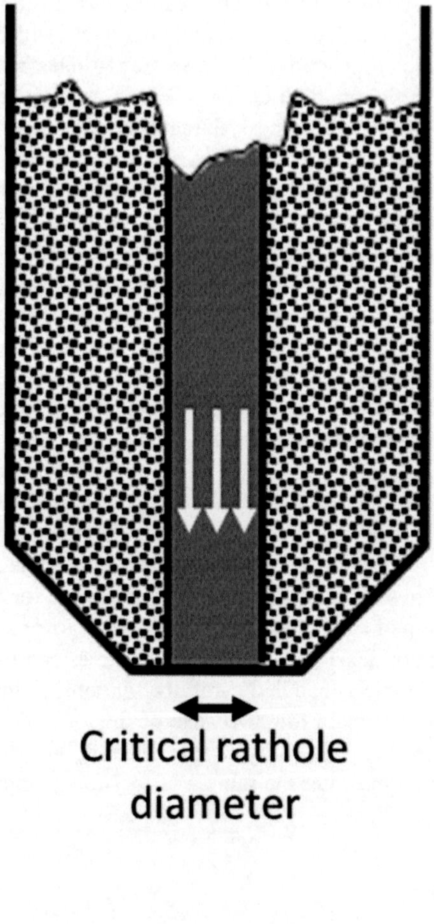


**Fig. 2.11** Funnel flow and formation of a "rathole" in a powder silo

To determine the critical rathole diameter of a powder, the powder is loaded into a cylindrical hole in a sintered metal base that has the ability to vent air from the sample. A piston is then placed on top of the powder and pressure is applied to compress the sample under a controlled and reproducible load. The base is then removed from around the powder, leaving a free-standing column of it. The piston is again applied on top of the column and the stress required to cause the column to fail is determined. This stress is a measure of cohesive strength. The critical rathole index is then calculated from the ratio of the cohesive strength of the powder to its bulk density, using a constant that is specific to the metal base and piston.

The compressibility of a powder is also measured in a piston press. In this case, the volume of powder at a certain load is determined by using the distance over which the piston is moved. A curve of density versus consolidation stress is then plotted, and from this, a relative compressibility is calculated.

## Oil Absorption

In general, we are deferring characterization techniques that are important to only a limited set of end-use applications to the chapters that describe formulation (Chaps. 16 for paint, Chap. 17 for plastics, and Chap. 18 for paper laminates). However, we will include here a discussion on oil absorption, which is a technique that is only important to coatings, because oil absorption values are determined by five of the physical properties that are described above—surface area, surface energy, intra-particle void volume, inter-particle void volume, and powder compressibility. For this reason, a description of oil absorption follows naturally from our discussions above.

The oil absorption test is used to determine the packing efficiency of particles in linseed oil [35, 36]. In this test, linseed oil is added to a single particle type or to a mix of particles. The particle mixes are typically the particles formulated into a paint (pigments and extenders), at the same relative proportions as in the paint. As the oil is added, it is incorporated into the particle mass by vigorously working the mass with a spatula or a rubber policeman. This process is often referred to as a "rub-out". The mix of oil and particles begins as a moist powder, but at a certain point, it converts into a paste—that is, a thick liquid. This is the defined endpoint of the test. At this point, all of the particles have been wetted with the oil (i.e., have a monolayer coating of oil on their surfaces) and there is just enough remaining oil to fill the voids between the particles.

Oil absorption is generally reported as the number of grams of oil needed to bring 100 g of particles to the test endpoint. In some contexts, this value is referred to as the "oil demand" of the particles, and the endpoint in the test is referred to as "satisfying the demand". The oil absorption value is important in paint formulation because it can be used to determine the minimum amount of organic resin required to form a solid film from a set volume of particles, a quantity known as the resin demand of the pigment. The conversion between oil absorption and resin demand is given in

Eq. (2.12). Resin demand is important as it determines the critical pigment volume concentration (CPVC) of the paint. This is discussed in greater detail in Chaps. 4 and 16.

$$\mathrm{CPVC} = \frac{100}{1 + \frac{(\mathrm{OA}) \bullet (\rho)}{93.5}} \tag{2.12}$$

As mentioned above, oil absorption values are affected by five particle parameters. Three of these are determined by the individual particles themselves—their surface area, surface energy, and intra-particle porosity. The effects of these are quite straightforward. The higher the surface area, the greater the amount of oil required to wet the particles, and so the higher the oil absorption value. Similarly, the greater the intra-particle void volume, the greater the amount of oil needed to fill the voids. Surface energy affects the wetting of the particles by the oil, and so also has a direct effect on oil absorption values.

Inter-particle voids are as important as intra-particle voids since all pores must be filled with oil. The filling of intra-particle voids, such as those found in diatomaceous earth (Fig. 2.7), is quite straightforward, and it is clear how the void volumes affect the amount of oil required to satisfy the particles. However, the filling of inter-particle voids is more complicated. When the oil/particle mix is worked, much of the energy applied to the system is used to compress the particle packing arrangement, decreasing the void volume of the particle mix. This is similar to the effect on bulk density of adding energy into a particle bed, both in terms of compressing the particles and in terms of being very dependent on the exact history of the particle mix. The compressibility of the powder is, therefore, the fifth particle property that affects oil absorption. The more energy introduced during the oil absorption test, the greater this compression and the lower the void volume of the resulting particle mix. This, in turn, decreases the measured oil absorption value of the mix. This dependence is a significant source of error for this test since different operators are likely to apply a different degree of force when incorporating the oil into the powder.

In addition to the uncontrolled application of energy, this test also suffers imprecision due to the need to interpret the true endpoint of the test. There is generally not a sharp change to a paste consistency as oil is incorporated into the particle bed, and different operators may have different definitions as to when the mass becomes a paste.

Finally, there is a sixth parameter that affects oil absorption but is not determined by the particle. This parameter is the exact identity of the oil used in the test. There are many types of linseed oil that vary in chemical composition, in particular, in the density of carboxylic acid groups in the polymer chain. The chemical composition is expressed as the acid number of the oil, and this value partially determines the ability of the oil to efficiently wet the particle. A comparison of oil absorption values for a variety of $TiO_2$ pigments using two different linseed oils, which differ in acid number, is shown in Fig. 2.12.

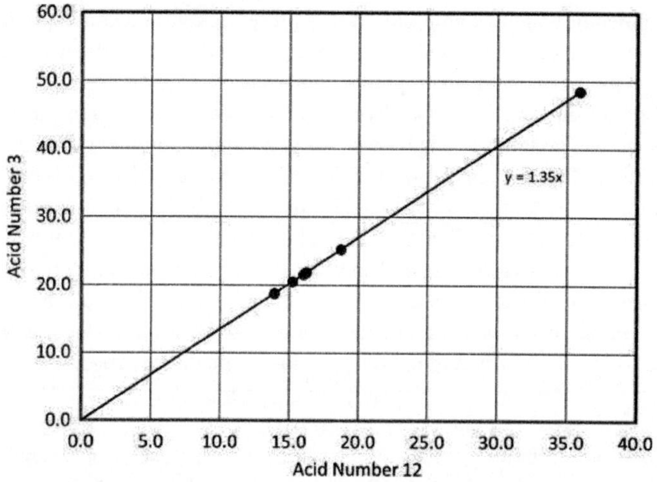


**Fig. 2.12** Comparison of oil absorption values for a variety of $TiO_2$ pigment grades using linseed oils with different acid contents. Values are the grams of oil per 100 g of particles

## Thermal Techniques

Different materials react to temperature changes in different, and often unique, ways. We are familiar with the change in spatial dimensions on heating (normally an expansion). In addition, heating can lead to a weight loss due to the volatilization of some (or all) components of a sample, and it can lead to changes in the heat capacity of the material (i.e., the change in temperature that occurs when a unit amount of heat is applied). Both weight loss and heat capacity change as a function of temperature is used to characterize particles.

### *TGA*

Thermal gravimetric analysis, or TGA, is a test in which weight loss is determined as a function of temperature. In this method, a small amount of power is placed on a balance that is itself in an oven. A gas is flowed over the sample as it is controllably heated and weight as a function of temperature is recorded.

Weight loss occurs for two reasons. The first is the burning of organic components when the analysis is done in air. Different organic materials or classes of organic materials have different ignition temperatures, and the onset of combustion can be used to identify these materials. The second cause of weight loss is the decomposition of materials that contain volatile constituents. The volatile product of such decompositions is typically water molecules from hydrated materials, but it can also be small molecules such as ammonia if it is present. A gas chromatograph or mass spectrometer is often used to analyze the composition of the evolved gases.

Many of the particles used in paints and plastics are hydrated minerals. These not only contain internal waters of hydration but also surface water adsorbed from the atmosphere. The latter normally evolves at a temperature near or slightly below 100 °C. Waters of hydration, on the other hand, can be released at significantly higher temperatures.[10] The temperature at which this occurs is dependent on the chemical composition of the sample. When the material is crystalline and well defined, all of the water is in the same environment and so volatilizes at the same temperature, giving a sharp weight loss in the temperature scan. On the other hand, in amorphous or poorly defined materials, each water molecule is in a slightly different local environment and so the weight loss is spread out over a wide temperature range.

A TGA scan of a mixture of amorphous hydrous alumina ($Al_2O_3{\cdot}nH_2O$) and crystalline hydrous alumina ($Al_2O_3{\cdot}3H_2O$) is shown in Fig. 2.13. Here, the weight loss due to the amorphous alumina is seen as a broad decrease over the entirety of the scan, while the weight loss due to the crystalline alumina is indicated by the relatively sharp feature between 250 and 300 °C. The amount of weight lost (13.4%) can be used to calculate the weight proportion of the crystalline material in this mixture (38.8%, since $Al_2O_3{\cdot}3H_2O$ is 34.6% water).

## *DSC*

Differential scanning calorimetry is used to characterize solids that undergo a temperature-dependent transition at the molecular level. In this test, the heat capacity of a material is measured as a function of temperature. The foundation of this test is that heat added to a sample can affect it in two ways. The first, and most obvious, is that it can increase the temperature of the material. The relationship between heat input and temperature rise for a unit mass of material (normally one gram) is the specific heat of the material. For most particles found in paints or plastics, at the temperatures of interest to us, specific heat generally varies only slightly with temperature (the exception is organic resin particles, for which specific heat changes sharply at certain temperatures, as discussed below). This results in a linear relationship between temperature and heat added to the sample, as shown in Fig. 2.14a.

The second effect of heat added to a sample is that it can cause a physical or chemical change within the material. In this case, heat enters the sample but does not increase its temperature, and so this is often referred to as latent heat. The most

[10] We might expect that all water would evolve at 100 °C—the boiling point of liquid water. While the end product is the same in all cases (gaseous water molecules), the starting point is different. The vaporization process—and the temperature at which it occurs—is dependent on how strongly a molecule is attached to its neighboring molecules. When the neighboring molecules are like molecules, as they are in liquid water, this process occurs at 100 °C. In other materials, the atomic environment surrounding the water molecules is different, and so, too, is the water evolution temperature.

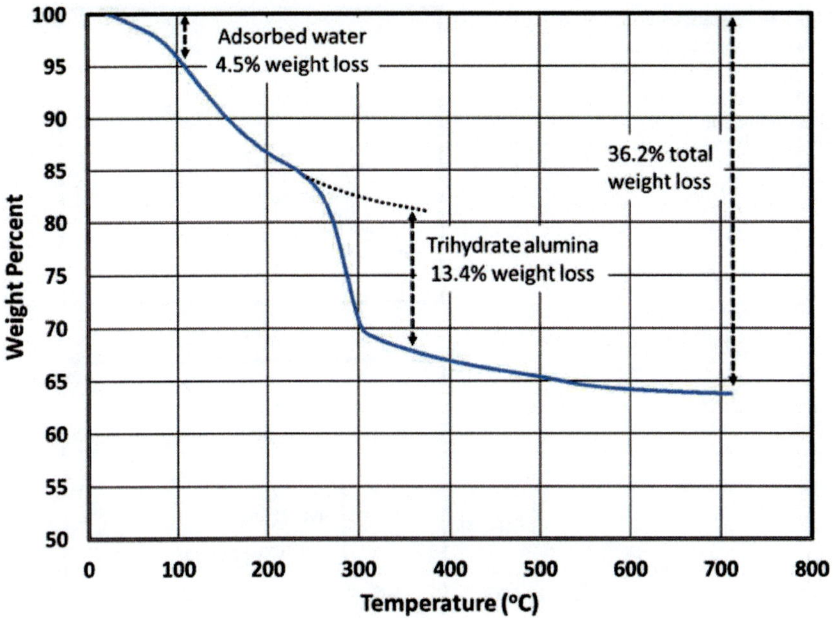


**Fig. 2.13** TGA weight loss scan of a mixture of amorphous hydrous alumina and crystalline hydrous alumina. Losses below approximately 100 °C are due to waters of adsorption. Broad losses between 100 and 700 °C are due to water loss from the amorphous alumina. Sharp loss between 250 and 300 °C is due to the decomposition of the crystalline alumina

obvious of these transitions are melting and boiling (Fig. 2.14b). Here, there are abrupt changes in the motion of molecules and these changes require energy.

Another example of such a transition is a change in the crystalline structure, or crystalline phase, of a material. For example, one solid phase of alumina may change into another when heat is added to the sample at a certain temperature, or an amorphous material may crystallize at a certain temperature (the crystallization temperature, or $T_c$). These transitions are not accompanied by any change in mass or temperature. Although crystalline changes are important in the plastics industry when a polymer is heated or cooled, the particles found within the plastic, or within a paint, do not typically undergo a phase change within the temperature range of most applications.

The final thermally induced change to a particle is its glass transition. This transition occurs in the polymer particles found in a latex paint, as well as in many plastics. At a specific temperature, termed the glass transition temperature and indicated with the abbreviation "Tg", there is a step-change increase in the mobility of sections of the polymer chains. This is accompanied by a change in the heat capacity of the material and is seen as a discontinuity in the curve relating heat added to a sample and its temperature (Fig. 2.14c). The glass transition and its importance to the resin found in waterborne coatings are discussed in more detail in Chap. 10.

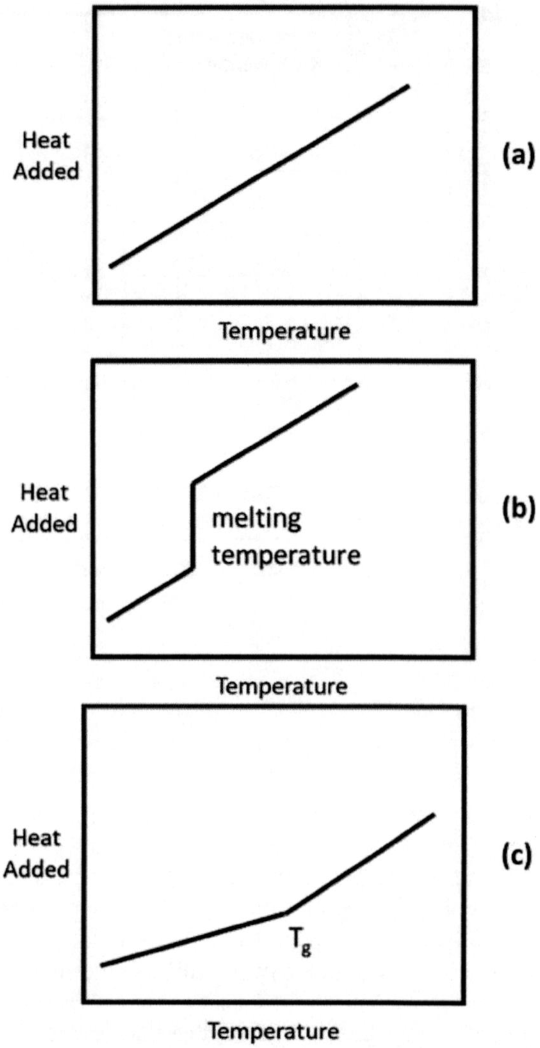


**Fig. 2.14** Relationship between temperature and heat added to a material. **a** Simple heating. **b** Melting. **c** Glass transition

The DSC experiment is simple in principle [37]. A known amount of heat is added to a sample, and the corresponding temperature change is determined as a function of this added heat. A control sample of known thermal characteristics is used as a reference for this measurement. There are two fundamentally different types of DSC instruments. Heat flow, or flux, instruments use a single furnace for the unknown and reference samples. The temperature difference between the samples is monitored and heat flow is calculated based on this difference. By contrast, a power-compensated DSC uses two furnaces, one for the unknown and the other for the reference. The two samples are kept at the same temperature during heating. The power required to heat the sample furnace to the same temperature as the reference furnace is monitored and the energy input per incremental increase in temperature is calculated from this.

## Elemental Analysis

A quantitative determination of the amounts of the elements found in a material is a powerful tool for determining the identity of any material, including powders or powder mixtures. This can be used for gross characterization of the sample as well as the detection of trace impurities. Such impurities often reveal the process used to make the particles (e.g., the sulfate and chloride routes for producing $TiO_2$ each have difficulties in removing certain impurities) or the geographical origin of a mineral sample (impurities are generally specific to a particular region).

There are a number of ways to measure the elemental composition of a powder, but we will focus on three—X-ray fluorescence spectroscopy (XRF), atomic absorption spectroscopy (AAS), and inductively coupled plasma atomic emission spectroscopy (ICP-AES).

In the XRF measurement, the material of interest is subjected to X-ray radiation. The atoms within the sample absorb X-rays with a concomitant ejection of a certain electron from the atomic inner core. This leads to a high energy excited state of the atom. An electron from the outer core of the atom will typically transfer to the inner core, replacing the ejected electron, and in the process will lose energy as a photon with a characteristic wavelength (also in the X-ray region of the light spectrum). The overall process of light absorption and re-emission at a lower energy wavelength is referred to as fluorescence.

The elemental composition of the material can be calculated by comparing the intensities of the fluorescent X-rays to those of known standards. This technique can be used to detect atoms between sodium and uranium on the periodic table, as the fluorescent X-rays for lighter elements are typically reabsorbed before they can leave the material.

AAS is based on the light absorption characteristics of gaseous atoms. In this analysis, a solution of the material of interest is heated (often by an argon plasma generated in an inductively coupled heating process) to a temperature that causes it to decompose. This generates individual atoms in the gas phase. A light is then applied to the gas and its intensity spectrum after passing through the sample is measured. Individual atoms of each element exhibit a characteristic absorption spectrum, and so the amount of a particular element can be determined by adsorption spikes in the spectrum. The wavelength of light can be modified to match the absorption features of a specific atom.

The final technique that we will consider is ICP-AES. In this technique, the sample of interest is again dissolved and subject to high enough temperatures to vaporize the material. In this case, an inductively coupled plasma is used to heat the sample and ionize the resulting vapor. These high-energy ions typically radiate, or emit, certain wavelengths of light. A detector determines the intensity of the emitted light and

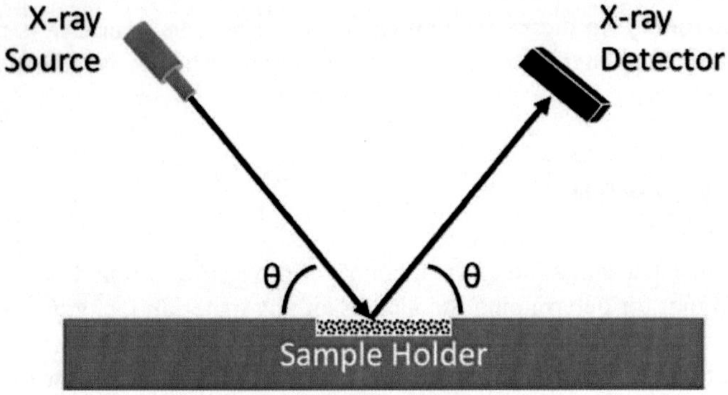


**Fig. 2.15** Geometry of the X-ray powder diffraction measurement

compares this to intensities for standards of the elements present. This technique is often used for trace metals analysis in addition to routine elemental analysis.

## Crystalline Phase Composition

When discussing the use of X-ray line broadening to determine particle size, we described the interaction of X-rays with crystalline materials. These interactions cause X-rays to scatter with certain intensities in certain directions (angles). The intensities are governed by the type of elements in the crystal and the directions are determined by the arrangement of these atoms within the crystal lattice (i.e., the crystal phase).

The combination of scattering angles and intensities is a fingerprint for a specific crystal type. These parameters are measured by scanning a powder sample with X-rays, simultaneously changing the angle of the incoming beam and the angle of the detector (Fig. 2.15). The results are typically reported as detected intensity as a function of angle (2θ). Different components for even very complex mixtures of crystals can normally be resolved and an approximate composition determined (Fig. 2.16).

## Microscopy

We mentioned microscopy earlier as a means for characterizing the size distribution of certain particles or particle mixtures. While it is quite capable of doing this, the overall usefulness of microscopy to particle characterization is significantly broader.

Two types of microscopes are available to the researcher—optical microscopes and electron microscopes. Optical microscopes are useful for imaging particles at the

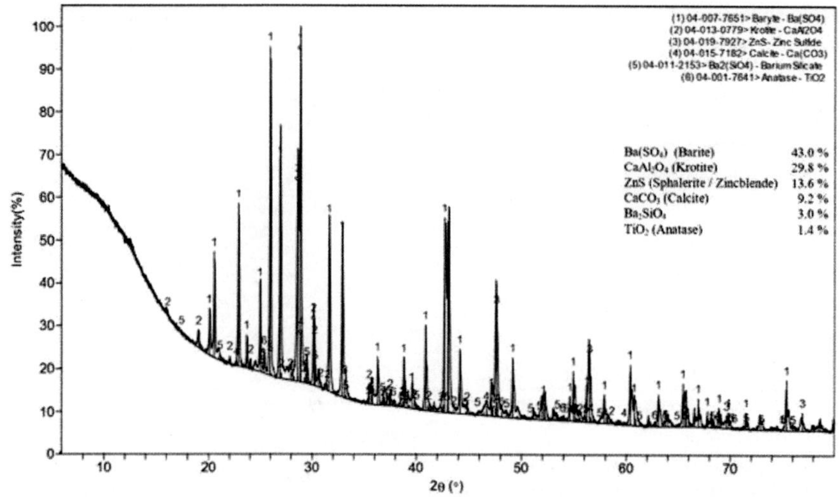


**Fig. 2.16** Powder X-ray diffraction pattern of a complex mix of many materials found in paints or plastics

larger end of the size range of interest to us, as well as agglomerates of particles, and this level of detection is adequate for many needs of the paint or plastics formulator. However, resolution in microscopy is theoretically limited to about one-half of the wavelength of the illuminating light,[11] although in practice it is impossible to resolve features smaller than a few microns using light microscopy. This is due to limitations on the depth of focus. Because of this, particles at the smaller end of the range of interest to us cannot be imaged using this technique.

This issue can be avoided by using probes of much smaller wavelengths. However, there are practical issues with this when the probe is light (X-rays). Instead, a high-energy electron beam is used for this purpose. Because of wave/particle duality, these electrons have associated with them a characteristic wavelength. The resolution limit of these electrons is about 0.1 nm—an improvement of several orders of magnitude compared to optical microscopes.

There are two fundamentally different ways to use electrons to image a sample, leading to two types of electron microscopes—scanning electron microscopes (SEMs) and transmission electron microscopes (TEMs). In both cases, an electron beam is focused on the sample. In the case of an SEM, the degree of interaction between the beam and sample is quantified in some way (i.e., a signal is created and processed), while in the case of a TEM the electrons that interact with the sample are projected onto a screen, directly creating an image of the particles without the need to create or process a signal.

[11] Light does not merely reflect from particles that are of similar size to the wavelength of light, but instead scatters from them (see Chapter 3).

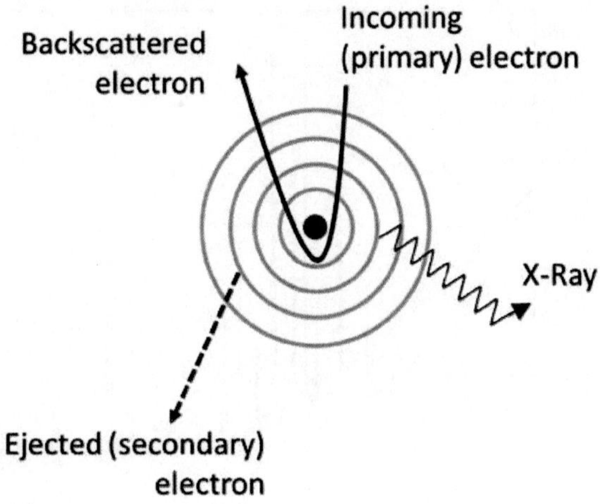


**Fig. 2.17** The three quantities detected in traditional scanning electron microscopy. The solid circle is the nucleus of an affected atom; concentric circles indicate electron shells

There are three main detection approaches for the SEM: detecting backscattered electrons, detecting secondary electrons (these are electrons that are ejected from atoms in the sample by the incoming electron beam), and detecting X-rays generated from the sample when the atoms that have ejected secondary electrons relax to a lower energy state (similar to X-ray fluorescence, discussed above, except using electrons rather than X-rays to eject a core electron from the atom). These three processes are shown schematically in Fig. 2.17.

For all three detection methods, the electron beam in an SEM is rastered across the sample of interest, and the image created from this scan is essentially a map of interaction intensity as a function of beam location. In effect, the intensity is measured for a pixel of area, and the pixels are combined to give an image. Note that this form of imaging is fundamentally different from optical imaging (or TEM imaging, discussed below), where a wave source (visible light or electrons) is applied to the entire sample at once, and the resulting waves are focused and imaged.

The three detection methods provide a somewhat different view of the sample. Because secondary electrons are easily absorbed, only those at the topmost region of the surface (1–10 nm) can escape the sample and be detected. In addition, these electrons are less likely to escape from valley or pore regions than they are from regions of prominence. These two factors enhance the topographical (surface) features of the sample, providing a three-dimensional appearance to these images (Fig. 2.18).

The sensitivity of secondary electrons to surface features can also be used for depth studies. In this case, the energy of the beam electrons is changed—higher energies allow for deeper penetration. An example of this is shown in Fig. 2.19. These show the same region of a paint film containing only $TiO_2$ and resin. The electron beam energy for one image was 3 keV and for the other was 10 keV. Film surface features and the topmost $TiO_2$ particles are seen at 3 keV while the top micron or so of the film is imaged at the higher beam energy.

**Fig. 2.18** SEM secondary electron image of a mixture of particles (curtesy of E. Najafi and H. Huezo, Chemours)

The intensity of electron backscattering is controlled by the atomic number of the atom. Heavier atoms have more electrons, and so there is a greater chance of an electron in the beam being scattered backward by the electrons in heavy atoms than is the case for lighter atoms. The contrast in these images, therefore, is determined by the elemental identities of the atoms in the sample, with the heavier atoms appearing brighter.

A comparison of secondary electron images and backscattered electron images for a paint sample is shown in Fig. 2.20. The surface-specific nature of the secondary electron technique, and the elemental contrast of the backscatter technique, are clearly seen in these images.

Finally, analysis of the X-ray energies can give information as to the elemental identity of the atoms present. This can either provide an element map of the entire image area or a spectrum for an individual region within the image. An example of X-ray fluorescence is given in Fig. 2.21. Here, four images are given for a sample mixture of hollow silica sphere microparticles (diameter 0.3 microns) and gold nanoparticles (40 nm). The secondary electron and backscatter electron images are shown in Fig. 2.21a and b. In the secondary electron image, there is no contrast between the two particle types, while in the backscatter electron image the gold nanoparticles are clearly highlighted in comparison to the silica microparticles. Figure 2.21c and d shows elemental maps for the two metal atoms found in the different particle types.

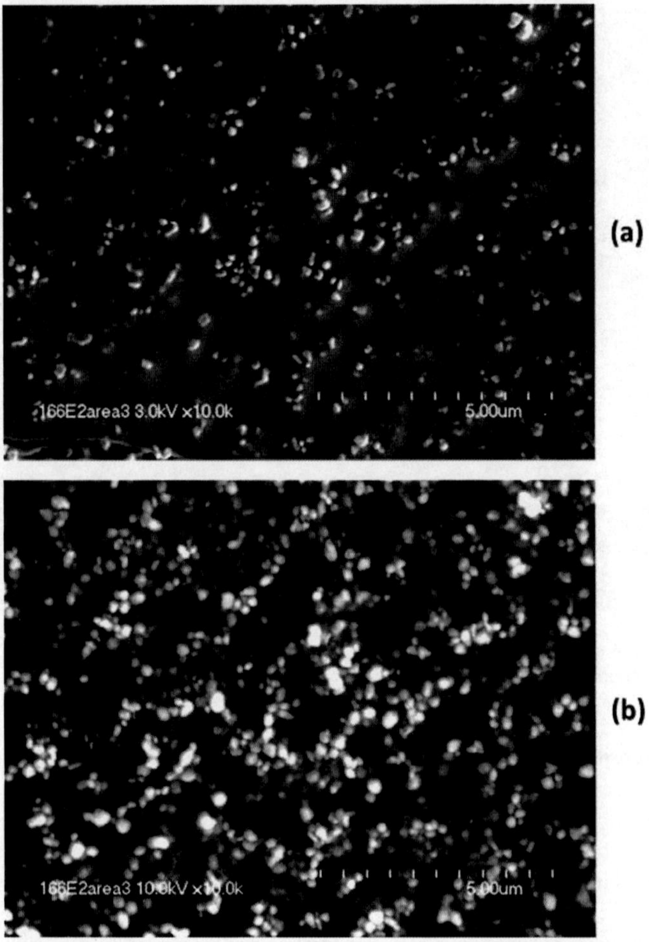


**Fig. 2.19** SEM secondary electron images of the same region of a paint film at different electron beam energies. **a** 3 keV. **b** 10 keV

SEM imaging of materials that are electrically insulating, which describes nearly all particles of interest to us, is complicated by the accumulation of electrical charge on the material surface. This electrical charge arises from electrons in the beam becoming trapped within the sample. Because the electrical fields created by this excess charge will deflect the electrons in the beam, the sample image becomes distorted. This problem can be solved by coating (sputtering) the sample with an electrically conductive material (e.g., gold, osmium, or carbon). However, this unavoidably alters the sample surface at the atomic level, and so surface features of only a few nanometers cannot be properly imaged when these coatings are used.

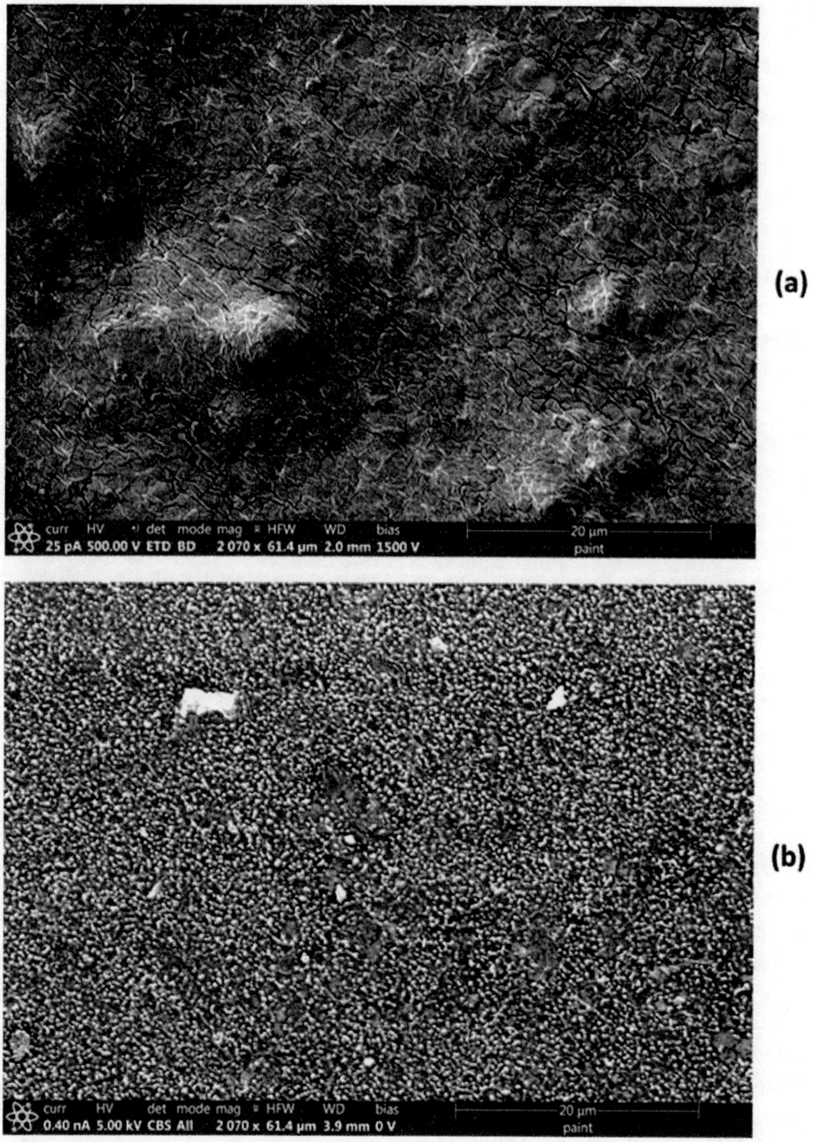


**Fig. 2.20** SEM images of a paint film using different detectors. **a** Secondary electron image. **b** Backscattered electron image. (curtesy of E. Najafi and H. Huezo, Chemours)

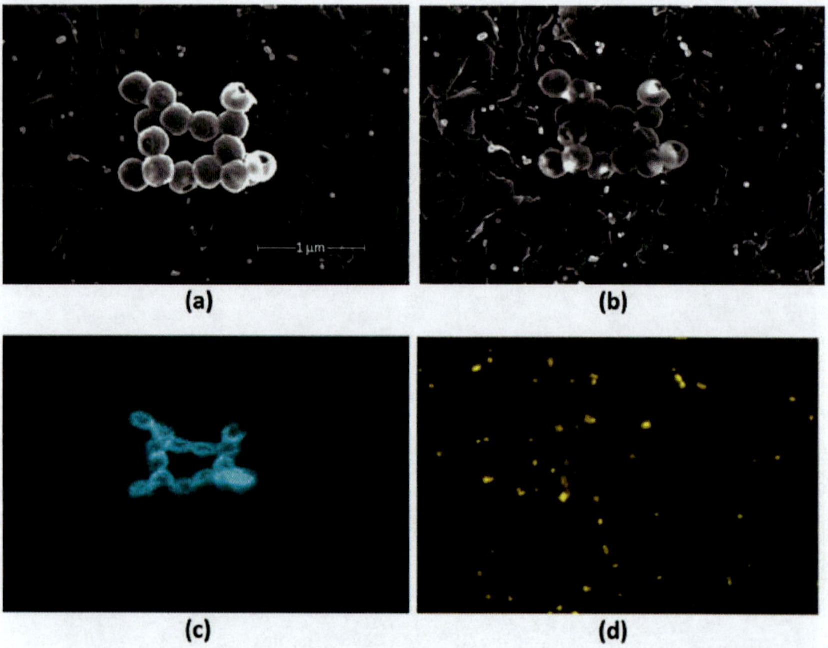


**Fig. 2.21** SEM images of a mixture of hollow silica sphere microparticles and nanogold particles. **a** Secondary electron image. **b** Backscatter electron image. **d** X-ray fluorescent image for silicon. **e** X-ray fluorescent image for gold. (curtesy of E. Najafi and H. Huezo, Chemours)

Unlike the SEM, the TEM, as the name implies, analyzes electrons that are transmitted through the sample. For this reason, the sample must be thin enough (typically less than 100 nm) to allow some of the electrons in the beam to pass through. After passing through the sample, beam is then expanded using a magnetic lens[12] and directed at a phosphor screen.

Electrons in the electron beam interact with the sample in two ways—through elastic scattering, which preserves the energy of the beam, and through inelastic scattering, during which the beam loses energy. Both types of interactions can be used to image the sample.

The TEM is capable of higher magnification than the SEM and can, in fact, resolve individual atoms or columns of atoms. This can be seen in Fig. 2.22, which is a TEM image of a pigmentary $TiO_2$ crystal coated with 3 wt% amorphous silica. Here arrayed columns of titanium atoms are clearly visible near the $TiO_2$/silica interface, whereas the amorphous coating shows no indication of regularity. The contrast seen in TEM images is controlled by the thickness of the sample, and the atomic identity of the sample—as with SEM, the electron beam in the TEM process interacts more intensely with heavier atoms than lighter atoms, making the amorphous silica in

[12] The ability of a magnetic lens to spread out an electron beam is far greater than the ability of a glass lens to spread out a light beam.

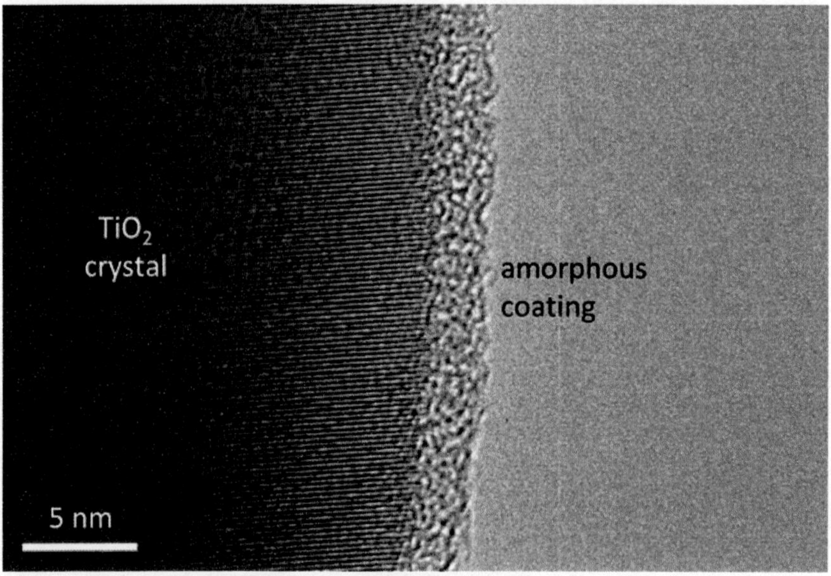


**Fig. 2.22** TEM image of a $TiO_2$ particle with a thin coating of amorphous silica

Fig. 2.22 lighter than the $TiO_2$ portion of the particle. In addition, resolution can be increased by using phase contrast.

Elemental analysis and mapping are also possible with the TEM. In this case, a technique known as electron energy loss spectroscopy (EELS) is used. Here, the electron beam causes an electron in a sample atom to eject, decreasing the energies of the beam electrons by a specific amount. The energies of the beam electrons are measured after passing through the sample, and from that, the elemental identity of the sample can be determined. This measurement is so precise that the oxidation state of the atom and even the crystalline phase of the material (which sets the local environment around the sample atoms) can be differentiated. An example of an EELS image is shown in Fig. 2.23 for a $TiO_2$ particle with a patchy alumina coverage.

## Surface Charge

The surfaces of particles dispersed in water typically carry an electrical charge. The sign of this charge, and its magnitude, are governed by dissolved ions in the water (especially $H^+$ and $OH^-$, the concentrations of which are determined by pH), particularly when the particle is a metal oxide.

Any un-neutralized or excess charge in a particle will be found at or near the particle surface. A charged particle surface in water will create an electrical double layer that surrounds that particle [38]. The inner electrical layer consists of ions

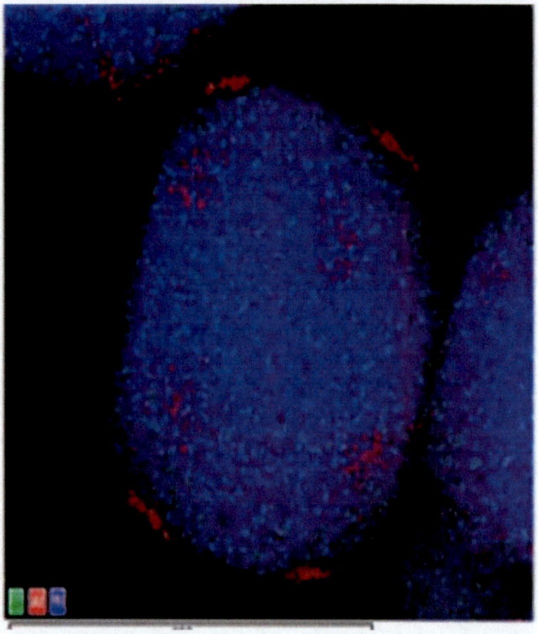

**Fig. 2.23** EELS image of a $TiO_2$ particle with patchy alumina coverage. Blue indicates titanium and red indicates aluminum (curtesy of E. Najafi, Chemours)

(either from water dissociation or any dissolved salts) that are tightly adsorbed onto the particle surface. These ions are not swept away if the solution around the particle moves relative to the particle. The outer layer is composed of counterions that remain in solution and screen the charge on the inner layer.

The exact distribution of ions around the charged surface, and the electrical potential as a function of distance from the particle surface, has been the subject of much experimental and theoretical attention since Helmholtz first reported his analysis of electrodes immersed in ion-containing solutions in 1853. Because this layer can extend into a relatively large region around the particles, it is often referred to as the diffuse layer. The amount to which it extends and the electrical potential as a function of distance are determined by the magnitude of the surface charge layer and the concentration of ionic species in the liquid phase (the diffuse layer is more compact if the ionic concentration is high).

The association of the ions within the diffuse layer with the particle is not strong and these ions can be swept away if the solution moves relative to the particle. The boundary between the mobile diffuse layer and the tightly held inner layer is referred to as the slipping plane. The electrical potential at this plane is called the zeta potential ($\zeta$) and is typically measured in millivolts. The surface charge, and its application to dispersion stabilization, are discussed in greater detail in Chap. 11.

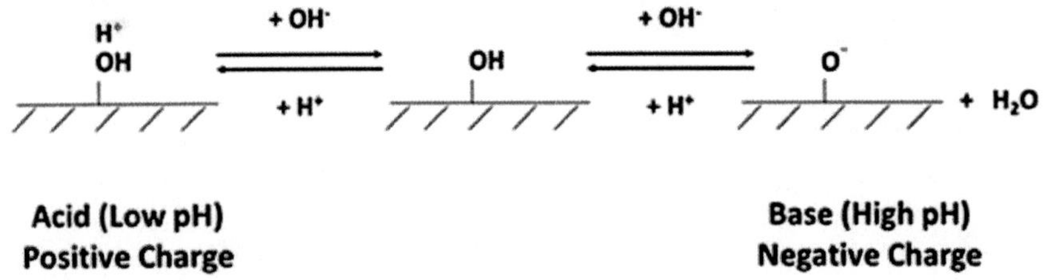


**Fig. 2.24** Modification of oxide surface charge by addition of acid or base

## *Controlling Surface Charge*

The surfaces of most metal oxides are terminated by oxygen groups (typically bridging two or more metal ions) and hydroxyl groups (typically bound to one metal ion, although bridging hydroxides are also seen). $H^+$ and/or $OH^-$ in the water phase of the slurry or paint react with these groups, bringing with them their electrical charge.

These reactions provide us with a convenient way to change the magnitude and the sign of the charge on a particle surface. We can give the surface a positive charge by adding acid (low pH), or a negative charge by adding base (high pH). In conditions between these (mid pH), the surface will have no net charge (Fig. 2.24).

We might assume that the pH value for no net charge would be 7.0, which is the value for which water has an equal number of dissolved hydroxide ions and protons. However, in the paragraph above we have carefully chosen the words describing this value as "mid pH", rather than as a pH value of 7.0. This is because the pH value for surface charge neutrality (called the iso-electric point (IEP) or point of zero charge (PZC)) varies from one type of particle to another, depending on the exact chemical nature of the particle.[13] This is an exception to the earlier statement that the ideal particle in a paint would have physical characteristics but would be chemically inert.

To understand why the IEP varies from one particle type to another, we will consider the family of chemical reactions that can be written as

$$\begin{gathered} \mathrm{X\text{-}OH + H_2O \rightleftarrows X\text{-}OH_2^+ + OH^-} \\ \mathrm{X\text{-}OH + H_2O \rightleftarrows X\text{-}O^- + H_3O^+} \\ \mathrm{X\text{-}OH + H^+ \rightleftarrows X\text{-}OH_2^+} \\ \mathrm{X\text{-}OH + OH^- \rightleftarrows X\text{-}O^- + H_2O} \\ \mathrm{X\text{-}OH \rightleftarrows X^+ + OH^-} \end{gathered}$$

For water, X = H. In this case, neutrality is achieved at a pH value of 7.0. When X is a metal atom at the surface of an oxide particle, however, the character of the

[13] There are still positively and negatively charged sites on the surface at the IEP, but these charges balance to give a net charge of zero. This situation is equivalent to the concept of zwitterions for discrete molecules.

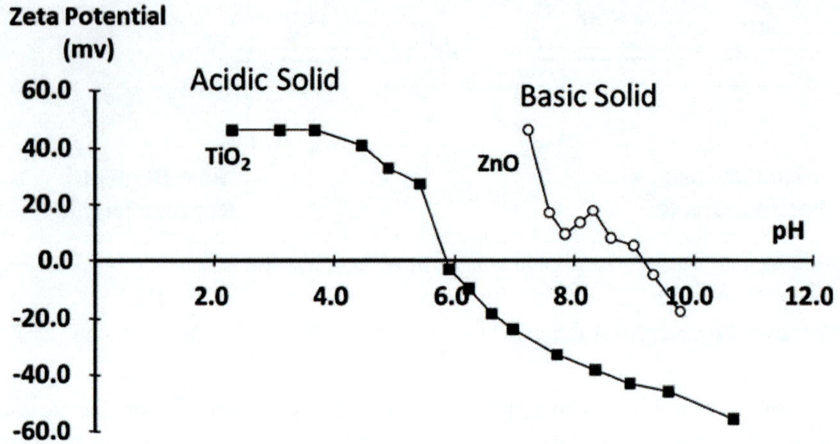


**Fig. 2.25** Zeta potential as a function of pH for an acidic solid ($TiO_2$) and a basic solid (ZnO)

surface O–H bonds vary depending on the identity of X. If X is electronegative, then the proton on the surface -OH group will be more acidic and when incorporated into water, the X–OH moiety will spontaneously disassociate to generate $H^+$ and $XO^-$—that is, the material will be an acid. If X is electropositive, then the oxygen in the hydroxide group will be more capable of supporting a second proton. This second proton is formed by the dissociation of water into $H^+$ and $OH^-$. After the reaction of the $H^+$ with the surface -OH group, the dissociated $OH^-$ ion remains in solution, raising the pH. In this case, the oxide is a base.

Since, by definition, the X group is different for different oxides, some oxide particles are acidic and some basic. We can see this by measuring the zeta potential on a particle in water as a function of pH using techniques described in the next section. The results for such measurements are shown in Fig. 2.25 for pigmentary titanium dioxide and zinc oxide. The intersections of these curves with the X-axis give the IEP values for these particles.

The curve shapes are similar for both materials. The charge on these surfaces is positive at low pH values and negative at high pH values, consistent with Fig. 2.24. However, the curves are horizontally shifted from one another. $TiO_2$ is a mildly acidic solid.[14] As a consequence, it is relatively easy to remove the surface protons, leaving a negative charge on the particle surface. This is favored enough that the surface protons will even dissociate under mildly acidic conditions (i.e. at pH values as low as 5.9). By contrast, ZnO is a basic solid. Protons from the water adsorbed onto it, releasing hydroxide, even under moderately basic conditions ($7.0 < pH < 9.2$), giving it a higher IEP value than that of $TiO_2$.

Most grades of pigmentary $TiO_2$ that are used in the paint industry, and many used in the plastics industry, are coated with hydrous alumina and, in some instances, a

[14] Pigmentary $TiO_2$ particles made by the chloride process invariably have some alumina on their surfaces, which shifts the IEP to a higher value than for pure rutile (IEP = 5.2).

combination of hydrous alumina and hydrous silica, or of hydrous alumina and zirconia. Since these coatings reside on the particle surfaces, they affect the acid/base characteristics of the particles. Silica is an acidic solid, while alumina is a basic solid, and the IEP values of these grades reflect contributions from all surfaces.

Ideally, the surface charge of a particle in a slurry or paint will not change over time. Such a change would indicate that the particle surface is chemically reacting with a dissolved species in the paint, replacing the original surface with a new surface composed of the reactive dissolved species. This type of undesired reaction is often seen in waterborne paints that contain both pigmentary $TiO_2$ and zinc oxide [39]. These paints often gel on storage. The cause of this gel formation was found to be the deposition of a layer of zinc ions onto certain surfaces in the $TiO_2$ pigment, which transforms the pigment surface into a zinc oxide surface. This deposition could be monitored by measuring the zeta potential of the $TiO_2$ as a function of pH as the paint ages. Because zinc oxide has a higher dispersant demand than pigmentary $TiO_2$, the paint becomes dispersant starved over time and gels (the effect of dispersant on paint formulation is discussed in Chaps. 11 and 16).

As noted above, different surface sites have different acid/base characteristics. This is an important consideration for particles that contain more than one type of metal. For example, clays typically have both alumina and silica (and some have additional cations). When a clay surface has both aluminum and silicon atoms exposed, then, in a slurry, that surface will have both positive and negative charges. This results in a near-neutral surface charge at the pH value of a typical paint. However, the faces of certain clays, such as laponite, can be terminated exclusively with either surface aluminum or surface silicon atoms. At pH values near neutral, this results in some faces being positively charged and others being negatively charged. The combination of positively and negatively charged surfaces leads to the formation of flocculates in a "house of cards" structure (Fig. 2.26). Such flocculation, when adequately controlled, can be used to modify the rheology of a slurry or paint.

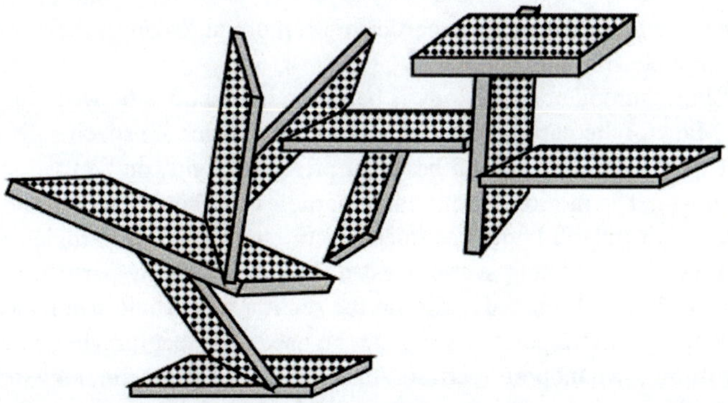

**Fig. 2.26** Face-to-edge house of cards structure seen in laponite. Patterned faces are negatively charged; gray faces are positively charged [40]

The surfaces of non-oxide solids can also be charged, but not with protons or hydroxides. As an example, barium sulfate consists of discrete barium cations and discrete sulfate anions. Both of these ions are exposed at the surfaces of barium sulfate particles. If sulfuric acid was added to a slurry of barium sulfate, we might expect the surface charge to become positive due to the adsorption of protons. However, the surface charge actually becomes negative.

The explanation for this behavior is that there are no hydroxyl or metal oxide groups present on the barium sulfate surfaces, so these surfaces do not react with the protons in sulfuric acid. Instead, the negative sulfate ions absorb onto the positive surface barium cations, in a sense extending the barium sulfate lattice by one sulfate anion. Because sulfate is negatively charged, the surface will become negatively charged. Similarly, the addition of a barium salt to a slurry of barium sulfate will result in positively charged surfaces. $Ba^{2+}$ and $SO_4{}^{2-}$ are referred to as the potential determining ions for barium sulfate. For metal oxide particles, the potential determining ions are typically $H^+$ and $OH^-$.

## *Measuring Surface Charge*

There are two technologies currently used to measure the zeta potential. In most forms of both technologies, the movement of particles within an electric field is monitored, a process referred to as electrophoresis [41]. Particle velocities are proportional to, among other factors, the zeta potential of the particles.

In these tests, the electric field is typically created using charged parallel plates. The slurry of interest is then placed between these plates, and the movement of particles is measured as a function of the voltage on the plates. Of particular interest is the movement of particles in an alternating field. In this case, not only is the velocity of the particle measured but also the ability of the particle to change direction. This ability is affected by the size of the particle as well as by the extent of the charged layer around it. In general, larger particles are more sluggish in their response to changes in the electric field.

The first measurement technology is based on interactions between the particles and light. Many of the particles used in paints and plastics are specifically designed to interact with light, so it should be no surprise that light can be used to monitor their movement. The motion of individual particles can be followed if the particles are large enough (this is known as microelectrophoresis). For particles below the imaging limit, dynamic light scattering can be used to follow particle movement. This technology was discussed earlier in the section on particle size measurement. Finally, tunable resistive pulse sensing can be used to detect individual particles as they move through a nanopore aperture. An advantage of this technology is that it can simultaneously determine the size of the individual particles as well as their charge when they traverse the nanopore.

A disadvantage of these light-based measurements is that they require the slurry or suspension to be significantly diluted when compared to the particle concentrations found in most commercial slurries and paints. The dilution process has the potential to significantly disturb the environment around the particles and change their properties, a process known as shocking. As such, it is often difficult to apply the results for the diluted material to the properties of the fully concentrated material. Shocking can be minimized by diluting the slurry or paint with the same liquid phase as is present in it. This can be done by separately making the liquid solution used in the slurry or paint, or by centrifuging a sample of the slurry or paint and using the separated supernatant as the diluent.

Low solids concentrations are not needed for the second measurement technology—electroacoustic detection. With this technology, there are two related but fundamentally different acoustical effects that can be used to monitor particle motion. The first is the colloid vibration current method. With this technology, sound waves are used to create a pressure gradient within the suspension. Particles move in response to this gradient. The ionic layers surrounding the particles move more slowly than the particles themselves, and this creates a dipole moment. This dipole moment creates an electric current, which is then measured and analyzed.

The second acoustical approach relies on the same principles but differs in how motion is created and detected. In this case, an alternating electric field is applied to the sample. The motion caused by the charged particles moving in response to this field creates an ultrasonic wave, which is then converted to an electrical signal using an audio transducer. This technique is in some ways the reverse of the colloid vibration current method.

All of these analytical techniques determine particle velocity under different electric fields, and this velocity data is analyzed in similar ways. Different analysis techniques can be applied to the data, with the most commonly used based on Smoluchowski's theory [42]. A variety of other theories have been developed to relate velocity to zeta potential under conditions where Smoluchowski's assumptions do not hold.

## pH and Acid/Base Capacity

In the section above, we showed that different solid materials show different acid and base characteristics. These characteristics can be quantified through titration of a slurry using a strong acid or base. The pH of a powder is typically determined by mixing the powder with deionized water (10% solid by weight) and simply measuring the pH of the resulting suspension [43]. This is an indicator of the acid or base strength of the solid, similar to the $pK_a$ or $pK_b$ value for a weak molecular acid or base. The total number of acid or base sites (or buffer capacity) can be measured by then titrating the slurry with the appropriate strong base or acid. The results are typically measured in milli-equivalents of acid (or base) per gram of material.

## Optical Properties

The interactions between visible light and particles can be quite complex, particularly for particles within the size range of visible light (approximately 380–700 nm). These interactions generally fall into two categories: light absorption and light scattering. Particles that interact with light strongly are referred to as pigment particles—colored pigments if the interaction is absorption, and white pigments if it is scattering.

Understandably, each type of interaction has very important consequences for the appearance of the paints or plastics that contain these particles. Because of their importance, these properties, and how they are measured, are discussed in depth in Chaps. 5, 6 and 15 (light absorption) and Chaps. 3 and 4 (light scattering) and so will not be discussed further here.

## Summary

A wide variety of particle properties affect an array of paint and plastics end-use performance attributes, and so particle characterization is an essential aspect of paint and plastics formulation. The particle properties of interest can generally be divided into two categories—chemical properties and physical properties.

Physical properties include particle size (and size distribution), particle surface area, intra- and inter-particle void sizes (and size distributions), thermal properties, shape, appearance, composition, and bulk behavior. Each of these properties can be measured or probed with an appropriate technique or instrument, and it is often the case where more than one analytical option is available for a particular property. In such cases, complementary information can often be generated using different methods or instruments.

The main chemical property of interest is the reactivity of the particle surfaces, particularly in a liquid carrier, such as in a paint, but also in air. In such environments, the particles can adsorb different species. In water, when the adsorbing species are ionic, an electrical charge will develop on the particle surface. We can measure this charge as a function of the species concentration. When the surface species are protons or hydroxyls, we can measure the acidic or basic strength of these species as well as their surface concentration.

In later chapters, we will show how these different analyses can be applied in ways that aid the understanding of the effect of different particles on the behaviors of paints and plastics and how they can assist in modified of existing products or the formulation of new ones.

# References


1. Heintzenberg, J.: Properties of the log-normal particle size distribution. Aerosol Sci. Technol. **21**(1), 46 (1994)
2. Zender, C.S.: Particle Size Distributions: Theory and Application to Aerosols, Clouds, and Soils. https://www.researchgate.net/publication/2393276_Particle_Size_Distributions_Theory_and_Application_to_Aerosols_Clouds_and_Soils
3. Kolmogorov, A.N.: On the logarithmically normal distribution law of particle sizes at the subdivision. Dokl. Akad. Nauk SSSR **31**(2), 99 (1941)
4. Söderlund, J., Kiss, L.B., Niklasson, G.A., Granquvist, C.G.: Log-normal size distribution in particle growth processes without coagulation. Phys. Rev. Let. **80**(11), 2386 (1998)
5. Wang, D., Fan, L.-S.: Particle characterization and behavior relevant to fluidized bed combustion and gasification systems. In: Scala, F. (Ed.) Fluidized Bed Technologies for Near-Zero Emission Combustion and Gasification. Woodhead (2013)
6. Ruzer, L.S., Harley, N.H. (Eds.): Aerosols Handbook: Measurement, Dosimetry, and Health Effects, 2nd Edn. CRC Press (2013)
7. Merkus, H.G.: Particle Size Measurements: Fundamentals, Practice, Quality. Springer (2009)
8. Allen, T.: Particle Size Measurement. Springer (2012)
9. Patterson, A.: The Scherrer formula for X-ray particle size determination. Phys. Rev. **56**(10), 978 (1939)
10. Stetefeld, J., McKenna, S.A., Patel, T.R.: Dynamic light scattering: a practical guide and applications in biomedical sciences. Biophys. Rev. **8**, 409 (2016)
11. Anon.: Acoustic spectroscopy for particle size measurement. Pig. Res. Technol. **29**(5) (2000)
12. Measurement and Characterization of Particles by Acoustic Methods – Part 1: Concepts and Procedures in Ultrasonic Attenuation Spectroscopy. ISO 20998-1 (2006)
13. Dunkin, A.S.: Characterization of Nanoparticles: Measurement Processes for Nanoparticles. In: Hodoroaba, V.-D., Unger, W.E.S., Shard, A.G. (Eds.). Elsevier (2020)
14. Fichera, O., Alpan, L., Laloy, J., et al.: Characterization of water-based paints containing titanium dioxide or carbon black as manufactured nanomaterials before and after atomization. Appl Nanosci. **9**, 515 (2019)
15. Giddings, J.C., Yang, F.J.F., Myers, M.N.: Sedimentation field-flow fractionation. Anal. Chem. **46**(13), 1917 (1974)
16. Allen, T.: Particle Size Analysis. In: Stanley-Wood, N.G., Lines, R.W. (Eds.). Royal Society of Chemistry (1992)
17. Standard Guide for Defect Detection and Rating of Plastic Films Using Optical Sensors. ASTM D7310
18. Pigments and Extenders—Methods of Dispersion and Assessment of Dispersibility in Plastics—Part 5: Determination by Filter Pressure Value Test. ISO 23900-5 (2015)
19. Thommes, M., et al.: Physisorption of gases, with special reference to the evaluation of surface area and pore size distribution. Pure Appl. Chem. **87**(9–10), 1051 (2015)
20. Brunauer, S., Emmett, P.H., Teller, E.: Adsorption of gases in multimolecular layers. J. Am. Chem. Soc. **60**(2), 309 (1938)
21. Rouquerol, J., Llewellyn, P., Rouquerol, F.: Is the BET equation applicable to microporous adsorbents? Stud. Surf. Sci. Catal. **160**(7), 49 (2007)
22. Rouquerol, J., Avnir, D., Fairbridge, C.W., Everett, D.H., Haynes, J.M., Pernicone, N., Ramsay, J.D.F., Sing, K.S.W., Unger, K.K.: Recommendations for the characterization of porous solids (technical report). Pure Appl. Chem. **66**(8), 1739 (1994)
23. Leofanti, G., Padovan, M., Tozzola, G., Venturelli, B.: Surface area and pore texture of catalysts. Catal. Today **41**(1), 207 (1998)
24. Thommes, M., Kaneko, K., Neimar, A.V., Olivier, J.P., Rodriguez-Reinoso, F., Rouquerol, J., Sing, K.S.W.: Physisorption of gases, with special reference to the evaluation of surface area and pore size distribution (IUPAC Technical Report). Pure Appl. Chem. **87**(9–10), 1051 (2015)
25. Washburn, E.W.: Note on a method of determining the distribution of pore sizes in a porous material. Proc. Natl. Acad. Sci. **7**(4), 115 (1921)

26. Barrett, E.P., Joyner, L.G., Halenda, P.R.: The determination of pore volume and area distributions in porous substances. I. Computations from nitrogen isotherms. J. Am. Chem. Soc. **73**(1), 373 (1951)
27. Seaton, N.A., Walton, J., Quirke, N.: A new analysis method for the determination of the pore-size distribution of porous carbons from nitrogen adsorption measurements. Carbon **27**, 853 (1989)
28. Landers, J., Yu, G., Neimark, A.V.: Density functional theory methods for characterization of porous materials. Colloids Surf. A **437**, 3 (2013)
29. Ravikovitcha, P.I., P. I., Haller, G. L., Neimarka, A. V.,: Density functional theory model for calculating pore size distributions: pore structure of nanoporous catalysts. Adv. Coll. Inter. Sci. **76–77**, 203 (1998)
30. Alghunaim, A., Kirdponpattara, S., Newby, B.-M.Z.: Techniques for determining contact angle and wettability of powders. Powd. Technol. **287**, 201 (2016)
31. Kirdponpattara, S., Phisalaphong, M., Min, B., Newby, B.-M.Z.: Application of washburn capillary rise for determining contact angles of powders/porous materials. J. Colloid Interface Sci. **397**, 169 (2013)
32. Castellanos, A.: The relationship between attractive interparticle forces and bulk behavior in dry and uncharged fine powders. Adv. Phys. **54**(4), 263 (2005)
33. Hassanpour, A., Hare, C., Pasha, M. (Eds.): Powder Flow: Theory, Characterisation and Application. Royal Society of Chemistry (2019)
34. Prescott, J.K., Barnum, R.S.: On powder flowability. Pharma. Technol. **24**(10), 60 (2000)
35. Standard Test Method for Oil Absorption of Pigments by Spatula Rub-out. ASTM D281-12(2021)
36. General Methods of Test for Pigments and Extenders—Part 5: Determination of Oil Absorption Value. ISO 787 Part 5 (1980)
37. Höhne, G.W.H., Hemminger, W.F., Flammersheim, H.J.: Differential Scanning Calorimetry. Springer, Berlin, Heidelberg (2003)
38. Measurement and Interpretation of Electrokinetic Phenomena. International Union of Pure and Applied Chemistry, Technical Report. Pure Appl. Chem. **77**(10), 1753 (2005)
39. Diebold, M.P., Bettler, C.R., Mukoda, D.M.: Mechanism of $TiO_2$/ZnO instability. J. Coatings Tech. **75**(942), 29 (2003)
40. Schofield, R.K., Samson, H.R.: Flocculation of kaolinite due to the attraction of oppositely charged crystal faces. Disc. Faraday Soc. **18**, 135 (1954)
41. Delgado, A.V., González-Caballero, F., Hunter, R.J., Koopal, L.K., Lyklema, J.: Measurement and interpretation of electrokinetic phenomena (IUPAC technical report). Pure Appl. Chem. **77**(10), 1753 (2005)
42. Smoluchowski, M.: Contribution to the theory of electro-osmosis and related phenomena. Bull. Int. Acad. Sci. Cracovie **3**, 184 (1903)
43. General Methods of Test for Pigments and Extenders—Part 9: Determination of pH Value of an Aqueous Suspension. ISO 787 Part 9 (2019)

# Part II
# Optical Properties

# Chapter 3
# Light Scattering 1—The Physics of Light Scattering

## Contents



## Introduction

The defining difference between pigment particles and extender (or filler) particles is whether the particle interacts meaningfully with light or not. In the paint or plastic matrix, pigment particles interact with light by either partially or completely absorbing it, scattering it, or doing both. For most pigment particles, one of the two types of interaction—scattering or absorption—dominates the other, and we can classify these particles as being white pigments (when light scattering dominates) or being colored pigments (when light absorption dominates).

---

**Supplementary Information** The online version contains supplementary material available at https://doi.org/10.1007/978-3-030-99083-1_3. The videos can be accessed individually by clicking the DOI link in the accompanying figure caption or by scanning this link with the SN More Media App.




M. Diebold et al., *Pigments, Extenders, and Particles in Surface Coatings and Plastics*, https://doi.org/10.1007/978-3-030-99083-1_3

An important aspect of these interactions is that each allows us to make a paint film or plastic object opaque. They do this in different ways. For light absorption, light is removed before it has the chance to pass through a paint film or a plastic article and return to the eye. For light scattering, light is redirected out of the film or article before it has passed through it. With either interaction, full opacity is achieved when light does not emerge from the back of the film or object, making the visual information of anything behind the film or article obscured. For a paint film, this visual information is the appearance of the substrate. For a plastic article, it is more generally anything behind that article.

There are many situations where opacity is desired, and so both types of interactions between light and particles are of high importance in paints, plastics, and papers. We will discuss light scattering in this chapter and in Chap. 4, and light absorption in Chaps. 5, 6, and 15.

## Light and Light Scattering

The interaction of light with a particle is determined by the nature of both interacting species—the light and the particle. While our interest is in the particle aspects of this interaction, we cannot understand this interaction without first understanding the nature of light. We will therefore review this briefly, before discussing the role of the particle in light scattering interactions.

In Chap. 1, we discussed the four fundamental forces of nature—electromagnetism, gravity, the weak nuclear force, and the strong nuclear force. Electromagnetism is the combination of the electrical and magnetic forces, which are grouped into one force because they do not occur independently of one another, but instead are inseparably linked. To see this, we will consider each aspect of electromagnetism individually, and then in combination, beginning with the electrical force.

Certain subatomic particles possess an electrical charge, which can be either positive or negative. While these charges are localized on the particles, their influence extends through nearby space. This influence occurs through the electrical field, which is defined as a property of a given region of space that affects the electrical energy or forces on an electrically charged particle located within that region. This is admittedly a somewhat circular definition, which is partially due to the fact that the exact mechanism by which the effects of electrical charge extend through space is, as yet, incompletely understood. Although the exact mechanism is not known, these effects can be described quite accurately using several physical laws and their associated mathematical equations.

Magnetism is in some ways similar to electricity and in other ways different from it. Like electrical charge, it is a property of some forms of matter, and like electrical charge, it has two different states (north and south). The influence of magnetism also extends through space, and magnetic objects create a magnetic field that is analogous to the electric field. Unlike electrical charges, however, the two magnetic states cannot exist in isolation and instead are always found in opposing pairs.[1]

The link between electricity and magnetism is that when one type of field changes, it creates a nearby change in the other. For example, flowing electrons create a magnetic field (this is the basis by which electrical motors operate), and moving magnets create an electric field (this is the basis by which the magnetic strip on credit cards conveys information). This link leads to the phenomenon of self-propagating fields, wherein a perturbation in one type of field causes a perturbation in the other type, and as this second type of field changes, it creates a change in the first field type. Since the fields extend in space, the changing fields can propagate through it. This traveling self-propagating field is light. Light can be characterized by speed, wavelength (distance over which a single perturbation spans), and frequency (number of perturbations per second). The three are related as shown in Eq. 3.1.

$$c = \lambda \cdot \nu \tag{3.1}$$

where

c is the speed of light,

λ is the wavelength of light, and.

ν is the frequency of light.

Einstein famously showed that the speed of light is a constant, but this is true only within a given medium. Light traveling through a vacuum does so at roughly 3 $\times$ $10^8$ m per second, but it travels appreciably more slowly through matter. Matter, in essence, creates a drag on light, in much the same way that knee-high water acts as a drag on movement when wading. Different materials affect the speed of light differently, universally by decreasing it. One way to quantify the amount by which a material decreases the speed of light is as the ratio of the speed through a vacuum divided by the speed through that material. This ratio is termed the refractive index of that material. Its value varies from one material to another, and from one wavelength to another, and is generally denoted by the lower case letter n.

Refractive indices are useful ways of expressing the effect of a material on light because many optical properties of a material can be affected by it. Importantly, these properties include the ways that light can change direction (scatter) when it moves from one material to another.

[1] Magnetic monopoles were originally predicted mathematically by Dirac, but as yet have not been detected.

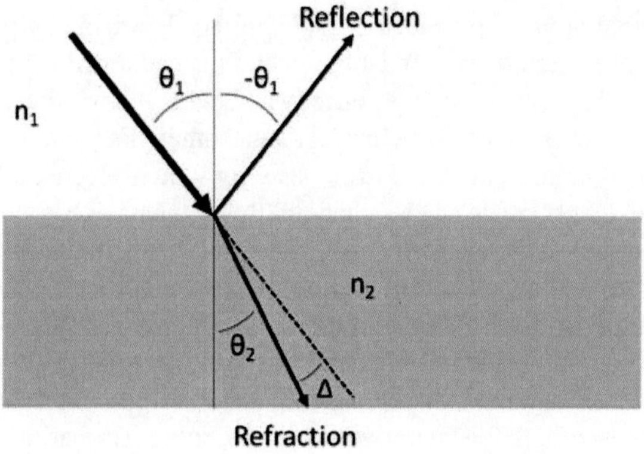


**Fig. 3.1** Reflection and refraction when light crosses a surface

## *Light Scattering Mechanisms*

Light scattering plays a crucial role in creating opacity in most paint, plastics, and paper applications. The term "scattering", when applied to light, refers to an interaction that changes the direction of the light, but not its intensity (brightness) or wavelength (color). The magnitude of this change in direction can be measured by an angle of deflection, and this angle may be large or small.

There are three means by which light traveling from one material to another changes direction. Two of these are familiar to us in our everyday experience—reflection and refraction (Fig. 3.1). These occur at the interface between two materials with different refractive indices (designated as $n_1$ and $n_2$). When one of these materials is a particle, this interface is the particle surface.

Reflection describes the light that rebounds from the surface at the mirror angle. Depending on this angle, reflection can occur with a large change in direction. While a large change in direction is desirable, not all the light that strikes a surface is reflected by it. The fraction that reflects is determined by the difference in refractive indices between the two materials, as given by Fresnel in Eq. 3.2[2] [1]. For reference, this fraction is shown for a particle in air, as a function of the refractive index of that particle, in Fig. 3.2.

$$\text{Reflected intensity} = \left[\frac{n_1 - n_2}{n_1 + n_2}\right]^2 \tag{3.2}$$

Light that is not reflected at the surface continues in the same general direction but is deflected by a relatively small angle. This redirection is termed "refraction". We are familiar with refraction when we see a straight rod partially submerged in water. The rod appears to bend at the water's surface, but, in fact, it is the light that bends at the surface (Fig. 3.3).

[2] This equation applies only to light that is incident normal to the surface. The relationship between reflected intensity and angle of incidence is complicated, but this does not alter our analysis.

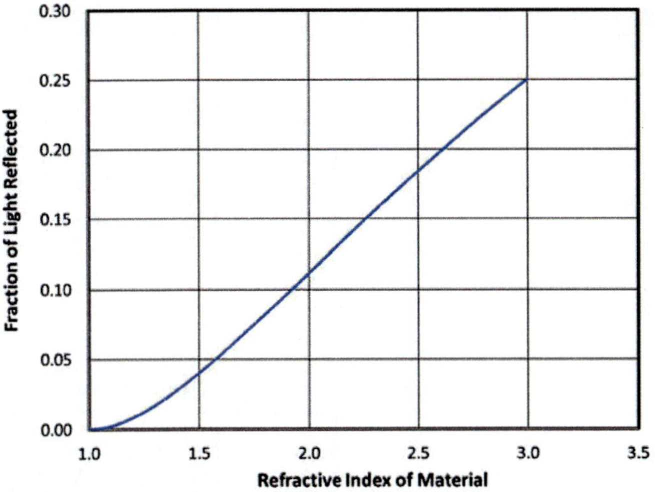


**Fig. 3.2** Fraction of light reflected at the interface of a material with air

**Fig. 3.3** Refraction at the air/water interface

As with reflection, the extent of refraction is determined by the difference in refractive indices between the two materials, but, for refraction, the angle of incidence ($\theta_1$ in Fig. 3.1) is also important. The mathematical relationship between the magnitude of the deflection ($\Delta$ in Fig. 3.1) and the refractive indices of the two materials is given in Eq. 3.3. This equation is often attributed to Snell, but others discovered it before him [2].

$$\text{Angle of Deflection} = \Delta = \theta_1 - \sin^{-1}\left[\frac{n_1 \sin(\theta_1)}{n_2}\right] \tag{3.3}$$

**Fig. 3.4** Wave diffraction. **a** Through a gap. **b** Around a barrier

The third mechanism by which light is scattered is diffraction. Diffraction is far less common in our everyday lives than reflection and refraction,[3] and so it merits closer analysis. Any wave that travels near an obstacle will be bent by it. This can be seen in ocean waves, which bend (diffract) around barriers such as rocks, piers, and even small islands. Interestingly, waves also diffract around openings in extended barriers. This is seen, for example, when waves enter a harbor through a small inlet. Babinet showed that the mathematics describing the diffraction of waves is symmetrical in this respect and so the effect of the obstructive object or the non-obstructive opening is identical [3]. These two forms of diffraction are shown schematically in Fig. 3.4. Similar considerations apply to sound waves and explain how sound can "bend" around a corner.

Diffraction also affects light waves. Here, an obstacle—for our purposes, a small particle—causes the light that passes near it to change direction. The magnitude of the change in direction is dependent on the wavelength in comparison to the size of the obstacle. If the particle is much smaller than the wavelength, it diffracts light poorly. However, when the particle size and light wavelength approach one another, diffraction can become a very important contributor to light scattering. Refractive indices, too, are important, and the greater the difference in refractive indices between the particle and the surrounding medium, the greater the magnitude of diffraction.

[3] There are few examples of light diffraction in our everyday lives because diffraction occurs at roughly the same scale as the wavelength of light being diffracted. Visible light spans the wavelength range of roughly 390 nm–700 nm, and features at this length scale are nearly impossible to see with the naked eye.

A commonality of the three light scattering mechanisms—reflection, refraction, and diffraction—is that the extent of scattering is dependent in part on the difference in refractive indices between the particle and the surrounding medium. The refractive indices of most polymers are near 1.5, and so to scatter light most strongly, the refractive index of a particle embedded in a polymer should be as far below this value or as far above it as possible. Air has the lowest refractive index (1.0), and so an air "particle" (i.e., a bubble or void in the polymer) is often used in paints, plastics, and papers to scatter light. At the other end of the refractive index scale, the rutile form of titanium dioxide has the highest refractive index for a colorless material (2.73 for light of wavelength 560 nm), and so it, too, is used in paints, plastics, and papers to scatter light. Light scattering is much stronger for rutile than for air, but air can cost nothing and so can be an economical way of achieving opacity.

## *The Scattering Cross Section*

An important consequence of light diffraction, particularly of visible light diffraction by a submicron particle, is that it causes light that passes close to the particle, but does not actually strike it, to be scattered by that particle, nonetheless. One way to quantify scattering is to do so based on the scattering cross-sectional area. The scattering cross section of a spherical particle is a circle centered at the particle center and perpendicular to the direction of the light. The size of the circle is such that any light striking the circle will be scattered by the particle—light that strikes the portion of the circle that is interior to the particle will be scattered by reflection and refraction, while light that passes through the portion of the circle outside the particle will be scattered by diffraction. The scattering cross-sectional concept is shown schematically in Fig. 3.5.

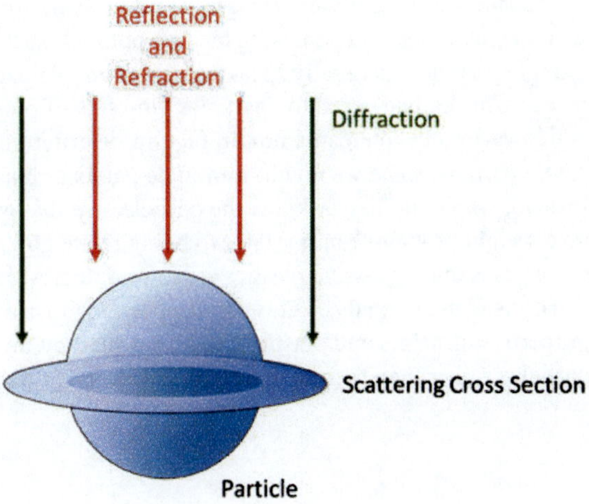


**Fig. 3.5** Scattering cross section of a spherical particle. Any light that passes through the circle will be scattered

## Light Scattering by a Single Particle—Mie Scattering

The phenomenon of light scattering by the particles formulated into paints, plastics, and papers is quite complex. In most of these materials, light is scattered by several particles as it travels into and out of the paint film or plastic object, with each scattering event redirecting the light in some way, large or small. Further complicating matters, the scattering strength of a particle within a polymer matrix is not invariant—instead, it is changed by the presence and identity of nearby particles. This has important implications in systems where the light scattering particles are agglomerated or where their concentration is high enough that particles are forced to be near one another.

Before we can consider light scattering in systems containing many particles, we must first understand light scattering by a single, isolated particle. This situation was approached theoretically by Mie in the first decade of the twentieth century [4]. Mie applied Maxwell's equations, which describe electromagnetic forces and fields, to the phenomenon of light scattering by a single spherical particle embedded in a matrix with a different refractive index. This analysis was further confined to particles having the same refractive index in all directions (many materials found in paints and plastics slow light to different extents in different crystalline directions). Within these constraints, Mie was able to solve Maxwell's equations exactly. The values calculated from the Mie analysis are generally referred to as "Mie Scattering" values.

Mie's analysis showed that the theoretical scattering strengths of particles are determined by the refractive indices of the particle and medium, the size of the particle, and the wavelength of the scattering light. We will use his equations to determine the values for these parameters that give the highest scattering efficiency under the above-listed constraints. Before doing so, however, we must be clear on the exact type of information that we desire from this analysis.

Our first inclination may be to determine the particle size that gives the greatest total scattering. This is, however, of little use to us. For example, a 10 cm glass sphere will scatter more light than a single 1 mm glass bead for the simple reason that it is much greater in size. There is, in fact, no optimal particle size for light scattering for a single particle—generally speaking, the larger the particle, the greater its scattering power. This extends even to "particles" the size of planets or larger.

However, our interest is not, in fact, in determining the greatest scattering for a single particle, since we do not formulate paints or plastics on a particle count basis. Instead, we would like to know the particle size that gives us the greatest scattering on a weight or volume basis (the two being related by particle density), since it is on these bases that paints and plastics are formulated. Although either weight or volume can be used in this analysis, it is more convenient to use volume because many optical properties of paints and plastics are determined by the volume concentration of the particles rather than their weight concentration.

## *Mie Analysis of Light Scattered by a Single Particle in a Polymer Matrix*

Our procedure for finding the optimal particle size for light scattering will be to first calculate the scattering cross-sectional area of a particle using the Mie equations and then normalize this value by dividing it by the particle volume. This gives a measure of scattering strength per unit volume of material, with units of area per volume, or one over length (most conveniently given in units of microns$^{-1}$). In the literature these values are often abbreviated as "S", but in this book, they will be abbreviated as "$S_{Mie}$" to avoid confusion with other measures of scattering that will be encountered in later chapters (in particular, the S constant in Kubelka–Munk theory).

The equations that Mie derived are quite complex, but one can find Mie calculators readily available online. These calculators use as inputs the particle size, the wavelength of the scattered light, and the refractive indices of the particle and the medium for light of that wavelength (recall that refractive index is wavelength dependent), and provide as output the calculated value of $S_{Mie}$. It is convenient to show this output graphically by holding three of the parameters constant and calculating $S_{Mie}$ as a function of the fourth. Because we are interested in the particle size that gives the greatest scattering per unit volume or weight, the parameter that will be varied in these graphs is particle size.

The results of the Mie calculations for different materials embedded in a polymer with a refractive index of 1.5 are shown in Fig. 3.6 for 560 nm light (this is the wavelength for which the human eye is the most sensitive—see Chap. 5). In this figure, we see the clear advantage that rutile has in scattering strength due to its

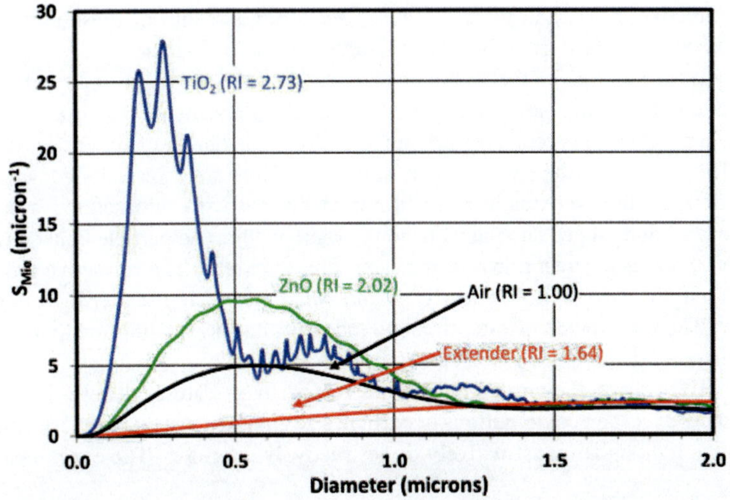


**Fig. 3.6** Mie calculation results for four materials embedded in a polymer resin (polymer refractive index 1.5; light wavelength 560 nm)

**Fig. 3.7** Opacities of paints made with different white particles. All paints are at the same thickness and have the same volume concentration of particles [5]

high refractive index, and we see that the scattering strength of these particles is maximized when they are roughly one-quarter micron in diameter.

We also see in Fig. 3.6 that although extender particles are white in air (where the refractive index of the surrounding medium is 1.0), they should be virtually transparent in polymer (where the refractive index of the surrounding medium is roughly 1.5). This is, in fact, the case, as seen in Fig. 3.7. Here are shown drawdowns of a series of paints that were made using different white particles, with refractive indices spanning a wide range. The paints have the same volumetric particle concentration (20 PVC) and are drawn down to the same film thickness. The refractive index of the resin in this case is 1.55. Clearly, extender particles do not scatter light in these paints.[4] The importance of the refractive index of the surrounding medium is also demonstrated in the video associated with Fig. 3.8.

The $TiO_2$ curve shown in Fig. 3.6 has many sharp features. These are due to resonances at certain combinations of particle size and light wavelength (560 nm) for which the light either constructively or destructively interferes. The exact locations

[4] By contrast, the paper industry relies heavily on extender particles to provide opacity. The key difference between this and paints or plastics is that the particles in a sheet of paper are surrounded by air rather than polymer.

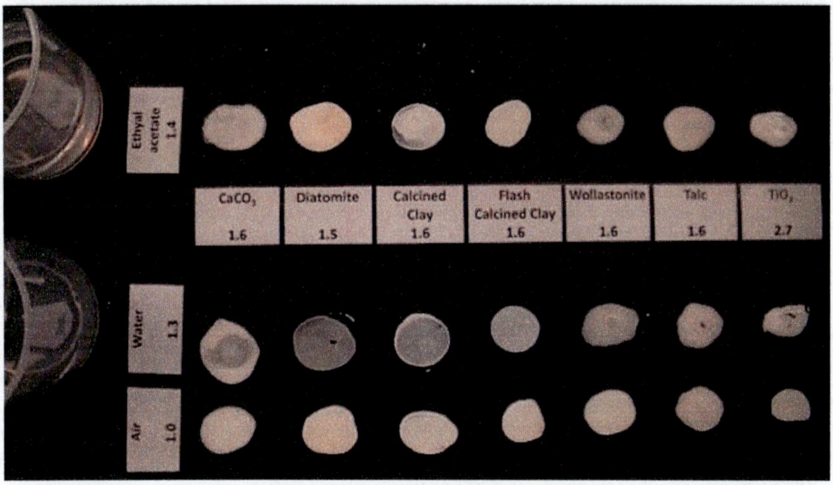


**Fig. 3.8** Video demonstration of the scattering power of various white powders when surrounded by materials with different refractive indices

of these resonances are different for different wavelengths of light, and over the entire range of visible light they average out to give a smoother curve (Fig. 3.9). As can be seen, there is in fact a broad maximum spanning the diameter range of about 0.20–0.30 microns, with a secondary peak around 0.77 microns.

Several decades after Mie's analysis, Weber tested the light scattering abilities of particles with varying refractive index and size. He developed an empirical equation, known as Weber's law, used to estimate the particle diameter that is optimal for light

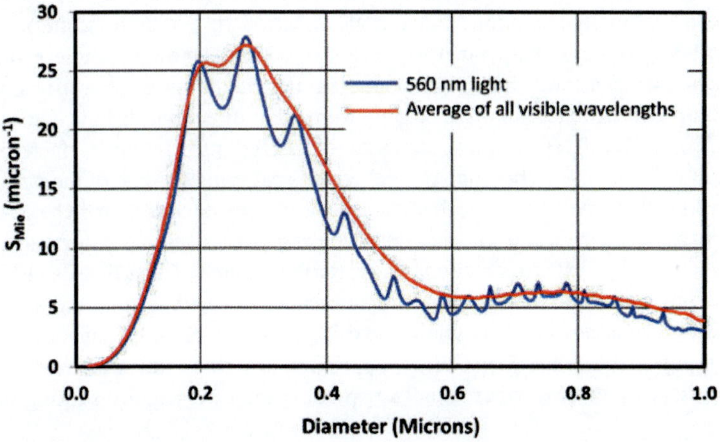


**Fig. 3.9** Average Mie scattering results for $TiO_2$ particles over the range of visible light wavelengths, as compared to the results for 560 nm light [5]

scattering for a given material [6]. This equation is shown as Eq. 3.4.

$$D_{optimal} = \frac{\lambda}{1.5(n_1 - n_2)} \tag{3.4}$$

where $n_1$ and $n_2$ are the refractive indices of the higher and lower refractive index materials, respectively.

We can test this equation by comparing its result for $TiO_2$ ($n_1 = 2.73$) embedded in a resin matrix ($n_2 = 1.50$) illuminated by 560 nm light, against the value calculated using the Mie theory (Fig. 3.6). The result from Eq. 3.4 is 0.30 microns, which is at the upper end of the range of that given by the Mie calculations, and slightly higher than the optimal size for an isolated $TiO_2$ particle as determined experimentally (about 0.23 microns).[5]

One will, on occasion, encounter paints that contain only resin and extender particles yet are white, or at least show some level of white opacity. This is not due to light scattering by the extender particles, as Fig. 3.7 clearly shows it cannot be, but is instead due to the presence of air voids that develop in films so crowded with particles that there is not enough resin to fill the regions between the particles. Such paints are discussed in detail in Chaps. 4 and 16.

## *The Scattering Volume*

The direction of light traveling through a paint or plastic that contains light scattering particles can, after enough scattering events, become random—a condition known as being "diffuse". This randomization of direction leads to the important concept of scattering volume. When discussing the scattering cross section of a particle, we did so by considering light coming from a single direction only, with the scattering cross section being oriented perpendicular to that direction. Through simple symmetry arguments, we can show that a particle has similar scattering cross sections, which are appropriately rotated, for light coming from other directions. If light comes from all angles (i.e., is diffuse), then the scattering cross sections will trace a sphere around the particle (Fig. 3.10). This sphere is called the scattering volume of the particle.

Any light that enters the scattering volume—from any direction—will be scattered by the particle. In the same way that the scattering cross section of a particle can be larger than its physical cross section, the scattering volume of a particle can be larger than its physical volume. As a point of reference, for visible light, the scattering volume of a quarter-micron titanium dioxide particle is roughly four times larger than its physical volume [5]. That is, approximately three times as much visible light scatters by coming close to a $TiO_2$ particle than by actually striking it. The

[5] While not obvious, the $TiO_2$ particle size for optimal light scattering is also affected by the concentration of the $TiO_2$ particles. Specifically, as concentration increases, the optimal size increases. This is discussed in more detail in Chap. 4.

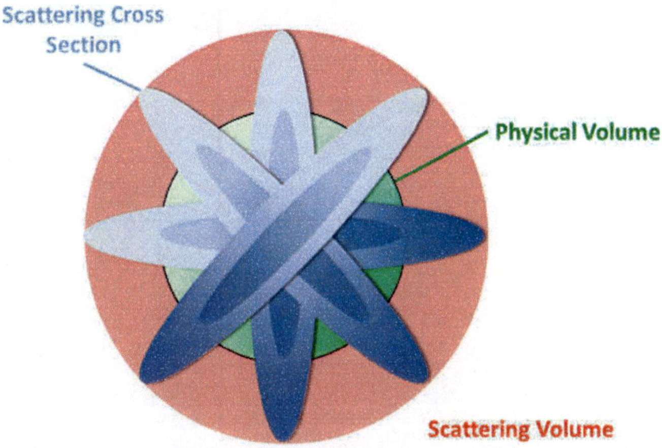


**Fig. 3.10** Generation of scattering volume by rotating the scattering cross section at all angles

implications of this for the scattering strength of a paint or plastic containing these particles are obvious.

## *Scattering Direction*

Earlier we defined light scattering as an event that causes the direction of the light to change, regardless of how large or small that change is. For maximum opacity, we obviously desire a large change in direction. However, we find that a single scattering event normally causes only a minor change in direction—certainly less than is needed to reverse the direction of the light. In his analysis, Mie developed an equation that calculates the theoretical intensity of the light deflected, as a function of deflection angle, for a single particle. The result of this calculation for 560 nm light scattered by a 0.25 micron diameter $TiO_2$ particle embedded in resin is shown in Fig. 3.11.

Clearly, the light is most likely to be deflected by a relatively small amount, and deflection in the backward direction has a very low probability. In order to return light back through the surface of a paint film or plastic article, the light must strike many particles. Even then, the light is not preferentially reflected back through the original surface, but instead, its direction is, at best, randomized (i.e., the light becomes diffuse), with the same amount of light proceeding through the film or object as being redirected out of it.

We will discuss light absorption in Chaps. 5, 6, and 15 in detail, but will here briefly compare the opacity efficiency of light-absorbing pigments to those of light scattering pigments. Even for a strong light scatterer such as $TiO_2$, dozens of photon/particle encounters are required for light to be scattered out of a paint or plastic. On the other hand, light absorption occurs in more than 99% of the encounters between a

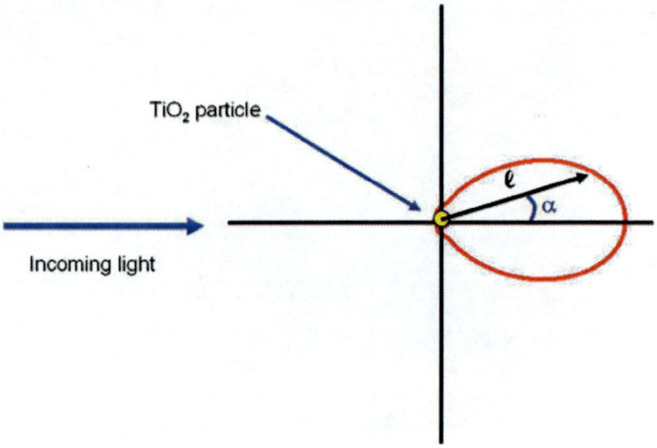


**Fig. 3.11** Scattering intensity versus scattering. Arrow length (l) is the relative scattering strength of light deflected at an angle α.[5]

photon and a strongly absorbing black pigment particle. For this reason, the weight of a black pigment in a high-quality opaque black paint is from one to two orders of magnitude less than that of a $TiO_2$ pigment in an opaque white paint.

## Scattering by Groups of Particles—Multiple and Dependent Light Scattering

Mie's analysis of light scattering by a single particle is useful for understanding how best to maximize light scattering—by maximizing the difference in refractive indices between the particle and surrounding resin while matching the particle size to the wavelength. However, as complex as this approach is, it is much too simple to adequately describe light scattering within a paint or plastic, where millions upon millions of particles are present.

We already described one shortcoming in using the Mie theory to quantify light scattering in a paint or plastic, which is that Mie's analysis only applies to a single light scattering event, whereas multiple scattering events are required to give opacity to a white paint film or plastic article. This limitation can be addressed through computer simulations of the paths of a large number of photons through an optically dense layer of the pigmented polymer [7].

A second, more consequential factor that is missing from Mie's analysis is the effect that one particle has on the scattering strength of nearby particles. Recall that, due to diffraction, the scattering volume of a particle can be larger than its physical volume. This is quite beneficial to opacity as it increases the scattering "footprint" of the particle, and so increases the amount of light scattered by it. However, there is a drawback associated with this property.

We can see this drawback if we consider the total light scattering from two particles. If the particles are far apart, separated by a few particle diameters or more, then each scatters light independently of the other, and the total amount of light scattering is simply twice that of a single particle. Imagine that we push the particles towards one another until they touch. While it is impossible for the physical volume of the particles to occupy the same space (i.e., to push one particle into another), this is not the case for the portions of the scattering volumes that reside outside of the particles. These regions are free to inter-penetrate, and, when they do, the total scattering volume of the system decreases by the amount of overlap volume. The overall effect is that the scattering strengths of particles decrease when particles come close to one another. We refer to the effect of the proximity of nearby particles on scattering strength as dependent light scattering and understand its cause to be scattering volume overlap [8–11].

It is quite common in paints, and even some plastics, for light scattering particles to approach one another closely enough that their scattering strengths are affected [12]. There are two causes for such close approaches. First, if the particles are attracted to one another, as small particles are (refer back to Chap. 1), then they are both difficult to initially separate (i.e., to disperse) and to remain separated (i.e., to stabilize). When the particles are not adequately dispersed during processing, we refer to the undispersed groups of particles as aggregates or agglomerates. In liquid paints, when dispersed particles join back together, we refer to these groups as flocculates. In either case, the practical result is the same—a loss of light scattering strength due to scattering volume overlap. Because of the importance of dispersion and dispersion stability to light scattering, these factors are discussed in detail in Chaps. 11 and 12.

The second cause for particles to approach is that their concentration is so high that it is simply impossible to keep them far apart. This effect is termed "crowding". We can estimate the particle concentration above which scattering volume overlap occurs by calculating the surface-to-surface distances[6] between perfectly arranged particles[7] as a function of their concentration. Such distances can be considered best-case scenarios—any deviation from these arrangements result in shorter distances between neighboring particles.

The results of these distance calculations are shown in Fig. 3.12 for 0.25 micron particles. Based on theoretical considerations, the surface-to-surface distance for which the scattering volume overlap of titanium dioxide particles of this size becomes important is about 0.5 microns [13]. This distance occurs at a $TiO_2$ particle volume

---

[6] Surface-to-surface distances are of greater importance than center to center distances because dependent light scattering is controlled by overlap of the portion of the scattering volume that is outside the particle volume—that is, in the region between the surfaces. For this reason, the extent of this overlap is best correlated to the distance between the particle surfaces.

[7] In three dimensions, there are two packing arrangements that do this—hexagonal closest packing and cubic closest packing.

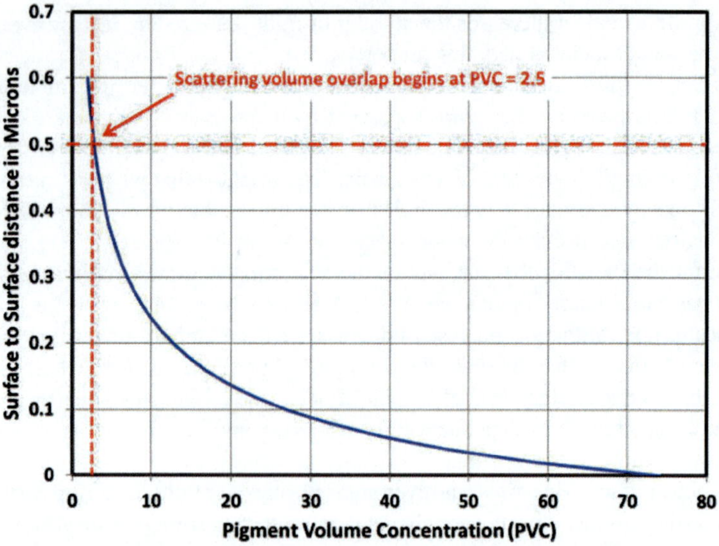


**Fig. 3.12** Surface-to-surface distances for perfectly spaced 0.25 micron particles as a function of particle concentration [5]

concentration (PVC)[8] of 2.5%. On a mass basis, this corresponds to 9.7 wt% $TiO_2$ in a typical polymer.

For comparison, the $TiO_2$ PVC values of most paints span the range from 6 to 25%, while those for plastics range from less than 0.5% for rigid objects to about 2% for thin films. For this reason, dependent light scattering is a much more important issue for paints than for plastics.

As an indicator of the potential magnitude of scattering loss, computer modeling has shown that two $TiO_2$ particles touching one another lose approximately 20% of their combined scattering power, while particles in the center of a 70-particle cluster lose approximately 70% of their scattering power [14]. The magnitude of these losses is consistent with the scattering volumes of the pigment particles being approximately four times larger than their physical volumes, as noted above.

As an aside, there is a proposition that has been made numerous times in the literature for increasing $TiO_2$ particle scattering strength by improving particle spacing at mid to high $TiO_2$ concentrations. The proposition is to cover the $TiO_2$ particles with a thick layer of material (up to a quarter micron) so that the $TiO_2$ portion of the pigment particles cannot approach one another closely enough to interfere with one another's scattering ability (or at least to keep the particles far enough apart to limit the amount of interference). Various proposals have included creating an air void from resin, with the $TiO_2$ particle inside the void ("nano-rattles") [15], placing a thick layer of polymer on the particle surface [16], and creating larger particles

[8] The use of PVC to measure particle concentration is discussed in detail in Chap. 4.

that are composed of appropriately spaced $TiO_2$ particles held in place by resin [17]. Often air voids are incorporated into the resin/particle composite as well [18].

The underlying premise of this proposition is that one can add high concentrations of $TiO_2$ to paints without the crowding normally associated with these concentrations of $TiO_2$ because the added particles are "frozen" in place at a set distance from one another. Unfortunately, there is a flaw in this reasoning.

We can see this flaw by looking at Fig. 3.12 from a different perspective. When we introduced this Figure, we did so as a means of describing the particle–particle distances as a function of $TiO_2$ concentration. However, we can also use this figure to determine the maximum concentration of particles as a function of the particle–particle distances. If we were to place, as an example, a 0.20 micron shell around the $TiO_2$ particles, then the closest distance these particles can approach one another is 0.40 microns. From Fig. 3.12, we see that the highest concentration of $TiO_2$ particles possible with this degree of spacing is 4.2 volume percent.

Thus, we cannot use these particles at $TiO_2$ concentrations above 4.2 volume percent. It is simply not possible to both space the particles well and to increase their concentration simultaneously—instead, we are limited to either having high concentrations of particles with minimal spacing (paints and paper) or distantly spaced particles at low particle concentrations (plastics).

## Summary

Particles embedded in a matrix can interact with light through either absorption or scattering. Each of these interactions can give opacity to paint, plastic, or paper, but with very different appearances: light absorption gives dark opacity (e.g., black paints), whereas light scattering gives light opacity (e.g., white paints).

Light scatters (changes direction) when it crosses the interface between materials with different refractive indices. There are three mechanisms by which this occurs: reflection, refraction, and diffraction. While all three occur when light is directed at any particle, their relative importance depends on the nature of the particle. Reflection and refraction of visible light dominate for macroscopic particles and objects while diffraction dominates at the micron scale.

Light scattering by a single spherical particle was first investigated by Mie. Using Maxwell's equations, Mie showed that scattering strength depends on the particle size, the wavelength of the scattered light, and the refractive indices of the particle and the surrounding medium. To maximize light scattering, the particle size must be matched to the light wavelength, and the difference in refractive indices must be as high as possible.

Two materials are typically used to scatter light in paints and plastics: $TiO_2$ particles, which have a refractive index greater than the polymer matrix, and air "particles" (bubbles or voids), which have a refractive index less than the polymer matrix. In both cases, the particle size for maximizing the amount of scattering per unit volume (or weight) of particle is somewhat less than a micron (about a quarter micron for $TiO_2$ and a half micron for air).

The scattering strength of a particle can be quantified by its scattering cross-sectional area. This is the area of a circle centered on the spherical particle and perpendicular to the direction of the light, and for which any light that it crosses will be scattered. Due to diffraction, the cross-sectional area can be significantly larger than the physical cross-sectional area of the sphere. The scattering cross-sectional concept can be extended to a scattering volume concept when light strikes the particle from random directions. For pigmentary $TiO_2$, the entire scattering volume is roughly four times greater than the physical volume of the particle itself.

Light scattering by more than one particle—as occurs in paints and plastics—is more complex than that by a single particle. One complication is that light must be scattered multiple times before it is returned through the surface of the paint or plastic. While multiple scattering complicates the analysis of light scattering, it can be modeled relatively simply and accurately by a computer.

A second, more consequential complication is the effect that particles have on the scattering efficiencies of their nearby neighbors. Because so much of a particle's scattering volume can reside outside the physical confines of that particle, the scattering volumes of two nearby particles can overlap, decreasing the entire scattering volume of the system. For pigmentary $TiO_2$ particles, which are roughly one-quarter micron in diameter, this undesirable effect occurs at surface-to-surface distances of roughly one-half micron. Such distances are unavoidable at $TiO_2$ volume concentrations greater than 2.5 volume percent, even when the particles are perfectly spaced. This concentration is greater than that typically seen in plastics applications but less than that seen in most paints. The implications of this effect on paint opacity are discussed in detail in Chap. 4.

## References

1. Fresnel, A.: Mémoire sur la loi des modifications que la réflexion imprime à la lumière polarisée (Memoir on the Law of the Modifications that Reflection Impresses on Polarized Light) (1866)
2. Kwan, A., Dudley, J., Lantz, E.: Who Really Discovered Snell's Law? Phys. World **15**(4), 64 (2002)
3. Babinet, A.: Memoirs d'optique m´et´eorologique (Memoirs of Meteorological Optics). Compt. Rend. Acad. Sci. **4**, 638 (1837)
4. Mie, G.: 'Beitrage zur Optik Truber Medien, Speziell Kolloidaler Metallosungen (Contributions to the Optics of Cloudy Media, Especially of Colloidal Metallic Solutions).' Ann. Phys. **25**, 377 (1908)
5. Diebold, M. P.: Application of Light Scattering to Coatings, Springer (2014)
6. Weber, H.: Lichtstreuung und Teilchengroflenverteilung kugelformiger Teilchen (Light Scattering and Particle Size Distribution of Spherical Particles). Kolloid-Zeitschrifl und Zeitschrift fiir Polymere **188**(1), 40 (1962)
7. Auger, J.C., Stout, B.: A Recursive Centered T-Matrix Algorithm to Solve the Multiple Scattering Equations; Numerical Validation. J. Quant. Spectrosc. Radiat. Transf. **79–80**, 533 (2003)

8. Fitzwater, S., Hook, J.W., III.: 'Dependent Scattering Theory: A New Approach to Predicting Scattering in Paints.' J. Coat. Technol. **57**(721), 39 (1985)
9. Wang, B.X., Zhao, C.Y.: Analysis of Dependent Scattering Mechanism in Hard-Sphere Yukawa Random Media. J. Appl. Phys. **123**, 223101 (2018)
10. Auger, J.C., Stout, B.: Dependent Light Scattering in White Paint Films: Clarification and Application of Theoretical Aspects. J. Coat. Technol. Res. **9**(3), 287 (2012)
11. Auger, J.C., McLoughlin, D.: Theoretical Analysis of Light Scattering Properties of Encapsulated Rutile Titanium Dioxide Pigments in Dependent Light Scattering Regime. Prog. Org. Coat. **77**, 1619 (2014)
12. Thiele, E.S., French, R.H.: Light-Scattering Properties of Representative, Morphological Rutile Titania Particles Studied Using a Finite-Element Method. J. Am. Ceram. Soc. **81**(3), 469 (1998)
13. Auger, J.C., Martinez, V.A., Stout, B.: Theoretical Study of the Scattering Efficiency of Rutile Titanium Dioxide Pigments as a Function of their Spatial Dispersion. J. Coat. Technol. Res. **6**(1), 89 (2003)
14. Thiele, E.S.: Scattering of Electromagnetic Radiation by Complex Microstructures in the Resonant Regime. PhD dissertation, University of Pennsylvania (1998).
15. Nguyen, D., Huynh, V., Lam, M., Serelis, A., Davey, T., Paravagna, O., Such, C., Hawkett, B.: Encapsulation by Directed PISA: RAFT-Based Polymer-Vesiculated Pigment for Opacity Enhancement in Paint Films. Macro. Rapid Commun. **42**, 2100008 (2021)
16. Lorimer, J.P., Mason, T.J., Kershaw, D., Livsey, I., Templeton-Knight, R.: Effect of Ultrasound on the Encapsulation of Titanium Dioxide Pigment. Colloid Polym. Sci. **269**, 392 (1991)
17. Korenkiewicz, S.M., Booth, K.A.: Method for Making Titanium Dioxide Pigment Grind Dispersion and Paint, US 9,487,670 (2016)
18. Stewart, D.M.D: Opacifiers for latex paints. In: Surface Coatings Volume 1, Raw Materials and their Usage; Third Edition, Editors: Oil and Colour Chemists' Association London, Springer (1993)

# Chapter 4
# Light Scattering 2—Light Scattering in Crowded Systems

## Contents





M. Diebold et al., *Pigments, Extenders, and Particles in Surface Coatings and Plastics*, https://doi.org/10.1007/978-3-030-99083-1_4

## Introduction

The complexity of light scattering is significantly greater in crowded systems than in dilute ones (here we define "crowded" systems as those having more than 2.5% volume concentration of $TiO_2$). In dilute systems, such as those typical of plastics applications, light scattering is proportional to $TiO_2$ concentration. However, in thin materials, such as paint films, particle concentrations must be proportionally higher than in thicker applications. Due to dependent light scattering, the response of scattering to $TiO_2$ concentration is not linear at these higher concentrations. The non-linear relationship between opacity and $TiO_2$ concentration has far-reaching consequences for the coatings industry and understanding light scattering in crowded systems is essential for effective paint formulation.

Similarly, $TiO_2$ particles are quite crowded in paper applications. Here the root cause is different: papers are generally composed of fibers that are only loosely intertwined. As such, these materials are mostly air, with large gaps (pores) between the fibers. Individual $TiO_2$ particles are far smaller than these gaps and so are prone to passing entirely through the paper during the wet stage of the papermaking process.

This issue is dealt with in two ways—first, retention aids are added into the process to affix the $TiO_2$ particles (and extender particles) onto the fiber surfaces. Second, the $TiO_2$ particles are intentionally flocculated into clusters (often hetero-clusters that include extender particles) that are large enough to be trapped by the fibers. Because of this, even though the volume concentration of the $TiO_2$ pigment may be low, the particles are crowded to the same extent as in highly concentrated paints. This is discussed in more detail in Chap. 18.

Our primary focus in this chapter is light scattering in paint systems. Many of the principles that are discussed here are demonstrated using paints that were developed solely for this purpose and without regard to other important paint properties. As such they are not meant to represent paints that would be suitable for commercial or practical purposes. The practical aspects of formulating such paints are discussed in detail in Chap. 16.

## Particle Volume in Paints—The PVC

Many important properties of paints, plastics, and papers are determined by the concentrations of the particles they contain. An important aspect of these concentrations is the way in which we measure them—on a weight basis, a volume basis, or something else. As a practical issue, using weight to define concentration is very convenient as we can measure weight directly and, in most types of manufacturing, we use weight to indicate how much of an ingredient to add to a production batch. However, when considering the properties of a material containing particles, it is often the volume of the particles that determine their effect on properties, rather than their weight. This is especially true in paint applications, where the particles typically occupy a significant portion of the paint film volume. For this reason, we quantify the concentrations of particles in paint films on a volume basis.

## *Measuring Particle Concentration*

There are two separate aspects of the volume concentration that are important to paint properties. The first has to do with the balance between the volume of resin and the volume of particles. Resin holds the particles together and attaches them to the substrate, and paints with similar balances between resin volume and particle volume often have similar mechanical and physical properties. For example, replacing one extender with another will normally result in similar paint properties if the replacement is done on an equal volume basis rather than an equal weight basis.[1]

The second reason we use volume to characterize concentrations in paints is that the scattering strength of $TiO_2$ particles decreases with decreasing surface-to-surface distances. This is because, as we saw in Chap. 3, the scattering volume of a $TiO_2$ particle is significantly greater than its physical volume. The fact that the scattering volume is much larger than the physical volume increases the total scattering strength of a particle considerably. However, at even modest concentrations of $TiO_2$, the particles are close enough together for their scattering volumes to overlap. This decreases the total scattering volume of the system and can be interpreted as a loss of $TiO_2$ scattering strength. As we will see, the rate of decrease in particle scattering strength is nearly constant with increasing concentration, and so volume concentration tells us how strongly the particles scatter light.

By contrast, in plastics applications, we typically use weight percent (also called "hundred weight" or "pounds per hundred"—pph) to characterize particle content. This is because at the low concentrations of $TiO_2$ that are typical of plastics applications (less than one volume percent), scattering loss due to scattering volume overlap is negligible. That is, the change in scattering loss with concentration, which makes the $TiO_2$ volume percent important in paints, is not an issue for plastics. In addition, particles are bought, sold, and used on a weight basis. Taken together these considerations support the use of weight as the basis for particle concentrations in plastics.

Particle concentrations measured on a volume basis are termed "pigment volume concentrations" and abbreviated as PVC. It is unfortunate that the word "pigment" is used here, rather than the word "particle", because this concentration is, in fact,

[1] Of course, the two are the same if the density of the particles are the same. This is true when replacing, for example, a calcium carbonate extender with one particle size with a calcium carbonate extender of another.

determined by the amounts of all particles, not just the pigment particles,[2,3] PVC is typically given as a percentage from 0 to 100, with the percent sign (%) omitted, but is occasionally given as a fraction from 0 to 1. It is usually clear from context which convention is being used.

The calculation of the overall PVC for a dry paint film is straightforward, as shown in Eq. 4.1. Because $TiO_2$ alone scatters light effectively, we often wish to calculate its relative volume separately. This quantity is called the $TiO_2$ PVC, and its calculation is shown in Eq. 4.2. When the particle concentration is low, the particles are said to be dilute, while, when it is high, they are said to be concentrated or crowded.

$$PVC = 100x \frac{(extender\ volume) + (TiO_2 volume)}{(extender\ volume) + (TiO_2\ volume) + (resin\ volume)} \tag{4.1}$$

$$TiO_2PVC = 100x \frac{TiO_2\ volume}{(extender\ volume) + (TiO_2\ volume) + (resin\ volume)} \tag{4.2}$$

Notice that in both equations the denominator is not the total volume of the paint film, but rather the combined volumes of the pigment, extender(s), and resin. In most paints, these combined volumes are, in fact, the total volume of the paint film. However, in paints that incorporate air pores for hiding purposes, we do not include the air volume in the PVC calculation. The implications of this are far reaching, as will be discussed later in this chapter.

The pigment-to-binder ratio (P/B) is sometimes used in the paint industry as an alternative to the PVC. This is a weight-based value that is defined as the mass of the pigments and extenders divided by the mass of the dried binder, and expressed as a fraction. Although the P/B ratio is a bit easier to calculate, we prefer to work with PVC as this is based on the physically more relevant volume composition of the film rather than the weight composition. The P/B ratio and PVC are related through Eqs. 4.3 and 4.4:

$$PVC = 100\ \mathrm{x} \frac{1}{\frac{\vartheta}{\mathrm{P/B}} + 1} \tag{4.3}$$

$$\mathrm{P/B} = \frac{\vartheta}{\left(\frac{100}{PVC}\right) - 1} \tag{4.4}$$

where $\vartheta$ is the ratio of the particle density to the resin density. If more than one type of particle is present, their weight average density is used in these equations.

[2] The origin of the term PVC dates to a time when it was common to call all particles added to a paint, including extender, "pigment". Today we restrict the word "pigment" to those particles that interact with light.

[3] It is also unfortunate that "PVC" is often used as the acronym for polyvinyl chloride, a plastic used in high volumes worldwide.

## *The Critical Pigment Volume Concentration (CPVC)*

The PVC tells us about the balance between the resin and particle contents. While the PVC concept appears straightforward, there is an important aspect of it that is somewhat complicated. This aspect is that above a certain PVC value—the critical PVC or CPVC—a third component enters the paint film: air.

Before discussing the implications of the air regarding the way that we view particle concentration, we will first describe how it is brought into the film through the following thought experiment. We begin with a cube, 100 microns on a side, that is filled with resin. The starting PVC is obviously zero, since there are no particles. Next, we wish to increase the PVC of the cube. We cannot do this by simply adding particles to the cube because the cube is already filled with resin. Instead, we must remove resin to make room for the added particles, that is, we must replace resin with an equal volume of particles. For our purposes, we will assume that the particles are added at random locations within the cube. If necessary, we will rearrange the particles already in the cube to create contiguous regions of resin large enough to accommodate new particles.

We can make these replacements, with the necessary particle rearrangements, only up to a point. At some concentration of particles, we find that while there is still resin present in the cube, no amount of particle rearrangement can create a region of pure resin that is large enough to fit an added particle. This concentration is the CPVC. At this concentration, there is just enough resin to coat each particle with a monolayer of resin and to fill the voids between the particle with resin [1]. We call the amount of resin needed to reach this state the "resin demand" of the particles.

We might expect the CPVC to be the upper limit of PVC, since no more particles can be fit into our 100 micron cube and so we cannot increase their concentration. This expectation is not correct, however. We can continue to increase the PVC because the PVC calculation uses only the volumes of the particles and the resin. We can continue to remove resin, while holding the particle concentration constant, by replacing resin with air. That is, we can continue to increase the PVC, as defined in Eqs. 4.1 and 4.2, by changing our thought experiment at the CPVC point from replacing resin with particles to replacing resin with air. The many implications of this change in our thought experiment are important and far reaching, as will be made clear below, but here we will observe that many of the properties of the paint film experience a discontinuity at the CPVC. Before discussing the implications of replacing resin with air, rather than with particles, we will consider two alternatives to use the PVC to express particle concentrations in systems above the CPVC.

The first of these is the pigment packing factor, $\phi$. This value gives the concentration of particles based on the entire volume of the paint film (or the volume of the cube in our thought experiment), including air, rather than the combined volumes of just the particles and resin. This calculation is shown in Eq. 4.5, which is similar to Eq. 4.1 except that the volume of air (when present) is included in the denominator. Unlike PVC, $\phi$ is typically reported as a fraction (from 0 to 1) rather than percentage. The pigment packing factor is identical to the PVC in paints below the

CPVC. Importantly, however, at and above the CPVC, the pigment packing factor is constant because we can place no more particles into our cube (i.e., neither the numerator nor the denominator of Eq. 4.5 change above the CPVC).

$$\phi = \frac{(extender\ volume) + (TiO_2\ volume)}{(extender\ volume) + (TiO_2\ volume) + (resin\ volume) + (air\ volume)} \quad (4.5)$$

The invariance of the pigment packing factor above the CPVC is an indicator of an important change that occurs at the CPVC. Below the CPVC, when we add particles, we increase their absolute concentration and so increase the degree of crowding (and the scattering volume overlap) that they experience. However, above the CPVC, particle crowding no longer increases. This means that there is no further increase in scattering volume overlap and so no further decrease in scattering strength as the PVC increases. That is, while more particles are present (the PVC is higher), the particles are no more crowded than at the CPVC. This seeming contradiction is explained by the total volume of the paint film being increased when air is added, while the combined volume of the particles and resin is held constant throughout. This will be explained in more detail later in this chapter.

The second alternative to the use of PVC to characterize particle concentrations is the reduced PVC. As we will see, paints do not all have the same CPVC value and, in fact, the CPVC values of two paints can differ by a significant amount. When comparing the properties of paints that differ in CPVC value, it is found that many paint properties are comparable when the degree to which the paints exceed (or, alternatively, are lower than) the CPVC is similar. This degree can be conveniently described by the reduced PVC, which is the ratio of the actual PVC to the CPVC and is typically indicated by the symbol $\Lambda$. The reduced PVC is less than one below the CPVC and greater than one above it.

## *Particle Packing in Crowded Systems*

Random packing of uniformly sized macroscopic spheres, for example, glass marbles or tennis balls, typically results in a volume density (CPVC) within the narrow range of $64 \pm 2\%$ [2]. By contrast, the packing density of particles found in paints, when at their CPVC values, span a much greater range, with values both above and below that of uniformly sized macroscopic spheres. There are three contributors to this difference in packing behaviors at the macroscopic and microscopic scales.

The first contributor is the relatively high attraction and frictional forces between small particles compared to their macroscopic counterparts. As described in Chap. 1, the packing of these particles in air is quite loose because their high surface areas give them high internal friction that "freezes" the particles in place and prevents the particles from settling or compressing into a denser state. The magnitude of this effect in air is such that the void fraction of nanosized particles can exceed 99%. A similar effect is seen when nanosized particles are packed in oil or resin, although to a lesser extent. As an example, nanoparticles of aluminum hydroxide (0.01 to 0.30 microns) have CPVC values from 13.0 to 33.2 [3].[4] Another way of saying this is that small particles have a high resin demand, and so a low CPVC value.

A second physical factor that can affect the CPVC of a particle is the particle shape. Packing efficiency generally increases when going from needles to plates to blocks. In addition, nanosized particles have associated with them a degree of "structure". This term applies to aggregates of these particles and refers to the openness of the three-dimensional arrangements of these particles. Quite often small particles form linear or spikey aggregates that cannot pack efficiently and so have low CPVC values. Examples of particles for which structure is important are fumed silica and carbon black, both of which have very low CPVC values (below 20).

Both frictional forces and a high degree of structure result in an open, low CPVC packing arrangement. By contrast, the third factor that affects CPVC, non-uniform particle size, results in a denser packing arrangement than is possible with densely packed, uniformly sized particles. This situation arises when a single particle type has a very broad particle size distribution or when two or more particle types with different particle diameters are mixed.

To demonstrate this principle as it applies to paints, we will first discuss the effect of multiple particle sizes on packing density in two dimensions. We begin by considering a field containing circles of diameter 10 units that are packed closely together (Fig. 4.1a). For closest packed arrangements of uniformly sized circles, the area coverage is 91%, but with the looser packing in Fig. 4.1a, the circles cover 75% of the area. Next, we place smaller circles (diameter 2 units) between the larger ones (Fig. 4.1b). These circles occupy 11% of the field, bringing the total area covered by all circles to 86%. Finally, we place still smaller circles (diameter 0.75 units) in the remaining void areas. This further increases the coverage to 93%, that is, the field is filled with loosely packed circles of three different sizes to a greater extent than is possible with efficiently packed circles of only one size.

The same principle applies to the filling of three-dimensional volumes. We can increase the CPVC of a particle mix by using particles of different sizes and nesting the smaller particles in the regions between the larger particles. This is demonstrated in Fig. 4.2, which shows the CPVC values of large extender particles (10 microns), small extender particles (0.55 microns), and mixtures of the two [4, 5]. Because small particles can displace resin in the regions between large particles, the total volume concentration of the particles in these mixtures (that is, their CPVC) can be greater than that of either particle packed efficiently by itself.

Note also that the CPVC value for the small particles (PVC = 41.6) is significantly reduced from the CPVC value of the large particles (PVC = 50.1), in accordance

[4] These values were determined by oil absorption measurements made using dioctyl phthalate oil (DOP) rather than linseed oil, which is used in a typical oil absorption test, as discussed later. This is relevant because DOP has a higher density (1.05 g/ml) than linseed oil (0.93 g/ml).

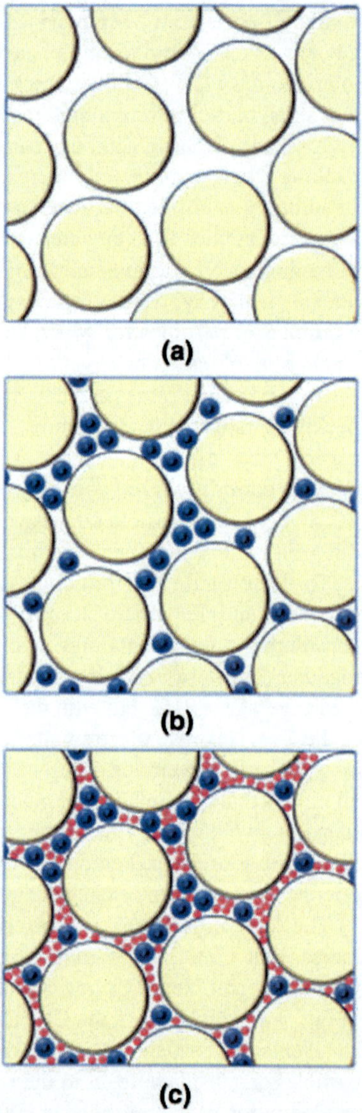

**Fig. 4.1** Increasing maximum coverage by nesting circles of different sizes

with the first two reasons for CPVC to vary that are outlined above. In addition to these reasons, particle surfaces are coated with a thin layer of resin, and the higher surface areas of small particles require more of resin to satisfy this need and so lowers the CPVC. This will be discussed in detail later in this chapter.

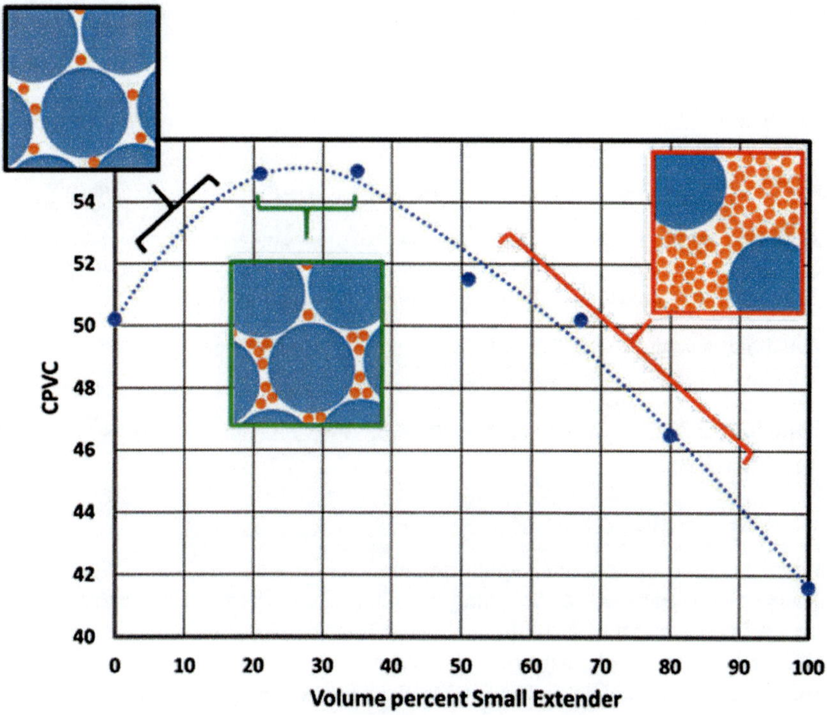


**Fig. 4.2** Critical pigment volume concentrations of a large extender particle (10 microns), a small extender particle (0.55 microns), and mixtures of the two. Adapted from [4, 5]

## The Opacity Versus PVC Curve

The relationship between a paint's composition and its opacity is complicated by fact that the scattering volume overlaps between $TiO_2$ particles, and with it their scattering strengths, are dependent on the concentrations of both the $TiO_2$ particles and the extender particles.[5] A practical way of addressing the formulation complexity that this causes is through a graph of paint opacity as a function of $TiO_2$ PVC. To use this graph as a formulation tool, we must first discuss the factors that determine it.

[5] Note that there is not a scattering volume overlap between $TiO_2$ particles and extender particles in contact with one another since the scattering volume of an extender particle embedded in a resin does not extend outside of its physical volume, due to it having a similar refractive index to that of the surrounding resin.

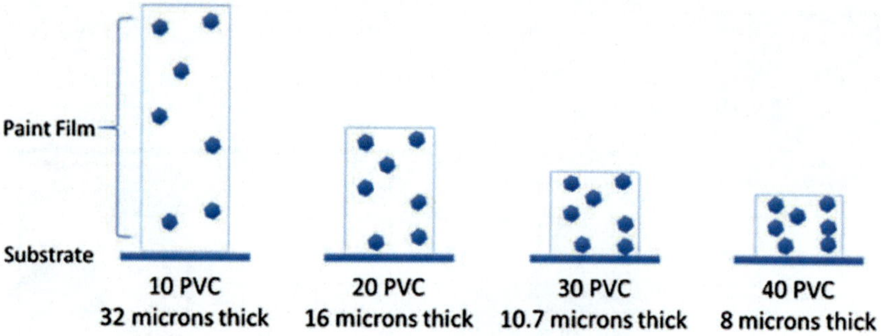


**Fig. 4.3** Relationship between PVC and dry thickness for the paints shown in Fig. 4.4

## *Visualizing the Loss of Scattering Efficiency Due to Dependent Light Scattering*

As was shown in Chap. 3, when the $TiO_2$ PVC exceeds about 2.5, the surface-to-surface distances between neighboring particles are small enough that even in an ideally spaced arrangement, the scattering volumes of the particles begin to overlap.[6] The detrimental effect of scattering volume overlap on paint opacity above this $TiO_2$ PVC can be seen visually in the following experiment: a series of paints were made at different $TiO_2$ PVC values and applied to black and white opacity charts at different film thicknesses. These thicknesses were chosen such that the total number of $TiO_2$ particles applied per square centimeter of dry film was the same (Fig. 4.3). Two grades of $TiO_2$ were investigated—a universal grade and a highly treated grade that is typically used in paints formulated above the CPVC (such paints are discussed in detail in Chap. 16).

Images for these two paint drawdown series are shown in Fig. 4.4. In both series, the opacity declines significantly as $TiO_2$ PVC increases, even though the same number of scattering centers cover each drawdown chart.

Close inspection of Fig. 4.4 shows that opacity generally decreases as $TiO_2$ PVC increases, as expected based on crowding arguments. However, there is a reversal of this trend at the highest PVC values for both paint series (the highest PVC paint made with the universal grade, and the three highest PVC paints made with the highly treated grade). This is due to these films being above the CPVC value for these pigments.[7] As such the films contain air voids that, as will be discussed in a later section of this chapter, are capable of scattering light on their own and also increase the scattering strength of the $TiO_2$ particles.

---

[6] Even below this PVC, scattering volume overlap typically occurs in paints because the particle arrangement is closer to random than to ideal, and so particles can, by chance, be close to one another.

[7] The significance of the difference in CPVC value for the highly treated grade (approximately 45) and the universal grade (approximately 60) will be discussed in Chap. 16.

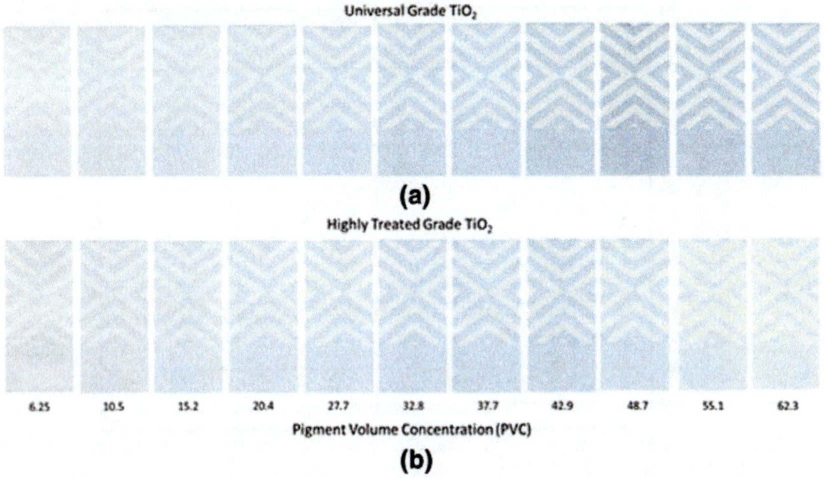


**Fig. 4.4** Effect of $TiO_2$ crowding on paint opacity for two grades of $TiO_2$ pigment. Each film has the same number of $TiO_2$ particles per square centimeter. **a** Universal $TiO_2$ grade. **b** Highly treated $TiO_2$ grade

## *Quantifying the Loss of Scattering Efficiency Due to Dependent Light Scattering*

The effect of $TiO_2$ PVC on film opacity is complex, since increasing the $TiO_2$ PVC increases the number of scattering centers, but at the same time reduces the scattering strength of these centers by increasing scattering volume overlap. To determine the overall effect of those opposing factors, we can experimentally measure the average scattering strength per $TiO_2$ particle as a function of PVC.[8]

The results of one such experiment, in this case a laboratory paint system,[9] are shown in Fig. 4.5. We see that the scattering strength of the $TiO_2$ particles decreases linearly with increasing particle concentration, and that the rate of decrease is relatively steep—the consequences of dependent light scattering can be quite significant between the two ends of the range of $TiO_2$ PVC values typically seen in paints (roughly 8 to 25). The scattering power difference is about 30%. Extrapolating this line to higher PVC values, we calculate that at the CPVC of this paint system (41.5), the total loss of scattering power is 75% compared to isolated $TiO_2$ particles.

[8] We do this by first measuring the scattering strength of an entire paint film (see Chap. 13), and then dividing that by the number of particles in the paint, using PVC as a proxy for the number of particles.

[9] We define a paint system as a series of paints, made with the same particle or particle mix, that spans a range of PVC values.

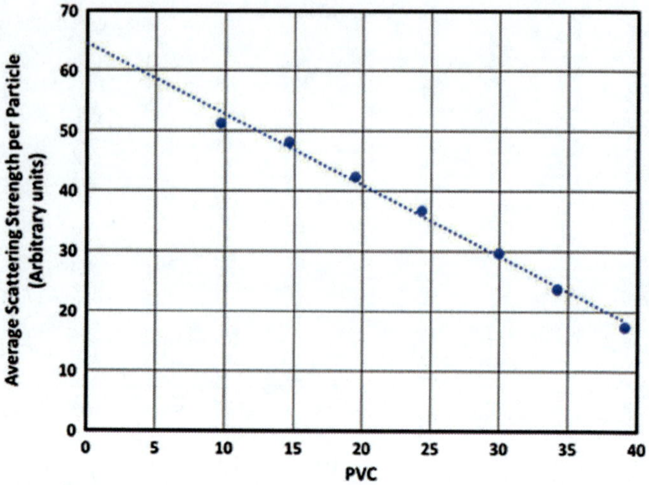


**Fig. 4.5** Scattering strength per particle for a series of paints made at different PVC values using the same resin and $TiO_2$ pigment

## Factors Affecting the Rate of Scattering Strength Loss at Increasing PVC Values

Studies similar to that shown in Fig. 4.5 for different paint systems reveal that the loss of scattering strength line varies according to the exact nature of the paint system. Some of the more important paint parameters that affect this relationship are the size of the resin particles, the amount of coalescing agent, the grade of $TiO_2$ pigment, and the size and concentration of any extender particles. This variability is shown for three pairs of paint systems in Fig. 4.6.

In the first pair of paint systems (Fig. 4.6a), the paints have the same grade of universal $TiO_2$ but differ in the size of the resin particles (the resin particles have the same chemistry). The second pair (Fig. 4.6b) shows paint systems made with the same universal grade of $TiO_2$ as in Fig. 4.6a, but with a different resin than the paints in Fig. 4.6a. In one case, $TiO_2$ is the only particle present while in the other a constant amount (15 PVC) of a 15 micron extender is also present. Finally, in the third pair (Fig. 4.6c), two types of $TiO_2$ pigments are compared (the same universal grade as shown in Fig. 4.6a, b, and a highly treated grade) in the same resin system as shown in Fig. 4.6b. Note that the blue lines in Fig. 4.6b, c are the same.

As is true for any line, those shown in Fig. 4.6 have two independent parameters—the y-axis intercept and the slope. Each of these is a direct measure of an important opacity attribute of the paints. The first, the y-axis intercept, is the scattering strength of an isolated $TiO_2$ pigment particle (that is, of a single particle for which there is no reduction in scattering volume due to the presence of close neighbors). We will refer to this as the intrinsic scattering strength of the particle.

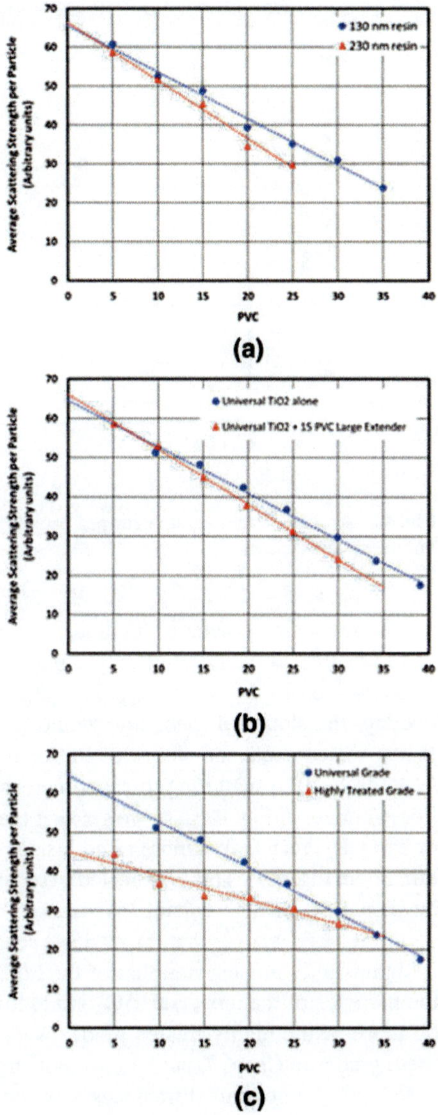


**Fig. 4.6** Average $TiO_2$ particle scattering strength as a function of PVC. **a** Universal $TiO_2$ pigment incorporated into paints with two different sizes of resins. **b** A universal $TiO_2$ pigment alone and with 15 PVC large extender. **c** Two different grades of $TiO_2$ pigment in the same paint as shown in (**b**)

We would expect that the intrinsic scattering strength of the $TiO_2$ particles to vary from one pigment grade to another, but to be independent of other aspects of the paint system (resin, extender, etc.). This should be the case for the four paint systems in Fig. 4.6a, b, since these are all made with the same universal grade of $TiO_2$. The lines for these paints are plotted together in Fig. 4.7, where we see that the y-axis intercepts (i.e., the intrinsic $TiO_2$ scattering strengths) for all four paint systems are, as expected, very nearly the same.

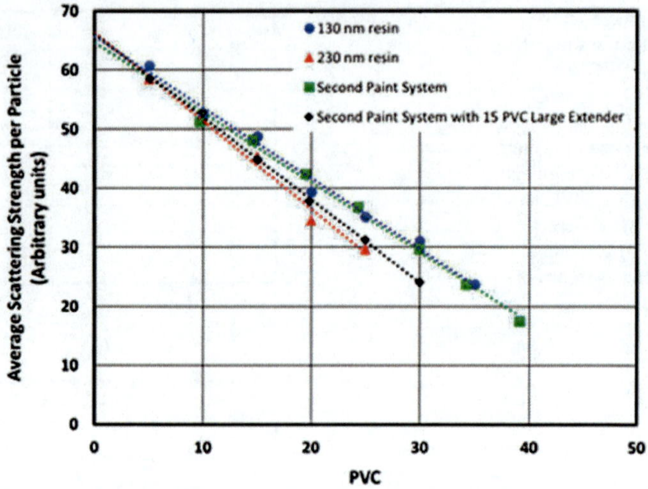


**Fig. 4.7** Average scattering strength per particle for the same universal grade of $TiO_2$ in four different paint systems

The second parameter of the lines shown in Fig. 4.6—their slope—is a measure of the degree to which scattering volume overlap increases with increasing $TiO_2$ concentration in the different paint systems. In the absence of scattering volume overlap, the slopes of these line would be zero and the line would be horizontal. When we compare the slopes of the lines in Fig. 4.6, we see that in some paint systems the $TiO_2$ scattering strength decreases more rapidly with PVC than in others. For example, in Fig. 4.6a, b, we see that the presence of large particles (either resin or extender) results in a more rapid loss of scattering power (this can be seen more directly in Fig. 4.7). This is expected based on the crowding effect that large particles have on $TiO_2$ particle spacing, as discussed earlier.

In Fig. 4.6c, we see that the intrinsic scattering strength of the universal pigment is significantly greater than that of the highly treated $TiO_2$ grade. We also see that the particles of the universal $TiO_2$ grade lose scattering strength more rapidly than do those of the highly treated grade. We will defer a more detailed discussion of these grades to Chap. 7, but here we note that these observations can be understood based on two important differences between these grades. First, as the name implies, the heavily treated grade has more surface treatment than the universal grade, and so a lower $TiO_2$ content (that is, there are fewer $TiO_2$ particles per kilogram of pigment). The reduced $TiO_2$ content of the pigment particles results in a reduction of their intrinsic scattering strength as particle concentration increases (i.e., their y-axis intercept).

Second, this heavy coating prevents the $TiO_2$ cores of the pigment particles from coming very close together (this is, in fact, its purpose). Because the surface coating acts as a barrier, preventing the close approach of the $TiO_2$ cores, it reduces the loss in opacity due to scattering volume overlap as the particle concentration increases.

The line for this grade, therefore, is less steep than that of the universal grade. In this case, it is roughly half as steep, indicating a nearly 50% reduction in the rate of loss of scattering strength.

## *Making the Opacity Versus PVC Curve*

Figure 4.5 shows the average scattering strength of a $TiO_2$ particle as a function of $TiO_2$ concentration for a laboratory paint system. While this information can be of great use to the paint formulator, paint consumers have little interest in the average scattering strength of a single particle. Their interests instead lie in the combined scattering of all particles, that is, the total scattering strength, or opacity, of the entire film.

The scattering strength of the entire film is simply the scattering power of an average $TiO_2$ particle (Fig. 4.5) multiplied by the number of particles in the film. It would be tedious to calculate the actual number of particles, so instead we multiply the scattering strength per particle by the $TiO_2$ PVC, which varies linearly with the number of particles. Doing this for the data points in Fig. 4.5 gives the curve in Fig. 4.8—the opacity versus PVC curve. This curve is among the more powerful tools available to the formulator of $TiO_2$-containing paints.

There is a sharp discontinuity of this curve at a PVC value of 41.5. This is the CPVC for this particular paint system. As was discussed in our thought experiment describing the process of increasing the PVC of a 100 micron cube of paint film, the

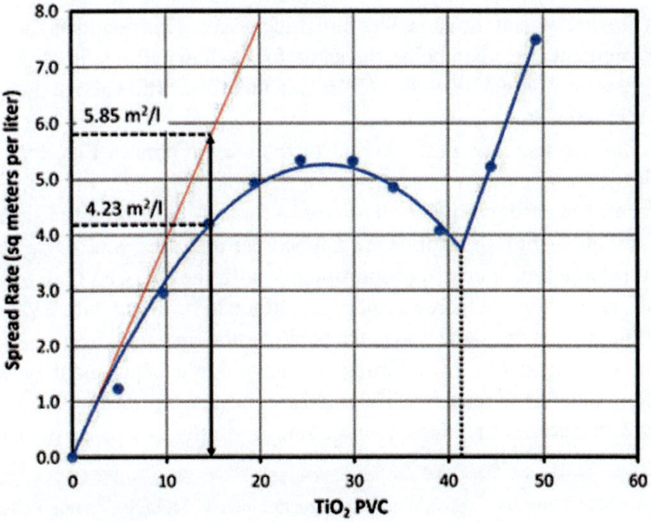


**Fig. 4.8** Opacity versus PVC curve for a paint containing only resin and a universal grade of $TiO_2$. This is the same paint formulation as used in Fig. 4.5

way PVC is increased below the CPVC is very different than the way it is increased above the CPVC—below the CPVC, resin is replaced with particles, while, above the CPVC, it is replaced with air. In that discussion we stated that this change in the thought experiment results in a sharp change in many paint properties at the CPVC point. Clearly film opacity is one of these properties.

There are many ways to characterize the scattering strength, or opacity, of a paint film. In opacity versus PVC charts, it is often convenient to do so using the spread rate of the paint, which is defined as the number of square meters that a liter of paint can cover at complete hide.[10] The benefit of using spread rate is that it directly displays the property of direct interest to the paint consumer, namely, how much paint is required for a specific job.

In practice, we construct the opacity versus PVC curve by measuring the spread rates of a series of white paints that are identical except for their $TiO_2$ PVC.[11] This is most easily done by first making large amounts of master paints at the two ends of the concentration range of interest and blending them in the appropriate ratios to give the target intermediate concentrations. If extenders are anticipated in the final formulation, they can either be held at a constant level throughout the paint series or varied in proportion to the paint PVC (that is, by using a constant ratio of $TiO_2$ and extender throughout the series). In addition, it is best (but not essential) that all the paints have the same volume percent solids. A bar chart representation of the relative volumes of the water, resin, and $TiO_2$ for the series of waterborne paints used to construct Figs. 4.5 and 4.8 is shown in Fig. 4.9a.

Next the paints are applied to black and white opacity cards (as shown in Fig. 4.4). Reflective measurements are made over the black and white backgrounds, and spread rate is calculated using the procedure in Chap. 13.

While not strictly necessary, it is conceptually easiest to consider the paints to have been drawn down at the same wet film thicknesses. This results in the combined volume of pigment and resin being the same for each dry film (Fig. 4.9b). The dry film thicknesses for paints below the CPVC are identical, while above the CPVC the thickness increases linearly with PVC.

The monotonic increase in the height of the orange bars in Fig. 4.9b indicates that the $TiO_2$ coverage rate, measured as, for example, the grams of $TiO_2$ per square meter, increases linearly with PVC. The reason for this linear relationship below the CPVC is obvious—the same solids are applied per unit area, and so the amount of $TiO_2$ per unit area will increase proportionally with the $TiO_2$ PVC. However, it is less obvious as to why this increase should continue to be linear with PVC, and why the rate of increase is the same above the CPVC as below.

We can understand this by returning to our thought experiment in which we sequentially replaced the resin in a 100 micron cube side with $TiO_2$ particles. Below the CPVC, the increase in $TiO_2$ coverage rate is simply due to an increase in the

[10] Most commercial paints have spread rate values greater than the laboratory paint used to generate Fig. 4.8. However, the laboratory paint was not optimized to give high levels of opacity, but rather was developed to best exemplify the concepts discussed here.

[11] The procedure for measuring the spread rate of a paint is given in Chap. 13.

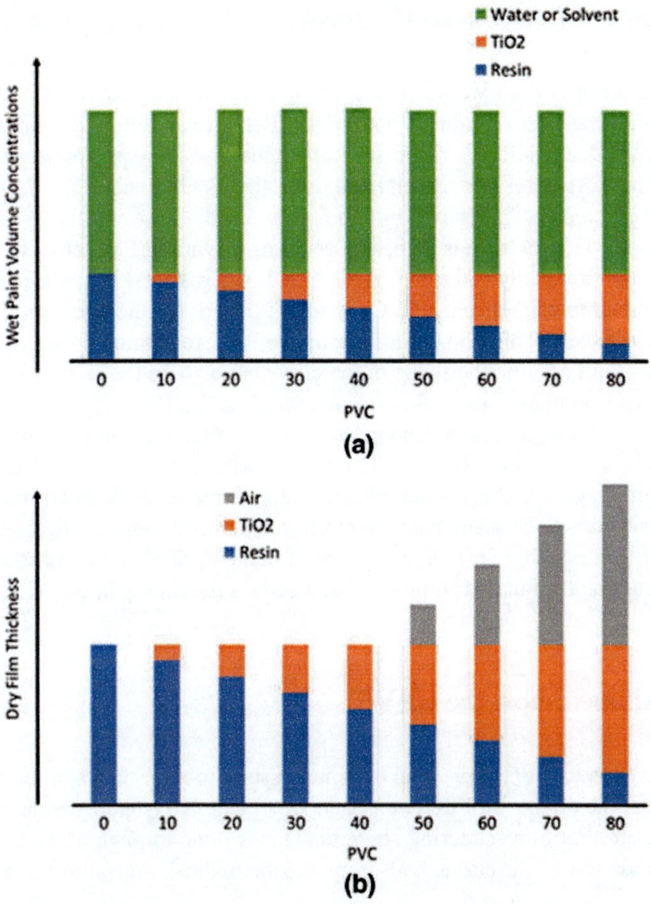


**Fig. 4.9** Relative proportions of **a** wet paint and **b** dried paint film for a PVC paint series. The CPVC for this paint is 41.5

number of $TiO_2$ particles in the cube. However, once we reach the CPVC value, we can no longer increase the number of $TiO_2$ particle in the cube, and so it seems that an increase in $TiO_2$ coverage with increasing PVC is no longer be possible. However, as shown in Fig. 4.9b, the volume of dry paint—that is, the number of 100 micron cubes in the film—increases above the CPVC due to the increase in dry film thickness (which, in turn, is due to the increase in air volume in the dry film). The increase in the number of cubes occurs at a rate exactly such that the relationship between the $TiO_2$ PVC and the $TiO_2$ coverage rate continues unabated as the paints pass across the CPVC threshold.

### *Significance of the Curve Maximum*

The significance of the PVC of the maximum in the opacity curve (PVC = 27.1 in Fig. 4.8) has often been confused both in the literature and in the coatings industry as a whole. It does indicate, of course, a maximum for some parameter of the paint system. This parameter is often claimed to be the PVC at which the $TiO_2$ scatters light most efficiently. This is in error—in Figs. 4.5 and 4.6 we saw that the scattering efficiency of a $TiO_2$ particle is highest at very large dilution (PVC close to zero) and decreases monotonically as the PVC is increased. Alternatively, it has been suggested that this maximum signifies the PVC for which the cost of the paint, on a coverage basis, is the lowest. This, too, is in error as the PVC for minimal cost of coverage is dependent not only on the shape of the curve but also on the relative costs of the resin and $TiO_2$ pigment, which do not enter into Fig. 4.8 [6].

The parameter being maximized in Fig. 4.8 is, in fact, the scattering efficiency of the entire film on a thickness basis, that is, the maximum in Fig. 4.8 gives the PVC for this paint system that gives the greatest hiding for a set thickness of film. This is an important consideration in many thin film paint applications, where only a limited amount of film is either desired or capable of being applied to the substrate. These paints should be formulated at the PVC that gives a maximum in the curve.[12]

## Curve Shape Below the CPVC

The opacity versus PVC curve is an extremely useful tool for the paint formulator as it can reveal the effects of different aspects of a paint ($TiO_2$ grade, resin, extenders if present, etc.) on film scattering strength. This is done through a careful analysis of various aspects of the curve, both from a mathematical viewpoint and a physical viewpoint.

### *Mathematical Analysis*

We will begin our analysis from a mathematical perspective and derive from this the expected shape of the curve below the CPVC (i.e., in the absence of air void scattering). To do this, we will return to our earlier description of the points on this curve as being calculated by multiplying, at each PVC value, the scattering power of an average particle at that PVC, by the number of particles at that PVC.

[12] Most paints are formulated at lower PVC values than the maximum in this curve for cost reasons—even though a lower PVC film must be thicker than the film at the PVC for the maximum in this curve, the $TiO_2$ in the film, by being at a lower PVC value, scatters light more efficiently and so less $TiO_2$ is required per unit area covered at complete hide.

From Fig. 4.5, we see that the scattering power of an average particle decreases linearly with increasing $TiO_2$ concentration. We can represent this with Eq. 4.6:

$$\text{Scattering per particle} = a \cdot (\text{PVC}) + b \tag{4.6}$$

As discussed when Fig. 4.5 was introduced, the slope of the line ("a" in Eq. 4.6) represents the rate of scattering loss as particle concentration increases. Because scattering power decreases with increasing PVC, the value of "a" is negative.

We can also write an equation for the number of particles in the paint as a function of the paint PVC:

$$\text{Number of particles} = c \cdot (\text{PVC}) \tag{4.7}$$

The number of particles, obviously, increases with increasing PVC, and so "c" in Eq. 4.7 is a positive number. Since there are no particles present when the PVC value is zero, there is no constant in this equation that would be analogous to the "b" term in Eq. 4.6.

To determine the total scattering strength of the paint film, we simply multiply the scattering power of an average particle by the number of particles:

$$\begin{aligned}\text{Total scattering} &= (\text{scattering per particle}) \times (\text{number of particles}) \\ &= (a \cdot c)(\text{PVC})^2 + (b \cdot c)(\text{PVC})\end{aligned} \tag{4.8}$$

Since Eqs. 4.6 and 4.7 are first order in PVC, their product (Eq. 4.8) is second order in PVC. That is, the total scattering (opacity) versus PVC curve is expected to be parabolic [7] as Fig. 4.8 shows it to be. Because there is no constant in Eq. 4.8 (or, more precisely, because the constant is zero), this function passes through the origin of the graph. The physical significance of this is that there is no film scattering when there are no particles present, which is a reasonable boundary condition to set for this curve. In addition, because the coefficient of the second-order term ("a·c") is negative, the parabola will face downward.

A second-order equation generally has associated with it three independent parameters—the coefficient of the second-order term ("a·c" in Eq. 4.8), the coefficient of the first-order term ("b·c" in Eq. 4.8), and a constant term (zero in this case). We saw that the constant term being zero had a physical significance; this is also true of the two other coefficients in this equation. The magnitude of the coefficient of the second-order term ("a·c") determines the degree of curvature of the parabola. When this coefficient is small, there is little curvature and the curve takes on a near-linear appearance.

The final parameter in a second-order equation—the coefficient of the first-order term (b·c in Eq. 4.8)—determines the location of the maximum of the curve (in terms of both the x-value and y-value at the curve maximum). When the other two parameters of Eq. 4.8 (the constant and the coefficient of the second-order term) are set, as they are here, the PVC of the maximum in the curve and the scattering power at this maximum cannot be set independently—instead they are linked through the coefficient of the first-order term.

The shape of the curve at low PVC values has important implications to materials with low concentrations of light scattering particles (such as most plastics). At low PVC, the second-order term in Eq. 4.8, $((a \cdot c)(PVC)^2)$, is quite small compared to the first-order term, $((b \cdot c)(PVC))$, and the curve in the vicinity of the origin is nearly linear. For this reason, the response of total scattering to PVC in most plastics applications, where the PVC value is quite low, is nearly linear. This is a mathematical restatement of the reason given for this earlier (that the surface-to-surface distances at low PVC values remain so large that there is very little increase in scattering volume overlap as the PVC increases).

## Examples of Opacity Versus PVC Curves

We saw in Fig. 4.6 that several parameters of a paint system can affect the relationship between $TiO_2$ scattering strength and PVC. In Fig. 4.10, we show the opacity versus PVC curves that are generated from the lines in Fig. 4.6.[13] As can be seen, these curves differ in (1) the tangent at the origin (shown as dotted lines in Fig. 4.10), (2) the degree of curvature below the CPVC, (3) the PVC of the curve maximum (i.e., the x-value at the maximum), (4) the spread rate at the curve maximum (i.e., the y-value at the maximum), (5) the CPVC value, and (6) the rate of opacity increase above the CPVC. In the previous section, we discussed the relationship between the first four of these differences and the mathematical equations describing these curves. We discussed the fifth difference—the PVC value of the CPVC—in the section on particle packing, and we will discuss the sixth difference—the shape of the curve above the CPVC—in the section on opacity above the CPVC.

## Quantifying the Opacity Loss Due to Scattering Volume Overlap

Returning to Fig. 4.8, the straight red line is the tangent of the curve at the origin. This line shows what the response of spread rate to increasing $TiO_2$ concentration would be if there were no reduction in $TiO_2$ scattering strength due to scattering volume overlap. We can use this red line to calculate scattering power loss due to scattering volume overlap. We see, for example, that the spread rate of the 15 PVC paint would be 5.85 $m^2/l$ in the absence of scattering volume overlap while the measured spread rate is only 4.23 $m^2/l$. This represents a scattering power loss of nearly 28%. At the CPVC, the reduction in scattering strength is 77%. Clearly no commercial paints

[13] In actual practice, the order of these steps is reversed—first we measure the opacity of the paint film (Fig. 4.10), and from that calculate the scattering strength of an average $TiO_2$ particle (Fig. 4.5).

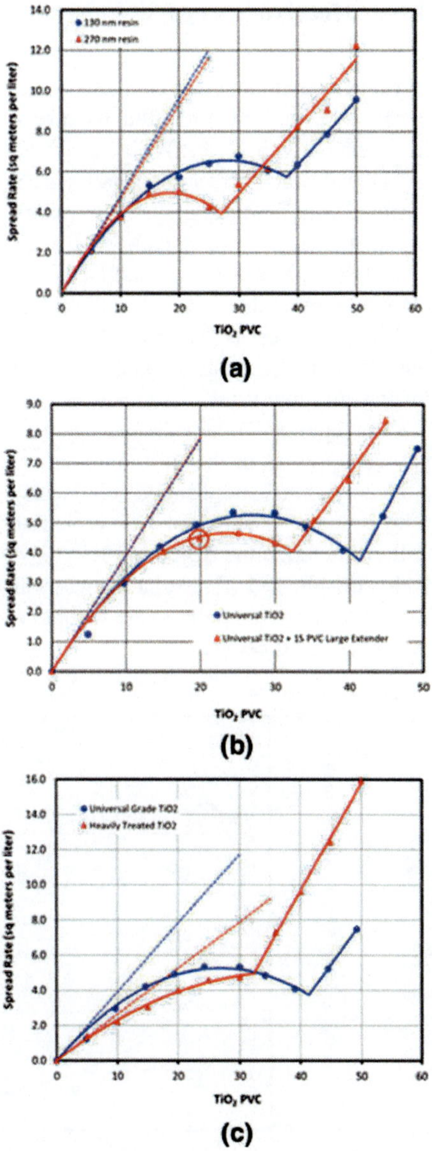


**Fig. 4.10** Opacity versus PVC curves for the paint pairs shown in Fig. 4.6. Dotted lines show expected scattering in the absence of scattering volume overlap (i.e., they are curve tangents at the origin). **a** Universal $TiO_2$ pigment incorporated into paints with two different sizes of resins. **b** A universal $TiO_2$ pigment alone and with 15 PVC large extender. **c** Two different grades of $TiO_2$ pigment in the same paint as shown in (**b**)

would ever be formulated with such a high $TiO_2$ content! Instead, when a high PVC is desired, high levels of extenders are used. It is important to note that the opacity losses seen here are unavoidable since they are due to the fundamental physics behind dependent light scattering, rather than to poor paint formulation.

## The Effective $TiO_2$ PVC

In paints containing only $TiO_2$ and resin, the PVC tells us not only how many $TiO_2$ particles are present in the film, but also the extent of $TiO_2$ crowding. As was shown in Chap. 3, Fig. 3.12, and reproduced here as Fig. 4.11, we can use simple geometry to calculate the surface-to-surface distances of ideally spaced 0.25 micron $TiO_2$ particles as a function of PVC. In paints that contain extender particles, however, the relationship between $TiO_2$ PVC and pigment crowding is altered. In particular, in the presence of extender particles, the $TiO_2$ particles are typically crowded more closely together than Fig. 4.11 would indicate.

When extender particles are present, it is useful to separate the information that the PVC tells us about the number of $TiO_2$ particles present in the paint from the information that it tells us about the degree to which these particles are crowded. We use $TiO_2$ PVC, as calculated in Eq. 4.2, to characterize the number of $TiO_2$ particles present in the film. To characterize the degree of $TiO_2$ crowding in the paint containing extender, we will develop a method to calculate the PVC value that reflects the same degree of $TiO_2$ crowding in the paint without extender. That is, we will characterize the degree of $TiO_2$ crowding in a paint containing extender, and with a certain $TiO_2$ PVC that we will call X, by saying it is the same as the degree $TiO_2$ crowding in a paint made without extender at a $TiO_2$ PVC of Y. This will be further clarified below.

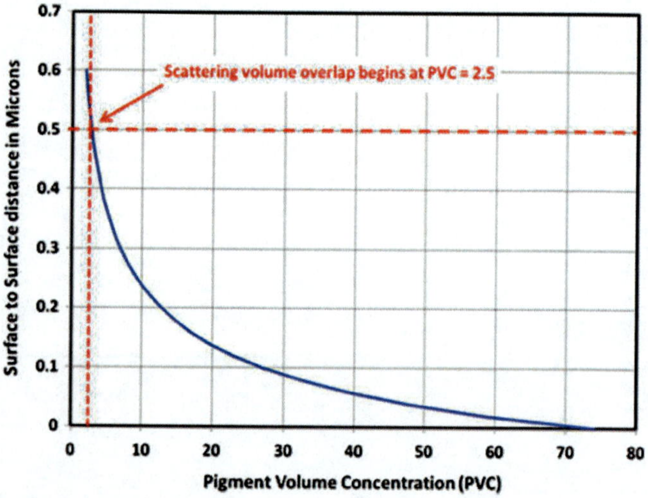


**Fig. 4.11** Surface-to-surface distances for perfectly spaced 0.25 micron particles as a function of particle concentration [6]

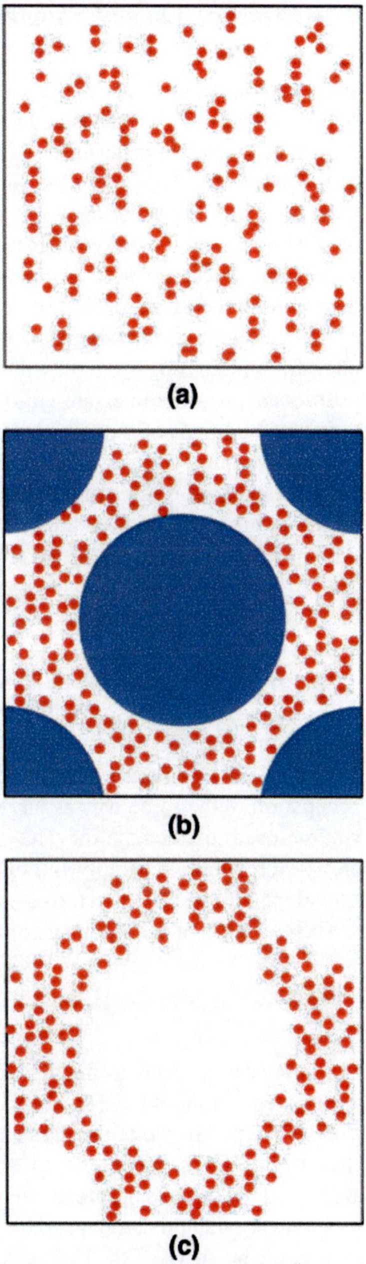


**Fig. 4.12** Crowding of $TiO_2$ particles (red) by the presence of large extender particles (blue). All three fields have the same number of $TiO_2$ particles

## *The Effect of Large Extender Particles on $TiO_2$ Crowding*

If we compare paints made with the same concentration of $TiO_2$ particles, one paint with large extender particles and the other without, we find that the degree of $TiO_2$ crowding is much greater in the paint made with large extender. The direct cause of this is that the $TiO_2$ particles cannot occupy the same regions in the film that are already occupied by the large extender particles. Instead, the $TiO_2$ particles must crowd into the interstitial voids between the extender particles. From a $TiO_2$ packing viewpoint, we can ignore the volume of film occupied by the extender particles when determining the extent of $TiO_2$ crowding.

This crowding is illustrated schematically in Fig. 4.12. In Fig. 4.12a, we see a paint for which $TiO_2$ is the only particle. As is apparent in this figure, the randomly positioned $TiO_2$ particles are entirely spread out within the entire region, and the scattering volumes of the particles overlap one another to only a minor extent. The situation is quite different when large extender particles are present (Fig. 4.12b). Here, despite the film having the same total volume and the same number of $TiO_2$ particles, the distribution of the $TiO_2$ particles is much different. This is more easily seen in Fig. 4.12c, where the $TiO_2$ particles are located in the same positions as in Fig. 4.12b, but the large extender particles have been removed for clarity. We see that the $TiO_2$ particles are much more crowded in the presence of large extender particles than in their absence.

Clearly the relevant volume for calculating the extent of $TiO_2$ crowding is not the entire volume of the paint film, but rather the volume available for $TiO_2$ particle occupation. Stieg noted that when the extender particles are of the same size as the $TiO_2$ particles or smaller, they do not restrict the volume available for $TiO_2$ particle occupation in the same way as large extender particles [8, 9]. He proposed that the volume used to calculate the $TiO_2$ PVC when we are interested in characterizing particle crowding is simply the entire volume less than the volume of the large extender particles. Stieg referred to the PVC value calculated in this way as the "effective PVC" of the pigment. His calculation for effective PVC is given in Eq. 4.9.

$$Effective\, TiO_2\, PVC = \frac{TiO_2\, volume}{(small\, extender\, volume) + (TiO_2\, volume) + (resin\, volume)} \tag{4.9}$$

The success of Stieg's effective $TiO_2$ PVC concept for paints containing relatively low concentrations of large extenders can be seen in Fig. 4.13. In Fig. 4.13a, the spread rates of paints are plotted as a function of the $TiO_2$ PVC as calculated by Eq. 4.2. Here we see the negative effect that the large particles have on $TiO_2$ crowding, as evidenced by the drop in spread rate with increasing extender content at a given $TiO_2$ PVC. We see that this drop is minimal at low $TiO_2$ concentrations, as expected, since even in the presence of the large extender particles there is plenty of room available for the $TiO_2$ particles to remain well separated at these low $TiO_2$ PVC values.

Using Stieg's effective PVC concept, we can make a model of expected opacity as a function of both extender PVC and $TiO_2$ PVC. The results from this model are plotted as dashed lines in Fig. 4.13a. Note that these are not best fit lines for the data

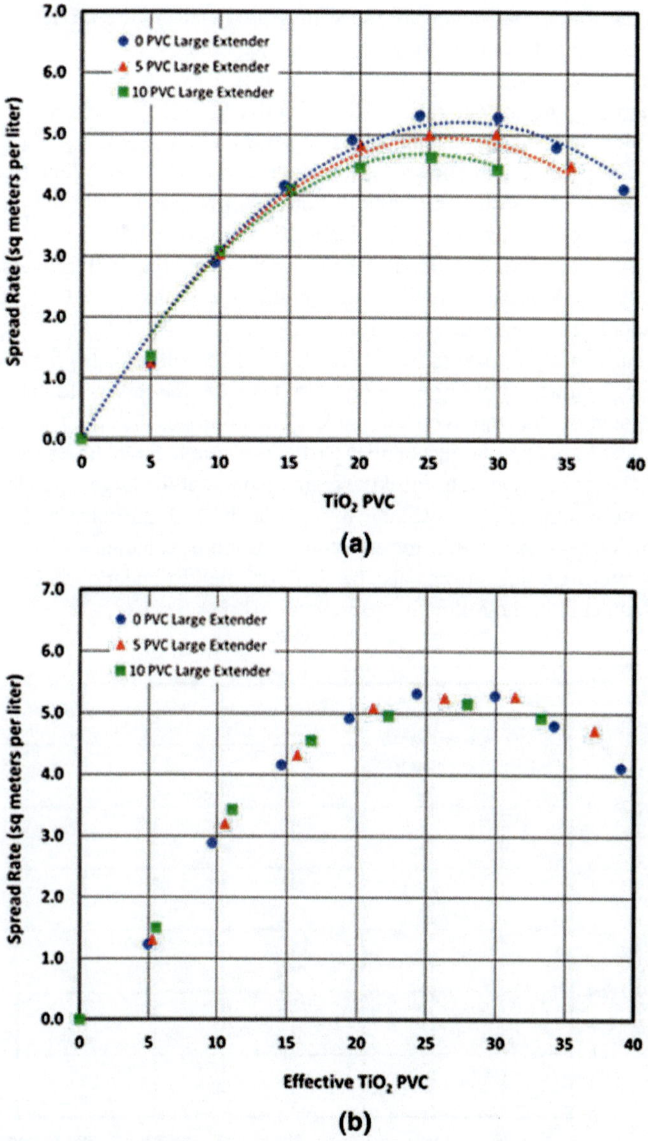


**Fig. 4.13** Response of spread rate to $TiO_2$ concentration in paints containing large extender particles. **a** Based on actual $TiO_2$ PVC. The dashed curves are from the effective PVC model, not best fit lines to the data. **b** Based on effective $TiO_2$ PVC

points on the graph, but rather are the values predicted by the model. The details regarding model construction are given in the appendix.

An alternative way to demonstrate the effective PVC concept is to multiply the spread rate of a paint by the ratio of the effective $TiO_2$ PVC over the actual $TiO_2$ PVC and plot the result against the effective $TiO_2$ PVC (rather than the actual PVC, as shown in Fig. 4.13a), as shown for this dataset in Fig. 4.13b. When this is done, the data points for the different paint systems will fall on a common curve if the assumptions associated with the effective PVC value are valid.

In Fig. 4.13, the concentrations of large extender are relatively modest (only as high as 10 PVC). At higher extender concentrations, the effective $TiO_2$ PVC concept fails to model the opacity data accurately. This can be seen in Fig. 4.14. Here we apply the same process used to make Fig. 4.13b to the spread rates for paints made at higher concentrations of large extender. In this case, the results do not agree with those predicted by the effective PVC model. The failure of the effective PVC model at high concentrations of large extender particles is due to the resin demand of these particles. The resin required to fulfil the resin demand of the large extender particles decreases the amount of resin available to separate the $TiO_2$ particles, and so the $TiO_2$ particles are closer together than the effective PVC model calculates. This same effect on $TiO_2$ scattering efficiency can also be seen for small extender particles, as will be discussed in detail in the following section.

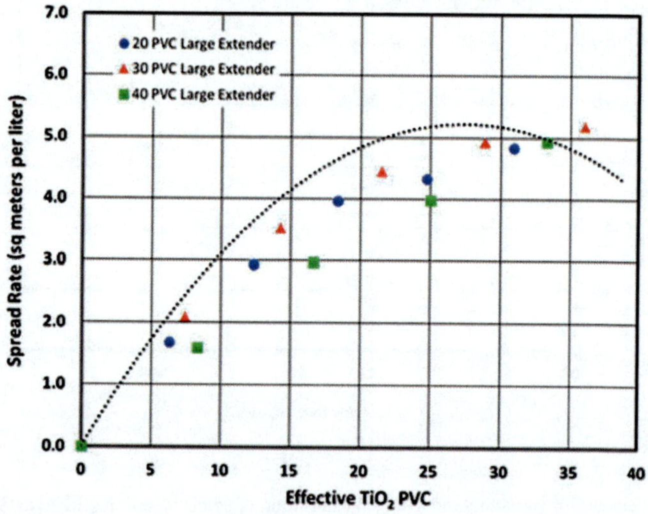


**Fig. 4.14** Effective PVC concept at high large extender concentrations. The dotted black line is the best fit line for the three series shown in Fig. 4.13b

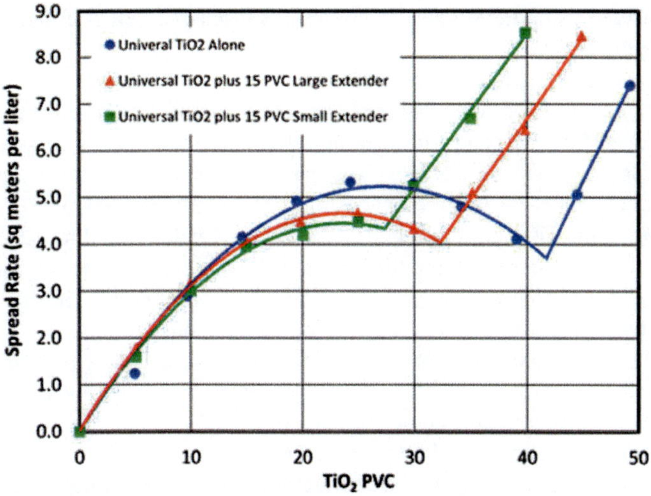


**Fig. 4.15** Effect of large and small size extender particles on paint opacity

## *The Effect of Small Extender Particles on $TiO_2$ Crowding*

Because well-dispersed small extender particles do not create the same large, contiguous regions of inaccessibility to $TiO_2$ pigment particles that an equal volume of large extender particles create, we might assume that the replacement of resin by small extender particles would not affect $TiO_2$ particle spacing. However, experiments show that while this assumption is often correct, it is sometimes incorrect. In fact, the presence of small extender particles can, in some cases, negatively affect paint opacity by roughly the same extent as an equal volume of large extender particles [10]. This is shown in Fig. 4.15, which compares the opacity versus PVC curves for paints made with a universal grade of $TiO_2$ alone and mixed with either a 15 micron extender or a 0.8 micron extender.[14]

A different mechanism applies to the reduction in $TiO_2$ scattering strength in the presence of small extender particles than that applied to large extender particles. This mechanism is related to that which explained the failure of the effective PVC concept at high concentrations of large extender particles, discussed earlier (Fig. 4.14). Both mechanisms are based on the resin demand of the extender particles.

Resin demand was discussed in Chap. 2 and is defined as the amount of resin required to bring a collection of particles to their CPVC point, that is, the amount of resin required to give each particle a thin (monomolecular) coating and to fill the spaces between the particles when they are packed at their highest density. This can also be described as the amount of resin that absorbs into the particle bed and is

[14] It is important to repeat that the laboratory paints used in these experiments were not formulated to be commercially viable, but rather to demonstrate certain principles. As such, in some cases, their behaviors cannot be directly extrapolated to those of commercial paints.

often measured using the oil absorption test, as discussed below in the section on measuring the CPVC.

To understand the role of extender resin demand in $TiO_2$ particle spacing, we will return to our analysis of the increase in $TiO_2$ particle crowding when large particles are introduced into the paint film (see Fig. 4.12), and the opacity loss due to this crowding. In the previous section, we approached this opacity loss from a geometric viewpoint. The physical volume of the film occupied by the extender particles is unavailable to the $TiO_2$ particles, and so the $TiO_2$ particles crowd together into the regions that are available to them (the interstitial voids between the large particles). The degree of crowding was quantified by calculating the effective $TiO_2$ PVC, which omits the large particle volume from the PVC calculation.

There is an alternative interpretation of the crowding effect. This interpretation focuses on the resin, rather than the extender particles. To understand this approach, we must first consider the different roles that resin plays in a paint film [6].

We begin by examining a film that is at the CPVC of the paint system. Here the particles are packed at their densest configuration. A portion of the resin coats the particles, and the remainder fills the voids between them. The combined amounts of resin are the resin demand of that particle mix. Paints formulated below their CPVC have a higher resin concentration than at the CPVC, and the resin in excess of the resin demand is situated between the particles, pushing them apart and decreasing their crowding (this is often referred to as diluting the particles). This resin improves the scattering efficiency of the individual $TiO_2$ particles.

For the sake of clarity, we will refer to the resin that coats the particles as "type 1" resin, which fills the voids between the particles at their densest packing as "type 2" resin, and the remaining resin, which pushes the particles apart from their densest packing, as "type 3" resin.[15] These designations are based on the role of the resin alone—the different types of resin are chemically and physically identical.

The amounts of types 1 and 2 resin determine the CPVC of the paint system[16] while the amount of type 3 resin determines the actual PVC—the greater the amount of type 3 resin, the greater the spacing between the particles and so the lower the PVC. Classifying the resin into these three types is also useful in explaining one reason for the CPVC of small particles to be lower than for large particles (Fig. 4.2). Because the surface areas of small particles are greater, on an equal volume basis, than large particles, their type 1 resin requirements are also greater. This increases the amount of resin at the CPVC conditions of the small particles (that is, their resin demand), and so decreases their CPVC value.

Returning to our alternative interpretation of the crowding effect, we next consider the addition of large extender particles to paints formulated with only $TiO_2$ and resin that are at a PVC value below the CPVC. Recall that we cannot simply "add" extender particles to this film—instead we substitute some of the resin with an equal volume of large particles. The resin that is substituted is not a random mix of the three types of

[15] Type 3 resin is often referred to a "free" resin, as it is not bound to the role of satisfying the resin demand of the particle mix.

[16] That is, the sum of the amounts of types 1 and 2 resin is the resin demand of the particles.

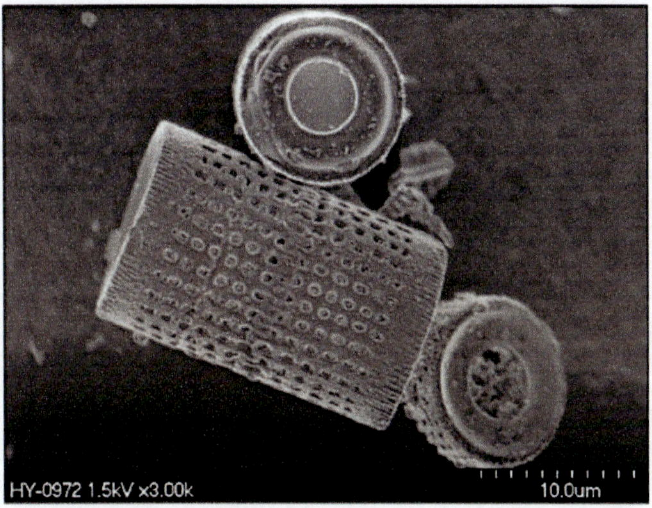


**Fig. 4.16** Example of diatomaceous earth particles [6]

resin—it is exclusively type 3 resin.[17] We are therefore replacing resin that separates $TiO_2$ particles with an equal amount of a material that is too large to get in between individual $TiO_2$ particles in the same way that type 3 resin can. From this viewpoint, the increase in $TiO_2$ particle crowding when large extender particles are formulated into a paint is due to a loss of type 3 resin.

If we extend this logic to the small particles, we would conclude that they should not cause a loss in opacity like their large counterparts because the small particles should be able to occupy voids between the $TiO_2$ particles and separate them in the same way that the type 3 resin they are replacing does. This is, in fact, an implicit assumption in the effective PVC model.

However, as seen in Fig. 4.15, this is not the case—the opacity for paints made with small extender (squares) is lower than for paints made in the absence of extender (circles). This analysis omits the fact that small particles reduce the amount of type 3 resin available in the paint system due to their resin demand [8]. That is, replacing a given volume of resin with small extender particles removes that volume of type 3 resin and converts some of the remaining type 3 resin into the types 1 and 2 resin needed to satisfy the small extender particles.

This effect can be illustrated by considering diatomaceous earth. This is a unique extender particle that is composed of large particles that are very porous (Chap. 9). These fascinating particles are the fossilized remains of tiny, one-celled organisms that encapsulate themselves in a porous shell (Fig. 4.16). These particles are in the size range of the large extender particles used in the paints discussed in this chapter, but they have a much higher resin demand because resin is needed not only to coat

[17] Above the CPVC, type 2 resin is exchanged. In the case of extremely high PVC values, when the reservoir of type 2 resin is depleted, then type 1 resin is exchanged.

the particles and fill the interstices between them, but also to fill the pores and open spaces within them.

As a result of their higher resin demand, the deleterious effect of diatomaceous earth on paint opacity is much more severe than the effect that extender particles of equal size (10 to 12 microns) have on opacity—in fact, the loss of opacity at equal PVC for diatomaceous earth is roughly double that of the large extender particles.[18] We can understand this readily—first, each particle of diatomaceous earth occupies the same volume as do large extender particles. Second, because their void fraction is so high (approximately 85%), a roughly equal volume of resin is required to fill the intraparticle voids. Thus, roughly twice the amount of type 3 resin is lost for diatomaceous earth as for solid extender on an equal particle volume basis.

We can now understand why the effective PVC concept fails at high concentrations of large extender particles. The effective CPVC calculation (Eq. 4.9) takes into account the amount of type 3 resin replaced based on the volume of the extender particles but does not include the type 3 resin loss due to the resin demand of the large particles, that is, the amount of type 3 resin that must convert to the type 1 resin that adsorbs onto the extender particles. Large particles consume relatively little type 1 resin, and so the error from omitting this source of type 3 resin loss from our effective PVC calculation is minor at low extender concentrations. However, at higher concentrations of large extender particles, the amount of type 3 resin lost due to conversion to type 1 resin is enough to negatively impact paint opacity (Fig. 4.15).

The fact that the opacity loss due to small particles is roughly equal to that of large particles in our laboratory paints can be understood by considering that the effective PVC calculation should include the volume of lost type 3 resin in the same way that it includes the volume of large extender particles. Equation 4.10 is the traditional way of calculating $TiO_2$ PVC (this is the same equation that was listed as Eq. 4.1 earlier in this chapter and is repeated here for convenience). To measure the effective PVC based on the volume of large extender particles, we modified the denominator of this equation by subtracting (removing) the volume of large particle extender (Eq. 4.11a). This equation simplifies to the effective PVC equation given earlier as Eq. 4.9 and repeated here as Eq. 4.11b.

$$\mathrm{TiO_2PVC} = \frac{\mathrm{TiO_2\,volume}}{(\mathrm{extender\,volume}) + (\mathrm{TiO_2\,volume}) + (\mathrm{resin\,volume})} \tag{4.10}$$

$$\mathrm{Effect\;TiO_2\;PVC} = \frac{\mathrm{TiO_2\;volume}}{\left[(\mathrm{TiO_2\;vol}) + (\mathrm{resin\;vol}) + (\mathrm{large\;extender\;vol})\right] - (\mathrm{large\;extender\;vol})} \tag{4.11a}$$

$$\mathrm{Effective\,TiO_2\,PVC} = \frac{\mathrm{TiO_2\,volume}}{[(\mathrm{TiO_2\,volume}) + (\mathrm{resin\,volume})]} \tag{4.11b}$$

[18] This is true when comparing paints below the CPVC value. Small amounts of diatomaceous earth can increase paint opacity by decreasing the CPVC to a level below the paint PVC. The air voids created by the CPVC dropping below the PVC will scatter light, as discussed elsewhere in this chapter and in Chap. 3.

For paints made with small extender particles, we can take into account the available volume loss due to the resin demand of these particles by subtracting their resin demand from the denominator of the $TiO_2$ PVC equation in a similar way that we subtracted the large extender particle volume from the denominator. This is done in Eq. 4.12a.

$$\text{EffectTiO}_2\text{PVC} = \frac{\text{TiO}_2\,\text{volume}}{[(\text{TiO}_2\,\text{vol}) + (\text{resin vol}) + (\text{small extender vol})] - (\text{resin demand vol})} \quad (4.12a)$$

$$\text{Effective TiO}_2\,\text{PVC} = \frac{\text{TiO}_2\,\text{volume}}{[(\text{TiO}_2\,\text{volume}) + (\text{resin volume})]} \quad (4.12b)$$

We can estimate the resin demand of most small extender particles as roughly the same volume as the particles themselves (this implies a CPVC value for the small extender particles of 50, which is consistent with those measured for most small particles). If the small extender volume and the resin demand volume for the small extender are equal, then we can simplify Eq. 4.12a to give Eq. 4.12b, which is exactly the same as the standard effective $TiO_2$ PVC equation when large extender particles are present (Eq. 4.11a). We can therefore understand how, in some paints, the effect of extender PVC on $TiO_2$ crowding is the same for small extender particles as large ones, as is borne out in Fig. 4.15.

The physical implications of this mathematical analysis are displayed schematically in Fig. 4.17. Here we consider, in two dimensions, the ideal spacing of the same number of small particles ($TiO_2$) in four particle scenarios.[19] The $TiO_2$ particles (0.25 micron diameter) are at the same concentration (20 PVC)[20] in all four scenarios. The first scenario (Fig. 4.17a) has $TiO_2$ as the only particle present, and the surface-to-surface distance between nearest neighbors (of which there are six) is 0.283 microns.

In the second scenario, a large hexagonal extender particle (side length 2.57 microns) is placed in the center of the field (Fig. 4.17b). This represents an extender PVC of 25. Again, the $TiO_2$ particles are positioned such that they pack as tightly as possible in the space available to them. Using Eq. 4.11b, we calculate the effective PVC of the $TiO_2$ to be 26.67. In this scenario, the surface-to-surface distance between nearest $TiO_2$ neighbors (of which there are again six) is only 0.212 microns, demonstrating the crowding effect of large extender particles.

The third scenario is somewhat more complex. Here we have replaced the large extender particle with small extender particles having the same total volume.[21] This collection of particles is packed together as tightly as possible, that is, at their CPVC

[19] A similar analysis applies to three dimensions, but this is more difficult to illustrate in figures.

[20] Note that we will refer to the concentration of the two-dimensional particles as if based on volume (PVC) although, technically, since this is a two-dimensional analysis, the concentration should be expressed on an area basis.

[21] The number of small extender particles is such that the volume of each is 25% greater than that of a $TiO_2$ particle. In this way, there are the same number of small extender particles in the film (their concentration is 25 PVC) as are $TiO_2$ particles (their concentration is 20 PVC).

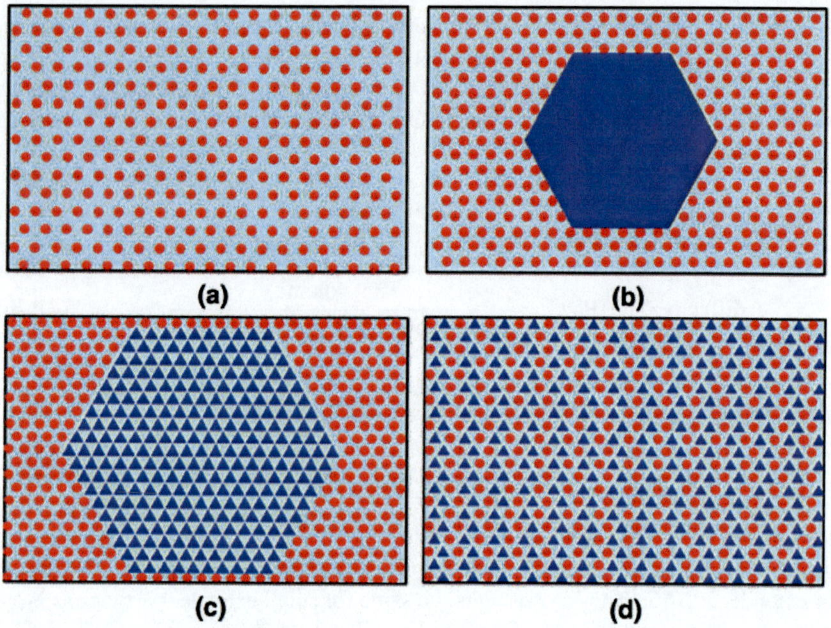


**Fig. 4.17** Particle spacing for different combinations of $TiO_2$ (red) and extender (blue) particles in a field of resin (gray). **a** $TiO_2$ particles (20 PVC) only. **b** $TiO_2$ particles (20 PVC) and large extender particle (25 PVC). **c** $TiO_2$ particles (20 PVC) and small extender particles (25 PVC) at their tightest packing (CPVC = 50). **d** $TiO_2$ particles (20 PVC) and small extender particles (25 PVC) distributed throughout the film

value, which we will define here as 50. The small particles occupy twice the space as the single large particle because of this and result in the removal of 25 PVC resin (the resin demand for 25 PVC of small particles). For the purposes of calculating the effective PVC of the $TiO_2$ particles, we use Eq. 4.11b and treat the collection of small particles, with their attendant resin, as a single large particle with twice the size of the large extender particle. Alternatively, we can use Eq. 4.11b, with the amount of small particle extender and excluded resin volume being equal (25 PVC each). In either case, we calculate an effective $TiO_2$ PVC of 40. The additional crowding due to the increase in inaccessible area between scenarios B and C leads to a reduction in the $TiO_2$ surface-to-surface distance to 0.127 microns.

Were we to terminate our analysis here, we would conclude that the crowding of the $TiO_2$ particles in the case of small extender particles (Fig. 4.17c) is twice that of an equal volume of large particles (Fig. 4.17b). We know, however, that this is not the case, as shown in Fig. 4.15. What is missing from the analysis is that there are two contributors to $TiO_2$ crowding in the case of the small particles (Fig. 4.17c). The first, as drawn in this image, is crowding due to the volume of the small particles being inaccessible to the $TiO_2$ particles (as is the case for the large particle in Fig. 4.17b). The second contributor is that the small particles have made an equal volume of type

**Table 4.1** Analysis of the four scenarios described in Fig. 4.17

| Scenario[a] | $TiO_2$ PVC | Effective $TiO_2$ PVC | Extender PVC | Surface-to-surface distance (μ) | Number of near neighbors |
|---|---|---|---|---|---|
| A | 20 | 20 | 0 | 0.283 | 6 |
| B | 20 | 26.7[b] | 25 | 0.212 | 6 |
| C | 20 | 40.0[c] | 50[d] | 0.127 | 6 |
| D | 20 | 40.0[c] | 50[d] | 0.127 | 2 |

[a] See Fig. 4.17 and the text for the description of each scenario
[b] Calculated using Eq. 4.9
[c] Calculated based on the resin available to the $TiO_2$ (see Eq. 4.12b)
[d] This includes the resin demand volume for the small extender particles

3 resin inaccessible to the $TiO_2$ particles (the resin demand of the small extender). The contribution to crowding for both of these components is, in this case, equal.

Figure 4.17c is quite artificial since it is a near impossibility that a random mix of the $TiO_2$ and small extender particles would result in all of the small extender particles positioning together as one mass. Instead, we would expect these particles to be interspersed with the $TiO_2$ particles. This is shown in Fig. 4.17d. In this case, the penalty due to exclusion of the $TiO_2$ particles from region occupied by the small particles is removed, but the penalty due to the resin demand of the small particles remains.

In this arrangement, the surface-to-surface distance between each $TiO_2$ particle and its nearest neighbors remains at 0.127 microns. However, the number of near neighbors has decreased from six (Fig. 4.17c) to two (Fig. 4.17d). This reduces the amount of scattering volume overlap. Based on the similarity in the measured scattering power of paints made with large and small extender particles (Fig. 4.15), we conclude that the total extent of scattering volume overlap must be roughly the same for large and small extender particles at equal extender concentrations.

The numerical details of this analysis are given in Table 4.1.

## *Using the Opacity Versus PVC Curve to Measure the Effective $TiO_2$ PVC*

The red line in Fig. 4.8 is a special case of a more general family of lines that we can place on the opacity versus PVC chart. This family comprises all lines that pass through the origin. The significance of these lines is that the scattering efficiencies of the individual $TiO_2$ particles for all paints that fall on the same line will be the same.

We can understand this by recalling that the scattering efficiency of an average particle in a paint is the scattering strength of the paint divided by the number of particles in it. In the opacity versus PVC curve, the y-axis value is the scattering strength of the film and the x-axis value is the number of particles, measured as the particle concentration. Any point on a straight line through the origin will have the same ratio of the two and therefore the same average scattering efficiency per individual $TiO_2$ particle.

We can use lines through the origin to experimentally determine the effective $TiO_2$ PVC of a given paint. Recall that the importance of the effective $TiO_2$ PVC concept is that it describes the local concentration of $TiO_2$ particles in a paint, omitting the regions of the film that are inaccessible to the $TiO_2$ particles (such as the regions occupied by individual large extender particles). As stated earlier, the scattering ability of a $TiO_2$ particle in a paint film is determined as if the concentration of $TiO_2$ particles was at their effective PVC value, rather than at their true PVC.

We will demonstrate how the lines of constant $TiO_2$ scattering efficiency can be used to determine the effective $TiO_2$ PVC using the paint made with 20 PVC $TiO_2$ and 15 PVC large extender (circled red triangle in Fig. 4.10b) and the series of paints made with the $TiO_2$ alone (blue line in Fig. 4.10b). We reproduce this information in Fig. 4.18. Because the large extender particles restrict the volume of film accessible to the pigment particles, the pigment particles scatter light less efficiently in the presence of extender than in its absence and so the red triangle in Fig. 4.18 is below the blue curve.

We can quantify the opacity cost of the extender by comparing the spread rate of the paint made in the absence of extender (4.93 $m^2/l$) to that in the presence of extender (4.47 $m^2/l$). This shows that the paint film loses roughly 10% of its scattering power by the exchange of 15 PVC resin with an equal volume of large extender.

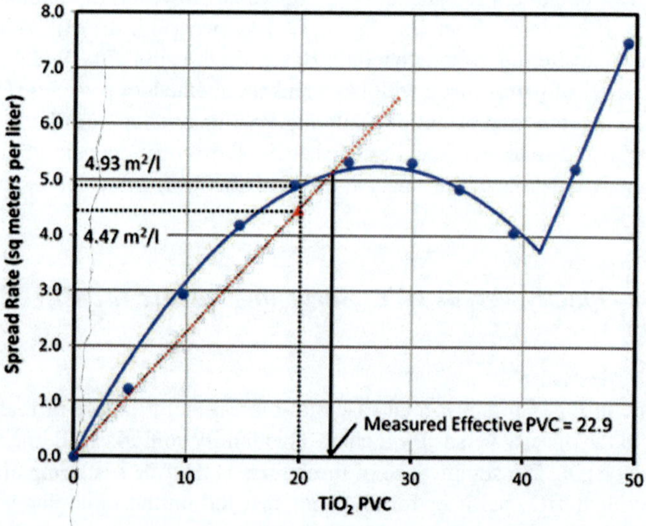


**Fig. 4.18** Comparing a paint made with 15 PVC large extender and 20 PVC universal $TiO_2$ (red triangle) to a series of paints made with the same universal $TiO_2$ in the absence of extender (blue line)

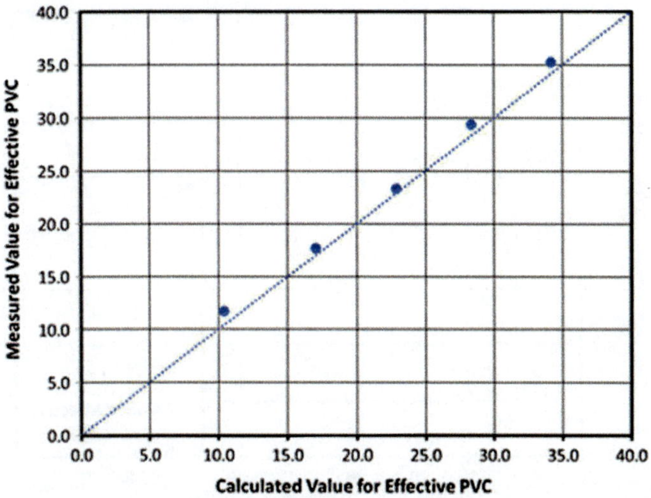


**Fig. 4.19** Comparison of effective PVC for the red curve in Fig. 4.10-B as calculated by Eq. 4.6 to the value experimentally determined as described in the text. The dashed line represents perfect agreement between the methods rather than the best fit line

In Fig. 4.18, we extend a red dashed line from the origin through the extended paint to the curve for the non-extended paint. The red line and blue curve intercept at a $TiO_2$ PVC value of 22.9. Since the average scattering efficiencies of the individual $TiO_2$ particles for all paints that are on this line are the same, the $TiO_2$ particles in the extended paint scatter light as if they were at 22.9 PVC rather than at their actual concentration of 20 PVC. That is, the measured effective PVC for this paint is 22.9. This is in good agreement with the value calculated from Eq. 4.9 (23.5 PVC).

We can extend this analysis to all paints made with 15 PVC large extender (shown in red in Fig. 4.10b) to determine the experimental values for the effective PVC for this paint series. In Fig. 4.19, we compare the experimentally determined effective PVC values to the calculated values. As can be seen, the values are in excellent agreement. The measured values are slightly higher than the theoretical values, possibly because of the slight resin demand of the large extender particles (this factor is not captured in Eq. 4.9).

For the extended paint shown in Fig. 4.18, we calculated a loss of the average $TiO_2$ scattering strength, due to crowding by the large extender particles, of roughly 10%.[22] This value is not uncommon for extended paints. In fact, many paints have significantly greater levels of extender, and in these paints the loss in average particle scattering strength can be much larger. As a worst-case example, a paint was made similar to the extended paint in Fig. 4.18, except with 25 PVC $TiO_2$ and 25 PVC

[22] Here we measure efficiency on the basis of the unextended paint, not on the basis of the scattering value of the $TiO_2$ pigment in the absence of scattering volume overlap (the basis for the analysis of Fig. 4.8).

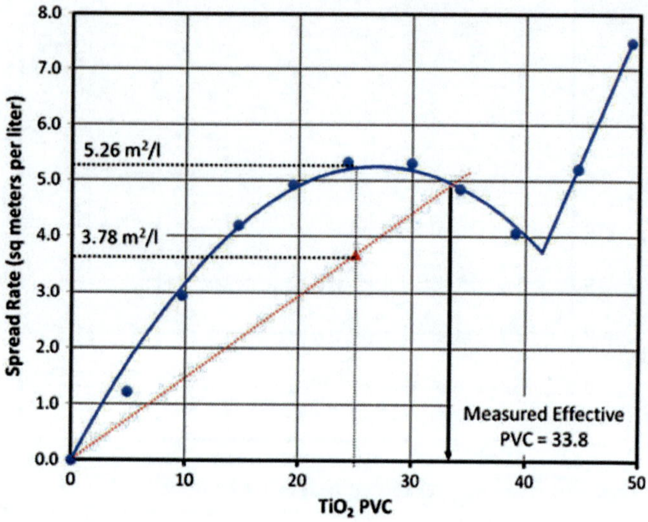


**Fig. 4.20** Comparing the paint made with 25 PVC large calcium carbonate extender and 25 PVC $TiO_2$ (red triangle) to a series of paints made with the same universal $TiO_2$ in the absence of extender (blue line). The red dashed line indicates the $TiO_2$ scattering efficiency of this paint

extender.[23] This is shown in Fig. 4.20. Here the paint loses nearly 30% of its scattering value due to crowding of the $TiO_2$ pigment by the large extender particles. The measured effective $TiO_2$ PVC of this paint, 33.8, is in good agreement with the calculated value or 33.3.

## Measuring the CPVC

The CPVC value for a specific single particle type or mixture of more than one particle types is typically measured in the laboratory using one of two strategies. In the first, paints are made over a range of PVC values while keeping the composition of the particle mix the same. This is most conveniently done by making master paints at the high and low ends of the PVC range of interest, then filling in paints at intermediate PVC values by mixing appropriate amounts of the two master paints.[24] Drawdowns are made of these paints and the paints tested for a property that shows a significant change at the CPVC.

[23] We consider this a worse case sample because the $TiO_2$ PVC—25 in this case—is at the maximum in the curve for the unextended paint (i.e., the blue curve in Fig. 4.18 and elsewhere). No reasonable paint would be formulated at a $TiO_2$ PVC above the maximum of this curve.

[24] Note that the two master paints must be made at the same total solid volume concentration, so that equal thicknesses of wet paint films will give equal thicknesses of dry films.

The film properties that change at the CPVC can be grouped into the following categories [11]:

- *Opacity*—Any of the opacity tests outlined in Chap. 13 will show a significant increase at the CPVC.
- *Gloss*—Air entrained in a film will cause the surface of that film to roughen. Gloss values are sensitive to roughness of this magnitude and so decrease at the CPVC. That said, this is among the least precise techniques of determining the CPVC since gloss generally decreases with PVC below the CPVC. The change in the rate of gloss loss can be similar in the two concentration regimes (below and above the CPVC). In addition, some paints below the CPVC are formulated to have low gloss (e.g., eggshell and flat sheens), which further compounds the uncertainty inherent in this method.
- *Profilometry*—This is an alternative technique to measure surface roughness. It is generally more accurate than gloss loss but requires a special instrument. Overall, it is less preferred than many other methods.
- *Density*—Below the CPVC, the measured densities of dry films agree quite well with the those calculated based on the densities of the constituents (i.e., the mass weighted average density). Above the CPVC, the measured density is lower than the calculated density due to the presence of air voids. This is a fairly accurate means of determining the CPVC, and also has the advantage that the air content of the film (that is, its porosity) can be determined from the difference in measured and calculated densities.
- *Pore Volume*—Pore volume of the paint film can be measured directly using mercury porosimetry. Below the CPVC there are no pores, and so this value will be zero. Like density, this method provides volumetric information about the pores. In addition to the total volume of the pores, it also reports the volumes on a size basis. This method can also be used to determine the void volume of hollow sphere polymer particles, as they collapse at specific pressures (this pressure changes from one grade of hollow sphere polymer to another). A specialized instrument is required for this test.
- *Elasticity and Tensile Strength*—These related properties drop sharply at the CPVC point, making them promising options for CPVC determination. However, they require free-standing films (i.e., films separated from the substrate), which makes sample preparation difficult and time-consuming. In addition, these properties can also be affected by the degree to which the latex particles in a waterborne paint coalesce, and so factors such as coalescent levels, drying time, and drying temperature must be carefully controlled.
- *Scrub Resistance*—This is an alternative method of determining film strength. This test differs from elasticity and tensile strength as it does not require a free-standing film (in fact, a durable substrate is required). Paints are submerged in a soap solution and scrubbed with a mechanical device. Scrub resistance is quantified as either the film loss after a certain number of scrub cycles or the number of scrub cycles to a predetermined failure point. As with elasticity and tensile

strength, the results of this test are affected not only by the presence of air, but also by how well the resin particles coalesce during film formation.

- *Stain Resistance*—Capillary forces pull a liquid deep into a film that has interconnected pores, and the degree to which a liquid stain that is applied to a film surface and then wiped off will discolor that surface changes at the CPVC value. However, there is some variability in this measurement when different liquids are applied, especially if the liquid carriers are chemically different (and so have different surface tensions) for the different stain materials.
- *Corrosion Resistance*—The ability of a paint film to prevent corrosion of a reactive substrate will obviously decrease substantially when the film contains pores. This technique is seldom used, however, because it, too, is sensitive to the quality of resin particle coalescence and because of the relatively long exposure times (weeks) required.
- *Outdoor Durability*—Intrusion of rainwater or dew into a porous film will significantly decrease the durability of that film and its ability to protect the substrate from weather. However, this test is seldom used because these tests require months to years of exposure.
- *Electrical Conductivity*—Paints are drawn on a metal panel and allowed to dry in an oven overnight. They are then soaked in salt water for a period of time. The panels are removed from the saltwater bath and, while still wet, one electrode of a resistance meter is applied to the painted surface and another electrode to the back of the panel. The resistance will be low (conductance high) if there are pore channels that travel from the film surface to the substrate surface [12].

The second strategy for determining CPVC is to use oil absorption values (refer back to Chap. 2). In this procedure, oil absorption is measured for a well-mixed sample of the particle blend of interest [13]. The end point of the oil absorption test occurs when there is just enough oil to make a thin coating on all of the surfaces and to fill the gaps between the particles at their tightest packing. This is the same description as the CPVC, except that oil, rather than resin, is the diluting material in the oil absorption method. Under the assumptions of this equivalence, the CPVC can be calculated from oil absorption using Eq. 4.13.

$$\mathrm{CPVC} = \frac{100}{1 + \frac{(\mathrm{OA})(\rho)}{93.5}} \tag{4.13}$$

In this equation, the CPVC is given on a percent basis (0 to 100%), OA is the oil absorption value, and $\rho$ is the density of the dry resin (in g/cc). The value 93.8 in this equation is one hundred times the density of linseed oil (the factor of 100 is included because oil absorption is reported as the grams of oil per 100 g of particles, and for Eq. 4.13 we use the grams of oil per gram of particles). Note that CPVC decreases as oil absorption increases, that is, there is a reciprocal relationship between them.

Although the oil absorption method generally predicts CPVC accurately for solventborne paints, it invariably over-estimates it for latex waterborne paints. The reason for this is that not all of the resin in the resin particles is free to flow into the crevices between the pigment and extender particles. Instead, resin particles can be considered as having two components—an outer component that can be solubilized by the coalescing agent and so can deform and flow, and an inner core that is shielded from the coalescing agent by the outer component and thus cannot deform or flow.[25] Since the resin in the core of the latex particles is unavailable for coating extender or pigment particles, or for filling the voids between them, more resin is required at the CPVC than for a solventborne paint (where all the resin is available for these roles). This results in a lower CPVC for a latex paint than for a solventborne paint, or than calculated using Eq. 4.13. This property of resin is referred to as its Binding Power Index (BPI) and will be discussed in detail in Chap. 10.

We will end this section with a note of caution. There are numerous examples in the literature of formulation experiments for which the oil absorptions values of particle mixtures (and the CPVC values calculated for them) are calculated by measuring the oil absorption values of the individual particle types and then using the weighted average of these values as the oil absorption for mixtures of them. While many mixture properties can be calculated as a weighted average of the properties of the individual particle types (e.g., density, composition, surface area, TGA weight loss, etc.), this is not the case for oil absorption or CPVC. As was clearly shown in Fig. 4.2 and discussed in the section on particle packing, the packing of particle mixtures can be very different than the packing of the individual constituents. For this reason, both oil absorption and CPVC must be measured for each particle mix and not inferred from the values for the individual mixture components.

## Opacity Above the CPVC

Inclusion of air voids in the dry paint film increases light scattering by two mechanisms. First, as noted above, air voids scatter light. An extreme example of light scattering by air voids is the bright white color of unpigmented polystyrene foams (e.g., Styrofoam®). These materials generally contain no $TiO_2$ or other light scattering particles. Instead, they achieve complete opacity exclusively through air void scattering (on a volume basis these materials are 98% air).

The second mechanism by which air voids increase light scattering arises from the refractive index of air (1.0) being significantly less than that of the resin (1.5), extender (1.5), and titanium dioxide pigment particles (2.73). Air inclusion therefore lowers the average refractive index of the film and so increases the difference in refractive indices between the $TiO_2$ particles and their surroundings [14]. As such, air voids increase the scattering strength of the titanium dioxide pigment.

---

[25] These two components of a resin particle should not be confused with the three types of resin described earlier. Resin type describes the role that a particular portion of resin plays in a paint film while the two components describe differences in the resin particle, rather than the paint film.

## *Dry Hide*

The contribution of air voids to the opacity of a dry paint film is referred to as dry hide. This designation is derived from the fact that this opacity is only apparent in the dry paint but is completely missing in the wet paint. This is because the wet paint does not contain air voids—these are only created when the liquid carrier of the paint (typically water) leaves the film. Aspects of wet hide will be discussed later, but first we will consider the nature of dry hide.

### The Opacity Versus CPVC Curve Above the CPVC

Referring back to Fig. 4.8, we see that the shape of the opacity versus PVC curve above the CPVC is quite different from that below the CPVC. More specifically, above the CPVC the opacity increases linearly with increasing PVC, while below the CPVC there is a decrease in the rate of increase in opacity as PVC increases, eventually leading to a net loss in film scattering power above a certain PVC value (for example, a PVC value of 27.1 for the paint system in Fig. 4.8).

As outlined above, the increase in film opacity with increasing PVC above the CPVC can be attributed to several factors. First, increasing the PVC increases the air void content of the film (see Fig. 4.9b). This provides the two benefits to light scattering and film opacity described above (the air voids scatter light themselves and they decrease the average refractive index of the region surrounding the $TiO_2$ particles).

As mentioned earlier, as the PVC is increased in paints above the CPVC, $TiO_2$ particle crowding does not increase. This is very different than increasing the PVC in paints below the CPVC, where an increase in PVC results in an increase in particle crowding. By definition, the particles cannot crowd any closer together than at the CPVC, and so when increasing the PVC in paints above their CPVC, we do not increase crowding or scattering volume overlap (instead we increase the number of particles by increasing film thickness).

Therefore, above the CPVC, increasing the PVC has the benefit of linearly increasing the number of light scattering centers (both $TiO_2$ particles and air voids) without the penalty of increasing the scattering volume overlap between the $TiO_2$ particles. As a result, we expect that paint opacity should also increase linearly. Figure 4.10 confirms this for a variety of paint systems.

## *Porosity Index*

When characterizing paints according to their PVC value (and their CPVC value), our focus is on the particles within the film. While this is an important determinant of many film properties, it is often useful to characterize films that are formulated above the CPVC from an alternative viewpoint—that of the air voids. This is because the concentration of air voids affects many important properties of the film. These include opacity, as discussed elsewhere, as well as film integrity, stain resistance, and substrate protection.

The most obvious way to characterize the pore content of a paint film would be to do so by volume based on the entire volume of the film (Eq. 4.14), in a manner analogous to using the packing factor calculation (Eq. 4.5) to characterize particle volume. This quantity is termed the film porosity.

$$\text{Film Porosity} = \frac{\text{airvolume}}{(\text{combined particle volumes}) + (\text{resin volume}) + (\text{airvolume})} \quad (4.14)$$

Film porosity compares the volume of air to the combined volumes of all film components. In doing so, we are differentiating between air, on the one hand, and the solid components (resin and particles), on the other. While this differentiation is useful in some situations, particularly with regard to understanding the contribution of air to light scattering, it is not the only way we can quantify the air content of the film. An alternative is to quantify the air void volume by the porosity index, which is based only on the volumes of the air voids and the resin (Eqs. 4.15 and 4.16):

$$\text{PI} = \frac{\text{airvolume}}{\text{airvolume} + \text{resin volume}} \quad (4.15)$$

$$= 1 - \frac{\text{CPVC}(100 - \text{PVC})}{\text{PVC}(100 - \text{CPVC})} \quad (4.16)$$

The usefulness of the porosity index is that it combines air and resin into a single component. That is, we can consider the film to be composed of solid particles that are held together by an air/polymer mixture. We are quite accustomed to mixtures of air and polymer in our daily lives—these are simply solid foams. The porosity index tells us how much air is in this foam mixture, and as such it is an excellent indicator of many important film properties, as listed above. As a general rule, paints made with the same resin particles and having the same porosity index have the same degree of "foaminess" and so similar mechanical and stain properties.[26]

## *Specialized $TiO_2$ Grades for Paints Formulated Above the CPVC*

$TiO_2$ spacing is of importance in all paints, especially in those formulated above the CPVC. Because of this, some $TiO_2$ grades have been specifically developed for high

[26] This is true only to a first approximation for scrub resistance. Pore size can also affect scrub resistance, as will be shown in Chap. 16.

PVC systems. We introduced these grades earlier as "highly treated $TiO_2$", a name that highlights their composition. We can also describe these same pigments as "high PVC grades", as they are used only in paints near or above their CPVC.

As will be discussed in Chap. 7, the particles in these grades are encased in a thick layer of alumina and silica. These two oxides are present in the form of a highly porous aluminosilicate. Because it is porous, the aluminosilicate occupies a high volume when compared to its weight. This is important because a denser material would unnecessarily decrease the $TiO_2$ content of the pigment (and with it, the number of scattering centers per unit weight of pigment).

These materials have long been thought to consist of individual $TiO_2$ particles that are enveloped in a thick layer of porous aluminosilicates (Fig. 4.21a). This layer would prevent the $TiO_2$ cores of the pigment particles from touching one another and so restrict the loss of opacity in crowded paint systems. However, there is an alternative morphology that explains the spacing benefit of the aluminosilicates. These particles could be composed of agglomerates of a few $TiO_2$ particles held at a fixed distance from one another (Fig. 4.21b). This is consistent with electron micrographs of these particles (Fig. 4.22) and also with the shape of the opacity versus PVC curves for these materials below the CPVC (discussed below).

Before comparing the shapes of the opacity versus PVC curves for the two types of $TiO_2$ pigment, we will first consider the particle spacing implications for the two different morphologies. If this pigment exists as individual particles with thick

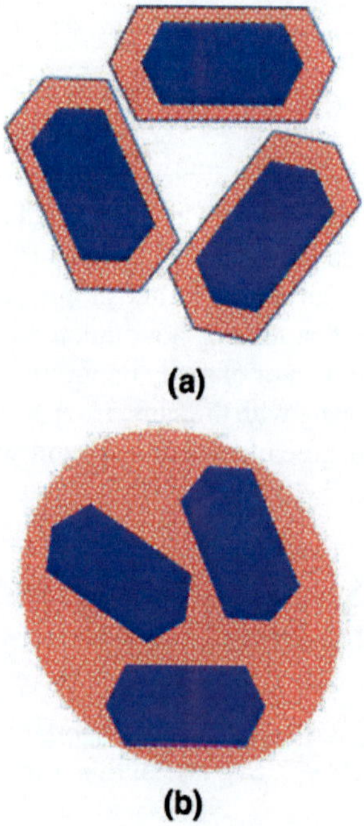


**Fig. 4.21** Different particle morphologies for highly treated $TiO_2$. Blue shapes are $TiO_2$, red regions are aluminosilicate. **a** Individually coated particles. **b** Small agglomerates

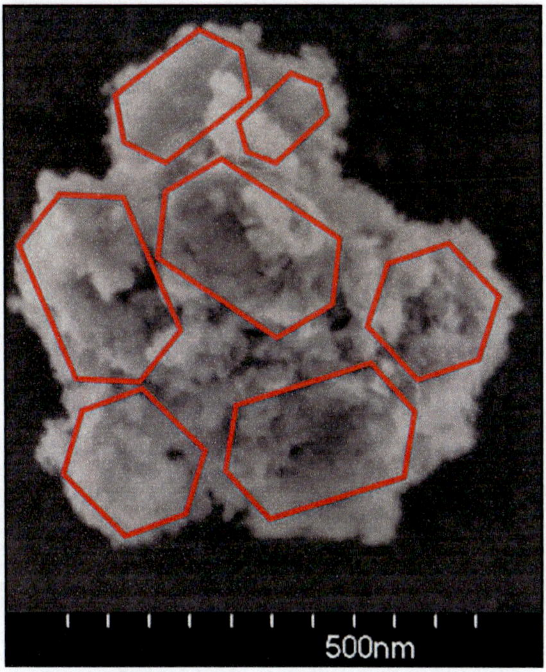


**Fig. 4.22** Electron micrograph of highly treated $TiO_2$ pigment. Red lines indicate the rough shapes of the $TiO_2$ particles within the aluminosilicate agglomerate

coatings (Fig. 4.21a), then the particle spacing will change as the PVC changes, just as it does for the universal grade. On the other hand, if the pigment exists as small agglomerates of spaced particles (Fig. 4.21b), then the distances between the agglomerates will change with PVC, but, to a first approximation, the distances between the individual particles will remain constant (because the $TiO_2$ particles in the agglomerates are frozen in place). This will not be exactly true, since the spacing between particles near the surfaces of two nearby agglomerates will change as the concentration of the agglomerates increases. Even so, the overall effect of PVC on crowding should be small.

For the agglomerate morphology, the opacity versus PVC curve in the region below the CPVC would be nearly linear, while for the individual particle morphology, there would be significant curvature in this part of the opacity versus PVC curve. Inspection of this curve for the highly treated grade, shown in red in Fig. 4.10c and reproduced in a larger format as Fig. 4.23, shows a nearly linear response of opacity to particle concentration for the highly treated $TiO_2$ pigment, consistent with the agglomerate morphology (Fig. 4.21b). In fact, a linear fit, with the constraint of passing through the origin, has an $R^2$ value of 0.986.

We will now shift our attention from comparing the two possible morphologies for the highly treated grade to comparing various aspects of the opacity versus PVC curves for the highly treated grade and a universal grade. As Fig. 4.23 shows, these curves cross one another at the CPVC value of the highly treated grade (PVC =

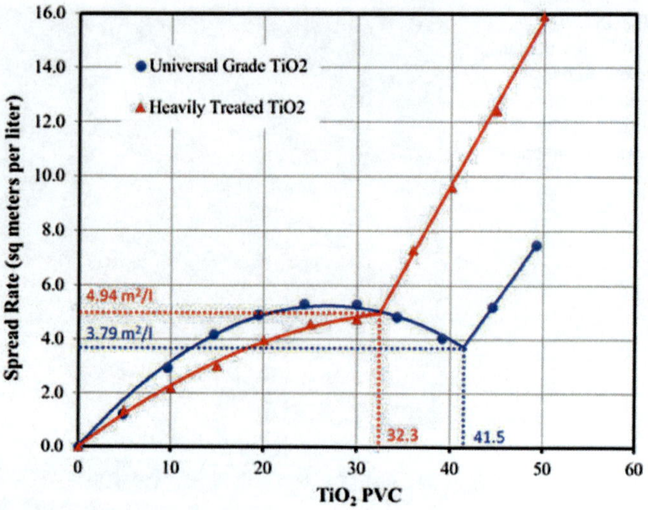


**Fig. 4.23** Comparison of the opacity versus PVC curves for a universal grade of $TiO_2$ and a highly treated grade of $TiO_2$

32.3). The CPVC for the highly treated grade occurs when the $TiO_2$ agglomerates are packed as tightly as possible. At this PVC, the two paint systems have the same number of particles and the same paint opacities (spread rate = 4.94 $m^2/l$), indicating that the particles in both paints must have the same light scattering strength and so must have the same $TiO_2$ spacing. This spacing is the fixed spacing of the individual particles in the agglomerate.[27] Although this is the CPVC for the highly treated grade, the universal particles do not yet touch, and so their CPVC value should be higher, consistent with Fig. 4.23.

A second important aspect of Fig. 4.23 is that, below the CPVC value of the highly treated grade, we see that, at any given PVC value, these particles scatter light less efficiently than those of the universal grade (the red line in Fig. 4.23 is below the blue line at PVC values below 32.3), while above that CPVC we see the opposite. This is because, by fixing the particle–particle distances in the agglomerate, we are preventing the particles from spacing further apart below the CPVC, where there is room for the particles to spread into. This decreases the hiding below the CPVC. That is, when we compare the spacing of the $TiO_2$ particles for the two grades, the constant spacing of the highly treated grade is closer together than the universal grade below the CPVC of the highly treated grade, while the spacing is the same at that PVC and is further apart for the highly treated grade at PVC values above that PVC.

Finally, we see in Fig. 4.23 that the opacities of the two paint systems at their respective CPVC value are quite different. At its CPVC value (PVC = 32.3), the opacity of the heavily treated grade (spread rate = 4.94 $m^2/l$) is roughly 30% higher

[27] We assume that at the CPVC condition, the distances between particles on the periphery of touching agglomerates are equal to the distances between particles within the same agglomerate.

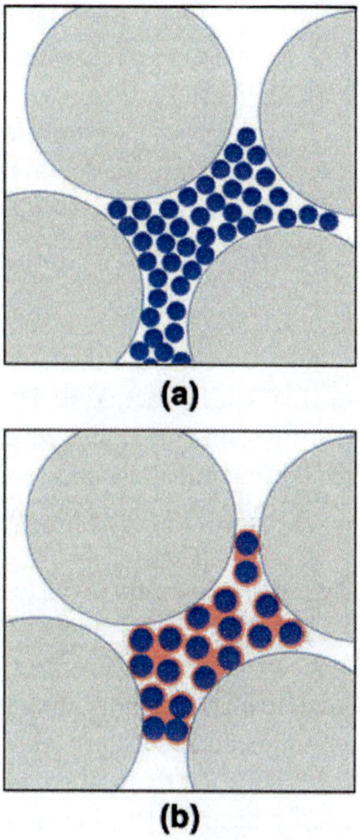


**Fig. 4.24** $TiO_2$ particle packing conditions in paints formulated above the CPVC. **a** Universal grade of $TiO_2$. **b** Highly treated grade of $TiO_2$. Gray circles are large extender particles. Red regions are porous aluminosilicates. Blue circles are $TiO_2$ cores

than that of the universal grade (spread rate = 3.79 $m^2/l$) at its CPVC value (PVC = 41.5). In paints formulated above the CPVC, the $TiO_2$ particles are typically packed in the voids between the large extender particles at particle densities that are at or near their CPVC condition, as shown in Fig. 4.24. Since the highly treated particles give better opacity at their CPVC than do the universal particles at theirs, there is a significant advantage to using highly treated $TiO_2$ grades in paints formulated above the CPVC. This will be discussed in more depth in Chap. 16.

An extension of this analysis can be made by referring again to the universal pigment curve in Fig. 4.23. We see that the CPVC for this grade (41.5) is above PVC at the maximum in this curve (27.1). This means that decreasing the CPVC of the universal grade from its measured value will actually improve the hiding power of the paint. By replacing the universal grade with the highly treated grade, we are decreasing the CPVC and so should expect the hiding power to improve.

## *Oiled Hide*

We can conceptually separate the factors that are responsible for increasing opacity as the PVC increases above the CPVC into two categories, both of which can be seen in Fig. 4.9b. The first are the two contributions from the air voids—scattering light in their own right and increasing the scattering strength of the $TiO_2$ particles. The second is the linear increase in the number of $TiO_2$ particles per unit area as the PVC increases.

It can be useful to separate these two contributions to increased opacity. In this way, the paint formulator can assess the effectiveness of the air voids. This effectiveness varies from one paint to another for several reasons, the most important of which is the size of the void [15]. This variability can be seen in the different slopes of the paint opacity versus PVC lines above the CPVC for different paint systems in Fig. 4.10.

We can separate these contributions by filling the air voids in the paint film of interest with mineral oil. The refractive index of mineral oil (1.47) is very similar to that of resin (typically from 1.48 to 1.55). This is not surprising since the two are similar chemically—they are both hydrocarbons, with the major difference between them being the length of the carbon chains, and this has little effect on refractive index. By replacing the air voids with mineral oil, we essentially return the film to the CPVC condition,[28] at least from a light scattering viewpoint.

The scattering strength of the paint film treated in this manner is termed the "oiled hide" of the film.[29] The average scattering power of the $TiO_2$ particles should be the same in all oiled films above the CPVC, regardless of the PVC, because the particles are in identical environments in all cases—packed as tightly as possible, and with the voids between them filled with hydrocarbon. The scattering power increases linearly simply because the film thickness and the number of $TiO_2$ particles per unit area increases as the PVC increases (Fig. 4.9b).

Because the average scattering power of the $TiO_2$ particles is the same in all oiled paints above the CPVC, we would expect them to be joined by a line that passes through the origin. This is in fact the case. In Fig. 4.25, we extend a line from the origin through the oiled spread rate values for the paints made with the highly treated $TiO_2$ grade in Fig. 4.10c. The correlation coefficient ($R^2$) for this line is 0.999, even with the constraint that it passes through the origin. Such excellent agreement between calculated and experimental values confirms the validity of the premise that the scattering volume overlaps between the $TiO_2$ particles in paints above the CPVC is constant.

## Wet Hide

The hiding power of the wet film immediately after it is applied to a substrate is termed "wet hide" [16]. Although the film remains in this state for only a short time,

---

[28] That is, the particles are packed as tightly as possible and the voids between them filled with hydrocarbon (resin in one case, a mix of resin and mineral oil in the other).

[29] In some regions of the world, this is instead called the "wet hide" of the film. We discourage this use of the term "wet hide" here as the same term is often used to define the hiding ability of the freshly applied wet paint, as discussed in the next section.

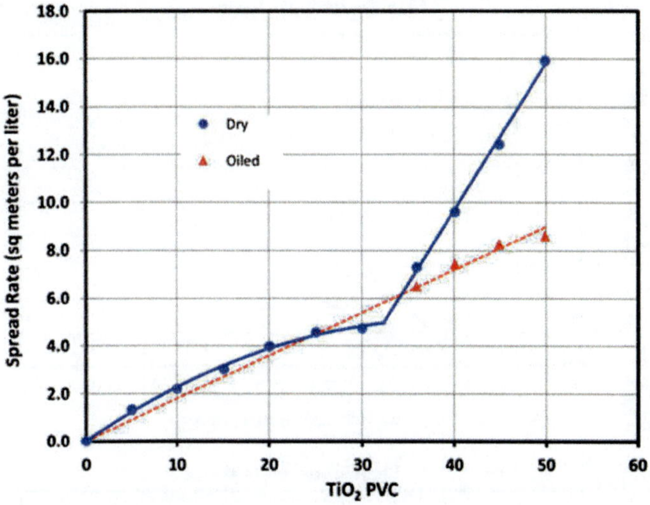


**Fig. 4.25** Red triangles are the oiled hide values above the CPVC for a heavily treated $TiO_2$ in paints made with the same resin as in Fig. 4.10b, c. The red dashed line intersects the origin and so indicates a line of constant $TiO_2$ scattering efficiency [6]

its opacity is nonetheless of great importance for architectural paints (also called wall or décor paints). These paints are typically applied by roller, sprayer, or, less commonly, brush. In each case, the painter can control, within reason, the thickness of the paint applied.[30] The painter uses the opacity of the wet film as a guide in judging how much paint to apply onto the substrate for complete hide.

Ideally, the wet opacity and dry opacity of the paint would be the same, in which case the wet hide would directly indicate the amount of paint to apply. However, these two can differ significantly from one another in some paints. If the wet opacity is weaker than the dry opacity, then the paint is likely to be overapplied, resulting in waste. If the wet opacity is stronger than the dry opacity, then the paint is likely to be underapplied, requiring time and labor for the later application of a second coat.

Wet and dry opacity can differ from one another for several reasons. First, the particle concentration in the wet paint is lower than that in the dry film. This dilutes the $TiO_2$ particles, lessening the scattering volume overlaps and so increasing the opacity of the wet paint compared to the dry film. Second, in waterborne paints, the average refractive index surrounding the particles is lower in the wet paints (where this environment is dominated by water) than dry films (where this environment is dominated by resin). In addition, extender and resin particles scatter light when surrounded by water, although this scattering is not nearly as efficient as that of the

[30] The lower end of the range of practical thickness is so thin that thickness variation due to the paint applicator (brush marks or, for roller application, stipple) results in visual non-uniformity across the wall. The upper end is dictated by the propensity of thick layers of wet paint to sag or run down vertical surfaces.

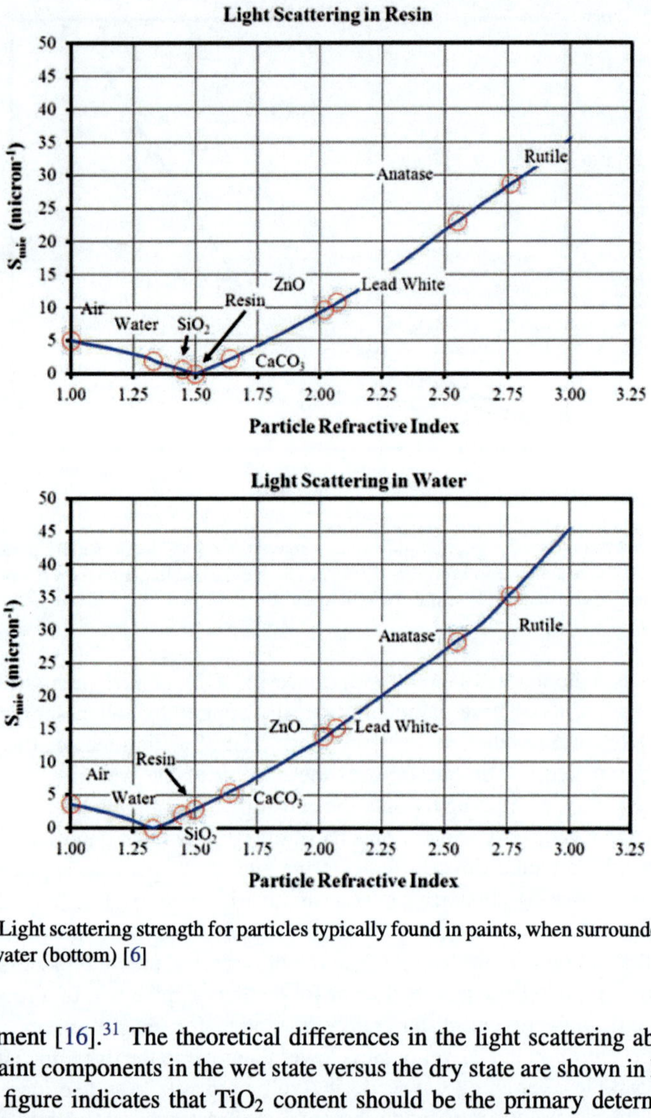


**Fig. 4.26** Light scattering strength for particles typically found in paints, when surrounded by resin (top) and water (bottom) [6]

$TiO_2$ pigment [16].[31] The theoretical differences in the light scattering abilities of various paint components in the wet state versus the dry state are shown in Fig. 4.26 [6]. This figure indicates that $TiO_2$ content should be the primary determinant of wet hide, with minor contributions from extender particles. This has been confirmed experimentally [16].

These opacity losses on drying are offset in paints formulated above the CPVC, which many interior architectural paints are, by the presence of air voids in the

[31] This is seen for clear lacquers, which are quite cloudy in the liquid state but completely transparent when dry, and with paints formulated below the CPVC with extenders only (deep base paints), which are also completely transparent when dry.

dry films.[32] This effect is so great that most paints formulated above the CPVC become more opaque on drying whereas paints formulated below the PVC become less opaque on drying.

## Particle Size for Optimal Light Scattering

Before leaving the topic of light scattering in crowded systems, we will revisit the topic of $TiO_2$ particle size for maximum scattering. In Chap. 3, we calculated the optimal particle size for a single, isolated $TiO_2$ particle using two methods—the Mie equation and Weber's law. The results spanned a range from about 0.21 microns to 0.31 microns.

These results apply to systems that do not experience scattering volume overlap between particles (i.e., systems with PVC values below about 2.5). In systems where scattering volume overlap does occur, such as those described in this chapter, it does so between the parts of the scattering volumes that reside outside of the particles themselves. For this reason, the degree of scattering volume overlap is determined primarily by the surface-to-surface distance between particles. This distance depends on both the PVC and on the size of the particles, with smaller particles being closer to one another than the larger particles at the same concentration.

This is shown in Fig. 4.27 for two diameters of ideally spaced particles—one with a diameter of 0.23 microns and the other with a diameter of 0.26 microns (compare with Fig. 4.11, which is for a 0.25 micron diameter particle). We chose these diameters for discussion because in Chap. 3, in Fig. 3.8, we saw, using the Mie equation, that the scattering strength of a 0.23 micron $TiO_2$ particle is slightly greater than that of a 0.26 micron particle under conditions where there is no scattering volume overlap and so the 0.23 micron particle is preferred for plastic applications, where the PVC is so low that there is no scattering volume overlap.[33]

In Fig. 4.27, we see that the surface-to-surface distances for these particles overlap at the two ends of the PVC range, but not in the center of the range, that is, they deviate at the PVC values typically of interest to paint formulators. This has important implications for the optimal $TiO_2$ particle size for paints, as follows.

Consider paints made with each particle size at 15.0 PVC. The surface-to-surface distance for the smaller $TiO_2$ particles, if ideally spaced, is 0.164 microns, while that for the larger particles is 0.183 microns. For the paint system shown in Fig. 4.5, we can calculate that for this paint system the loss of opacity due to scattering volume overlap at 0.164 microns is 5% greater than at 0.183 microns, and so in this case we

[32] The opacity of paints containing hollow sphere polymer pigments (described in Chaps. 7 and 10) also increases on drying because these spheres contain water in the wet state (and so do not scatter light) but contain air in the dry state.

[33] The 0.23 micron particle is also preferred because its undertone, which is bluer than that of the larger particle, offsets the slight yellow color of many polymers (see Chap. 2 for information on pigment undertone).

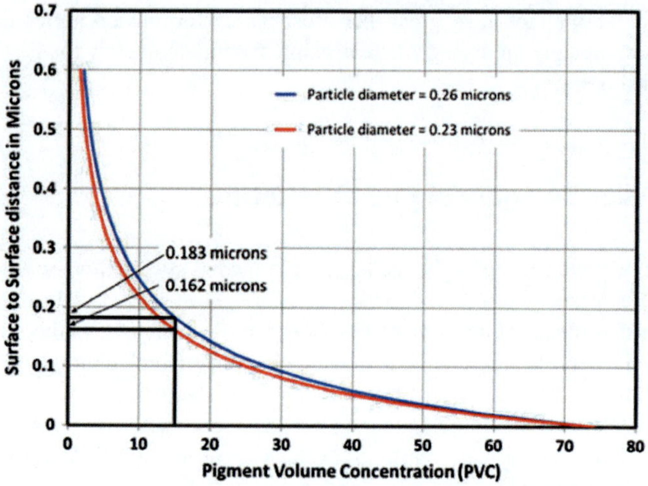


**Fig. 4.27** Surface-to-surface distances as a function of particle concentration for two sizes of particles

would expect the larger particles to scatter light 5% more efficiently at 15.0 PVC than the smaller particles.

Overall we see, therefore, that the optimal particle size for $TiO_2$ light scattering is dependent on the concentration of the intended application, and that a larger particle is preferred at $TiO_2$ concentrations typical of paints than for those concentrations typical for plastics. In addition, we see that using the wrong particle size can result in a roughly 5% loss in opacity. Large $TiO_2$ producers are capable of controlling their particle size within this range and so produce different primary particles for different end-use applications.

## Light Scattering in Paper Laminates

Paper laminates require a high level of opacity, and the papers used in them typically contain roughly 40% $TiO_2$ by weight. That said, because of the low densities of most paper ingredients, as well as the extensive pore structure of paper, this amounts to only about 10% $TiO_2$ by volume. As such we would expect particle crowding and scattering volume overlap to be relatively low. This is not the case, however, because the particles in paper are not distributed evenly throughout the sheet, but rather are found in small regions, or pockets, of high particle density. The effective PVC value within these pockets can readily exceed 50 and these particles are typically packed as tightly as at their CPVC condition.

There are two important consequences of this non-homogenous packing. The first is that the response of opacity to $TiO_2$ loading is generally linear, unlike what is seen in paints (e.g., Fig. 4.8). This is because increasing the loading of $TiO_2$ does not alter the amount of scattering volume overlap of the existing $TiO_2$—it alters the number of $TiO_2$ pockets, but not their local density. The second is that there is a significant opportunity to improve opacity by better spacing the particles when they are packed at their highest density. This will be discussed in more detail in Chap. 18.

## Summary

Light scattering in systems containing more than 2.5 volume percent $TiO_2$, which includes most paints, is complicated by the negative effect that near neighbor particles have on one another's scattering strength. These effects are due to overlapping scattering volumes, which reduce the total scattering of the system. As particle concentration increases, the distances between particles decrease and so the scattering loss due to scattering volume overlap increases. At the same time, as the $TiO_2$ particle concentration increases, there are more scattering centers in the paint. As a result of these two opposing effects, the relationship between $TiO_2$ concentration and paint opacity is non-linear.

This relationship is most usefully expressed through a chart of paint opacity as a function of $TiO_2$ PVC. In such charts, the opacity increases steeply at low $TiO_2$ volume concentrations, where the particles are well separated and so do not suffer from scattering volume overlap. At roughly 2.5 PVC, scattering volumes begin to overlap, and the opacity benefit from increasing the number of scattering particles begins to decrease. This loss of effectiveness grows as more and more particles are added, and, at a certain point, increasing the particle concentration actually leads to a net loss of opacity.

This loss in opacity with increasing particle concentration continues until the paint arrives at its CPVC value, where the particles are crowded as densely as possible and there is just enough resin to coat the particles and fill the voids between them. As the PVC continues to increase, air becomes incorporated into the paint film. Air voids increase the film opacity because they can scatter light in their own right and because they improve the scattering efficiency of the $TiO_2$ pigment. However, these opacity improvements come at a cost in terms of paint properties—the air voids join together into a network of pores that reduce the strength of the film and degrade the protective properties of the paint on the substrate.

The presence of extender particles in a paint further complicates the relationship between $TiO_2$ content and film opacity. Both large and small extender particles cause the $TiO_2$ particles to crowd one another, but for different reasons. Large extenders exclude the $TiO_2$ particles from large regions in the film, while small extender consumes resin that would otherwise be used to space $TiO_2$ particles.

It is convenient to quantify these crowding effects through the effective $TiO_2$ PVC concept. To do this the equations used to calculate the $TiO_2$ PVC are modified to remove the volume of the paint that is inaccessible to the $TiO_2$ particles. In this way, we can compute the amount of scattering volume overlap in terms of the local concentration of the $TiO_2$ particles.

Although extender particles cause the $TiO_2$ particles to crowd together, and so decrease their scattering strength, they can have a beneficial effect on opacity as they can be used to bring a paint above its CPVC value. This could be done using $TiO_2$ pigment alone, but the particle concentrations needed to reach and exceed the CPVC value more than offset any economic benefit of improving opacity through the incorporation of air, which is free.

Using extender particles to exceed the CPVC results in cost savings beyond the opacity boost due to bringing air into the film—extender particles also replace relatively expensive resin with an inexpensive substitute. This benefit is maximized by formulating the paint to have a high CPVC value, which can be done by using mixtures of extender particles with different particle sizes. In such mixtures, the larger particles are packed as efficiently as possible, which results in the replacement of much resin, and the smaller particles pack in the voids between the larger particles, replacing the resin that would otherwise be there and further increasing the amount or resin replaced by extender.

## Appendix

In this appendix, we will show the process for modeling the strength of $TiO_2$ scattering in paint films that contain large extender particles based on the scattering that we see for the same $TiO_2$ particles in an unextended paint. This allows us to predict the expected opacities of extended paints based on the measured opacities of unextended paints.

Basis for the model:

The opacity of a paint containing only $TiO_2$ and resin (which we will refer to as an "unextended paint") is invariably greater than that of a similar paint for which a portion of resin has been replaced by large extender (which we will refer to as an "extended paint"). The opacity penalty for the extended paint is due to the loss of scattering efficiency per $TiO_2$ particle because the extender particles crowd the $TiO_2$ particles together, as discussed in the text.

It is useful to have a means of describing the degree of $TiO_2$ crowding in a paint. An obvious way would be by the $TiO_2$ concentration—the higher the $TiO_2$ concentration, the greater the crowding. We can therefore describe the degree of $TiO_2$ crowding in an extended paint with a certain PVC value based on the PVC value that would give the same degree of crowding in an unextended paint. As discussed in the text, the effective PVC describes this degree of crowding. For example, a paint with an actual $TiO_2$ PVC of 20 and large extender PVC of 40 would have an effective $TiO_2$ PVC of 33.3, as calculated by Eq. 4.9 in the text. This means that the scattering efficiency of the $TiO_2$ particles in the extended paint (at 20 PVC $TiO_2$) is the same as the scattering efficiency that the $TiO_2$ particles would have in an unextended paint at 33.3 PVC

$TiO_2$. Note that the opacity of the extended paint at 20 PVC $TiO_2$ will not be the same as the opacity of the unextended paint at 33.3 PVC $TiO_2$ (i.e., at the effective PVC) because, although the $TiO_2$ particles experience the same amount of crowding, there are more of them present in the unextended paint (since it is at a higher PVC).

In our model, we will develop a way to modify the opacities of the unextended paints by decreasing the scattering strength of each $TiO_2$ particle to reflect the opacity penalty of extender crowding in the extended paints.

Developing the model:

To demonstrate how this model can be developed, we will use the opacities of an unextended paint system to estimate the opacities of a paint system with 20 PVC large extender.

The model is based on experimentally determined opacity data for the unextended paint system. The first step in this process is to measure the spread rate (or another opacity indicator) of a series of unextended paints that span a range of $TiO_2$ values. Data for the paint system with the universal grade of $TiO_2$ that are shown in this chapter are given in the second column of Table 4.2 and plotted in Fig. 4.28.

Next, a measure of the scattering strengths per individual $TiO_2$ particle in these paints is calculated by dividing the spread rate (a measure of total scattering) by the $TiO_2$ PVC (a measure of number of $TiO_2$ particles). This is shown for the unextended paint in the third column in Table 4.2 and as the blue line and symbols in Fig. 4.29.

The best fit equation for this particular set of data is

$$\text{Scattering per unit concentration (unextended paint)} = -0.0063 \cdot \text{PVC} + 0.3634$$

We next calculate the expected line for the paint series with 20 PVC large particle extender. To do this we will define a factor F that is the ratio of the reciprocal of one minus the extender PVC when expressed as a fraction. In this case, F equals 1.25 ($=\frac{1}{1-0.2}$). Multiplying the actual $TiO_2$ PVC of the extended paints by factor F gives

**Table 4.2** Scattering parameters for unextended paints below the CPVC

| PVC | Measured spread rate ($m^2/l$) | Scattering per unit concentration[a] |
|---|---|---|
| 0 | 0 | – |
| 4.9 | 1.55 | 0.313 |
| 9.6 | 2.89 | 0.300 |
| 14.6 | 4.15 | 0.284 |
| 19.4 | 4.91 | 0.253 |
| 24.3 | 5.32 | 0.219 |
| 29.9 | 5.29 | 0.177 |
| 34.2 | 4.80 | 0.140 |
| 39.1 | 4.11 | 0.105 |

[a] spread rate divided by PVC

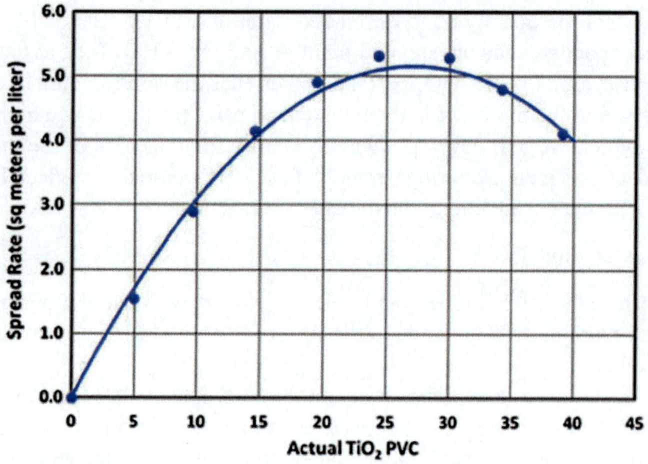


**Fig. 4.28** Measured spread rate values for unextended paint system

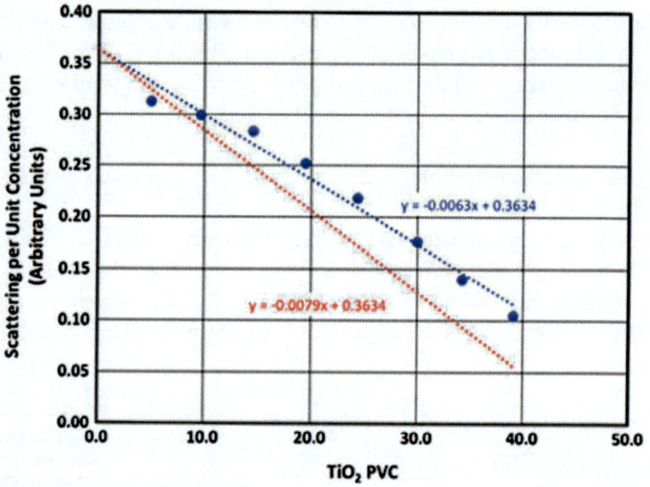


**Fig. 4.29** Scattering per unit volume for the unextended paint series (blue) and calculated values for the 20 PVC extended paint series (red)

the effective $TiO_2$ PVC:

$$\text{Effective PVC} = \text{F} \cdot \text{PVC}$$

Because the scattering per unit concentration is linear with PVC, we can calculate the scattering per particle for the extended paint series as simply the scattering per particle at the effective $TiO_2$:

$$\begin{aligned}\text{Scattering per unit concentration (extended paint)} &= -0.0063 \cdot (\text{effectivePVC}) + 0.3634 \\ &= -0.0063 \cdot (\text{F} \cdot \text{PVC}) + 0.3634 \\ &= -0.0079 \cdot \text{PVC} + 0.3634\end{aligned}$$

This equation is shown as the red line in Fig. 4.29. We can use this line to predict the opacity values for the extended paints as a function of PVC by multiplying the scattering per unit concentration for the $TiO_2$ particles in the extended paints, as expressed in the equation above, by the actual $TiO_2$ PVC (that is, reversing the procedure used to develop Fig. 4.29 from Fig. 4.28). The results of this calculation are shown in Fig. 4.30.

This model is useful for calculating the expected opacity of paints made with extenders based on the opacity of paints made without extenders. In some situations, we may wish to do the reverse, that is, we may have measured the opacity versus $TiO_2$ PVC curve for a paint system that has extender in it and wish to calculate the opacity that the paints would have without the extender. One way of doing this is the reverse of the procedure discussed above, that is, to calculate the best fit line for the scattering per unit concentration for the extended paints, then adjust the slope of the line based on the effective $TiO_2$ PVC.

There is a second way to accomplish this task. To demonstrate this, we begin by plotting one data point from the opacity versus PVC curve for the extended paint that has 20 PVC large extender and 25 PVC $TiO_2$. This is shown as a red triangle in Fig. 4.31. We can extend a line from the origin, through this data point, and to the $TiO_2$ PVC of 31.25 (blue circle in Fig. 4.31). This PVC is the effective PVC value

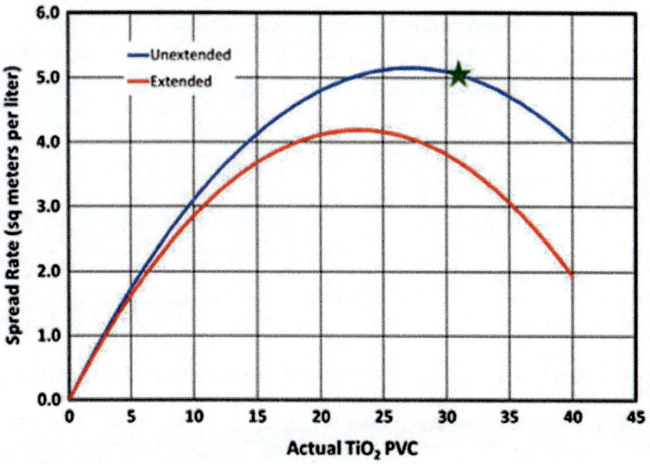


**Fig. 4.30** Modeled opacity for extended paint system based on the opacity of the unextended paint system

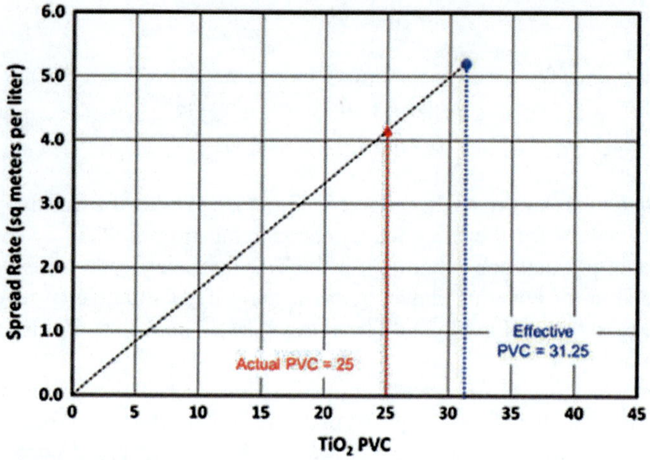


**Fig. 4.31** Spread rate data points. Red triangle—extended paint at an actual $TiO_2$ PVC of 25. Blue circle—opacity of this paint at the effective $TiO_2$ PVC

of the paint and was arrived at by multiplying the actual PVC (PVC = 25) by factor F (1.25 in this case).

As explained in the text, all data points on the line going through the origin have the same $TiO_2$ scattering strength. The blue circle, therefore, indicates the same $TiO_2$ scattering strength as the unextended paint would have at a PVC of 31.25, and so this point will be a point on the curve for the unextended paint. For reference, this circle is indicated as a green star in Fig. 4.30.

We can repeat this process for any paint in the extended paint series to recreate the blue line in Fig. 4.30. However, in this case, the X-axis will be the effective PVC of the extended paint, rather than its actual PVC (note that the effective PVC values for the unextended paints equal their actual PVC). This is how Fig. 4.13b was created.

## References


1. Asbeck, W.K., Van Loo, M.: Critical Pigment Volume Relationships. Ind. Eng. Chem. **7**, 1470 (1949)
2. Berryman, G.: Random Close Packing of Hard Spheres and Disks. Phys. Rev. A **27**, 1053 (1983)
3. Takemura, K: Gibbsite Type Aluminum Hydroxide Particles having High Oil Absorption", US 7,438,977 (2008)
4. Stieg, F.B.: Are Conventional Trade Sales Formulating Practices Wasteful? J. Coat. Technol. **48**(612), 51 (1976)
5. Stieg, F. B., "The Coatings Industry Looks at Fillers", Am. Chem. Soc. Div. Org. Coatings Plast. Chem., **33**(3), 5 (1973)
6. Diebold, M. P.; Application of Light Scattering to Coatings, Springer (2014)

7. Auger, J.C., Stout, B.: Dependent Light Scattering in White Paint Films: Clarification and Application of Theoretical Aspects. J. Coat. Technol. Res. **9**(3), 287 (2012)
8. Stieg, F. B., "The Effect of Extenders on the Hiding Power of Titanium Pigments", Off. Digest, 52 (January, 1959)
9. Stieg, F.B.: Effect of extender on crowding of titanium pigment. J. Coat. Technol. **61**(778), 67 (1989)
10. Cutrone, L.: Influence of Fine-Particle Size Extenders on the Optical Properties of Latex Paints. J. Coat. Technol. **58**(736), 83 (1986)
11. Bierwagen, G.P., Rich, D.C.: The critical pigment volume concentration in latex coatings. Prog. Org. Coat. **11**, 339 (1983)
12. Asbeck, W.K.: Critical pigment volume concentration measurements, a very fast method. CoatingsTech, **2**(12) (2005)
13. "Standard Test Method for Oil Absorption of Pigments by Spatula Rub-out", ASTM D281 (2016)
14. Stieg, F.B.: Air as a Dispersion Medium. Ind. Eng. Prod. Res. Develop. **13**(1), 41 (1974)
15. De Backer, S.; Diebold, M. P., "Influence of Paint Quality on the Environmental Footprint of Architectural Paints", Euro. Coat. J., 30 (Jan 2022)
16. Dollani, H., Elton-Legrix, A.: Coverage Uncovered. Europ. Coat. J. **10**, 40 (2021)

# Chapter 5
# Color 1—Seeing Color

## Contents



## Introduction

Vision is arguably the most important of our five senses: when our eyes are open, approximately two-thirds of the electrical activity in our brain is dedicated to the vision process [1]. The complex process of seeing begins when light is generated by a source and ultimately ends after complex processing, within both the eye and the brain, of the nerve impulses generated by light that enters the eye.

With vision comes color. Color is an important, and sometimes vital, part of our lives. Color is one of the first concepts we learn as children, and most of us could identify the different colors before we could say their names. We use color to provide aesthetics (the red color of a car), to convey information (the red color of a stop light), to convey a message or idea (the red color identifying a certain brand of soft drink),



M. Diebold et al., *Pigments, Extenders, and Particles in Surface Coatings and Plastics*, https://doi.org/10.1007/978-3-030-99083-1_5

or for pure pleasure (the red color of the sky during sunset). Nearly all coatings are colored, for these or other reasons, and this coloring is almost always provided by small pigment particles in the coating.

The primary goals in this chapter and the next two are to develop a system for quantitatively measuring color and to apply that quantification to the mixing and/or matching paint colors. We will do this by first discussing the visual process of sensing colors, in this chapter, and then developing several frameworks that can describe, precisely, any color, in Chap. 6. In Chap. 15, we will apply this knowledge to colors in paints and plastics. In Chap. 8, we will discuss the pigment particles that generate these colors.

## The Nature of Light

Before discussing color, we must discuss light. Light is a repeating, self-propagating disturbance of the electric and magnetic fields. This disturbance propagates through a vacuum at slightly less than $3 \times 10^8$ m/s—what is typically quoted as the speed of light. Light propagates more slowly through matter, such as glass or the resin in a paint film. The repeat time of the disturbance is the inverse of the light frequency, and the distance over which the light propagates during one repeat time is its wavelength. The disturbance itself is referred to as a photon. Energy is contained within a photon, with the amount of energy being linearly proportional to the frequency of the light and inversely proportional to wavelength. Higher frequency, shorter wavelength photons are higher in energy than lower frequency, longer wavelength photons. Light below a certain wavelength is energetic enough to break chemical bonds, an important topic that we will discuss in Chap. 14.

In principle, light of any wavelength is possible, and while we have not created or detected light of very long or very short wavelengths, we have done so over a wavelength range of approximately 16 orders of magnitude. Visible light spans a much narrower range—from approximately 400 to approximately 700 nm (Fig. 5.1). Different wavelengths of visible light appear to us differently from one another—this is color. Moving across the range of visible light, beginning at the shortest wavelength, color progresses from violet to blue to green to yellow to orange and finally to red—a progression familiar to us in rainbows.

Often we discuss light intensity as a function of wavelength. We refer to this as a spectrum. When the light comes directly from a light emitting source (rather than being reflected from an object), we call this the spectral power distribution of the source.[1] Separation of light into its individual wavelengths is done through refraction, the bending of light as it travels through a transparent media. In the lab, we use a glass

[1] In some literature, the term "spectral power distribution" is used to describe the reflectance spectrum of an object. To avoid ambiguity, we will use this term only with respect to the emitted spectrum from some light source. This is in agreement with the definition of spectral power distribution given by the Illuminating Engineering Society of North America.

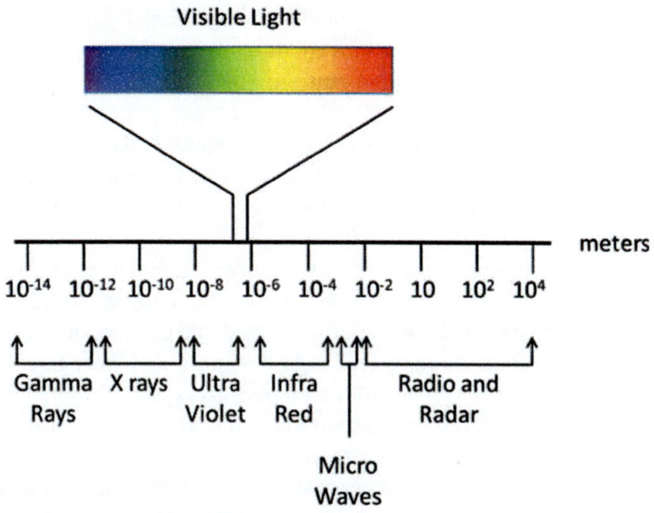


**Fig. 5.1** Range of accessible wavelengths

prism as the transparent media that bends the light; in nature, raindrops spread the light according to wavelength. This separation of colors occurs because the refractive indices of materials change with wavelength. This affects the angle at which the light is bent by a prism or drop of water. Since different wavelengths are bent at different angles, the light spreads out by wavelength and, therefore, by color.

## The Nature of Color

Color is a sensation that we experience when looking at a lighted object. This sensation is distinct from shape and brightness. To understand why we experience this sensation (the topic of this chapter), how we can measure this sensation (the topic of Chap. 6), and how we can control and manipulate this sensation (the topic of Chap. 15), we must first understand the fundamental nature of the colors themselves.

### *Primary Colors and Color Mixing*

When we separate white light by wavelength with a prism, we find that each wavelength appears as a certain color. We call these the spectral colors, as they are the components of the visible spectrum. They are also referred to as monochromic colors since they consist of a single wavelength. Combining light of different wavelengths gives us new colors that are not found in the spectrum (for example, brown and pink).

We will call colors that consist of more than one wavelength "complex colors". The spectra for these colors range from relatively simple (made from as few as two spectral colors) to quite complex (with no nonzero components over the entire visible range).

That we can combine colors to form new colors leads to the concept of primary colors. Because the human eye has three types of cells that respond to color (discussed below), we can create most—but not all[2]—colors from just three starting colors. These starting colors are called primaries.

Over the last 200 years, several workers attempted to find a set of three primary colors from which all other colors can be made. These attempts failed for reasons that are discussed in Chap. 6. However, it was found that most shades of color could be made by combinations of red, green, and blue lights, and these three colors are called the additive primary colors and are familiar to us as the colors of the lights found in our television screens and computer monitors. Similarly, for printing or painting, where color arises from light absorption rather than emission, most colors can be made from the combination of three light absorbing pigments—yellow, cyan, and magenta. These three colors are called the subtractive primary colors and are familiar to us as the colors of inks used in printers.

Additive primaries are usually associated with colored light sources. We have all seen colored lights by observing light directly from its source, typically a lamp. In color science, it is more convenient to study colored lights by reflecting them onto a white screen, rather than by observing them directly from their sources. This is because colored lights can be easily combined by projecting them onto the same portion of a screen. This combination usually creates a third color, and so we see that we can create new colors by adding light from different colored sources, as noted above.

For additive mixing, we begin with an absence of light (black) and then add colored lights to this. This is often depicted by overlapping circles of the three additive primaries, red, green, and blue, on a black background to make yellow, cyan, magenta, and white (Fig. 5.2a). When created in this fashion, color is considered to be a property of the light.

Subtractive primaries are associated with pigments, and color is manifested by illuminating an object containing pigment with white light. Here the object appears colored because it selectively absorbs specific wavelengths of light and reflects the others. As was the case with colored lights, we can combine color pigments to produce new colors. For subtractive color mixing, we do this by beginning with a white background that reflects all colors, and then add pigments to selectively remove certain colors or wavelengths. This is often depicted by the combination of yellow, cyan, and magenta overlapping circles on a white background to produce red, green, blue, and black, as shown in Fig. 5.2b. Colors produced in this way are considered properties of the object, rather than the light.

---

[2] It may be more accurate to say that not all shades of colors can be created from any set of three primary colors.

(a)

(b)

**Fig. 5.2** Mixing of primary colors. Top—additive mixing. Bottom—subtractive mixing

In both cases, we can define "complimentary" colors as those colors that are diametrically opposite on the color mixing images in Fig. 5.2. For example, green and magenta are on opposite sides on these mixing images. For additive mixing, they combine to give white, while for subtractive mixing, they combine to give black.

An important aspect of the primary color concept is that the colors formed by combining primary colors depend on whether they are being additively combined (using colored lights) or subtractively combined (using pigments). For example, mixing green and red lights produces yellow (additive mixing), while mixing green and red pigments produces gray or black (subtractive mixing).

We might expect that our main interest would be in subtractive color mixing, since we color paints by adding pigments that selectively absorb light. However, we sense and interpret colors by an additive process, additively combining the nerve signals from the various color detecting cells in the eye. In addition to sensing colors on an additive basis, we measure colors on this basis. Because of this we will use the properties of additive mixing to develop color theory and follow this in Chap. 15 by a discussion on subtractive color mixing to understand and control the colors of paints and plastics.

## *Color as a Property*

A white card illuminated by red light and a red card illuminated by white light will look identical, yet in the first case we say that the light is colored while in the second case we say that the card is colored. That is, we might say that color is a property of light or of an object.

Not all color scientists agree with this assertion. An argument can be made that color is in fact a sensation that we perceive only after our brains have processed nerve impulses from the eye created when light strikes it, that is, it only exists in our brains.[3] Consider that the combination of roughly equal amounts of light with wavelength 460 nm ("blue") and light of wavelength 570 nm ("yellow") produces white. How can we say that 460 nm light is blue if blue cannot be seen in this simple combination? Similarly, in our example above, how can we say that a card is a certain color (white, in this case) if it appears to be a different color when illuminated by different wavelengths of light? In general, how can we say an object or light has a set color when the color that we perceive changes depending on viewing circumstances?

This is, to a certain extent, an argument of definition, and as such we are free to define "color" any way we wish. That said, there is a logic to the concept that color exists only as a sensation within our brains. This logic will be further supported later in this chapter, where we will see that very different combinations of lights can produce identical colors. We will make the distinction between the definition of color as a property of objects and lights, and the definition of color as a perception created within our brains, by referring to the latter as a "color sensation". Thus, we would say that the white card illuminated by red light and the red card illuminated by white light give the same color sensation, without needing to distinguish whether this sensation arises from the spectral distribution of the light striking the card or the light absorption properties of its surface.

## *Hue, Lightness, and Saturation*

The simple names of colors—red, blue, purple, etc.—are referred to as "hues". While there is no exact number of hues (definitions may vary somewhat from one person to the next), it generally numbers between 11 and 16 [2]. As an example, Newton counted seven hues among the spectral colors he saw when sunlight passed through a prism.[4] We can increase the number of color names by adding descriptors to the

---

[3] This topic is analogous to the age-old question of whether a tree falling in the woods, with no one nearby to hear it, would make a sound. Everyone should agree that this would cause vibrations in the air (sound waves), but the debate is whether we define sound as these vibrations or define it as the sensation our ears give us due to these vibrations.

[4] Newton based his belief that there should be seven spectral colors by analogy to the musical scale having seven notes before repeating itself (i.e., the musical octave). It was his belief that there was

hues—cherry red, brick red, blood red, wine red, etc.—but even these are limited and typically unique to one hue (e.g., there is no such thing as "blood yellow").

While these descriptors may increase color names to a few dozen, this is clearly far too few to uniquely describe the millions of colors that our eyes can distinguish. Two additional factors have been developed to add to the hue designations—lightness (sometimes referred to as brightness[5] or luminosity) and saturation (also referred to as purity or intensity).

Lightness is well understood by the average person. It is the property of an object that is determined by how much incident light is reflected from it. Perceived lightness is dependent on both the intensity of the reflected light and the sensitivity of the eye to the wavelengths being reflected (the eye is not equally sensitive to all visible wavelengths).

Saturation describes how bold or intense a color is. The most saturated colors are the spectral colors, which consist of only a single wavelength of light. As other colors are added to a spectral color, the visual intensity diminishes. This is particularly true when white is blended into a spectral color. The color shifts towards grayness, and, if taken to the extreme, the color becomes the gray color of equal lightness to the original color.

This process is shown in Fig. 5.3. Here we see two spectra—the first (a) is for a highly saturated green light (nearly monochromatic), and the box inset in this plot shows, approximately, its color. The second spectrum (b) has the same peak, but in this case white light has been added to it (as a uniform amount at each visible wavelength). The color is still green, but with a washed-out appearance, that is, it has lower saturation (compare the insets).

Figure 5.4 gives an example of the ways that lightness and saturation alter a color's appearance. All nine squares are the same hue (blue), but their lightness and saturation differ as shown. At the extreme of no lightness we see black, while at the extreme of no saturation we see gray. At the other ends of these scales, we see white at complete lightness and a vibrant blue at complete saturation.

A visit to a décor/architectural paint store will provide an example of distinguishing colors by hue, lightness, and saturation. In these stores, colors are usually displayed on cards, with each card showing the same hue at several different lightness values (Fig. 5.5). These cards are typically displayed in rectangular arrays where hue changes in one direction (left to right or top to bottom), and brightness and saturation change in the other.

---

a logical or higher order to Nature that would manifest similar behaviors for different fundamental properties.

[5] Some authors use the words "lightness" and "brightness" interchangeably. However, in this chapter the word "lightness" will be used to describe a property of the color of an object. Color lightness can vary from complete (white) to none (black). We will use the term "brightness" to describe the intensity of a light. Brightness can vary from dazzling (full sun) to dim (faint star).

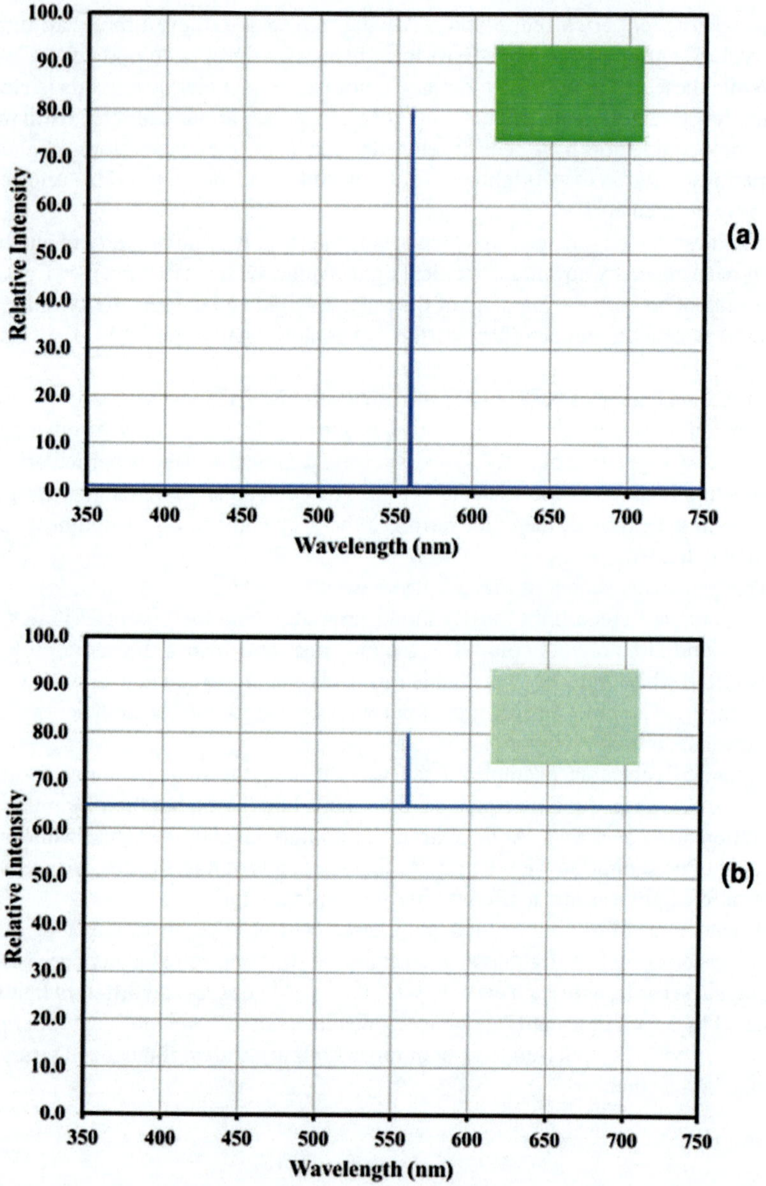


**Fig. 5.3** Reflectance of a saturated (**a**) and desaturated (**b**) green color

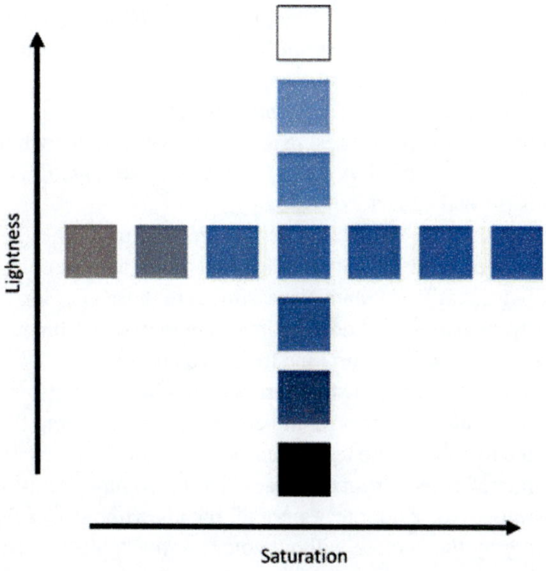


**Fig. 5.4** Example of changes in lightness and in saturation

**Fig. 5.5** Color cards arranged by hue, lightness, and saturation

## *The Achromatic "Colors"—Black, White, and Gray*

Most colors are conceptually equal. That is, while red and blue are different colors, we would not say that red is more "colorish" than blue, or that blue is much more of a color than red is. While this applies to most colors, there are three colors that are fundamentally different from the others. These are black, white, and gray.

Black and white are unique and entangled in the following way: combining all colors of light gives white, while combining all colors of pigment gives black. In addition, all colors except black can be produced with colored lights, and all colors except white can be produced with color pigments.[6]

Because of these unique properties, many would contend that white is not a color, that is, instead it is a "mix of all colors".[7] Likewise, many would say that black is not a color, but is instead a "lack" of color.

So, are black and white colors? Since we are free to define the word "color" any way we wish, we could agree that black and white are not colors. Yet they are appearance attributes that are in many ways similar to those of colors, and, if they are not colors, then what are they? And what about gray, the combination of the black and white—can two non-colors combine to form a color?

In color science, we make a distinction between black, white, and gray, on the one hand, and the remaining colors on the other hand, by defining these three as the neutral or achromatic colors. The term "achromatic colors" is a contradiction, since the word "achromatic" comes from the Greek word chroma—color—with the prefix "a", meaning without (for example, "amoral" means without morals). This makes white, black, and gray the "colors without color" or the "colorless colors". While it is difficult to define what, exactly, is meant by "colorless colors", we can intuitively understand why this term is used to describe white, gray, and black, and how these colors are fundamentally different from the others.

## Components of Color Vision

As described previously, the process of seeing an object begins when light is radiated by a source. The radiated light then strikes the object and is modified by it—certain wavelengths are reflected from the object, certain wavelengths are absorbed by it, and the remaining wavelengths are transmitted through it. The modified light, referred to as the stimulus, is then captured by our eyes, and through the interaction of the light with optically sensitive chemicals (opsins) in our vision cells, nerve impulses are created, partially processed within the eye, and then sent on to our brains for further processing. Finally, from the fully processed information, we perceive the color sensation of the object.

There are four factors that determine how colors are perceived. They are the sensitivity of the eye to the different wavelengths of light, the spectral power distribution of the light source, the light absorption characteristics of the object, and certain physiological factors.

---

[6] Following longstanding convention in the pigment industry, we use the term "color pigments" to refer to any color except white.

[7] If we consider black a color, then technically speaking a mix of all colors would be gray, that is, white to which black has been added.

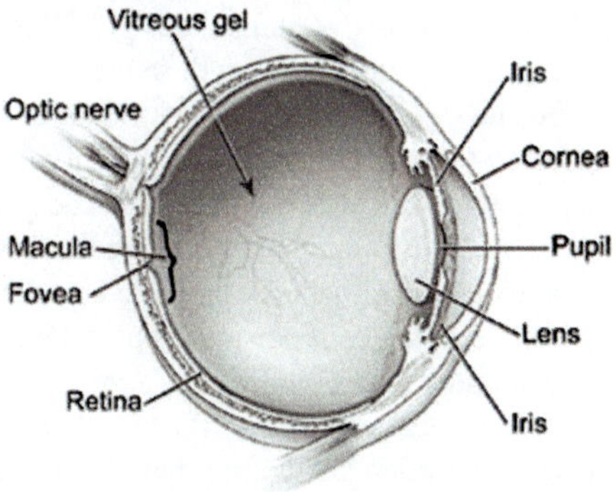


**Fig. 5.6** Components of the eye. *Source* National Eye Institute, National Institutes of Health. License found at https://creativecommons.org/licenses/by/2.0/

The first three of these can be described mathematically or, more commonly, graphically, as spectra. The fourth factor is subjective—it varies from one person to another and from one circumstance to another—and is therefore not readily quantifiable. We will discuss these four factors in turn, but before doing so we must understand how the eye and brain work together to detect and process light.

## *The Eye*

The eye is a complex and highly sensitive organ consisting of many parts that work in unison (Fig. 5.6). Light enters the eye through the pupil and is focused by the lens. It then passes through the vitreous gel and strikes the back surface (the retina). Here specialized cells absorb different wavelengths of visible light (ranging from approximately 400 to 700 nm) and generate the nerve impulses that are then processed into an image.

The image processing aspect of vision is very complex. Surprisingly, it begins within the eye.[8] Most individual light detecting cells are not in direct contact with the brain. Instead the stimulus inputs from groups of cells are combined and partially processed before the signal leaves the eye. These partially processed signals are fed to the cortex, via the optic nerve, to be further processed into images.

There are four types of cells that absorb light. These cells are classified based on their shape and function. Rod-shaped cells are sensitive to most visible wavelengths while cone-shaped cells—of which there are three types—are sensitive to different regions of the visible spectrum. Rod cells are very sensitive and can even respond to a single photon. By contrast, significantly higher brightness is needed to activate the cone cells. The cone cell types differ in the range of the visible spectrum over

[8] One theory for the development of eyes proposes that they evolved from brain cells that were sensitive to light. If correct, this would explain why the eye can process the nerve impulses before passing them on to the vision center of the brain. There is paleological evidence for this aspect of evolution.

which they are active and are designated S, M, and L cones (abbreviations for short, medium, and long wavelength sensing).

Because of the different sensitivities of the rods and cones, we rely on rod vision (referred to as scotopic vision) in dimly lit environments and cone vision (photopic vision) in well-lit environments. Rods do not contribute to vision in well-lit conditions because they are completely saturated at these light levels, and thus there is no contrast between the rods that are stimulated by different parts of the scene.

Light sensing cells can only report the total light flux over their active wavelength ranges, rather than brightness as a function of wavelength. That is, each independent cell absorbs light over a specific wavelength range, but cannot distinguish among the different wavelengths within that range. The output from these cells is, therefore, a single value that is a sensitivity weighted average of the intensities of light over the wavelength range.

Because there is only one type of rod cell, we have only one parameter available to us to describe the information that these cells give us. That parameter is lightness. We therefore cannot distinguish the color of an object in dim light; under these conditions we only see an object achromatically (white, black, or gray). This gives rise to the adage that "all cats are gray in the dark" [3].

We see the non-achromatic colors when the setting is bright enough for significant light absorption by the cone cells. As noted above, the three types of cone cells are sensitive to different regions of the visible spectrum, and color sensation is made through a complex comparison of the intensities among these three different cells, as will be detailed in the section on opponent color theory below.

Rod and cone cells are not evenly distributed on the retina. Cone cells are concentrated in the fovea, a narrow area directly behind the lens. Due to its location directly behind the lens, light falling on the fovea is more finely focused than light falling on other areas of the retina. In dim light, the cone cells are inactive, and vision comes from the rod cells, which are outside the fovea and so are more poorly focused. This is one reason that finely detailed activities, such as reading and sewing, are difficult under dim lighting.[9]

Just as visual acuity diminishes with age, so, too, does the ability to distinguish colors. This is due to a number of reasons, the most common of which are clouding of the lens (cataracts) and yellowing of the tissues and liquids along the light path. For this reason, manual color testing is normally done using youthful color matchers.

## Sensitivity of the Eye to Light

Our eyes are not equally efficient at detecting all visible wavelengths of light. We are more sensitive to wavelengths in the middle of the visible range, such as green, and less sensitive to those at the edges, such as blue and red. Over the course of the

[9] The prevalence of rod cells in the periphery of the fovea also explains why it is easier to detect faint stars by looking at the off-center, that is, by positioning them in the "corner" of our eyes, rather than directly in the center of our view.

twentieth century, several workers measured the apparent brightness of a light as a function of light wavelength. In these tests, a subject was shown a circular field which could be lit alternately by two light sources. The first was a constant intensity white light, and the other was a source that projected the entire range of spectral colors, one wavelength at a time. The latter light could be adjusted both in terms of intensity and wavelength. The two lights were rapidly flashed in succession, resulting in a noticeable flicker. The subject adjusted the intensity of the spectral light until the flicker could no longer be seen. This occurred when the perceived intensity of the two lights were equal.

These experiments were run twice—once under bright conditions (for which the cone cells were active, and the rod cells saturated) [4] and once under low light conditions (for which the cones cells were inactive and the rod cells were activated) [5, 6]. Plots of sensitivity versus wavelength for these two conditions, each scaled to a maximum value of 1.0, are shown in Fig. 5.7. These curves are known as the scotopic (rod cells) and the photopic (cone cells) luminosity functions, and are designated $V'(\lambda)$ and $V(\lambda)$, respectively. Note that there is a slight offsetting of these curves. Because of this, red objects appear black in dim light (rod cells do not detect red light) while blue objects appear gray (rod cells do detect blue light).

Overall, our eyes are sensitive to light intensities that span roughly five orders of magnitude, ranging from bright sunlight to the faintest of stars. However, we can only discriminate roughly two orders of magnitude at any given time. We cannot, for example, see a faint star at night if we are under a bright streetlight. Because we can only see a subrange of light intensities at any given time, we find that changing from a bright environment to a dim one, or vice versa, requires several minutes for the eye to adjust to the average radiance it is receiving.

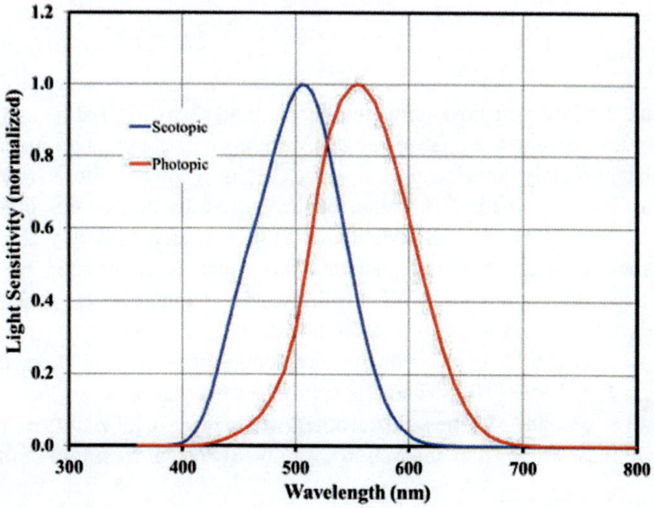


**Fig. 5.7** Relative sensitivities of the rod cells (scotopic vision) and the cone cells (photopic vision)

**Fig. 5.8** Brightness scales. Top: Linear with true lightness. Bottom: Linear with perceived lightness

An aspect of being able to detect brightnesses over such a large range is that we do not interpret brightness on a linear scale [7]. Instead we do so by a power law equation. This is demonstrated in Fig. 5.8. In this figure, there are two progressions from black to white. In the first progression (Fig. 5.8a), the lightness of the gray squares increases linearly with light intensity. This is in contrast with our visual interpretation of these squares—it seems that dark shades dominate the scale.

By contrast, the progression of gray shades in Fig. 5.8b appears much more uniform than in Fig. 5.8a. This is, however, an illusion. In Fig. 5.8b, the lightness of the squares increases each time by a power of 2.2, rather than by a constant amount. This is the basis of the gamma correction on electronic screens [8].

This finding is consistent with Weber's law, which states that a just noticeable difference between two stimuli—perceived light intensities in this case—is a constant proportion of the original stimuli. For vision this means that the smallest increment in added brightness that is perceivable by the eye is a percent of the original brightness, rather than an absolute amount, and so is not the same for all levels of brightness.

## Color Sensing

A common thread running throughout this chapter and Chap. 6 is that we can describe color using three parameters. As we will see, different sets of three parameters can be used, but to completely describe any color, each set must consist of three independent parameters. The reason for this is that our eyes give us three color inputs—one from each cone type (L, M, or S). Mathematically, we can completely describe any situation where there are three independent values, such as color vision, by a system of three independent parameters. We saw this earlier in our ability to describe color qualitatively using hue, lightness, and saturation.

The sensitivity ranges of each of the three cone types, for the average person, are shown in Fig. 5.9 [9, 10].[10] There is significant overlap between the responses, particularly between the M cones and the other two. In fact, for all visible wavelengths except the longest, there is an absorption overlap between at least two of the cones.

[10] Interestingly, much was already known about the existence and spectral absorptivities of these cells long before their spectra were determined experimentally [11–13].

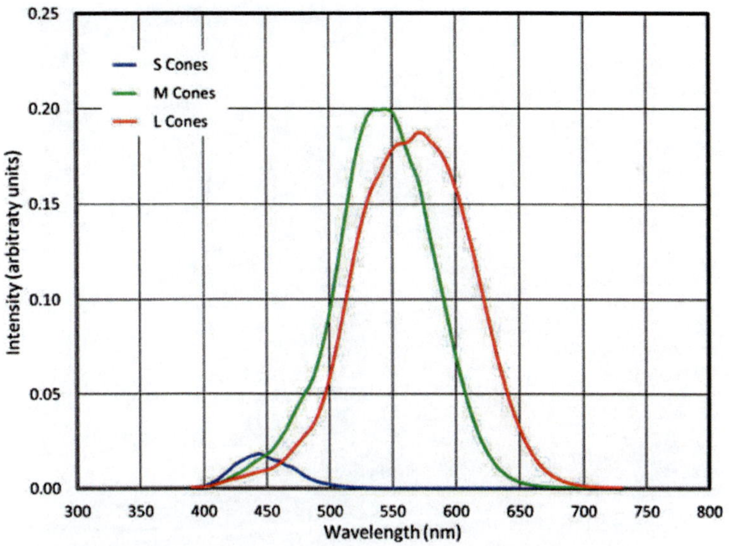


**Fig. 5.9** Visible light absorption of the three cones

Although the absorption strength of the S cones is much less than the other two, the information from this cone is amplified when processing light and so is of equal importance as the information from the other two cones.

An obvious way to describe colors using three parameters is to measure the intensities of the nerve impulses from each of the three cones when observing a given color. However, despite the logic of this, we do not use cone absorptivities as the basis for color science. This is for two reasons. First, these absorptivities are quite difficult to measure and were not determined until relatively recently, by which time color science had already developed alternative, equally valid ways to quantify colors. Second, this information cannot guide us as to how to reproduce colors using three primary lights (a very important purpose of color science is to tell us how to accurately reproduce colors using only a combination of red, green, and blue light sources). As we will see in Chap. 6, the original method of quantifying colors (RGB color space) did just that.

## *Sources and Illuminates*

The creation and projection of light is the starting point of any visual sensation. While we can purchase lights of virtually any color, in our day-to-day lives we interact primarily with white lights. When we examine different white light bulbs, however, we find that they do not all have the same appearance. Some "white" lights are somewhat blue, while others are somewhat yellow (Fig. 5.10). If we examine the

**Fig. 5.10** Two "white" lights of various tones. Left—"soft warm" (color temperature 2700 K); right—"daylight" (color temperature 5000 K)

intensities of these lights over the range of visible wavelengths, that is, their spectral power distribution, we would find that they can differ quite significantly, even though they are all considered "white" lights.

Daylight is similar. While we might expect light from the sun would have a constant color appearance, it, in fact, varies during the day. When the sun is near the horizon it has a distinctive red appearance, while when it is overhead it has a yellowish cast. Daylight is different on a cloudy day than a clear one, and can be different in the different compass directions (north, south, etc.).

That there are "white" lights of differing color is important because the spectral power distribution of a light that strikes an object has an effect on the perceived color of that object (as discussed earlier). An example of this can be seen in Fig. 5.11. In Fig. 5.11a, we see a white and a purple patch over a gray background, as illuminated by a white fluorescent light. In Fig. 5.11b, we see the same scene, but this time illuminated under yellow light. Here we see the purple patch and gray background appear as nearly identical, despite being very easy to differentiate under white light.

Because of the diversity of the actual color of "white" lights, we must clearly specify the light source used for any color analysis. In this way, the results from different laboratories or different instruments can be adjusted for spectral differences between them, and so measured spectra can be compared independent of the light

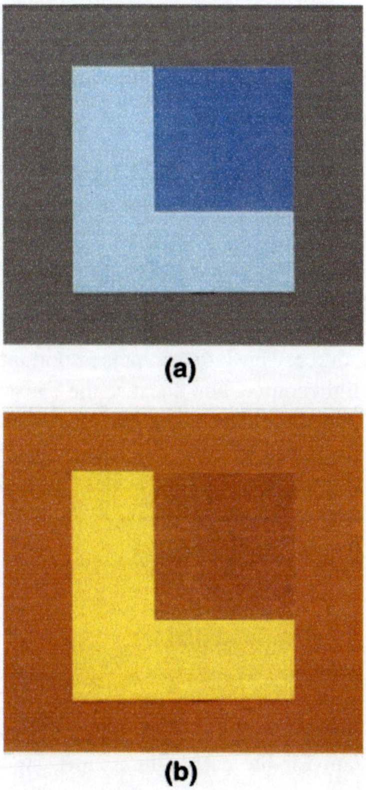


**Fig. 5.11** Photographs of blue square on white paper laid over gray background. **a** Illuminated by white light. **b** Illuminated by yellow light

source. The spectrum of each light source is typically characterized by a table of intensity values as a function of wavelength at 5 or 10 nm intervals.

Several standards societies and organizations define specific spectral power distributions of lights that are used in color science. In this book, we will focus on spectral compositions specified by the International Commission on Illumination, also known by the French name "Commission Internationale de l'Elcairage", and most commonly referenced by the initials CIE. This is the international authority under which much work in color science has been codified, beginning in the early twentieth century and continuing to the present day.

The CIE makes a distinction between a light source and an illuminant. A light source is a physical object that creates and projects light. The most common light sources in everyday life are light bulbs and the sun. An illuminant, on the other hand, is a set of spectral intensities that may or may not be producible by a physical light source. Illuminants are typically given in table form (intensity as a function of wavelength), and illuminant (and source) spectral power distributions are designated as $S(\lambda)$. The primary benefit of illuminants is that they are constant, whereas the emissions of a light source can change over time. Thus, by specifying illuminants, we use constant values for the spectral power distribution of the light.

Ideally there would be a source that replicates each illuminant. While some illuminants can be approximated with certain lamps (or combinations of lamps and color filters), others cannot. This is a problem for color matching using the eye, but, as we will see, is not an issue for color measurement using a spectrophotometer.[11]

Obviously, any color measurement device must use a real light source. The direct output from color measurement devices is a table of reflected light intensities at each wavelength tested (normally in 5 to 10 nm increments). If the spectral power distribution of the source is known, the measured intensity data can be manipulated mathematically to simulate the results expected for any illuminant or other light source. This latter point is important because not all optical instruments use the same light source, and because the emission spectrum from a given light source can shift over time. Such mathematical manipulations are generally done by weighing the measured spectral power distributions, one wavelength or wavelength interval at a time, by the ratio that scales the spectral power distribution of the actual source to the spectral power distribution of the illuminant or alternative source at that wavelength or wavelength interval.

## Blackbody Radiators

An important consideration of the CIE when creating standard illuminants was that they approximate a natural or an artificial light source.[12] This was especially important for the early illuminants, since all color matching was done by eye at that time.

When the CIE defined the first three standard illuminants (A, B, and C), all important sources of light were incandescent, that is, were sources that generated visible light through heat. These sources include both sun and filament (incandescent) light bulbs. This type of light is referred to as blackbody or thermal radiation and is familiar to us as the red glow of hot objects.

The theory behind blackbody radiation was worked out by Max Planck in 1900. Early that year he discovered an equation fitting the spectra of this radiation, and then later in the year developed the underlying theory to explain it. Interestingly, one implication of his work is that the energy of radiation is not continuous, but rather that this energy is quantized into discrete values, or quanta (individual photons). This is the basis of quantum physics.

Planck derived an equation that showed that the total intensity of radiation increases with temperature while the peak in the radiation distribution curve shifts to lower wavelengths (Fig. 5.12). At temperatures of about 800 K, objects begin to

[11] A spectrophotometer is a device that measures intensity as a function of wavelength of the light that is reflected from an object. Spectroradiometers are related devices that measure the intensity as a function of wavelength for light that is emitted directly by a source.

[12] This is not true for Illuminant E, the equal power spectrum, which is discussed below.

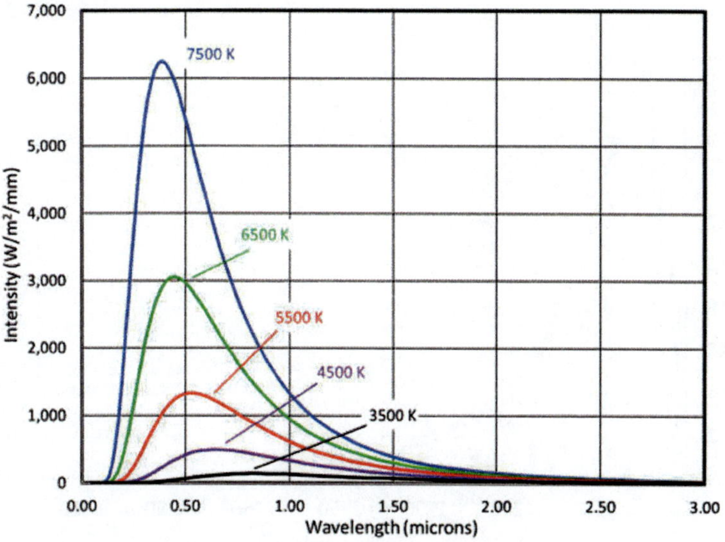


**Fig. 5.12** Spectral intensity of blackbodies at different temperatures

glow at the red edge of the visible spectrum. With increasing temperature, the color shifts from red to orange, yellow, white, and, eventually, blue, although blue is seen only at temperatures significantly hotter than the sun.[13]

Since the sun generates light through thermal radiation, it is not surprising that sunlight can be well approximated by an ideal blackbody curve. In fact, the extraterrestrial solar spectrum (that is, radiation as it is emitted by the sun and unmodified by the Earth's atmosphere) is well fit by a Planckian distribution for a 5800 K blackbody (Fig. 5.13). Note that the solar spectrum at the Earth's surface (also shown in Fig. 5.13) is less intense and much more jagged than the extraterrestrial spectrum. This is because many wavelengths of light are selectively absorbed by different molecules in the atmosphere.

Although many illuminants do not strictly follow Planckian distributions, color temperature is a useful means to characterize them. Color temperatures for non-Planckian spectral power distributions are the temperature for which the visible light portion of the Planckian spectrum most closely matches the spectral power distribution of the source or illuminant. Such temperatures are known as the correlated color temperature of the light.

Describing illumination based on correlated color temperature is a convenience, but it is not meant to be literal. For example, the true temperature of our atmosphere averages below 0 °C, yet the correlated color temperature of blue sky is approximately

[13] It may be surprising that such a high temperature is necessary for blackbody radiation to appear blue, since we are accustomed to seeing blue in flames, and flames are much cooler than the sun. Flames are blue not due to blackbody radiation, but instead because chemicals within the flame selectively emit blue light when hot.

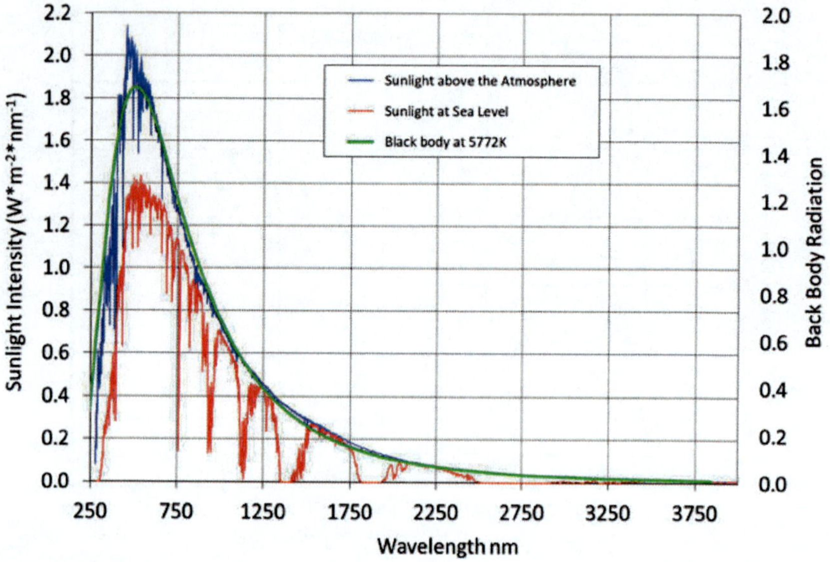


**Fig. 5.13** Intensity of extraterrestrial solar radiation, solar radiation at the Earth's surface, and blackbody radiation at 5772 K

10,000 K—over 30 times its actual temperature of the sky and nearly double the temperature at the sun's surface![14]

## Illuminants A, B, and C

The first three illuminants, Illuminants A, B, and C, were defined by the CIE in 1931. The spectral power distributions of these illuminants are shown in Fig. 5.14.[15] Illuminant A was chosen to match incandescent light bulbs that were in wide use at that time. The light from these sources, typically tungsten-filament bulbs, was closely approximated by a 2856 K blackbody radiator (Fig. 5.15).

Illuminants B and C were chosen to match different aspects of sunlight. The solar spectrum as we experience it at the Earth's surface is not constant. Instead it varies with weather conditions, time of day, season, and compass direction. Illuminant B

[14] The correlated color temperature of the sky is so high is not because it is hot, but rather because it is blue. The Earth's atmosphere separates the blue and red components of sunlight, the result being that blue light is visible in all regions of the sky while red light is restricted to regions of the sky very close to the sun (giving rise to red sunrises and sunsets). This color separation biases the spectral power distribution of the blue regions of sky to lower wavelengths, with the result being that the best fit Planckian distribution is shifted to a much higher temperature than it would be in the absence of this color separation.

[15] Note that there is a convention to scale S(λ) distributions for illuminants such that the value at 560 nm is 100.0.

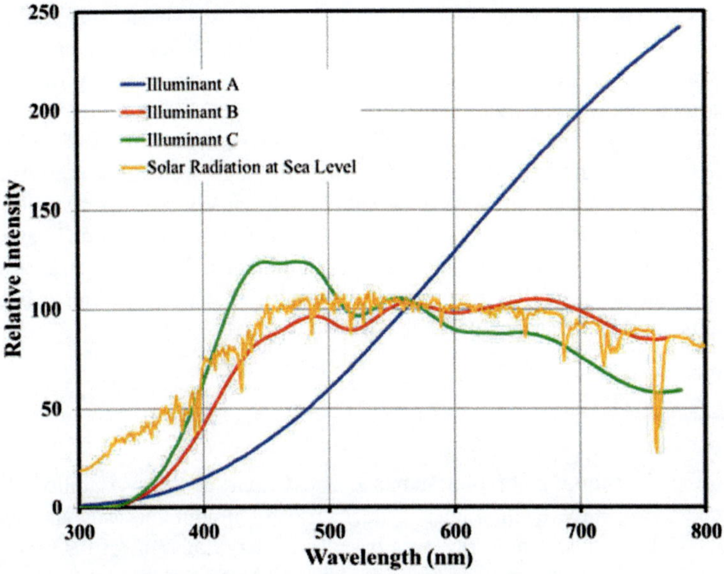


Fig. 5.14 CIE illuminants A, B, and C, and sunlight at sea level

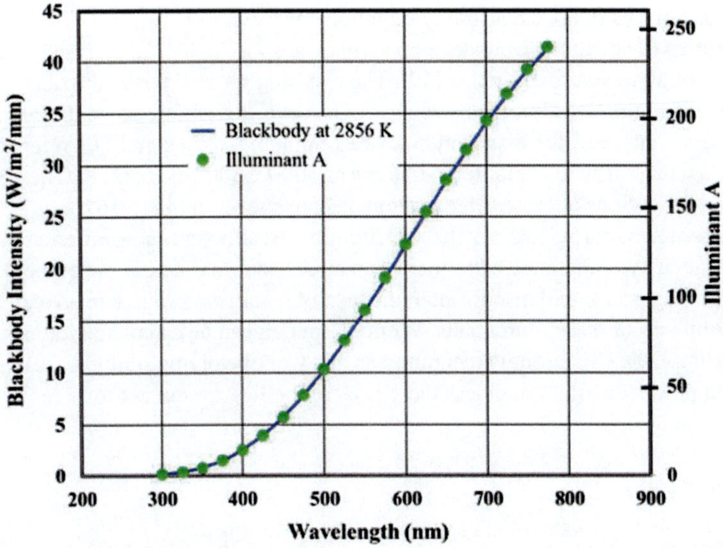


Fig. 5.15 Comparison of blackbody radiation to illuminant A (scaled)

was chosen to match daylight illumination for the noon sun (blackbody temperature 4874 K) while illuminant C matched "average" daylight (blackbody temperature of 6774 K). In practice, these illuminants were approximated by passing incandescent light through appropriate light filters.

Unfortunately, Illuminants B and C did not match sunlight well, particularly near the edge of the ultraviolet region (Fig. 5.14). Although we cannot see ultraviolet light, it is an important component of the spectrum because many materials fluoresce, converting ultraviolet light into visible light. Both Illuminants B and C have since been replaced by the D series illuminants. Illuminant B is no longer used at all, while Illuminant C is still used occasionally, and its spectral power distribution is still maintained and distributed by the CIE.

### D Illuminants

Due to the shortcomings of Illuminants B and C, and to the availability of more advanced light source technologies, a new series of illuminants was approved by the CIE in the 1960s. This series was based on a large amount of data measured on sunlight at different times of the day, days of the year, latitudes, and weather conditions. Judd and others defined a spectrum that matched overhead sunlight well (this spectrum was described by a table of values, rather than as a mathematical equation such as Plank's blackbody radiation equation) [14–16].

A series of equations were developed that transpose this dataset from one correlated color temperature to another [17]. The resulting spectral power distributions are designated as the $D_{xx}$ illuminants, where xx is a two-digit reference to the correlated color temperature of the distribution. For example, $D_{50}$, $D_{55}$, and $D_{65}$ refer to illuminants with correlated color temperatures of 5000, 5500, and 6500 K. The spectral power distributions of these three illuminants are shown in Fig. 5.16.

There are no sources that match the D illuminants, although using different sources simultaneously, sometimes with filters, can closely approximate some of them. That said, for reasons we will discuss later, the lack of a matching source in no way limits the usefulness of these illuminants. While D spectra can be calculated for any color temperature, the CIE strongly recommends that work involving sunlight uses the $D_{65}$ spectral power distribution unless there is a compelling reason not to.

### F Illuminants

The D series of illuminants approximate sunlight very well, but much of our day is spent indoors. Interior lighting is often provided by fluorescent light sources, and the spectral power distributions of these sources are very different from sources based on thermal radiation. Fluorescent lights do not operate under the principles of blackbody radiation, but instead rely on the conversion of ultraviolet light into visible light.

Fluorescent bulbs typically contain mercury. High-voltage electricity is discharged in the bulb, causing the mercury atoms to emit ultraviolet radiation. This

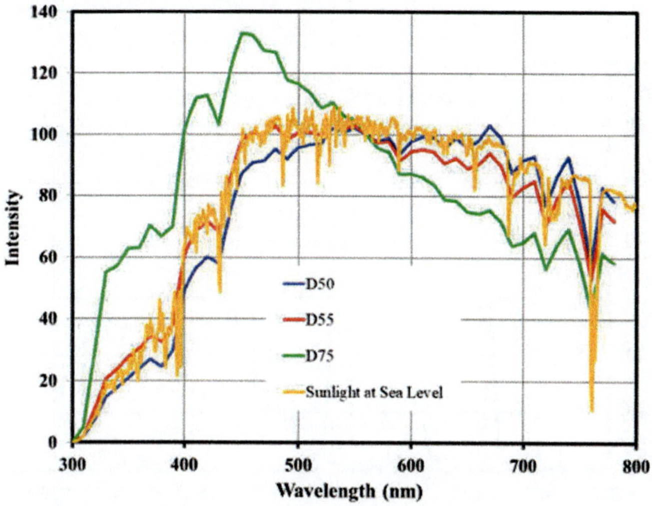


**Fig. 5.16** Relative intensities of the illuminant D series

emission is not thermal—it instead is due to an energy transition among the electrons on the energized mercury atoms. As such it is much more efficient than incandescent illumination, where much energy is wasted generating light outside the visible range.[16] The ultraviolet radiation generated by the excitation of mercury atoms is absorbed by phosphors that are coated on the interior surface of these bulbs, and these phosphors emit specific wavelengths of visible light. By choosing the correct combination of phosphors, the color sensation of white can be generated.

The CIE has designated 12 fluorescent illuminants that differ in their spectral power distributions. As we will see later, it is not necessary for all visible wavelengths of light to be present in order to see white. In fact, Newton famously showed that white light can be created by the combination of just two colored lights—blue and yellow. The result of this is that the spectral power distributions of fluorescent illuminates are quite jagged and discontinuous, unlike blackbody radiation, and yet these sources produce the color sensation of white. An example spectrum of one such illuminant, F12, is shown in Fig. 5.17.

## Illuminant E

Illuminant E is fundamentally different than the other CIE illuminants. All other CIE illuminants were specifically developed to closely match different light sources, so that we can approximate the effect of the spectral power distributions on perceived

[16] As little as 2% of the energy used in incandescent lighting is converted to visible light. Most of the remainder is heat.

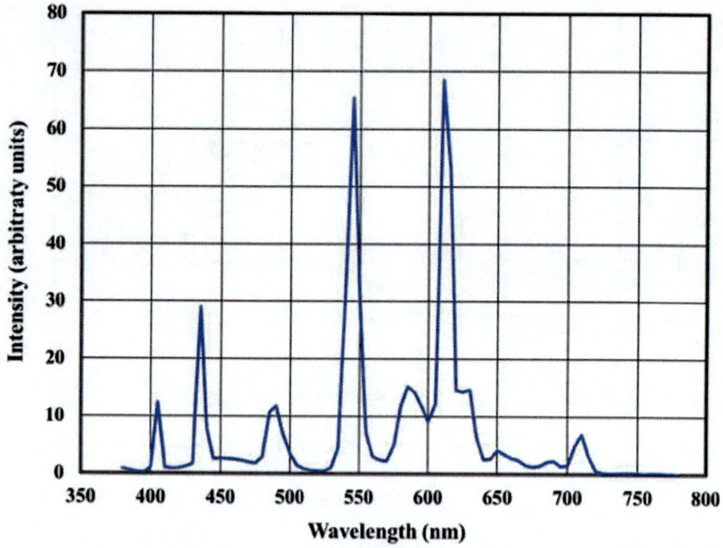


**Fig. 5.17** Spectral power distribution of illuminant F12

color. Illuminant E, on the other hand, was developed to eliminate, rather than approximate, this effect. Illuminant E is also known as the equal-energy radiator, giving it a constant intensity over the visible light range. Since no portion of the visible spectrum is favored against another portion, Illuminant E is achromatic. Although other illuminants are "white" lights, white objects typically have a slight color when illuminated by them.

There are no known sources for illuminant E, nor are there meant to be. The value of illuminant E is as a reference that gives all visible wavelengths equal importance. This illuminant was used extensively in the development of instrumental methods of characterizing colors, and we will refer to it often later in this chapter as well as in Chap. 6.

## *The Reflectance Spectra*

The light absorption properties of an object determine that object's color. We typically characterize an object's absorption properties by measuring the fraction of light, as a function of wavelength, that is reflected from it. The spectra of the light illuminating the object partially determines the object's color, as we saw in Fig. 5.10. Ideally, we would like to measure the intensity of light reflected, as a function of wavelength, using either a source that is important for the application at hand (e.g., using the solar spectrum for outdoor applications) or a source that will not bias the color of the reflected light.

The ideally unbiased source is described by Illuminant E, for which all wavelengths of light are present at equal intensities. There are, however, no known sources that duplicate the Illuminant E spectrum. That said, we can measure the spectral power distribution of a light source and the reflectance spectrum of the object illuminated by it, and mathematically (through division of the reflected intensity by the source intensity) remove the contribution of the source intensity to that object's reflectance. This gives the percent reflectance as a function of wavelength for that object, which we call the reflectance spectrum and denote as $R(\lambda)$.[17]

## The Stimulus Spectrum

As described earlier, there are two ways that we can make an object appear, for example, red. The first is to paint the object red and observe it under white light. Alternatively, the object could be painted white and observed under red light. The visual sensations of these two situations are identical and, based solely on observations, it is not possible to distinguish the two. This is because the spectrum of light entering the eye is the same in both circumstances.

Because it is the combination of light source intensity and light absorption strength that determines the perceived color of an object, it is useful to combine these two into a single spectrum. We call this the stimulus spectrum and denote it as $S_t(\lambda)$. This is simply the product of $S(\lambda)$ and $R(\lambda)$ for each wavelength at a given interval (most commonly every 5 nm). We can either measure the stimulus spectrum directly, using a real light source, or calculate it for a specific combination of object reflectance and light source or illuminant. This process is illustrated in Fig. 5.18.

A foundation of color science is that if two stimulus spectra are the same, then they give the same color sensation. However, the converse of this, that if two spectra give the same color sensation, then the spectra must be the same, is not true. That is, we can have two spectra with different compositions give identical color sensations. This is the basis for all color reproduction technologies, such as colors on the printed page, or on a computer or video monitor. In these cases, three primary colors are combined to give cone absorptions (S, M, and L) identical to the cone absorptions of the original color, regardless of how complex the stimulus spectrum of the original color is. This condition is known as metamerism.

## Metamerism

It is estimated that the average person can distinguish as many as ten million colors. One might reasonably expect that this would mean that we need ten million different phosphors or LED colors on our computer monitors, or ten million different color

[17] Note that this is different from tristimulus R, which we will discuss in Chap. 6.

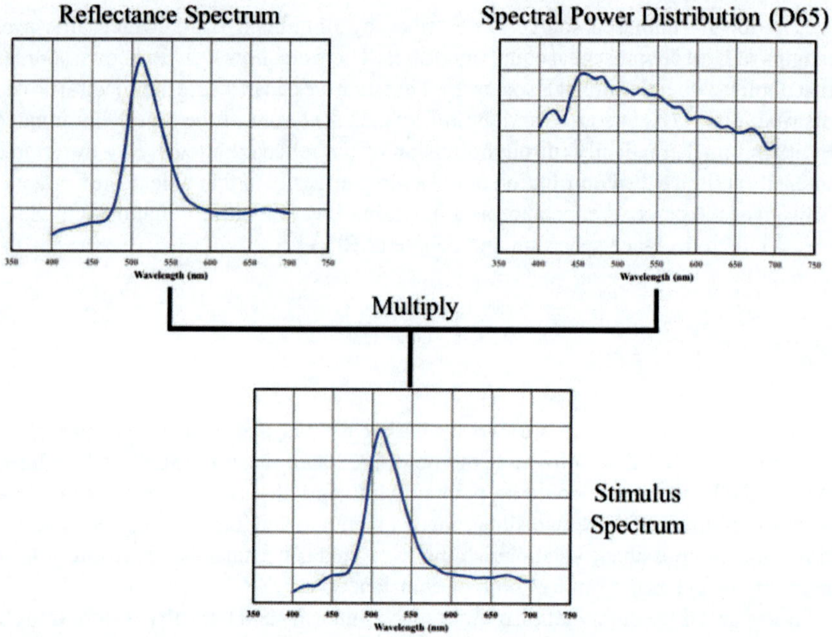


**Fig. 5.18** Combination of reflectance spectrum and spectral distribution to give stimulus spectrum

pigments in our inks, to accurately depict the world as we see it. Happily, this is not the case, and as discussed above, we can create most colors through a mixture of just three primaries—red, green, and blue. Those colors that we cannot create with these primaries can be closely approximated by those that we can.

Because we encounter it every day, we take for granted that, by using just three colored light sources, we can recreate colors accurately on a screen or, by using just three color pigments, recreate colors accurately on the printed page. Yet this is quite an astonishing situation! We can appreciate this by comparing vision to other senses. For example, we cannot play a symphony with just three musical notes, that is, by using three acoustical "primaries". Similarly, we cannot replicate every taste using just three spices, nor every smell using just three perfumes. For these, the rich range of sensations requires a continuous array of "primaries". Yet vision, the most complex of our senses, requires only three!

The phenomenon by which different stimulus spectra give identical color sensations is called metamerism. This arises because there are just three outputs from the eye to the brain (the intensities of the three cone cell signals), yet there is an essentially infinite number of inputs that determine these outputs (the intensities of light at all visible wavelengths). With so many more inputs than outputs, we might well expect that any specific color could be formed by a number of different stimulus spectra.

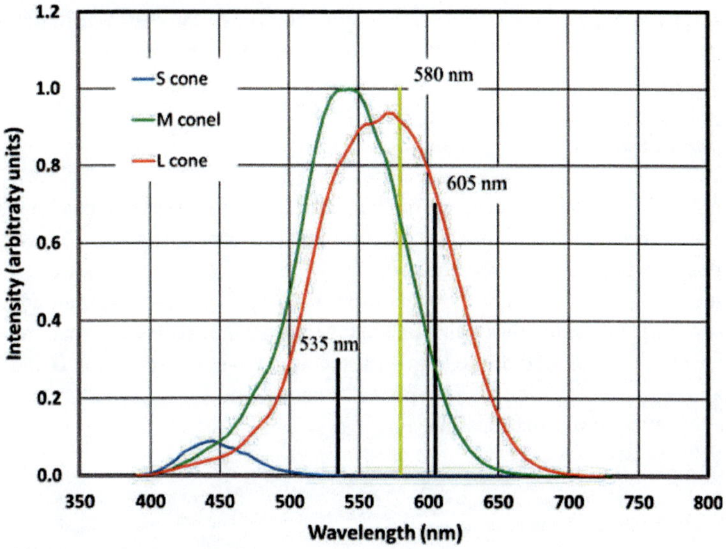


**Fig. 5.19** Combination of monochromatic green light and red light (black vertical lines) gives the same color sensation as monochromatic yellow light (yellow vertical line)

An example of metamerism is given in Fig. 5.19, which is similar to Fig. 5.9. Here we create an identical yellow color sensation in two ways—first by using spectral yellow light (580 nm) and second by mixing 3 parts of 535 nm green light with 7 parts of red 605 nm light. The color formed by combining these amounts of 535 and 605 nm light is impossible to distinguish, by eye, from the yellow light at 580 nm because in both instances we stimulate the three types of cone cells to the same extent relative to one another. This is in fact the way that our monitors and screens create yellow light. There are no yellow phosphors or LEDs in monitors or screens—instead the intense yellow that we can create on these displays is a mixture of green and red lights.

We may wonder the degree to which a set of metameric pairs truly match—can we, with the unaided eye, find a viewing condition for which the pair no longer match? For lights, screens, and monitors, the answer to this question is no. No matter how we change the viewing condition (for example, adding other lights to both lights in the metameric pair, or diminishing the intensity of the lights equally), the colors will look the same as one another.[18] This is also true for any of the physiological factors that will be described later. However, as we will see in a Chap. 6, when the colors

[18] Of course, it would be possible to differentiate the visual stimuli if we are aided by a prism or diffraction grating, either of which separates the light into its spectral components, or if we passed the light through a colored filter that screened portions of the spectra that are different from one another, such as, for the yellow metamers under consideration, a filter that removed light in the range of 500–550 nm.

in question are object colors (rather than light colors), we can distinguish metameric pairs by changing the spectral power distribution of the light source.

## Grassmann's Laws

The implications of metamerism are quite far reaching and serve the basis for what are known as Grassmann's laws of color mixtures. These laws are named after Hermann Grassmann, who first reported them in 1853 [18]. Grassmann's laws are a series of equalities that dictate how metameric pairs of lights interact with an additional light stimulus. Consider two pairs of metamers, the first pair with spectral power distributions **A** and **B**, and the second with spectral power distributions **D** and **E**. We will call the color sensation $\mathbf{C}_1$ for spectra **A** and **B**, and $\mathbf{C}_2$ for spectra **D** and **E**.

We can show these relationships as

$$\mathbf{A} \equiv \mathbf{B} \equiv \mathbf{C}_1$$

$$\mathbf{D} \equiv \mathbf{E} \equiv \mathbf{C}_2$$

Here the bold capital letters signify a certain color sensation, and the "≡" sign signifies that the color sensations match one another, including both the color itself (e.g., red, green, yellow, etc.) and its intensity (dark, bright, etc.).

Grassmann found that these color sensations obey the rules of algebra. For example, if we add light spectrum **D** to both **A** and **B**, the resulting stimuli continue to give the same color sensation as one another:

$$\mathbf{A} + \mathbf{D} \equiv \mathbf{B} + \mathbf{D}$$

Likewise, we can add **D** to **A** and **E** to **B** and retain equal appearance:

$$\mathbf{A} + \mathbf{D} \equiv \mathbf{B} + \mathbf{E}$$

We can also move a color from one side of the equation to the other through subtraction:

$$\text{If } \mathbf{C}_3 \equiv \mathbf{A} + \mathbf{D}, \text{ then } \mathbf{C}_3 - \mathbf{D} \equiv \mathbf{A} \text{ where } \mathbf{C}_3 \text{ is a third color.}$$

This implies that we can have a "negative" color—a seemingly nonsensical entity. We will discuss in Chap. 6 the ways that a negative color can manifest,[19] and the importance of the concept of negative colors to color science.

[19] We may suppose that we can subtract a color using an appropriate light filter, but this is not so. There is no physical process of truly subtracting a color—this is purely mathematical. That said, the concept of negative colors is extremely useful and quite analogous to the concept of imaginary numbers based on the square root of negative one.

Finally, we can increase the intensities of both **A** and **B** equally and maintain appearance equality. For example, if we increase the intensity of **A** by 30% and the intensity of **B** by the same 30%, the resulting stimuli are still equivalent to one another:

$$\mathbf{A} + 0.3 * \mathbf{A} \equiv \mathbf{B} + 0.3 * \mathbf{B}$$

$$1.3 * \mathbf{A} \equiv 1.3 * \mathbf{B}$$

This final equation shows that two metameric stimuli do not shift away from one another as light intensity changes. Note that this equality is only applicable to intensities for which photopic vision applies (that is, for intensities strong enough that cone vision is dominant). Since the photopic and scotopic sensitivity curves are shifted from one another (Fig. 5.7), stimuli that are metameric under bright lights (cone vision only) can often have different lightnesses under dim lighting conditions (rod vision only).

As we will see later, the concept of metamerism, and the fact that colors follow Grassmann's laws, is used extensively in color matching experiments and in the application of these experiments to color theory and reproduction.

## *Physiological Factors*

It has been said that the eye receives light while the brain perceives it. This insight brings us to the final factors in observing colors—those that arise from the way that the brain interprets nerve impulses from the eye. These factors are referred to as the physiological factors of color vision and in some cases can be even more important to our perception of color than the other factors discussed above. As described earlier, the eye partially processes signals from the individual cone and rod cells before sending this information on to the brain for further processing. The details of this processing are quite complex and are still the subject of active study, but even in the current incomplete state we can make some observations about them.

### Opponent Color Theory

We will begin by detailing the process by which the outputs from the light sensitive cone cells are turned into the sensation of color. The most obvious procedure for doing this would be to have the brain interpret the nerve impulses from each cone type separately. This way of sensing colors can be linked to a color display monitor that has knobs for separately adjusting the intensities of three primary colors (red, green, and blue). We call this color processing model the "trichromatic theory" [11, 13].

While this is the simplest model for color sensation, it is not the only possibility by which this process can occur. Since there are three types of cones, any color sensing process must involve three inputs, but the inputs do not have to be the direct intensities of the three cone types. In 1892, Ewald Hering, an early worker in color science, proposed an alternative explanation that is more complex than the trichromatic theory for processing each cone type separately, but that explains many observations that are not explainable with the simpler theory [19].

Hering's proposition, referred to as opponent color vision, is that the raw nerve signals from the cone cells are processed by the eye, prior to being sent to the brain. It is these semi-processed signals that are then further processed and interpreted by the brain as color sensations. This preprocessing of the cone signals is done by combining the raw nerve signals from the three cone cell types in three different ways. Each of these combinations reflects a balance between two "opponent" colors. These balances are light/dark, red/green, and blue/yellow, and are conceptually calculated from the S, M, and L cone cell outputs as follows:

$$\text{Light/dark} = \text{L} + \text{M}$$

$$\text{Red/green} = \text{L} - \text{M}$$

$$\text{Blue/yellow} = \text{S} - (\text{L} + \text{M})$$

There is an important property of the last two balances (red/green and blue/yellow). While we can have, for example, a reddish orange or an orangish red, we cannot have a greenish red or a reddish green. Instead these two colors combine to form yellow—a completely different color sensation than red or green. A combination of red and green that is stronger in red will give a reddish yellow, while a combination that is stronger in green will give a greenish yellow, but we do not perceive these as reddish green or greenish red.

In the same way, we cannot have a yellowish blue or a bluish yellow. These two colors combine to form white or gray, the achromatic colors. A combination of blue and yellow that is stronger in blue will give a bluish gray, and a combination that is stronger in yellow will give a yellowish gray, but we do not see bluish yellow or yellowish blue.

These two ways of processing color—trichromatic and opponent—are not in conflict with one another. Rather, they explain different parts of the color perception process. The trichromatic theory correctly postulates that there are three types of color cone cells, and that it is the relative intensities of the output from these three cone cells that determine color. The opponent theory explains how the output from these cells are subsequently combined in a later step of color processing.

## Other Factors

The brain is very active in interpreting what we see, not only by processing the data from our eyes in an absolute sense, but also by processing the data in a relative or comparative sense. By this we mean that the brain processes a color sensation in part by comparing the color in one area of the visual field to the colors in other areas of the visual field. This is very different from how color detection instruments function. The color detector will faithfully report the wavelength intensities of an area of an image completely independent of the colors in adjacent parts of the image. The brain is not this impartial; instead it relies on the entire field of view when interpreting light signals.

This is easily demonstrated by various optical illusions that trick our brain into seeing something different than what an electronic detector would see. Once such illusion, which affects our interpretation of lightness, is shown in Fig. 5.20. In the top image (a), we see a game board with alternating light and dark squares illuminated from the side. Between the board and the light is a cylinder that creates a shadow on some squares. Two squares on the game board are indicated with letters. It is clear that

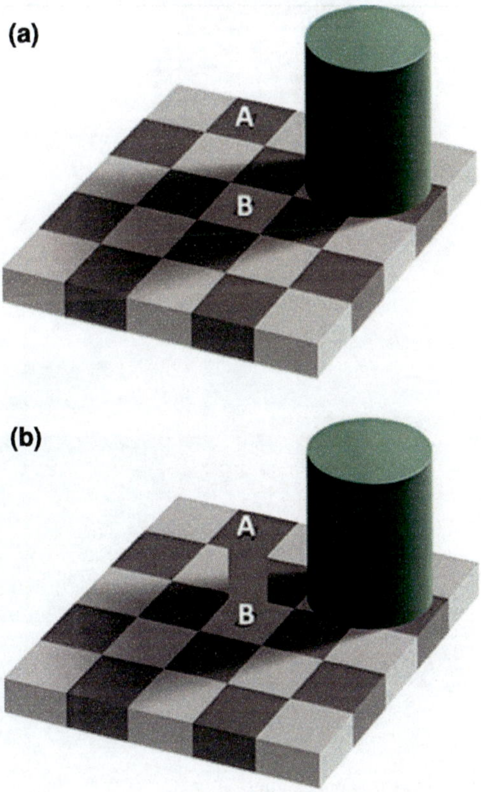


**Fig. 5.20** Lightness optical illusion. *Source* Edward H. Adelson, Wikipedia Commons (license found at https://creativecommons.org/licenses/by-sa/3.0/deed.en)

the square in the illuminated section of the board (square A) is a dark square, while the square in the shadowed area (square B) is a light square. However, by adding a uniformly gray rectangle to the image, we see that, in fact, the two squares have equal lightness! Because we are so accustomed to the way that objects appear differently in light and shadow, our brains automatically take this information into account when interpreting this image. An electronic light detector would not be fooled by this illusion and would faithfully report the correct light intensities regardless of the visual information in the remainder of the image.

Figure 5.21 shows an illusion in which we misinterpret color rather than lightness. To view this illusion properly, first cover the lower image (Fig. 5.21b]), leaving a pair of striped boxes composed of two alternating colors (Fig. 5.21a). In each striped box, one of the colors is green, but the shades of green appear different. Above each striped box is a rectangle of green color that appears to match the green in the striped box.

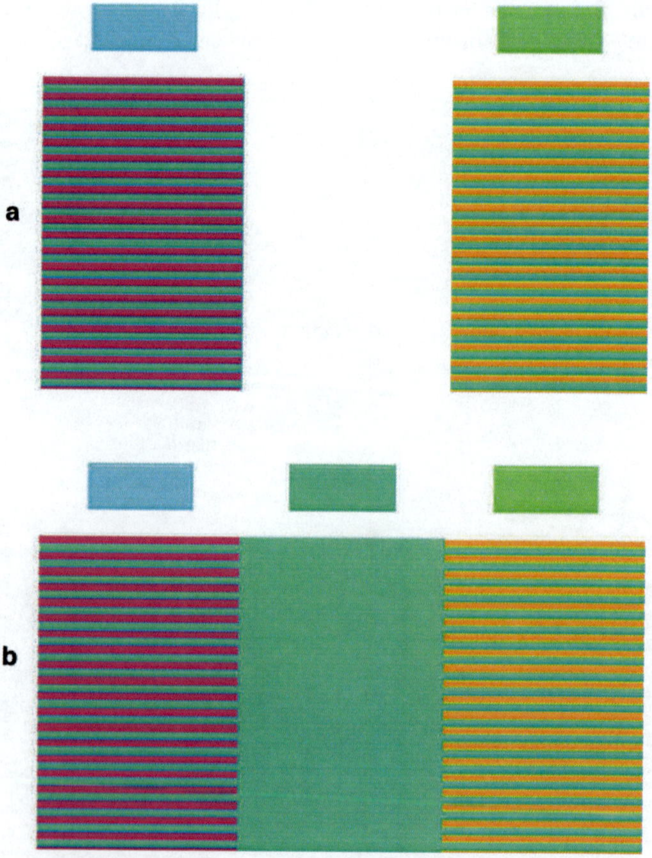


**Fig. 5.21** Color optical illusion

Next we move our attention to the lower half of this figure (Fig. 5.21b). Here we see the same two striped boxes, but now with a block of green color between them. By careful comparison of the areas of the striped boxes closest to this central green area, we can see that, in fact, the green stripes in both striped boxes are the same shade as the green in the region between them, and so the same shade of green as one another! Here, again, an electronic detector would not be fooled by the proximity of one color to another.

Another example of the effect of physiology on our interpretation of colors occurs in our everyday life. Imagine being with a friend inside a room that has no windows and is illuminated solely by an artificial light. Then go outside with that friend. We would perceive the friend's shirt as being the same color indoors as out. Although the shirt appears, to us, to be the same color indoors as out, its appearance has actually changed because the source of light falling on it has changed. That is, the stimulus spectrum has changed because the source spectrum has changed. This perception that the shirt color has not changed is known as color constancy [20].

In this situation, an electronic detector would measure these colors—colors that are considered the same to us—as being different. So our eyes can be confused both into thinking that one color is in fact two (Fig. 5.21) as well as thinking that two colors are one (the color of the shirt indoors and out).

Finally, we note that it is often, if not always, the case that we have incomplete visual information. We do not notice this because our brains are adept at seamlessly "filling in" this missing information. For example, consider the three partial circles in Fig. 5.22a. In our mind's eye, we perceive a white triangle joining the circle centers, but it would be incorrect to say that we *imagine* this. Imagination takes effort, and here it takes effort *not* to see a triangle, and to instead see three partial circles, as shown in Fig. 5.22b (here it *does* take imagination to see a triangle joining the circle centers).

This illusion occurs because the brain fills in what seems to be missing visual information—in the case of Fig. 5.22a, the sides of a triangle. Such "filling in" of missing or incomplete visual information is done all the time, in different ways, as

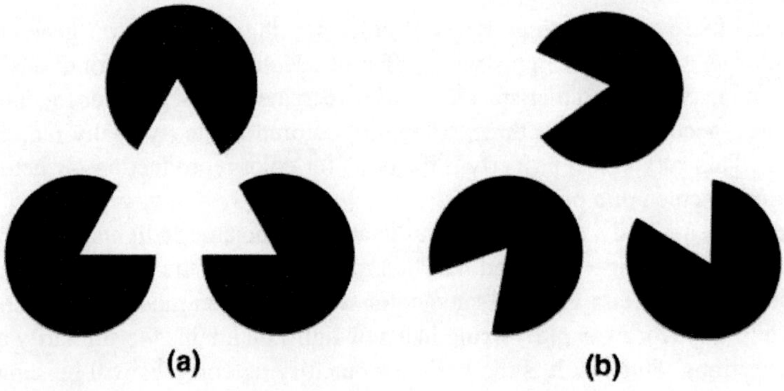


**Fig. 5.22** Aligned (**a**) and non-aligned (**b**) circles with missing wedges

we observe the world around us. While this processing serves an invaluable purpose, it can also mislead us in certain situations, such as those in Fig. 5.22 as well as in Figs. 5.20 and 5.21.

These physiological factors, as well as many others, were certainly known to early color scientists. While they are just as important as the other factors that determine color, there is not, unfortunately, an accurate way to quantify them. They occur within our brains and so are hidden from external measurement. Because of this, in the early development of color measurement and matching, scientists took strict precautions to avoid or minimize these effects in their studies.

We may be concerned that these precautions are so restrictive that they limit the ability to apply the results of these studies to real-life situations, where physiological factors can play a dominating role in the perception of color. While this is true, the intention of these early experiments was to determine ways to uniquely identify a color using an optical instrument, and to determine when two colors match (i.e., are metameric). As laid out in this section, such an instrument would not suffer from these physiological effects, and so the highly controlled nature of these studies does not in any way detract from the utility their results and conclusions. We will discuss these early experiments, and their subsequent modifications, in Chap. 6.

## Summary

Color vision is a complex process comprising many steps—the creation of light, its modification when reflected from an absorbing surface, the absorption of the modified light by special cells within our eyes, and the processing of the nerve impulses from these cells, both within the eye and within the brain.

The color sensing cells in our eyes fall into three categories, depending on the range of wavelengths that they absorb. These are the S, M, and L cell types, corresponding to the absorption of short, medium, and long wavelengths of visible light. The sensation of color is determined by the balance between the intensities of the light absorbed by these different cells.

Because there are only three types of color sensing cells, we can generate the same color sensation with lights with different spectral power distributions. This phenomenon, termed metamerism, allows us to recreate color sensations for complex reflectance spectra using just three differently colored lights (typically red, green, and blue). This very useful property is the basis for color reproduction—whether on an electronic screen or a printed page.

Grassmann devised a set of laws that dictate how metameric lights appear when mixed with other lights. He found that this mixing follows the laws of algebra. In particular, if two spectra give the same color sensation, then modifying the spectra in the same way (for example, mixing in a new light) maintains the similarity of the color sensations. This result is the basis for quantifying color, as will be shown in Chap. 6.

Finally, while there are many physical properties that affect color sensations, there are physiological aspects of color sensing that can be of equal importance. For example, we interpret one color based on the visual information of other colors in nearby areas of the visual field. Because these physiological factors occur within the brain, we cannot adequately measure them and, therefore, cannot anticipate their effects on color sensation. Experimenters typically control these factors tightly so that they do not contribute to the appearance or sensation of different colors.

## References


1. Sells, S.B., Fixott, R.S.: Evaluation of research on effects of visual training on visual functions. Am. J. Ophthalmol. **44**(2), 230 (1957)
2. Berlin, B., Kay, P.: Basic Color Terms: Their Universality and Evolution. University of California Press (1969)
3. Schiff, S.: A Great Improvisation: Franklin, France, and the Birth of America, p. 236. Henry Holt (2005)
4. Commission Internationale de l'Eclairage Proceedings, 1924. Cambridge University Press, Cambridge (1926)
5. Crawford, B.H.: The scotopic visibility function. Proc. Phys. Soc. **B62**, 321 (1949)
6. Wald, G.: Human vision and the spectrum. Science **101**, 653 (1945)
7. Fechner, G.T., Boring, E.G., Howes, D.H.: Elements of Psychophysics. Holt, Rinehart and Winston (1966)
8. Poynton, C. A.: A Technical Introduction to Digital Video. Wiley (1996)
9. Stockman, A., MacLeod, D.I.A., Johnson, N.E.: Spectral sensitivities of the human cones. J. Opt. Soc. Am. A **10**(12), 2491 (1993)
10. Wald, G.: The receptors of human color vision. Science **145**, 1007 (1964)
11. Young, T.: Bakerian lecture: on the theory of light and colours. Philos. Trans. R. Soc. Lond. **92**, 12 (1802)
12. von Helmholtz, H.L.F.: Uber die Theorie der zusammengesetzten Farben. Ann. Phys. **163**(9), 45 (1852)
13. Hecht, S.: The development of Thomas Young's theory of color vision. J. Opt. Soc. Am. **20**(5), 231 (1930)
14. Henderson, S.T., Hodgkiss, D.: The spectral energy distribution of daylight. Br. J. Appl. Phys. **14**(3), 125 (1963)
15. Henderson, S.T., Hodgkiss, D.: The spectral energy distribution of daylight. Br. J. Appl. Phys. **15**(8), 947 (1964)
16. Condit, H.R., Grum, F.: The spectral energy distribution of daylight. J. Opt. Soc. Am. **54**(7), 937 (1964)
17. Judd, D.B., MacAdam, D.L., Wyszecki, G.: Spectral distribution of typical daylight as a function of correlated color temperature. J. Opt. Soc. Am. **54**(8), 1031 (1964)
18. Grassmann, H.: Zur Theorie der Farbenmischung. Ann. Phys. Chem. **165**(5), 69 (1853)
19. Hering, E.: Outlines of a Theory of the Light Sense. Harvard University Press (1964)
20. Foster, D.H.: Color constancy. Vis. Res. **51**(7), 674 (2011)

# Chapter 6
# Color 2—Measuring Color

## Contents



## Introduction

Characterizing colors in a quantitative way is important to industry and, more generally, to society at large. This gives us the ability to specify exactly what color we wish for a certain application. An international beverage company may want, for the sake of brand recognition, the same exact red color in their advertisements in all regions of the world. How does this company specify to printers in each region what this exact shade of red should look like? How can the beverage company protect its trademark for this particular color against use by other beverage companies? These problems can only be solved by developing a system by which colors can be specified in precise and quantifiable terms.

 
M. Diebold et al., *Pigments, Extenders, and Particles in Surface Coatings and Plastics*, https://doi.org/10.1007/978-3-030-99083-1_6

Historically, color measuring systems were based on the premise that three primary colors (red, green, and blue) could be combined to match any other color. The relative intensities of the primaries would form the basis for describing a color, and we call these intensities the tristimulus values for that color. Tristimulus values can be plotted in two or three dimensions to express the color space for a particular measurement system.

Color matching studies were first done in the 1920s, and the first means of specifying color quantitatively was published in 1931. In the years since there has been much additional research into the subject of color quantification. The fruits of this research go beyond merely describing colors—they go to the heart of how we create and interpret colors. Over the last century, several systems for characterizing color have been developed, each addressing specific shortcomings of their predecessor. In this chapter we describe three of these systems: the RGB, XYZ, and L*a*b* color spaces.

## Characterizing Colors

Before discussing methods of color characterization, it is important to specify what, exactly, we mean by this somewhat ambiguous term. Our interests in this chapter are not in the process or procedure for measuring color—things such as how to calibrate a color measuring instrument, the time required to warm up a light source, or the best geometry for optical measurements. We will instead address the fundamental underpinnings of such characterizations; that is, the underlying strategy for measuring color, rather than the tactical means for doing this.

### *Systems Based on Cataloging Colors*

The first efforts to quantify colors were based on color matching and discrimination by eyes. These efforts began in the late part of the nineteenth century to the early part of the twentieth century. Several methods were developed, and many of these have been refined in the time since. These are best understood as being qualitative or, at most, semi-quantitative, and generally involve the development of a physical catalog of colors that can be used as references for comparison purposes. The three dimensions typically chosen to describe color are hue, a lightness property, and a saturation property.[1]

[1] Saturation, also called cleanness, chromaticity, or chroma, refers to the visual intensity of a color. Complete saturation is seen in monochromatic (spectral) colors. Adding other wavelengths to this decreases the visual intensity and saturation value for the resulting color.

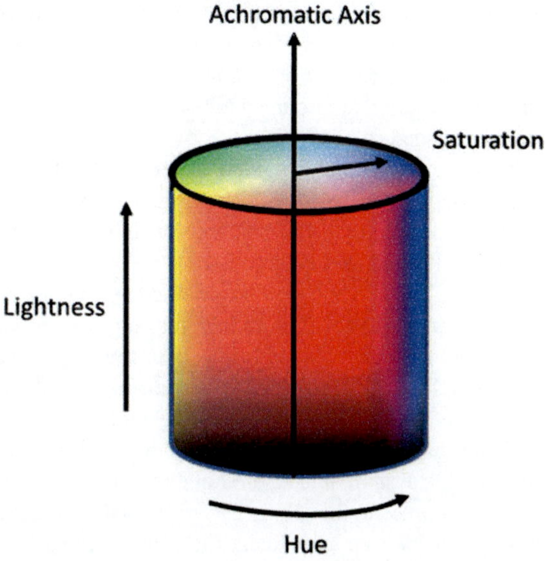


**Fig. 6.1** Cylindric color space showing saturations, lightness, and hue

Because color systems use three parameters, early systems for cataloging colors often did so in terms of three-dimensional objects, such as spheres, cones, or cylinders. One such example is shown in Fig. 6.1. Here we see hue around the circular cross-section of a cylinder, with saturation going from completely unsaturated along the central axis to completely saturated on the cylinder surface. Lightness follows the central axis, with black at the bottom of the cylinder and white on top.

In the late nineteenth century, A. H. Munsell, a professor of arts at the Massachusetts Normal Art School in Boston, realized that the regular solid shapes typically used by these systems had a flaw. That flaw is that all of these systems were not perceptually uniform. What is meant by this is that if, for example, the colors were described by a cylinder (as they are in Fig. 6.1), then pairs of colors separated by the same distance in color space would not have the same degree of perceived color difference. For example, in the green region of color space, two colors separated by a certain distance might look very similar to one another while two colors in the red region that are separated by the same distance might look very different. Regardless of the form, all color spaces mapped to regular solids showed this lack of perceptual uniformity.

This realization that color space does not need to fit into an ideal solid freed Munsell to develop a color space that was irregular, but perceptually more uniform. His space is shaped like a very irregular oblate or cylinder. The circumference of the solid is described by ten main hues, each of which is given a unique letter designation. Lightness, which Munsell called value, varies along the central, or neutral, axis, in a way similar to Fig. 6.1. Finally, radiating out from the central axis is the measure of chroma or saturation.

The main ten hues that Munsell employed, along with their letter designation, are the principal hues **R**ed, **Y**ellow, **G**reen, **B**lue, and **P**urple, along with intermediate hues between each pair (e.g., **YR**, **YG**, etc.). The gaps between each of the ten hues are then subdivided into ten additional hues, numbered 1 through 10, that create an equally spaced color gradient (i.e., is perceptually uniform). This generates a total of 100 unique hues. Any given hue is denoted by its principal hue letter and a number describing the further subdivision. For example, **5R** denotes the intermediate red color; **4R** is slightly away from this hue in the **RP** direction, and **6R** is slightly away from this hue in the **YR** direction.

Each of the 100 hues can be modified by changing its value and/or chroma. Value can vary from 0.5 to 12, with low values representing dark colors, and chroma can vary from 1 to 16, with low values representing unsaturation. Not all combinations of hue, value, and chroma exist. For example, only a few **R**, **Y**, and **RY** hues have chroma as high as 16 while essentially all hues have a chroma as low as 1.

Note that white, gray, and black—the "achromatic" colors—are not considered hues in this system, and so are not explicitly mentioned in this nomenclature. They are in fact present, though, in each of the color designations. This is because adding white or black to a hue changes its value (giving black and white at the extremes) while adding both white and black together (i.e., gray) changes the chroma (giving gray and complete saturation at the extremes of low and high chroma).

By being perceptually uniform, changing the value of any color attribute (hue, value, or chroma) by one unit results in the same apparent change in appearance. Munsell chose these colors by eye, rather than using any sort of instrument, and a major benefit of his system—which is still used widely today—is its high degree of perceptual uniformity. Munsell originally published his color space in 1905 as an atlas. Today labeled color chips are available that can be easily removed from a holder and held up to any color for comparison.

## *Quantifying Colors Instrumentally*

The development of the Munsell color system and others like it was a first step towards placing color measurements on a firm scientific footing. The next step, which occurred in the late 1920s and early 1930s, was the development of a technology that could measure and quantify any reflectance spectrum instrumentally.

The method of doing this consists of first measuring the stimulus spectrum ($S_t(\lambda)$) of the color of interest. This spectrum is calculated by multiplying the source spectrum, $S(\lambda)$,[2] by the reflectance spectrum $R(\lambda)$[3] of that color, wavelength by wavelength. Once the stimulus spectrum is measured, the spectrum is analyzed in a way that ascribes three parameters for that color—in the early studies, this was the amounts of specific red, green, and blue light that can be combined to match the color of

[2] Also called the spectral power distribution of the source.

[3] Note that this is not tristimulus R, which we will discuss later in this chapter.

interest. These amounts are called tristimulus values of that color—"tri" because there are three of them, and "stimulus" because they describe how the lights will stimulate the eye. Any colors with the same tristimulus values will appear identical to the eye, even if the stimulus spectra are different. This situation, where more than one spectrum gives the same color sensation, is referred to as "metamerism" and was discussed in Chap. 5 and will be discussed further in Chap. 15.

Before discussing the experiments from which a tristimulus measurement system was developed, we must first discuss how we quantify brightness.

## Light Detectors

What would it mean if we asked an experimenter whether two differently colored lights had the same brightness? This request seems quite straightforward, but there is, in fact, an ambiguity to it, and that is the question of how we define brightness. One way would be to define it as the energy or power of the light, measured in, for example, Watts per unit area or per solid angle. However, we saw in Chap. 5 that the eye is not equally sensitive to all colors of light, and so we may wish to define the brightness of a light based on human perception rather than an instrumental response.

These two ways of defining brightness differ in the type of detector used to measure it. Defining brightness according to an energy measurement requires an instrumental detector while defining it based on perception uses the eye as the detector. These two brightness measurements are different for virtually any colored light. While we can convert between these two types of brightness using the photopic spectral curve $V(\lambda)$, which shows the sensitivity of the eye as a function of wavelength under well-lit conditions (Fig. 6.2) it is important that, when we discuss light intensity, we specify the basis of measurement.

The convention in color science is to refer to instrumental intensities with words that begin with the prefix "rad", such as radiance, radiant intensity, and radiant flux. These parameters are measured on an energy basis. When we refer to intensity as perceived by the eye, we use words with the prefix "lum", including luminance, luminous flux, and luminosity. These parameters are characterized by the perceived brightness of the light.

## The Color Matching Experiment

Much of our understanding of colors comes from experiments that involve the matching of colored lights by trained observers. In a typical color matching experiment, a box is constructed such that an observer can peer through a circular hole at a white screen at the far end (Fig. 6.3a). This screen is split in the viewing direction by a divider that prevents lights projected on one side of the screen from mixing with those projected on the other side. One side of the screen is illuminated by a light of a certain color. This area is referred to as the test field. The other side of the screen

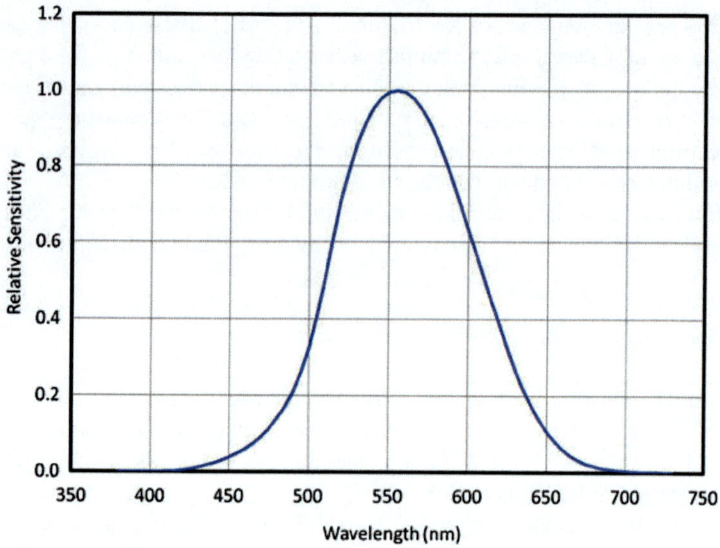


**Fig. 6.2** Photopic absorption curve V(λ)

is illuminated by three lights, typically red, green, and blue, that have independent intensity controls. This area of the screen is the reference or primary field.

In a typical experiment, the observer adjusts the intensities of the three lights in the reference field until they match the color and brightness of the light in the test field. The power settings for the reference lights needed to reproduce the sensation of the test field are the tristimulus values for that color and form the basis for quantifying color. Note that here the eye is the light detector.

## The $\bar{r}(\lambda)$, $\bar{g}(\lambda)$, and $\bar{b}(\lambda)$ Color Matching Functions

Early work in color matching began in the late 1920s when Wright and Guild independently conducted color matching experiments using three primary light sources to match the spectral colors from 380 to 780 nm at 5 nm intervals [1–3]. Both investigators used red, green, and blue primary lights, but the exact wavelengths used by these two workers were different. However, through matrix algebra Wright's results were transformed to the primary set that Guild used: 700 nm (red), 546.1 nm (green), and 435.8 nm (blue). Spectral (monochromatic) lights at these three wavelengths are designated as **R**, **G**, and **B**. The intensities of the three primaries needed to match the color sensation of a specific wavelength of visible light are called the spectral tristimulus values for this wavelength—"spectral" because they match a spectral color—and are notated as R, G, and B.[4]

[4] Note that these R, G, and B values here are *not* the same as the RGB values used to describe color on a computer monitor or screen.

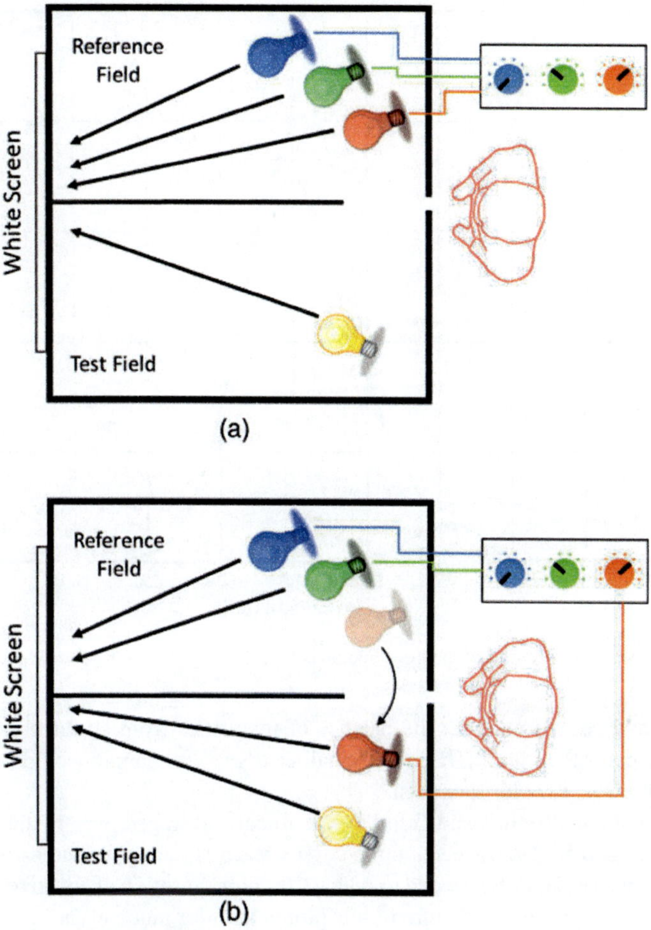


**Fig. 6.3** Color matching experiments

This choice of these primary light wavelengths was not dictated by Nature—it does not correspond to the maximum absorptivities of any of the three types of cone cells, which were not even known at the time, nor is it special to the eye in any other way. The choice was instead based on the practical requirements of convenience and accuracy. The green and blue wavelengths are strong lines in the mercury-vapor lamp emission spectrum, and so are ideal for providing accurately controlled high-intensity green and blue primary lights. The choice of the red wavelength (700 nm) was based on the finding that the eye is relatively insensitive to color differences, as a function of wavelength, in this region of the visible spectrum. This can be understood from Fig. 5.9 in Chap. 5, reproduced here as Fig. 6.4, where the sensitivities of the S and M cone receptors are shown to be essentially zero at and above 700 nm. Any small variation in wavelength around 700 nm will change the intensity of only the

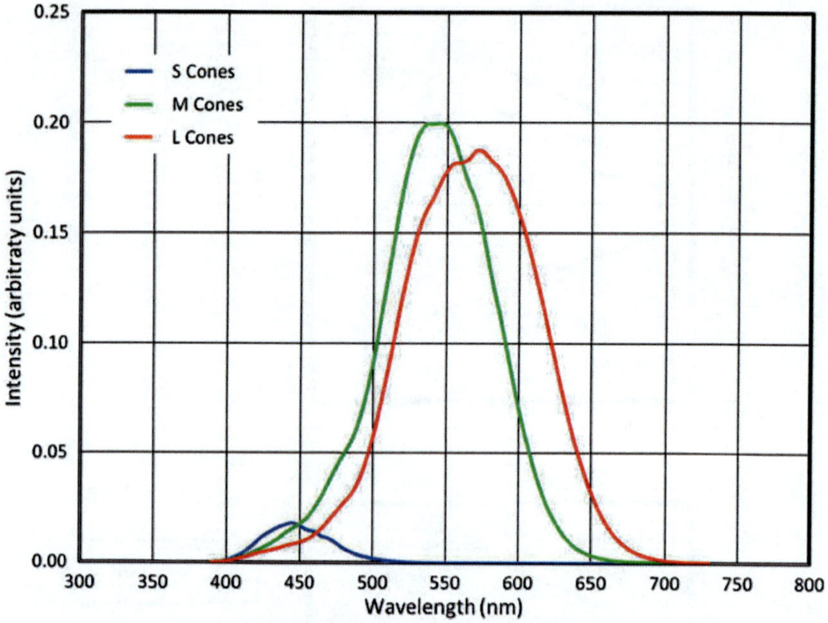


**Fig. 6.4** Spectral sensitivities of the three human cone cells

L cones, and so will not alter the balance of intensities from all three cones (i.e., the hue remains the same). Therefore, small errors in wavelength would not lead to significant errors in color sensation.

The results of Wright's and Guild's experiments, averaged over 17 subjects, are shown in Fig. 6.5.[5] Here we see three curves, each showing the intensities of the primary lights (R, G, or B) needed to match the color sensation of any given spectral color. The most prominent feature of this plot is that, for much of the spectrum, the power of the red primary overwhelms those of the green and blue primaries (so much so that the y-axis in Fig. 6.5 is truncated to allow the G and B curves to be seen). This is because, of the three primary wavelengths, the eye is least sensitive to red (refer back to the photopic luminosity function shown in Fig. 6.4). Although the red lamp is emitting high energy in the red region (high radiance), the eye does not see a high brightness (low luminance).

To place all three curves on an equal basis, the CIE scaled the data such that the areas under the three curves are equal. The ratio of scale factors was 72.0962:1.3791:1 for **R**, **G**, and **B**, respectively. The resulting curves are plotted in Fig. 6.6 and a table of values is illustrated in Table 6.1. The individual curves, and the data table describing them, are called the CIE 1931 $\bar{r}(\lambda)$, $\bar{g}(\lambda)$, and $\bar{b}(\lambda)$ color matching functions. Here the bars over the letters indicate that these are average curves (averaged over the

[5] Wright's experiments were done using ten observers while Guild's work was based on seven. All observers were young men with normal vision who had been extensively trained in color matching.

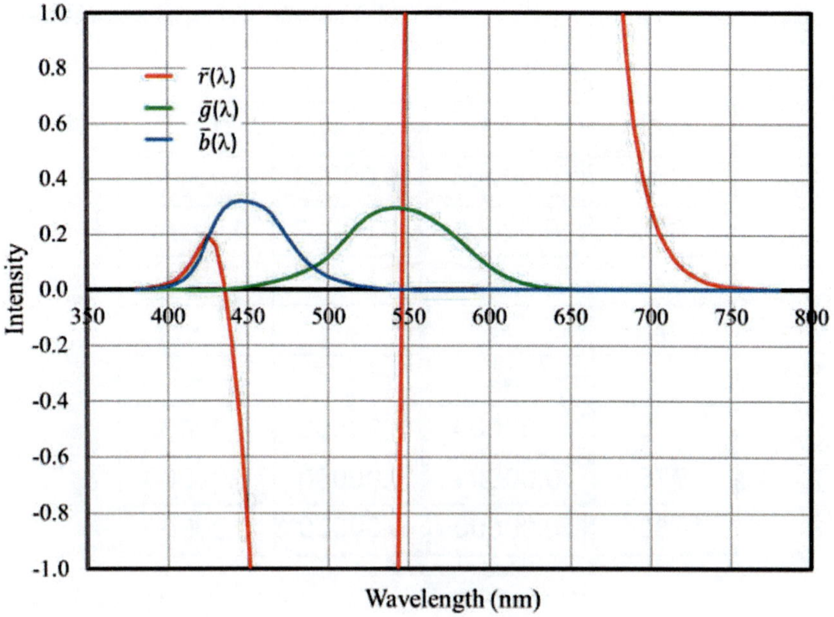


**Fig. 6.5** The raw (unscaled) intensities for the $\bar{r}(\lambda)$, $\bar{g}(\lambda)$, and $\bar{b}(\lambda)$ color matching functions

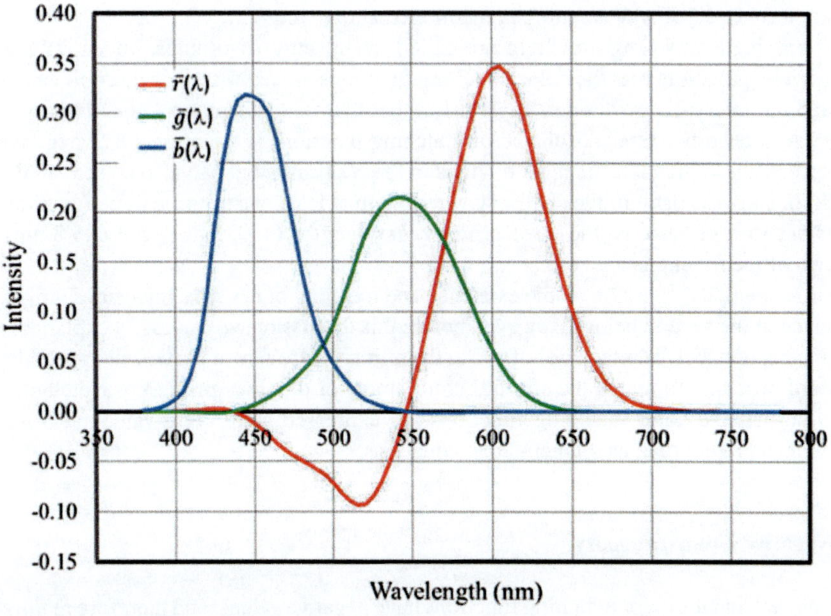


**Fig. 6.6** The normalized intensities for the $\bar{r}(\lambda)$, $\bar{g}(\lambda)$, and $\bar{b}(\lambda)$ color matching functions

**Table 6.1** CIE 1931 $\bar{r}(\lambda)$, $\bar{g}(\lambda)$, and $\bar{b}(\lambda)$ color matching functions. Blue outline designates a color matching function; red outline designates a spectral tristimulus value

| $\lambda$, nm | $\bar{r}(\lambda)$ | $\bar{g}(\lambda)$ | $\bar{b}(\lambda)$ |
|---|---|---|---|
| 380 | 0.00003 | -0.00001 | 0.00117 |
| 385 | 0.00005 | -0.00002 | 0.00189 |
| • | • | • | • |
| • | • | • | • |
| • | • | • | • |
| 770 | 0.00003 | 0.00000 | 0.00000 |
| 775 | 0.00001 | 0.00000 | 0.00000 |
| 780 | 0.00000 | 0.00000 | 0.00000 |

17 individuals tested by Wright and Guild), and the "($\lambda$)" term indicates the curves are a function of wavelength. As mentioned above, the $\bar{r}(\lambda)$, $\bar{g}(\lambda)$, and $\bar{b}(\lambda)$ values for a single wavelength of light are called the spectral tristimulus values for that wavelength. Note that the color matching functions are expressed as discrete values at 5 nm intervals, as found in Table 6.1, rather than as mathematical equations.

A noteworthy aspect of the color matching functions is the presence of negative tristimulus values. This is most obvious in the $\bar{r}(\lambda)$ curve but can also be seen in the $\bar{g}(\lambda)$, and $\bar{b}(\lambda)$ data. In fact, at every wavelength at least one of these three functions is negative or zero. At the three primary wavelengths (700, 546.1, and 435.8 nm), two of the functions are zero.

A detailed discussion on the existence and meaning of negative intensities will be given in the section below. However, before this discussion we can easily explain the observation that the intensities of two of the primaries are zero at the wavelength of the third primary. To match the spectral color at any of the three primary wavelengths. Note that the color matching functions are expressed as discrete values at 5 nm intervals rather than as mathematical equations.

### Negative Color Intensity

The fact that the color matching functions have negative values, and therefore require negative light intensities, merits some explanation. To understand this phenomenon, we will consider the intensities of the primaries required to match 490 nm light, which is an equal mix of blue and green. In Wright's and Guild's experiments, the subject would begin the match by adjusting the combination of luminances of the three primary colors to give the sensation of white light. The subject would then begin the process of adjusting the intensities of the primary lights to move their balance towards the color of the test light. For 490 nm light, this would be done by decreasing the intensity of the red primary light.

In this process, as the intensity of the red light decreased, the color of the primary field approached that of the test field. This was continued until the red primary light intensity was reduced to zero. At this point the test and primary fields still did not match, but it appeared that if more red light could be removed from the primary field (i.e., if the change effected by decreasing the content of the red light could be continued), a color match could be made.

Of course, once the red lamp intensity is brought to zero, no more red light could be removed from the reference field. However, an equivalent process could be used—*adding red light to the test field*. To do this, the subject would move the red lamp from the reference field to the test field, a procedure called "desaturation" of the test color (Fig. 6.3b). The subject would adjust the intensities of the three primary lights to achieve a match.

The result of this trial was that the subject matched the test color with red added to it, to a mix of green and blue primary lights. This can be written in equation form as

$$\text{Sen}(490\ \text{nm}) + \bar{r}(490\ \text{nm})\mathbf{R} \equiv \bar{g}(490\ \text{nm})\mathbf{G} + \bar{b}(490\ \text{nm})\mathbf{B}$$

where Sen(490) denotes the color sensation of 490 nm light and the symbol "≡" means the colors on the two sides of the equation appear the same [4]. That is, they have equivalent color sensations.

While the subject had made a color match, it is not the one in which our interests lie. The purpose of these experiments was to determine the amount of the three primaries needed to match a given spectral (test) color, not the amount of two of the primaries needed to match a *mix* of that spectral color and the third primary.

We can, however, accomplish our goal through the application of Grassmann's laws (see Chap. 5). These laws state that color sensations follow the rules of algebra. Using these laws, we remove the value of $\bar{r}(490)$ on the left side of this equation by subtracting this value from both sides. This results in a negative red light intensity on the right side of the equation:

$$\text{S}(490\ \text{nm}) \equiv \bar{g}(490\ \text{nm}) + \bar{b}(490\ \text{nm}) - \bar{r}(490\ \text{nm})$$

We now see the meaning of a negative primary light intensity. It is the intensity of a primary light required to be mixed into the test field light in order to achieve a color match with the other two primaries.

There is an opportunity for incorrectly interpreting the meaning of a negative light intensity as being the equivalent of removing certain wavelengths of light through

**Table 6.2** Definitions of the symbols used in color science

| RGB Space | XYZ Space | Property | Description |
|---|---|---|---|
| R, G, B | X, Y, Z | Primary light | Source or illuminant used for color space |
| R, G, B | X, Y, Z | Tristimulus value | Amount of primary required to match any color |
| $\bar{r}(\lambda), \bar{g}(\lambda), \bar{b}(\lambda)$ | $\bar{x}(\lambda), \bar{y}(\lambda), \bar{z}(\lambda)$ | Color Matching Function | Amount of each primary needed to match spectral color |
| $\bar{r}(\lambda'), \bar{g}(\lambda'), \bar{b}(\lambda')$ | $\bar{x}(\lambda'), \bar{y}(\lambda'), \bar{z}(\lambda')$ | Spectral tristimulus value | Tristimulus values at wavelength λ', where λ' is a specific wavelength |
| | $\bar{x}(\lambda), \bar{y}(\lambda), \bar{z}(\lambda)$ | CIE 1931 Standard Observer | Basis for quantifying colors |
| | $\bar{x}_{10}(\lambda), \bar{y}_{10}(\lambda), \bar{z}_{10}(\lambda)$ | CIE 1964 Standard Observer | Updated standard observer for wider viewing angle |
| $r, g, b$ | $x, y, z$ | Chromaticity coordinate | Tristimulus values normalized to sum to a value of one |

light absorption. For example, we might expect, based on the above equation, that S(490) could be matched by adding certain amounts of green and blue primary lights and then filtering out some amount of the red primary light from this mixture. This is not the case. The mixture of green and blue lights does not contain any light of the same wavelength as the red primary (700 nm), and so a light filter that selectively removed 700 nm light would have no effect on the appearance of the green and blue mixture.

Although negative light was introduced here as a mathematical construct, we can, in a sense, experience it under certain conditions. If we were to stare at a uniformly colored screen for 30 seconds or longer, and then shift our gaze to a white screen, we would briefly see an after-image color on the white screen that is the complementary color to the original color.[6] This new color is seen because, during the 30 seconds spent looking at the first color, our eyes became insensitive to that color due to the depletion of the chemicals in the cone cells (opsins) that respond to this color. Alternatively, we can say that the brain partially lowers the perceptual impact of a persistent color. When we shift our gaze to the white screen, our brains continue to filter the first color out, and we are left with the complimentary color (i.e., the screen appears as white with the original color removed). Another way of saying this is that when we look at the white screen, we are seeing white with the original color subtracted from our color sensation.

In Table 6.1, and other tables that we have not yet encountered, there are several parameters that can be easily confused because the same combinations of letters are used to describe them. To avoid confusion, we list in Table 6.2 the specific names and definitions of these various quantities, with the understanding that we will not encounter some of these parameters until later in this chapter.

[6] Complimentary colors are pairs of colors that, when additively mixed, produce a white color sensation. For example, blue and yellow are complimentary colors, as are green and magenta.

## *Describing Spectral Colors*

The goal of the color matching experiments was to provide a framework for specifying any color using three numbers based on three specific parameters—the R, G, and B tristimulus values. They are the intensities of the **R, G,** and **B** primaries needed to match any specific color sensation, regardless of the reflectance spectrum of that color.

Specifying a spectral color in terms of the amount of each primary needed to match that color is straightforward. This is, in fact, exactly what Wright and Guild did in their color matching experiments, where the $\overline{r}(\lambda)$, $\overline{g}(\lambda)$, and $\overline{b}(\lambda)$ values were measured for each spectral color at 5 nm intervals. The RGB tristimulus values for a spectral color of specific wavelength $\lambda'$ are, therefore, $R = \overline{r}\,(\lambda')$, $G = \overline{g}\,(\lambda')$, and $B = \overline{b}\,(\lambda')$.

## *Describing Complex Colors*

We will next consider a thought experiment in which we determine the R, G and B values for the relatively simple case of a combination of two wavelengths. In this thought experiment we will combine light of wavelength $\lambda_1 = 540$ nm (spectral green), with an intensity of 100, with light of wavelength $\lambda_2 = 620$ nm (spectral red), with an intensity of 80. The $\overline{r}(\lambda)$, $\overline{g}(\lambda)$, and $\overline{b}(\lambda)$ values for these wavelengths, as well as for $\lambda_3 = 540$ nm (spectral yellow), which we will discuss later, are:

| $\lambda$ | $\overline{r}(\lambda)$ | $\overline{g}(\lambda)$ | $\overline{b}(\lambda)$ |
|---|---|---|---|
| 540 nm | -0.03152 | 0.21466 | 0.00146 |
| 575 nm | 0.20715 | 0.15429 | -0.00123 |
| 620 nm | 0.29708 | 0.01828 | -0.00015 |

In this analysis we will illuminate a white screen using projectors that are capable of producing light of any individually wavelength or combination of individual wavelengths. Using different projectors, we will illuminate the screen with various intensities of light at wavelengths 540 nm, 620 nm, and the wavelengths of the **R**, **G** and **B** primaries (700 nm, 546.1 nm and 435.8 nm, respectively).

Two of Grassmann's Laws (Chap. 5) can be summarized by the following equations:

$$\text{If } \mathbf{A} \equiv \mathbf{B}, \text{ then } c\mathbf{A} \equiv c\mathbf{B}, \text{ where c is a constant}$$

$$\text{If } \mathbf{A} \equiv \mathbf{B}, \text{ and } \mathbf{D} \equiv \mathbf{E}, \text{ then } \mathbf{A} + \mathbf{B} \equiv \mathbf{D} + \mathbf{E}$$

Where **A, B, D,** and **E** are color sensations.

By the definition of the spectral tristimulus values, the sensation of monochromatic light at 540 nm can be replicated by the **R**, **G** and **B** primaries with intensities $\overline{r}(540)$, $\overline{g}(540)$ and $\overline{b}(540)$:

$$\text{Sen}(540\ \text{nm}) \equiv \bar{r}(540\ \text{nm})\mathbf{R} + \bar{g}(540\ \text{nm})\mathbf{G} + \bar{b}(540\ \text{nm})\mathbf{B}$$

Setting the c value in the first Grassmann Law listed above to 80, we see that:

$$80 \cdot \text{Sen}(540\ \text{nm}) \equiv 80 \cdot \bar{r}(540\ \text{nm})\mathbf{R} + 80 \cdot \bar{g}(540\ \text{nm})\mathbf{G} + 80 \cdot \bar{b}(540\ \text{nm})\mathbf{B}$$

In our thought experiment, we will use one projector to project light with an intensity of 80 and the wavelength 540 nm onto the screen. The R, G and B values for the light reflected from the screen are:

$$R = S_t(540\ \text{nm}) \cdot \bar{r}(540\ \text{nm})$$

$$G = S_t(540\ \text{nm}) \cdot \bar{g}(540\ \text{nm})$$

$$B = S_t(540\ \text{nm}) \cdot \bar{b}(540\ \text{nm})$$

$S_t(540\ \text{nm})$ is simply the intensity of the light source (80) multiplied by the reflectance of the screen (1.0), and so the R, G and B values for this color are −2.5216, 17.1728, and 0.1168. Another way to produce the same color sensation would be to illuminate the screen with light of the three R, G and B primary wavelengths using these **R**, **G** and **B** values as their corresponding intensities.[7] This situation is shown in Fig. 6.7a.

We can repeat this analysis for the projection of light with an intensity of 100 and the wavelength 620 nm. This is shown in Fig. 6.7b.

Finally, we consider the situation where the two spectral lights are combined (Fig. 6.7c). We do this by adding the sensations of the individual lights expressed as both the individual wavelengths and as the **R**, **G** and **B** wavelengths, using the second Grassmann Law as summarized above (if **A** ≡ **B**, and **D** ≡ **E**, then **A**+**B** ≡ **D**+**E**) :

$$\begin{aligned} 80 \cdot \text{Sen}(540\ \text{nm}) &\equiv -2.52 \cdot \mathbf{R} + 17.17 \cdot \mathbf{G} + 0.12 \cdot \mathbf{B} \\ + 100 \cdot \text{Sen}(620\ \text{nm}) &\equiv 29.71 \cdot \mathbf{R} + 1.83 \cdot \mathbf{G} - 0.02 \cdot \mathbf{B} \\ \hline 80 \cdot \text{Sen}(540\ \text{nm}) + 100 \cdot \text{Sen}(620\ \text{nm}) &\equiv 27.19 \cdot \mathbf{R} + 19.00 \cdot \mathbf{G} + 0.10 \cdot \mathbf{B} \end{aligned}$$

And so the R, G and B tristimulus value for this combination of two wavelengths are 27.19, 19.00 and 0.10. Generalizing this example, we see that the R, G and B tristimulus values for any mixture of two wavelengths is simply:

$$R = S_t(\lambda_1) \cdot \bar{r}(\lambda_1) + S_t(\lambda_2) \cdot \bar{r}(\lambda_2)$$

$$G = S_t(\lambda_1) \cdot \bar{g}(\lambda_1) + S_t(\lambda_2) \cdot \bar{g}(\lambda_2)$$

[7] Of course, it is physically impossible to project light with a negative intensity, as must be done here for the **R** primary wavelength, but for the purposes of our thought experiment we can ignore this physical impossibility.

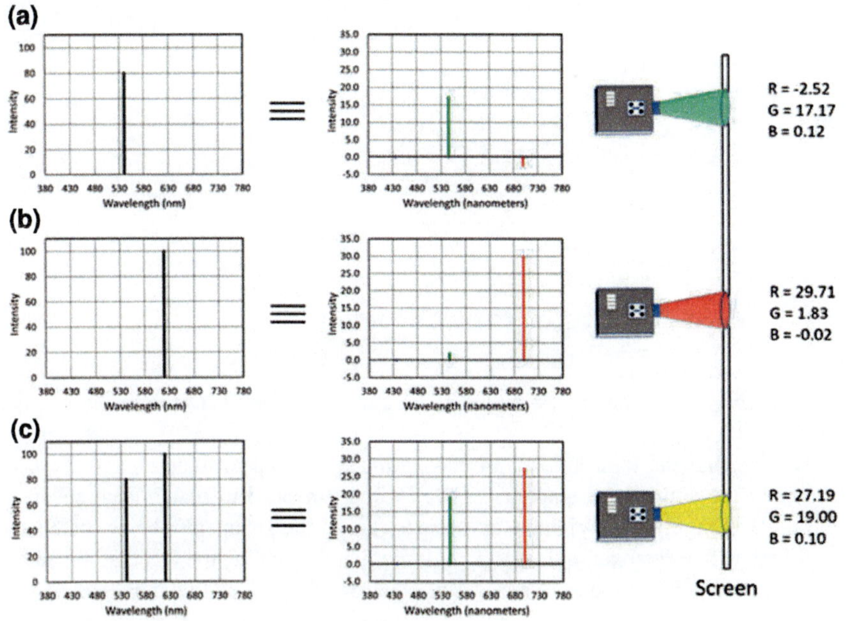


**Fig. 6.7** Mixing two spectral colors using light projectors

$$B = S_t(\lambda_1) \cdot \bar{b}(\lambda_1) + S_t(\lambda_2) \cdot \bar{b}(\lambda_2)$$

By extension, if we mixed a third wavelength, $\lambda_3$, into our reflectance spectrum, then the R, G, and B values for this mix would be

$$R = S_t(\lambda_1) \cdot \bar{r}(\lambda_1) + S_t(\lambda_2) \cdot \bar{r}(\lambda_2) + S_t(\lambda_3) \cdot \bar{r}(\lambda_3)$$

$$G = S_t(\lambda_1) \cdot \bar{g}(\lambda_1) + S_t(\lambda_2) \cdot \bar{g}(\lambda_2) + S_t(\lambda_3) \cdot \bar{g}(\lambda_3)$$

$$B = S_t(\lambda_1) \cdot \bar{b}(\lambda_1) + S_t(\lambda_2) \cdot \bar{b}(\lambda_2) + S_t(\lambda_3) \cdot \bar{b}(\lambda_3)$$

We are now in a position to consider any color with any reflectance curve. We do this by extending the above procedure. As stated earlier, the $\bar{r}(\lambda)$, $\bar{g}(\lambda)$, and $\bar{b}(\lambda)$ functions are reported at 5 nm intervals. At each of these wavelengths we multiply the $\bar{r}(\lambda)$, $\bar{g}(\lambda)$, and $\bar{b}(\lambda)$ values by the intensity of the stimulus at that wavelength ($S_t(\lambda)$), and then sum these values over all wavelengths. This gives us the following equations for determining the tristimulus values for any color[8]:

[8] Note that if the color matching functions were continuous, the summations ($\Sigma$) in these equations would be replaced by integrations ($\int$). Recall also that the $\bar{r}(\lambda)$, $\bar{g}(\lambda)$, and $\bar{b}(\lambda)$ values have been scaled so that the areas under these curves are equal.

$$R = k \sum_{\lambda=380}^{780} S_t(\lambda) \cdot \bar{r}(\lambda)$$
$$G = k \sum_{\lambda=380}^{780} S_t(\lambda) \cdot \bar{g}(\lambda) \tag{6.1}$$
$$B = k \sum_{\lambda=380}^{780} S_t(\lambda) \cdot \bar{b}(\lambda)$$

Where

$$K = 5.6507 \cdot 100/ \sum_{\lambda=380}^{780} S(\lambda) \cdot V(\lambda) \cdot \Delta(\lambda) = 565.07/ \sum_{\lambda=380}^{780} S(\lambda) \cdot V(\lambda) \cdot \Delta(\lambda) \tag{6.2}$$

Note that Eq. 1 uses the stimulus spectrum ($S_t(\lambda)$) and Eq. 2 uses the source spectrum ($S(\lambda)$).

We see that the total amount of each primary needed to match a given color is simply the sum of the amounts of that primary needed to match each spectral color, wavelength by wavelength, in the spectrum of the given color being matched, modified by a normalization factor (k). Although not intuitive, the fact that the areas under the $\bar{r}(\lambda)$, $\bar{g}(\lambda)$, and $\bar{b}(\lambda)$ curves are equal results in equal quantities of R, G, and B giving achromatic colors (white, gray, and black). Said differently, the R, G, and B values for Illuminant E are equal to one another.

We can display this process for each primary graphically as the product of two curves (the stimulus curve and the color matching curve), followed by a determination of the area under the resulting curve. These three areas are the tristimulus values of the color in question and represent the amount of each primary light that is needed to match that color, regardless of the actual stimulus spectrum. It is important to recognize that we are not matching the *stimulus spectrum* of the color in question but are instead matching the *sensation* of this color using the three primary colors.

## Chromaticity Coordinates *r*, *g*, *b*

We pause here to summarize our method of quantifying the color of an object. We first determine the stimulus spectrum of that object. We can do this either by illuminating an object with the light source of interest and measuring the resulting stimulus spectrum, or by multiplying, wavelength by wavelength, the spectrum of the illuminant by the reflectance spectrum of that object, to give a calculated stimulus spectrum.[9] This is shown at the top of Fig. 6.8.

The next step is to calculate the relative amounts of the three primary wavelengths needed to reproduce the color sensation of the stimulus spectrum. To do this, we multiply, again wavelength by wavelength, the stimulus spectrum with the $\bar{r}(\lambda)$,

[9] The difference between source spectra and illuminant spectra is that sources physically generate light with a specific spectral power distribution while illuminants are theoretical spectral power distributions that may or may not be physically realized by a light source.

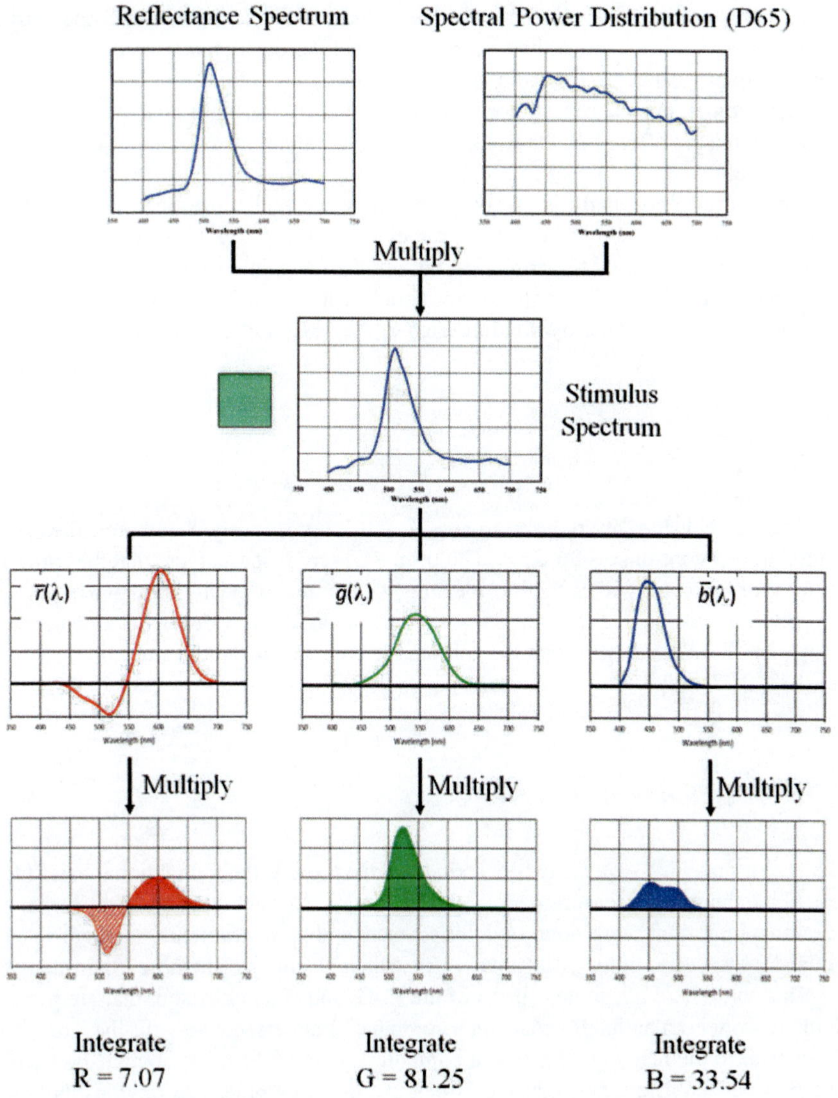


**Fig. 6.8** Calculations of tristimulus values for a sample reflectance spectrum. The color of this stimulus is shown next to the stimulus spectrum

$\overline{g}(\lambda)$, and $\overline{b}(\lambda)$ color matching functions (middle portion of Fig. 6.8). This produces three graphs (one for each primary). For each we calculate the area under the curve by simply adding together the curve values from the lowest wavelength to the highest (bottom of Fig. 6.8).

The resulting areas, when multiplied by the normalizing factor k, are the R, G and B tristimulus values for that color and represent the amount of each primary light that is needed to match that color, regardless of the actual stimulus spectrum. It is important to recognize that we are not matching the *stimulus spectrum* of the color in question but are instead matching the *sensation* of this color using the three primary colors. For the reflectance spectrum and source spectral power distribution shown at the top of Fig 6.8, the R, G, B tristimulus values are 7.07, 81.26 and 33.54, respectively.

Returning to our analysis of complex colors, we find it useful to describe the intensities of the primaries in relative, rather than absolute, terms—that is, in terms that show the balance of the primaries, but not their absolute intensities. In this way we separate the chromaticity information about a color from the intensity information (brightness). This is done by dividing each of the R, G, and B values by their sum:

$$\begin{aligned} r &= \frac{\mathrm{R}}{\mathrm{R+G+B}} \\ g &= \frac{\mathrm{G}}{\mathrm{R+G+B}} \\ b &= \frac{\mathrm{B}}{\mathrm{R+G+B}} \end{aligned} \tag{6.3}$$

These new values are referred to as *r*, *g*, and *b*, respectively,[10] and are called the chromaticity coordinates for a given color in RGB color space. Note that the sum of values for any color will equal one, meaning we are free to specify only two of them. In addition, we noted above that the achromatic colors arise from equal amounts of R, G, and B. We see, then, that the achromatic colors have equal values of *r*, *g*, and *b* (0.33, 0.33, and 0.33).

## *The 1931 Standard Observer*

The color matching experiments described above provide the foundational basis for the first quantitative measurement method for color. As such they represent perhaps the most important work done in color science, and all instrumental methods used today for measuring color can ultimately trace their origins back to these experiments.

That said, there are some aspects of the R, G, and B color system that are problematic from both an intellectual and a computational perspective—for the latter, at least prior to the ready availability of computers. One of the more obvious issues is the prevalence of negative light intensities, but there are other less obvious, but no less objectionable, aspects of the RGB color space. These prompted the CIE to seek an alternative set of primaries and color matching functions—what we now know as the Standard Observer.

[10] These are *not* the same as $\bar{r}(\lambda)$, $\bar{g}(\lambda)$, and $\bar{b}(\lambda)$ color-matching functions. The *r*, *g*, and *b* values describe any color, including those with contributions from multiple wavelengths, while the $\bar{r}(\lambda)$, $\bar{g}(\lambda)$, and $\bar{b}(\lambda)$ color-matching functions describe only the spectral colors.

## The $\bar{x}(\lambda)$, $\bar{y}(\lambda)$, and $\bar{z}(\lambda)$ Color Matching Functions

To overcome difficulties associated with the RGB color space (detailed below), and to make chromaticity coordinates more intuitive, the CIE, in 1931, developed a second set of primaries, with associated color matching functions, to describe color. This set is comprised of three linear combinations of the $\bar{r}(\lambda)$, $\bar{g}(\lambda)$, and $\bar{b}(\lambda)$ color-matching functions.[11] The new primaries are **X**, **Y**, and **Z**, the new color-matching functions are $\bar{x}(\lambda)$, $\bar{y}(\lambda)$, and $\bar{z}(\lambda)$ and the new chromaticity coordinates are $x$, $y$, and $z$. As we will see, these parameters are, in many ways, analogous to those of the **R**, **G**, and **B** color scheme, although there are some fundamental differences, particularly in the nature of the primaries themselves.

The specific requirements that the CIE placed on the new color scheme are as follows:

- Each function must be positive or zero for all visible wavelengths.
  - This avoids negative values, which simplifies calculations and decreases the opportunity for computational mistakes, an important consideration in pre-computer days.
- The areas under the $\bar{x}(\lambda)$, $\bar{y}(\lambda)$, and $\bar{z}(\lambda)$ curves must be equal.
  - This is equivalent to stipulating that the X, Y, and Z values for the constant power white light source (Illuminant E) must be equal to one another, and, therefore, that the x, y, and z values must each equal $\frac{1}{3}$ for this illuminant, as is the case for $r$, $g$, and $b$ under Illuminant E.
- One of these functions ($\bar{y}(\lambda)$) must be identical to the photopic luminance function (V($\lambda$)).
  - This reduces the number of calculations needed to fully characterize a color, since the lightness of a paint, as determined by the photopic luminance function, is an important aspect of a color sensation.
  - In addition, this gives an easy interpretation of the $\bar{y}(\lambda)$ color-matching function—it is the film lightness. All hue and saturation information are carried in the $\bar{x}(\lambda)$ and $\bar{z}(\lambda)$ functions, and all lightness information in $\bar{y}(\lambda)$.
- One of these functions ($\bar{z}(\lambda)$) must be zero over as much of the spectral range as possible.
  - In practice, $\bar{z}(\lambda)$ is zero above 650 nm.
  - This simplifies chromaticity coordinate calculations over the range where $\bar{z}(\lambda)$ is zero since here only X and Y values need to be calculated.

[11] As a consequence of Grassmann's laws, any linear combination of valid color-matching functions generates a valid new color-matching function. Here we will create three new color-matching functions, $\bar{x}(\lambda)$, $\bar{y}(\lambda)$, and $\bar{z}(\lambda)$, from linear combinations of the $\bar{r}(\lambda)$, $\bar{g}(\lambda)$, and $\bar{b}(\lambda)$ functions.

- The area enclosed by the spectral colors must fill as completely as possible the triangle joining the three primaries when plotting $x$ against $y$.

A matrix transformation was developed that satisfied all of these requirements.[5] The equations used to convert the $\bar{r}(\lambda)$, $\bar{g}(\lambda)$, and $\bar{b}(\lambda)$ color matching functions into the $\bar{x}(\lambda)$, $\bar{y}(\lambda)$, and $\bar{z}(\lambda)$ color matching functions are shown as Eq. 6.4 [5, 6].

$$\begin{aligned}
\bar{x}(\lambda) &= 5.6507 \cdot (0.49000\,\bar{r}(\lambda) + 0.31000\,\bar{g}(\lambda) + 0.20000\,\bar{b}(\lambda)) \\
\bar{y}(\lambda) &= 5.6507 \cdot (0.17697\,\bar{r}(\lambda) + 0.81240\,\bar{g}(\lambda) + 0.01063\,\bar{b}(\lambda)) \\
\bar{z}(\lambda) &= 5.6507 \cdot (0.01000\,\bar{g}(\lambda) + 0.99000\,\bar{b}(\lambda))
\end{aligned} \tag{6.4}$$

The equations for the reverse transformation, from the $\bar{x}(\lambda)$, $\bar{y}(\lambda)$, and $\bar{z}(\lambda)$ color matching functions to the $\bar{r}(\lambda)$, $\bar{g}(\lambda)$, and $\bar{b}(\lambda)$ color matching functions, are shown as Eq. 6.5.

$$\begin{aligned}
\bar{r}(\lambda) &= 0.41847\,\bar{x}(\lambda) - 0.15866\,\bar{y}(\lambda) - 0.08284\,\bar{z}(\lambda) \\
\bar{g}(\lambda) &= -0.09117\,\bar{x}(\lambda) + 0.25243\,\bar{y}(\lambda) + 0.01571\,\bar{z}(\lambda) \\
\bar{b}(\lambda) &= 0.00092\,\bar{x}(\lambda) - 0.00255\,\bar{y}(\lambda) + 0.17860\,\bar{z}(\lambda)
\end{aligned} \tag{6.5}$$

We can calculate the X, Y and Z tristimulus values directly from the R, G and B tristimulus using a set of equations similar to those used to translate the color matching functions, as given by Eq. 6.6.

$$\begin{aligned}
X &= 0.49000\,R + 0.31000\,G + 0.20000\,B \\
Y &= 0.17697\,R + 0.81240\,G + 0.01063\,B \\
Z &= 0.01000\,G + 0.99000\,B
\end{aligned} \tag{6.6}$$

Finally, we can reverse transform X, Y and Z tristimulus values into R, G and B tristimulus values using Eq. 6.7.

$$\begin{aligned}
R &= 5.6507 \cdot (0.41847\,X - 0.15866\,Y - 0.08284\,Z) \\
G &= 5.6507 \cdot (-0.09117\,X + 0.25243\,Y + 0.01571\,Z) \\
B &= 5.6507 \cdot (0.00092\,X - 0.00255\,Y + 0.17860\,Z)
\end{aligned} \tag{6.7}$$

Alternatively, the X, Y and Z tristimulus values can be calculated directly from a stimulus spectrum using the $\bar{x}(\lambda)$, $\bar{y}(\lambda)$, and $\bar{z}(\lambda)$ color matching functions in a similar way to that used to calculate R, G and B tristimulus values from the $\bar{r}(\lambda)$, $\bar{g}(\lambda)$, and $\bar{b}(\lambda)$ color matching functions and a stimulus spectrum, as shown in Eqs. 6.8 and 6.9 [7].

$$\begin{aligned}
X &= k \sum_{\lambda=380}^{780} S_t(\lambda) \cdot \bar{x}(\lambda) \cdot \Delta\lambda \\
Y &= k \sum_{\lambda=380}^{780} S_t(\lambda) \cdot \bar{y}(\lambda) \cdot \Delta\lambda \\
Z &= k \sum_{\lambda=380}^{780} S_t(\lambda) \cdot \bar{z}(\lambda) \cdot \Delta\lambda
\end{aligned} \tag{6.8}$$

$$\text{where k} = 100 / \sum_{\lambda=380}^{780} S(\lambda) \cdot \overline{y}(\lambda) \cdot \Delta\lambda \tag{6.9}$$

Note that Eq. 6.8 uses the stimulus spectrum $S_t(\lambda)$ and Eq. 6.9 uses the source spectrum $S(\lambda)$.

Equations 6.4 and 6.6 are related to one another by a factor of 5.6507 (often denoted as 1/0.17697 in the literature), as are Eqs. 6.5 and 6.7. In addition, this same factor relates the k normalization factor used to calculate R, G and B tristimulus values from reflectance data (Eq. 2) to the k normalization factor used to calculate X, Y and Z tristimulus values from reflectance data (Eq. 6.9)[12].

The origin of this factor of 5.6507 is the desire to make the $V(\lambda)$ and $\overline{y}(\lambda)$ equations exactly equal one another (rather than proportional to one another) [6]. The areas under the $\overline{r}(\lambda)$, $\overline{g}(\lambda)$ and $\overline{b}(\lambda)$ curves are equal to one another but are smaller than the area under the $V(\lambda)$ curve by this factor of 5.6507. This is because the basis for scaling the $\overline{r}(\lambda)$, $\overline{g}(\lambda)$ and $\overline{b}(\lambda)$ color matching functions to the $V(\lambda)$ function was that the values of $\overline{r}(\lambda)$ and $V(\lambda)$ be equal at the wavelength 700 nm. By using the scaling factor 5.6507 when transforming from the $\overline{r}(\lambda)$, $\overline{g}(\lambda)$ and $\overline{b}(\lambda)$ color matching functions to the $\overline{x}(\lambda)$, $\overline{y}(\lambda)$ and $\overline{z}(\lambda)$ color matching functions, the areas under the $V(\lambda)$ and the $\overline{x}(\lambda)$, $\overline{y}(\lambda)$ and $\overline{z}(\lambda)$ curves become equal and the X, Y and Z values for a perfect diffuse reflector under Illuminant E are each 100. Note that under these same conditions the R, G and B values are also each 100, since the k normalization factor used in these calculations includes the factor of 5.6507 (see Eq. 6.2).

Mention should be made of a correction factor that is commonly omitted in routine color measurements. When light crosses an interface, such as the air/coatings interface at the top surface of a paint film, it is partially reflected. This crossing occurs twice—first when the light travels from air to the film, and second when the redirected light travels from the film back into air [8]. The exact nature of this correction is discussed in Chap. 15. In practice, we find that omitting this correction creates little error when comparing paints that are similar to one another, particularly in $TiO_2$ content and in gloss.

The $\overline{x}(\lambda)$, $\overline{y}(\lambda)$, and $\overline{z}(\lambda)$ color-matching functions over the visible wavelengths of light are shown in Fig. 6.9. These functions, taken together, are known as the 1931 Standard Observer. The word "Observer" here does not refer to a person, but rather to the table of data comprising the $\overline{x}(\lambda)$, $\overline{y}(\lambda)$, and $\overline{z}(\lambda)$ values (Table 6.3) [9]. The development of the Standard Observer in 1931 was a giant step forward for the nascent field of color science, and the Standard Observer is still used nearly universally among color scientists.

Note that the $\overline{x}(\lambda)$, $\overline{y}(\lambda)$, and $\overline{z}(\lambda)$ color-matching functions were not based on new measurements of color, but instead were derived mathematically from the color matching data that Guild and Wright had already reported. The mathematical transformation of the $\overline{r}(\lambda)$, $\overline{g}(\lambda)$, and $\overline{b}(\lambda)$ color-matching functions into a new set of

[12] When comparing Eqs. 2 to 9, recall that the $V(\lambda)$ and $\overline{y}(\lambda)$ functions are equal to one another, as stipulated in the requirements of the $\overline{x}(\lambda)$, $\overline{y}(\lambda)$ and $\overline{z}(\lambda)$ color matching functions.

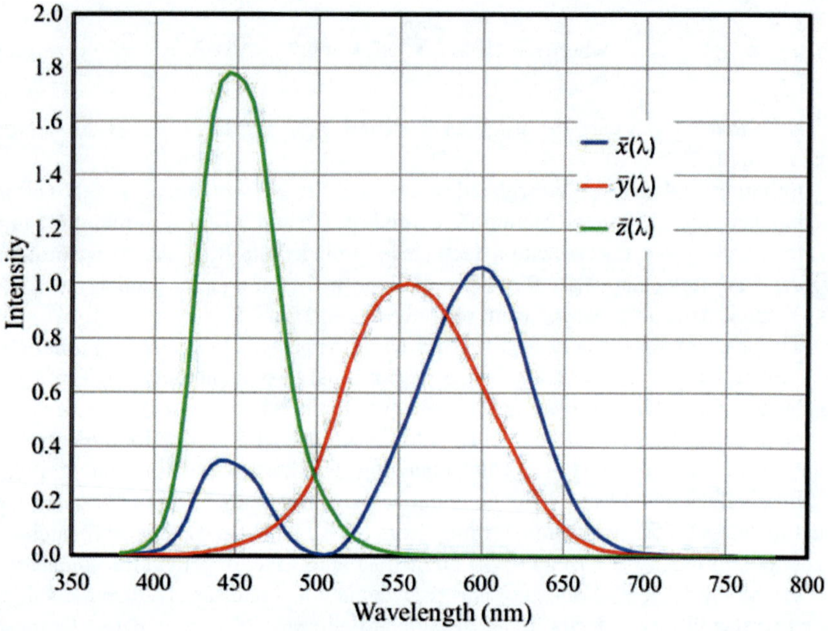


**Fig. 6.9** $\bar{x}(\lambda)$, $\bar{y}(\lambda)$, and $\bar{z}(\lambda)$ color-matching functions

**Table 6.3** CIE 1931 $\bar{x}(\lambda)$, $\bar{y}(\lambda)$, and $\bar{z}(\lambda)$ color matching functions. Blue outline designates a color matching function; red outline designates a spectral tristimulus value; green outline designates the standard observer

| $\lambda$, nm | $\bar{x}(\lambda)$ | $\bar{y}(\lambda)$ | $\bar{z}(\lambda)$ |
|---|---|---|---|
| 380 | 0.001368 | 0.000039 | 0.006450 |
| 385 | 0.002236 | 0.000064 | 0.010550 |
| • | • | • | • |
| • | • | • | • |
| • | • | • | • |
| 770 | 0.000083 | 0.000030 | 0.000000 |
| 775 | 0.000059 | 0.000021 | 0.000000 |
| 780 | 0.000042 | 0.000015 | 0.000000 |

color-matching functions is made possible by Grassmann's Laws, which state that color sensation follows the laws of algebra.

## x, y, *and* z *Chromaticity Coordinates and the 1931 CIE* xy *Chromaticity Diagram*

Just as was the case with the R, G, and B tristimulus values, we can increase the usefulness of the X, Y and Z tristimulus values by placing them on a relative basis. This is done by dividing each coordinate by the sum of the three tristimulus values, X, Y, and Z to give *x*, *y*, and *z* chromaticity coordinates; in the same way, we derived the *r*, *g*, and *b* chromaticity coordinates from R, G, and B:

$$\begin{aligned} x &= \frac{X}{X+Y+Z} \\ y &= \frac{Y}{X+Y+Z} \\ z &= \frac{Z}{X+Y+Z} \end{aligned} \tag{6.5}$$

Note that $x + y + z = 1.0$ for any color.

As was the case with RGB color space, we calculate three parameters (*x*, *y*, and *z*) from our tristimulus values (X, Y, and Z), but only two of these parameters are independent (e.g., *z* can be determined from the *x* and *y* values). We require three independent parameters to uniquely identify a color since we have three types of color cells (the S, M, and L cone cells). We cannot use all three chromaticity coordinates to identify a color, since *z* can be calculated from *x* and *y*. The third parameter must be derived from the X, Y, and Z values. Traditionally, colors have been described by *x* and *y*, which combined determine hue and saturation, and Y, which, by virtue of being identical to the photopic luminance function (V(λ)), defines lightness. By this convention, we separated the chromaticity aspects of color from the achromatic/lightness aspect of the color. In this system colors are characterized using the format *xy*Y.

It is quite useful to display XYZ color space using a plot of *x* versus *y*. We begin the construction of this plot with the placement of the spectral colors on a Cartesian plane, starting with the chromaticity coordinates for 780 nm light. From Table 6.3 we see that the color-matching functions, which are also the spectral tristimulus values, at 780 nm are X = 0.000042, Y = 0.000015, and Z = 0.0. Using Eq. 6.5, we calculate the chromaticity coordinates *x* and *y* to be 0.735 and 0.265, respectively. We plot this as a dark red point in Fig. 6.10.

We can continue adding spectral colors to our plot. In Fig. 6.10 we see the placement of selected wavelengths, down to 380 nm, indicated by points colored as closely as possible to the spectral colors they represent. We can join these points together to form the spectral locus. We can join the two endpoints together (380 and 780 nm) with what is known as the purple line. This generates a geometric figure that is entirely convex.

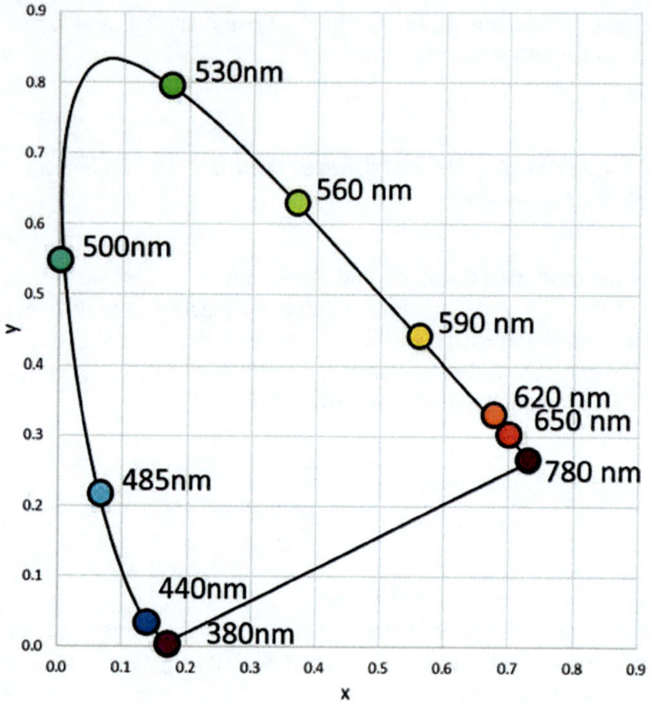


**Fig. 6.10** *x* and *y* locations of spectral colors

There are two noteworthy features of the spectral locus. First, we find that the *x* and *y* values for all wavelengths above 680 nm are approximately equal to one another, which is consistent with the fact that only the L cone is active over this wavelength range (Fig. 6.4), and so there is no differentiation of color perception over this range—the color sensation is the same, although the perceived intensity changes with wavelength.

The second notable aspect of this diagram is that between 550 and 780 nm the spectral locus curve is linear, or nearly so. This is because the *z* coordinate for these wavelengths is at or near zero (Fig. 6.9), and so the spectral locus follows the line $x + y = 1.0$ over this wavelength range.

Complex colors can also be placed on this diagram. For reasons that will soon be apparent, the complex colors all reside in the interior of the geometric figure formed by the spectral locus and the purple line. In Fig. 6.11 we see the interior of this figure colored at each point according to the color represented by those *x* and *y* values. Note that these colors are only approximate since no color reproduction technology, whether electronic or print, can exactly reproduce all visible colors—an issue we will discuss shortly.

The totality of this plot is known as the CIE 1931 *xy* Chromaticity Diagram. The region enclosed by the spectral locus and the purple line is often referred to as the

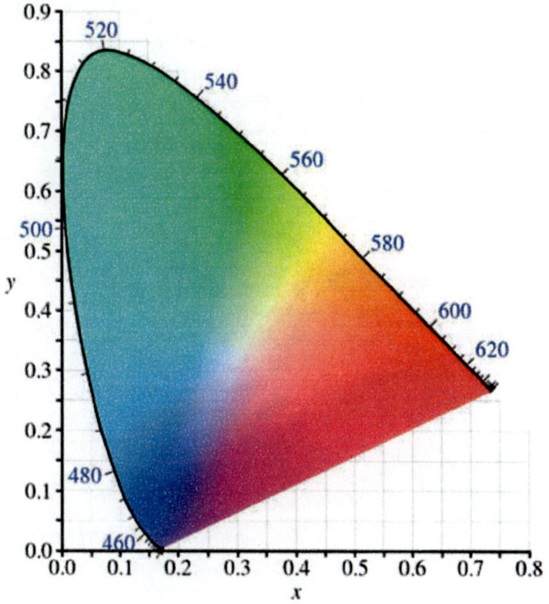


**Fig. 6.11** *xy* chromaticity diagram

color shoe, due to its supposed resemblance to a horseshoe. However, here we will refer to it as the visible gamut. Points outside this gamut cannot be created by any combination of visible wavelengths and so cannot be physically realized (discussed below).

Because of the nature of the mathematics involved (specifically, the equations describing Grassmann's Laws), if we pick any two colors within the visible gamut and draw a line between them, then the colors on that line can be made by additively mixing the two original colors. The inverse is also true—any colors that can be made by mixing the two original colors must fall on the line between them. We can now understand the designation of the borderline joining the two ends of the locus as being the "purple" line. All points on this line are combinations of red and blue light, making them purple (this color is often referred to as magenta, especially when the amount of red sensation is high compared to the blue sensation). Colors on this line are often called the non-spectral colors, although some assign this name to the colors that we call here the complex colors.

This property of color mixing can be extended to more than two colors. For example, if we place three colors in the visible gamut, then any mixture of these colors will be found within the triangular area bounded by these three colors—that is, the triangle represents the gamut of colors that can be made by mixing these particular three colors. We designate these three colors as the primaries for that

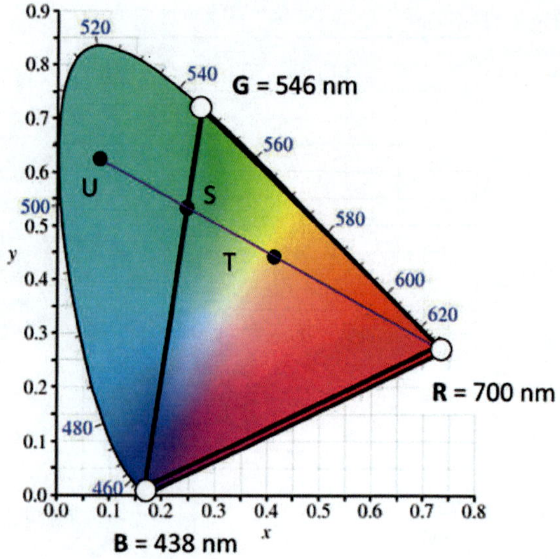


**Fig. 6.12** Creating colors from the R, G, and B primaries

gamut. Note that primaries do not need to be spectral, and any point on the xy chromaticity diagram can be used as a primary.[13]

To demonstrate how we can form colors within the interior of a particular color gamut, we have placed the **R, G,** and **B** primaries on the visible gamut in Fig. 6.12. To create color **T** from the **R, G,** and **B** primaries, we first generate color **S** by combining the proper ratio of primaries **G** and **B**. We then combine colors **S** and **R** to give color **T**. Of course, we could also get color **T** by first combining **G** and **R**, or by first combining **B** and **R**, and then adding the third primary to that combination.

In Fig. 6.12, color **U** lies outside the triangle formed by the **R, G,** and **B** primaries, and so cannot be created by combining these three primaries. However, we can express color **U** in terms of these primaries by using negative light intensities. Color **U** can be created by first adding the proper ratio of primaries **G** and **B** to give color **S**, as before, but then *subtracting* a certain amount of primary **R** from **S**, by moving along the **RS** line from **S** and away from **R**. In this case, the magnitudes of **R** for both **T** and **U** are similar, but the sign of **R** is positive for **T** and negative for **U**.

The gamut for the **R, G,** and **B** primaries covers only about half of the visible gamut. The colors missing from it are mainly the spectral colors between **B** and **G** primaries (435.8 and 546.1 nm) and the complex colors that can be made from them. These colors are mostly shades of green. Because they lie outside the **R, G,** and **B** triangle, the spectral colors in this large section of the spectral locus all have negative values for the **R** primary, some quite large, which is consistent with Fig. 6.6. Similarly, the spectral colors between the **G** and **R** primaries have small negative

[13] Although any three colors can be used as primaries, inspection of Fig. 6.11 shows that the largest gamut is formed when the primaries are red, green, and blue—which explains why this particular set of primaries is used in additive mixing applications such as monitors and electronic screens.

values for **B**, and spectral colors at wavelengths lower than the **B** primary will have small negative **G** values. We can now see why at least one of the $\bar{r}(\lambda)$, $\bar{g}(\lambda)$, and $\bar{b}(\lambda)$ spectral tristimulus values is zero or negative at all visible wavelengths.

We can increase the size of a color gamut by increasing the number of primaries. When using more than three primaries, the color gamut is the polygon whose perimeter is formed by joining neighboring colors with lines. Adding primary colors that are not within the original gamut will therefore expand the gamut, allowing for more color possibilities than the original gamut.[14] The use of more than three colors to broaden the gamut is quite common in the paint industry, where as many as a dozen color tints are used to form a large color gamut. In this case, however, color comes from light absorption. Although additional primaries will increase the color gamut, in situations where colors come from light generation, such as monitors and screens, in practice only three primaries are used.

Figure 6.12 also shows us why it is not possible to create all visible colors from any set of real primaries. There is a geometric principle that states that any polygon with all its vertex points located within a convex shape, such as the visible gamut, will not be able to enclose the entirety of that convex shape—some part of that shape will always reside outside of the polygon. Since all visible colors reside within the visible gamut, no polygon made by joining any finite set of these colors, regardless of the number of them (i.e., the number of primary colors), can encompass the entire visible gamut region.

## Points Outside the Visible Gamut

The visible gamut fills roughly half the area of the *xy* chromaticity diagram. We know that points within the gamut represent visible colors, but how do we interpret the points residing outside of the gamut? We call these points "imaginary colors", because such points cannot be made by any combination of visible light. This is true because the visible gamut is convex or straight along its entire perimeter. We can understand the implications of points that are outside the visible gamut both mathematically and physiologically.

Mathematically, it is impossible to create a stimulus spectrum that produces $x$ and $y$ values outside of the visible gamut. This is easiest to see for $x$ and $y$ pairs that lie to the right of the $y = 1 - x$ line. Recall that $x + y + z$, by definition, must be equal to one, and that a stipulation of the XYZ color space is that all tristimulus values (and, by extension, all chromaticity coordinates) are positive. The sum of the $x$ and $y$ values for any points to the right of the $y = 1 - x$ line is greater than one, and so, in order for $x + y + z$ to equal one in this region of the chromaticity diagram, the $z$ values must be negative, which is not possible.

Points outside the visible gamut but to the left of the $y = 1 - x$ line are also not possible, although the reasoning here is less obvious than for points to the right

[14] Note, however, that adding primaries located within the existing gamut does not increase the size of the gamut.

of the $y = 1 - x$ line. For points outside the visible gamut but on or to the left of the $y = 1 - x$ line, there are simply no stimulus spectra, no matter how carefully constructed, for which $x + y + z = 1.0$.

As an example of this statement, consider the **Y** primary itself. Here $x$ and $z$ are zero and $y$ is one, satisfying the equation $x + y + z = 1.0$. However, inspection of Fig. 6.9 shows that there are no wavelengths for which $y$ is non-zero and both $x$ and $z$ are zero. By extension we can say that there are no combinations of wavelengths—that is, no spectra—that can be constructed that satisfy the requirement that both $x$ and $z$ are simultaneously zero and $y$ is not. We can generalize this to all points outside the visible gamut by saying that these points are mathematically impossible to obtain and physically impossible to produce.

We can understand why we cannot see points outside the gamut physiologically by referring back to Fig. 6.4. In this figure we see the absorption spectra of the three color sensing cells in our eyes (the short wavelength, mid wavelength, and long wavelength cone cells). We perceive color sensations based on the degree to which each of these color sensing cells is activated by light. Note that the M curve in Fig. 6.4 overlaps entirely with the L and S cell responses. We can therefore never get a color sensation that is determined solely by the M cells—that sensation will always have contributions from one or both of the other cone cell types. If the sensation produced by activation of the M cells alone could be plotted on the *xy* chromaticity diagram, it would fall outside of the visible gamut.[15]

We saw earlier that, due to the fully convex nature of the visible gamut, the color matching functions for any set of real primaries (i.e., primaries contained within the visible gamut) will, by necessity, have some negative values (i.e., some portion of the visible gamut will be outside the triangle formed by any set of real primaries). A corollary of this result is that any triangle that encompasses the entire visible gamut must have corner points (primaries) that fall outside the visible gamut. This is true for the **X**, **Y**, and **Z** primaries, which are found in the *xy* chromaticity diagram at $x$, $y$ coordinates (1,0), (0,1), and (0,0).

## Imaginary Primary Colors

Points that fall outside of the visible gamut, such as the **X**, **Y**, and **Z** primaries, are impossible to create physically and, as a result, are called imaginary colors. We may feel uncomfortable that the **X**, **Y**, and **Z** primaries, which are so important to color science, cannot be physically made or seen. However, the purpose of XYZ color space is to uniquely characterize, or quantify, all color sensations (i.e., the entire visible gamut) using convenient (all positive) values to do this. This is in contrast to color matching experiments and the color spaces derived from them (such as RGB color space). The purpose of those color spaces is to specify how a certain color can be duplicated using only three primary lights (such as is done with electronic screens and monitors). In this case it is imperative that the three primaries be real, otherwise

[15] Actually, it would be more proper to say that the gamut would be extended to include this point.

we could not physically create colors from them. This requirement is irrelevant to XYZ color space since its purpose is to quantify, rather than create, the colors of the visible gamut.

The use of imaginary primaries in XYZ color space is very similar to the use of negative light intensities in RGB color space. In both cases we are using Grassmann's laws to construct a mathematical framework to describe colors. This is also similar to the concept of imaginary numbers, which are based on the square root of negative one and are used to facilitate certain mathematical and engineering operations. The square root of a negative number has no physical meaning, just as imaginary primaries and negative light intensities have no physical manifestation. This does not, however, diminish in any way the value or utility of any of these concepts.

Finally, we note that XYZ color space is but one of an infinite number of color spaces that can be derived from linear combinations of the $\bar{r}(\lambda)$, $\bar{g}(\lambda)$, and $\bar{b}(\lambda)$ color-matching functions. Combinations that give real primaries will, in all cases, have some negative light intensities, whereas combinations for which the values of all color matching functions are positive will have imaginary primaries. The factor that distinguishes XYZ color space from all other color spaces based on imaginary primaries is that the area of the visible gamut, relative to the area mapped out by a triangle joining the primaries, is a maximum (this was the fifth stipulation of the CIE for the XYZ color space, mentioned above). By spreading the visible gamut out over the largest area possible, we improve the ability to distinguish between colors that are very close to one another.

## White Point, Dominant Wavelength, Complimentary Color, and Excitation Purity

In Chap. 5 we noted that no white light sources or illuminants are truly white, with the exception of Illuminant E, which has a constant spectral power distribution and so is achromatic. Instead, real sources and the other illuminants have a weak color component—usually blue or yellow. We can measure the tristimulus values for these sources by measuring the reflectance spectrum of the source from an ideal white screen (i.e., a screen that reflects all light that falls on it); for illuminants, we can calculate their tristimulus values directly. From these tristimulus values, we can calculate the $x$ and $y$ chromaticity coordinates of the sources or illuminants. These $x$ and $y$ chromaticity coordinates are called the white points of the source or illuminant.

The white points for several sources and illuminants are shown in Fig. 6.13. The black line on this figure represents values for ideal black body radiators, as a function of temperature (numerals in the figure). This line is called the Planckian locus, recognizing that the equations describing black body radiators were first derived by Max Planck. We often refer to the "temperature" of an illuminant or source—this is the temperature on the Planckian locus that is closest to the white point of that illuminant or source and is termed the correlated color temperature for that illuminant or source.

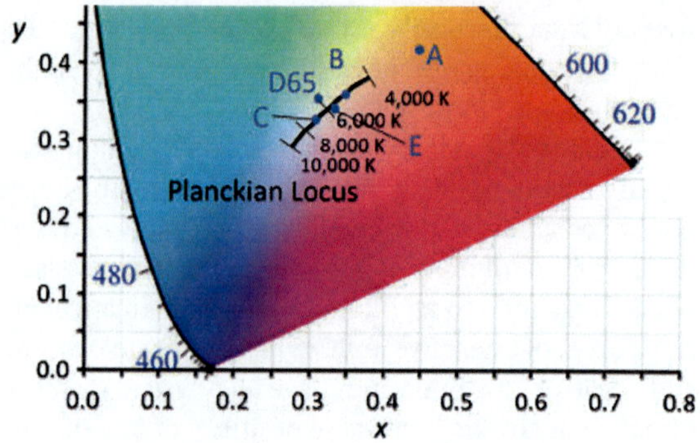


**Fig. 6.13** White points for different illuminants

When discussing ways to categorize colors, we saw the usefulness of the concept of saturation. We defined the spectral colors as being fully saturated, and any combination of colors as being unsaturated. In that discussion we described the degree of saturation on a purely qualitative basis—for example, we could arrange colors by degree of saturation, and pick out colors that formed a uniform progression of saturation levels, but we did not in any way quantify saturation.

We are now in the position where we can assign numeric values to saturation using the *xy* chromaticity diagram. In this case we will quantify the degree of saturation as "excitation purity", and to do so we will make use of the concept of dominant wavelength.

The dominant wavelength of a color describes the spectral color that can be modified (desaturated) by the addition of white light to match the color of interest. We show this in Fig. 6.14. We begin with point D, a spectral orange color, and Point W, the white point (in this case $x = y = z = 1/3$ for Illuminant E). We join these points with a line segment and place Point X on that segment. Color X can be made by mixing Color D with white (color W) since Color X falls on the line between Colors D and W. By definition, mixing a color with white desaturates that color. We call Color D the dominant color for Color X, and the wavelength of D (586 nm) is the dominant wavelength of Color X.

We can extend the line between Points D and W to the left, again intersecting the spectral locus, this time at Point C. Color C is the complimentary color to Color X, meaning a mix of the proper proportion of the two will give white, since point W falls on the line between Points D and C. The complimentary wavelength of Color X is the wavelength of Color C (485 nm).

We can give a numeric value of the degree of saturation of Color X by taking the ratio of the distance between the white point (W) and Color X and the distance between the white point (W) and the spectral value for the dominant wavelength of Color X (Point D). That is, the degree of saturation, which is a value between 0 and 1, is the ratio of the length of A to the length of B in Fig. 6.14. This is also referred to as the excitation purity of Color X.

Not all colors have a dominant wavelength. Extension of the line between Point W and any color inside the triangle formed by the white point, the spectral locus at

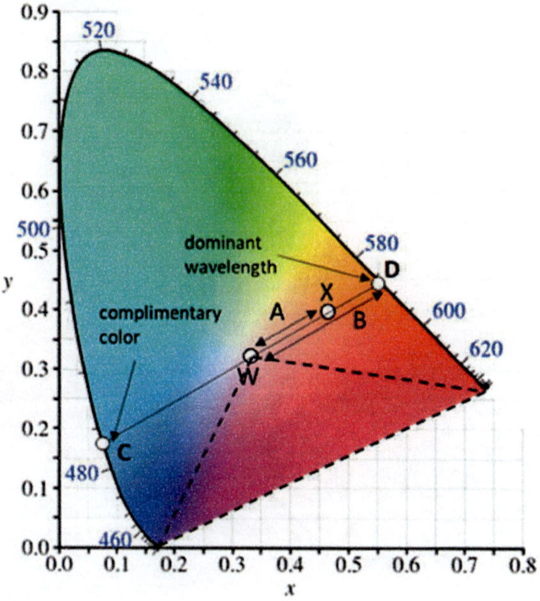


**Fig. 6.14** Dominant wavelength, complimentary color, and saturation value parameters for color X

380 nm, and the spectral locus at 780 nm, as depicted by the dashed lines in Fig. 6.14, will intersect at some point on the purple line, rather than a point on the spectral locus. As such they do not have a dominant wavelength. Colors in this triangle are called "non-spectral" colors and we describe them by their complimentary wavelengths.

## Visualizing the X, Y, and Z Primaries

Although we cannot physically create the **X, Y,** and **Z** primaries, since they are imaginary, we can gain an understanding of their relationship to the visible colors as follows:

First, we draw a line from the white point (Point W in Fig. 6.15) to each of these primaries. For **Y** and **Z,** these lines intersect the spectral locus at 520 nm and 475 nm, respectively. These are the dominant wavelengths for the points on the lines that are within the visible gamut. The line for **X** crosses the line of purples rather than the spectral locus.

Next, we focus on the line from the white point to the **Y** primary. Beginning at the white point and following this line, the color sensation shifts from white, to very light green (Point A), to a more intense green (Point B). The progression to greater

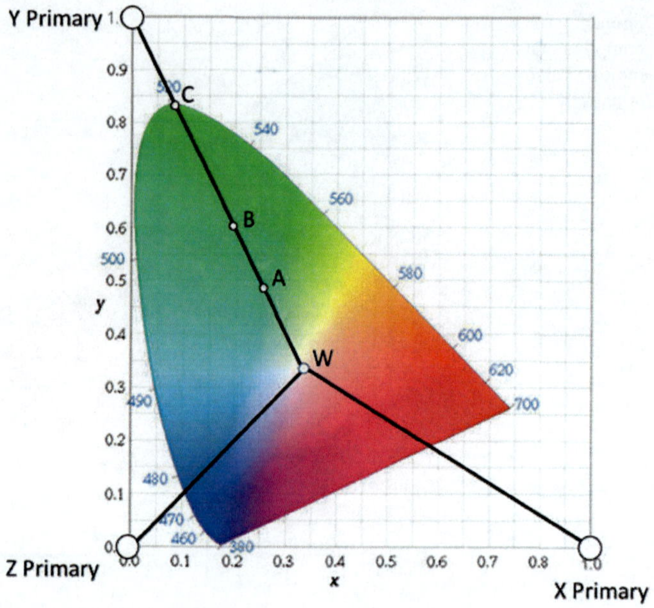


**Fig. 6.15** Extrapolating the colors of the X, Y, and Z primaries

saturation ends when we reach the point $x = 0.063$ and $y = 0.71$ (Point C). Here we intercept the spectral locus at a wavelength of 520 nm and the color is saturated.[16]

The spectral locus is a physical endpoint of this process, as we cannot remove any additional white light to continue on the line towards the **Y** primary. However, our sensations are somewhat different. Based on the shift that we saw in going from the white point to the spectral locus, we can imagine that, if we could continue, the green color would become even more intense. That is to say, our vision does not tell us that this is an obvious stopping point in the appearance of the color.[17]

If our perception could continue beyond the spectral locus, we would continue to see a green color, but one that is more saturated than is physically achievable. Eventually, we will reach the **Y** primary, and it, like all other points on this line, will be a shade of green based on the green color at 520 nm.

We can repeat this process to show that the **Z** primary is a color more intense than a saturated blue color. Our analysis of the **X** primary is slightly different since the line between it and the white point intersects on the purple line, rather than the spectral

[16] Note that the colors along the line are not, strictly speaking, different saturations of a single hue. Instead, the hue varies slightly along the line. This variability does not, however, alter the conclusions of this thought experiment.

[17] This is quite different than a somewhat analogous progression. As we travel on the achromatic scale from white to black, we see a progressively darkening gray. This progression ends at black when our eyes no longer receive any stimulus. We cannot imagine an increasing darkness beyond black—there are no degrees of blackness in the same sense that there are degrees of greenness.

locus. The position of the intersection with the purple line is considerably closer to red than to blue, and we can say that primary **X** would represent a reddish-purple color more intense than any reddish-purple colors that can be made by combining real colors.

## Perceptual Uniformity

An inspection of the *xy* chromaticity diagram shows that areas for the different colors are not equal. This is most obvious with green—over half of the diagram is green or a greenish color. This abundance of green does not match the distribution of the 10 million or so colors that we can discriminate—half of these are not green. This issue is not with our eyes, but rather with the *xy* chromaticity diagram.

In Fig. 6.16 we see two pairs of points chosen on the *xy* chromaticity diagram. The points in each pair are separated by 0.10 units in both x and y. A semi-circle is extracted from each region. Each pair of semi-circles is then joined to make a complete circle for which the top and bottom halves are from different points in the visual gamut.

The two green colors are nearly indistinguishable from one another while the red colors are quite distinct. There are two reasons for this: Technical limitations of color

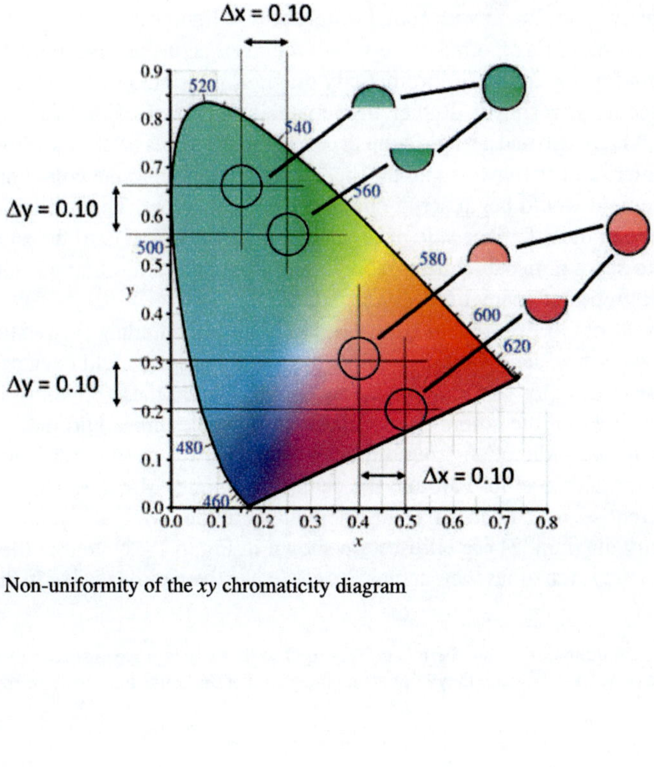


**Fig. 6.16** Non-uniformity of the *xy* chromaticity diagram

**Fig. 6.17** Brightness scales. Top: Linear with true lightness. Bottom: Linear with perceived lightness. All numerals are the same shade of gray

rendition,[18] and a difference in perceptual uniformity. Here we are concerned with the difference in perceptual uniformity.

Perceptual uniformity is based on the degree to which all pairs of colors a certain distance apart in a color space appear different. Ideally, we would want the same distance between pairs of points on the *xy* chromaticity diagram to have the same degree of difference in appearance. While this is not a requirement of a color space, it would be a convenience, as there would be no guesswork as to the tolerances needed to reproduce a certain color regardless of where in color space that color fell. We saw a similar situation in our discussion in Chap. 5 of the sensitivity of the eye to differences in the lightness in a series of gray colors (Fig. 5.8 in that chapter; reproduced here as Fig. 6.17). There we saw that our perception of differences in gray intensity is not linear with actual differences in lightness.

In 1942, MacAdam reported the results of a study that further highlighted the issue of visual uniformity in the *xy* chromaticity diagram [10]. This study was a modified color-matching experiment. Rather than matching spectral colors, as Wright and Guild did, MacAdam had a single subject color match a series of 25 colors distributed over the interior of the visible gamut. In a typical experiment, the colors on the test side of the field would begin set at one of the 25 test colors. The intensities of the three colors on the reference side of the field were then varied until the subject was just able to see a noticeable difference between the colors on the two fields. This process was repeated many times for each color.

The results of this experiment are a series of points surrounding each of the 25 test colors showing how far from these colors the *x* and *y* values could deviate and still present the same color sensation as the test color. Ideally these points would form a circle centered on the colors under investigation since this would indicate equal sensitivity in every direction around the test color. However, MacAdam found that the points formed an ellipse around the central test point. Moreover, these ellipses were different sizes and had different orientations in the different regions of the *xy* chromaticity diagram. These ellipses are shown in Fig. 6.18. Note that the ellipses are shown here at ten times their actual sizes, to make them more visible in the figure.

[18] The technical limitation is that there is no way to produce these two green shades on either the printed page or a screen because they fall outside the gamut of the colors used in these applications

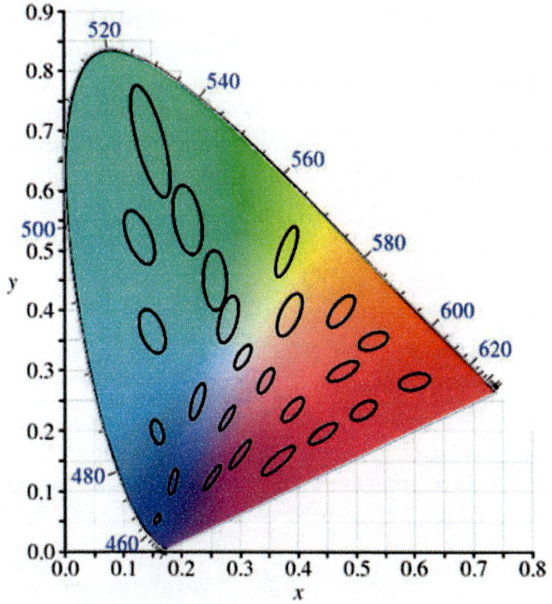


**Fig. 6.18** *xy* chromaticity diagram showing ellipses of undiscernible color differences. The ellipses are drawn at 10 times their true size to enhance visibility

As we might expect based on Fig. 6.16, larger ellipses were found in the green region of the *xy* chromaticity diagram and smaller ellipses in the regions for other colors. This confirms the fact that the *xy* chromaticity diagram shows a significant deviation from perceptual uniformity and is skewed heavily towards green.

## *Perceptually Uniform Color Spaces*

There is an obvious benefit to developing a perceptually uniform color space—color tolerances would not need to be specific to any given color, and a color difference of a certain magnitude would be equally perceptible across all color spaces.

Ever since the 1931 CIE Standard Observer was first developed, a number of alternative color systems were investigated. A common feature of these investigations is that the color of an object is measured in the same way as for the RGB color space (i.e., measuring the R, G, and B tristimulus values of the object). The differences among these investigations are how these tristimulus values are transformed into the

(inks or LED pixels). That said, most media stretch their colors to regions outside their gamut and so these colors will appear different.

new color space. That is, the color spaces differ in the set of equations applied to the R, G, and B tristimulus values to give the tristimulus values for that color space.[19]

Attempts to develop a uniform color space met with varying degrees of success, but none were determined to be completely uniform. This was not due to a lack of effort by the various researchers, but instead more likely reflects the fact that color sensations shift with conditions, as discussed in the section in Chap. 5 on physiological factors, as well as from person to person, and so there might not be, after all, a color space that is uniform under all conditions and to all people.

## *CIE L*a*b**

While many color spaces have been developed, we will focus here on the system that today is commonly used to characterize the colors of coatings. This system is called CIE L*a*b* and is also referred to as CIELAB. CIE L*a*b* is calculated from X, Y, and Z according to these equations[20]:

$$\begin{aligned} L^* &= 116\left(\frac{Y}{Y_n}\right)^{1/3} - 16 \\ a^* &= 500\left(\left(\frac{X}{X_n}\right)^{1/3} - \left(\frac{Y}{Y_n}\right)^{1/3}\right) \\ b^* &= 200\left(\left(\frac{Y}{Y_n}\right)^{1/3} - \left(\frac{Z}{Z_n}\right)^{1/3}\right) \end{aligned} \tag{6.6}$$

In these equations, $X_n$, $Y_n$, and $Z_n$ are the ratios between the measured tristimulus values and the tristimulus values for full reflectance from a white panel for the particular light source used, with $Y_n$ being scaled to a value of 100.

An attractive aspect of the CIE L*a*b* color space is that it characterizes colors in an opponent process, as the brain does. As discussed in Chap. 5, the outputs from the eye are transformed by the brain into three parameters—the balance between darkness and lightness, the balance between green and red, and the balance between yellow and blue. In the CIE L*a*b* color space, L* is the brightness component

[19] Mention should be made here of the 1964 CIE Standard Observer. This was developed by color-matching experiments similar to the 1931 Standard Observer, except the experiments were done using a wider view angle (10°, rather than 2°) and incorporating corrections that were identified after the 1931 Standard Observer was created. At 2°, only the central area of the retina (the macula) is activated. This region has few rod cells, and so the wider field altered the proportion of rod cells to cone cells, resulting in an a slight shift in color perception. The color matching functions for this new color space are written as $\bar{x}_{10}(\lambda)$, $\bar{y}_{10}(\lambda)$, and $\bar{z}_{10}(\lambda)$, and the tristimulus values are known as $X_{10}$, $Y_{10}$, and $Z_{10}$. In this case the motivation was to improve color matching, rather than to increase perceptual uniformity. However, the difference between the two-color systems is slight. Even many decades after its introduction, the 1964 Standard Observer has yet to gain popularity, and the 1931 Standard Observer still sees wide use.

[20] Note that alternative equations are used in the limited regions of color space where $X/X_o$, $Y/Y_o$, or $Z/Z_o$ is less than 0.01.

and has a scale from 0 to 100, with higher numbers representing brighter colors. a* represents the red/green balance of the color, with negative numbers indicating green, and positive numbers indicating red; a value of zero indicates neutrality with respect to red and green. b* is similarly scaled, but in this case with the blue and yellow. Negative values indicate blue while positive values indicate yellow. Achromatic colors have a* and b* values near zero.

CIE L*a*b* values allow for an intuitive way to characterize color differences between two materials. This method is to calculate the Euclidean distance between colors using their L*, a*, and b* coordinates:

$$\Delta E = \sqrt{(\Delta L^*)^2 + (\Delta a^*)^2 + (\Delta b^*)^2} \tag{6.7}$$

where ΔL*, Δa*, and Δb* are the differences in L*, a*, and b* values between the colors being compared, and ΔE is the color difference.[21] Note that this calculation assumes that a unit change in any direction (L*, a*, or b*) will give the same degree of perceived difference in color (i.e., this calculation treats L*, a*, and b* equally and so implicitly assumes perceptual uniformity).

The original intent of the CIE L*a*b* color space was that a ΔE value of 1.0 would represent a just noticeable difference between two colors. While the perceptual uniformity of the CIE L*a*b* color space is a definite improvement over the RGB and XYZ color spaces, it is, in fact, not the case that a ΔE value of 1.0 is always equal to the just noticeable difference between two colors. Instead, values for the just noticeable difference have been reported as little as 0.2 ΔE units [11] to as high as 2.3 ΔE units [12], depending on where in L*a*b* space the colors are found. Although other color spaces have been proposed since the introduction of L*a*b* space, in the hopes of developing a more uniform space, none have gained favor in the field [12].

## Miscellany

The field of color science is both broad and deep, and this chapter has only scratched its surface. While the intention of this book is not to exhaustively describe this subject, there are a few additional aspects of color science that may prove of interest to the reader. We will consider these now.

---

[21] Here E stands for "Empfindung", German for "sensation".

## *Color Blindness*

Our trichromatic vision gives us the ability to distinguish spectral colors from one another. However, about 8% of the population is unable to do this, to one extent or another.[22] We refer to such individuals as being color blind and say that they cannot see certain colors. This description is somewhat misleading—a person with red/green color blindness can, for example, see a green object; the issue is whether this person can distinguish green from red. We will broadly refer to the inability to distinguish colors normally as "anomalous color perception". We call people with normal vision "trichromats" and, for reasons that will be made obvious below, we refer to those with many types of anomalous color perception as being "dichromats".

Much of what was first learned about color theory was based not only on what the normal person perceives but, just as importantly, on what a person with anomalous color perception perceives.[23] For example, most color deficiencies could be aggregated into two related groups of three, correctly suggesting that there are three types of color-sensing cells.

The term "anomalous trichromacy color blindness" is applied to the condition where one of the three cone cell types is abnormal. Here all three types of cone cells are present (L, M, and S), but one of the cell types is mutated. These mutations lead to three types of anomalous trichromacy color blindness—protanomaly, which is the mutation on the L-type cells, deuteranomaly, which is the mutation of the M-type cells, and, more rarely, tritanomaly, which is the mutation of the S-type cells.

These mutations cause either the diminution of the nerve signal output of these cells or a shift of the absorption curves of these cells towards the absorption curve for one of the other two cell types. This diminishes, but does not entirely remove, the ability to distinguish colors based on three color primaries (i.e., people with this form of anomalous vision are still trichromats).

For people experiencing protanomaly, for example, the L cone is shifted to lower wavelengths. As a result, those who experience this anomaly possess a shorter range of visible wavelengths than normal trichromats—because light residing at the high wavelength end of the visible spectrum can only be detected by the L cone. When the light absorption character of this cone is shifted to lower wavelengths, light of these wavelengths cannot be detected at all.

Individuals with deuteranomaly have color distinction difficulties similar to those of individuals with protanomaly because in both cases the M and L cones are shifted towards one another. The opponent nature of color perception, as discussed in Chap. 5, is based, in part, on the brain's interpretation of the balance between red and green light. This balance is determined by the relative strengths of the signals from the L

[22] Approximately eight times as many men as women have anomalous color perception. Most forms of anomalous color perception are recessive traits caused by a mutated gene on the X chromosome. Since men have only one copy of this chromosome, while women have two, it is much less common for women to display anomalous color perception than for men.

[23] In this regard the rare individual with one normal eye and one abnormal eye was especially useful to color scientists [13].

and M cones. For both protanomaly and deuteranomaly the absorption spectra for the L and M cones nearly overlap, and so regardless of the stimulus spectrum presented to individuals with either type of anomaly, the outputs from the two types of cells will be equal to one another, or nearly so. This leads to the inability to distinguish between red and green light since these wavelengths cause the same responses for the two cone cells.

Individuals with tritanomaly, which is quite rare and not gender biased, have mutated S cells. In this case the absorption of these cells shifts towards green. This results in the inability to distinguish blue and yellow, and to distinguish color combinations containing them (e.g., yellow from pink and red from purple).

For these three conditions, the absorption of one type of color cell is shifted from its normal position. There are three other, more severe, types of anomalous color perception that arise when one of the three cone cell types is absent altogether. These are protanopia (lack of L cone cells), deuteranopia (lack of M cells), and tritanopia (lack of S cells). The symptoms of these conditions are similar, but more severe, than their similarly named anomalous trichromacy counterparts.

All of these forms of anomalous color perception affect the ability of an individual to perceive differences in color that are obvious to those with normal color vision. This can be interpreted as an extreme form of observer metamerism—some color pairs that look different from those with normal cone cell function are metameric to those with anomalous color perception. In this way we understand that this is not a true blindness to certain colors—objects with those colors can still be seen. However, they will look identical to objects that, to the normal trichromatic eye, appear differently colored.

While we focus on the inability to distinguish colors when discussing individuals with anomalous color perception, there are some instances where the opposite is true. For example, people with protanopia can experience a color sensation that comes solely from the M cone. For normal trichromats, the M cone cannot be excited in isolation, since its absorption spectrum overlaps with those of the L cone at longer wavelengths and the S cone at shorter wavelengths (refer back to Fig. 6.4). In addition, certain colors that are metameric to the normal trichromat will look different to those with protanopia. For this reason, individuals with anomalous color perception have allegedly been used during wartime to identify camouflaged areas—here the camouflage is metameric to foliage to the normal trichromat but not to a dichromat [14, 15].

## *Color Vision in Non-humans*

Color reproduction using just three primaries can so accurately recreate a color that, in some cases, is formed by a complex absorption spectrum, that we may take for granted that such reproductions will appear similarly lifelike to all animals. After all, if a painting and a picture of that same painting look identical to us, it seems clear that they should look identical to any observer, human or otherwise.

However, this is not the case. In this chapter we reviewed the color theory and color spaces that have been evolving since the early twentieth century. All of these color systems can ultimately be traced back to the original color-matching experiments reported by Wright and Guild in the 1920s. These experiments measured the color vision of humans and the results of this research were ultimately based on the specific color-sensing cone cells that are common to most people. The color cells of other animals do not absorb at the same wavelength of light that those of humans, and most mammals are dichromats, possessing only two types of cone cells.

For this reason, both printed and electronic images that appear to accurately reproduce the colors of a scene to humans will appear significantly off-color to non-humans. These reproductions are accurate to humans because they were developed to be, with these developments ultimately based on human color-matching functions. Note that these reproductions do not reproduce the reflectance spectrum of the scene, but rather they produce the same sensation as the original scene, using three primary colors. The colors in the scene and the image will be different from animals with different color-sensing cell light absorption characteristics and color-matching functions.

## *Are There Colors That We Cannot See?*

At different times throughout this chapter, the reader might have wondered whether there are colors that we cannot see with our trichromatic vision. One trivial example is light of a wavelength outside the visible range. We assign color names to some regions of the electromagnetic spectrum outside this range—for example, ultraviolet and infrared—even though we cannot see them. The existence of electromagnetic radiation outside the visible gamut is not surprising, and we should have no expectation that our eyes can detect all light wavelengths. There are of course examples elsewhere in the animal kingdom of organisms with the ability to readily detect wavelengths beyond the accepted visible range of human vision.

A second possibility would be colors that fall outside the gamut of visible colors in the *xy* chromaticity diagram—the so-called imaginary colors. As discussed previously, these points cannot be made from any combination of visible light. For example, we saw there that there was no wavelength or combination of wavelengths that gave a positive value for **Y** with the values of the **X** and **Z** primaries being zero.

This same result can be derived differently, based on Grassmann's laws as follows: Since we can only see colors within the visible range of wavelengths, any visible color must consist of one or more of the spectral colors. Because the visual gamut is convex, the line segment connecting any pair of spectral colors must fall within the color gamut. There are, therefore, no combination of wavelengths that would not be visible to us.

That said, there are, in a sense, colors that we cannot see, or, perhaps more accurately stated, that we cannot differentiate with our trichromatic vision. These are metameric colors. We saw earlier that we perceive yellow both when we are presented with 570 nm light (this is the yellow seen in the rainbow) and when we are presented with a combination of green (520 nm) and red (650 nm) light (this is the yellow on our computer monitors). If we had a fourth cone cell that adsorbed light in this region of the spectrum, we would be able to differentiate between these two stimulus spectra—that is, they would no longer be metameric—because the response of this fourth cone cell would be different for each of these two ways of creating yellow.

This reasoning is related to the earlier discussion on color deficiencies. To an individual with protanopia (i.e., with only active S and M type cones), red and green are metameric—they look the same. By adding the missing L cones, this individual would be able to easily discriminate red from green since the response of the L cone is different from these two colors. Similarly, if we could add a new cone to our trichromatic vision, we would be able to discriminate more colors, including differentiating between the metameric pair of yellow color sensations referenced above.

Finally, we note that we can see certain color sensations under special conditions that we cannot see under normal viewing conditions. We do this by saturating (or fatiguing) certain of our cone cell types by first looking at a colored field for several tens of seconds. When we shift our gaze from that field to a field of a different color, we stimulate our cone cells in a way that cannot be replicated in normal vision.

For example, we can create a green afterimage by looking at a red computer display for about 30 seconds, and then changing the color of the display to white. We see green, rather than white, because during the 30 seconds we gazed at the red screen, the vision processing occurring in our brains compensated for the overwhelming redness by shifting its perception towards white and, at the same time, the opsins in the L cones become depleted, making these cones incapable of contributing to color sensation. Since green is the complimentary color of red, that meant mentally adding green to the red screen. When we then change the field to white, we fleetingly see the green that has been mentally added to the scene.

What if, rather than shifting our gaze from red to white, we shift it from red to green? In this case, we see what can be regarded as a super-saturated green. We mentally add the green due to red saturation to the green on the screen. This super-saturated green color is called a "forbidden" color and can only be seen in this fashion—we could not, for example, formulate a paint with this color sensation.

## Summary

The first successful efforts to quantify colors were color-matching experiments performed separately by Wright and Guild in the early part of the last century. In these experiments, subjects would recreate spectral colors by additively mixing the light from three monochromatic sources—one red, one green, and one blue, which are designated as the **R**, **G** and **B** primaries. The relative intensities of the three lights needed to match each spectral color are called the spectral tristimulus values for these colors, and the values over the entire range of visible wavelengths are termed the red, green, and blue color-matching functions.

The amounts of the three primary colors needed to reproduce any complex color (i.e., those colors with spectra that included light intensity from more than one wavelength) can be calculated based on the color-matching functions. In this process the amounts of each primary needed to match the complex color, at each wavelength, are calculated and the values for each wavelength throughout the visible spectrum are added together. The resulting amounts are called the RGB tristimulus values for that color and are unique to each color sensation.

The color-matching functions derived using the red, green, and blue primaries represented a major advance in placing color on a firm scientific basis. However, there were some aspects of these functions that made them difficult to use, particularly before the ready availability of computers. At that time all calculations were done by hand or using a slide rule, and streamlining the procedure to calculate tristimulus values would save much time and reduce the likelihood of error. In addition to calculating the three tristimulus values, there was a value in calculating the brightness of the color. Also, over the spectral gamut one or more of the red, green and blue tristimulus values are negative, which created another opportunity for making mistakes in the calculation of the tristimulus values.

To overcome these and other problems associated with RGB color space, the CIE, in 1931, mathematically transformed that color space into a new color space—the XYZ color space. All tristimulus values in this color space are positive, and paint brightness is automatically calculated in this space because the Y tristimulus value matched the photopic brightness function ($V_\lambda$). Collectively the XYZ color-matching functions are known as the Standard Observer. To have only positive values for the XYZ color-matching functions, the **X**, **Y**, and **Z** primaries had to be imaginary—that is, there is no light source that can generate these primary colors. This is not important, however, since our goal is not to reproduce a color with primaries, but rather to assign a unique set of XYZ values to each color sensation.

Colors can be conveniently plotted and organized on the *xy* chromaticity diagram. Additive mixtures of any two colors are found on the line joining these two colors, and color gamuts for three (or more) primaries are located within the triangle (or polygon) created by joining these colors together.

## References

1. Wright, D.W.: A re-determination of the trichromatic coefficients of the spectral colours. Trans. Opt. Soc. Lond. **30**, 141 (1928–1929)
2. Wright, D.W.: A re-determination of the mixture curves of the spectrum. Trans. Opt. Soc. Lond. **31**, 201 (1928–1929)
3. Guild, J.: The colorimetric properties of the spectrum. Trans. R. Soc. Lond., Ser. A **230**, 149 (1931)
4. Wright, D.W.: The Measurement of Colour, 4th edn. Van Nostrand Reinhold (1960)
5. Fairman, H.S., Brill, M.H., Hemmendinger, H.: How the CIE 1931 color-matching functions were derived from wright-guild data. Color Res. Appl. **22**(11), 11 (1997)
6. Oleari, C.: Standard Colorimetry: Definitions, Algorithms and Software. Wiley (2016)

7. Standard practice for computing the colors of objects by using the CIE system. ASTM E308-15, ASTM National (2015)
8. Saunderson, J.L.: Calculation of the color of pigmented plastics. J. Opt. Soc. Am. **32**, 727 (1942)
9. Hunter, R.S., Harold, R.W.: The Measurement of Appearance, 2nd edn. Wiley (1987)
10. Macadam, D.L.: Visual sensitivities to color differences in daylight. J. Opt. Soc. Am. **32**(5), 28 (1942)
11. Tr´emeau, A., Konik, H., Lozano, V.: Limits of using a digital color camera for color image processing. In: Proceedings of the IS&T/OSA Optics & Imaging in the Information Age, Rochester, New York, p. 150 (October 1996)
12. Mahy, M., Van Eycken, L., Oosterlinck, A.: Evaluation of uniform color spaces developed after the adoption of CIELAB and CIELUV. Color Res. Appl. **19**(2), 105 (1994)
13. Judd, D.B., Wyszecki, G.: Color in Science, Business and Industry, 3rd edn. Wiley Series in Pure and Applied Optics (1975).
14. Colour-blindness and camouflage. Nature **146,** 226 (1940)
15. Morgan, M.J., Adam, A., Mollon, J.D.: Dichromats detect colour-camouflaged objects that are not detected by trichromats. Proc. R. Soc. Lond. Ser. B: Biol. Sci. **248**, 291 (1992)

# Part III
# Particle Types

# Chapter 7
# White Pigments

## Contents



## Introduction

Chapters in this book discuss entire classes of materials—color pigments (Chap. 8), extenders (Chap. 9), and resin particles (Chap. 10). Yet here we devote nearly an entire chapter to only one material class—titanium dioxide pigments. This seeming imbalance is fully justified by a number of factors. The first is the volume of $TiO_2$ pigments sold annually. It is estimated that on a weight basis $TiO_2$ accounts for about two-thirds of all pigments used annually. On a value basis, it accounts for roughly 44% [1].



M. Diebold et al., *Pigments, Extenders, and Particles in Surface Coatings and Plastics*, https://doi.org/10.1007/978-3-030-99083-1_7

In addition, it is used in nearly all opaque paints and plastics (the exception being black), and at levels significantly higher than that of the pigments that provide color. For example, a high-quality light blue paint may contain 300 g of $TiO_2$ per liter (2.5 pounds per gallon) but only a few grams of blue pigment per liter. This disparity is due to the different ways that these particles interact with light. Light scattering by the $TiO_2$ is much weaker than light adsorption by color pigment particles, as discussed in Chap. 3.

Finally, unlike extenders and color pigments, for which there are a multitude of choices for any given paint or plastic, there are only two materials that provide whiteness, brightness, and opacity—$TiO_2$ and air. Other white pigments were used in the past, but today none can compete with $TiO_2$ in respect to its high scattering strength and non-toxicity.

$TiO_2$ can be considered a specialty chemical in the sense that it must be customized for each specific end-use application. One size does not fit all, and using the $TiO_2$ grade in an application for which it was not designed will cause a significant reduction in its opacity performance, as well as likely resulting in the degradation of other end-use properties.

There are two processes for making the $TiO_2$ particles themselves—the sulfate process and the chloride process. Once made, the particle surfaces are modified to enhance certain attributes of the pigment or the material into which the pigment is incorporated. In this chapter, we discuss these aspects of manufacture as well as the end-use properties that can be controlled by the pigment producer.

## Pigment Manufacture

The physical and chemical properties of small particles discussed in Chap. 1 are dictated by the chemical nature, surface properties, and dimensional characteristics of the particles. Properties such as size, shape, and purity of pigmentary $TiO_2$ are controlled by the production process. There are two different commercial processes for producing pigmentary $TiO_2$: the sulfate and the chloride processes.

In both cases, the process for its manufacture begins with mining the ore, and upgrading its value by doing physical separations from other minerals that are present with it. These separations are typically made on a size, density, magnetic, or electrostatic basis. In addition, the iron content of the ore, often present as an iron titanate rather than separate iron species, can be reduced by slagging or other chemical means. Together these processes are often referred to as ore beneficiation and can increase concentration of the desired particles from a few percent to upwards of 90%. The benefits of beneficiation of the ore are that it decreases the amount of material that is shipped to the ultimate manufacturing site and the amount of material that must be reacted and disposed of at that site. There is also environmental advantage that the removed impurities are still in their natural state and so can be directly returned to the environment.

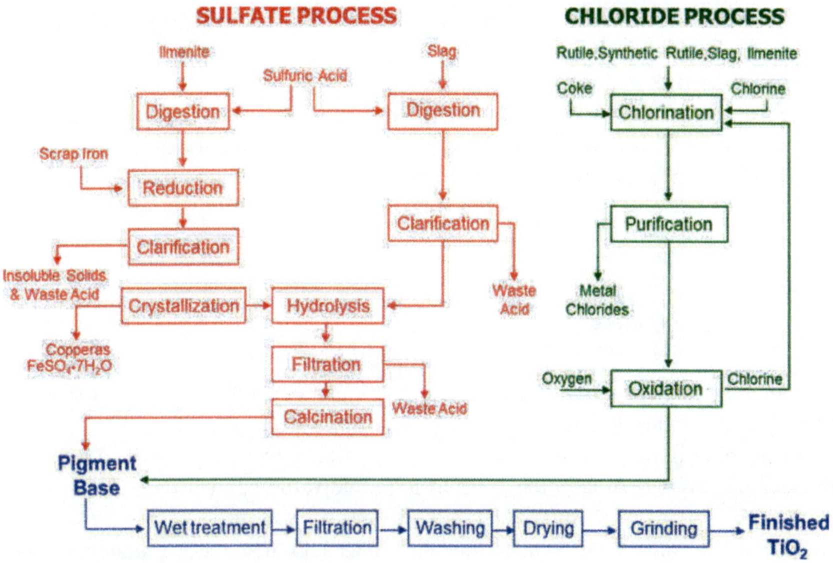


**Fig. 7.1** Commercial routes to pigmentary $TiO_2$

## *Sulfate Process*

In the sulfate process (see Fig. 7.1), the upgraded ore is digested in concentrated sulfuric acid to produce titanyl sulfate. To keep impurities in their more soluble lower oxidation states, reducing conditions are maintained. The titanyl sulfate solution is then purified by first removing the insoluble contaminants and then selectively crystallizing the titanyl sulfate. This material is hydrolyzed and then calcined to produce pigmentary $TiO_2$ particles. Straight hydrolysis will produce only anatase when calcined, but rutile can be obtained through the addition of rutile seed crystals during the hydrolysis and subsequent calcination steps. The size of the particles is determined by the calcination conditions, and, often, by the incorporation of small amounts of other oxides.

## *Chloride Process*

In the chloride process, the titanium ore is reacted with chlorine gas at high temperatures in the presence of a carbon source. The various minerals present in most ores are oxides, and under these conditions the oxygen content of the oxides is liberated as carbon dioxide while the metal portions are transformed into metal chlorides. Titanium tetrachloride is a liquid under ambient conditions whereas other metal chlorides are typically solids (e.g., ferrous and ferric chlorides) or gases (e.g., aluminum and silicon chlorides). The titanium tetrachloride is purified from these chlorides by fractional distillation and other associated processes.

The purified titanium tetrachloride is then reacted with pre-heated oxygen in a flame reactor, generating pigmentary size $TiO_2$ particles. The by-product of this reaction, chlorine gas, is recycled to the chlorination reactor. Note that this process does not create a closed loop for the chlorine gas, since the impurities of the chlorination reactor are metal chlorides, which remove a portion of the chlorine from the process.

## *Property Differences*

The main benefits of the sulfate route are that it can generally use a wider range of ores than the chloride route and that it is simpler from a production viewpoint, allowing new producers to enter the market with a relatively simple and unsophisticated process. In addition, both anatase and rutile can be made through this route.

Working against these benefits are the drawbacks that rutile production from the sulfate process can contain anatase impurities, and that non-titanium impurities are more easily removed in the chloride process, generally giving chloride grades a higher brightness than their anatase counterparts. In addition, particle size can be better controlled in the chloride route. The sulfate process also produces high amounts of spent (diluted) sulfuric acid, which must be disposed of or concentrated for recycle in the process. Finally, the chloride route is a continuous process whereas the sulfate route is batch, and so product consistency tends to be better in the chloride route.

## *Surface Treatment*

Once the particles are made, the two process routes converge (Fig. 7.1), and the particles are collected as a water slurry. The next process step is to modify the "raw" particles to optimize performance in specific end-use applications. This is required because the particles in their as-made condition degrade many important properties of paints or plastics, such as opacity, gloss, and durability. This degradation is due to interactions between the particle surface and its surroundings, whether that is water or solvent (a liquid paint), molten resin (plastics during manufacture), or the resin matrix in the final end-use application (paint film, plastic article, or paper laminate). Without modifying the particles in some way, they are generally incompatible with the other ingredients in the paint or plastic.

Since these interactions occur at the particle surface, and since the refractive index of the particle is a bulk, rather than a surface, property, we can modify the pigment surface to improve particle interactions while leaving the most important aspect of the pigment—its light scattering ability—unchanged.[1] This modification is accomplished through the deposition of a material onto the surface, and for this reason most $TiO_2$ pigment grades are not pure $TiO_2$. The pigment user should not regard the non-$TiO_2$ components as impurities—they are quite the opposite. Even though present at only a few percent, these materials greatly improve the value of the pigment and without them it is unlikely that $TiO_2$ could be used effectively in most applications [2].

In principle, we can deposit any material we wish onto the particle surface. However, if the particle is to maintain many of its important features, we must restrict the characteristics of these surface materials. The important features include colorlessness, non-toxicity, and insolubility in both water and solvents. In addition, these deposited materials, and their precursors, must be easy to handle, relatively inexpensive, and effective at low concentrations so as to limit dilution of the $TiO_2$ content of the pigment. When we apply these restrictions to the large number of potential particle surface coatings, we arrive at a very short list of practical candidates. This list includes hydrous silica, hydrous alumina, and, at low levels, zirconia. In addition, an organic is often added to the particles near the end of the process (in the grinding stage, discussed in the next section).

The process for depositing these materials is referred to as "surface treatment" or "wet treatment", the latter designation referencing that these treatments are normally deposited onto the pigment when it is a water slurry. The strategy for this deposition is to add a water-soluble precursor of the desired surface coating to a water slurry of the freshly made particles, thoroughly mix the precursor, and then change the slurry conditions (temperature and/or pH) in a way that precipitates the precursor. The exact nature of the change in slurry conditions is important—the same precursor can be precipitated in different ways to alter different properties of the pigment and the end-use materials made from them.

Silica is deposited on the particles using sodium silicate as its precursor. Sodium silicate is a viscous, high pH material that normally contains an excess of NaOH to stabilize against precipitation before use. When the sodium silicate is added to the $TiO_2$ slurry, the slurry pH increases. The silicate is then neutralized with a strong acid, typically hydrochloric acid or sulfuric acid. During this precipitation the pH remains high because most of the added acid is consumed by the silicate and excess NaOH, rather than remaining in solution and lowering the pH. The pH drops rapidly at the end of the deposition when there is no longer free silicate with which the acid can react.

The silica can be deposited in two forms that differ in the local environment around the individual silicon atoms and in the morphology of the coating. In each case, the silicon atoms are bound to four oxygen atoms, and each oxygen atom is bound to two other atoms. The forms differ in the identity of the second atom bound to the oxygen—it can be a second silicon atom (in which case the oxygen atom bridges two silicon atoms), or it can be a hydrogen atom (in which case the oxygen atom is part of a terminal hydroxide).

---

[1] Except by a small reduction due to the slight dilution of the $TiO_2$ by the surface modifier(s).

The first form of silica is referred to as "dense" silica or "Iler" silica, the latter designation reflecting the first researcher to deposit this form of silica onto $TiO_2$ [3, 4]. In this form, bridging oxygen atoms dominate, making most of the silicon atoms internal to the structure. This form is typically deposited under high pH and temperature conditions, and with a slow addition of the neutralizing acid [3]. Under these conditions the silicon atoms are more likely to deposit onto existing surfaces (the $TiO_2$ surface or already-deposited silica), rather than as a separate particle, when they are neutralized. This results in a tight, coherent coating of silica on the $TiO_2$ surface.

The silica formed in this way deposits as an amorphous material, and as such it has no shape preference (morphology) of its own. Such a coating is desired to mitigate the formation of chemical radicals on the surface of the $TiO_2$ when exposed to UV light. As will be discussed in Chap. 14, these radicals can leave the surface of the pigment particle and react destructively with the polymer matrix of the paint film or plastic article. Electron micrographs showing this tight, amorphous layer at three magnifications are shown in Fig. 7.2. Because it coats the existing $TiO_2$ surface, rather than forming new particles (with new surfaces), this form of silica does not change the surface area of the pigment appreciably.

Ideally a single layer of silica would suffice for encapsulating the $TiO_2$ surface and preventing free radical formation. This would amount to approximately 0.4% $SiO_2$ by weight. However, in practice, higher levels are required for maximum performance, and grades that contain silica will typically have an amount between 1.5% and 7.5% [5–7]. These higher levels are needed because the silicate moieties deposit randomly on the surface and are just as likely to cover an already covered portion of the surface as they are to cover a uncoated portion of the surface.

The second form of silica has an abundance of Si–OH groups and is referred to as "ionic" silica because these groups deprotonate easily, ionizing the surface of this material. The Si–OH groups are, by their nature, only found at a surface, and because they are predominant in this form of silica, silica deposited in this form has a high surface area and so greatly increases the surface area of the pigment. This form is optimized by precipitating the silica at low temperature, low pH, and quickly [8]. This form of silica is heavily hydrated and so deposits as a thick, porous layer on the particle surface. When used, it is typically present at between 5.0% and 10.0% as $SiO_2$ by weight, and it greatly increases the surface area of the pigment particles. The raw particles have a surface area of roughly 7–8 $m^2/g$, while particles treated with ionic silica at these levels have surface areas between 45 and 55 $m^2/g$.

Because of its appearance under an electron microscope (Fig. 7.3), this form of silica is also called "fluffy" silica. It is used in heavily treated grades made for highly loaded paints, as discussed below.

Alumina, too, can be deposited in two forms, but in this case the local environments around the aluminum atoms are the same (octahedral coordination by bridging oxygen atoms), but the long-range order is different. These forms are pseudo-boehmite, which is crystalline, and amorphous alumina. Each of these forms of alumina provides a different functionality to the pigment.

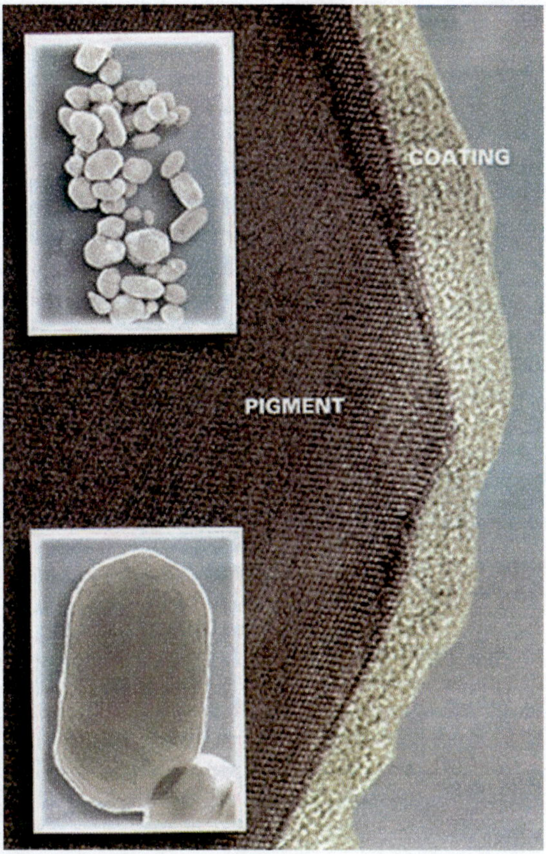


**Fig. 7.2** Dense layer of silica on $TiO_2$ particles at three magnifications. Note the contrast between the lattice fringes of the crystalline $TiO_2$ core of the particles and the unstructured nature of the amorphous silica surface

The pseudo-boehmite form exists as small (10–20 nm) particles that decorate the particle surface (Fig. 7.4). The function of these particles is to improve the attachment of dispersant molecules onto the pigment surface (see Chap. 11). Although they are present at low amounts (typically 1.5–3.5% measured as $Al_2O_3$ by weight), they roughly double the pigment surface area to between 15 and 20 $m^2/g$.

The pseudo-boehmite particles do not entirely coat the surface in the same way as dense silica or amorphous alumina (see below), and so there the surface is still mostly $TiO_2$. As such it does not improve paint or plastics durability as do the amorphous oxides. This does not decrease their effectiveness, however, at improving pigment dispersion.

The amorphous form of hydrous alumina is similar in morphology and function to the dense form of silica. Its primary role is to increase paint or plastics durability. It has an advantage over silica in that it does not harm the gloss potential of the pigment, as does silica. However, unlike hydrous silica, which is kinetically stable, amorphous alumina can rearrange to the crystalline pseudo-boehmite phase under

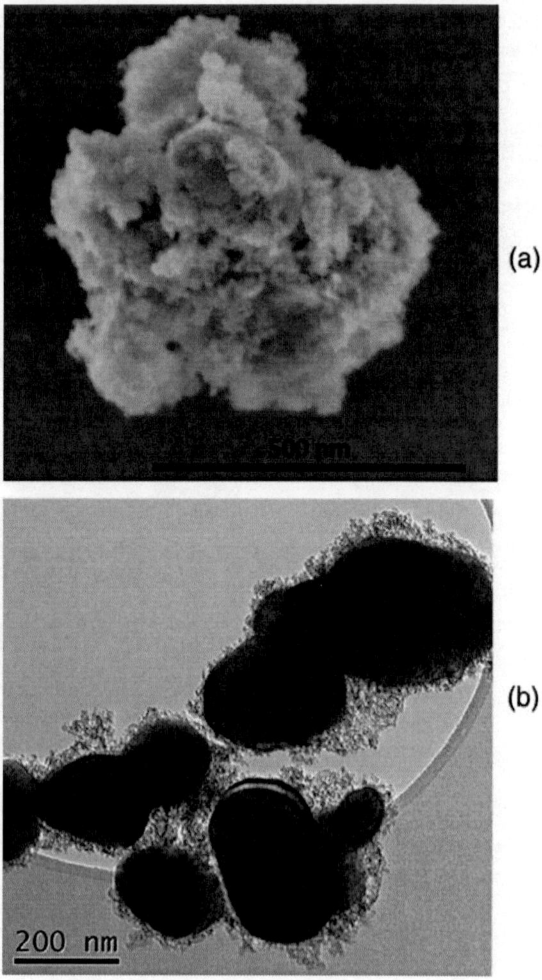


**Fig. 7.3** Electron micrographs of ionic silica on $TiO_2$. **a** SEM. **b** TEM

surface treatment conditions. It is essential that the deposition conditions be closely controlled in order to prevent crystallization. A heteroatom is often incorporated into the amorphous structure in order to stabilize it against crystallization. This atom is most commonly zirconium that is co-precipitated with the alumina [9], although other atoms such as molybdenum, sulfate, phosphate, tin, or fluoride can also be used [10–16].

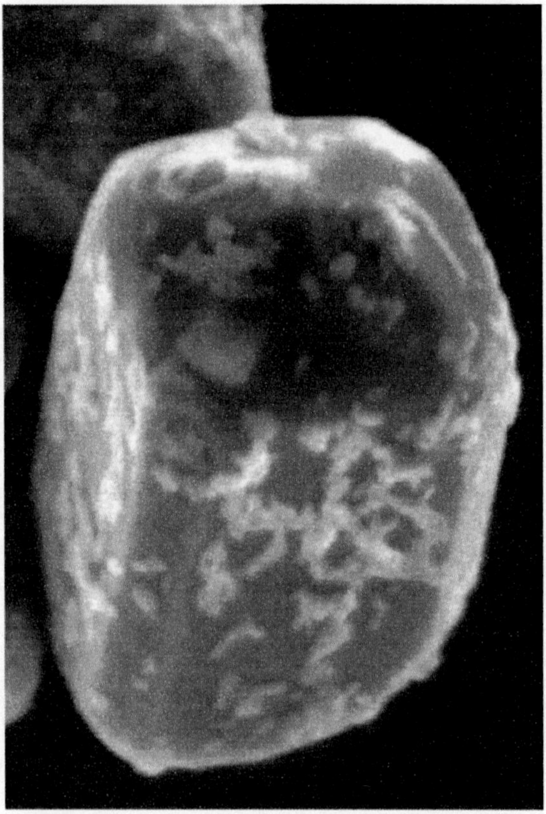

**Fig. 7.4** SEM image of pigmentary $TiO_2$ coated with pseudo-boehmite particles

## *Finishing*

After surface treatments are complete, the $TiO_2$ must still be filtered, washed, dried, ground, and packaged or made into a slurry. These last steps are referred to as the finishing process.

In the filter and wash steps, the pigment from wet treatment is dewatered, washed free of soluble impurities, and dried. Several types of filters can be used in this step. Although, for example, filter presses are generally more costly to install and operate, they have the advantage of producing a higher solid discharge that requires less heat to dry. This results in a cost savings and a reduction in $CO_2$ emissions.

The filter discharge can be dried in a number of oven types. It is important to match the dryer type to the form of the filter discharge. The mass exiting the dryer is then passed through a particle/gas separator which removes the $TiO_2$ from the gas stream.

A certain degree of particle aggregation is inevitable during $TiO_2$ production. These agglomerates can be formed during $TiO_2$ particle formation, during the surface treatment step when touching particles are encapsulated together by the surface treatment agent, and during drying when the surface tension of the water exiting the slurry droplet or cake fragment pulls the particles together. It is therefore necessary to grind the pigment prior to final packaging.

The most effective way to achieve this grind is with a fluid-energy mill ("micronizer"). This type of mill has a disk-shaped chamber (Fig. 7.5). Steam and pigment are tangentially injected into the chamber at supersonic speeds. The clusters of particles collide with the outer wall of the chamber, other particles moving at high speeds, and particles slowed by previous collisions. The centrifugal force generated keeps larger particles towards the outer edge of the fluid-energy mill, and only the desired smaller particles can travel to the center where the steam is discharged. The pigment/steam mix is then separated in a bag house or cyclone. The exiting steam can be used in other aspects of the process. Note that this process separates touching particles, but it does not actually grind the individual particles into smaller sizes.

A uniform particle flow is required for the particles to grind with optimal efficiency. Organic "grind aids" can be applied to the pigment surface prior to injection into the fluid-energy mill to facilitate a uniform flow. These aids are typically low weight, high functionality organic molecules. In addition, organic hydrophobic materials can be applied to enhance dispersibility of the $TiO_2$ pigment in plastics applications [17].

While fluid-energy mills are effective for pigment grinding, they also require high energies. Because of this the pigment is often pre-ground in the slurry state with media, sand, or ball mills earlier in the process (either before or after surface treatment). This reduces the very large agglomerates into smaller clusters that are more efficiently ground in a fluid-energy mill to achieve primary particle sizes.

Finally, the resulting powder is either packaged in bags (typically 25 kg), semi-bulk containers (½ or full metric ton sacks), or bulk containers (20–40 metric tons) destined for silo storage. The pigment may also be made into a water slurry ready for incorporation into a latex paint. This has the advantage that it allows the paint manufacturer to skip a step in the paint making process, and also allows for a more exact metering of the pigment into the paint batch (that is, the minimum increment of pigment added is less than for dry material). However, care is required to assure

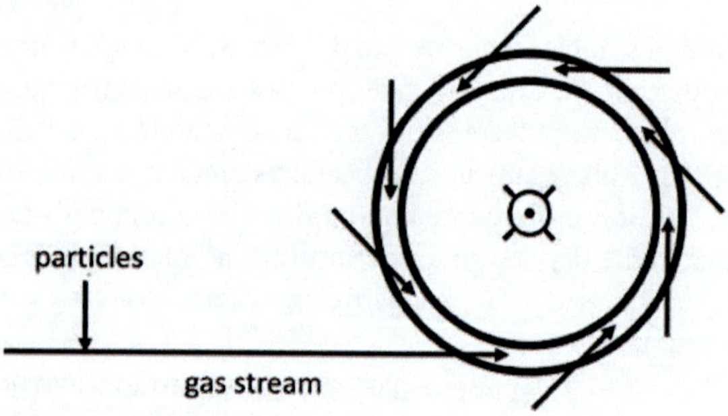


**Fig. 7.5** Overhead view of a fluid-energy mill. Gas exits out of the page at the center of the chamber

that the slurry remains stable for extended periods of time, and the slurry must be actively acted on during storage to reduce settling.

## Pigment Design

A wide assortment of $TiO_2$ pigment grades is available for use in the coatings, plastics, and paper laminate industries. Each grade is customized with a specific application in mind. For example, grades are available that optimize paint gloss, durability, processibility, electrical resistance, and so on. Similarly, grades are available for plastics that are optimized for processibility, durability, and color properties (e.g., undertone).

In general, we can divide most $TiO_2$ pigment grades into one of two categories: those for specific applications (e.g., superdurable paints, electrodeposition coatings, etc.) and those that are meant to satisfy the needs of multiple paint or plastics types. The latter are referred to as "universal" or "multipurpose" grades, and their value to the paint or plastics manufacturer is raw material rationalization—by using a universal grade, a single pigment can be inventoried, rather than a separate pigment for each paint or plastic type.

$TiO_2$ pigment grades typically differ in two ways: the exact particle size distribution (in particular, the mean size) and the surface treatment (the identity, amounts, and deposition conditions for the surface coatings). The former determines the undertone of the pigment—blue or neutral. In regard to the latter, it is remarkable that so many different product properties can be altered and optimized by using only a few surface treatment materials. This reflects the ingenuity and creativity of $TiO_2$ product developers in making the most out of limited materials available to them.

### *Particle Size*

The importance of $TiO_2$ particle size to light scattering strength was discussed in Chap. 3. There we saw that the optimal particle diameter for $TiO_2$ to scatter visible light is roughly 0.25 microns. We also saw, in Chap. 4, that the exact optimal size depends on the concentration of the particles in the polymer matrix. We can divide these concentrations into two categories—low concentrations (typically less than 1% by volume) in plastics and high concentrations (typically between 8 and 20% by volume) in coatings. The optimal particle size at low concentrations is roughly 0.23 microns while that at the higher concentrations is roughly 0.26 microns.

Control of size over such a small range can be a challenge to the pigment producer. However, those producers with a long history of manufacture have generally optimized their processes to allow for such fine particle control. Although differentiating grades by the typical particle size measurement techniques is difficult (see Chap. 2), differences can be detected by eye. The smaller particles preferentially scatter blue light while the larger particles scatter all wavelengths more evenly. As a result, a mixture of the $TiO_2$ particles of interest with a black pigment (generally carbon black or black iron oxide) will have a blue undertone for grades destined for plastic and a more neutral undertone for those destined for coatings.

## *Durability*

The term "durable" is often used in connection with a particular grade of $TiO_2$ pigment. This term describes the effect of the $TiO_2$ on the outdoor durability of the paint or plastic into which it is incorporated, rather than the durability of the $TiO_2$ pigment itself. Unlike the organic matrices of paints and plastics, $TiO_2$ and the oxides used to coat its surfaces are fully oxidized and so do not react with oxygen from the air.

As described above, pigment grades are made durable by the application of an amorphous coating that completely separates the $TiO_2$ surface from its surroundings. This enhances paint and plastic durabilities because the $TiO_2$ surface becomes reactive when irradiated by UV light. Details of this reactivity are described in Chap. 14.

Surface-deposited silica gives a higher durability than amorphous alumina [18], and a combination of the two gives better durability than either alone [6, 7]. However, silica-coated $TiO_2$ has traditionally been viewed as difficult to disperse and an inhibitor of high gloss in paint applications. For this reason, many durable grades of $TiO_2$ are based on amorphous alumina coatings rather than silica coatings.

There is an additional consideration regarding the use of hydrous oxides to improve the durability of a plastic. The temperatures in certain plastics processing steps can be high enough to drive a significant portion of the waters of hydration out of these pigments. This causes gas bubble to form, which, in thin film applications, leads to thin spots or holes. This process is referred to as lacing. Alternative coatings that contain fewer waters of hydration (and so do not lace) yet still provide a beneficial effect on durability have been reported [14, 19].

## *Dispersibility*

The dispersion of particles into a liquid paint is covered in detail in Chap. 11 and into a molten polymer in Chap. 12. There are specific considerations for optimizing $TiO_2$ dispersion in both paint and plastic processing. Ideally these particles would disperse quickly and with minimal energy. In the case of paints, the dispersion must remain stable for extended periods of time.

The ease of dispersion can be controlled using suitable surface modifications. For coating grades of $TiO_2$, this consists mainly of depositing the proper form of alumina as the outermost pigment surface layer. Doing so not only improves the wettability of the $TiO_2$ pigment, which can be a slow step in dispersion, but can also improve dispersion stability by supplying a surface to which dispersant molecules attach strongly and permanently. In addition, certain additives present during the alumina precipitation can enhance the dispersibility of the pigment in water [7, 8].

Pigment dispersion in plastics can be improved by making the pigment surface hydrophobic (lipophilic). The hydrous oxide surfaces of $TiO_2$ pigment are hydrophilic and as such are incompatible with the molten polymer into which it is incorporated, making dispersion difficult. This issue can be reduced significantly by adding an appropriate surface agent during the final stages of pigment finishing [20]. This also has the advantage of reducing the amount of surface water adsorbed on the pigment, and thus reducing the possibility of lacing.

## *Gloss*

The gloss of a paint film is the visual impression given from the light reflected at the film surface. It is determined by the quantity of the light that is specularly reflected from a surface (i.e., reflected at the mirror angle) [19]. As such it is a function of the surface smoothness at the optical scale (0.5 microns and above).

Surface smoothness is affected by a number of factors. Obviously, a major factor is the size of the particles in the paint or plastic. Extender particles are generally very large compared to the wavelength of visible light, and as such their presence will typically decrease gloss. This occurs by near-surface particles perturbing the surface resin, that is, the particles themselves do not protrude through the surface, but nonetheless have an effect on it.

Although near-surface $TiO_2$ particles will also affect the surface smoothness, when the particles are in an individual and separated state, the surface perturbations are too small to impact gloss. However, groups of particles (agglomerates of flocculates) are large enough to affect gloss (Fig. 7.6). This effect can be seen, as an example, by

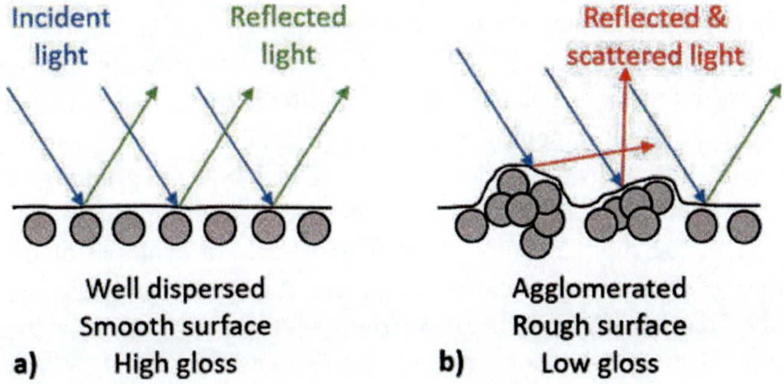


**Fig. 7.6** Comparison of the impact of agglomeration and surface roughness of two paint films on gloss

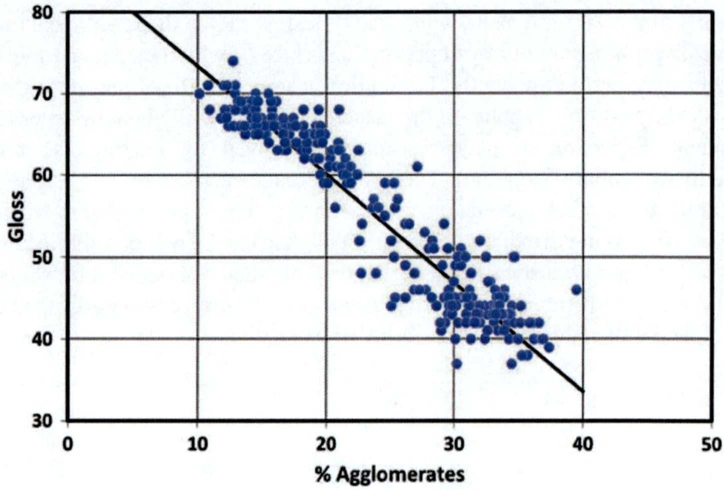


**Fig. 7.7** Effect of number of $TiO_2$ agglomerates on the gloss of a standard emulsion paint

comparing the gloss of a standard emulsion paint made with different $TiO_2$ samples as a function of the percent of particle agglomerates in those samples (Fig. 7.7).

As mentioned above, not all grades of $TiO_2$ give the same degree of gloss. This can be attributed to three factors. The first is the degree to which the particles maintain their separate identities as they are formed. In both the sulfate and the chloride processes, particle formation takes place at relatively high temperatures. In the sulfate process, this is during the calcination step, and in the chloride process the reaction of $TiCl_4$ with oxygen occurs in a flame reactor. Small particles become sticky at high temperatures and are prone to attaching to one another. For this reason, the exact conditions of the calcination or flame reaction, in particular, the time at temperature and the cooling rate, affect the number of agglomerates in the final product.

The second factor that affects the relationship between $TiO_2$ grade and gloss is the nature of the oxides deposited in the surface treatment step. In particular, the silica that is deposited to increase the durability of the paints or plastics into which a $TiO_2$ pigment is incorporated has a deleterious effect on gloss. In some cases, this issue can be overcome by replacing one oxide with another, although this can cause negative effects on other important end-use properties.

Finally, grind intensity affects gloss. This applies to both the grinding that is done both in the fluid-energy mill at the final stages of pigment production and during incorporation of the pigment into paints or plastics. An example of the latter is shown in Fig. 7.8. Here we show the development of distinctness of image (one form of gloss) as a function of grind time in a paint.

In general, the gloss of a paint can be readily decreased by adding a flatting agent, but it is not possible to add a material that will increase the gloss of a paint. Because of this, universal or multipurpose grades of $TiO_2$ pigment will typically have a high gloss potential.

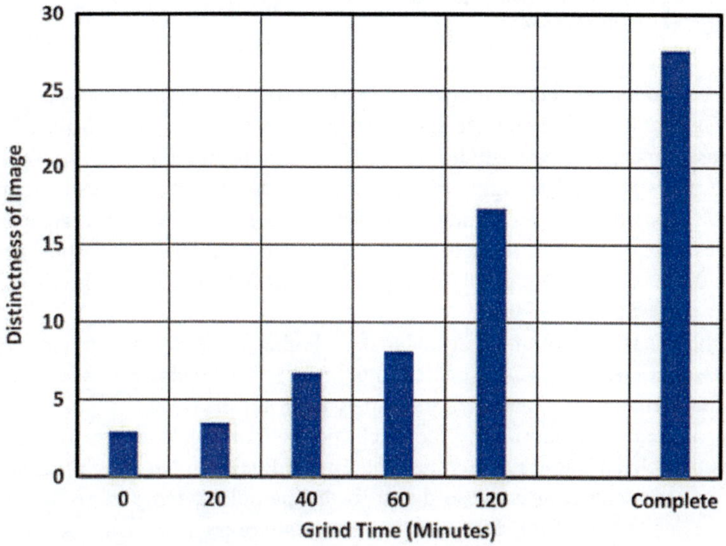


**Fig. 7.8** Evolution of distinctness of image during high shear dispersion $TiO_2$ paint grind. Complete dispersion was achieved by subsequent media milling

## *Opacity*

The light scattering ability of $TiO_2$ is determined by two factors: particle size and degree of particle dispersion. In Chap. 3, we saw the importance of particle size to light scattering strength. Light scattering is optimized when the particles are roughly 0.25 microns. However, as discussed in Chap. 4, there are slight differences in the optimal size based on the concentration of the particles in the polymer matrix. At the low concentrations, typical of plastics applications, slightly smaller particles are the most efficient. At the higher concentrations seen in paints, slightly larger particles are optimal.

In Chap. 4, we also saw the importance of dispersion on light scattering. Particles that are close together interfere with one another's ability to scatter light due to scattering volume overlap. As a result, good dispersion of the $TiO_2$ particles in the paint or plastic is necessary for their optimal efficiency, in the same way that good dispersion is necessary for high paint gloss. Chapters 11 and 12 describe the dispersion processes, and how they are optimized, for particles incorporated into paints and plastics, respectively.

### *Highly Treated Grades*

As described earlier, $TiO_2$ pigment grades tend to fall into the categories of those optimized for a specific end-use and those intended for multiple end-uses. Perhaps the most important of the grades optimized for a specific end-use are the highly treated pigments used in highly crowded paint systems. These paints contain high concentrations of extender, which are typically present as a mix of large and small particles. In these paints, there is not enough resin to fill the inter-particle regions, and so these regions are partially occupied with air. The effect of these air voids on opacity will be discussed later.

Highly treated grades are preferred in these highly crowded paint systems because the $TiO_2$ particles, which are crowded into the interstitial regions between the larger extender particles, are spaced farther apart than universal $TiO_2$ pigment in these regions. Because of dilution by the thick, porous coating, there are fewer highly treated particles in these regions than there would be an equal weight of universal $TiO_2$, but the scattering increase due to better particle spacing more than offsets decrease due to fewer $TiO_2$ particles per unit weight of pigment. This gives crowded paints made with the highly treated particles better opacity than those with the universal particles, as was discussed in depth in Chap. 4.

## Alternative White Pigments

Because of its unmatched refractive index, titanium dioxide is the preeminent white pigment used in the coatings, plastics, and paper laminate industries. However, other white pigments are sometimes used in these industries, particularly in niche applications where $TiO_2$ brings an unwanted property. As an example, $TiO_2$ particles scratch the reinforcing glass fibers used in certain plastics applications, causing them to break during processing. Here a less abrasive white pigment must be used, despite the loss of opacity. Another example is the continued use of some older pigments due to the cost or lack of availability of $TiO_2$.

### *Alternative Particles*

Particulate alternatives to $TiO_2$ pigments include zinc oxide, zinc sulfide, and lithopone. In the past, oxides of lead and antimony have been used for this purpose, but due to their toxicity and expense,[2] these pigments have all but disappeared from use. The relative opacities of these materials, as reported by Stieg [21], are given in Table 7.1. These values underscore the vast hiding power advantage that $TiO_2$ enjoys over its alternatives.

[2] This refers to expense on a coverage basis—that is, based on the cost to make a unit area of paint film or plastic opaque—rather than on a weight basis.

**Table 7.1** Relative hiding strengths of white pigments in paints [21]

| Material | Relative opacity |
|---|---|
| Titanium dioxide | 100.0 |
| Zinc sulfide | 50.4 |
| Lithopone | 23.5 |
| Zinc oxide | 17.4 |
| Carbonate white lead | 13.0 |
| Basic lead sulfate | 11.3 |

Zinc oxide had been used extensively in the Southern USA and other regions having persistent high humidity because it is a mildewstat. It achieves this property through its slight solubility in water—dissolved zinc ions discourage the growth of mildew and fungus. The use of zinc oxide in paints has mostly disappeared because it is an aquatic toxin. This is directly linked to its mildewstatic properties—while it is desirable for it to be toxic to micro-organisms on a painted surface, it is undesirable for it to be toxic to micro-organisms in a river or lake. It is still sometimes used at low levels as a bactericide for the liquid paint during storage.

Zinc sulfide is sometimes used by itself in paint primers, where it can indicate where the paint has been applied without the need to completely hide the substrate [22]. In addition, it has high thermal conductivity (which helps paints and plastics stay cool), lower UV absorption (an advantage in UV-cured paints and applications using optical brighteners), low abrasivity (making it preferred in glass fiber reinforced plastics and in situations where metal marking during plastics production is an issue), and is relatively easy to disperse in plastics.

Zinc sulfide is more commonly used as lithopone, which is an equal molar mix of zinc sulfide and barium sulfate (28% ZnS and 72% $BaSO_4$ on a weight basis). This material is conveniently made by mixing zinc sulfate and barium sulfide in water. Both of these materials dissolve in water, but their reaction products do not. Pure lithopone was substantially replaced by $TiO_2$ in the mid part of the last century. As a point of comparison, the current annual market for lithopone, on a monetary basis, is about 1% that of titanium dioxide. Modified lithopone, which was fortified with an excess of zinc sulfide or with titanium dioxide, enjoyed a brief popularity during the transition from lithophone to titanium dioxide [21]. Roughly three-quarters of current lithopone production goes into the coatings market, where it is still used in some applications due to its low cost, its ability to protect the paint from UV light (although the ZnS component of lithopone can discolor on UV exposure), and its fungicidal properties. It is used in the plastics industry when lacing (water vapor release during plastic film processing) would otherwise be an issue.

## *Entrained Air*

In addition to the materials described above, air can be used as a $TiO_2$ alternative for white opacity. This is because the refractive index of air is different from that of resin—in this case, lower than the resin matrix, rather than higher, as is the case with the materials described above. In order to scatter light effectively, air must be present in the form of discrete voids (bubbles) or pores (networks of joined pores). The functional difference between these forms is that pores allow for the ready transport of liquids and gases through the polymer matrix while discrete air units do not. In addition, because pores are larger than individual voids, they weaken the polymer to a greater extent than voids. These two forms of air also differ in the way they are created.

Discrete voids can be created in a number of ways. In certain plastics, such as polystyrene, a blowing agent can be added to the molten polymer [23]. This agent is a material that volatilizes at process temperatures, either through decomposition or boiling, and releases gas bubbles that are then trapped within the polymer when it solidifies. Alternatively, an inert gas can be dissolved in the molten resin under pressure. When the pressure is released, for example, by letting the mixture down to atmospheric pressure, bubbles are formed.

An excellent example of this form of air void is expanded polystyrene (EPS), a material familiar to many by the brand name "Styrofoam®". EPS is typically bright white but contains no white pigments. Instead light is scattered from a multitude of voids within the polymer. The voids in EPS materials are created by vaporizing small pockets of pentane that are incorporated into molten resin.

Air pockets in plastic films can also be formed through cavitation [24, 25]. In this process, small particles, often calcium carbonate, are incorporated into a plastic film, and then the film stretched. When the adhesive strength at the polymer/particle interfaces is low, stretching the film pulls the polymer away from the particles, creating elongated voids with a particle at the center.

The waterborne coatings industry also incorporates air voids into paint films. One way of doing this is by the use of hollow sphere opaque particles or HSOPs (these are also discussed in Chaps. 10 and 16) [26–28]. These are individual spherical polymer shells with an empty center. In the liquid paint, the center is filled with water, but in the dry film it is filled with air. These particles are formed by encapsulating a swellable polymer sphere with a rigid polymer that has an accessible softening point. The internal polymer is swelled at a temperature where the outer polymer is soft. The temperature is then lowered to make the shell rigid, and then the internal polymer is de-swelled to create a void.

Due to the limited difference in refractive index between air and resin, these particles allow for only a partial replacement of $TiO_2$, and only in paints for which the $TiO_2$ pigment volume concentration (PVC) is close to 20. Although the use of HSOPs is currently restricted to paints and limited paper laminate applications, research is currently underway to apply this technology to plastics [29].

Flash calcined clay particles are an alternative means by which encapsulated air can be brought into a paint film. Clay is a mineral composed of alternating layers of silica and alumina that are joined together by waters of hydration (see Chap. 9). When clay particles are rapidly heated, their outer surfaces quickly become glassy, sealing the interior of the particle. The interstitial water molecules then vaporize, resulting in the formation of internal, sealed voids.

As mentioned in the section on highly treated grades of $TiO_2$, a different type of air scattering is commonly used to opacify flat architectural (décor) paints. Here the paint is formulated with a high content of particles—mostly extender particles, but lesser amounts of pigmentary particles are typical—and a low content of resin. The balance between particle and resin volumes is so great that there is not enough resin to completely fill the voids between the particles. Instead much of the inter-particle voids are filled by air. We refer to such paints as being formulated above CPVC, a subject that is discussed in depth in Chaps. 4 and 16.

The air regions formed in this way are fundamentally different than those described above—those described above are discrete, self-contained voids or bubbles while the voids found in paints formulated above their CPVC join together into an extensive network of pores. This has implications on the strength, protectiveness, and stainability of paints employing this strategy.

Many older technologies, which included pigment composites and various ways of incorporating air into a paint or plastic [21], are no longer in use [30].

## Summary

Titanium dioxide pigments are the highest volume pigment used in the paint and plastic industries. This is because of its high whiteness, brightness, and opacifying abilities. Most paints, even those that are not white, contain $TiO_2$ as an opacifying agent, often in high amounts.

The opacity of pigmentary $TiO_2$ has two origins. The first is its high refractive index, which maximizes scattering on a volume basis (refer to Chap. 3). The second is the $TiO_2$ industry's ability to tightly control the $TiO_2$ particle size, as well as modify the particle surfaces in order to enhance other end-use properties.

There are two processes to form the individual $TiO_2$ particles. In each process, the titanium containing ore is reacted to liberate the titanium content. In one case, this is done with sulfuric acid, to make titanyl sulfate, and in the other with chlorine gas to make titanium tetrachloride. In both cases, the titanium containing compound is separated from impurities, and then reacted to form $TiO_2$. Process conditions are critical for achieving the correct particle size.

Regardless of how the individual particles are made, they are then surface modified to improve the interactions between the particles and the other components of the paint or plastic. These modifications consist of the precipitation of hydrous alumina and/or silica under carefully controlled conditions so as the surface agent precipitates in the proper form. In some cases, a third metal oxide, such as zirconia, is co-precipitated with the alumina. Finally, a thin organic coating is often placed on the particles immediately before they are ground.

The size of the $TiO_2$ particle determines its undertone. The optimal particle size is dependent on the concentration of the $TiO_2$ pigment in the end-use application. Applications with low $TiO_2$ concentrations, such as most plastics, benefit from slightly smaller particles, and applications with relatively high $TiO_2$ content, such as paints, benefit from slightly larger particles.

The surface treatments applied to the particles affect the dispersibility of the pigment as well as the gloss, opacity, and durability of the paint or plastic into which it is incorporated. In general, good dispersion is desired as this optimizes opacity and allows for a high paint gloss. This is desired even in paints with mid-to-low gloss because the high gloss grades of $TiO_2$ can be used in these paints, and a flatting agent can be added to reduce gloss to the desired level. As such the $TiO_2$ grades can go into multiple paint applications, earning them the designation of "universal" $TiO_2$ pigment.

## References

1. Schonbrun, Z.: The quest for the next billion-dollar color. Bloomberg Businessweek (2018)
2. Veronovski, N.: Titanium dioxide—material for a sustainable environment. IntechOpen **421**, (2018)
3. Iler, R.K.: Product comprising a skin of dense, hydrated amorphous silica bound upon a core of another solid material and a process of making same. US 2,885,366 (1969)
4. Iler, R.K.: The chemistry of silica: solubility, polymerization, colloid and surface properties and biochemistry of silica. Wiley-Interscience (1979)
5. Werner, A.J.: Titanium dioxide pigment coated with silica and alumina. US Re. 27,818 (1973)
6. Diebold, M.P., Bettler, C.R.: Easy to disperse, high durability $tio_2$ pigment and method for making same. US 6,783,586 (2004)
7. Diebold, M.P., Bettler, C.R.: Easy to disperse, high durability $tio_2$ pigment and method for making same. US 8,105,432 (2012)
8. Diebold, M.P., Kraiter, D.K., Rusnak, E., Shih, A.: Treated inorganic pigments having improved dispersibility and use thereof in coating compositions. US 9,539,557 (2017)
9. Craig, D.H., Elliott, J.D., Ray, H.E.: Process for manufacturing zirconia-treated titanium dioxide pigments. US 7,238,231 (2007)
10. Diebold, M.P., Baidins, A.: High gloss durable $TiO_2$ pigment. US 5,554,216 (1994)
11. Juergens, V., Siekman, J., Blumel, A.B., Schmitt, V.: Method for surface treatment of a titanium dioxide pigment. US 8,641,870 (2014)
12. Wachi, H., Egami, Y., Maekawa, H.: Surface coated particles and use of same. WO2014104225 (2014)
13. Drews-Nicolai, L., Bluemel, S.: Method for the post-treatment of titanium dioxide pigments. US 2006/0034739 (2006)
14. Jernakoff, P., Bolt, J.D.: Preparation of lacing resistant, titanium dioxide particles for use in photodurable thin film production. US 2020/0332074 (2020)
15. Jacobson, H.W.: $TiO_2$ pigment bearing a coatings with fluoride ions and laminate and coatings based thereon. US 4,460,665 (1984)
16. Tooley, P.A., Niedenzu, P.M., Reid, A.H.Jr.: Organosilicon treatment of $TiO_2$ pigment bearing a coating with fluoride ions. US 5,562,990 (1996)

17. Tooley, P.A., Holtzen, D.A., Musiano, J.A.: Processibility and lacing resistance when silanized pigments are incorporated in polymers. US 5,959,004 (1999)
18. Kinniard, S.P., Campeotto, A.: Improved method for manufacturing high opacity, durable pigment. EP 1 373 413 (2002)
19. Birmingham, J.N. et al.: Preparation of organic additive-treated, pyrogenic silica-encapsulated titanium dioxide particles. US 7,795,330 (2010)
20. Braun, J.H., Baidins, A., Marganski, R.E.: $TiO_2$ pigment technology: a review. Prog. Organic. Coat. **20**, 105 (1992)
21. Stieg, F.B.: Opaque white pigments in coatings. In: Tess, R.W., Poehlein, G.W. (eds.) Applied Polymer Science (Am. Chem. Soc.), vol. 285 (1985)
22. Pfaff, G.: Zinc sulfide pigments. Phys. Sci. Rev. **6**(8), 369 (2021)
23. Jin, F.-L., Zhao, M., Park, M., Park, S.-J.: Recent trends of foaming in polymer processing: a review. Polymers **11**(6), 953 (2019)
24. Breil, J.: Biaxially oriented films for packaging applications. In: Ebnesajjad, S. (ed.) Plastic Films in Food Packaging. William Andrew Publishing (2013)
25. Brunner, M., Tinkl, M., Brookes, D., Schulz, K.: Calcium carbonate as cavitation and voiding agents in biaxially oriented films. POLYMERS Communiqué **87** (2017)
26. McDonald, C.J., Devon, M.J.: Hollow latex particles: synthesis and applications. Adv. Colloid. Inter. Sci. **99**, 181 (2002)
27. Hungenberg, K.-D., Jahns, E.: Trends in emulsion polymerization processes from an industrial perspective. In: Polymer Reaction Engineering of Dispersed Systems. Springer (2017)
28. Jiang, S., Van Dyk, A., Maurice, A., Bohling, J., Fasano, D., Brownelf, S.: Design colloidal particle morphology and self-assembly for coatings applications. Chem. Soc. Rev. **46**(12), 3792 (2017)
29. Silva, J.F.A., et al.: Use of multi-hollow polyester particles as opacifying agent for injection-molded polyethylene. Polymers **12**(6), 1331 (2020)
30. Braun, J.H., Dickinson, J.G.: White pigments. In: Craver, C., Carraher, C. (eds.) Applied Polymer Science: 21st Century. Elsevier (2000)

# Chapter 8
# Color Pigments

## Contents





M. Diebold et al., *Pigments, Extenders, and Particles in Surface Coatings and Plastics*, https://doi.org/10.1007/978-3-030-99083-1_8

## Introduction

Appearance and aesthetics are of high importance in most paint and plastics applications. While there are several aspects of appearance, such as opacity and gloss, perhaps the most important is color. Consumer color preferences are often the deciding factor in a purchase, whether for a room paint, vinyl siding, or an automobile.

Color, in turn, is controlled by color pigment particles, and many formulation aspects of paints and plastics that use color pigment particles can be related directly to particle properties such as particle size and shape, crystalline form, and dispersibility. In this chapter we will first discuss the phenomenon of selective light absorption, and follow this with discussions as to the types of color pigments that are available to the formulator, aspects of their more important properties, and how they are made. We will end with a discussion of color pigment naming. Although there are several ways to name color pigments, in this chapter we will use the Colour Index™ system. Formulating with colors is covered in Chap. 15.

## Light Absorption

The defining feature of color pigments is that they absorb visible light.[1] It is therefore important that we understand the light absorption phenomenon if we are to understand how color pigments can be used to their greatest advantage.

Light is a form of energy, and because energy cannot be destroyed, light absorption requires the conversion of light energy into the second type of energy. There are many possibilities for the form that this second type of energy can take—electrical energy, kinetic energy, rotational energy, vibrational energy, electronic energy, and so on. The energy of visible light is most closely matched to the electronic energy within an atom or molecule, and for this reason, most colored pigments function by transforming light energy into electronic energy. To understand how this occurs, we must first discuss what, exactly, we mean by the term "electronic energy", and how this energy can be created from light energy.

### *The Electronic Nature of Atoms and Molecules*

Molecules are composed of atoms, which are, in turn, composed of protons, electrons, and, except for hydrogen, neutrons. Because they possess electrical charge, the proximity of protons and electrons to one another in an atom or a molecule results in the existence of an electrical energy associated with that atom or molecule. The exact amount of energy depends on the configuration of the electrons and protons,

[1] In the case of interference pigments, color is generated by selective reflection of light according to wavelength and viewing angle. This is discussed in a later section of this chapter.

and, under certain circumstances, this configuration can be changed by adding (or removing) energy to (or from) the atom or molecule. To understand this process, we must first understand the relationship between the configuration of the electrons and protons and their energy.

## The Electronic Structure of Atoms

The electrons, protons, and neutrons in atoms are not found in a random or unstructured way within the atom. Instead, the atom consists of a small, dense nucleus containing all constituent protons and neutrons, with the electrons forming a diffuse "cloud" surrounding this nucleus. These clouds are referred to as atomic orbitals, and their shapes and energies are understood on the basis of quantum mechanics.

A revolution in our understanding of atomic structure occurred in the early part of the last century. Physicists demonstrated at that time that the electrons within an atom followed certain rules. The first of these is that electron energies are quantized, meaning each electron has a discrete amount of energy, and intermediate energies are forbidden. In addition, the electrons are not stationary but instead move quickly around the nucleus in the atomic orbitals mentioned above. These orbitals have specific shapes, and the energy of an electron is determined by which orbital it occupies.[2] We classify atomic orbitals based on quantum numbers. There are three quantum numbers that describe the size, shape, and energy of an orbital, and these are designated as n, $\ell$, and $m_\ell$.

$n$ is the primary quantum number and is allowed integer values from 1 to infinity. $\ell$ is the orbital angular momentum quantum number,[3] and it determines the shape of the orbital. Shapes can vary from spherical ($\ell = 0$) to symmetrical lobe structures and more complex geometries. $\ell$ values are integers running from 0 to $(n - 1)$ and are designated by letters. The sequence of letters for $\ell = 0$–5 are *s*, *p*, *d*, *f*, *g*, and *h*. Figure 8.1 shows the shapes of *s*, *p*, and *d* orbitals.

$m_\ell$ is the magnetic quantum number, and it determines the orientation of the orbital. Its values run from $-\ell$ to $+\ell$, again as integers. For example, for a given n quantum number, there are three *p* orbitals ($\ell = 1$) that have the same shape but are at right angles to one another. These three orbitals correspond to the $m_\ell$ quantum numbers of $-1$, 0, and 1.[4] The shapes of *d* ($\ell = 2$) and higher $\ell$ orbitals are more complex.

There is a fourth quantum number, $m_s$, that does not determine the shape or energy of an orbital under normal conditions, but instead describes a property of electrons known as spin angular momentum. The values for $m_s$ are ±½. Each atomic orbital can contain no more than two electrons, one with a positive spin and one with a negative spin. This rule is known as the Pauli exclusion principle, and as a result of

---

[2] The energy of an orbital is always given as a negative number, indicating energy is needed to remove the electron from the atom or molecule.

[3] $\ell$ is sometimes called the azimuthal quantum number.

[4] Strictly speaking, the orbitals shown in Fig. 8.1 are mixtures of the $m_\ell = -1$, 0, and 1 orbitals.

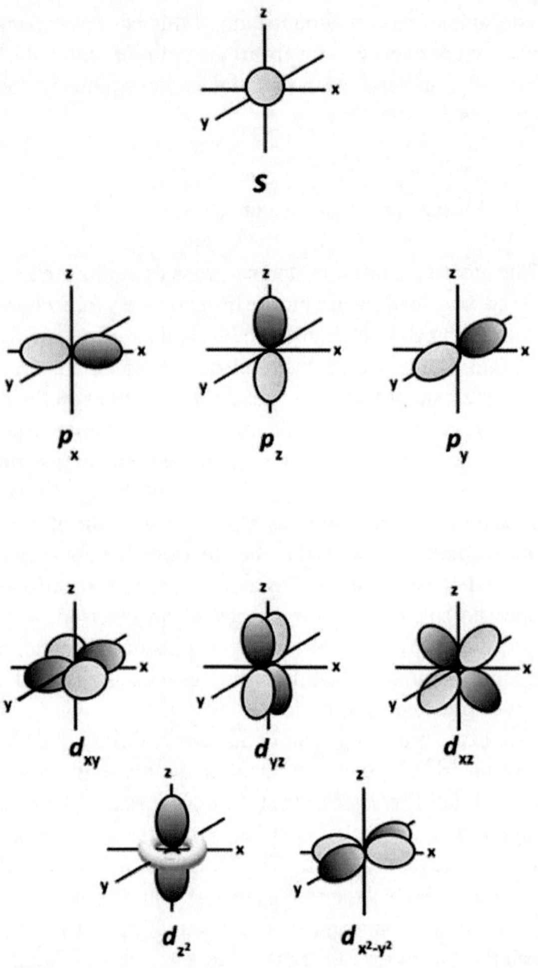


**Fig. 8.1** Shapes of the *s*, *p*, and *d* atomic orbitals

it, each electron in an atom can be uniquely identified by its four quantum numbers—the three that describe shape, orientation, and energy of the orbital, and the one that describes the electron spin.

The origin of an atom's energy is electrostatic and arises from the attraction between the oppositely charged electrons and protons. When only one electron is present, the energies of the atomic orbitals depend only on the n quantum number, with lower n values being the most stable (the lowest, or more negative, energy). In this case many orbitals have the same energy, a situation referred to as degeneracy. As $n$ increases, the orbitals become more diffuse, placing the occupying electrons further from the nucleus and decreasing the interactions between them and the positively charged nucleus.

The situation is more complex for atoms (or ions) that have two or more electrons. In addition to electron–proton attractions, we must contend with electron–electron repulsions. We can account for electron–electron repulsions through the following thought process: we begin with a bare nucleus, then add electrons to it, one at a time. Based on thermodynamic arguments, the electrons will occupy orbitals in order of their energy, with the lowest energy orbitals being occupied first (this is known as the Aufbau principle). These orbitals are the ones that are closest to the nucleus, and we refer to the electrons in these orbitals as the core electrons.

As more electrons are added, the lowest energy (core) orbitals fill, and the additional electrons must be added to higher energy orbitals (called the valence orbitals[5]). The core electrons, being closer to the nucleus, shield the valence electrons from the nucleus. This decreases the attractive energies between the valence electrons and the protons. Now the shape of the orbital (determined by the $\ell$ quantum number) becomes important because different orbital shapes penetrate the core electrons to different extents. This leads to an energy separation between orbitals with the same n number but different $\ell$ numbers. The ordering of the orbitals, from low energy to high, is 1*s*, 2*s*, 2*p*, 3*s*, 3*p*, 4*s*, 3*d*, 4*p*, 5*s*, and so on.[6]

A convenient way to depict the energies and shapes of atomic orbitals is through an energy level diagram, such as that shown in Fig. 8.2 for a fluorine atom. Here orbitals are placed according to energy, with the most negative at the bottom. Occupation of an orbital by an electron is shown as an arrow, the direction of which (up or down) indicates the spin quantum number $m_s$. Note that there are theoretically an infinite number of atomic orbitals, but only those occupied or readily available for occupation are shown.

## The Electronic Structure of Molecules

The concept of chemical bonds between atoms is a fundamental building block of science [1]. These bonds create molecules, and reactions that occur between atoms or molecules do so by the creation of or, in some instances, the destruction of, chemical bonds. Bonds are formed between atoms when they share electrons. This sharing of electrons can lower the total electronic energy of the system, which is a thermodynamically favorable process.

Individual atoms within a molecule retain much of their electronic nature. This is especially true for the core orbitals, which are physically closest to the nucleus and so overlap (and interact) very little with orbitals on other atoms. However, the highest energy occupied orbitals of the different atoms (the valence orbitals) are more diffuse and so can interact with one another to a degree. It is the perturbation of valence atomic orbitals that leads to the formation of bonds between atoms.

[5] These are sometimes called the frontier orbitals.

[6] In this orbital designation, the numeral refers to the principal quantum number n and the letter to the angular momentum quantum number $\ell$.

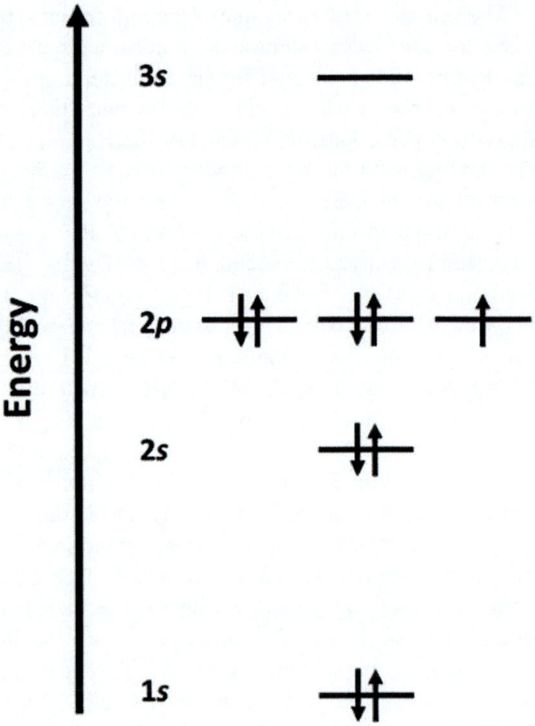


**Fig. 8.2** Energies and ground-state occupation of the atomic orbitals for fluorine

When atomic orbitals of similar energy on different atoms are spatially close to one another, they combine with one another to form new orbitals. These are termed molecular orbitals, and they follow the same quantum mechanical rules as atomic orbitals. They have specific shapes and energies, and they can hold no more than two electrons each (one of each spin state). In addition, there is a conservation principle for orbitals—one molecular orbital is formed for each participating atomic orbital.

The energies of molecular orbitals depend on the way that the atomic orbitals combine. The shapes and energies of the simplest molecular orbitals—those made from the combination of *s* orbitals on two like atoms—are shown in Fig. 8.3. In this molecular energy diagram, the energies of the atomic orbitals on the two isolated atoms are shown on either side, and the energies of the molecular orbitals made by their combination are shown in the center. The diagonal lines indicate which atomic orbitals are used to form a given molecular orbital in this figure. Only one atomic orbital is shown for each atom for the sake of clarity.

In Fig. 8.3 we see that one of the two molecular orbitals has a lower energy than the constituent atomic orbitals, and the other has a higher energy. We call the former "bonding" orbitals and the latter "anti-bonding" orbitals. When only two electrons are present in the atomic orbitals (one electron from each atom), the bonding molecular orbital fills, and the net energy of the molecule is decreased. This creates a chemical

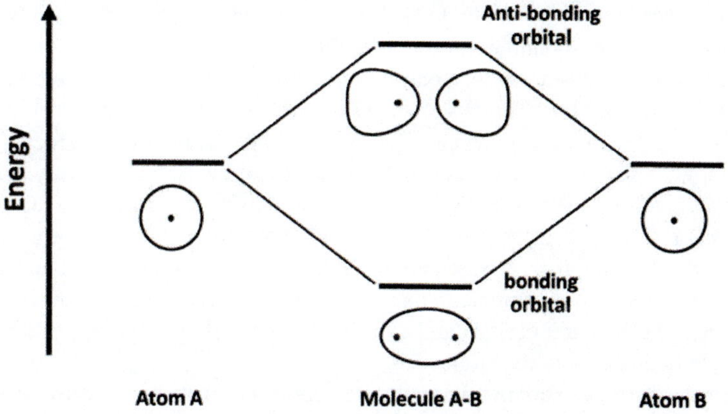


**Fig. 8.3** Combination of *s* atomic orbitals on two atoms to form molecular bonding and anti-bonding orbitals

bond between the different atoms, stabilizing the molecule. If the atomic orbitals each contained two electrons, then the anti-bonding orbital would also be occupied and the energy benefit of the electrons moving from the atomic orbitals into the bonding molecular orbital is slightly more than offset by the energy penalty of the electrons moving into the anti-bonding orbital. In this case no chemical bond will be formed.

The molecular orbitals shown in Fig. 8.3 have equal contributions from each atomic orbital, and so the electrons in the molecular orbital will be shared equally by the two atoms. If the two atoms bonding together are not the same, such as in the case of carbon monoxide, then the atomic orbitals will not participate equally in each of the molecular orbitals. In particular, the low energy (bonding) molecular orbitals will be composed of a higher proportion of the lower energy atomic orbital, and the high energy (anti-bonding) molecular orbital will have a higher proportion of the higher energy atomic orbital. When only the bonding orbitals are filled, this results in a charge separation with a partial negative charge accumulating on the atom with the lower energy atomic orbital. Depending on the symmetry of the molecule, a dipole moment can result from this partial charge separation.[7] In the extreme case, the "bonding" orbital can be composed of nearly 100% of the lower energy atomic orbital, leading to complete charge separation and the formation of a salt. In a salt the different atoms are held together through simple electrical charges (ionic bonding) rather than by the sharing of electrons (covalent bonding).

The energy difference between the bonding and anti-bonding orbitals is determined by the degree of spatial overlap between the atomic orbitals and by the energy difference between them. The greater the spatial overlap between the atomic orbitals,

[7] A dipole is the separation of electric charges. A dipole moment is the measure of the electric field created by a dipole.

and the closer they are in energy to one another, the greater the difference between the bonding and anti-bonding molecular orbitals.

Because atoms have more than one atomic orbital, molecules have more than one set of molecular orbitals. This is shown in Fig. 8.4, where we see two fluorine atoms combine to give a fluorine molecule. The energies of the 1*s* and 2*s* orbitals on the two atoms align with one another. Because they are core orbitals, they are very compact and so there is little spatial overlap between the orbitals on different atoms. Therefore there is very little energy separation between the resulting molecular orbitals. The 2*p* orbitals have significantly greater spatial overlap and so the molecular orbitals derived from them show a greater range of energies. In this diagram we see that there are three filled bonding orbitals and two filled anti-bonding orbitals, resulting in a net single bond between the atoms.

We have so far considered diatomic molecules, but the vast majority of matter that we interact with is composed of molecules containing more than two atoms. In this case molecular orbitals generally form from the atomic orbitals on adjacent atoms. When there are more filled bonding orbitals than filled anti-bonding orbitals, a chemical bond will exist between the atoms, just as is the case for diatomic molecules. These bonds are generally localized—that is, they are located around and between the pairs of nearby atoms.

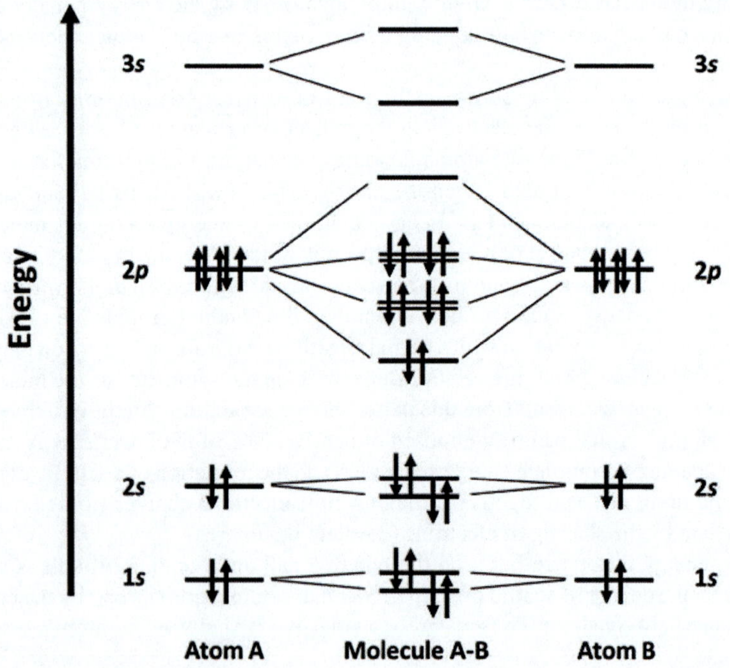


**Fig. 8.4** Molecular orbitals for diatomic fluorine. Y-axis not to scale

Most discrete molecules, both organic and inorganic are held together by these localized bonds (also referred to as "discrete" bonds). The main exceptions are salts, which are held together by electrostatic attractions, and certain organic molecules, which will be discussed below.

The most important orbitals in a molecule are the valence orbitals because they determine the chemical nature and reactivity of the molecule. These are the highest occupied molecular orbital (the HOMO) and the lowest unoccupied molecular orbital (the LUMO). These orbitals are often responsible for the color, if any, of a molecule. The mechanism for this is discussed in the next section.

## Electronic Transitions

The Aufbau principle specifies that electrons fill orbitals in order from lowest energies to higher energies, and this gives us the most stable state, or the ground state, of the atom or molecule. However, electrons can be moved (or promoted) to empty (or partially empty) higher energy orbitals if the atom or molecule is given the amount of energy that exactly matches the difference in energy between the ground state orbital and the higher energy orbital. The promotion of an electron therefore leads to a higher energy, electronically excited state of the atom or molecule.

The amount of energy necessary for electron promotion is usually greater than that available from ambient thermal energy. Energy high enough to promote an electron is often supplied as light energy. As an example, light in the visible region of the spectrum can be absorbed by some atoms and molecules quite readily, putting the absorbing species into an electronically excited state. Most often, the electronically excited species quickly shed its excess energy in the form of heat. Overall, this results in the transformation of light energy into heat energy.

The exact wavelength of light needed to promote an electron in a molecule (which determines the color of that molecule) and the probability that that wavelength will be absorbed (which determines color intensity) are determined by the electronic structure of that molecule. There are certain classes of molecules that have common electronic structures, and so absorb light of similar (but not identical) energies.

We might expect materials to have extremely sharp absorptions since the energy of the absorbing light must exactly match that needed for an electronic transition. However, this is not generally the case because the absorbing energy can simultaneously change the electronic, rotational, and vibrational energies of the molecule. Like electronic energies, these two latter forms of molecular energy are quantized. Rotational energies are the lowest of these, and on their own their states can be changed by absorbing microwave radiation. Vibrational energies are higher and fit in the infrared region of the light spectrum. When the light of a wavelength is close to, but not exactly equal to, the energy needed for an electronic transition, it can be absorbed by simultaneously changing a rotational and/or vibrational energy state.

When the HOMO is a bonding orbital and the LUMO an anti-bonding orbital, as is the case for diatomic fluorine (Fig. 8.4), the promotion of an electron from the HOMO to the LUMO results in the bond between the atoms being broken. The energy benefit of the electron in the partially occupied bonding molecular orbital (compared to the energy of an electron in the atomic orbital) is completely offset by the energy penalty of the electron in the anti-bonding molecular orbital, and so there is no net bonding interaction between the two atoms.

This condition—the HOMO being a bonding orbital and the LUMO and anti-bonding orbital—occurs routinely in organic compounds, particularly those with localized chemical bonds. The absorption of light by these compounds generally results in bond breaking. The energies of the bonds between the atoms in most organic compounds (i.e., carbon–carbon bonds, carbon–hydrogen bonds, carbon–oxygen bonds, etc.) correspond to the energy of ultraviolet light photons.

This is consistent with our everyday experiences—most organic compounds, such as solvents or pigment-free plastics, are colorless, and they decompose when exposed to ultraviolet light. Some inorganic compounds, such as many oxides and carbonates, are also colorless, but are not affected by ultraviolet light. This is because the energy difference between the HOMO and LUMO is greater than the energy of ultraviolet light. A notable exception to this is titanium dioxide, which is colorless but strongly absorbs UV light [2]. The implications of this absorption on the durability of paints and plastics are discussed in Chap. 14.

## Delocalized Bonds and Aromaticity

In addition to localized bonds between pairs of adjacent carbon atoms, some organic compounds have delocalized bonds that extend over many carbon (and sometimes other) atoms. Such bonds occur when there is a chain or ring of coplanar carbon atoms. Localized single bonds form between adjacent carbon atoms, but each of these atoms also has a *p*-type orbital projecting out of the plane. The *p*-type orbitals combine to form bonds that include all of the atoms within the chain or ring, and electrons found within the resulting molecular orbitals are delocalized over these atoms.[8] This delocalization provides additional stability (lower energies) to these orbitals.[9]

A simple example of delocalization is seen in benzene. Benzene is a hexagonal ring of six coplanar carbon atoms that are each bonded to a hydrogen atom and to two neighboring carbon atoms. The *p*-type orbital that projects from each carbon atom join together into six molecular orbitals. The shapes and energies of these molecular orbitals are shown in Fig. 8.5. Here we see that there are three bonding orbitals, all

---

[8] Note that only those ring systems that satisfy the Hückel rule ($4n + 2$ electrons in the *p*-type orbitals, where $n$ is an integer) can benefit energetically from these systems. For benzene, $n$ equals one and there are six electrons in the delocalized orbitals.

[9] Based on quantum mechanical principles (more specifically, the "particle in a box" analysis), the spatial confinement of electrons generally increases their energies. By delocalizing molecular orbitals over a number of atoms, electrons have a greater region of occupancy and so their energies decrease. This added energetic stability generally increases with the number of atoms participating in the delocalized network, since this increases the occupancy region of the electrons (i.e., the size of the "box").

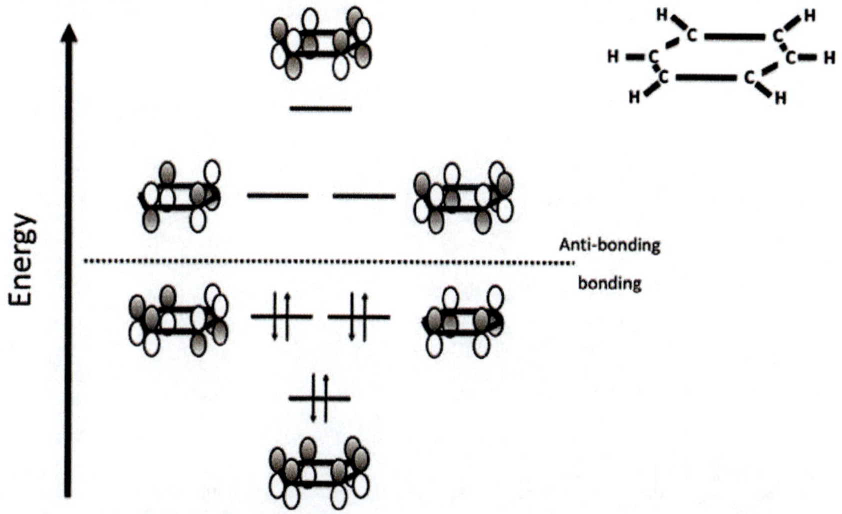


**Fig. 8.5** Delocalized molecular orbitals on benzene. Inset shows a stick structure of this molecule

filled, and three anti-bonding orbitals, all empty. We also see that the total energy of this system is lower than that of a series of alternating single and double bonds between the carbon atoms.

Organic compounds that contain delocalized bonds such as those in benzene are termed "conjugated" or "aromatic", although the latter term is based on an early misconception these materials have a stronger odor than non-aromatic compounds with similar compositions (it was, in fact, impurities in these materials that gave them their distinct odor). When the aromatic rings are found within a large molecule, they are referred to as aryl groups.

Molecules with two (naphthalene), three (anthracene), or more edge-sharing six-membered rings can also be aromatic, as can ring molecules for which some carbon atoms are replaced with heteroatoms such as nitrogen (e.g., pyridine) or oxygen (e.g., furan). Aromaticity is seen in straight-chain hydrocarbons for which there are alternating single and double bonds (e.g., β-carotene). Finally, molecules containing aromatic rings joined by a single atom or an atom chain with alternating single and double bonds can have molecular orbitals that are delocalized over the entire molecule, rather than systems of isolated delocalized regions (this is the case for crystal violet and aryl azo compounds, both of which are discussed later).

Delocalized bonds are sometimes depicted using "resonance structures". These show a molecule as a superposition of localized double bonds. Examples are shown in Fig. 8.6 for benzene (which has two resonant structures), naphthalene (which has three), and anthracene (which has four). These individual representations do not indicate stable isomers that contain alternating double and single bonds. Instead, the actual structure is an average of the individual resonant structures, and all of the bonding interactions are equivalent (i.e., each carbon is joined to its neighbors with a

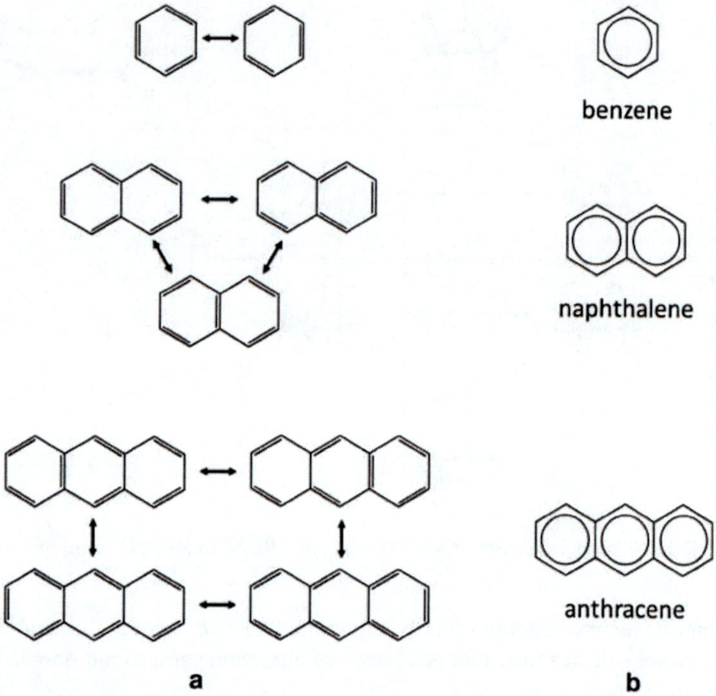


**Fig. 8.6** Depictions of electronic structures for benzene, naphthalene, and anthracene. **a** Resonance structures. **b** Averaged structures

bond order of 1½). In aromatic ring compounds, the average structures are typically indicated by circles within the rings (Fig. 8.6b).

While the HOMO and LUMO energy differences for most carbon–carbon localized bonds correspond to ultraviolet light energies, the HOMO and LUMO energy differences in aromatic compounds span a wider range of values and can extend into the visible spectrum. In general, the shift towards visible light increases with an increasing number of atoms that participate in the network of delocalized bonds. For example, benzene derivatives are generally not colored whereas many anthracene derivatives are.

One way of increasing the number of atoms participating in delocalization is to make larger rings. However, it is much more common to do this by making polycyclic compounds that consist of linked or fused rings (such as naphthalene and anthracene, Fig. 8.6) [3]. For this reason, nearly all organic color pigments are polycyclic aromatic compounds (discussed below).

A crucial aspect of visible light absorption by an aromatic compound is that, although it promotes an electron from a bonding orbital into an anti-bonding orbital, it does not break a chemical bond. Instead, the effect of the promotion from a bonding orbital to an anti-bonding orbital is spread through the entire delocalized system.

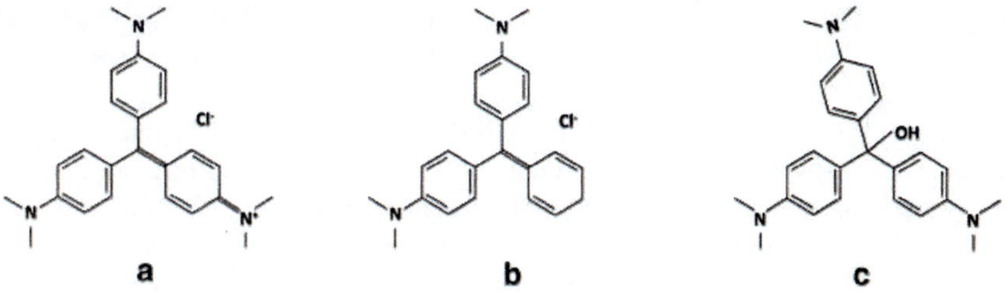


**Fig. 8.7** Chemically similar molecules. **a** Crystal violet. **b** Malachite green. **c** Crystal violet carbinol base. Only one resonance structure is shown for each molecule

Light absorption therefore decreases the average bond order of all of the bonds by a proportional amount, rather than destroying an individual bond between two atoms. For example, the promotion of an electron into the LUMO of benzene (see Fig. 8.5) lowers the average bond order between the carbon atoms from 1.50 to 1.33, rather than lowering the bond order between a single pair of atoms by one (i.e., rather than breaking an individual bond).

This is very important because when the excited aromatic molecule sheds its excess electronic energy as heat, it returns to its initial ground state and so is unchanged by the absorption process. Rather than being consumed by light absorption, the molecule is able to absorb light over and over again. This can be contrasted with the destruction of localized bonds via ultraviolet light absorption. Here individual bonds are broken and new chemical species, which do not have the same light absorption properties, are formed. In this case, ultraviolet light absorption is an irreversible, one-time event.

For aromatic pigment molecules, both the wavelength of their light absorption (color) and the strength of this absorption (color intensity or tint strength, as described in Chap. 13) can be quite sensitive to small changes in molecular structure. An example of this sensitivity can be seen in crystal violet, also known as C.I. Methyl Violet 10B,[10] and its close relatives—malachite green (also known as C.I. Basic Green 4) and crystal violet carbinol base. These three compounds are shown in Fig. 8.7a through c, respectively.

Crystal violet is composed of three benzene rings, each with an amine group, linked by a central carbon atom. The central carbon atom plays a dominant role in the electronic structure of this molecule. This atom has a *p*-type orbital that projects out of the plane of the molecule, and this orbital links together the *p*-type orbitals on each of the three substituted benzene rings, resulting in extensive delocalization of the valence molecular orbitals over the entire molecule.[11]

When dissolved in water, crystal violet is intensely purple, but the substitution of one amino group by a hydrogen atom results in a molecule (malachite green,

[10] C.I. stands for Colour Index™. This system of naming pigments is described later in this chapter.

[11] Ideally the three benzene rings in crystal violet and malachite green would be coplanar, as this maximizes delocalized bonding, but due to steric constraints these molecules have a propeller or paddlewheel structure.

Fig. 8.5b) that, when dissolved in water, is an intense green. Attaching a hydroxyl group to the central carbon atom of crystal violet breaks the uninterrupted network of *p*-type orbitals and results in the colorless material (crystal violet base, Fig. 8.5c). This transforms the delocalized network containing 19 carbon atoms into three smaller, independent networks of only six carbon atoms—that is, the electrons on each aryl ring remain delocalized, but the sets of delocalized bonds on the three rings are isolated from one other. As is the case for benzene itself, the difference in energy between the HOMO and LUMO orbitals in the three isolated six-membered rings is too high for visible light absorption, resulting in a colorless compound.

An important implication of the high degree of sensitivity of light absorption to the exact structure of the molecule is that colored pigments can quickly lose their intensity if they are attacked by a single reactive species. This results in color fade occurring early in exterior exposure, at the beginning of the degradation process, as opposed to later in exterior exposure after a significant change in molecular structure occurred. This is discussed below and in Chap. 14.

## Simultaneous Light Absorption and Light Scattering

As a prelude to our discussion of color pigments, we will point out that we often consider light absorption and light scattering by particles as mutually exclusive phenomena. For example, Chaps. 3 and 4 were devoted solely to light scattering, and our focus in this chapter is on light absorption. However, both occur simultaneously, to different degrees, in any interaction between a particle and light. This results in a distinction of color pigments as being either opaque or transparent. Opaque particles absorb and scatter light in roughly the same proportions, while light scattering in transparent particles is negligible. The degree of light scattering desired for a given application will determine which type of color pigment is used.

## Pigment Families

There are many ways to categorize pigments, the most obvious being by color. This is, in fact, the convention normally followed for naming a pigment, as described in the section on pigment nomenclature at the end of this chapter. A second useful way to classify pigments is on the basis of chemical type. The broadest such classification is by chemical nature—organic or inorganic. Within each of these types are specific chemical families; for example, azo organic pigments or complex inorganic color pigments. We will review a few members of each of these broad families in the following sections. This review will be brief since our interest is in the particulate properties of these materials, rather than their chemical form. That said, understanding the chemical nature of these materials is important for understanding many of their particulate properties. Comprehensive information about these families of pigments is available elsewhere [4–10].

## *Comparing Pigments to Dyes*

Before discussing the various pigment families, we must define exactly what we mean by the designation of a material as being a color pigment, and how a color pigment differs from other colored materials (specifically, dyes). Both dyes and pigments are classified as "colorants" since both confer color. Color is typically achieved through light absorption (discussed above), although other optical phenomena are used in some special-effects pigments (discussed below).

The property that distinguishes dyes from pigments is solubility. Simply stated, dyes are soluble in the media in which they are incorporated while pigments are not. Solubility is actively avoided in paint applications because dyes leach from paint films when wetted. Plastics applications can use dyes that are compatible with the polymer type [11]. These are generally used when a bright and transparent appearance is desired since dyes are invariably too small to scatter visible light.

Dyes can be made insoluble (i.e., converted into pigments) in three ways. First, ionic dyes can be precipitated as insoluble salts using certain counterions. For basic dyes, metal cations such as $Na^+$, $Ba^{2+}$, $Ca^{2+}$, etc., are used, while for acidic dyes, polyoxo anions (e.g., phosphomolybdate) are used. In some cases, dyes that are not ionic can have charge-conferring groups attached to them (such as carbonate or sulfate), making them ionic. These materials are commonly referred to as toners.

The second way of transforming dyes into pigments is to adsorb them onto a colorless, insoluble particle. The dye adsorbing solid is typically barium sulfate (blanc fixe) or a hydrated aluminate such as alum. These pigments are referred to as lakes. They are little used as commercial pigments today but are of historical importance.

Finally, vat dyes can be used as pigments in their own right [5]. A vat dye is used to color a textile fabric but is, in fact, a pigment, being an insoluble colored material. These materials can be used as dyes by first chemically reducing them in water into soluble, typically colorless or near colorless, forms. These dissolved molecules are then attached to the textile, followed by re-oxidization to their original colored state.

## *Organic Pigments*

Organic color pigments are very important in coatings and plastics applications [5]. Organic chemicals are generally defined as those chemicals that contain carbon–hydrogen bonds (In most cases there is more than one carbon atom, with the carbon atoms joined together through single, double, or triple carbon–carbon bonds). Additional atom types may also be present, most commonly oxygen, nitrogen and/or sulfur, in order of prevalence. Each of these atoms can be bonded to the carbon and hydrogen atoms, or to each other, in a number of ways, each of which forms a different chemical functionality. Compounds with similar functionality are grouped into families such as alcohols, aldehydes, ketones, amines, acids, etc.

The majority of organic pigments fall into three broad families based on their chemical functionalities—azo pigments, phthalocyanine pigments, and other polycyclic pigments. Each of these families can be further divided into closely related compounds. In each case the basic molecules can be modified by substituting different groups onto the molecular skeleton. These groups are called auxochromes.

## Azo Pigments

Azo pigments [12–14] are quite numerous and account for significantly more than half of all organic color pigments. The defining characteristic of an azo compound is that it contains a pair of nitrogen atoms that are joined by double bonds. Each nitrogen atom is also bound to another organic moiety, or "R" group, resulting in an R–N = N–R′ segment in the molecule. This linkage is bent at each nitrogen atom, due to a filled atomic orbital projecting in the same plane as the R and R′ groups:

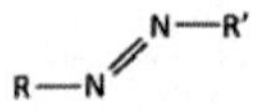


A fourth atomic orbital projects above and below each nitrogen atom in the R–N = N–R′ segment and, when R and R′ are aryl groups, these projecting orbitals participate in electron delocalization, combining the delocalization present in the separate R and R′ groups into a single, extensive aromatic molecular orbital system that has many resonant forms (similar to crystal violet, discussed above). Azo pigments can have a single azo group (monoazo pigments), or two azo groups (diazo pigments). A few representative azo pigments are shown in Fig. 8.8.

The exact structure of these molecules has been debated over the years, as they can exist as tautomers [13]. A tautomer is a form of isomer that differs in the type of organic functionality. "Azo" pigments can be in the azo/enol form or the hydrazone/keto form. These forms are shown in Fig. 8.9 for C.I. Pigment Red 112.[12] Historically, these pigments were assigned the azo/enol form, thus their names, but it is now thought that most exist in the hydrazone/keto form [5, 13]. For this reason, they are more appropriately called hydrazone pigments. That said, the azo designation, both in terms of pigment name and structural representation (e.g., Fig. 8.8), is still more common than the hydrazone designation.

Many azo colorants are water-soluble dyes, but these can be converted into insoluble pigments through the attachment of ionic groups onto the molecule, such as

[12] Note that this is fundamentally different from the resonance structures of aromatic molecules. In that case a single true structure exists that is an average of the resonance structures, while each form of tautomer is a unique molecule. That said, interconversion of tautomers is often facile.

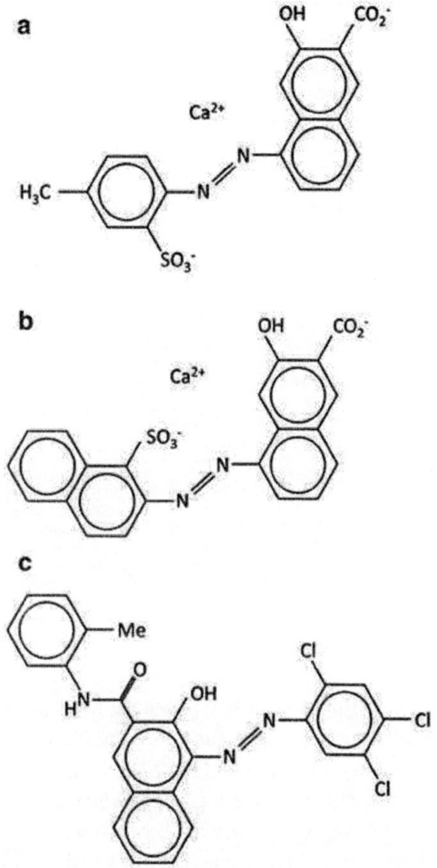


**Fig. 8.8** Examples of monoazo pigments. **a** C.I. Pigment Red 57:1. **b** C.I. Pigment Red 63:1. **c** C.I. Pigment Red 112. Note in all cases the azo/keto form is shown

sulfate (–$SO_3^-$) or carboxylate (–$CO_2^-$), followed by their precipitation as a metal salt (discussed above). The identity and location of the ionic groups, and the metal used to form the salt, affect the exact color of the pigment. Although azo pigments can be made for nearly any color, only those in the yellow–orange–red color range are typically produced commercially—the green and blue color spaces are better served by phthalocyanine pigments, which have lower cost, high color strength, and better lightfastness than most azo pigments.

## Phthalocyanine Pigments

Phthalocyanines are the second most important class of organic pigments, accounting for between a quarter and a third of all color pigment production. In many ways phthalocyanines are the ideal pigment for paint and plastics applications. They are lightfast, have strong tint strength (so much so that they are normally combined

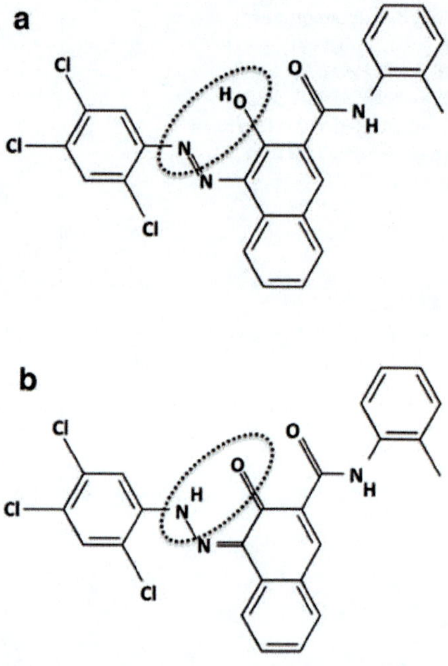


**Fig. 8.9** Tautomers of C.I. Pigment Red 112. **a** Azo/enol form. **b** Hydrazone/keto form

with white or yellow pigments), have high thermal and chemical stability, are non-toxic, are relatively easy to disperse (particularly surface treated grades), and are relatively inexpensive [15–17]. Because of this combination of desirable properties, these pigments have replaced nearly all other organic blue and green pigments.

The defining property of these pigments is the phthalocyanine molecule (Fig. 8.10a), which can be modified in two ways to make different hues. First, as is the case with other organic molecules, substitutions can be made to one or more atoms that are attached to an aromatic ring. This is normally the substitution of one or more hydrogen atoms with chlorine and/or bromine atoms. This shifts the color from blue to green, or even yellow green. Second, the two hydrogen atoms in the center of the molecule are acidic and can be replaced with cationic metals [18]. Over 66 metal complexes are known [19], but the only metal phthalocyanine pigment of significant commercial interest is copper phthalocyanine (Fig. 8.10b).

As noted above, the colors of phthalocyanine pigments vary from blue to yellow green and are controlled by the replacement of the ring hydrogen atoms with different amounts of chlorine or bromine atoms. In general, the shift to green increases with increasing replacement and is greater for bromine than for chlorine. However, particle size and crystal phase are also important. Particle sizes below 0.15 microns are available and provide the greatest color strength. These molecules can assemble in solid particles in more than one way (Fig. 8.11), leading to polymorphism, which is the occurrence of more than one crystal structure (discussed below). These different crystalline forms have different color, tint strength, and dispersion properties [16].

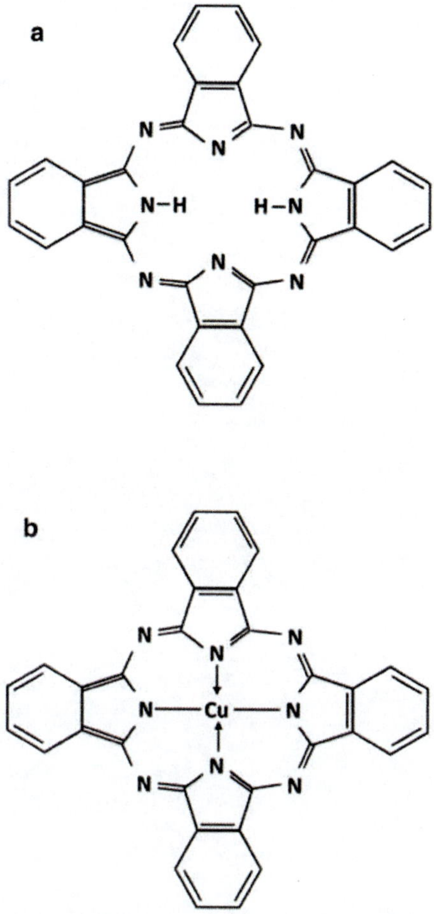

**Fig. 8.10** Structure of phthalocyanine pigments. **a** Metal free phthalocyanine (C.I. Pigment Blue 16) and **b** copper phthalocyanine (C.I. Pigment Blue 15). The arrows in the copper phthalocyanine structure indicate dative (electron donor) bonds

## Other Polycyclic Pigments

Nearly all organic pigments contain more than one aromatic ring (i.e., are polycyclic), including the azo and phthalocyanine pigments. Because of the importance of these two pigment families as color pigments, each was covered separately above. Here we will discuss some of the other, less prominent, polycyclic aromatic pigments.

The first of these is the anthraquinone [20]. Anthraquinones are used extensively as dyes, where they are second in importance only to azo dyes. However, they are sometimes used as pigments (vat dyes). Anthraquinone itself is colorless, but derivatives containing electron donor groups such as hydroxy or, more commonly, amino groups, are colored. An example of such a pigment is indanthrone (C.I. Pigment Blue 60), which is a fusion of two 2-aminoanthraquinone molecules (Fig. 8.12).

a

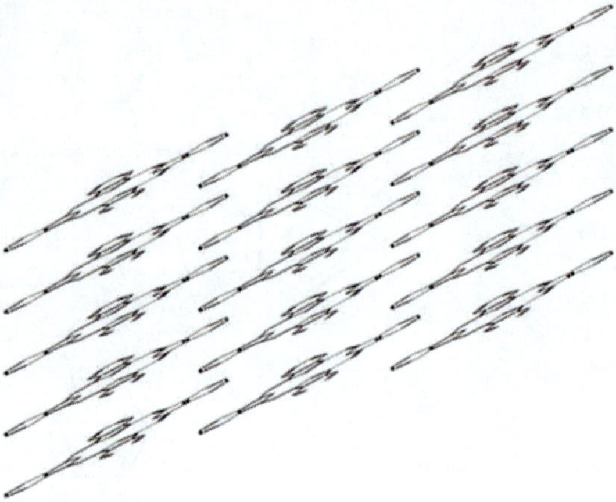

b

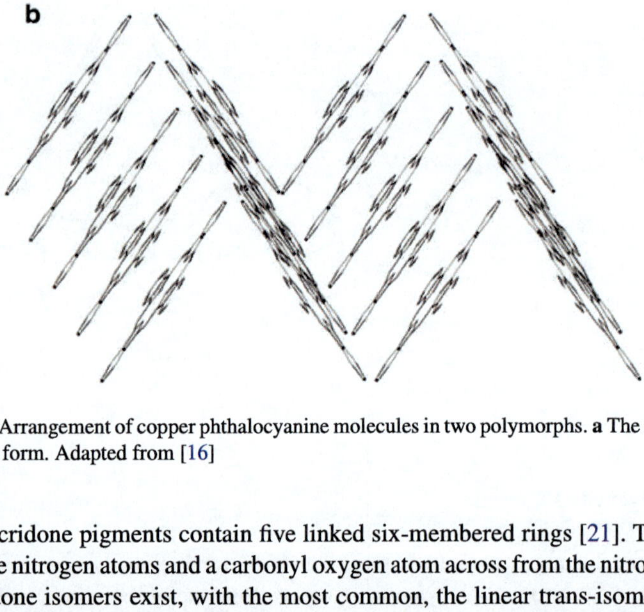

**Fig. 8.11** Arrangement of copper phthalocyanine molecules in two polymorphs. **a** The alpha form. **b** The beta form. Adapted from [16]

Quinacridone pigments contain five linked six-membered rings [21]. Two of the rings have nitrogen atoms and a carbonyl oxygen atom across from the nitrogen. Four quinacridone isomers exist, with the most common, the linear trans-isomer, shown in Fig. 8.13. Various quinacridone derivatives are known for their lightfastness and as such are found in high-performance exterior applications such as automotive and exterior paints.

Perylenes are also used as high-performance pigments [20, 22]. Perylene itself is colorless, but many of its amine derivatives are colored. Perylene and a representative pigment made from it are shown in Fig. 8.14. Most perylene pigments are derivatives based on nitrogen-containing rings on either end of the molecule, such as seen in Fig. 8.14b for C.I. Pigment Red 179.

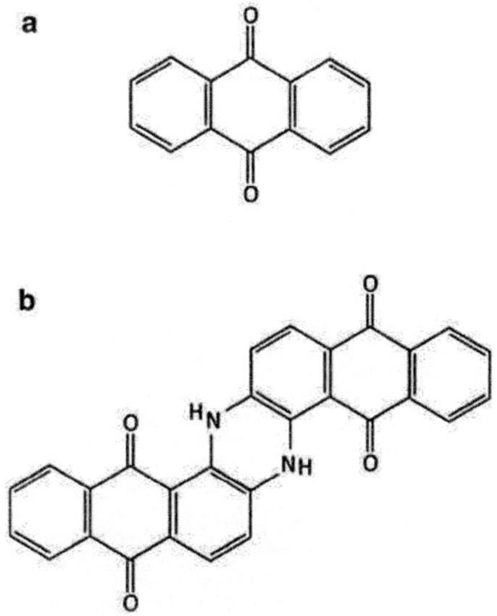


**Fig. 8.12** Molecular structure of **a** anthraquinone and **b** C.I. Pigment Blue 60

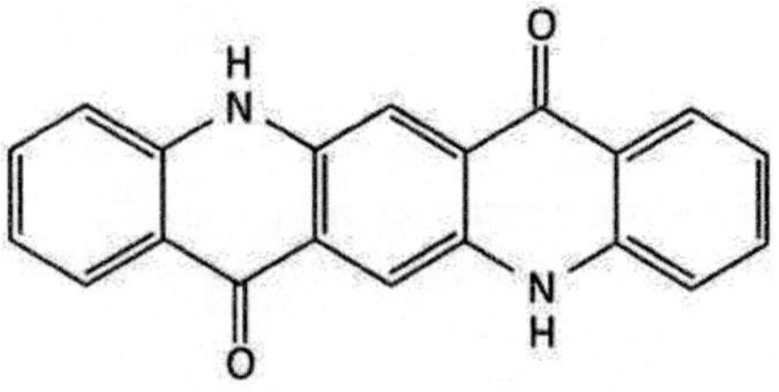

**Fig. 8.13** Quinacridone (C.I. Pigment Violet 19)

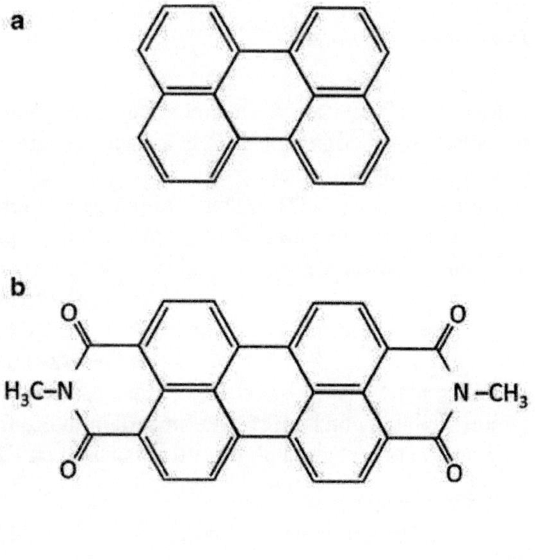


**Fig. 8.14** **a** Perylene and **b** C.I. Pigment Red 179

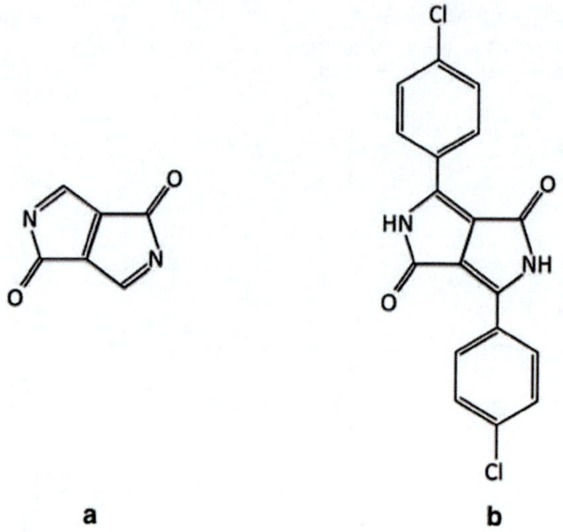


**Fig. 8.15** **a** DPP structure. **b** C.I. Pigment Red 254. This was the first DPP pigment commercialized

The final polycyclic pigment class that we will discuss is the relatively new family of pyrrolo[3,4-c]pyrrole-1,4-dione derivatives (also known as diketopyrrolopyrrole, or DPP, pigments) [23]. The molecular structure of DPP and one of its derivatives, CI. Pigment Red 254, are shown in Fig. 8.15. These materials were developed as high-performance pigments in the 1980s [24]. They have good lightfastness and stability, high chromaticity, and cover a gamut of colors centered in the red/orange region of color space.

## *Inorganic Pigments*

Inorganic pigments are characterized as having one or more metal ions in their structures. Most commonly this is a transition metal, which are those metals that have a partially filled set of *d*-orbitals. As is the case for organic pigments, the electrons in the HOMOs of standard inorganic pigments can be promoted by the absorption of visible light.[13] There are two main types of promotion: simple *d* to *d* promotion, and charge transfer between the metals and atoms close to them.

Simple *d* to *d* promotion occurs when the promoted electron both begins and ends on the same metal atom. These reactions are formally forbidden by the rules of electron promotion, one of which is that the $\ell$ quantum number, which is 3 for all *d* electrons, must change by one for electron promotions within an atom, at least for symmetric compounds. However, in certain molecules these promotions do occur. An example of a colored pigment that is based on a *d* to *d* promotion is potassium

[13] Color is produced in some inorganic pigments through light interference effects, as discussed later.

cobalt nitrite (also known as cobalt yellow, Indian yellow, or Aureolin), with the chemical formula $K_3[Co(NO_2)_6]$. Ruby, while not a pigment, also gets its color from a *d* to *d* promotion; this one occurs in isolated chromium atoms that are doped into an alumina matrix. Some colored oxides, such as chrome green (chromium(III) oxide), also derive their color from *d* to *d* promotion.

The electron involved in *d* to *d* promotions remains within the excited metal atom. In charge transfer, on the other hand, a promoted electron is exchanged between the metal atom and an atom (or chemical species) to which it is attached. When the metal is present as an oxide, hydrous oxide, or sulfide, the attached atom is oxygen or sulfur. These atoms bridge the metal cations. When the metal is present instead as part of a discrete molecule, the attached species are referred to as ligands, and these can include cyanide, ammonia or amines, water, and certain organic species. The electron transfer can go in either direction (from metal to ligand or from ligand to metal), depending on the accessible oxidation states of the metal. In most instances the transfer is from the attached species to the metal cation, resulting in a partial (and temporary) reduction in the metal oxidation state.

In systems with more than one type of metal atom (or with the same type of metal atom present in different oxidation states), strong light absorption can occur through the transfer of an electron from one metal atom to another. An example of this is iron blue (also called Parisian Blue and Berlin Blue), which has the molecular formula $(Fe_4{}^{3+})(Fe^{2+}(CN)_6)_3$. The intense blue color of this pigment is due to the transfer of an electron from a $Fe^{2+}$ atom to a $Fe^{3+}$ atom. Note that the quantum rules restricting *d* to *d* transitions do not apply in this case since the *d* atomic orbitals are on different metal atoms.

## Natural versus Synthetic Inorganic Pigments

The use of inorganic pigments predates recorded history. Wall paintings in Spain that date back approximately 36,000 years were created using both charcoal (a rudimentary form of carbon black) and various shades of red iron oxides.

Although various colored minerals have been ground for use as pigments for many centuries, the most prevalent natural pigments are various iron oxides that range in color from red to orange to black. These include umbers (which contain between 5 and 20% manganese dioxide) and sienna. These pigments have traditionally been used as "earth tones" in paints as well as ceramics and cement. Due to their inferior performance compared to synthetic pigments, their use today is limited to inexpensive, low-value paints, cements, and several other minor applications.

In contrast, the use of synthetic inorganic pigments is widespread in paint and plastics. These pigments can be synthetic analogs of colored minerals or new materials that are not found in nature. Synthetic pigments are preferred over their natural counterparts due to their cleaner hues and their consistent color and physical properties. Product consistency is achieved through purification of the raw materials and the ability to control particle size, shape, and crystalline structure during manufacture.

Inorganic color pigments can be classified by using a number of criteria. Distinctions are usually made based on whether the material is an oxide or hydroxy oxide, or by the metal(s) involved (except for carbon black, which is essentially pure carbon). Our discussion of the different inorganic pigments is organized in the latter manner (i.e., by metal content).

## Iron Oxides

Iron oxides are the most widely used inorganic color pigments. Several forms of iron oxide can be used as pigments, differentiated primarily by the oxidation state of the iron (i.e., the ferrous +2 state or the ferric +3 state), the presence of water or hydroxide, the crystalline structure, and particle size. Most iron oxide pigments are synthetic versions of known minerals, including hematite ($\alpha$-$Fe_2O_3$), goethite ($\alpha$-FeO(OH)), lepidocrocite ($\gamma$-FeO(OH)), and magnetite ($Fe_2O_4$), the latter being used as both for its pigmentary and magnetic properties.

The hues of various iron oxides vary from red (hematite) to orange (lepidocrocite) to yellow (goethite) and black (magnetite), all being relatively pure [25]. Brown can also be formed by mixing various proportions of the red, yellow, and black forms, or by introducing low concentrations of another metal (e.g., manganese) into $Fe_2O_3$. Hue is affected by particle size, with goethite changing from green yellow to brown yellow and lepidocrocite from yellow to orange with increasing size, and by particle shape (rods are preferred for yellow goethite).

Transparent iron oxides are the only true inorganic transparent color pigments.[14] They are typically yellow, orange, or red, depending on the exact chemical form. The red oxide is important as it is the only transparent red pigment that can withstand the temperatures of typical plastics processing. These materials can be made through a precipitation route (discussed below) or by the decomposition of iron pentacarbonyl ($Fe(CO)_5$). The precipitation route initially produces a hydrated yellow oxide (goethite structure) that can be transformed into the anhydrous yellow oxide (hematite structure) by heating.

Iron oxides typically have good opacity (except for the transparent forms) and high tint strength. All iron oxide pigments have high light stability.

## Chromium(III) Oxide

Chromium(III) oxide, $Cr_2O_3$, is an example of a strictly synthetic pigment, as it does not exist in nature in concentrations high enough for pigmentary purposes [7]. This is an important green pigment for paints and is valued for its high stability. Its use in plastics is more limited due to its low color intensity. It is used in military camouflage because it reflects infrared light as well as visible light.

[14] Colorless transparent inorganic pigments can be made from colorless materials such as $TiO_2$ and ZnO. These are often used as UV light absorbers in applications such as sunscreens.

## Cadmium Sulfide

Cadmium sulfide pigments span the color range from yellow to red. Color differences are due to differences in particle size as well as the presence of zinc or selenium atoms. When selenium is added at levels below approximately 10%, an orange color is formed, whereas higher selenium levels produce a red color. Both the orange and the red pigments have particle sizes on the order of 1.0 micron. Pure CdS pigment is yellow and has a particle size of roughly 0.4 microns.

## Complex Inorganic Color Pigments

Inorganic color pigments often contain more than one metal atom type. Such materials are called mixed metal oxide pigments or, more commonly, complex inorganic color pigments (CICPs). CICPs are not simply physical mixtures of two pigments that have different metals, but instead are single compounds containing the two (or more) metals combined at the atomic level.

CICPs typically take one of three crystalline structures, each of which is commonly found as minerals [26]. These are the hematite structure ($M_2O_3$, where M is a metal), rutile structure ($MO_2$), and spinel structure ($M_3O_4$). Note that the average oxidation states of the metal atoms in these three structures are different. The metals used include zinc, iron, chromium, nickel, titanium, and cobalt, and their colors cover a wide range from yellow to green, blue, and black.

As a class, CICPs show high stability to chemicals, ultraviolet radiation, and heat. This stability can be understood based on the manufacturing process for these materials. These pigments are made by heating mixtures of the appropriate starting materials to temperatures between 800 and 1,300 °C [26]. Clearly, materials produced at such temperatures will be stable at even the most aggressive temperatures seen in paint or plastic manufacture. Similarly, CICPs are stable towards reactions with other chemicals because, after being heated to such temperatures, they are typically in their thermodynamically lowest energy state. These materials also have the benefit of high infrared and solar reflectivities (see Chap. 19), which differentiate them from most organic pigments.

## Carbon Black

Carbon black particles are produced through the incomplete combustion of organic molecules such as oil, natural gas, or acetylene. They are typically much smaller than other inorganic pigment particles, and even smaller than many organic pigments. Primary particle size is the principal means of differentiating grades, ranging from a few nanometers to half a micron. The size and structure of agglomerates are of equal importance to the size of the primary particles themselves because these agglomerate structures can survive the dispersion process. While the property of agglomerate size is obvious, the concept of "structure", as applied to carbon black pigments, is also important.

Structure refers to the degree of openness of the agglomerate. Some agglomerates are compact, with an aciniform, grape-like cluster, shape, while others have a linear, usually somewhat branched, arrangement. Pigments with the latter condition are referred to as having "high structure". Structure affects dispersibility and viscosity in liquid carriers (i.e., paints), as well as the color sensation of the pigment.

There are three color-related aspects of carbon black pigments. The first is jetness, which refers to the degree to which the pigment deviates from achromaticity. This can be characterized by the b* value for the material (see Chap. 6). High jetness indicates a lack of chromaticity or a "pureness" of black. The second measure is the degree of blackness. This indicates how dark the pigment is and is a measure of the reflectance of the material under complete opacity conditions. Finally, the tint strength of the pigment is a measure of how much light a unit amount of material absorbs. This is measured as the reflectance value of a paste consisting of a mix of carbon black and zinc oxide and is made on a comparative basis to the reflectance value of a standard or reference with a defined tint strength of 100.

Due to their small size and intrinsic hydrophobicity, carbon black pigments can be notoriously difficult to wet and disperse into paints and plastics. Dispersion can be enhanced by introducing a small number of oxygen functional groups (alcohols, aldehydes and ketones, and carboxylic acids) onto the particle surfaces. This is done either chemically or through partial oxidation of the particle surfaces.

## *Comparison Between Organic and Inorganic Pigments*

There are many general trends and properties that differentiate most organic pigments from most inorganic pigments. Several of these are listed below. The reader is cautioned, however, that these are generalizations and not strict rules, and as such there are exceptions to each.

In general, organic pigments have a lower hiding power (or opacity) and higher transparency than inorganic pigments. There are two reasons for this. First, the refractive indices of organic pigments are closely matched to those of the polymer surrounding them, since these are all hydrocarbons. This reduces the light scattering strength of the particles. In contrast, the refractive indices of inorganic color pigments can be quite high—even exceeding that of $TiO_2$. Second, the particle sizes of organic pigments are typically smaller than those of inorganic pigments (the notable exception being carbon black, which is considered inorganic and includes some of the smallest sizes for all pigments). Organic pigments are generally too small to effectively scatter visible light while inorganic pigments are often of the correct size to maximize light scattering.

The smaller sizes of organic particles give them higher surface areas and, in coatings applications, higher oil absorption values, greater dispersant demands, and higher liquid paint viscosities [22]. These can all limit their concentrations in paints.

The smaller sizes also make these particles more difficult to disperse in either paints or plastics and more difficult to stabilize against flocculation in paints.

Pigment density is lower with organic pigments than inorganic pigments. Coupled with their larger sizes, this makes inorganic pigments more susceptible to settling during periods of quiescence.

The saturation, or color purity, of organic pigments is generally greater than inorganic pigments. While there are inorganic pigments with high chromaticity, most have dull or muted colors.

Inorganic pigments are generally more stable against heat, oxygen, and light. This is because organic pigments, like all hydrocarbons, are fundamentally unstable in an oxidative environment such as the atmosphere, particularly when exposed to ultraviolet radiation. The thermodynamically most stable forms of carbon and hydrogen are carbon dioxide and water. In contrast, the atoms in most inorganic pigments, particularly the oxides or hydrous oxides, are already in their fully oxidized state and so do not react with atmospheric oxygen. In addition, many inorganic pigments are produced through calcination at temperatures in excess of several hundred degrees, and so are inert under ambient conditions and at temperatures encountered in paint of plastics production. Finally, some inorganic pigments do not absorb ultraviolet light, so they are immune to its effects.

Finally, there are certain differences between organic and inorganic pigments that are not physical or chemical in nature. The costs of organic pigments are generally higher than inorganic pigments because of the more extensive processing required to produce them (discussed below). There are also more color choices for inorganic pigments than organic pigments, particularly when stability is required. That said, over the past few decades there has been an overall shift away from inorganic pigments and to their organic counterparts due to the generally higher saturation and greater tint strength of organic pigments.

## *Special Effects Pigments*

In our previous discussions on light scattering (Chaps. 3 and 4), we assumed that scattering would ultimately randomize the direction of the light (i.e., would render it diffuse). In most applications this is desired as it makes the object appearance constant with respect to the viewing angle. This avoids optical effects that would be distracting or otherwise objectionable.

In some applications, however, such effects are desired as they add a visual dimension to appearance, often described colloquially as "pop". Examples include a metallic appearance, or a change in color with viewing angle, either at a small scale (pearlescence) or at a large scale (flip or flip-flop). In many cases the color is not formed through selective light absorption, as it is for a conventional color pigment, but rather by light interference, which is a special case of selective light reflection. The formation of color by this mechanism is familiar to us when we view a thin layer of oil on water, which generates a rainbow effect that changes with both viewing angle and oil thickness.

Pigments that provide these unconventional color sensations are referred to as "special effects" (or simply "effects") pigments. There are two major classes of these pigments, which differ in the way they interact with light and the appearance characteristics that they confer. These are metallic pigments and pearlescent, or interference, pigments.

## Metallic Pigments

The metallic effect is created by light reflection below the paint or plastic surface [8]. This effect is created by using thin platelets of metal, usually stabilized aluminum [27], that are aligned with one another. In paints, this alignment is achieved as the film contracts during drying (Fig. 8.16), and is enhanced by a low solids content of the paint, since low solids paints experience greater loss of volume (greater contraction) on drying than high solids paints. A hydrophobic organic material, such as stearic acid, is typically coated onto these particles to bring them near the dry film surface, which further enhances the uniformity of particle alignment. This hydrophobic effect is greatest in coatings with a low polarity liquid carrier (i.e., solvent-borne paints or inks) and is more difficult to achieve in highly polar water-borne systems. Pigments with these hydrophobic coatings are known as leafing pigments.

In plastics, particle alignment is achieved along the flow lines developed as the molten polymer moves through the process system. Due to the significantly higher viscosity of a polymer melt compared to a liquid paint, there is no differentiation in alignment between leafing and non-leafing pigments.

The metallic effect is seen when the reflectance from the object (paint film or plastic article) is not entirely specular (i.e., not all at the mirror angle of the surface). Due to a slight misalignment of the different platelets, light scatters over a small range of angles centered at the mirror angle. This produces a "flop" effect, where the brightness of the surface changes significantly with viewing angle.

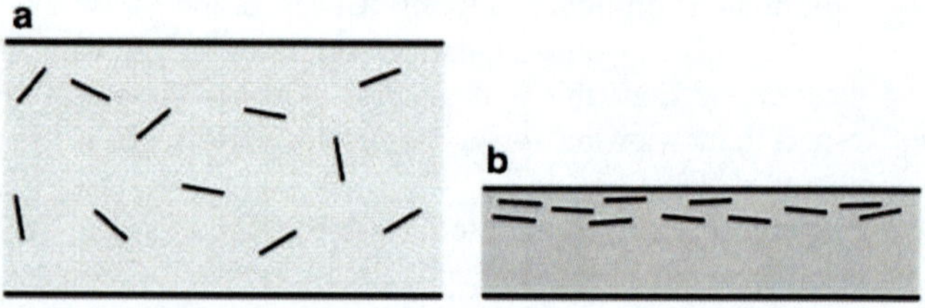


**Fig. 8.16** Metallic platelet alignment due to film contraction during paint drying. **a** Wet paint with randomly aligned platelets. **b** Dry film with aligned particles

Colored metallic paints use transparent pigments (see below). This is because non-transparent (opaque) pigments randomize the direction of light through scattering, while the metallic effect depends on the light all traveling in nearly the same direction.

## Pearlescent and Interference Pigments

Pearlescent, or interference, pigments confer two appearance attributes to a material [9]. First, they preferentially reflect different colors of light at different angles. Second, they can add a sensation of depth to the material, unlike conventional pigments, which have a solid appearance. Combined these create a pearl-like or nacreous appearance.

The defining attribute of pearlescent pigments is their ability to separate and preferentially reflect individual colors from white light. This is achieved through interference, which describes the interactions of light that are partially reflected from two parallel surfaces separated by a distance of similar dimensions to (but greater than) the wavelength of light. This phenomenon is shown in Fig. 8.17. Here light is partially reflected at the surface of a platelet. The portion of light not reflected is transmitted to the lower interface of the platelet, where a portion is reflected. When light rays 1 ($r_1$ in Fig. 8.17) and 2 ($r_2$) are parallel and in phase with one another (i.e., are from the same distant source), and are offset from one another by a certain distance, then the reflection of light ray 2 from the bottom surface meets the reflection of light ray 1 from the upper surface at point P in this figure. When they meet at point P, light ray 2 has traveled farther than light ray 1.

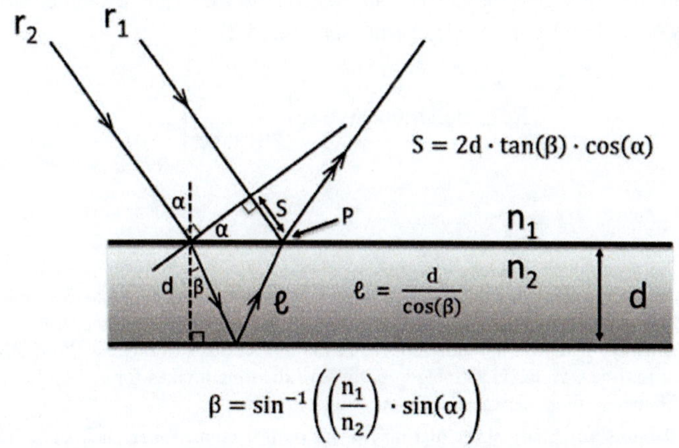


**Fig. 8.17** Interference of light reflected from parallel surfaces. $n_1$ and $n_2$ are the refractive indices of the two materials on either side of the surfaces

When the following condition is met,[15,16]

$$\frac{2n_2d \bullet \cos(\beta)}{\lambda} = m \tag{8.1}$$

where

- $n_2$ is the refractive index of the layered material,
- $d$ is the thickness of the layered material,
- $\beta$ is the angle indicated in Fig. 8.17,
- $\lambda$ is the wavelength of light in a vacuum (i.e., a medium with a refractive index of 1.0), and
- $m$ is an integer (that is, an even multiple of ½),

then the reflections of $r_1$ and $r_2$ are in phase when they meet; that is, they combine constructively and so the intensities of the two combining streams reinforce one another and there is a strong reflection for that wavelength of light ($\lambda$) and at that angle of reflection ($\alpha$). If, on the other hand, $m$ is a half integer (i.e., an odd multiple of ½), then the two light streams combine destructively and so cancel one another, resulting in the reflected intensity of that wavelength of light and at that angle being zero.

The overall effect of this interference is that the light reflected from the platelet will appear colored, with the wavelength of the reflected light (i.e., its color) depending on the refractive indices of the two materials,[17] the distance between the surfaces, and the angle of observation.

The amount of light reflected at a surface, in relation to the total light intensity, depends on the refractive indices ($n_1$ and $n_2$) of the materials on either side of the surface, as described by the Fresnel equation (Eq. 8.2)[18]:

$$\text{Reflected intensity} = \left[\frac{n_1 - n_2}{n_1 + n_2}\right]^2 \tag{8.2}$$

---

[15] Note that, to derive Eq. 8.1, the change in light wavelength that occurs when the light crosses from the medium into the platelet must be taken into account. It is, in fact, not the absolute distances of the two pathways that are important, but rather the time it takes to traverse those distances. If the difference in time traveled is an integer multiple of the time it takes for one cycle (that is, the inverse of frequency), then the waves combine constructively.

[16] Equation 1 omits any phase shifts that may occur on reflection. When such shifts occur, they must be account for in this equation. In particular, the time required for the light to undergo such a phase shift must be included in the travel times referred to in the previous footnote.

[17] Although $n_1$ does not explicitly appear in Eq. 8.1, it is a part of the relationship between the angles $\alpha$ and $\beta$ and so does play a role in this determination.

[18] Strictly speaking, Eq. 8.2 applies when the angle to the normal ($\alpha$ in Fig. 8.17) is zero. However, its deviation when the angle is greater than zero does not affect this argument.

As can be seen, the proportion of light reflected increases with an increase in the difference in refractive indices, and so platelets with high refractive indices show the highest level of pearlescence.

Pearlescent pigments are similar to metallic pigments as both are platelets with a preferred orientation relative to the object surface and both reflect the light that strikes them. Importantly, however, not all of the light that strikes an interference pigment is reflected from it. This allows the light to penetrate more deeply into the object and to reflect from different depths. This difference is shown schematically in Fig. 8.18. The deeper penetration of light into the object containing a pearlescent pigment, and its reflection from different depths within the object, confer a three-dimensional characteristic or appearance of depth to the object, which contributes to the pearlescent effect.

Since the wavelength of visible light is centered at roughly one-half micron, we would expect platelets with characteristic thicknesses on the order of a few microns to be most effective at creating this effect, and this is the case. It is not possible to manufacture these platelets with identical thicknesses; instead, a statistical distribution of thicknesses is seen. The platelet materials commonly used in these applications, and their refractive indices, are shown in Table 8.1.

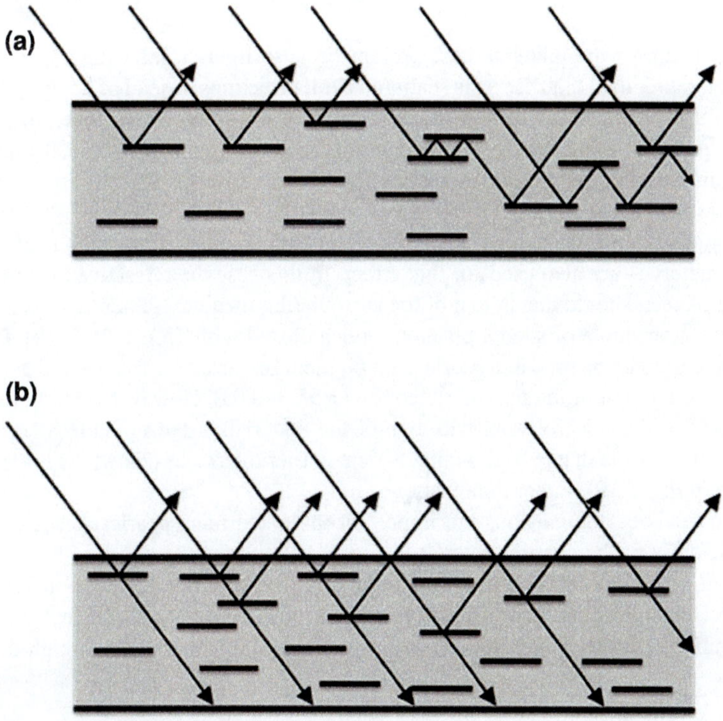


**Fig. 8.18** Reflection from platelet particles. **a** metallic pigments. **b** pearlescent pigments

**Table 8.1** Uncoated materials commonly used as pearlescent pigments

| Material | Refractive index |
|---|---|
| Pearl essence | 1.85 |
| BiOCl | 2.15 |
| $TiO_2$ | 2.73 |
| $Fe_2O_3$ | 2.80–3.01 |

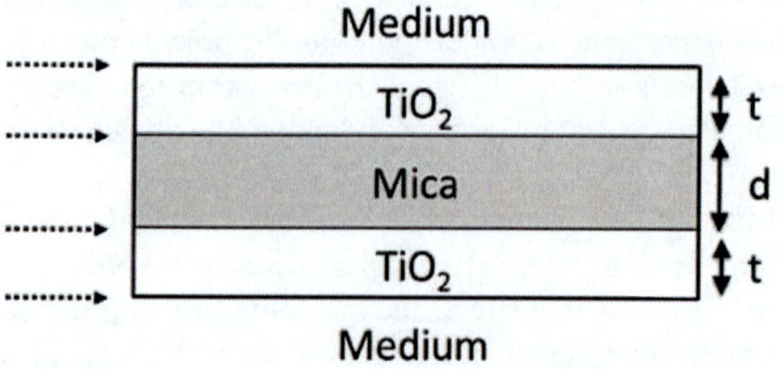


**Fig. 8.19** A pearlescent pigment composed of $TiO_2$ (refractive index 2.73) coated onto mica (refractive index 1.59) embedded in an organic medium (refractive index 1.55). Horizontal dashed arrows indicate interfaces between materials with substantially different refractive indices

The width of the platelets is important because light striking the edges of the platelet, rather than their faces, is scattered in all directions instead of being reflected at the mirror angle. The consequences of this scattering are greatest when the width of the particles is the least. Widths can vary within the range of 1–200 microns, depending on the desired appearance. Larger particles favor luster[19] while smaller particles favor opacity [28].

In addition to uncoated platelets, platelets that are coated with high refractive index materials are also used for this effect. In this case the refractive index of the coated platelet should match that of the surrounding medium, rather than be greater than it. An example of such a platelet is mica coated with $TiO_2$ (Fig. 8.19). On its own, mica cannot be used as a pearlescent pigment because its refractive index (1.59) is too close to that of the polymer medium ($1.55 \pm 0.05$). However, coating the top and bottom of a mica platelet with one of the materials listed in Table 8.1 gives a composite particle that partially reflects light at four interfaces (shown by horizontal arrows in Fig. 8.19), rather than just two.

Two thickness dimensions are important in coated mica pearlescent pigments: the thickness of the coating layer (*t* in Fig. 8.19) and the thickness of the platelet substrate ( d in Fig. 8.19). As mentioned above, the substrate thickness (d) cannot be precisely controlled and instead typically follows a statistical distribution. This distribution typically spans thicknesses from 0.3 and 0.6 microns, although thicker

---

[19] Luster is defined as the contrast between the light specularly reflected from one part of a surface and the light diffusely reflected from an adjacent area [28]. That is, high luster is seen when light is reflected strongly at the mirror angle (specularly), but weakly slightly off of this angle (diffusely).

particles (up to a few microns) are sometimes used. On the other hand, the thickness of the surface layer (t) can be tightly controlled. This thickness affects the color of the pearlescent pigment, and different grades of pigment are available with different coating thicknesses. For example, in the case of $TiO_2$ on mica (Fig. 8.19), as the layer thickness increases, the reflected color shifts from white or silver (thickness ~ 0.05 micron) to yellow, red, blue, and green (thickness ~ 0.14 micron) while the transmitted color shifts from white to blue, green, red, and yellow (i.e., the complement of the reflected color).

Synthetic silica, alumina, and borosilicate can be used in place of mica. They generally show less variation in thickness than mica, are clearer (mica has a yellow tone), and have lower refractive indices, enhancing the interference effect.

Interesting color effects can be seen when the coating layer is colored (e.g., $Fe_2O_3$). In this case the color of the platelet is determined by both light interference and absorption, and a bronze and copper appearance can be generated. In dual effect or combination mica pigments, a five-layer sandwich is made in which the $TiO_2$ coated mica is further coated with a colored (light-absorbing) layer. These complicated pigments are used primarily in cosmetics, where relatively small quantities can confer high value to the product, justifying the costs of the extensive production process required to make these pigments, but they are also found in high-value paints and plastics.

Interesting color effects are also seen when pearlescent pigments are used in the presence of conventional color pigments. In this case the phenomenon of color-flop, in which the color of an object changes with viewing angle, can be exhibited. Such an effect is often used in automotive coatings, where it both changes the color of a vehicle as it drives past the observer and it enhances the curves on the vehicle when at rest.

## Important Pigment Properties

As is true for all small particles found in paints, plastics, and paper laminates, there are many physical and chemical properties of color pigments that affect their behavior in their end-use application. The more important of these are discussed below, beginning with the defining aspect of these particles—color.

### *Color*

It is important to differentiate between the color of a pigment and the color of a paint or plastic. In Chap. 6 we described methods to measure and quantify the color of a paint or plastic, and we can apply these methods to pigments themselves. But there is a complication when doing so: In what state do we measure the pigment color? For example, we could press a pellet of the pigment and measure the color of that.

However, we would find this to be very dark, even for pigments that make a bright paint or plastic. In addition to low brightness, there can be a marked shift in hue when concentrating the pigment in this way.

To understand the various ways that pigment color can be measured, and how the results can differ from one another, we first need to discuss the light absorption process.

## Light Absorption

We noted earlier that the energy needed to promote an electron determines the color of a molecule, and that the probability with which that energy is absorbed by that molecule determines its color intensity or tint strength. This probability is characterized as the absorption efficiency of the material at that wavelength of light and is quantified as an absorption coefficient.

Pursuit of a scientific understanding of light absorption, and its quantification, goes back to the early 1700s. Bouguer reported that the amount of light transmitted through a translucent object was dependent on that object's thickness. This relationship between thickness and percent transmittance was not, however, linear (blue circles, Fig. 8.20). This is not surprising—a linear relationship would give a negative percent transmittance value for objects above a certain thickness.

Bouguer found that the decrease in transmittance was instead logarithmic, and so plotting the log of transmittance against thickness results in a linear fit (red triangles,

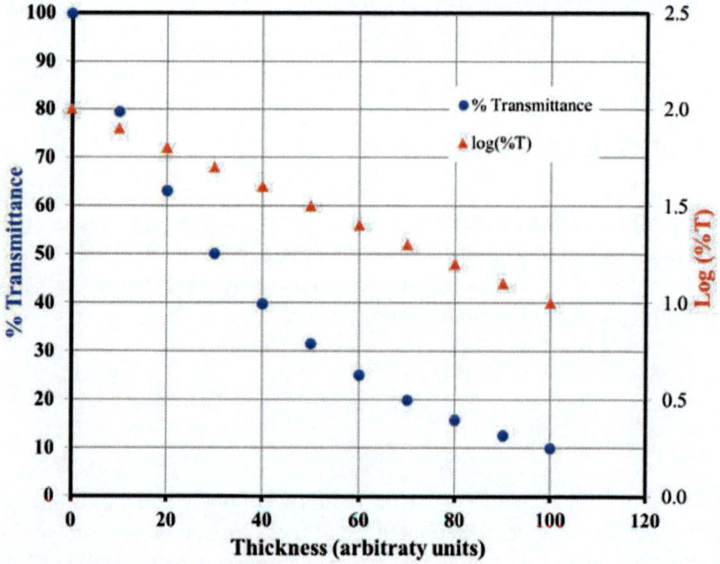


Fig. 8.20 % Transmittance and log (% transmittance) as a function of object thickness

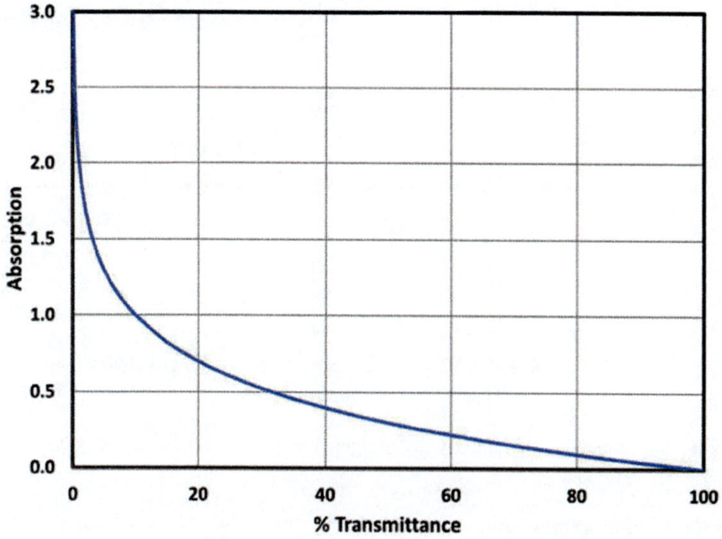


**Fig. 8.21** Relationship between A and %T

Fig. 8.20). This leads us to a different way of quantifying the light removed by atoms or molecules—absorption, rather than transmittance.

Absorbance and percent transmittance are related by the following equations[20]:

$$\%\mathrm{T} = 10^{(2-\mathrm{A})} \tag{8.3}$$

$$\mathrm{A} = 2 - \log(\%T) \tag{8.4}$$

If transmittance is expressed as a fraction rather than a percent, the number "2" is removed from these equations.

A chart of absorbance as a function of percent transmittance is given in Fig. 8.21.

Later, Beer showed that absorption was dependent on the concentration of the absorbing species in the same way that it is dependent on sample thickness—that is, absorption is linearly related to concentration. We can combine the findings that absorption increases linearly with pathlength and concentration, and depends on the absorption strength of the absorbing species, into the Beer–Lambert law:

$$A = \varepsilon bc \tag{8.5}$$

where

[20] Somewhat confusingly, the term "% absorbance" is sometimes used to describe the percent of light that is not transmitted (i.e., 100 − %T). This is not what we are referring to here and elsewhere in this chapter.

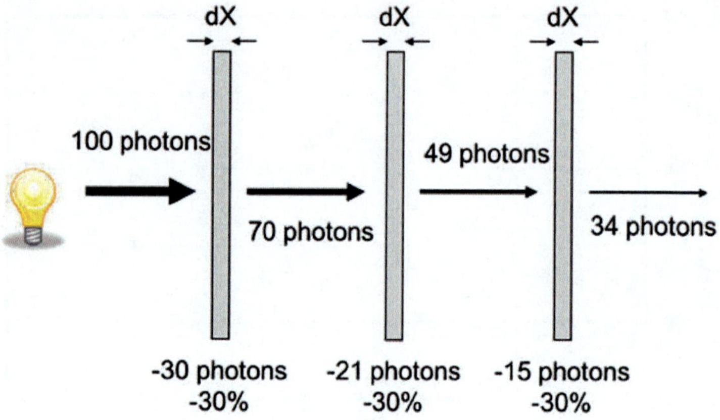


**Fig. 8.22** Proportional loss of intensity due to light absorption

*A* is the total amount of light absorbed,
*ε* is the absorption coefficient, sometimes called the extinction coefficient or molar absorptivity,[21]
*b* is the pathlength of the light through the materials, and
*c* is the concentration of the absorbing specials.

These relationships are consistent with the proposition that the amount of light absorbed by an object is a *proportion* of the incoming light, rather than an *absolute number* of photons. For example, in Fig. 8.22, we see that each thickness of the sample absorbs the same fraction of incoming photons (30%) while the absolute number of absorbed photons decreases as fewer and fewer photons remain in the beam.

The foundational basis for the Beer–Lambert law is that absorbance is determined by how much material the light passes through (said differently, by the number of photon/absorber species interactions). We can control this amount either through concentration or through the distance through which the light travels. This finding supports the proposition that species absorb light independently of one another.[22] This is very different from light scattering, where the scattering strength of a particle is a strong function of its proximity to other scattering particles, and, therefore, on its concentration (Chap. 4).

The non-linear relationship between A and %R (Fig. 8.21) can result in the color of a material shifting if the concentration of the colorant changes (*c* in Eq. 8.5) or if the thickness of the material changes (*b* in Eq. 8.5). This color shift occurs if the

[21] The units of ε depend on the units of b and c and are chosen such that A is unitless. b is typically given in units of cm and c is typically given in units of mol/l.

[22] The linear response between absorbance and concentration can break down for strongly absorbing systems and in situations where paint components chemically interact with one another.

light is absorbed in different regions of the visible spectrum with different absorption strengths.

This is best demonstrated by an example. Consider drawdowns of two paints for which the same color pigment is present, but at two different concentrations—in Paint X the concentration is one-tenth that in Paint Y. In this thought experiment, we will assume that the absorption coefficient of the pigment in the red region of the spectrum is ten times that in the blue region, and that the absorption coefficient in the green region is so high that essentially all green light is absorbed in both paints.

We will also, for the purposes of this example, assume that, at the thickness of the drawdown, the total light absorbed, A, in Paint X in the red region is 0.1, resulting in an A value for this paint in the blue region of 0.01 (one-tenth that in the red region). These correspond to %R values of 79.4% and 97.7% for the red and blue regions, respectively, in Paint X, giving a ratio of blue light to red reflected light of 1.23. Under these conditions the color would be considered a somewhat blue shade of purple.

Next, we consider Paint Y, with ten times the concentration of pigment as paint X, drawn down at the same thickness as above. This increases the A values to 1.0 and 0.1 for red and blue light, respectively. These A values correspond to %R values of 10.0% and 79.4%, respectively. Now the amount of blue light is 7.5 times that of the red light. Paint Y is darker than Paint X, but it also has a distinctly different color—it is blue, while Paint X is blue/purple. The overall effect is that by increasing the concentration of pigment, we have shifted the color balance, and therefore the color itself, without having changed the absorbance properties of the pigment.[23] This analysis underscores the importance of concentration and thickness when describing the color of a pigment. Examples of color changes that can occur for these and similar reasons are discussed in the next section.

## Masstone and Tint Tone

Because the perceived color of a pigment changes with concentration, pigment color is generally determined under two conditions. The first condition is a paint film or plastic object that has no other pigments and that is thick enough to be completely opaque. This color is referred to as the masstone of the pigment and is generally quite dark. The second condition for determining pigment color is to dilute a paint or plastic made with that pigment with a high concentration of white pigment. There is no standard ratio of color pigment to white pigment, so this must be specified when describing the color mix. Ratios of 10:90 or 5:95, color pigment to white pigment, are often used. The white pigment uniformly increases reflectance across the entire visible light range with an effect similar to decreasing the color pigment concentration. The resulting color is referred to as the tint tone of the pigment and is, not surprisingly, quite bright compared to the masstone.

[23] In Chap. 15 we will quantify the effect of colorant concentration on color.

**Fig. 8.23** **a** masstone and **b** tint tone (10:90 with $TiO_2$) of Nickel Azo Yellow (C.I. Pigment Yellow 150) [29]

In most instances there is a clear visual relationship between a masstone and tint tone. However, for reasons discussed in the previous section, this is not always the case. An example of this is the masstone and tint tone of nickel azo yellow pigment (C.I. Pigment Yellow 150), as shown in Fig. 8.23 [29]. Here the masstone has a distinctive green appearance whereas the tint tone is a pure yellow.

Masstone and tint tone can be conveyed through physical samples or by using the color scales typically used for paints and plastics (e.g., L* a* b*; see Chap. 6). Standard techniques for making these measurements have been developed by ASTM [30].

A hue shift is also sometimes seen when a black colorant is added to a paint containing a color pigment and a white pigment (this is referred to as "shading" the paint). This is in many ways the opposite of diluting a masstone with a white pigment to give a tint tone. Here we are increasing the total absorption of the paint over the visible light spectrum uniformly, whereas with the tint tone we are decreasing the total absorption of the paint over the visible light spectrum uniformly. Subsequently, the color shifts for the same reason as described above for masstone. This is shown in Fig. 8.24 for a different yellow pigment (C.I. Pigment Yellow 184) than shown in Fig. 8.23.

## *Transparency and Opacity*

When discussing white pigments (Chap. 7), our focus was on their ability to provide opacity through light scattering, while when discussing color pigments, we focus on light absorption. However, some color pigments both absorb and scatter light—in fact, the refractive indices of many inorganic pigments actually exceed that of $TiO_2$. For example, the refractive indices of iron(III) oxides vary from 2.80 to 3.01,

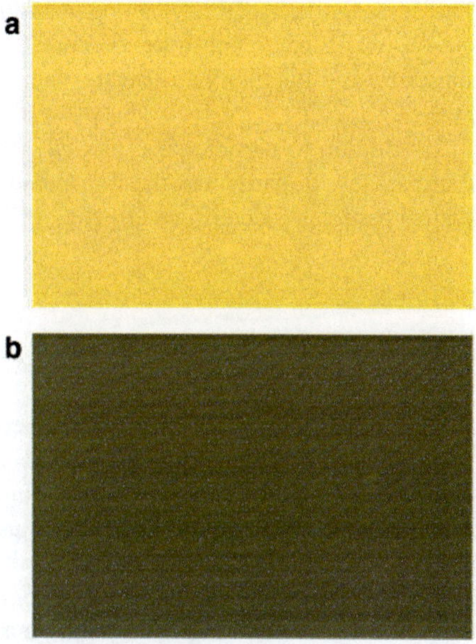


**Fig. 8.24** C.I. Pigment Yellow 184 in **a** a paint containing a white pigment (tint tone) and **b** a paint containing a black pigment (shade)

depending on the form.[24] Color pigments that also scatter light are referred to as opaque pigments, while those that do not are referred to as transparent pigments.

While light scattering benefits opacity, it is not always desired in a color pigment. In certain applications, it is advantageous for the pigment to allow the unabsorbed light to continue on its original path. Inks are an example of such an application. Inks are typically applied to an opaque surface, most usually the white surface of the paper, so do not themselves need to provide opacity. That is, it is the white paper, rather than the colorant in an ink, that scatters light back to the eye of the observer. By using transparent pigments, new colors can be created by applying layers of different color inks (an opaque ink would conceal any layers of ink beneath it and so would not create a new color).

Transparent pigments are also used in translucent or tinted plastic applications [31]. This allows for visibility through the object while at the same time imparting a color appearance. Transparent pigments are also often used in conjunction with effects pigments to achieve a metallic effect (discussed above).

Transparency is achieved by particle size control—transparent pigment particles are too small to scatter visible light (typically these particles are between 2 and 15 nm) [7]. Size control within this regime is more commonly seen for organic pigments, and the only important transparent inorganic color pigments are the transparent iron oxides and transparent cobalt blue. Phthalocyanines and quinacridones are the most transparent of the organic color pigments, while pyrroles are among the most opaque.

[24] For reference, germanium metal has perhaps the highest refractive index in the visible spectrum, with a value of 4.05.

The degree of transparency can be quantified based on the reflectivity of a paint or plastic over a black substrate, compared to the reflectivity of the uncoated black substrate, as a function of either the path length or the color pigment concentration [32].

In addition to particle size, particle shape and orientation are also a factor for transparency. In particular, the transparency (or, alternatively, the opacity) of rod-shaped particles is determined by their orientation with respect to the light source.

## *Lightfastness*

In the earlier discussion on electronic structure, we noted that ultraviolet light has the energy required to break the single bonds (primarily C–C and C–H bonds) that are found in organic molecules. As such, organic molecules are susceptible to breakdown when exposed to natural sunlight. The ample availability of atmospheric oxygen hastens this breakdown. As described earlier, the final dispositions of the carbon and hydrogen atoms in a degraded hydrocarbon are carbon dioxide and water, although fading of organic color pigments begins much earlier in the degradation process. The ability of an organic molecule to resist these degradation reactions is referred to as durability when discussing resins (see Chap. 14) and as lightfastness when discussing pigments.

Certain classes of organic pigments are more lightfast than others. Phthalocyanines, for example, have a high lightfastness whereas the lightfastness of many azo pigments is low. This reflects not only the rate at which the pigment degrades but also the sensitivity of the color to the change in the molecular structure of the pigment. In some cases, changing the bonding of a single atom can give a large change in both the color and the color intensity (tint strength) of a pigment. An example of this was shown in Fig. 8.7, where a change in the bonding of the central carbon atom of crystal violet (from being bonded to three other atoms to being bonded to four) resulted in a total loss of color. Another example of this is shown in Fig. 8.25, where

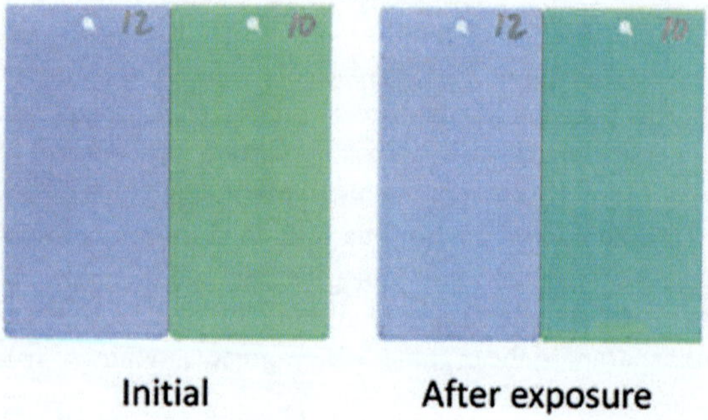


**Fig. 8.25** Color shift after exposure. The blue paint becomes lighter in color, reflecting the appearance of "chalk" on the paint surface, while the green paint becomes lighter and bluer, the shift in color reflecting the partial degradation of the pigment in this paint. This is discussed in more detail in Chap. 14

there is a clear shift in color of a yellow-green paint towards blue. This example is also discussed in Chap. 14.

As a general rule, lightfastness is most important in paints and plastics that are used in exterior applications and least important in inks, where exposure to sunlight is limited (e.g., in the pages of a closed book) or the application is not intended for long-term use (e.g., in a newspaper or advertisement).

## *Crystal Structure*

Atoms and molecules form solid particles when they benefit energetically from close interactions with one another. In general, there is one arrangement of atoms or molecules that is thermodynamically the most stable, and to optimize this stability the atoms and molecules repeat this arrangement throughout the solid particle. Particles exhibiting a regularly repeating arrangement of atoms or molecules are called crystals.

While there is typically one arrangement for which the energy is minimized for a given material, in some cases there are multiple arrangements for which the energy differences are small, leading to more than one crystal structure (phase) of that material. This is analogous to the creation of different two-dimensional patterns from a set of tiles. When this occurs, the different crystalline phases are referred to as polymorphs. Examples of polymorphism are the two phases of $TiO_2$ (anatase and rutile) and the existence of multiple phases of various iron oxides.

Polymorphism is common for planar aromatic materials, including many organic pigments because there can be strong interactions between the aromatic rings on different molecules. The extensive network of delocalized bonds in aromatic compounds is positioned above and below the plane of the atoms (refer back to Fig. 8.5). Inter-molecular stabilization can occur when the delocalized molecular orbitals stack on top of one another, and stacking is quite common for these materials. Polymorphs often exist that differ in the way that individual molecules arrange with respect to one another in each stack, and in the way that the stacks arrange with respect to one another. This was discussed briefly in the section on phthalocyanine pigments above and was shown schematically in Fig. 8.11 for two different arrangements of copper phthalocyanine stacks [16].

The alignment of delocalized orbitals that occurs when aromatic molecules are stacked on one another perturbs the molecular orbital energies of these delocalized bonds and so adjusts the wavelength of light absorbed by these materials. This accounts for color shifts that are often seen between different crystalline forms of a pigment.

Polymorphism is important in pigment applications because many properties of pigment particles are different for different crystal forms. These properties are both physical, such as density, solubility, lightfastness and particle shape, and optical, such as color and tint strength. Crystal phase can be controlled during particle formation by the presence of certain additives that promote or inhibit certain crystalline phases and by the presence of seed crystals.

Like the crystal phase, crystal size and shape can be controlled by certain additives. This occurs when the additive promotes or inhibits the growth of certain particle faces [16]. A molecule that adsorbs strongly onto a crystal face will inhibit the growth of that face. When this molecule adsorbs on all faces, the resulting particles are small since growth is completely hindered (this is used for particle size control—see the next section). If absorption is preferential to only one type of face, then the shape of the particle will be changed since this will slow the growth rate of that face while not affecting the growth rate of the other faces.

Particle shape is important because it can affect the hue of certain pigments as well as particle packing efficiencies in highly concentrated conditions, such as seen in colorant concentrates (for paints) and masterbatches (for plastics), as discussed below. Shape can also affect the overall hydrophilicity or hydrophobicity of the particle if one face is more hydrophilic than another. This is sometimes seen in copper phthalocyanine pigments [16].

Solubility is also affected by crystalline form since the strength of the interactions between individual molecules in a particle is an important factor in determining the solubility of that particle. Even slight solubility can be an issue because this can lead to the formation of new, more stable, crystalline phases of the pigment. This occurs because thermodynamically less stable crystalline phases are more soluble than thermodynamically more stable phases, and so a dynamic equilibrium is established for which molecules are preferentially dissolved from the less stable phase and redeposited onto the more stable phase.

Slight solubility can also lead to an increase in the particle size of existing crystalline phases. This occurs because the solubility of small particles is greater than the solubility of large particles of the same crystalline phase.[25] Under dynamic conditions, molecules on the particle surfaces continually dissolve and reprecipitate, with the rate of dissolution and reprecipitation depending on the particle solubility. Because solubility is greater for smaller particles than for larger particles, molecules preferentially dissolve from the smaller particles and preferentially deposit onto the larger particles. This results in the disappearance of small particles and the growth of large particles—a process known as Ostwald ripening [33].

Particle growth or phase change that occurs during paint preparation or storage are undesirable because they cause uncontrolled changes in the physical and optical properties of the paints. This leads to paint instability and to the color of a paint shifting over time.

Finally, we note that certain particles can affect the crystallization rate of molten polymer, which is a slow step in polymer processing, by increasing the rate of crystal nucleation of the molten polymer. Different crystalline forms of a pigment can affect this rate to different extents [34, 35].

---

[25] This is a manifestation of surface energy. The thermodynamic driver for this behavior is a decrease in overall surface area, with the attendant decrease in surface energy.

## *Particle Size*

Particle size affects many properties of color pigment particles. As is the case with any particle, particle size can affect the wettability, dispersibility, and dispersion stability of particles in a polymer medium (see Chaps. 11 and 12). In addition, the light absorption properties of particles, which in turn affect both the color and the color intensity (tint strength) of color pigments, are affected by particle size [25], as will be discussed next.

### Effect of Particle Size on Light Absorption

Before discussing the effect of particle size on light absorption, it is useful to review, for comparative purposes, the effect of particle size on light scattering (Chap. 3). In our discussion of white pigments, we saw that there is a complex relationship between particle size and the light scattering power of a unit weight of pigment. For particles larger than the wavelength of light, scattering occurs when light strikes the particle surface. Since the surface area of a unit weight of particles increases with decreasing particle size, the amount of light scattered on a mass basis increases with decreasing size in this size regime.

This relationship changes when particles are near in size to the wavelength of light. Under these conditions scattering can also occur when the light comes close to the particle but misses it (diffraction). This effectively increases the scattering cross-section of the particle beyond its physical cross-section and results in an even greater increase in scattering power per unit weight as the particle size decreases. However, when the particle size decreases to a value significantly below the wavelength of light, the fundamental nature of the scattering interaction changes, and not all of the light that strikes a particle is scattered by it (i.e., its scattering cross-section becomes smaller than its physical cross-section). In this size region, scattering strength per unit weight decreases rapidly with decreasing particle size.

This complex relationship between particle size and light scattering strength per unit mass of particles leads to there being an optimal particle size for light scattering (roughly half the wavelength of light). Particles above this size have a lower surface area per unit mass, and so intercept less light. Particles below this size have a fundamentally weaker interaction with light, and so, on a per mass basis, also intercept less light.

The factors that determine the effect of particle size on light absorption are in some ways similar to those that determine the effect of particle size on light scattering, and in other ways different. Light absorption occurs when light strikes a particle surface, which is similar to light scattering for particles significantly larger than the wavelength of light. However, unlike light scattering, the light absorption cross-section of a particle is nearly always the same as the physical cross-section, and so we would expect light absorption, on a unit mass basis, to increase linearly with decreasing particle diameter without limit. Unlike in light scattering, however, the amount of light absorbed by a particle also depends on the pathlength of the light through it (i.e., the particle thickness), and this decreases with decreasing particle diameter.

Overall, then, there are two opposing effects of particle size on the amount of light absorbed by a unit mass of material, as was the case for light scattering. However, the details of these effects are fundamentally different for the two forms of light/particle interaction. For light scattering, the two opposing effects are the amount of surface area and the inability of small particles to interact with light in a scattering fashion, while for light absorption the two effects are the amount of surface area and the distance traveled through the particle. We may reasonably wonder which of these opposing effects is more important for optimizing light absorption, or if a balance between required, as was the case with light scattering, which would lead to an optimal particle size for absorption.

We can address this issue through the following thought experiment. We will consider two blue pigment samples that differ only in particle size—the linear dimensions of one sample are three times greater than the other. The size of the smaller particles is such that one-half of the red light that strikes them is absorbed, and one-half passes through. Since the larger particles are three times as thick as the smaller particles, they will allow only one-eighth of the light striking them to pass through. This is because half the light striking the particle is absorbed by the top third of the particle, half of the remaining light is absorbed by the middle third, and half of that remaining light is absorbed by the bottom third.

In our thought experiment, we will disperse the same weight of these particles into two films. For simplicity, we will specify a film concentration such that the smaller particles occupy one half of the film surface, and we will focus on a region of the film that contains 27 of the small particles (Fig. 8.26a) or one of the large particles (Fig. 8.26b).[26] Since the large particle has only one-third of the cross-sectional area as the combined cross-sectional areas of the small particles, it only covers one-sixth of the film surface.

One-quarter (25%) of the red light directed at the film with the smaller particles is absorbed—half of the light strikes the particle surfaces, and half of that light is absorbed by them. However, only 7/48 (14.6%) of the light striking the surface of the film with the large particle is absorbed—one-sixth of the light strikes the particles, and seven-eighths of that light is absorbed.

The overall result is that, for this particular size, increasing the particle side length by a factor of three decreases the amount of light absorbed by 58.3% (0.583 = 0.146/0.25). This analysis can be extended to other particle sizes, and in all cases the light absorption strength of a mass of small particles is greater than for the same mass of large particles. Taken to the extreme, we can think of individual molecules

[26] Both of these regions are the same size, since they contain the same volume of particles and have the same concentration.

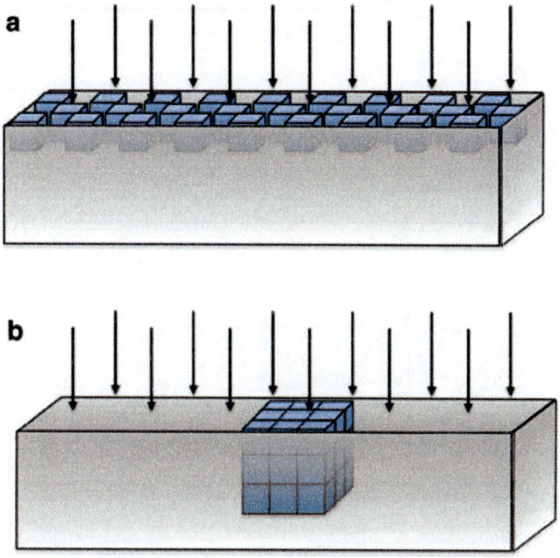


**Fig. 8.26** Light absorbed by an equal weight of **a** small particles and **b** large particles in films. Arrows indicate incident light. The total volumetric concentration is the same in these two films

that are dissolved in a liquid as being "particles" of vanishingly small dimensions, and they generally absorb light accordingly.[27]

The effect of particle size on pigment absorption strength, on a unit mass basis (e.g., the absorption per gram of pigment) is shown in Fig. 8.27. The Y-axis in this figure is the absorption strength of a particle relative to the absorption strength of an infinitely small particle (i.e., a dissolved molecule). The X-axis is the diameter of the particle relative to the diameter of a particle thick enough to absorb half the incident light (i.e., the size of the particles in Fig. 8.26a). For reference, this size is roughly 10 nm for copper phthalocyanine. As indicated by the thin black lines, a particle of this size absorbs only 72% as much light as an infinitely small particle (i.e., as a dissolved molecule), and a particle three times this size (i.e., similar to the particle size in Fig. 8.26b) absorbs only 42% as much light as a dissolved molecule (and, as discussed in context with Fig. 8.26, absorbs only 58.3% as much light as the smaller particles).

An example of the effect of particle size on light adsorption strength can be found in carbon black [4]. High color channel blacks, which are made by impinging a gas flame onto a metal plate, have particle sizes in the range of 5–15 nm. The absorption intensities, referred to as "jetness" for black pigments, of high color channel blacks are high. Furnace blacks, with particle sizes in the range of 50–200 nm, are made by the combustion of oil or gas, and have intermediate jetness, while lampblacks, made by burning oils or tars in a deficit of oxygen, have particle sizes in excess of 500 nm.

[27] This is not true when molecules within the particles affect one another's electronic structure, as was described in the previous section on crystal structure, or when the dissolved molecules interact electronically with solvent molecules.

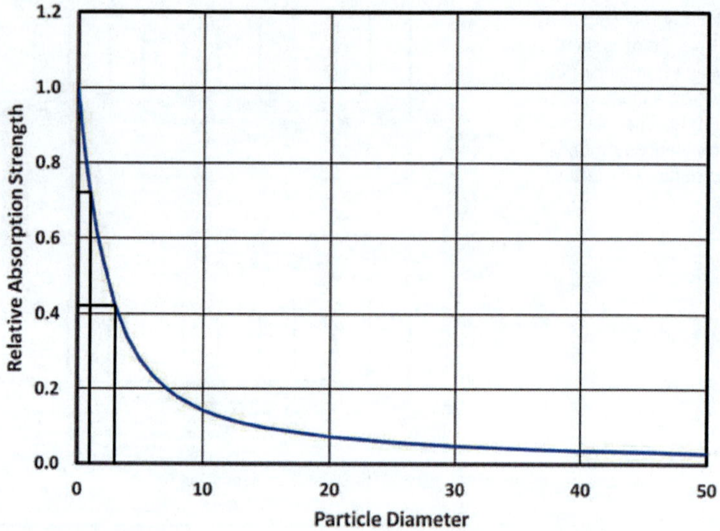


**Fig. 8.27** Theoretical effect of particle size on the amount of light absorbed by a unit weight of pigment

The jetness of lampblacks is so low that they are primarily used in gray, rather than black, paints or plastics.

The thought experiment described by Figs. 8.26 and 8.27 is strictly theoretical and there are many practical considerations that can alter these conclusions. In particular, although absorption efficiency increases with decreasing particle size, there are practical factors that limit the fineness of the particles. The large surface areas of small particles make them very difficult to disperse in either a plastic or a paint, and difficult to keep dispersed in a paint. This is particularly true for organic pigments, which often have low energy surfaces that are difficult to wet (see Chaps. 16 and 17). Small size particles also greatly increase the viscosity of any liquid into which the particles are incorporated when compared to large particles. This restricts the maximum concentration of small particles that can be used in a paint [22]. A practical balance of these different effects is reached for most color pigments in the submicron to nano particle dimensions.

## Controlling Particle Size

There are two fundamentally different ways of controlling particle size: by starting with large particles and creating smaller particles from them (grinding), or by starting with small particles (or dissolved molecules) and building larger particles from them (deposition). Both of these technologies are used for color pigments.

Grinding, or comminution, has been used for centuries to decrease the size of natural pigments, and today is also used to control the particle sizes of some synthetic pigments as well. This method of particle size control is generally less desirable than size control via deposition because it consumes much energy. It can discolor the pigment particles by abrasion with the sides of a metal wall or with the dispersing media, when present. It can chemically alter the particles due to the creation of very high local temperatures and pressures, and, finally, it can introduce particulate impurities from abraded media or beads. In addition, there is typically a diminishing return on grinding, and many particles cannot be ground below a certain size.

Alternatively, particle size can be controlled during the particle formation process. This strategy generally applies to particles formed by precipitation, although not always, as the above example for carbon black shows. Crystal size can be controlled by the concentration, temperature, pH, and rate of reagent addition during the reactions that produce the particles [33]. These factors affect the relative rate at which dissolved species precipitate as new crystals (i.e., the rate of particle formation) compared to the rate at which the dissolved species adds on to an existing crystal (i.e., the rate of crystal growth).

In general, particle growth is favored when the concentration of particles is high and when the degree of supersaturation is low. High particle concentrations favor existing particle growth over new particle creation because there are always particle surfaces close to a molecule when it deposits. A low degree of supersaturation favors particle growth because it decreases the likelihood of two molecules that are precipitating to encounter one another and create a new particle before they contact the surface of an existing particle and deposit there. A low degree of supersaturation is favored by long precipitation times, slow rates of reactant addition, and high temperatures. The degree of supersaturation is also affected by the order of addition and the manner in which the starting materials are combined. In addition to particles growing while they are initially being formed, they can also grow after they are formed by Ostwald ripening, as discussed above.

## *Dispersibility*

The dispersity of a pigment is of high importance because the light absorption efficiency of agglomerates or flocculates is less than that of individually dispersed particles [25]. This can be understood by considering the agglomerates or flocculates as being the optical equivalent of larger particles. As discussed above, large particles absorb light less efficiently than small particles.

Groups of undispersed particles can exist in a paint film or plastic if the particles are not adequately dispersed during the grinding operation (see Chaps. 11 and 12, respectively). These groups of particles are referred to as agglomerates. For paints, groups of undispersed particles can also form during storage or drying if the particles are not properly stabilized. Groups of particles that were initially dispersed in a paint, but then re-combine, are referred to as flocculates. When poor dispersion is seen in a dry paint film, it is important for the paint formulator to know whether it is due to inadequate initial dispersion (agglomeration) or to dispersion instability (flocculation), as the corrective actions are very different for the two (see Chap. 11).

For both paints and plastics, initial dispersion is determined by the characteristics of the particle surface (e.g., its wettability and surface energy), the size of the particle (small particles are more difficult to separate than large particles due to their higher surface areas) and the degree to which energy can be rapidly and efficiently used to pull the particles apart, rather than, say, to simply overcome drag due to high viscosity, or to heat the dispersion by mixing rather than grinding.

Even at low concentrations, slurries made with particles that have high oil absorption values (and high surface areas) can be highly viscous. This can make it difficult to adequately disperse small pigment particles within the time constraints of paint production. Low dispersion rates are also seen if the particle surfaces are difficult to wet. This is often true of organic pigments, since they generally have inert, low energy surfaces that do not wet easily. Because of this, many paint manufacturers prefer to purchase their color pigments in the form of pre-dispersed liquid concentrates. Similarly, many plastics manufacturers purchase their color pigments in the form of plastics concentrates, or masterbatches. Both of these forms of concentrated pigment are discussed in the section on pigment manufacture below.

Both the dispersibility and, for paints, the dispersion stability, of a pigment can be significantly enhanced by surface modifications. These are in some ways similar to those applied to $TiO_2$ pigments, as discussed in Chap. 7, and in some ways different.

## Surface Modification

Dispersion improvement through surface modification is an important aspect of many pigments. Many of the materials added during the production of pigments to control their particle size also assist in their wettability. Surface modifications with other materials can assist in both ease of dispersion and dispersion stability. These surface modifiers are typically applied to organic pigments, since they are typically the most difficult to wet and disperse, although certain inorganic pigments, most notably $TiO_2$ (see Chap. 7), are also treated in this way. In all cases they are generally applied at the end of the production process, in a step referred to as "conditioning" in color pigment manufacture and "finishing" for $TiO_2$.

The surface modifications for organic pigments generally fall into three categories. The first is the addition of rosin during particle synthesis [16]. This controls particle size and often improves particle wettability. Rosin is often used on azo pigments, although other classes of pigments also use rosin to a lesser extent.

The second category of modification is a limited, tightly controlled oxidation of the pigment surface. This is generally applied to carbon blacks and to some organic red pigments. Here the intent is to react the particles with enough oxygen to create surface oxygen functional groups, such as alcohols, aldehydes, ketones, ethers, esters, and carboxylates. These functional groups increase the polarity at the surface and therefore the wettability of the particle.

The third type of modification is the surface deposition of a chemical derivative of that pigment that can improve particle wettability by increasing the polarity of the particle surface [36]. An example of such a derivative is a sulfonated version of the original pigment molecule. Since these derivatives have the same overall shape and chemistry as the original pigment molecule, they generally fit onto the particle surface in a fashion similar to the original molecules, and so these attachments are strong.

The surfaces of inorganic color pigments can be modified by deposition of an inert, colorless material. The modification of $TiO_2$ surfaces using silica and alumina was described in Chap. 7, and similar materials can be applied to inorganic color pigments such as C.I. Pigment Yellow 184 (bismuth vanadate) and C.I. Pigment Black 29 (mixed cobalt iron oxide). In addition, the surface energies of inorganic pigments that are incorporated into plastics can be lowered, if desired, by the deposition of a hydrophobic organic material onto the particle surface [37]. This not only increases plastics processing rates but also allows for higher concentrations of particles in masterbatches (discussed below).

## Pigment Manufacture

Entire books are dedicated to the subject of pigment manufacture [5–7, 10, 23], and we will not exhaustively review these methods here. Our intent is to give a flavor for the types of processes involved in color pigment manufacture.

The two types of pigments—organic and inorganic—are generally produced in very different ways. Organic pigments are produced by the reactions of petroleum distillates and their derivatives, typically in an organic solution and at temperatures no higher than 150 °C. Inorganic pigments are produced by either high-temperature calcination or by precipitation.

### *Organic Pigments*

Broadly speaking, most organic pigments are made following a similar series of operations, as shown in Fig. 8.28. This schematic is meant to be quite general—some steps are omitted or repeated for some pigments, and the precise details of each step vary greatly for the different pigment families.

The raw materials for organic pigments are referred to as primaries. These are separated from petroleum through distillation and include substituted benzene, naphthalene, and anthracene derivatives. These are then functionalized by the addition of reactive or simple substitution groups such as nitrate, sulfate, halogens, etc., to give substituted primaries. These are often purified using, for example, distillation or precipitation. The substituted primaries are then reacted together to assemble the final pigment particle in a crude form.

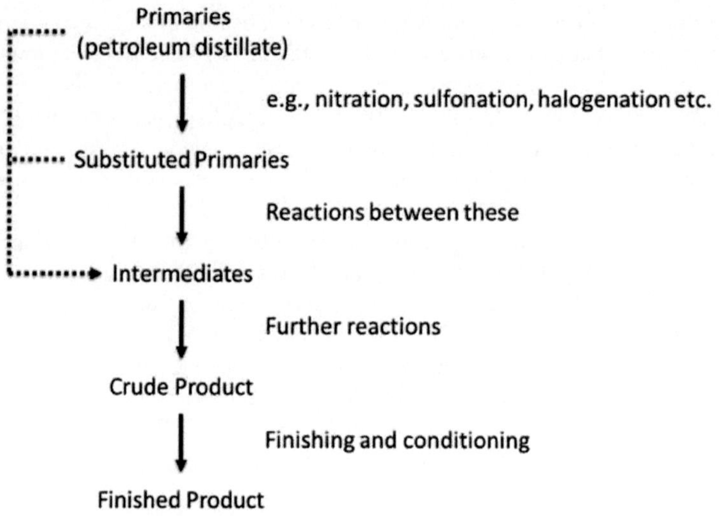


**Fig. 8.28** Generic reaction scheme for organic pigments [10]

The crude product must undergo a number of operations to produce the final pigment. Purification is necessary and can be achieved by filtration and washing. Optical and other properties can be optimized by adjusting the particle size. Particles that are too large can be wet or dry ground, or dissolved and recrystallized (see below). Particles that are too small can be enlarged through Ostwald ripening, as discussed above [33]. In some instances, adjusting particle size by either process can change the crystalline form (phase) of the particle.

In addition to altering particle size, recrystallization can remove crystal defects that are harmful to optical properties, as well as alter the particle shape and crystalline phase. Recrystallization is often achieved by dissolving the material in one solvent and then adding a second solvent to effect precipitation. In the various recrystallization processes the exact concentrations, solvents, addition rates, agitation conditions, temperatures, and reaction times are critical for optimizing the various particle properties. Even the geometry of the reaction vessels can be important [33].

Once the particles are of the proper size and crystalline phase, the final step is their conditioning. In this step the particle surfaces are modified, primarily to enhance dispersibility and dispersion stability, as discussed above, but also in some cases to confer better durability to the particle. This is typically accomplished by applying a material to the particle surfaces. Rosin and its derivatives are often applied to azo pigment particle surfaces for dispersibility, particle growth inhibition, and handling purposes. Fatty acids are also used to condition surfaces, for much the same reasons. Derivatives of the parent pigment molecule, specifically designed to improve dispersibility or stability, can also be applied to the surfaces. The use of such derivatives is common for phthalocyanine pigments.

Finally, particles can be weakly (and reversibly) agglomerated to enhance handling (particle flow) and minimize dusting.

## *Inorganic Pigments*

Most inorganic pigments are produced either by high-temperature solid-state methods (calcination) or by controlled precipitation. Certain pigments can be produced by other methods that are specific to that material—for example, transparent iron oxides are typically made via the thermal decomposition of iron pentacarbonyl ($Fe(CO)_5$).

In the solid-state method, pure, finely divided solid raw materials are combined and then calcined at high temperatures (500–1,400 °C) to produce the desired compound. The raw materials can be metal oxides, hydroxides, carbonates, nitrates, etc. The calcination step can be done in either an oxidizing or a reducing atmosphere, depending on the desired oxidation state of the metal. In some instances, the reaction mass is cooled, crushed, mixed, and then returned to the calciner to ensure complete reaction. Calcination is typically restricted to oxides and sulfides. Complex inorganic color pigments are produced exclusively using this method.

The material made by calcination is typically too coarse for pigmentary purposes and so is ground to the required fineness. In some cases, surface treatments are applied to the particles to enhance pigmentary properties, as discussed elsewhere.

The precipitation method of inorganic pigment production is straightforward in concept. Solutions of various starting materials are combined and the desired product is precipitated. This precipitate is then separated from the mother liquor and purified, typically through filtration and washing. Surface treatments can also be applied if desired. In many cases (e.g., cadmium yellow, bismuth vanadate molybdate, etc.) the precipitate is an intermediate that is subsequently calcined to produce the final pigment.

## *Pigment Concentrates*

Concentrated forms of pre-dispersed pigments are available to both the paint and plastics manufacturer [22, 38]. Using these materials frees the paint or plastic manufacturer from the difficulty of dispersing the pigments, which is particularly challenging for organic pigments, and also ensures product consistency. In addition, using concentrates free paint or plastics manufacturers from the difficulties of handling dry powders.

In the coatings industry, concentrated colorants are often used for "point of sale", or POS, paints. These are paints that are colored at the paint store, as opposed to "factory colors" that are colored at the manufacturing site. POS paints have the dual advantages that they are available in a much broader range of colors than factory colors[28] yet require minimum inventory at the paint store. Rather than stocking separate containers for each combination of color and sheen, only a limited number of white base paints, with various sheen levels, are kept on hand. Each sheen level will typically come in three varieties, depending on the brightness level of the final paint (the base paints for bright or pastel colors have high levels of $TiO_2$, while base paints for dark colors use little to no $TiO_2$). Color paints are made by accurately dispensing liquid colorant into the base paints. Depending on the paint product line, from six to nine liquid colorants are used. In addition to having a large color palette, POS paints can match virtually any color using the techniques that are described in Chap. 15.

The material demands on liquid colorants are quite significant. In addition to color and optical consistency, these materials must remain stable during storage and must be compatible with the other ingredients in the paints—including other colorants. In addition, universal colorants must be compatible with both solvent-borne and water-borne paints. This is usually accomplished using steric dispersants that have both hydrophilic and hydrophobic arms. These arms are capable of extending into water and organic solvent, respectively (dispersants are discussed in depth in Chap. 11).

The plastics industry uses color masterbatches for the same purpose as liquid colorants are used in the coatings industry. In this case a highly concentrated colored plastic is produced in a polymeric carrier material. These plastics are granulated and sold to the final plastics manufacturer, who can then meter (let down) the differently colored masterbatches into the melted polymer as needed to produce the desired final color. This same concept is also used to meter other additives into the plastic. In addition to particulate masterbatches, liquid masterbatches are available that can be pumped into the screws of plastics compounding machines.

## Pigment Nomenclature

A widely accepted system for classifying and naming colorants (both dyes and pigments) is jointly maintained by the American Association of Textile Chemists and Colorists and the Society of Dyers and Colourists.[29] This system consists of a database of colorants identified by a hue-based Colour Index™ Generic Name and a five or six-digit Colour Index™ Constitution Number. The system is referred to as the Colour Index™ International [39]. When describing a colorant, both the generic name and the constitution number are generally preceded by the letters "C.I.", standing for "Colour Index". We have used this designation when discussing individual pigments throughout this chapter.

[28] Factory colors are typically limited to a dozen or so while POS paints can be made in hundreds, if not thousands, of colors.

[29] Note that this is fundamentally different from systems that define colors (see Chap. 6). Here we are defining the materials that create colors, not the colors themselves.

Generic names are organized by hue. These typically take the form of a hue name[30] followed by a number—for example, C.I. Pigment Red 48. The number simply reflects the chronological order in which the pigments were added to the index. Often the names are abbreviated, for example, PR 48 for the above pigment.[31]

In contrast, the constitution numbers are organized by pigment chemistry, with only indirect categorization by color. For example, phthalocyanine pigments all have constitution numbers between 74,000 and 74,999, with no regard for their actual color.

Both the generic names and constitution numbers for certain pigments can be modified with a colon number—for example, C.I. Pigment Red 48 and C.I. Pigment Red 48:1 are related (but different) pigments, as are C.I. 42,535 and C.I. 42,535:1. In each case, pigments differentiated by colon numbers are "similar" to one another. As a general rule, "similarity" can refer to minor differences in the chemical identity of the pigment (e.g., organic pigments for which one halogen has been substituted for another, or ionic pigments that differ in counterion) or to differences in crystal structures.

The use of colons is generally consistent between the generic names and the constitution numbers. For example, C.I. Pigment Yellow 62 and C.I. Pigment Yellow 42:1 have corresponding constitution numbers of C.I. 13,940 and C.I. 13,940:1. That said, there are occasional inconsistencies in the use of colon numbers, both within and between the two types of index (generic name and constitution number). As an example, C.I. Pigment Yellow 36 and C.I. Pigment Yellow 36:1 have constitution numbers that are not related by a colon—instead, they are C.I. 77,955 and C.I. 77,956. On the other hand, the pigments with constitution numbers C.I. 77,283:1 and C.I. 77,283:3 have different generic names that are not related by a colon—they are C.I. Pigment Orange 75 and C.I. Pigment Red 265.

These inconsistencies are partially responsible for the current assignment of six-digit constitution numbers to newly added colorants (historically only five digits were used for constitution numbers), with the intent of avoiding the assignment of colon numbers in favor of giving newly added materials unique designations.

Finally, note that some materials have two or more generic names, reflecting their use as a colorant in more than one application. For example, C.I. Pigment Blue 60 is identical to C.I. Vat Blue 4.

## Summary

Color is produced when a material absorbs part or all of the visible light spectrum. Light absorption occurs when light strikes a molecule with the exact energy needed to

---

[30] There are ten hues used for generic names—red, orange, yellow, green, blue, violet, black, white, brown, and metal.

[31] Because blue, brown, and black all begin with the letter b, they are abbreviated as B, Br, and Bk. For example, PBr is a brown pigment whereas PB is a blue pigment.

promote an electron from an occupied molecular orbital to an unoccupied molecular orbital. The wavelength of light that is absorbed, and the intensity with which it is absorbed, determine the color and tint strength of the molecule, respectively. When the molecule is in the solid state, the resulting particle is a color pigment particle. When it is in the dissolved state, it is a dye molecule.

All color pigments have the common purpose of conferring color to an object, which, for our interests, is a paint film or plastic article. That said, they represent a near infinite diversity with respect to chemical identity and manufacturing process. This diversity can be conveniently organized into organic and inorganic pigments, based on the chemical nature of the constituent molecules.

Organic pigments are generally aromatic, polycyclic materials that have an extensive system of delocalized molecular orbitals. The energy differences between these orbitals match the energy of visible light photons. This can be compared to the energy differences between localized orbitals, which are higher and correspond to the energy of ultraviolet light. The absorption of visible light by molecules with delocalized orbitals does not result in chemical bonds being broken, and so these molecules are not consumed in the absorption process.

In broad terms, organic pigments can be divided into three families. The first, azo pigments, are perhaps the most important class of organic pigments. These pigments are characterized by a pair of nitrogen atoms, each attached to an aromatic group, that are joined together by a double bond. Industrially important azo pigments are generally red, orange, or yellow.

The second family of organic pigments, the derivatives of copper phthalocyanine, are the most important organic blue and green pigments. They are preferred to other types of organic blue and green pigments because of their high absorption strength, high chromaticity, high stability, and low production costs. In many ways these are the ideal color pigment.

All other organic pigments fall under the general class of polycyclic pigments. These include many chemical families that consist of only a limited number of commercially important pigments. These pigments are generally valued over azo or copper phthalocyanine pigments due to their enhanced performance properties, most notably their pure colors and lightfastness.

Regardless of organic pigment family, the raw materials used for organic pigment manufacture are aromatic components of petroleum. Individual chemicals are separated via distillation and then are reacted with different agents to append functional groups on the starting materials. The purpose of these functional groups is to adjust the color of the final pigment, improve the physical properties of the final pigment, and/or react with other aromatic molecules to produce more complex pigments. These reactions form crude particles that are then refined and, in some cases, surface treated to give the finished product.

Inorganic pigments are typically oxides, hydrous oxides, sulfides, or salts of metal atoms. Like organic pigments, they span the entire range of colors. They are typically made by either a high-temperature solid-state reaction (calcination) or by precipitation. Some are made in two steps—precipitation followed by calcination. These particles tend to be larger than organic pigments and have lower chromaticity or saturation (i.e., are muddier), but they are also easier to disperse, provide higher levels of opacity, and are more stable against light and many chemicals.

Regardless of the chemical type, all pigment particles share the same set of important properties. Chief among these, obviously, is color, but opacity, crystal structure, particle size, lightfastness, and dispersibility are also important. These affect the end-use applications for which a specific pigment can be used, as well as the type of equipment required to incorporate that pigment into the paint or plastic.

## References


1. Pauling, L.: The Nature of the Chemical Bond and the Structure of Molecules and Crystals. Cornell University Press (1960)
2. Qu, Z.-W., Kroes, G.-J.: Theoretical study of the electronic structure and stability of titanium dioxide clusters $(TiO_2)^n$ with $n = 1$–9. J. Phys. Chem. B **110**(18), 8998 (2006)
3. Hunger, K., Schmidt, M.U.: Polycyclic pigments. In: Industrial Organic Pigments: Production, Crystal Structures, Properties, Applications, 4th edn. Wiley (2018)
4. Wicks, Z.W.Jr., Jones, F.N., Pappas, S.P.: Organic Coatings Science and Technology, 2nd edn. Wiley-Interscience (1999)
5. Hunger, K., Schmidt, M.U.: Industrial Organic Pigments: Production, Crystal Structures, Properties, Applications, 4th edn. Wiley (2018)
6. Smith, H.M. (ed.): High Performance Pigments. Wiley (2001)
7. Buxbaum, G., Pfaff, G. (eds.): Industrial Inorganic Pigments, 3rd edn. Wiley (2005)
8. Wissling, P. (ed.): Metal Effects Pigments. Vincentz (2006)
9. Pfaff, G. (ed.): Special Effects Pigments. Vincentz (2008)
10. Abel, A.G.: Pigments for Paint. In: Lambourne, R., Strivens, T.A. (eds.) Paint and Surface Coatings: Theory and Practice. Woodhead Publishing (1999)
11. Sabreen, S.R.: How coloring plastics affects secondary processes. Plast. Decor. (2014)
12. Berrie, B.H.; Lomax, S.Q.: Azo pigments: their history, synthesis, properties, and use in artists' materials. In: Studies in the History of Art: 57—Conservation Research 1996/1997, 9. National Gallery of Art, Washington (1996)
13. Azopigmente, https://second.wiki/wiki/azopigmente#google_vignette
14. Hunger, K., Schmidt, M.U.: Hydrazone pigments (formerly called azo pigments). In: Industrial Organic Pigments: Production, Crystal Structures, Properties, Applications, 4th edn. Wiley (2018)
15. Gregory, P.: Industrial applications of phthalocyanines. J. Porph. Phthalo. **4**(4), 432 (2000)
16. Aravindakshan, A.: Copper phthalocyanines. Paint Coat. Indust. (2005)
17. Christie, R., Abel, A.: Phthalocyanine blue pigments. Phys. Sci. Rev. **6**(9), 391 (2021)
18. Davidson, A.T.: The effect of the metal atom on the absorption spectra of phthalocyanine films. J. Chem. Phys. **77**, 168 (1982)
19. Wöhrle, D., et al.: Practical applications of phthalocyanines—from dyes and pigments to materials for optical, electronic and photo-electronic devices. Macroheterocycles **5**(3), 191 (2012)
20. Christie, R., Abel, A.: Anthraquinonoid pigments. Phys. Sci. Rev. **6**(8), 299 (2021)
21. Jaffe, E.E.: Quinacridone pigments. In: Smith, H.M. (ed.) High Performance Pigments. Wiley (2001)
22. Greene, M.: Perylene pigments. In: Smith, H.M. (ed.) High Performance Pigments. Wiley (2001).
23. Christie, R., Abel, A.: Diketopyrrolopyrrole (DPP) pigments. Phys. Sci. Rev. **6**(7), 281 (2021)

24. Bao, W.W., Li, R., Dai, Z.C., Tang, J., Shi, X., Geng, J.T., Deng, Z.F., Hua, J.: Diketopyrrolopyrrole (DPP)-based materials and its applications: a review. Front. Chem. **8**, Article 679 (2020)
25. Guy, A.: Coatings beyond binders. In: Marrion, A. (ed.) The Chemistry and Physics of Coatings, 2nd edn. Royal Society of Chemistry (2004)
26. Biller, K.: The use of mixed metal oxide pigments in industrial coatings. Paint Coat. Indust. (2010)
27. Men, P., Lianf, H., He, J., Chen, J., Geng, B., Li, W.: Preparation of alkali-resistant aluminum pigment encapsulated with fluoropolymer by in situ polymerization. J. Coat. Technol. Res. **18**, 1227 (2021)
28. Hunter, R.S.: The Measurement of Appearance. Wiley (1975)
29. Spengeman, W.F.: Pigments. In: Sward, G.G. (ed.) Paint Testing Manual. ASTM (1972)
30. Standard Test Method for Color and Strength of Chromatic Pigments with a Mechanical Muller. ASTM D387 (2008)
31. Pfaff, G.: Transparent pigments. Phys. Sci. Rev. **6**(6), 218 (2021)
32. Determination of Indices for the Transparency of Pigmented and Unpigmented Systems—Colorimetric Method. DIN 55988:2019-01 (2019)
33. Vicum, L., Massotti, M., Iggland, M.: Precipitation and crystallization of pigments. In: Myerson, A.S., Erdemir, D., Lee, A.Y. (eds.) Handbook of Industrial Crystallization, 3rd edn. Cambridge University Press (2019)
34. Mubarak, Y., Martin, P.J., Harkin-Jones, E.: Effect of nucleating agents and pigments on crystallization, morphology, and mechanical properties of polypropylene. Plast. Rubber. Comp. **29**(7), 307 (2000)
35. Pfaff, G.: Colorants in plastic applications. Phys. Sci. Rev. **6**(2), 20190104 (2021)
36. Gooch, C., Aravindakshan, A.: The anatomy of multipurpose solventborne colourants. Paint Coat. Indust. (2011)
37. Kostelnik, R.J., Weber, L., El-Shoubary, M.: Processes for preparing hydrophobic inorganic oxide pigments, WO 01/34711 (2001)
38. Muller, B.: Colorants for thermoplastic polymers. In: Kutz, M. (ed.) Applied Plastics Engineering Handbook: Processing and Materials, 1st edn. Elsevier (2011)
39. The Colour Index™ is published by Society of Dyers and Colourists and American Association of Textile Chemists and Colorists online at colour-index.com.

# Chapter 9
# Extender Particles

## Contents



## Introduction

Extenders can be defined as colorless, or nearly colorless, particles with refractive indices below 1.7—that is, so low that the particles do not scatter light meaningfully when embedded in a polymer matrix [1–3]. These materials were introduced to the paints and plastics industries as simple fillers to replace resin or polymer with a less expensive substitute. While many extenders are still used in this passive manner, others take on a more active role, providing a means for the formulator to adjust and improve the performance attributes of their product [4–8].

The evolving importance of extenders can be attributed to two factors. First, paints and plastics have become more sophisticated over time, and as such, every opportunity is taken by the formulator to enhance product performance. Extenders offer

 
M. Diebold et al., *Pigments, Extenders, and Particles in Surface Coatings and Plastics*, https://doi.org/10.1007/978-3-030-99083-1_9

one such opportunity. Second, over the years the mining and extender manufacturing industries have undergone significant technological improvements, allowing for enhanced extenders to be produced economically. These enhancements apply to extender properties such as particle size, whiteness, and processibility in paint or plastics applications. The resulting particles are often referred to as "functional fillers", highlighting the active part that these materials can play in polymer matrices. In this chapter and throughout this book, we will refer to these materials as "extenders" rather than "fillers", as the latter term does not accurately convey the value that these particles bring to the paint and plastics industries.

Extender production is linked intimately to the mining industry. With few exceptions, the raw materials for extenders are minerals that are mined out of the ground. In some cases, these are simply cleaned and sold in essentially the same condition as they came out of the ground, while in other cases, they are processed more extensively. This processing can include more rigorous purification, grinding, and particle size separation (classification). In some cases, the natural minerals are reacted chemically to provide an intermediate material that is then back-reacted to the original mineral. This allows for improved purity and greater control of particle size.

The extender business is typically organized at the regional level. This is because the cost of transporting extenders over distances of more than a few hundred kilometers would add significantly to their overall costs, and because the raw materials for extenders are well distributed geographically, making them locally available in most cases. This can be contrasted to the pigments business, which is typically global and can involve intercontinental transportation.

## Important Extender Properties

Many paints and plastics' properties can be affected by extender particles. Table 9.1 summarizes the most important of these properties [7, 9, 10]. Many of the effects that extenders have on the properties listed in this table are based on the physical properties of the extender particles themselves. Many of these effects are similar for extenders that share certain properties. For example, the effect that a certain size of platelet particles will have on paint rheology is similar for all platelet extenders of that size, regardless of their chemical or mineral composition. The extender's physical properties that affect paint and plastic performance include, in rough order of importance, particle size (and size distribution), particle shape (and aspect ratio, for non-uniform particles), particle porosity, particle packing ability, and particle hardness. In addition, some chemical properties of extenders can affect paint and plastic performance. These are the pH of the particle surface, as well as the reactivity of the particle to water, acids, bases, salts, and solvents.

Particle size and size distribution are important for a number of reasons. In general, extenders cover the range of 0.01 to 45 microns, although most are in the range of 0.5 to 10 microns. This range is limited for practical reasons. The use of particles at the high end of the range can cause defects in paint or plastic films, while the use

**Table 9.1** End-Use Properties of Paints and Plastics that can be Affected by Extender Particles

| Paints | Plastics |
|---|---|
| • Flow and Viscosity | • Process aid |
| • Settling | • Shorter fusion time |
| • Strength/toughness | • Melt flow and extensibility |
| • Reinforcement | • Nucleation for crystallization |
| • Cracking and shrinking | • Moldability |
| • Scrub and abrasion resistance | • Strength |
| • Burnish | • Impact |
| • Adhesion (to substrate and inter coat) | • Modulus |
| • Gas and liquid permeability | • Stiffness |
| • Weather resistance | • Tensile |
| • Corrosion resistance | • Reinforcement |
| • Dry Hide | • Shrinkage/thermal expansion |
| • Gloss | • Gas and liquid permeability |
| | • Weather resistance |

of particles at the low end of the range can cause excessively high viscosity in the liquid paint or polymer melt.

There are many times more particles in a gram of small size extender than there are in a gram of large size extender. For identically shaped particles, the number of particles decreases as the third power of particle size. Likewise, a gram of small particles has a greater surface area than a gram of large particles. In this case, the change is linear with particle size. Higher surface areas increase the amount of material that a unit weight of particles can adsorb, which increases oil absorption[1] and binder demand in paints. Higher surface areas also increase the magnitude of particle–particle attractions, since these occur at particle surfaces. This affects the dispersibility of the particles in both a liquid paint and a polymer melt.

Particle size can influence the color or brightness of an extender. In their pure form, the minerals used as extenders are colorless. However, there are invariably colored impurities present in those that are mined. These colored impurities can be reduced through purification, but it is difficult to entirely eliminate discoloration. However, since small particles scatter light more efficiently than an equal weight of large particles, as the particle size is reduced, the material becomes brighter and whiter.

This effect can be demonstrated using macroscopic glass particles. Figure 9.1a shows four particle sizes of the same brown glass (refractive index 1.52). As can be

[1] Oil absorption refers to the grams of linseed oil needed to "liquify" one hundred grams of particles (see Chapter 2).

**Fig. 9.1** Brown glass particles of different sizes. **a** Particles in air. **b** Particles immersed in mineral oil. Adapted from [10]

seen, in air (refractive index 1.0), the smallest particles are merely off-white, while the largest particles are intensely colored. This brightening effect is significantly reduced when the particles are embedded in a resin or polymer matrix, since the higher refractive indices of these matrices (1.50–1.60) reduce the light scattering ability of the particles. This can be seen in Fig. 9.1b, where mineral oil (refractive index 1.49) has been added to the particles.

Particle size distribution is important because particles with a broad size distribution can pack more densely than those with a narrow distribution. This is because the smaller particles can pack into the voids between the larger ones (refer to Chaps. 4 and 16). Related to this, small extender particles (< one micron) are capable of situating themselves between pigmentary particles, which typically have sizes between 0.2

and 2.0 microns. These small extender particles improve the spacing of the pigmentary particles in comparison to the same volume of large extender particles, since the latter crowd the pigmentary particles into confined regions. Particle packing also affects the critical pigment volume concentration (CPVC) of paint, as discussed in Chap. 4, as well as the melt flow viscosity of plastics. There is still much detail to learn about the effects of particle size distribution on the properties of paints and plastics, and this is an area of active research [11, 12].

Particle shape affects a number of physical properties of the paint or plastic. As discussed in Chap. 1, particle shapes can be categorized into three groups, although the demarcation between these is often blurry. These categories are the acicular one-dimensional particles (needles, rods, and fibers); the plate-like two-dimensional particles (platelets); and the solid three-dimensional particles (nodules or blocks). As a general rule, when different particle shapes behave differently in an application, two-dimensional particles behave more similarly to one-dimensional particles than to three-dimensional particles.

Particle shape is often determined by the type and arrangement of atoms in the solid. In this regard, there are two basic types of extenders—insoluble salts and oxides. Both can be hydrated with water of crystallization. The most common salt extenders are calcium carbonate and barium sulfate. The structure of salts is such that discrete cations and anions alternate in all directions, leading to a three-dimensional, blocky particle.

The atomic arrangements in oxide extenders are more complex. In general, these consists of networks of metal cations (typically aluminum or silicon) bridged by oxygen or hydroxide anions. These networks can be one-dimensional chains, as in wollastonite, two-dimensional sheets, as in kaolin, talc, and mica, or three-dimensional space-filling structures, as in feldspar and silica. The bonds within networks are typically quite strong (ionic or covalent bonds), while the bonds between the networks are much weaker (hydrogen or van der Waals bonds). Particles generally extend the greatest in the direction of the strongest bonds, and so wollastonite particles are needles, kaolin, talc, and mica particles are platelets, and feldspar and silica particles are blocks.

Each particle shape confers certain properties to the material in which they are embedded. Perhaps the most important effect of needles is their ability to provide strength to a paint film or plastic object. When oriented randomly, they reinforce these materials in all three dimensions. This reinforcement is manifested in a number of ways, particularly in plastics, where several strength parameters can be affected (bending strength, impact resistance, stiffness, etc.). In paints, this can lead to impact resistance (which is important for automotive paints) as well as scratch resistance.

Needles also affect the viscosity of a liquid paint or a polymer melt through their formation into weak networks. In paints, these networks are designed to quickly form under conditions of no shear, such as during storage or immediately after application to a substrate, and to break down under the shearing action of paint application (by brush, roller or spray device). This allows the paint to be easily applied to the substrate yet also resist running or sagging after application. It also prevents sedimentation of other particles in paints while they are stored, since these particles will be trapped in the network of needle-shaped particles.

Platelets also affect paint viscosity and polymer melt flow in the same ways. In addition, should settling of paints occur, the consistency of sediments that contain platelets is typically such that they can be easily reincorporated into the paint through simple stirring. When platelets are aligned and oriented parallel to a painted surface, they increase substrate protection by reducing the permeability of the paint to air, water vapor, liquid water, and corrosive chemicals. This reduced permeability is due to the tortuous path that an intruding molecule must travel to reach the substrate surface.

Both needles and platelets can decrease crack formation during the paint drying process. Paints will "mud crack" during drying if stresses due to film contraction exceed a certain threshold. This is a common concern for dilute paints, which lose much volume during drying, as well as for thickly applied paints, from which solvent or water often evaporates unevenly. This temporarily forms regions with different physical properties. Needles and platelets can relieve these stresses as they are created, preventing crack formation.

Platelets, like needles, reinforce plastics and increase many of their strength attributes. However, platelets tend to orient along melt flow lines, which results in strength improvements in only two directions, rather than in all three dimensions. In addition to this reinforcement effect, platelets and needles can improve strength by acting as concentration and relief points for stress.

Blocky particles have a lesser effect on the properties of paints and plastics because they do not form extensive networks. As such their effect on viscosity is minimal except at high concentrations. In addition, they do not reinforce paints and plastics in the same way as needles and platelets, although they can act as concentration points for stress. This allows for some level of strength improvement.

There are also several important effects of extenders that are not related to their size or shape. Hard particles improve the abrasion and scratch resistance of plastics and paint films, but also increase the abrasion of equipment during processing. Particles near a surface can roughen it, providing a means to control the sheen in paints and plastics. In addition, these particles can provide a "tooth" to a paint—that is, increase the adhesion between the paint and the substrate, and the inter-coat adhesion between layers of paint. When used at high concentrations in paints, extenders improve opacity through dry hide, which is light scattering by air voids within the paint (see Chaps. 4 and 7). Finally, low levels of some extenders can be used as anti-blocking agents in plastic films, preventing sheets of film from sticking to one another.

The surface character of extender particles plays an important role in both paint and plastic applications. It is important that the surface is compatible with its surrounding matrix. In paint films and plastics, this matrix is a polymer, while for liquid paints it is water or a solvent. Most extender particles are hydrates, which makes them hydrophilic and, as a general rule, easily dispersed and stabilized in water. Particle surfaces can be modified by the application of a hydrophobic material, such as a stearate or silane, leading to stronger binding of the polymer to the particle surface. This improves processibility in plastics [13] and corrosion protection in paints [14].

**Table 9.2** Important Properties of Extenders

| Mineral | Chemical Composition | Mohs Hardness | Density g/ml | Refractive index | Particle Shape |
|---|---|---|---|---|---|
| Calcium Carbonate | $CaCO_3$ | 3 | 2.6–2.8 | 1.59 | Nodular |
| Dolomite | $Ca/Mg(CO_3)$ | 3.5–4 | 2.8–2.9 | 1.62 | Nodular |
| Kaolinite | $Al_4(OH)_8Si_4O_{10}$ | 1–2 | 2.6 | 1.56 | Platelet |
| Calcined Kaolin | $Al_4(OH)_8Si_4O_{10}$ | 5–6 | 2.6 | 1.62 | Platelet |
| Talc | $Mg_3(OH)_2(Si_2O_5)_2$ | 1 | 2.6–2.8 | 1.57 | Platelet |
| Mica (muscovite) | $KAl_2(AlSi_3O_{10})(OH)_2$ | 2.5 | 2.82 | 1.56 | Platelet |
| Feldspar | $KAlSi_3O_8$ | 6 | 2.5–2.6 | 1.53 | Nodular |
| Wollastonite | $CaSiO_3$ | 4.5–5.5 | 2.84 | 1.6 | Acicular |
| Precipitated aluminum silicate | Various | – | 2.1 | 1.46 | Nodular |
| Quartz | $SiO_2$ | 7 | 2.65 | 1.55 | Nodular |
| Diatomaceous earth | $SiO_2$ | 5.5 | 2.0–2.34 | 1.45 | Complex |
| Barite | $BaSO_4$ | 3.3 | 4.48 | 1.65 | Nodular |
| Calcium sulfate | $CaSO_4$ | 3.5 | 2.97 | 1.58 | Nodular |

Many of the effects of extender particles on paints and plastics properties are not linear with extender concentration. For example, properties that are strictly surface related, such as anti-blocking, respond rapidly to increases in extender concentration up to a point, that point being when the surface region is saturated with particles. Other properties vary with particle concentration in a more complex way. When testing or characterizing an extender in a paint or plastic, it is best to do so on a full series of particle concentrations, rather than to extrapolate from the results at only one or two concentrations.

For convenience, many of the most important physical properties of the extender types that will be discussed in this chapter are given in Table 9.2. Note that this list excludes some factors that are important in only limited applications, such as electrical impedance and magnetic characteristics.

## Classification of Extenders

Given the wide variety of extenders used in the paint and plastics industry, it is useful to classify and organize these materials in some logical way. Since most commercial extenders originate from natural mineral deposits, the first level of classification can be made according to mineral origin. There are four broad classes of minerals used as extenders: carbonates, silicates, different forms of silica, and others. In each case, there are several important sub-types. The majority of these are discussed below, starting with the carbonates.

## *The Carbonates*

The term "carbonate" refers to any material that contains the $CO_3^{2-}$ anion. There is a multitude of cations that can counter the charge of the carbonate ion, including $Ca^{2+}$, $Mg^{2+}$, $Na^+$, $K^+$, and $H^+$. Of these, calcium carbonate and a mix of calcium and magnesium carbonate (dolomite) are the most commonly used extenders in the paint, plastics, and paper industries. This is due to their high brightness and commercial affordability. The high demand for these materials results in a broad choice of carbonate extender products [15].

Calcium carbonate, also known as calcite or limestone, forms under a number of geological conditions. Most calcium carbonate originates biogenically through the deposition of clam or other shells, although some are also formed by direct precipitation from fresh or ocean waters. When a carbonate sediment is formed, it is gradually compacted under its own weight, resulting in the formation of chalk. Chalk particles are typically a few microns in size, and even after grinding, the remains of seashells can be observed under the electron microscope.

Under increasing pressure from the overlaying of more sediment or other rock, the pores in chalk fill with precipitated calcite, forming limestone. When the temperature increases to between 200 and 500 °C and the pressure rises above 1000 bar, the limestone melts [2]. It then crystallizes slowly as marble, which is a metamorphic form of calcium carbonate. The purity of the calcium carbonate typically increases as a result of this crystallization. Because even small amounts of certain impurities (for example $Fe^{2+}$) can cause strong discoloration, the increased purity of marble gives it higher brightness and lower yellowness than chalk or limestone.

Calcium carbonate deposits are widespread, and each deposit has its own characteristic particle size, shape, and impurity profile. Because they originate as surface deposits, calcium carbonates are generally found close to the surface and so are relatively easy to mine. Both limestone and marble are used extensively to make calcium carbonate extenders. Chalk is less favored as an extender due to its higher concentration of colored impurities.

Calcium carbonate extenders process very well in both water-borne and solvent-borne coatings. Their surface areas are typically low, with values between 1 to 10 $m^2/g$, but can go up to as high as 20 $m^2/g$ for the most finely ground natural particles as well as those formed synthetically by precipitation. As a consequence of their low surface area, the oil absorption and binder demand are both relatively low, with oil absorption values generally varying between 12 and 25 g oil per 100 g particles [2].

Due to their blocky shapes, calcium carbonate particles do not form extensive networks in slurries or liquid paints, and so they have only a moderate thickening effect on them. The lack of network formation allows for calcium carbonate levels as high as 60% in paints, in contrast to many other extenders, such as talc and mica, that can only be used at much lower levels.

Calcium carbonate dissolves in acids, releasing $CO_2$. Since most water-borne coatings are formulated under alkali conditions, this is seldom a problem. However, some paints, such as those that are used for cathodic electrodeposition, and wood coatings that are based on cationic resins, are formulated at pH values of 6.5 or less, and so cannot use calcium carbonate as an extender. Calcium carbonates are often used in anti-corrosion applications, where they buffer the coating at relatively high pH values.

Calcium carbonate can be used to improve paint opacity. There are two different ways that this can be accomplished. The first is to raise the volume concentration of particles in the dry film to a level above the critical pigment volume concentration, or CPVC. As was discussed in Chap. 4, this results in the incorporation of air voids into the film that, due to the low refractive index of air, provide light scattering and opacity.

Second, particles that are close in size to $TiO_2$ are claimed to act as spacers, separating the $TiO_2$ particles and thereby increasing their scattering efficiency. This is most often seen when larger extender particles, which tend to crowd the $TiO_2$ particles together, are replaced with smaller extender particles. In this case, the improvement in opacity is not due to active spacing on the part of the small extender, but instead to the reduction of active crowding on the part of the large extender [16].

Calcium carbonate accounts for more than half of the total amount of extenders used by the plastics industry. Its extensive use can be attributed to many of the same factors that lead to its wide use in paints. In addition, calcium carbonate has no water of crystallization, which makes it ideally suitable for high-temperature plastics processing, where water volatilization can be problematic.[2] This is particularly true for plastic films. Here water vapor bubbles can form holes in the film, which are then enlarged as the film is pulled during processing (Fig. 9.2), creating severe defects in the film product. This process is known as lacing, since in the more extreme cases the resulting film strongly resembles lace fabric.

Calcium carbonate also reduces plastics shrinkage, has good anti-plate out properties, and can be conveniently surface treated with stearic acid (or a stearic acid salt) for improved processing. It does not strongly reinforce plastics in the same way as platelets or needles, although it does improve impact resistance. Thermoplastics are the largest single application for calcium carbonates in plastics, with lesser amounts used in elastomers and thermosets [17]. Of the carbonate types, ground calcium carbonate (see below) is used predominantly in plastic applications.

Calcium carbonate extenders are generally available in two forms that differ in manufacture, both of which can be further surface modified with fatty acids. The simpler form, ground calcium carbonate (GCC), is made by physically processing material that is directly extracted from the earth. Dolomite extenders, which are closely related to calcium carbonate extenders, are also processed in this way. The

[2] Other extenders lacking waters of crystallization are available, but these typically require the costly and energy consuming calcination of hydrated materials.

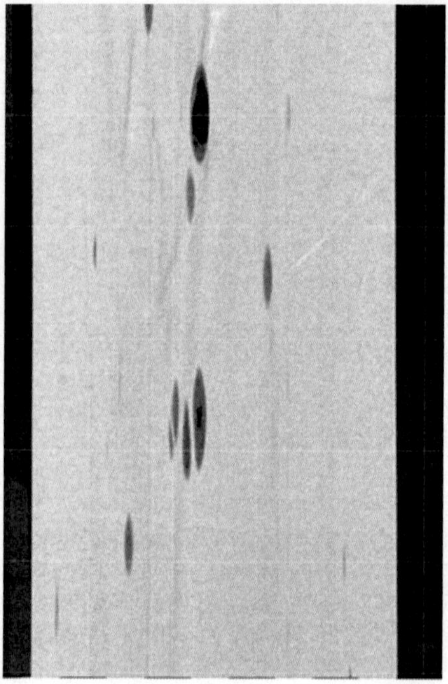

**Fig. 9.2** Lacing of a polymer film due to water vaporization during processing

second form of calcium carbonate extender is precipitated calcium carbonate (PCC), which is manufactured by heating calcium carbonate to form lime (CaO), then reacting the lime with carbon dioxide to remake calcium carbonate (this is further detailed in the section below on the production of synthetic extenders). Purity and particle size can be more finely controlled by this process than by the GCC process.

We will discuss each of these materials in turns, beginning with GCC.

## Ground Calcium Carbonates (GCC)

The ground calcium carbonate grades used as paint and plastics extenders typically have median particle diameters ($D_{50}$)[3] between 2 and 15 microns, although smaller particles ($D_{50} < 0.5$ microns) can also be made. Both wet grinding and dry grinding can be used to reduce particles. Fatty acids are often used to enhance grinding.

The wet grinding process allows for smaller and more uniformly sized particles but requires the additional steps of water separation (typically by filtration) and drying. In addition, wet grinding allows for colored impurities to be removed via flotation, and it is more cost-efficient than dry grinding when grinding to small particle size. This is partly because high electrostatic charges build up in the dry grinding

[3] $D_{50}$ is the diameter for which half the particles (by weight) are larger and half are smaller. See Chapter 2 for additional details on particle size determination.

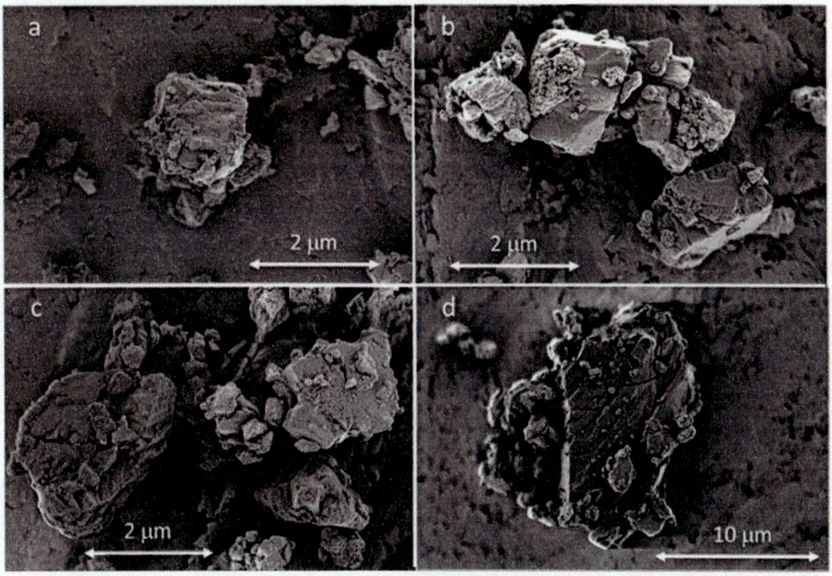


**Fig. 9.3** Micrographs of ground $CaCO_3$ with different median particle diameters **a** wet ground $CaCO_3$ ($D_{50}$ particle size 0.7 micron). **b–d** dry ground $CaCO_3$ with $D_{50}$ particle sizes of 0.9 microns, 5 microns, and 15 microns, respectively (courtesy of E. Najafi and H. Huezo, Chemours)

process once the particles are reduced below a certain size. On the other side of the size spectrum, particle sizes up to several millimeters are used in putties, plasters, mineral adhesives, and sealants. Examples of typical GCC materials processed by these different grinding techniques are shown in Fig. 9.3.

Ground calcium carbonate is known for its blocky morphology, which distinguishes it from the more common platelet and needle structures of other extenders. As can be seen in Fig. 9.3, the distribution of particle sizes within a sample can be quite broad and can include particles with diameters as low as 0.2 microns, even in the largest average particle size sample. (Fig. 9.3d) Grinding decreases particle size not only by shifting the size distribution to smaller particles, but also by increasing the population of very small particles. That is, grinding not only alters the median particle size; but also alters the width of the particle size distribution [2].

In plastics applications, GCC materials are used both in their traditional capacity as extenders as well as to create opaque micro-voids in polypropylene and polyethylene films. To do this, small calcium carbonate particles (1 to 2 microns), often pretreated with a fatty acid such as stearic acid, are incorporated into these films, which are then stretched. These particles do not adhere strongly to the polymer, and when the films are stretched, the polymer pulls away from the particle surface. This creates elongated holes around the particles that, as described in Chap. 7, provide for opacity through light scattering.

## Precipitated Calcium Carbonate (PCC)

While GCC is characterized by a nodular shape, precipitated calcium carbonates can have several different particle shapes depending on the manufacturing conditions. As shown in Fig. 9.4, PCC particle sizes can be much smaller than those of ground $CaCO_3$. This is because small particle sizes are attained when the particles form, rather than through a grinding step. In addition to a lower overall particle size, the size distributions of PCC extenders are narrower than those of GCC extenders. PCC extenders are also brighter and less yellow than their GCC counterparts. These factors combine to give a very good overall performance of PCC extenders in paint and plastics applications. However, this good performance comes at a cost—the selling prices for PCC extenders are significantly higher than those of their GCC counterparts, due to the costs of the additional processing steps.

Historically, PCC extenders have been favored in plastics applications over their GCC counterparts because of their finer particle size—polymer grade PCCs are typically between 0.02 and 0.1 microns [17]. However, their higher cost has been an issue for use in these applications. Over time advances in GCC processing have improved the performance of these materials in plastics, though, resulting in a shift to GCC extenders in many applications.

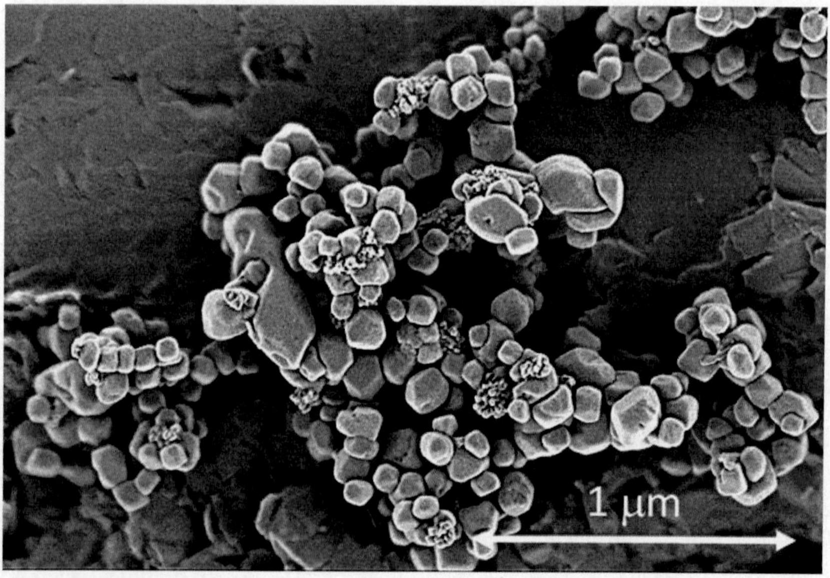


**Fig. 9.4** Micrograph of precipitated $CaCO_3$ (courtesy of E. Najafi and H. Huezo, Chemours)

### Dolomite

Dolomite is a mixed carbonate with roughly equimolar amounts of calcium and magnesium. Extenders made from dolomite are not as widespread as those based on calcium carbonate, due to their lower abundance and slightly higher cost. The properties of dolomite are generally in line with those of calcium carbonate, except it is slightly higher density, hardness, whiteness, and reactivity. The higher density of dolomite gives it a slightly higher bulk density and better flow than calcium carbonate. Like calcium carbonate, dolomite is hydrophilic and disperses quickly and easily in water. Also, like calcium carbonate, every deposit has its own set of unique properties. Overall, dolomites can be used in paints nearly interchangeably with GCC.

## *Silicates*

Silicates are among the most diverse class of minerals. They are also quite widespread, making up over three quarters of the Earth's crust. Despite their large variety, all silicates have one structural feature in common—they all contain silicon atoms that are surrounded by a tetrahedron of four oxygen atoms. The various silicates differ in the ways that these tetrahedra are arranged, both with respect to one another and with respect to tetrahedra or octahedra formed by other atoms, most typically aluminum, magnesium, or alkali metals. The oxygen atoms are shared between these polyhedra, typically bridging two or three silicon or other metal atoms. In some cases, the oxygen atoms are bound to hydrogen atoms, forming hydroxyl groups that can bridge two or more metal atoms or can form hydrogen bonds to other oxygen atoms. Hydroxyl groups are also present at the particle surface.

A number of silicates are used as extenders in paints and plastics. We will discuss six of the most common, beginning with kaolin.

### Kaolin

Kaolin, also called china clay or simply clay, is an important extender for coatings and, to a lesser extent, plastics. Kaolin is composed primarily of the mineral kaolinite ($Al_2O_3 \cdot 2SiO_2 \cdot 2H_2O$), with lesser amounts of other minerals such as feldspar, quartz, and talc. It is produced in three forms—hydrous kaolin, partially dehydrated kaolin (metakaolin), and fully dehydrated (anhydrous) kaolin. The fully dehydrated kaolin can be made by two processes, each giving different properties. These processes are conventional calcination and flash calcination. The advantage of flash calcination is that it produces particles with sealed air voids. These voids increase the opacity in paints and plastics via light scattering. Representative electron micrographs of these various forms of kaolin are shown in Fig. 9.5.

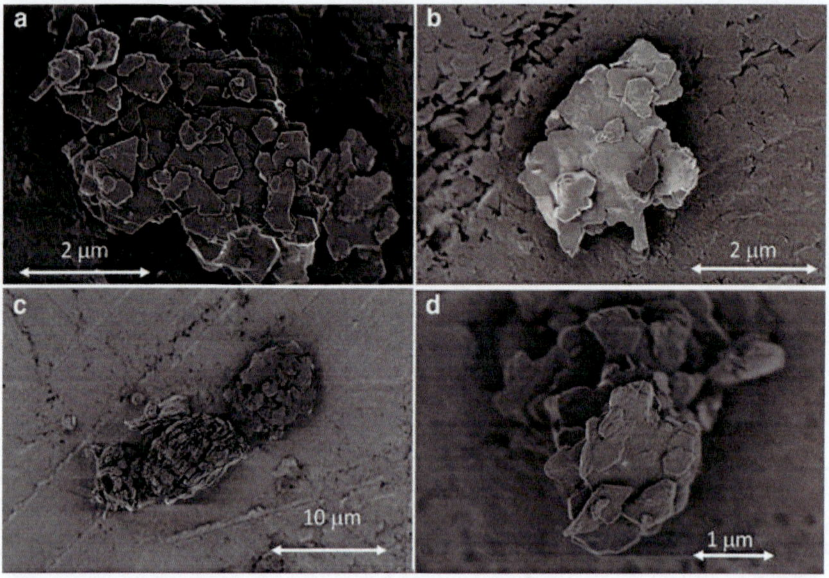


**Fig. 9.5** Electron micrographs of kaolin extenders (courtesy of E. Najafi and H. Huezo, Chemours)

Kaolin deposits are distributed globally and are formed by the weathering of aluminosilicate minerals such as feldspar. Kaolin is typically found as a soft clay (2.5 on the Mohs scale) that can be extracted through either wet or dry mining. In its native state, it is usually discolored, but brightness can be increased to relatively high levels through purification and/or calcination. Calcination also increases particle hardness, bringing it to a value between 4 and 5 on the Mohs scale.

At the atomic level, kaolin can best be described as sheets of $Si_2O_5$ layers bound to sheets of $Al_2(OH)_4$ layers (Fig. 9.6). The sheets are held together by hydrogen bonding, and as such crystals of kaolin are typically found as platelets with aspect ratios (width to height) between 5:1 and 20:1. The widths of materials used as extenders can be as high as 10 microns [18]. However, the widths of the particles used in paints and plastics are more typically between 0.2 and 5 microns. The smallest particles that are used as extenders can have aspect ratios as high as 50:1. Sheet-like silica arrangements such as those in kaolin are quite common in nature, and minerals containing them are termed phyllosilicates.

Calcination of hydrous kaolin results in an increase in particle size due to the bonding together of the faces of different particles [18, 20–22]. That said, small size calcined kaolin products are widely used—these are made by calcination of especially small hydrous kaolin particles. The merging together of particles during calcination gives these particles a blockier shape than the hydrated particles.

Kaolin is mostly used in paper applications, where it supplies brightness and opacity. However, the hydrated and anhydrous forms are also used extensively in paints, and the anhydrous form is sometimes used in plastics. Metakaolin, which is

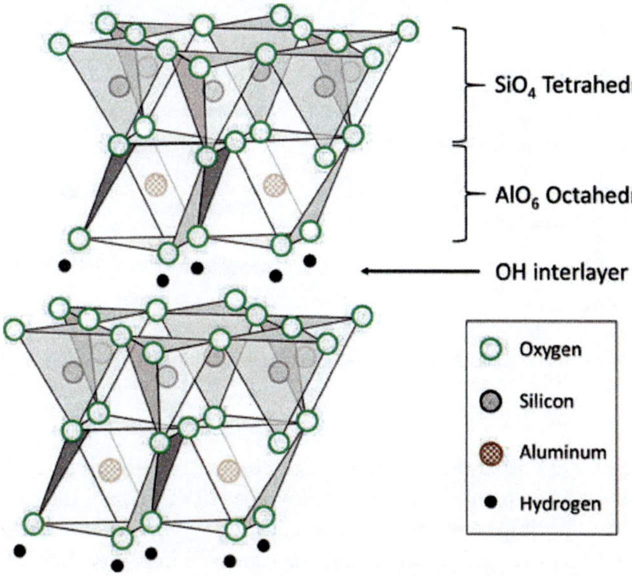


**Fig. 9.6** Atomic structure of kaolinite. Adapted from [19]

incompletely dehydrated kaolin, is used in cement but seldom found in paints. It can be used in PVC cable [18] and has been investigated as an extender for thermoplastics [23].

Kaolin is hydrophilic and so disperses easily in water. Its platy structure affects the rheology of liquid paints, giving them body and improving flow properties such as improved brushability, sag resistance, and leveling. They also minimize crack formation during drying and sedimentation during storage.

Due to the platy nature of these particles, they reinforce dry paint films, increasing their toughness and burnish resistance (burnish is an increase in gloss when dry paints are rubbed). They also improve film adhesion to the substrate and to any overcoats of paint. Kaolin improves the weatherability of paints by reducing water permeability, and for this reason, it is sometimes found in highly durable industrial paints.

Under conditions of very low shear (for example, during storage, when gravity is the only shear force), the faces and edges of kaolin platelets interact, organizing the particles into a "house of cards" structure (Fig. 9.7a). This occurs because the edges of the particles are silica rich, and thus negatively charged, while the faces are alumina rich, and thus positively charged. This structure can be easily broken down in paints by stirring or by application using a brush or roller, both of which generate shear. Under shear conditions, the platelets align in a different fashion (Fig. 9.7b). The organized structure quickly reforms after paint application, preventing the paint from dripping or sagging. The structure can also be either further stabilized or destabilized by the addition of ions that can alter the charge on the particle edges [18].

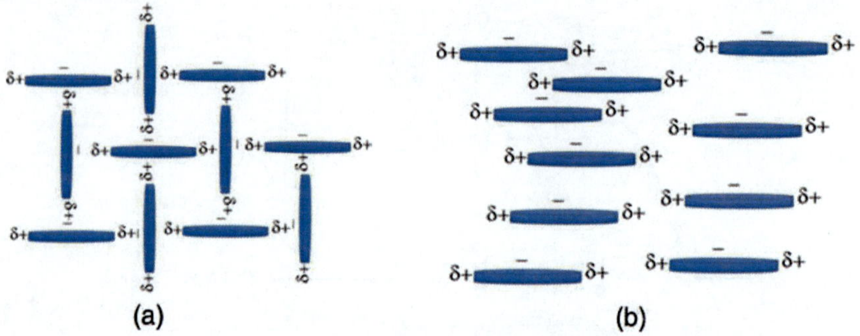


Fig. 9.7 Schematic representation of the self-organizing structures of clays at **a** low shear and at **b** high shear [24]

Hydrous grades, due to their small size, can be used in high gloss paints. They are commonly used in paints formulated below the CPVC, where they are claimed to improve the spacing between $TiO_2$ particles, allowing for some reduction of $TiO_2$ without a loss in opacity. Calcined grades are used in paints above the CPVC, where they increase opacity through dry hide (see Chap. 4). While they are considered to be more efficient at providing dry hide than hydrous kaolin or calcium carbonate, their larger size relegates them to matt or flat finishes. Brightness and opacity improvements are the main factors dictating the choice of kaolin above other extenders.

Kaolin is less prevalent in plastic applications due to its water of crystallization, relatively poor color, and poor heat aging in use. Calcined grades are more commonly used than hydrous grades, and these materials are often surface modified with stearic acid or silanes for better dispersibility and compatibility with the polymer matrix [18]. The finest grades (<2 microns) improve the tensile strength of plastics. They are used at low levels as anti-blocking agents for plastic films.

## Talc

Talc (or talcum) is a common mineral with the chemical formula $Mg_3Si_4O_{10}(OH)_2$. It is perhaps best known for its softness and is, in fact, the softest mineral, with a defined Mohs value of 1. This limits scratch and scrub resistance in paints and plastics, and it also lessens the wear of production equipment and minimizes discoloration due to this wear. It is formed when calcium is washed from other minerals by carbonated groundwater. As is true of most mined extenders, many properties, such as degree of discoloration, particle size, and impurity profile, will vary depending on the source.

The atomic structure of talc consists of oxide sheets stacked on top of one another [25]. These sheets, in turn, are composed of a two-dimensional layer of octahedral $MgO_4(OH)_2$ sandwiched between a pair of two-dimensional layers of tetrahedral $SiO_4$ (Fig. 9.8). The structure of these sheets is common in mineralogy, and they

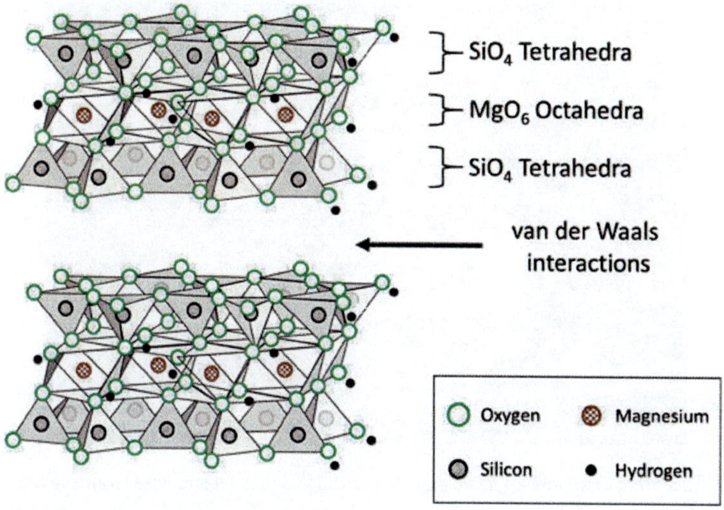


**Fig. 9.8** Atomic structure of talc. Adapted from [25]

are referred to as "TOT" sheets (Tetrahedron–Octahedron–Tetrahedron), based on the arrangement of the layers. Each sheet is electrically neutral and so there are no water of crystallization or ions between them. This results in the sheets being held together by only weak van der Waals interactions. The sheets can slide easily over one another, in a similar way to sheets of graphite, and this makes talc an excellent dry lubricant.

The lamellar atomic structure of talc results in most particles being platelets, although fiber forms are also known [26]. An image of a typical talc particle is shown in Fig. 9.9. Particle sizes range from below 1 micron to 15 microns. The smaller particles are referred to as "microcrystalline" and the larger particles as "macrocrystalline" [2, 27]. The particle aspect ratio is partially dependent on the size of the particle, with macrocrystalline particles generally having aspect ratios as high as 35:1 and microcrystalline particles having aspect ratios below 10:1 [2].

Talc affects paint properties in many of the same ways as other platelet extenders. Talc particles orient during drying, lessening mud cracking, decreasing film permeability, and increasing adhesion. It also has a strong effect on paint rheology, increasing viscosity during storage and shear thinning during application. The amount of talc used in paints is commonly below 20%, being limited by its effect of increasing viscosity.

Talc is hydrophobic, which differentiates it from most extender minerals. This hydrophobicity imparts water repellency to paint films. This, in turn, enhances the benefit it brings to anti-corrosion paints. Small-sized talc materials have relatively high oil absorption demand values, which decreases paint CPVC and enhances opacity through dry hide. Despite being hydrophobic, talc is not difficult to disperse if the appropriate wetting agents and dispersants are used.

**Fig. 9.9** Electron micrograph of a talc particle showing its platelet structure (courtesy of E. Najafi and H. Huezo, Chemours)

In polypropylene and other thermoset plastics, talc shows a good balance between strength and stiffness. It also provides good gas barrier properties and increases heat distortion temperatures [27]. Due to its lubricating nature, it can require extensive melt processing to assure good wet-in and dispersion. Particles with a diameter of less than 5 microns are preferred, although particles with diameter as high as 12 microns can be used. Talc is also valued as a good nucleating agent in polypropylene and nylon, which results in decreased processing times and higher plant throughput.

## Mica

The term "mica" refers to a group of 37 similar minerals of which only two, muscovite ($KAl_2(AlSi_3O_{10})(OH)_2$) and phlogopite ($KMg_3(AlSi_3O_{10})(OH)_2$), are of commercial importance. However, only the muscovite form is used by the paints and plastics industries, due to its lower degree of discoloration [28]. It is well known for its sheet-like structure and is typically found as flaky sheets, or books, that can have widths of one meter or more. Raw mica is typically processed by subjecting the mined particles to air jets. This both delaminates (exfoliates) the particles and decreases their median size. For extender applications, these flakes are then wet or dry ground into particles with diameters of 5 microns or greater, making mica one of the coarser extenders [26]. The aspect ratios of particles typically used in paints and plastics are similar to those of talc, although, for some materials, these values are very high (150:1). Mica has a relatively low hardness (2.5 on the Mohs scale—although this can be higher when silica impurities are present) and is basic in water slurries [2].

The atomic structure of mica is based on sheets of tetrahedrally coordinated silicon atoms with intervening octahedral layers of other atoms, similar to talc. Muscovite

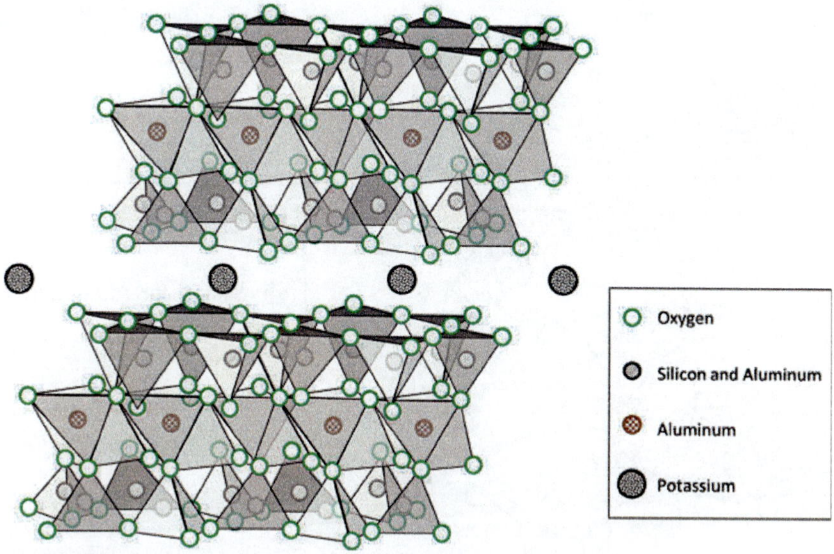


**Fig. 9.10** Atomic structure of muscovite mica. Note periodic absences of aluminum atoms in the octahedral mid-layer of each sheet. Adapted from [28]

and phlogopite have different atomic arrangements in the octahedral sheet, due to the different number of cations. In addition, in some forms of mica (including muscovite), some of the central atoms in the oxygen octahedra can be absent. Finally, varying amounts of silicon atoms in the tetrahedral layer can be replaced by aluminum atoms. In all cases, the sheets have a TOT-c structure, where "c" indicates the presence of a counterion between the TOT (Tetrahedron–Octahedron–Tetrahedron) sheets. The structure of muscovite is shown in Fig. 9.10.

The physical differences between mica and talc can be attributed to the presence of counterions in mica. These are required because the sheets in mica are electrically charged, while those in talc are not. Although the counterions in mica increase the bonding strength between the sheets (electrostatic forces are stronger than van der Waals forces), the bond strengths between the sheets are weaker than those within the sheets. This allows the sheets to be easily delaminated, although not to the same extent as those in talc.

The properties that mica confer to paints are similar to those of other platelet particles. Mica modifies rheology, minimizes settling, reinforces both the drying paint and the dry paint film, promotes corrosion resistance by the paint, improves film strength (for example, scrub and abrasion resistance), enhances film durability, and improves film adhesion. Mica has a distinctive sparkle appearance. Although this property is used advantageously to create special effect pigments (see Chap. 8), it is not always welcomed in paints. Overall, mica tends to be used only at relatively low concentrations in paints (5% or less of the total extender quantity).

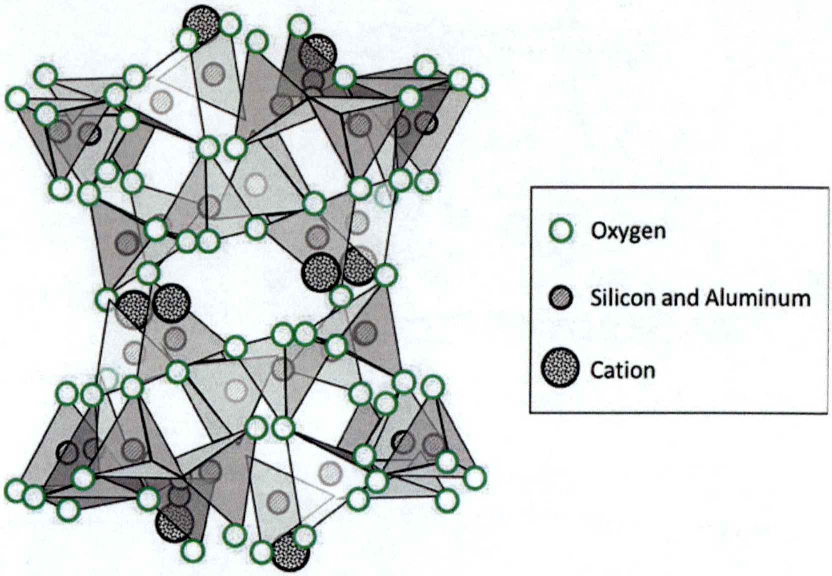


**Fig. 9.11** Atomic structure of feldspar. Tetrahedra are centered around either silicon or aluminum atoms. In anorthite, half of these are aluminum atoms, while in albite and microcline, a quarter are aluminum atoms. Adapted from [31]

Mica is valued in the plastic industry for its reinforcing properties [29], as well as its ability to increase the heat deflection temperature, decrease warpage, and reduce shrinking [30]. It also attenuates sound, which is desirable in automotive applications. The most common plastic application of mica is in polypropylene, where it is valued for its strong reinforcement and high dimensional stability. Mica has a high surface polarity, which decreases its compatibility with many polymers. This can cause poor adhesion between the polymer and particle, resulting in void formation at these interfaces, accompanied by a reduction in film strength. This incompatibility also affects the dispersibility of the particles. Surface treatments with silanes can minimize these deleterious effects [30].

### Feldspar

Feldspars are the most abundant family of minerals, accounting for more than half of the earth's crust. There are over 25 feldspar mineral types, but only three are used as fillers—orthoclase ($KAlSi_3O_8$), albite ($NaAlSi_3O_8$), and anorthite ($CaAl_2Si_2O_8$). Nepheline syenite is a related, but less pure, feldspar-like material that is often classified with the feldspars. Deposits of nepheline syenite are much less common than feldspar. Feldspars are generally known for their blocky structure and high hardness (6.0 on the Mohs scale). This level of hardness limits the degree to which these particles can be ground and can cause excessive equipment wear, both when it is processed and when it is used. However, this hardness translates to better strength in plastics and provides surface toughness to paints. Feldspars generally have adequate brightness for most paint and plastics applications. Due to their blocky structure, feldspars have lower surface areas, oil absorption values, and binder demands than non-blocky extenders with the same particle size.

The blocky nature of feldspar is a consequence of its three-dimensional, space-filling structure. This can be contrasted to the platelet shape of kaolin, talc, and mica, and is similar to calcium carbonate and silica. The atomic structure of feldspars consists of silicon and aluminum oxide tetrahedra joined together, through bridging oxygen atoms, into four-membered rings (Fig. 9.11). These rings are arranged in chains that join other chains that are oriented in different directions. This structure is similar to that of quartz (discussed below).

The oxide network in feldspar is negatively charged. This is due to the difference in electrical charge between the silicon (+4) and aluminum (+3) atoms—each aluminum atom in feldspar must have an associated counterion. The excess negative charges are balanced by different cations in the different feldspar minerals. In many cases, these cations are present as a mix, creating solid solutions. The counterions are located in cavities between the intersecting chains. The presence of counterions produces cleavage planes in feldspar.

Feldspar offers many benefits to paints. Its blocky shape allows it to be used at high concentrations, and its hardness confers good scratch and abrasion resistance to the paint film. Because fledspar is transparent in the ultraviolet, it can be used in radiatively cured paints. In addition, it generally improves paint brightness and durability. Balanced against these advantages are the limited degree to which these particles can be ground—typically the finest particles are 2 to 3 microns [2]—and their hardness, which is undesirable in many paint applications.

The hardness of feldspar also restricts its use in plastic applications. Its refractive index (between 1.50 and 1.55) gives it greater transparency in most polymers than any other filler at an equal concentration [32]. It can be used in films at low levels (0.1 to 1% by weight) as an anti-blocking agent [33]. It also has been used for greenhouse glass as it readily passes visible and ultraviolet light but reflects infrared light. This prevents greenhouses from overheating during the day and overcooling at night [34].

## Wollastonite

Wollastonite is a calcium silicate with the chemical formula $CaSiO_3$[35]. Deposits are not as widespread as those of the other extenders considered here—for example, there are only two mines in the United States, and they are located only a few hundred kilometers from one another [36]. Unlike the other silicates discussed so far, though, wollastonite particles are acicular. Their aspect ratios fall into two groups. The "powder" grades have aspect ratios between 3:1 and 5:1 while the "HAR" (High Aspect Ratio) grades have ratios that vary between 12:1 and 20:1. Grades used as paint or plastic extenders can be up to ten microns in their long direction.

Wollastonite is a relatively new mineral extender, having first been commercialized in the mid-twentieth century [37]. Its importance rose in the 1970s and 1980s when the negative health effects of asbestos became fully understood. Due to its acicular shape, wollastonite became a substitute for short fiber asbestos during that time [36]. Wollastonite is available in relatively pure form (97 to 99%), giving it good brightness and low discoloration. Wollastonite is the only white acicular extender currently available in industrial quantities.

The pH values of wollastonite slurries can be as high as 11, which is quite unusual for an extender. This is because in water a coating of highly caustic calcium hydroxide forms on the particle surface [36]. Wollastonite is moderately hard, with a Mohs value ranging from 4.5 to 5.5. Since it is anhydrous, it evolves little water when heated (weight loss at 1,000 °C is between 0.5 and 2.0%). Although it is anhydrous, overgrinding this material causes it to partially hydrate due to water absorption onto the freshly reated surfaces. In addition, overgrinding reduces the needles to nodular particles, resulting in a loss of many of the desirable properties that are due to its acicular shape.

Wollastonite's acicular particle shape is a consequence of its atomic structure. Wollastonite is composed of oriented one-dimensional, corner-sharing $SiO_4$ tetrahedra (Fig. 9.12). Calcium counterions reside in cavities between these chains. An electron micrograph of a wollastonite extender particle is shown in Fig. 9.13.

Wollastonite is used as a flattening agent and suspension aid in paints. In comparison to other silicates, it has a low binder demand and minimal effect on paint viscosity. These properties allow it to replace greater amounts of paint resin than is possible with the other silicates. Wollastonite is used to impart roughness, scrub and scratch resistance, hardness, adhesion, and durability to paints. Its high alkalinity gives it good corrosion protection properties, and it can synergistically enhance the anti-corrosion performance of other inhibitor pigments [38, 39]. Corrosion resistance is perhaps its most important property for the paint industry.

The largest industrial use of wollastonite is as an extender in plastics [36]. Due to its acicular shape, it can replace glass fibers in some fiberglass applications. Here it is valued for its lower hardness and abrasivity than glass, which has a Mohs hardness of 7. Fine particle size grades can be used for imparting scratch resistance. Wollastonite is valued in automotive plastics applications for giving strength, impact resistance, and dimensional stability. Wollastonite grades used in plastics, and in some paint applications, are often surface treated with bifunctional silanes. This imparts excellent plastic reinforcing performance, since it strengthens the interactions between the particles and the surrounding matrix.

### Precipitated Aluminum Silicate

As nature shows, there are a very large number of ways that aluminum and silicon can combine to form stable oxide structures. The energetic favorability of solid aluminum silicates is so great that mixing solutions of sodium aluminate and sodium silicate results in a violent release of heat and the near-instantaneous formation of a

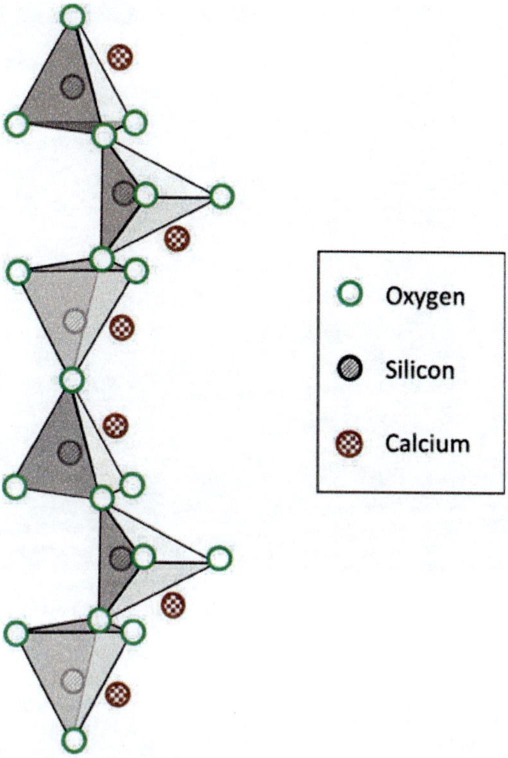


**Fig. 9.12** Atomic structure of wollastonite. Adapted from [37]

voluminous, bright white solid mass. Many forms of precipitated aluminum silicates are useful as extenders. We will not review the entire gamut of them here, but will mention some commonalities that they have.

Like all synthetic extenders, the precipitated aluminum silicates are characterized by their high purity and the resulting high brightness and low yellowness. The mean particle size varies from type to type, but is generally low, with some materials being less than 0.10 microns in diameter. Size is often difficult to measure because these materials tend to be at the lower limit of many types of size measuring technologies (see Chap. 2). Most precipitated aluminum silicates are nodular because their chemical bonds extend in three dimensions, similar to feldspar and silica.

Like other small particles, these materials have high surface areas, high oil absorption values, and high binder demands, and so they strongly influence paint and polymer melt viscosity. This limits their use to 5% or less in paints. Due to their small size, precipitated aluminum silicates allow for more optimal spacing of $TiO_2$ pigment particles than most extender particles (see Chap. 16). This increases opacity

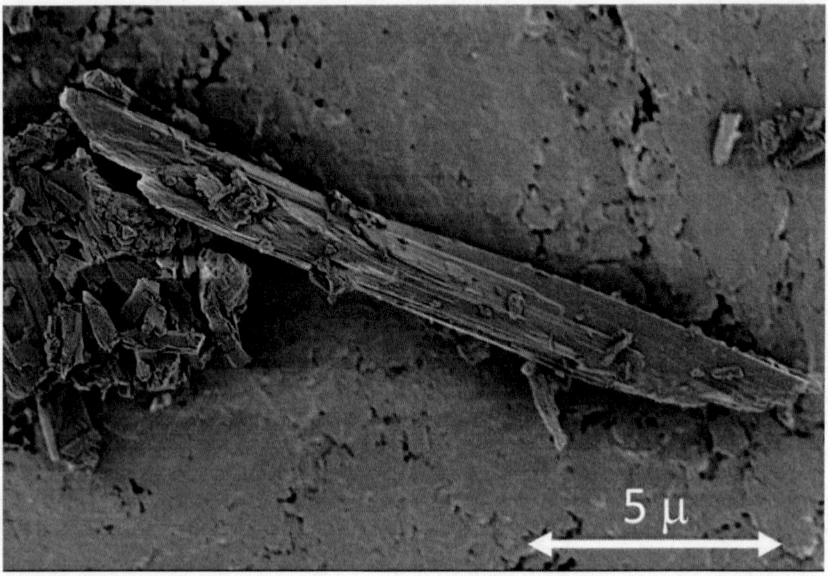


**Fig. 9.13** Micrograph of wollastonite showing its acicular structure (courtesy of E. Najafi and H. Huezo, Chemours)

**Fig. 9.14** Ball and stick schematic representation of the three-dimensional network of $SiO_2$ tetrahedrons found in quartz. The white balls are silica; the black balls are oxygen

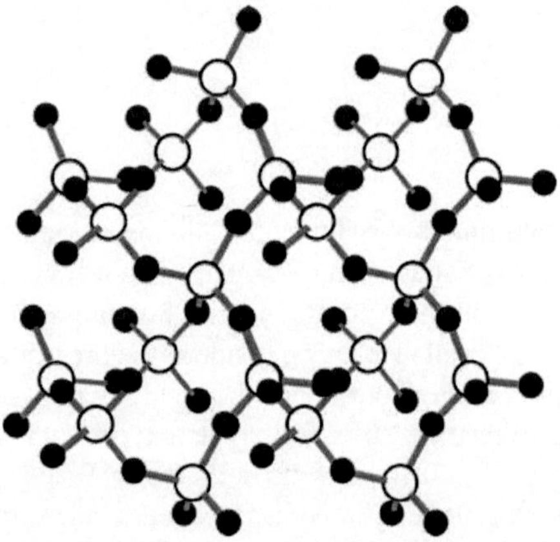

both below and, especially, above the CPVC of a paint, which contributes significantly to dry hide.

## *Silicas*

Pure silica has the chemical formula $SiO_2$. It is most commonly found in nature as the mineral quartz and, less commonly, the mineral cristobalite. It is also found in a variety of amorphous forms. Impurities in amorphous silica can impact its color and structure, such as seen in opals and obsidian. The different forms of silica are often found as sands, and together they comprise roughly 10% of the earth's crust. The distinguishing characteristics of silica are its high resistance to heat and chemicals, and its extreme hardness (7 on the Mohs scale).

A wide variety of silica extenders are available to the paint and plastics industries. These differ in particle shape, particle size, agglomerate structure, purity, and means of manufacture (natural, precipitated or fumed). The major types of silica extenders are quartz and diatomaceous earth, both of which are natural, and fumed and precipitated silicas, both of which are manufactured.

The atomic structure of silica is quite simple, consisting of a three-dimensional array of silicon atoms that are tetrahedrally bound to four oxygen atoms. Each oxygen atom is bound to two silicon atoms, resulting in bridges between the silica-centered tetrahedra (Fig. 9.14). In amorphous silica, the local tetrahedral structure surrounding the silica atoms is unchanged, but the tetrahedra are not aligned in the regular arrangement as they are in quartz and cristobalite. In all forms, the surface terminates with a mixture of Si–O–Si bridges and Si–OH hydroxyl groups. The surface hydroxide groups bond strongly to silanol- or siloxide-based organic surface treating agents.

### Quartz

Quartz is the main component of many sands and as a result is a very abundant mineral. It is typically found as particles that are too large for extender use, although some deposits of microcrystalline silica are known. Quartz can be comminuted to extender sizes using typical grinding processes, but due to its hardness, it is not ground to sizes below 2 microns [2]. Care must be taken, both in manufacture and use, to avoid discoloration (graying) due to the wear of metal equipment.

Quartz particles are typically blocky or nodular, as expected based on their three-dimensional atomic structure (Fig. 9.14). The nodular shape and relatively large particle size give quartz low surface areas and low oil absorption values. This allows for high loadings in paints without significantly altering viscosity. Quartz is often found in water-borne paints that use silicate binders in place of organic resin. Here, the dissolved silicate bonds strongly with the quartz surface as the paint dries . Quartz is also used in thick coatings to give high wear resistance [2], and in anti-corrosive paints to enhance substrate protection. The noticeable yellow cast of natural quartz often prevents its use in bright or white coatings.

Quartz is not typically used in plastic applications.

## Diatomaceous Earth (Kieselguhr)

Diatomaceous earth, also known as kieselguhr or diatomite, is a unique material. While its composition is amorphous silica, which is not unusual, its shape is not like any other mineral (Fig. 9.15). Diatomaceous earth is formed by the fossilization of the skeletons, or shells, of single-celled phytoplankton called diatoms. The structure of the original shell is preserved in the fossilization process. These structures are defined not only by their extraordinary shapes, but also by their extreme porosity (note the different pore structures and sizes in the different particles shown in Fig. 9.15).

Over 12,000 thousand species of diatoms are known, leading to an enormous array of shapes and sizes. The particle size is unique in the sense that it is determined biogenically, rather than through geological processes or comminution during processing. Grinding to smaller particles, in fact, leads to a loss in many of the unique properties for which diatomaceous earth is used.

Each deposit of diatomaceous earth is unique, both in the mixture of diatom structures and the nature of any impurities. The shape of a diatom is determined by its species, which, in turn, is dependent on the geographical conditions when the diatom was living and the age of the deposit. Different deposits can vary in brightness

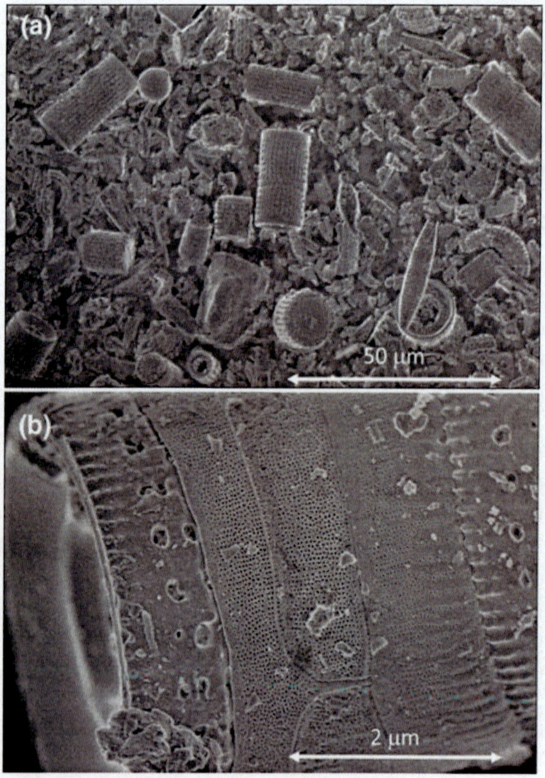


**Fig. 9.15** Electron micrograph of diatomaceous earth. **a** Assortment of shapes. **b** Assortment of pores (courtesy of E. Najafi and H. Huezo, Chemours)

as well as age, and darker materials are often calcined to improve brightness. This also reduces the surface area and converts the initially amorphous silica into crystalline silica (quartz).

Regardless of their shape and composition, diatomaceous earth particles always have a characterized by its very low density, which is attributable to their very high porosity. In addition, being composed of silica, it is very abrasive, with a Mohs hardness value of 6 (somewhat less than quartz). This unique blend of properties allows diatomaceous earth to be used in a wide range of applications, including toothpaste, catalyst supports, and liquid absorbent in addition to its broad use as an extender in paints and, to a lesser extent, plastics.

The combination of large particle size (2 to 20 microns) and high oil absorption (100 to 150 g oil per 100 g particles) make diatomaceous earth a useful flattening agent in paints. It is typically used for this purpose at levels of only a few weight percent [26]. Higher levels do not further decrease paint gloss but instead result in a large increase in paint viscosity, as a result of the high oil absorption of the material. Its high oil absorption also results in a strong reduction of the CPVC of a paint, which is advantageous in paints formulated above the CPVC for dry hide (Chap. 4).

Diatomaceous earth also helps maintain other particles in suspension and gives good brushability to the liquid paint. Its porosity allows paints to dry quickly and uniformly. It enhances the adhesion of paints to the substrate as well as improves tooth in undercoats (i.e., the interlayer adhesion). However, unlike many extenders, diatomaceous earth generally weakens the paint film. In addition, it is generally too yellow to be used in bright or white coatings.

In plastic film applications, diatomaceous earth is used at low levels as an anti-blocking agent.

### Fumed Silica

Fumed or pyrogenic silica, as the name suggests, is produced by a flame process in which an aerosol of silicon tetrachloride is hydrolyzed with water. This process is similar in some ways to the chloride process for titanium dioxide, where titanium tetrachloride is reacted in a flame to produce the oxide (see Chap. 7). However, for fumed silica, the oxygenation species is water, resulting in hydrochloric acid as a by-product, whereas in titanium dioxide, the oxygenation species is oxygen, resulting in chlorine gas as a by-product. This gas is returned to the earlier stages in the titanium dioxide process.

The primary particle size of fumed silica ranges from 7 to 40 nm. However, individual primary particles do not survive the flame process and are instead joined into various structures. These structures consist of aggregates, with sizes of around 120 nm, which, in turn, join together into looser agglomerates, which are typically greater than 1 micron [40]. The primary particles and agglomerate structure of fumed silica are shown in Fig. 9.16. In many ways, the structure found in fumed silica is similar to that of carbon black, which is also made by a flame process (see Chap. 8).

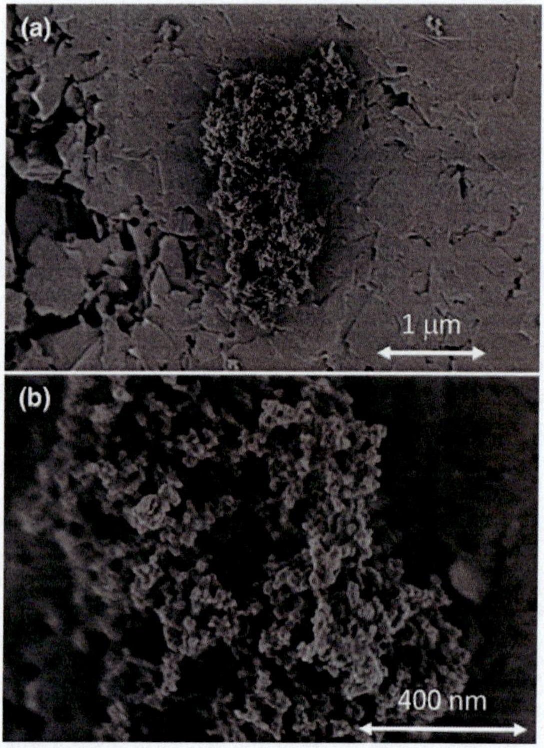


**Fig. 9.16** Electron micrograph of fumed silica under **a** magnification of 25,000 × and **b** magnification of 100,000x (courtesy of E. Najafi and H. Huezo, Chemours)

Fumed silica has an extremely high surface area (between 50 and 400 $m^2/g$) and an extremely low bulk density (0.03 g/ml). This density is so low that fumed silica must be compressed or granulated for handling purposes. This can be a challenge since over-compression can result in difficulties in dispersing the particles into a liquid paint or plastic melt. The surface silanol groups give fumed silica a hydrophilic character, although various organic post-treatments can make the particles hydrophobic.

In coatings, the very high surface area and the propensity to form large, three-dimensional networks via weak hydrogen bond linkages make fumed silica an effective rheological control agent at levels between 0.4 and 2.0 percent. Its high hardness improves the scratch and abrasion resistance of dry paint films, and its chemical inertness is an advantage in anti-corrosion paints.

Fumed silica also finds use in paints as an effective flattening agent [41]. Due to its small size, fumed silica particles readily migrate to the paint film surface during drying. There the particles reduced surface smoothness. Fumed silica also enhances dirt pick-up resistance due to its location on the film surface.[40] It is also used at levels below 1% as a flow aid in powder coatings. In this capacity, it shortens dispersion and melt times.

Fumed silica is used in plastics to adjust rheology and to provide anti-blocking for plastic films. It is also used in some niche plastics applications [42]. It is not used in plastics as much as precipitated silica is (discussed next), due to its higher cost.

### Precipitated Silica

Precipitated silica is produced by the neutralization of a silicate salt solution. The resulting particles are in many ways similar to fumed silica particles, having high surface areas, high oil absorption values, high binder demands, and low bulk densities. An important difference between these forms of silica is the higher amounts of water in the precipitated material—up to 7%, with as much as half of that being surface hydroxyl groups [42].

Precipitated silica is formed as round or blocky particles with a median particle diameter between 5 and 100 nm. However, agglomerates as large as 100 microns can form when the particles are dried. These large particles can be ground to diameters between 1 and 40 microns. Like fumed silica, precipitated silica has a very low bulk density and must be compressed or granulated for handling purposes.

The high levels of surface hydroxyl groups make these particles quite amenable to silane surface treatments. These alter the hydrophobic/hydrophilic balance of the particles, allowing them to be optimized for specific organic matrices. They are also sometimes treated with waxes for the same reason. Surface treatments are typically done at levels between 5 and 15% [43].

Precipitated silica is employed in paints for many of the same reasons as fumed silica. It is, however, more likely to be used as a flattening agent, and less likely to be used as a rheology modifier, as fumed silica is. In plastics, it is used to enhance strength (its largest single use is to enhance the strength of rubber tires). It is also used to modify the rheologic properties of polymer melts. Recent advances in surface treatment have significantly improved the dispersibility of precipitated silica in plastics [42].

## *Barium Sulfate*

Barium sulfate can be used as an extender in both its natural form (barite) and in synthetic form (blanc fixe). It is a salt with a structure consisting of discrete ions extending into three dimensions. This results in a nodular or blocky particle. Barium sulfate is relatively soft, with a Mohs hardness of 3.

Barium sulfate has two features that distinguish it from other extenders. First, it has a significantly higher density (4.4 g/ml) than all other common extenders (these vary from roughly 2.5 to 3.0 g/ml—refer to Table 9.2). Second, it has a refractive index of 1.64, which is higher than other extenders and of resins or polymers. While this is not high enough for barium sulfate to be classified as an opacifying agent, it does offer a limited degree of light scattering.

Natural barium sulfate extenders are typically comminuted to sizes between 1 and 15 microns through either wet or dry grinding processes [26]. The particles are nodular, as mentioned above, and the combination of relatively large size and high density gives barium sulfate extenders lower surface areas (1 to 4 $m^2/g$) than most other extenders. This also gives barite a low oil absorption value (roughly 15). Synthetic barium sulfate, known as blanc fixe, has a medium particle size ranging from 0.7 to 4 microns, depending on the synthesis conditions. The process for making blanc fixe is discussed in the section on synthetic extenders below.

Barium sulfate is used in coatings for similar reasons as other nodular particles. Although the high density of barium sulfate is an advantage in some ways, it is a disadvantage in others. It settles easily, and it significantly increases the density of paint. This is economically unfavorable in most paint markets, since paint ingredients are purchased on a weight basis and typically sold on a volume basis. Unlike many extenders, blanc fixe can be used in gloss paints as it does not roughen the film surface. It improves the toughness of a paint film, and its chemical resistance is valued in anti-corrosion paints.

Barium sulfate is used in the plastics industry, where its density confers good vibrational mass dampening. Its high density also suggests solidity and good quality to plastic consumers. Blanc fixe is also used in high concentrations in polypropylene or polystyrene for X-ray shielding applications.

### *Calcium Sulfate*

Calcium sulfate was at one time used extensively in the coatings industry. A blend of calcium sulfate and titanium dioxide was also popular, both to the $TiO_2$ manufacturer (as a way of disposing of sulfate by-product from the sulfate manufacturing process [44]) and to the paint manufacturer (as a way of decreasing the cost of opacity). However, this was at a time when paints were predominately solventborne. Calcium sulfate is soluble enough in water to create problems in water-borne paints [43], and the production of these calcium titanate pigments, as they were known, was discontinued in the early 1970s [45].

Calcium sulfate is used as an extender in plastics, where its main value is as a simple economic substitute for polymer.

## Production of Mineral Extenders

Extenders originate from natural minerals and rocks that can be mined either in the open air or underground. Once the materials of interest are removed from the earth, they are broken apart into ever-finer particles, initially by mechanical crushers, and then through wet or dry grinding. In some cases, the material is separated on a size basis during or after this process. A small fraction of extenders (the synthetic extenders) undergoes additional processing. This typically involves, in this order, the reaction of the raw material to a new material, purification (in some cases), then reaction to the desired material. These reactions are run under very tightly controlled conditions, resulting in particles with high purity, controlled size, and enhanced functionality. In general, the smaller size extenders, such as precipitated calcium carbonate, barium sulfate, and silica, are made synthetically, while larger size extenders are manufactured directly from the mined minerals [3].

## *Natural Extenders*

Mining is an essential aspect of extender production. The first step in the mining process is prospecting. During this phase, a number of aspects of the potential mine site are evaluated, including the extent and quality of the deposit, the thickness and composition of any overburden that must be removed, and the probable equipment needs and process steps to extract and purify the desired material. In addition to these technical and economical evaluations, consideration must be given to local, regional, and national government regulations, as well as existing regional infrastructure.

The installation of the mine begins once the prospecting phase has been successfully completed. The most appropriate mining techniques are evaluated and selected to match the specific site details. Typical equipment includes bucket-wheel and shovel excavators, conveyor belts, heavy-duty trucks, etc. The mining of highly compacted materials may require the use of explosives. All in all, several years can typically elapse between initially identifying a potential mine site and finally extracting the first shovelful of mineral from the ground.

Once mining commences, the material is removed and mechanically crushed to particles with diameters from 100 to 300 mm. After the physical removal of loose soil, detritus, and substandard material, the remaining material may be washed, ground, filtered, wet or dry milled, and classified, resulting in the final extender product. During the purification process, ferrous impurities, which often discolor the final product, may be removed via a magnetic separator.

Both dry and wet grinding can be used to reduce particle size. Dry grinding, which is typically done in ball mills, can cost-effectively reduce particles to a median diameter as low as 1.5 microns, depending on the material. For smaller particles, wet grinding becomes more economical despite the extra energy required to separate water from the material and then dry it. In the wet grinding process, an aqueous slurry with a high solid content is fed into ball mills that are designed to work with slurry materials. These mills can grind the particles to sizes below one micron. When the particles have the desired diameter, the material is dewatered, dried, and packaged. Alternatively, it can be delivered as a slurry, which avoids the water separation and drying steps.

Air classification is an essential aspect of the dry grinding process in many extender mining operations. There are several variants of air classifiers, but they are all based on the premise that small particles, having a large surface area on a

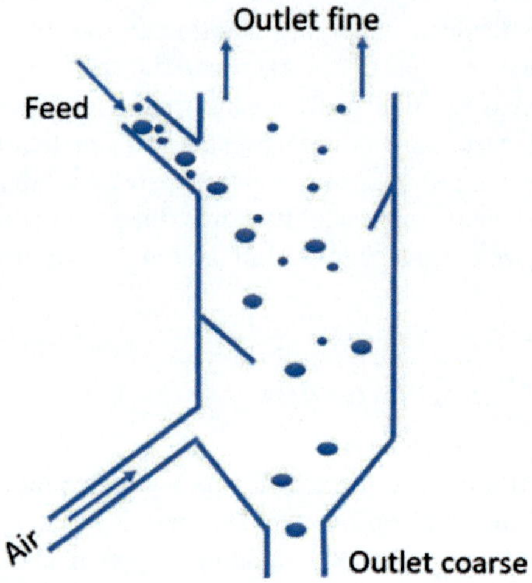


**Fig. 9.17** Schematic representation of an air classifier

weight basis, have greater air drag than large particles. The mix of ground particles is gravity fed through the side a chamber with a current of air blowing upwards. The smaller particles are entrained in the rising air stream and exit through the top of the chamber, while the larger particles fall to the base of the chamber and are returned to the mill (Fig. 9.17). The entrained particles are collected vertically from the air stream that exits through the top of the chamber. These particles can be classified by size according to the height they reach in the exiting air stream.

Some extenders undergo thermal processing, which can significantly improve many properties. This can be illustrated with the heat treatment of kaolin. Calcined, or anhydrous kaolin is produced by heating wet-ground kaolin to high temperatures. The dehydration process occurs in stages [46]. Between 450 and 650 °C, approximately 90% of the waters of crystallization are removed, converting the kaolin into an amorphous material called metakaolin. Metakaolin is used extensively in the cement industry, but not as an extender for paints or plastics. Further heating to between 900 and 955 °C removes the remaining water and converts the material to a spinel phase that can be used as an extender in paints and plastics. The hot material is cooled, then ground to undo the agglomeration that occurs during the calcination process. The calcination process increases purity, decreases discoloration, and alters the particle size and shape.

Flash calcination is a special variant of the calcination process that is unique to kaolin clay. In this process, ground kaolin is heated at rates between $10^3$ and $10^5$ degrees per second over a very short time (typically a few tenths of a second), then quenched at similar rates of cooling [47]. Under these conditions, dehydration is not complete, but the material is in a different form than when it is incompletely dehydrated through conventional calcination. Instead of creating metakaolin, the flash calcination process expands the platy structure of the kaolin particles and, at the same time, anneals the surfaces of the particles, which are heated the most, into a glassy structure. This seals the voids that are formed by the exiting water vapor and result in an amorphous particle that contains very small internal air cavities. These air cavities scatter light in end-use applications, improving opacity and allowing for a decrease in $TiO_2$ content. They also contribute to the wet hide of paint.

## *Synthetic Extenders*

Many minerals that are formed naturally can also be synthesized artificially. The starting material for this process is often the same same mineral that is being synthesized. For example, mined calcium carbonate is used as the starting material for synthetic calcium carbonate. The reason for doing this process is that the synthetic mineral can have higher purity (and less discoloration), a lower average particle size (and tighter size distribution), or both.

Although the synthetic details differ from one extender to another, nearly all follow the same strategy. The raw mineral is first reacted chemically to form a new species that contains the element of interest. This new species is, optionally, purified, and then is reacted a second time to return it to its original composition. This conversion process is well controlled in order to form the proper crystalline phase, the desired median particle size, and the optimal particle size distribution. This is similar in concept to the manufacture of pigmentary titanium dioxide (Chap. 7), where the titanium content of the ore is transformed into titanyl sulfate or titanium tetrachloride, purified, and then reacted to form pigmentary $TiO_2$ particles.

### Precipitated Calcium Carbonate

Precipitated calcium carbonate (PCC) is typically made by a three-step process. During the first step, $CaCO_3$ (limestone) of high purity is heated above 900 °C, where it decomposes into calcium oxide (lime) and $CO_2$. The carbon dioxide is collected to be used in the final step of the process, while the lime is immersed in water as the second process step. This produces a calcium hydroxide (slaked lime) suspension. In the third step, the $CO_2$ that was formed in the first step is added to this suspension, resulting in the deposition of the precipitated calcium carbonate particles. By adjusting reaction variables such as pressure, temperature, time, and the presence of chemical additives, the particle shape and size distribution can be controlled. Grinding or separation via classification may be necessary to produce the final extender material.

### Precipitated Barium Sulfate

Blanc fixe, or synthetic barium sulfate, is produced in a two-step process from the mineral barite ($BaSO_4$). First, barite is reacted with coke (carbon) at high temperatures, producing barium sulfide (BaS) and carbon monoxide. The barium sulfide is purified, dissolved in water, and reacted with a sulfate source such as sulfuric acid.[4] Barium sulfate is completely insoluble in water[5] and precipitates as a fine white powder.

Physical properties such as median particle size, particle size distribution, and shape are controllable using production process parameters such as concentration, temperature, agitation, pH, and aging time. Once the desired precipitate is formed, it is purified by filtration and washing, often multiple times. In the last step of production, the material is dried and ground to break up the agglomerates that form during drying. Blanc fixe particles are controlled in the range of 0.7 to 4 microns, making them smaller, and their particle size narrower, than the original barite particles [1]. They are also significantly brighter than their natural counterparts, allowing them to be used in bright paints.[6]

## *Surface Treatment of Extenders*

The surfaces of most natural extenders are hydrophilic, allowing for easy dispersion in water-borne paints. However, a hydrophobic surface can be advantageous in plastics and in some paint applications. In plastics, a hydrophobic surface would be more compatible with the polymer matrix, similar to the principle that "like dissolves like". In the same way, a hydrophobic particle would be more compatible in a non-polar paint system. In addition, a hydrophobic particle can make the paint film more water-resistant. The use of hydrophobic particles is significantly more common in plastics applications than paint applications.

For these reasons, extenders are often treated with an organic, hydrophobic agent as a final stage in their production. These agents modify both the particle–particle and particle–matrix interactions. While the particle–matrix interactions are usually the more important, the change in particle–particle interactions affects the bulk flow and bulk density of the extender, as well as its dustiness.

The modifiers used for plastics applications can be classified into two groups—coupling surface modifiers and non-coupling surface modifiers [48]. This distinction

[4] Lithopone is a related a white pigment that was used extensively before pigmentary $TiO_2$ was developed commercially. It is made by combining a barium sulfide solution with a zinc sulfate solution. Barium sulfate and zinc sulfide simultaneously precipitate from this mixture.

[5] The insolubility of barium sulfate can be demonstrated by its use as an X-ray contrast agent. Although soluble barium is quite toxic, barium sulfate can be safely ingested or otherwise entered into the body with no ill effects.

[6] The low hardness of barium sulfate also contributes to paint and plastics brightness because these soft particles do not abrade production equipment, which would discolor the paint or plastic.

is made according to the type of modifier/polymer interaction. Non-coupling modifiers are too short to enter into the polymer phase and so affect only the wettability and dispersibility of the particles. The effect of wetting has been the subject of debate [6, 48], but it appears to be negative—the surface modifier decreases the surface energy of the extender particles [49], while wetting is greatest on high energy surfaces [50]. The non-coupling modifiers include fatty acids such as stearic acid and its derivatives, and also some silanes with short alkane chains. Most of the surface-treated extenders that are used in the paint industry are coated with fatty acids.

Coupling modifiers, on the other hand, partially enter into the plastic matrix. An interphase region is believed to exist within the polymer that is near the particle surface. This region is especially important for the incorporation of particles with high surface energy (e.g., non-surface treated particles) into the polymer melt [6]. This region has properties that differ from the bulk polymer, and it is believed that coupling modifiers on the particle surfaces can diffuse into this region, strengthening the interaction between the particle and resin. The thickness of this region varies with the identity of the particle and the polymer, and is believed to range from one nanometer to over one micron.

Coupling modifiers are very similar in concept to the dispersant molecules used to stabilize particles in liquid paints (Chap. 11). They consist of two parts—a head and a tail [48]. The function of the modifier head is to bond to the surface of the particle. There are two types of heads—reactive and non-reactive. The reactive heads chemically bond to the particle surface, while the non-reactive heads do not. Reactive heads are often composed of silanols or alkoxysilanes, which can react with a surface hydroxide, releasing a water molecule. Carboxylic acid groups can also be used for this function. The tails are typically long alkyl chains, sometimes with reactive moieties that can form chemical bonds with the polymer.

## Summary

Extenders exist in a wide range of chemical compositions, shapes, particle sizes and size distributions, and physical and chemical properties. This results in an equally wide range of interactions with and uses in paints and plastics, going beyond the traditional role of extenders as particles that simply fill a paint or plastic, lowering its cost. The functionalities that these particles can bring to paint or plastic include enhancing processability, altering the rheology of liquid paints and molten polymer, preventing sedimentation of paints, improving various strength attributes (crack resistance, scratch resistance, tensile strength, stiffness, etc.), increasing weatherability, improving corrosion resistance, and enhancing certain optical properties (particularly dry hide and gloss in paints).

Extenders can be conveniently categorized according to their mineral composition. The major categories of extenders are the carbonates (which are the most commonly used extenders in paints and plastics), the silicates, the silicas, and finally insoluble salts. The particle shape is a function of the atomic arrangement within the crystal lattice. Particles with bonding interactions extending in all three dimensions are nodular or blocky. Those for which strong bonding interactions extend only in two directions, with weak hydrogen or van der Waals bonding in the third direction, are platy or lamellar, while those for which strong bonding interactions occur through one-dimensional chains are needle-like or acicular.

The origins of all mineral extenders are geological deposits. Extenders are extracted from the ground, then physically separated from other materials, such as loose soil or other detritus. Most extenders undergo grinding and size classification. Impure extenders can be processed by reactions that convert them to an alternate chemical, then back reaction to the original mineral. This allows for additional purification and for control of median particle size, particle size distribution, and, in some cases, particle shape. While this additional processing comes at a cost, in some applications, this cost is more than offset by the improved properties.

Additional details about the utility of extenders in paints and plastics formulations can be found in Chap. 16 (for paints) and Chap. 17 (for plastics).

## References


1. Parker, D. H., *Principles of Surface Coatings Technology*, Wiley (1966).
2. Gysau, D., *Fillers for Paints 2nd Revised Edition*, Vincentz (2011).
3. Roelle, W.; Mannheim, K. G., "Extenders", in *Paints, Coatings and Solvents, 2d Edition*, Wiley (1998).
4. *Functional Fillers for Plastics*, Ed. Xanthos, M., Springer (2005).
5. Anon., "Extender Particles", *Pig. Res. Technol.*, **27**(2) (1998).
6. Móczó, J.; Pukánszky, B., "Particulate Fillers in Thermoplastics", in *Fillers for Polymer Applications, Polymers and Polymeric Composites: A Reference Series*, Ed. Rothon, R., Springer (2017).
7. Wypych, G., *Handbook of Fillers, 4th Edition*, ChemTec (2016).
8. DeArmitt, C.; Rothon, R., "Particulate Fillers, Selection, and Use in Polymer Composites", in *Encyclopedia of Polymers and Composites*, Springer (2015).
9. Abel, A. G., "Pigments for Paints", in Paint and Surface Coatings: Theory and Practice, Eds. Lambourne, R.' Strivens, T. A., Woodhead Publishing (1987).
10. Hunter, R. S.; Harold, R. W., *The Measurement of Appearance, 2nd Edition*, Wiley (1987).
11. Werner, R.: Effect of Extenders with Narrow and Broad Particle Size Distributions on the Properties of Coatings. J. Coat. Technol. **72**(903), 71 (2000)
12. Krolikowask, A., Komorowski, L., Bonora, P.L.: The Effect of Size and Distribution of Inert Pigment on the Performance of Organic Coatings. Corr. Eng. Sci. Technol. **56**(2), 137 (2021)
13. Tooley, P. A.; Holtzen, D. A.; Musiano, J. A., "Processibility and Lacing Resistance when Silanized Pigments are Incorporated in Polymers", US 5,959,004 (1999).
14. Shan, G.; Diebold, M. P., "Aqueous Corrosion Resistant Coatings with Surface-Hydrophobic Inorganic Particles", US 20200157358 (2020).
15. Ertekin, B.: Chalking Up the Advantaged of $CaCO_3$. Polym. Paint. Col. J. **201**(4556), 32 (2011)
16. Diebold, M.P.: A Monte Carlo determination of the effectiveness of nanoparticles as spacers for optimizing $TiO_2$ opacity. JCT Res. **8**(5), 541 (2011)
17. Rothon, R.; Paynter, C., "Calcium Carbonate Fillers", in *Fillers for Polymer Applications, Polymers and Polymeric Composites: A Reference Series*, Ed. Rothon, R., Springer (2017).
18. Rothon, R., "China Clay or Kaolin", in *Fillers for Polymer Applications, Polymers and Polymeric Composites: A Reference Series*, Ed. Rothon, R., Springer (2017).

19. Kotal, M., Bhowmick, A.K.: Polymer Nanocomposites from Modified Clays: Recent Advances and Challenges. Prog. Polym. Sci. **51**, 127 (2015)
20. Thomas, R. E., "High Temperature Processing of Kaolinitic Materials", thesis, The University of Birmingham (2010).
21. Güneyisi, E.; Gesoglu, M.; Özturan, T.; Mermerdas¸, K., "Microstructural Properties and Pozzolanic Activity of Calcined Kaolins as Supplementary Cementing Materials", *Can. J. Civ. Eng.*, **39**, 1274 (2012).
22. Sennett, P., "Changes in the Physical Properties of Kaolin on Exposure to Elevated Temperature", *Proceedings of the 9th international Clay Conference, Strasbourg, 1989. Vol V*, 71 (1990).
23. Raj, M.M., Thummar, A., Raj, L.M., Patel, H.: Studies on Effects of Metakaolin as a Filler in Some Commercial Thermoplastics Polymers. Arch. Appl. Sci. Res. **7**(1), 1 (2015)
24. Schofield, R.K., Samson, H.R.: Flocculation of Kaolinite due to the Attraction of Oppositely Charged Crystal Faces. Disc. Faraday Soc. **18**, 135 (1954)
25. Martin, F., et al.: A Review of Ni and Co Incorporation During Talc Synthesis: Applications to Crystal Chemistry, Industrial Compounds and Natural Ni- and Co-rich Ore. J. Geochem. Expl. **200**, 27 (2019)
26. Ralston, H. P.; "Extender Pigments", in *Print and Coating Testing Manual*, ASTM (1995).
27. Rothon, R., "Talcs", in *Fillers for Polymer Applications, Polymers and Polymeric Composites: A Reference Series*, Ed. Rothon, R., Springer (2017).
28. Anon., "Strength and Sheen", *Paint Coat. Tech.*, May (2003).
29. Luis, J.; Woodhams; R. T., Xanthos, M., "The Effect of Flake Aspect Ratio on the Flexural Properties of Mica Reinforced Plastics", *Poly. Eng. Sci.*, **13**(2), 139 (1973).
30. Zilles, J. U., "Micas", in *Fillers for Polymer Applications, Polymers and Polymeric Composites: A Reference Series*, Ed. Rothon, R., Springer (2017).
31. Pakhomova, A., et al.: Polymorphism of Feldspars Above 10 Gpa. Nat. Commun. **11**, 2721 (2020)
32. Katz, J. V.; Milewski, H. S., *Handbook of Fillers and Reinforcements for Plastics*, Van Nostrand Reinhold Co (1978).
33. Zilles, J. U., "Feldspar and Syenites", in *Fillers for Polymer Applications, Polymers and Polymeric Composites: A Reference Series*, Ed. Rothon, R., Springer (2017).
34. Schwartz, H.; Shachar, A.; Bar, A., "Absorbent Go for Agricultural Films", $4^{th}$ European Congress of Additives and Colors, March 2005.
35. Zilles, J. U., "Wollastonites", in *Fillers for Polymer Applications, Polymers and Polymeric Composites: A Reference Series*, Ed. Rothon, R., Springer (2017).
36. "Wollastonite - A Versatile Industrial Mineral", U.S. Geological Survey Fact Sheet 0002–01 (2016).
37. Anon., "Wollastonite: A Versatile Functional Filler", *Paint Coat. Tech.*, November (2002).
38. Hare, C.J.: The Evolution of Calcium Metasilicate in Paint and Coatings. Mod. Paint Coat. **83**(12), 43 (1993)
39. Wolfe, M., "Calcium Metasilicate Maintains Performance, Minimizes Cost", *Paint Coat. Tech.*, May (2012).
40. Reader, C.J., Nargiello, M.: The Use of Engineered Silica to Enhance Coatings. CoatingsTech **17**(6), 36 (2020)
41. Lin, B. T.; Biesiada, C., "Novel Synthetic Silica Matting Agents for Polyaspartic Coatings", *Proceedings of the 2016 Waterborne Symposium* (2016).
42. Rothon, R., "Precipitated and Fumed Silicas and Related Products", in *Fillers for Polymer Applications, Polymers and Polymeric Composites: A Reference Series*, Ed. Rothon, R., Springer (2017).
43. Braun, J.H., Baidins, A., Marganski, R.E.: $TiO_2$ pigment technology: a review. Prog Organic Coat. **20**, 105 (1992)
44. Sullivan, R. W., "Calcium Sulfate Extended Titanium Dioxide Pigments", US 2,549,261 (1948).
45. Stieg, F.B.: Calcium-Based Titanium Pigments: Death of a Giant. J. Paint Technol. **44**(565), 63 (1972)

46. Khokhani, A.; Folmar, K. W., "New Generation Kaolin Based Paint Pigment Extender", EP 2,744,863 (2013).
47. Bridson, D., Davies, T.W., Harrison, D.P.: Properties of Flash-Calcined Kaolinite. Clays Clay Miner. **33**(3), 258 (1985)
48. DeArmitt, C.; Rothon, R., "Surface Modifiers for Use with Particulate Fillers", in *Fillers for Polymer Applications, Polymers and Polymeric Composites: A Reference Series*, Ed. Rothon, R., Springer (2017).
49. Móczó, J., Fekete, E.: László, K; Pukánszky, B, "Aggregation of Particulate Fillers: Factors, Determinations and Properties." Macro. Symp. **194**(1), 111 (2003)
50. Fox, H.W., Hare, E.F., Zisman, W.A.: Wetting Properties of Organic Liquids on High-Energy Surfaces. J. Phys. Chem. **59**(10), 1097 (1955)

# Chapter 10
# Resin Particles

## Contents



## Introduction

Since their introduction by Dow Chemical in the mid-1940s [1], latex paints have grown from novelties into the dominant technology in many market segments, particularly architectural/décor coatings. Today water-borne coatings account for roughly 40% of the total coatings market, making this the largest single technology for coatings. By comparison, solvent coatings account for roughly 25% of the total market, with powder coatings, UV cured coatings, high solids coatings, and specialty coatings making up most of the remaining volume.

The dominance of water-borne coatings is due to a combination of factors, the most important of which is the reduced volatile organic compound (VOC) emissions compared to the solvent-borne paints that they replaced. Legislative pressures worldwide are continuously becoming tighter on VOC emissions, creating an ever-increasing drive to produce paints with very low or no VOC discharge [2]. In addition to this, water-borne paints do not carry the flammability risk of solvent-borne paints,



M. Diebold et al., *Pigments, Extenders, and Particles in Surface Coatings and Plastics*, https://doi.org/10.1007/978-3-030-99083-1_10

have shorter drying times, are easier to clean up after application, and do not carry the health risks or the objectionable odor of many solvent-borne paints.

Most application research on the resin particles that are used in water-borne paints has focused on the polymer chemistry of these materials. This is appropriate given the importance of not only the selection of raw materials but also the exact reaction conditions used to make these particles. As such, numerous review articles and books have been written on the polymeric aspects of latex particles, and the wide variety of paint end-use performance attributes that these polymers affect [3–7].

Our interest here, however, is quite different. We will focus on the nature of the resin particles themselves, as they exist in the liquid paint, and on their coalescence into the continuous phase of the dry paint. While resin particles share many properties with the other particles in the liquid paint, their coalescence into the continuous phase is unique to them—all other particles in the liquid paint maintain their individuality after film formation. In addition to describing the coalescence process and the factors that affect it, we will discuss resin particles that contain air voids (for air void hiding below the CPVC) and reactive resin particles that attach to the $TiO_2$ pigment particles, enhancing the spacing between these particles and so optimizing their light scattering potential.

## Comparison of Solvent-Borne and Water-Borne Coatings

A reasonable starting point for understanding the properties of resin (latex) particles in water-borne paints[1] is a simplified comparison of these paints to their solvent-borne counterparts. After application and drying, the two paint types create similar cohesive films consisting of high molecular weight organic polymer molecules intertwined with one another and holding any particles in the film in place. However, the film-forming mechanisms for these two paint types are quite different.

In solvent paints, the long polymeric chains that will form into the continuous phase of the dry paint film are dissolved in a chemically similar organic solvent. As the solvent evaporates, the long strands of resin intertwine and chemically crosslink, both of which lock these polymeric molecules into place as a solid film. The non-film forming particles, too, are locked into place within the matrix of the resin as the film dries. A thin layer of resin coats the non-film-forming particle surfaces, preventing them from directly touching one another or the substrate, or from being exposed to the film surface.

The resin in water-borne paints, on the other hand, is not dissolved in the paint liquid. Instead, it is present as discrete particles with size ranges similar to those of the non-film-forming particles. As the paint dries, the surfaces of the resin

[1] These paints are sometimes referred to as "emulsion" paints, but this is incorrect. The resin particles are made by first generating a suspension of an immiscible liquid monomer in water, which is the definition of an emulsion, but these suspended liquid droplets are then polymerized into solid particles, which is the definition of a latex.

particles partially solubilize due to interactions with coalescing agents (discussed below). The partially dissolved outer shells of the resin particles intertwine with one another, allowing them to coalesce into a coherent film as the last of the water and the coalescing agent evaporates. The partially dissolved resin in the drying water-borne paint film is relatively easily deformed and squeezes onto and between the non-film-forming particles in a way similar to the resin in a drying solvent-borne paint.

A second important difference between solvent-borne and water-borne paints is the amount of polymer that can be dissolved in the paint solvent. For solvent-borne paints, this amount is restricted by paint viscosity, which increases greatly with polymer concentration (as well as with molecular weight). The rate of increase is especially high at higher solids levels. At high polymer concentrations, the intertwining polymer molecules cause a drag or resistance to flow (the same phenomenon occurs as a solvent-borne paint dries). High solids solvent-borne paints (>65% solids) can be made, but their viscosities are roughly three to four times that of conventional lower solids solvent-borne paint and they must be applied by a spray gun.

This can be contrasted to the response of water-borne paint viscosity to latex concentration. Because the polymer molecules on different resin particles do not intertwine in a water-borne paint in the same way that the polymer molecules in a solvent-borne do, there is a much lower dependence of viscosity on resin concentration. In this respect the latex particles can be as highly concentrated as any other particles in the liquid paint.

An obvious yet important aspect of water-borne resins is that they are used as water slurries. As such the particles must be stable—that is, they must not agglomerate during paint production or storage. However, an explicit requirement of these particles is that they interact so strongly during drying that they not only touch one another but also intimately join together. This is accomplished by taking advantage of the chemical and physical changes that occur during paint drying. Physically, surface and capillary forces grow to the point of overcoming the electrostatic repulsions that stabilize the particles in the initial paint. Chemically the concentration of the coalescing agents, if present, increases, increasing their effectiveness at allowing the individual particles to join into a coherent solid phase.

## Coalescence

Film formation in water-borne paints is quite complex and with the particles involved going through a series of chemical and physical changes. However, we can grossly simplify the formation process by considering it to consist of three distinct steps, as outlined schematically in Fig. 10.1 [8]. Initially, there is enough water present that the resin particles are well separated. As water evaporates, the latex particles become more concentrated and begin to deform into a honeycomb arrangement.

Upon further drying the polymer molecules on the surfaces of different particles diffuse into one another and intertwine, creating in the drying film a single mass from

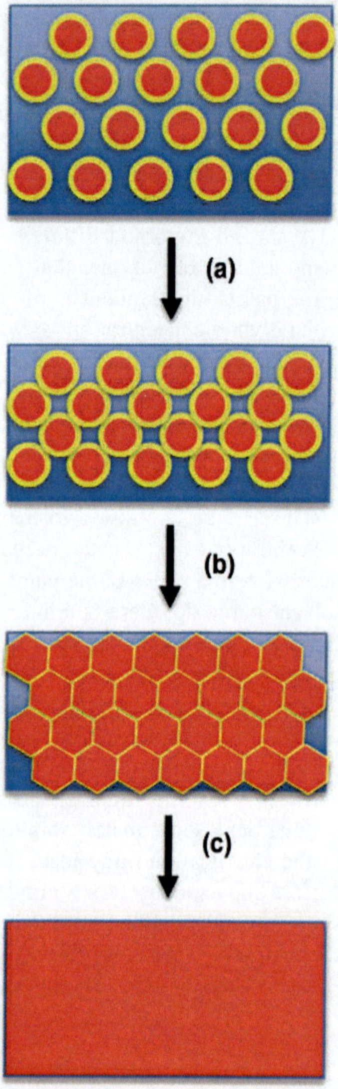


**Fig. 10.1** The three steps of film formation in a water-borne paint. **a** Initial evaporation. **b** Particle alignment and deformation. **c** Diffusion and coalescence

the individual particles. Often this intertwining of the polymer molecules is enhanced by chemically bonding to one another through crosslinking. The quality of a paint film is strongly affected by the degree to which the resin particles coalesce, which is, in turn, determined by the properties of both the resin and the certain aspects of the paint system.

## Factors Affecting Coalescence and Film Formation

The nature of resin polymers varies greatly according to the chemistry of the monomer(s) and the exact conditions of manufacture. This allows the polymer scientist to create resins that satisfy often highly demanding end-use requirements for properties such as paint weatherability, stain and dirt pick-up resistance, scrubbability, gloss, impact resistance, and so on.

The chemistry and manufacture of the polymers used in water-borne paints are detailed in numerous books and papers, as noted earlier. For our purposes, however, we will limit our attention to the primary function of the latex particles, which is to coalesce into a coherent film. It is only after the film forms that the paint properties listed above become pertinent. As such the successful transformation of latex particles into a coherent film is a critical aspect of virtually all latex paints.

The successful coalescence of individual latex particles into a continuous film requires the careful formulation of the paint. Often a factor that optimizes one aspect of film formation will have a negative impact on another. As is ever the case, a balance between competing requirements is necessary to arrive at a paint that satisfies all end-use demands in the most cost-effective manner. The primary parameters that can be varied to achieve such a balance are certain thermal properties of the latex resin and the type and amounts of certain additives (plasticizer or coalescing aids) that promote the intertwining of resin molecules on different latex particles.

### *Tg and MFFT*

When heated, polymers can undergo a variety of different thermal changes. A familiar one is melting, which is the complete transformation from a solid to a liquid. A sharp melting temperature is normally seen only in crystalline polymers, while most resin used in paints is amorphous, or a mixture of crystalline and amorphous domains. Rather than having a sharp melting point, these materials typically show a more nuanced physical change when heated to a certain temperature known as the glass transition temperature, or Tg.[2] [9, 10] At this temperature there is a change from a hard, brittle, glassy state to a soft, sticky, pliable state. Polymers below their Tg are often referred to as "hard" while those above their Tg are referred to as "soft".

This change that occurs at the Tg and above is physically caused by an increase in the internal space available to the resin molecules (i.e., their free volume). As is true of most materials, the volume of latex particles increases with increasing temperature. At the Tg this increase in free volume is enough to allow the polymer chains a greater freedom of motion, increasing their mobility and allowing them to deform when placed under stress. Near and above their Tg, the surfaces of latex

[2] This is not a phase change, since the material is solid both before and after, but it nonetheless is a sharp and significant change in properties.

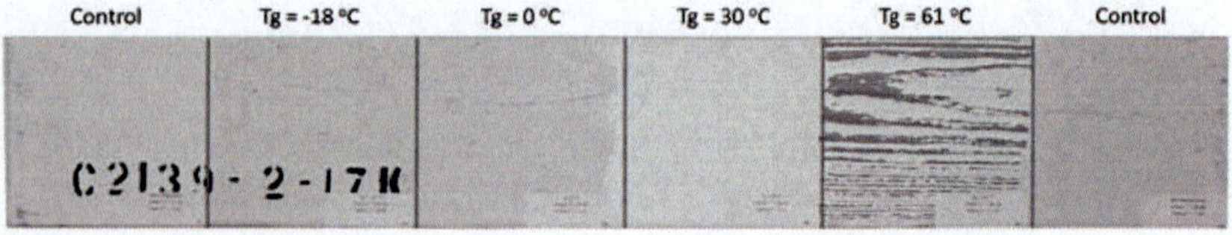


**Fig. 10.2** Effect of Tg on dirt adhesion and film flexibility (Courtesy P. Lutz)

particles are fluid enough that polymer chains from different particles can intertwine spontaneously, allowing these particles to coalesce.

An important difference between the Tg of an amorphous polymer and the melting point of a crystalline polymer is that the change in properties that occur on melting do so at a single temperature, whereas the softening of an amorphous polymer continues to increase as the temperature increases. For this reason, the coalescing abilities of most latex particles improve at higher temperatures.

The coldest temperature for which latex particles will coalesce is called the minimum film-forming temperature (MFFT), below which there is simply not enough intermingling of the surface polymer molecules on different latex particles for coalescence to occur.[3] Not surprisingly, the MFFT is strongly linked to the Tg—more specifically, it is typically a few degrees cooler than the Tg. The difference between Tg and MFFT can be increased to some extent using coalescing agents, which are discussed below.

Tg values for resins are important both in the particle state (especially for the coalescence step of the drying paint) and in the final dry film. Our focus in this chapter will be on the effect of Tg on coalescence, but we will briefly review the effect of Tg on film properties.

Films above their Tg are more pliable and less prone to cracking or flaking, which is desirable if the substrate expands or contracts with changing temperature or humidity. In addition, these films typically have good impact resistance.

Balanced against the advantages seen in films above their Tg temperature, however, are many disadvantages. Because the surface molecules in a polymeric film above its Tg are fluid, these films are often tacky. This leads to increased dirt pick-up and to blocking, which is the sticking together of two freshly painted surfaces when pressed against one another. In some cases, the adhesion of the film surfaces can be stronger than the adhesion between the film and substrate, and separating the surfaces after pressing them together can result in the film being pulled from one of the surfaces. Other problems with films used at temperatures above their Tg are that they can be easily scratched or marred.

The effect of Tg on film flexibility (as seen by paint flaking), and on film tackiness (as seen by darkening due to dirt adhesion), is shown in Fig. 10.2. Here we see a

[3] The MFFT of a resin particle is measured by applying a suspension of resin on a bar for which there is a temperature gradient and allowed to dry (this gradient being created by subjecting the ends of the bar to different temperatures). The MFFT is determined from the position on the bar where the film transitions from being clear (coalesced) to cloudy (uncoalesced).

pine panel divided into six sections. The first and last sections were painted with the same control paint, and the four interior sections were coated with paints that were similar to one another except for their Tg value. The panel was exposed for one year in the vicinity of a heavily used highway and under high ambient temperatures (daily high temperatures between 25 and 35 °C for much of the year). The darkening of the first three of these paints lessens as Tg increases, as expected since the films should become less tacky as Tg is increased. While the final film in the Tg series remained mostly dirt-free, it had unacceptable cracking and peeling due to its inflexibility.

Not surprisingly, the Tg of a latex depends strongly on its chemical composition. This includes aspects such as type of resin (thermal plastic or thermal setting), degree of crystallinity, crosslink density, degree of branching, chain length, and bonding interactions between molecules. Typical Tg values span a range of roughly −80 °C to over 100 °C, with any intermediate value being possible through co-polymerization of hard and soft monomers. For most water-borne paints, the Tg is somewhere between 10 and 20 °C, while some oven-dried paints can have Tg values closer to 100 °C.

There is no single ideal Tg for a paint. A low Tg is desirable from a film-forming standpoint, but many end-use properties of the paint film are also affected by Tg, as described above. The ideal Tg for a particular paint depends on its application temperature and its intended use. For example, exterior paints used in dirty environments should have Tg values above ambient, to prevent excessive dirt pick-up. On the other hand, elastomeric roof coatings must be flexible and so are formulated with a Tg below ambient.

## *Coalescing Agents*

As outlined above, there are conflicting factors that the formulator faces when choosing the Tg of the resin used in a paint. On the one hand, in most applications a relatively high Tg will reduce the disadvantages of a “soft” film (blocking, dirt pick-up, etc.). On the other hand, the MFFT must be lower than the application temperature in order for a true film to form, and the MFFT is typically only a few degrees different than the Tg. These conflicting requirements usually result in an unacceptably narrow window of temperature over which the paint can be applied, if the application temperature is near the in-use temperature.

One resolution to these conflicting requirements is to dry the paints at temperatures above the intended in-use temperature (i.e., to oven dry). In this way a high Tg (and high MFFT) resin can be used while at the same time achieving a hard film during service. This strategy works for paints that are applied at the factory (i.e., OEM paints), but the majority of water-borne paints are used on substrates that cannot be placed in an oven (e.g., the wall of a room). In these cases a second resolution is available. This option is to add a low level of a material that will soften the surface of the particle, lowering its MFFT while the material is present in the drying film [1, 11, 12].

These additives are collectively called plasticizers, as they increase the plasticity, or deformability, of the particle surfaces. Plasticizers can be divided into two types: those that are intended to remain in the film after drying, and those that are intended to leave the film. The latter is typically referred to as film-forming agents, coalescents, coalescing aids, or coalescing agents, and these are the more common type of plasticizers used in paints because permanent plasticizers result in permanent film softness.

Coalescing agents are typically organic molecules with limited water solubility. Because of their limited solubility, these molecules tend to segregate on the organic resin surfaces. This segregation increases as water evaporate from the drying paint and the molecular concentration increases. These molecules provide a dual role at the latter stages of drying—they increase the plasticity of the surface layer, enhancing the deformability step of film formation, and, since they are organic, they partially solubilize the polymer molecules on the resin surfaces, allowing polymer molecules on separate particles to intermingle, enhancing the coalescence step of film formation[4] [13].

Coalescing agents are typically used at concentrations of 3% or lower and at these levels can decrease the MFFT of a resin by as much as 15 °C. Coalescing agents cover a range of hydrophobicity/hydrophilicity balances and evaporation rates. Ideally, the evaporation rate will be lower than that of water, to ensure the material remains at the latter stages of film formation, but high enough that the molecules do not remain in the film for extended times after drying. This avoids the initial poor blocking performance and water sensitivity of the films.

## Binding Power Index

Coalescing agents soften only the top layer of the resin particle, which creates a hard particle core. The relative amounts of soft and hard resin, and the size of the resin core, has an important effect on the distribution of pigment and extender particles in a water-borne paint film. This is because, unlike in solvent-borne paints, only the soft surface layer of the resin particles, and not its entire polymer content, is available for film formation. To understand the implications of this, we will first classify resin in the final paint film into three categories, or types, as was explained briefly in Chap. 4.

The first category, termed "type 1" resin, adsorbs as a thin layer on the particle surfaces. In nearly all paint films, whether originating from a solvent-borne paint or a water-borne paint, all particles are coated with a monolayer of resin polymer.[5] As such there are seldom direct surface-to-surface contacts between particles, and few if any exposed particle surfaces on the film surface.

---

[4] That is, they increase the free volume of the polymer chains.

[5] The exceptions being films with very low levels of resin (i.e., films at very high PVC), where there is no types 2 or 3 resin and insufficient type 1 resin to coat all particle surfaces.

The second category of resin, "type 2" resin, is the resin that fills the voids between the coated particles when they are packed at their densest arrangement—that is, at the CPVC condition.[6] If there is enough type 2 resin to completely fill these voids, then the film is below CPVC. If there is not enough type 2 resin to completely fill the voids, then the voids are partially filled with air. The total volume of types 1 and 2 resin at the CPVC is the "resin demand" of the non-resin particles present in the paint (pigments and extenders) and is similar in concept to the oil demand, as measured in the oil absorption test, for these particle mixtures. In fact, the oil demand of a given particle mixture in the OA test is essentially equal to the resin demand of that mixture in solvent-borne paints, and OA measurements are often used to calculate CPVC values of solvent-borne paints, as discussed elsewhere.

The final category of resin, "type 3" resin, is any resin in excess of the resin demand of the particles. This resin pushes the particles apart, increasing their separation and, for $TiO_2$ particles, increasing their scattering strengths. This resin is often referred to as "free resin". In our thought experiment from Chap. 4 we increased the PVC of a film that was below its CPVC by replacing resin with particles. The resin being replaced in this thought experiment is exclusively type 3 resin. Above the CPVC we increased film PVC by replacing resin with air. Such a replacement is initially done only using type 2 resin. Once all of the two resins are replaced, we increase PVC by exchanging type 1 resin with air.

With this in mind, we will next consider the availability of the polymer in the latex particles to transform into each of these three types of resin. In the presence of a coalescing agent, the surface of the latex particle is deformable and on film formation can become any of the three types of resin. However, the rigid core beneath this layer cannot deform, and so it cannot transform into type 1 or 2 resin.

There are significant ramifications of the solid core being unable to transform into type 1 or 2 resin. These can best be understood using the concept of resin "efficiency". We will define any portion of the resin polymer that can transform into any of the three resin types as being fully efficient. This includes all resin in a solvent-borne paint but is limited to the deformable surface layers of the resin particles in a water-borne paint. We define the efficiency of a resin particle as the fraction of the polymer that can satisfy the resin demand of the other particles in the paint (i.e., of the extender and pigment particles) by transforming to type 1 or 2 resin. This quantity is referred to as the "binder efficiency" (symbol $e$) or the "binding power index", BPI [14]. In this text we will use the latter term.

A consequence of the BPI is that the total amount of resin required to satisfy the resin demand of a mixture of pigment and extender particles is greater for water-borne paints than it is for solvent-borne paints (for which the resin is 100% efficient) [15]. A comparison of the measured CPVC values for a universal grade of $TiO_2$ in a solvent-borne paint and two water-borne paints made with chemically identical but differently sized resin particles is shown in Table 10.1. Note that we can consider the size of the resin molecules in a solvent-borne paint as the lower limit of resin

[6] The CPVC is defined as the state where all the particles are coated with a thin layer of resin, packed in their tightest configuration, and the voids between them filled with resin.

**Table 10.1** CPVC values and deformable layer thicknesses as a function of resin particle radius

| Paint type | Resin particle radius (nanometers) | CPVC value | BPI | Core radius (nanometers) | Deformable layer thickness (nanometers) |
|---|---|---|---|---|---|
| Solvent-borne | 0 | 54.6 | 1.0 | 0.0 | – |
| Water-borne | 65 | 38.2 | 0.514 | 51.0 | 14.0 |
| Water-borne | 135 | 27.1 | 0.309 | 119.4 | 15.6 |

particle size (i.e., as particles with a size of zero), and for this "particle" the BPI is at the top limit of its range (1.0). Resin particles at the other limit of the BPI range (0.0) are rigid hard particles that are completely non-coalescible.

We can calculate the BPI of a specific latex resin by comparing the amount of resin required to reach the CPVC value of a particle mix if all of the resin was available for transformation into types 1 and 2 resin (i.e., the CPVC calculated from an OA measurement of the particle mixture, or the CPVC measured in a solvent-borne paint), to the amount of latex resin required to reach the CPVC value for this particle mix. For the former we will use the volume of oil required to reach the oil absorption end-point for the non-resin particle mix. This definition of BPI is shown mathematically in Eq. 10.1:

$$\mathrm{BPI} = \frac{V_{\mathrm{Oil}}}{V_{\mathrm{Latex}}} \tag{10.1}$$

where $V_{\mathrm{Oil}}$ is the volume percent of the oil at the oil absorption endpoint and $V_{\mathrm{Latex}}$ is the volume percent of latex resin at the paint CPVC (calculated as 100 - CPVC).

Equation 10.1 is useful as a way of understanding the meaning of the BPI, but it is not ideal for calculating it because the OA measurement is made on a weight basis rather than a volume basis. As such, Eq. 10.2 is more useful for the BPI calculation:

$$\mathrm{BPI} = \frac{\mathrm{CPVC}_{\mathrm{Latex}}(100 - \mathrm{CPVC}_{\mathrm{Oil}})}{\mathrm{CPVC}_{\mathrm{Oil}}(100 - \mathrm{CPVC}_{\mathrm{Latex}})} \tag{10.2}$$

The logic behind this equation is best illustrated through an example. Consider a pigment/extender particle mix that has a solvent-borne CPVC value, calculated from its oil absorption, of 54.6, and a measured CPVC for a particular latex resin of 38.2 (these are the values for the first two entries in Table 10.1).

For one liter of dry solvent-borne paint film at its CPVC value (54.6) we have 0.546 L of particles and, by difference, 0.454 L of resin. The resin demand of the particles, therefore, is 0.832 L of types 1 and 2 resin per liter particles (0.832 = (0.454/0.546)).

We next consider the content of one liter of dry latex paint film at its CPVC value (38.2). Here we have 0.382 L of the pigment/extender particle mix and 0.618 L of resin. The resin demand of the 0.382 L of the pigment/extender particle mix is 0.318

L (0.318 = 0.382 × 0.832). The binding efficiency, or BPI, of this resin is therefore 0.514 (0.514 = (0.318/0.618)). In other words we say that the latex paint requires 0.618 L of latex resin in order to obtain the 0.318 L of types 1 and 2 resin needed to satisfy the resin demand of 0.382 L of the pigment/extender particle mix. We arrive at the same BPI value of 0.514 by substituting the CPVC values measured for the latex paint (38.2) and calculated based on the oil absorption value (54.6) in Eq. 10.2.

We can use the results of the BPI calculation to determine the thickness of the deformable outer layer on the latex particle for the two different latex particles in Table 10.1. We do this by first calculating the total volume of a particle, then multiplying this volume by the BPI of the resin. This gives the volume of the deformable shell. By difference we can calculate the volume of the rigid core. From this volume we can calculate the radius of the solid core, and finally the thickness of the outer layer is simply the particle radius minus the core radius.

The calculated thickness for the deformable layers of the two latex resins is given in the last column of Table 10.1. Here we see good agreement between the thicknesses of the deformable layers (within 10%), which implies the coalescing agent penetrates the resin particles by the same distance, regardless of their size. This is consistent with the fact that these binders have the same chemistry and the paints use the same coalescing agent at the same concentration.

A second consequence of the BPI concept for latex particles is that the spacing of the pigment and extender particles is dependent on the size of the resin particles. This is because the pigment and extender particles cannot penetrate the rigid cores of the resin particles. This is analogous to the effect of large extender particles on the distribution of $TiO_2$ particles that was discussed in Chap. 4. There we saw that large extender particles crowd small $TiO_2$ pigment particles into the voids between them, which resulted in a decrease in scattering power of the $TiO_2$ particles.

The effect of latex particle size can be seen in the electron micrographs shown in Fig. 10.3. Here we compare the images of paints made with the same universal grade of $TiO_2$ at the same PVC (20). The top two images are of paints made with different size resin particles (note that these are not the paints described in Table 10.1). The mean resin particle size for these paints is shown as a red circle in each image. The final image shows the same $TiO_2$ pigment in a solvent-borne paint. The distributions of particles are different and the particle spacing decreases with increasing resin particle size. The consequences of these spacing differences on opacity are quite significant—Paint A has only 82% of the opacity of Paint B, which in turn has only 94% of the opacity of Paint C (the solvent-borne paint).

The effect of resin particle size on the scattering strength of the $TiO_2$ particles is shown in Fig. 10.4, which was generated using the two latex particles shown in Table 10.1. Here we see the significant effect of resin particle size on $TiO_2$ crowding, as was also seen in Fig. 10.3. In addition, we see that the CPVC value for the larger resin particle is much lower than that of the smaller particle. This is expected since the resin efficiency (BPI) of the larger particle is lower than that of the smaller particle.

BPI is a property of a specific latex resin and varies from one latex to another, but, for a given resin, does not vary from one paint to another. The main parameters

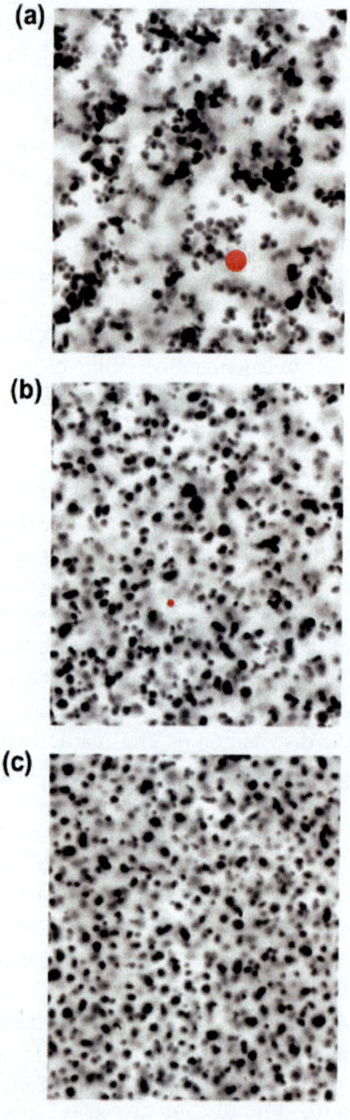


**Fig. 10.3** Effect of resin particle size on $TiO_2$ Dispersion. Paints (**a**) and (**b**) are latex paints (average particle sizes are indicated by red circles) while Paint (**c**) is a solvent-borne paint [16]

determining the BPI are the size of the latex particle, the amount and type (or effectiveness) of the coalescing agent, the Tg of the resin, and the temperature [17]. The effects of these parameters are straightforward. Because coalescing agents can only penetrate to a certain depth, large resin particles have a larger rigid core than small resin particles and so a lower BPI [18]. Higher amounts of coalescing agent will increase the amount of deformable resin, as will a lower Tg and a higher application temperature. The overall effect of these parameters on BPI is shown in Fig. 10.5.

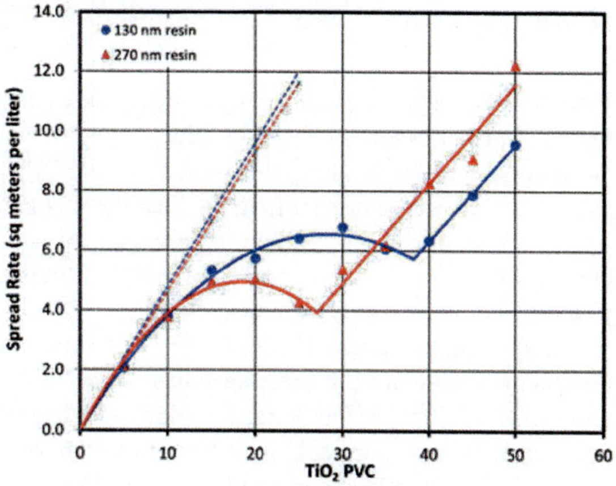


**Fig. 10.4** Effect of resin particle size on the opacity versus $TiO_2$ PVC curve

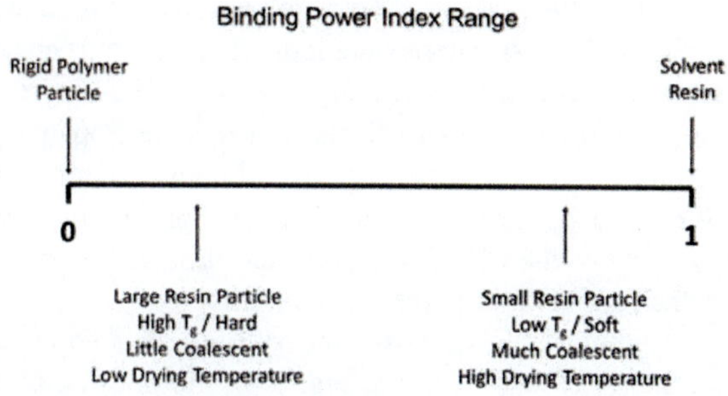


**Fig. 10.5** Factors affecting the BPI

As will be discussed in Chap. 16, the BPI is an important consideration in the formulation of a latex paint.

## Specialized Resin Particles for Improved Opacity

On a volume basis, $TiO_2$ particles are the most efficient scatterers of visible light and so $TiO_2$ is the strongest white opacifier. However, on a weight basis, $TiO_2$ pigment is often one of the more expensive ingredients in a paint. Much effort has been spent over the last 60 years to decrease the cost of white opacity by either finding a less efficient but lower-cost $TiO_2$ alternative that gives higher opacity per unit cost or by increasing the scattering power of the $TiO_2$ pigment (primarily through better spacing in the paint film as discussed in Chap. 4). Both of these strategies have been pursued through the use of specialized resin particles.

## *Hollow Sphere Opaque Polymer*

Paint or plastics opacity can be improved by the inclusion of air voids in the polymer matrix. In plastics air voids can be added using blowing agents, while in paints this is often done by formulating the paint above its CPVC value—that is, by replacing so much resin with extender particles that there is no longer enough resin to fill the inter-particle voids. While this is an inexpensive way to improve paint opacity, it comes at the cost of a significant reduction in film integrity as these voids generally join together into an extensive pore network.

An alternative method of adding air voids to paint is by using air encapsulated in polymer resin—that is, discrete air voids or bubbles. In the past this was done using large (>10 microns) polymer particles into which air voids and, in some cases, $TiO_2$ particles were incorporated. These "vesiculated beads" were commercially available in the 1970s and 1980s but eventually could not compete with $TiO_2$, which itself underwent significant opacity improvements over this time [19].

In the 1980s, a second form of encapsulated air was introduced to the coatings industry [20–22]. These were individual air voids encapsulated within a single resin particle of roughly pigmentary size. These have variously been called voided latex particles, hollow latex particles, hollow sphere opaque polymers (HSOPs), or simply opaque polymers. In this text we will refer to them as HSOPs. The critical advantage of employing air in this form, rather than in the form created when formulating a paint above the CPVC value, is that these voids remain discrete and so do not lead to the formation of pores.

Although light scattering theory suggests an optimal air void size of roughly 0.5 microns (refer to Chap. 3, Fig. 3.6), the optimal size of voids for these particles has, in practice, been found to be roughly 0.30 microns. This difference may reflect the fact that as the particle size of the non-$TiO_2$ components of a paint increases (e.g., resin size, and extender size), the $TiO_2$ particles become crowded and lose scattering efficiency and so the scattering gains of increasing the void size to 0.5 microns are offset by the losses due to increased $TiO_2$ crowding. These materials are commercially available with ranges of pore diameter from roughly 0.25–0.35 microns and shell thicknesses from 0.05 to 0.10 microns, and a total particle size of 0.4–0.5 microns.

HSOPs can be created by a number of related processes. At its simplest, there are three steps involved in making these particles. First, cores of soft, hydrophilic particles in water are synthesized. The material used must be capable of swelling when certain environmental conditions, such as pH, are changed.

Next, a shell of a hard, typically more hydrophobic resin is placed around these cores. This shell must have a Tg that is significantly higher than ambient temperatures but be capable of being exceeded during particle manufacture. It must also be permeable to water and small ions, such as simple bases, in order to allow for the inner core to swell.

Finally, the slurry of core/shell particles is heated to a temperature above the Tg of the shell, typically 50–100 °C, although higher temperatures are possible in pressure vessels, and the conditions of the slurry are changed in such a way as to swell the inner core. On cooling, the outer shell becomes rigid. When these particles are dried, their inner cores, which, on a volume basis, consist mostly of the water that swelled into the particle in the third step of the synthesis, fill with air.

There are many important reaction parameters that control the exact nature of the HSOP particles. These include composition (both the chemical identity and relative balance of the different species), number of shells,[7] thermal properties or hardness of the shell (i.e., Tg), softness of the core, method of expansion, particle size of core, and thickness of shell(s).

It is worth noting that the desired particle morphology—with the softer, more hydrophilic polymer as the core and the harder, more hydrophobic polymer as the shell—is thermodynamically unstable against the polymers being in the reverse order [23]. However, through a careful balance of kinetic and thermodynamic variables, the desired morphology can be attained.

Core swelling is most commonly achieved by incorporating carboxylic acid groups into the core polymer. Base is then added during the final step of production. This base penetrates the shell and reacts with the acid groups of the core polymer, increasing the ionic strength within the core. Osmotic pressure then forces water through the shell, decreasing the concentration of the ions within the core. Enough water infiltrates the particle to create a substantial increase in core volume. Other mechanisms can be used for swelling—for example, incorporating basic groups into the core polymer and acidifying the slurry to swell the core, or using organic molecules for swelling (although in this case the VOC content of the slurry is relatively high).

The particle shells have several important requirements. They must be rigid at paint application temperatures; otherwise, they would collapse under the high compressive forces present during film drying. Styrene or styrene co-polymers are typically used and can be crosslinked to increase strength against collapse.

The shells also must be permeable. Permeability is often achieved by incorporating a small amount of polar co-polymer into the shell resin, or by adding plasticizers (which must be later removed so as to keep the VOC content of the material low). Finally, they must not be coalescible, as this would cause them to self-coalesce and collapse when water is removed from the interior void. For this reason, the HSOPs are considered as particles, and not resin, when calculating paint PVC (i.e., HSOPs are considered extender particles in Chap. 4, Eq. 4.1). That said, modified HSOP particles have been reported for which a coalescible outer layer is placed over the hard shell layer [24]. In this case the particles can be considered to contribute volume to both the resin and the particle terms in the PVC equation.

Unlike $TiO_2$, the scattering strength of the HSOP particles does not decrease with particle concentration—that is, dependent light scattering is not an issue with these particles (Chap. 3). This is thought to be because the shells of touching particles are

[7] In particular, a "tie" coat is often deposited after core formation but before shell creation, especially when the two polymers are substantially different in terms of chemical identity and physical properties.

thick enough so as to prevent scattering volume overlap of the inner air voids. In addition, at the sizes used in commercial applications, the particles are small enough that they do not negatively impact the spacing of the $TiO_2$ particles.

Although initially developed in the 1970s, these materials have undergone a number of changes over the years and remain an area of active research interest [25]. These materials are typically used to decrease the $TiO_2$ content of a paint. In this regard they are not 1:1 replacements of $TiO_2$, but instead a higher volume of HSOP particles is required.[8] Replacement rates as high as 20% are reported for semi-gloss and satin paints and up to 50% for flat paints [26]. Replacement levels are typically limited by wet hiding (these particles do not scatter light in the wet paint) and by burnish resistance. A new version of this strategy has been reported that can be used in polyethylene plastic [27].

## *Reactive Resin and Resin/$TiO_2$ Composite Particles*

A second resin-based technology for improving light scattering in paints is the use of reactive resins. These are resin particles that attach to the surface of the $TiO_2$ particles, creating $TiO_2$/resin composite particles. The attached resin particles provide a steric barrier layer that prevents the close approach of the $TiO_2$ particles to one another and so improves their spacing. In this regard they are similar to the heavily treated $TiO_2$ particles discussed in Chap. 4. However, they do not dilute the $TiO_2$ content of the pigment, do not significantly change the resin demand of the pigment, and do not have some of the other detrimental effects that the heavy treatment can have on film properties.

There are two requirements for these resin particles. The first is that they must be small, preferably smaller than the $TiO_2$ particles so as to properly space the $TiO_2$ particles.[9] Second, they must interact with the $TiO_2$ particles strongly enough to bind to them but, at the same time, weakly enough so as to avoid extensive $TiO_2$ flocculation through the formation of $TiO_2$/resin/$TiO_2$ chains or agglomerates [24]. This second requirement also dictates the order of addition of the resin particles and $TiO_2$ particles dispersed in water. Specifically, the $TiO_2$ particle must be added to the resin. In the early stages of the reverse addition—that is, the addition of resin to $TiO_2$—the slurry is $TiO_2$ particle rich and the excess $TiO_2$ particles will attach to already deposited resin particles, resulting in the creation of bridges between particles and the formation of the extensive floc structures described above.

[8] Note that there are two synergistic effects that these particles have when replacing $TiO_2$—they scatter light themselves, and by decreasing the $TiO_2$ content of the paint, they increase the scattering strength of the remaining particles (Chap. 4).

[9] Large resin particles will more distantly space the $TiO_2$ particles, but there are far fewer large particles than small particles at the same particle volume. Because of this, if enough large particles are present to space each $TiO_2$ particle, the resin concentration would be so high as to limit the $TiO_2$ concentration below that necessary for single or double coat coverage.

The interactions between the resin particles and the $TiO_2$ surfaces are generated by the incorporation of up to a few percent of functional groups within the polymer particles. A number of different functional groups have been investigated, with the most successful being phosphates [28, 29]. Note that the surfaces of the $TiO_2$ pigment particles are actually dominated by different combinations of hydrous alumina, hydrous silica, and hydrous zirconia that are deposited on the bare $TiO_2$ particles to improve various aspects of their performance (refer to Chap. 7). Several different reactive resin grades are available to optimize the interactions with the different $TiO_2$ surface compositions.

Because these composite particles are present in the wet film; they can, in some cases, slightly improve wet opacity (unlike HSOP particles). In addition, they are coalescible and so can partially or entirely replace the conventional resin. Because they prevent direct $TiO_2$ particle–particle interactions in the dry film, they improve the quality of the film. This is manifest as an improvement in barrier properties (e.g., corrosion and humidity resistance), stain, scrubbability, dirt pick-up resistance, tannin stain blocking, and efflorescence resistance [24].

As is the case with HSOP particles, reactive resins are reported to allow for a $TiO_2$ reduction of up to 20% at high $TiO_2$ PVC values (i.e., above 20), where there is a significant loss of $TiO_2$ scattering strength due to crowding, although lower reduction amounts are possible at lower $TiO_2$ PVC. In addition, the HSOP and reactive resin technologies improve opacity by different mechanisms and their effects are sometimes additive [28].

## Summary

The resin in water-borne paints is present as particles rather than as discrete polymeric molecules, as is the case in solvent-borne paints. The function of these resin particles is to coalesce into the continuous phase of the drying paint film. This process, and the factors that affect it, are the focus of this chapter.

Coalescence is defined as the intertwining of polymer molecules residing in touching resin particles. This process converts resin from a particulate into the coherent phase necessary for film formation. Several parameters control the ease and extent of coalescence. In this regard the most important property of the particles themselves is their minimum film formation temperature (MFFT), which is determined, in part, by their glass transition temperature (Tg). The MFFT is the temperature below which coalescence cannot occur. Typically the MFFT of a polymer will be a few °C lower than the Tg.

In general, a high Tg is desired for the final paint film, as films below their Tg are soft and tacky. However, this is at odds with the need for a low MFFT for good film formation. This conflict can be resolved by adding a coalescing agent to the liquid paint. This is an organic molecule that will be driven to the resin particle surfaces during the paint drying process. Once there, these materials partially solubilize the surface polymer molecules. This reduces the MFFT of the polymer by up to 15 °C, which allows polymer molecules on touching surfaces to intertwine at temperatures significantly below the normal MFFT. The coalescing aid must leave the film in order to restore the Tg of the polymer to above ambient. It does this by evaporating from the film during the final stages of drying.

The central roles of the resin in a paint film are to coat the pigment and extender particles and to fill the voids between them. In a solvent-borne paint, all of the resin is available for these purposes. However, in a water-borne paint only the portion of the resin that is partially solubilized by the coalescing agent is available for these purposes. Because of this a higher amount of resin is necessary to coat the pigment and extender particles and fill the voids between them when they are packed at their tightest—that is, to satisfy their resin demand and meet their CPVC point—than is needed for solvent-borne paints.

The binding power index (BPI) is used to quantify the loss in effectiveness of the resin in fulfilling its roles in the dry film. This index is simply the ratio of the concentration of the resin at the CPVC of a solvent-borne paint (or linseed oil, if using oil absorption to determine CPVC) to the concentration needed to reach the CPVC for a similar waterborne paint resin. We will encounter this important parameter again when we discuss paint formulation in Chap. 16.

Finally, specialized resin particles are available that improve the hiding power of the paint film. They do this by either scattering light themselves or by improving the $TiO_2$ particle spacing—and so their scattering strength. Resin light scattering is accomplished with hollow polymer spheres, which bring air into the paint without the disadvantages associated with the air voids that are found in paints formulated above their CPVC. In the case of HSOP particles, the air is present as discrete voids, while when formulating above the CPVC air is present as extensive pores. Resin that improves $TiO_2$ spacing does so by permanently attaching small resin particles to the $TiO_2$ surfaces while still in the liquid paint, preventing a close approach of these particles in the dry film. Such materials are called reactive resins, and the $TiO_2$/resin species formed in this process are called composite particles.

## References


1. Taylor, J.W., Klots, T.D.: An applied approach to film formation – the glass transition temperature evolution of plasticized latex films. PCI Mag. (2002)
2. Burns, M.E.: A Comparison of Solvent and Water-Borne Alkyd Coatings and the History of VOC Regulation in the United States. Masters Thesis, California Polytechnic State University San Luis Obispo (2016)
3. Freitag, W., Stoye, D.: Types of Paints and Coatings. Paints, Coatings and Solvents Second, Completely Revised Edition, 11 (1998)
4. Padget, J.: Polymers for Water-Based Coatings - A Systematic Overview. J. Coat. Technol. **66**(839), 89 (1994)
5. Klier, J., Bohling, J., Keefe, M.: Evolution of Functional Polymer Colloids for Coatings and Other Applications. AIChE J. **62**(7), 2238 (2016)
6. van Herk, A.M.: Chemistry and Technology of Emulsion Polymerisation, Second Edition. Wiley (2013)

7. Bandiera, M., Balk, R., Barandiaran, M.J.: One-pot synthesis of waterborne polymeric dispersions stabilized with Alkali-Soluble resins. Polymers **10**(1), 88 (2018)
8. Keddie, J.L.: Film formation of latex. Mater. Sci. Eng. R **21**(3), 101 (1998)
9. Curtzwiler, G., et al.: The world of surface coatings is centered around the glass transition temperature, but which one. Part 1. Coatings Tech, **11**(8) 28 (2014)
10. Curtzwiler, G., et al.: The world of surface coatings is centered around the glass transition temperature, but which one. Part 2. Coatings Tech, **11**(9) 40 (2014)
11. Berce, P., Skale, S., Razborsek, T., Slemnik, M.: Influence of Coalescing Aids on the Latex Properties and Film Formation of Waterborne Coatings. J. Appl. Polym. Sci. **134**(31), 45142 (2017)
12. Ludwig, L., Schabel, W., Kind, M., Castaing, J.-C.: Drying and Film Formation of Industrial Waterborne Lattices. AIChE Journ. **53**(3), 549 (2007)
13. Winnik, M.A., Yang, Y.: Latex Formation at the Molecular Level: The Effect of Coalescing Aids on Polymer Diffusion. J. Coat. Technol. **64**(811), 51 (1992)
14. Berardi, P.: Parameters Affecting the CPVC of Resins in Aqueous Dispersions. Paint Technol. **27**(7), 24 (1963)
15. Bierwagen, G.P., Rich, D.C.: The Critical Pigment Volume Concentration in Latex Coatings. Prog. Org. Coat. **11**, 339 (1983)
16. Diebold, M.P.: Application of Light Scattering to Coatings, Springer (2014)
17. Schaller, E.J.: Critical Pigment Volume Concentration of Emulsion Based Paints. J. Paint Technol. **40**(525), 433 (1968)
18. del Rio, G., Rudin, A.: Latex Particle Size and CPVC. Prog. Org. Coat. **28**, 259 (1996)
19. Woodbridge, R.: Dry hiding – an alternative to Titanium Dioxide. In: Woodbridge, R., Blackie (Ed.)., Principles of Paint Formulation (1991)
20. McDonald, C.J., Devon, M.J.: Hollow Latex Particles: Synthesis and Applications. Adv. Colloid. Inter. Science **99**, 181 (2002)
21. Hungenberg, K-D., Jahns, E.: Trends in emulsion polymerization processes from an industrial perspective. In: Polymer Reaction Engineering of Dispersed Systems, Springer (2017)
22. Jiang, S., Van Dyk, A., Maurice, A., Bohling, J., Fasano, D., Brownelf, S.: Design Colloidal Particle Morphology and Self-assembly for Coatings Applications. Chem. Soc. Rev. **46**(12), 3792 (2017)
23. Schuler, B., et al.: Structure and Properties of Multiphase Particles and their Impact on the Performance of Architectural Coatings. J. Appl. Polym. Sci. **40**, 139 (2000)
24. Brown, W.: Hollow Latex Particles: Binders that Provide Opacity. Paint Coat. Ind. **24**(9), 56 (2008)
25. Eckenrode, H.M., Fasano, D.M.: Shining new light on opaque polymer. PCI Magazine, 40 (2012)
26. Adamson, L., Fasano, D.: Polymeric hiding technologies that make $TiO_2$ work smarter. PCI Magazine, 22 (2011)
27. Silva, J.F.A., et al.: Use of Multi-Hollow Polyester Particles as Opacifying Agent for Injection-molded Polyethylene. Polymers **12**(6), 1331 (2020)
28. Wildeson, J., Smith, A., Gong, X., Scriven, L.E., Davis, H.T.: Understanding and Improvement of $TiO_2$ Efficiency in Waterborne Paints through Latex Design. JCT CoatingsTech **5**(7), 32 (2008)
29. Bohling, J., Fradkin, D., Gao, W., Hook, J.W., Wang, T., Wang, T.: Process for improving hiding efficiency in pigmented paints. US 9,518,192 (2016)

# Part IV
# Aspects of Formulation

# Chapter 11
# Dispersion of Small Particles in Liquid Paints

## Contents



## Introduction

The dispersion of particles into a liquid medium is a multistep process that is vitally important for the manufacture of coatings. The quality of the dispersion directly determines whether the optimal performance is achieved for a coating formulation. When discussing particle dispersion in a liquid, we are referring to the dispersion of pigments (white or colored) and extenders in water or solvent with the aid of a dispersant (dispersed resin particles are discussed in Chap. 10). As discussed in previous

**Supplementary Information** The online version contains supplementary material available at https://doi.org/10.1007/978-3-030-99083-1_11. The videos can be accessed individually by clicking the DOI link in the accompanying figure caption or by scanning this link with the SN More Media App.



M. Diebold et al., *Pigments, Extenders, and Particles in Surface Coatings and Plastics*, https://doi.org/10.1007/978-3-030-99083-1_11

chapters, the pigments and extenders in a coating formulation contribute important properties to the final product, but only if the particles are properly dispersed. For example, pigments like $TiO_2$, which are prized for their light scattering, (see Chap. 3), contribute optimal scattering only when they are properly dispersed. Because of their size, these particles can easily agglomerate, and it is necessary not only to initially separate them, but also to keep them separated. If the particles are not dispersed properly in the final paint film, the quality of the film will suffer regardless of the quality of the particles that went into it. The primary properties of the paint that are affected by the quality of the particle dispersion are gloss, color, tint strength, and hiding power.

To better understand dispersion, we can divide the process into three major steps: wetting, separation, and stabilization. The degree of wetting is impacted by the surface energy of the particles and the surface tension of the liquid. Once wetted-in, the particles are then separated through a combination of spontaneous capillary forces and applied mechanical forces, specifically impact and/or shear forces. Finally, the separated particles are stabilized against re-agglomeration or flocculation by the introduction of electrostatic or steric repulsive forces adequate to overcome the innate attractive forces that pull the particles together. This final step is typically carried out during grinding, when the particles are being separated.

The goal of this chapter is to introduce the importance of a high-quality dispersion, discuss the dispersion process, and explore the practical decisions that result in proper dispersion and optimal paint performance. We will do this by discussing the three steps of the dispersion process in turn. We will then apply these principles to practical aspects of paint making, such as equipment selection, particle design, processing conditions, and formulation.

## The Importance of Good Particle Dispersion

Before discussing the dispersion process, we must first establish the benefits of good dispersion, and conversely, the consequences of poor dispersion.

### *Impacts on Paint Quality*

The important properties and contributions of pigments and extenders to coatings have been described in previous chapters. One of the most important is light scattering by pigments such as $TiO_2$. The fundamentals of light scattering were discussed in detail in Chap. 3. There we saw that good particle spacing is necessary for optimal light scattering. Even those particles that are ideal in other ways, such as scattering size and high refractive index, do not scatter light optimally if they are not well separated. Other paint properties, such as color, gloss, and fineness of grind are also negatively impacted by poor dispersion.

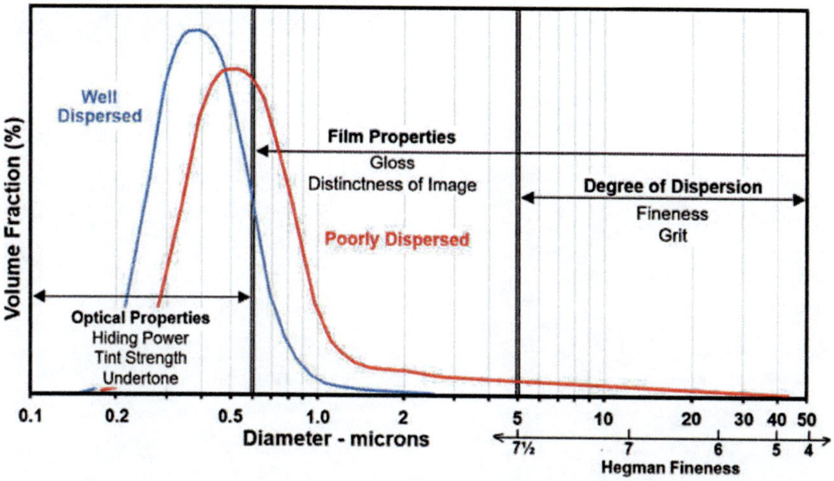


**Fig. 11.1** Particle size distribution of two different $TiO_2$ dispersions and the effect on optical and film properties. Primary particle sizes for the two dispersions are similar (0.25 microns)

The effects of poor dispersion on film properties are determined by the size distribution of the particles. In Fig. 11.1, we show two idealized particle size distribution curves for dispersed $TiO_2$ particles. The curves differ in degree of dispersion, which can be measured by their width or by the location of their maxima. Obviously, better dispersion results in fewer larger particles or particle agglomerates, leading to a sharper curve with a maximum size at a smaller particle diameter. For reference, the size of the primary particles is centered around 0.25 microns, and most particles with a diameter greater than 0.3 microns are, in fact, agglomerates of these primaries rather than large individual particles.

Overlaid on this figure are the optical properties affected by particles or agglomerates in a given size range, including various manifestations of gloss, hiding power, tint strength, undertone, and grit or texture/fineness. The effects of dispersion specifically on the gloss and scattering efficiency of $TiO_2$ in a paint film are visualized in Fig. 11.2. A well-dispersed system has a smoother surface leading to greater specular light reflectance and higher gloss. The separation of $TiO_2$ particles reduces scattering volume overlap, increasing the scattering efficiency of the paint. As can be seen, the degree of dispersion impacts many important coating properties, and dispersion issues can have significant negative impacts on paint performance. On the other hand, improvements in dispersion or particle dispersibility can lead to improved properties and reduced costs of manufacture.

Finally, we note that since the quality of the dispersion impacts these properties, these properties can be used to quantify the degree of dispersion and can be compared to the expected values for a well dispersed sample. Of particular value is the measurement of these properties as a function of dispersion energy or time, since this is of direct relevance to batch turn-over time and energy costs per unit of paint.

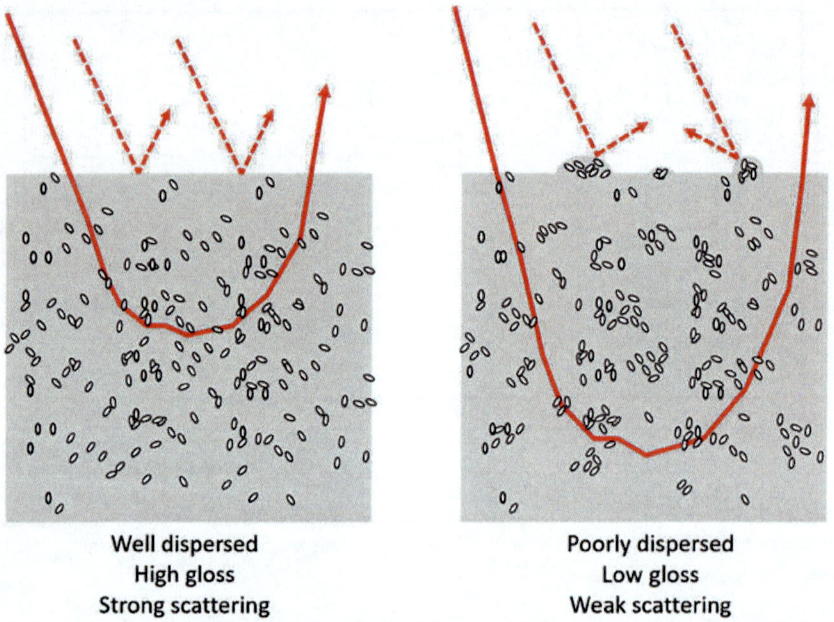


**Fig. 11.2** Impact of the degree of dispersion on paint film gloss and light scattering. Reflected incident light (dashed lines) and scattered light pathlength (solid lines)

## The Dispersion Process

The dispersion of particles in a liquid carrier is a complex process. The properties of the surfaces of both the pigment and extender particles, of the liquid medium (water or organic solvent), and of the interfaces between them dictate the ease and quality of dispersion. We not only want to achieve a good dispersion, but also want to achieve it quickly and efficiently. We accomplish this by fast particle wet-in, rapidly breaking apart any loosely held clumps, efficiently separating the agglomerates into primary particles, and stabilizing the separated particle against any reformation of agglomerates. Each of these steps is described below.

### *Wetting*

The first of the three major steps of dispersion is the wetting process. Powders consisting of small dry particles are, by volume, mostly air, with the air content increasing as particle size decreases. As an example, $TiO_2$ particles (0.25 microns) pack with approximately 75% air voids. In the wetting process, all of this air and

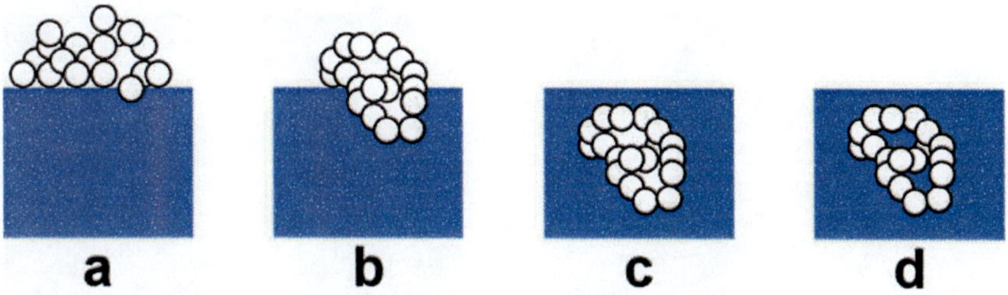


**Fig. 11.3** Illustration of the steps of wetting a dry powder, where the liquid first adheres to the surface of the particles and then penetrates the pores of any agglomerates

any surface or inter-particle moisture must be displaced by the liquid carrier. The wetting process proceeds through four stages, as illustrated in Fig. 11.3.

First, the dry particles, which are invariably present as loose agglomerates of either primary particles or particle aggregates, are added on top of the liquid. The agglomerates are then submerged into the liquid, partially wetting the surfaces of the particles on the periphery of the agglomerate. Once fully submerged, the air within the inter-particle voids is displaced by the liquid and all remaining particle surfaces are wetted. These steps will sometimes occur spontaneously, if given enough time. However, they are greatly accelerated by mixing during particle addition.

## Surface Tension Versus Surface Energy

The wetting process is dictated by the surface properties of the particles and the nature of the liquid carrier. More specifically, the wetting step is determined by the energetics of the particle–air, particle–particle, and particle–liquid interfaces and by the surface tension of the liquid. The surface tension of the liquid arises from an imbalance of the attractive forces felt by molecules on the liquid surface versus those in the bulk of the liquid (Fig. 11.4).

The molecules in the bulk of the liquid are acted on equally in all directions by the attractive forces of the molecules that surround them. However, the forces on the molecules at the liquid surface are unbalanced since they are only partially surrounded by other liquid molecules. This creates an inward pull on the surface molecules and explains why liquid droplets contract to the shape that gives them the minimum surface area possible, which is a perfect sphere. Surface tension is measured as the energy required to increase the liquid surface area, and its units are Joules per meter squared, $J/m^2$, or Newtons per meter, N/m.

Wetting involves the interfaces between three phases of material—air, the wetting liquid (water for most paints), and the solid to be wetted [1]. During the wetting process, we replace particle/air and liquid/air interfaces with particle/liquid interfaces. The ease or difficulty with which wetting occurs depends on the relative strengths of these interactions—if the particle/liquid interface is more favorable than the particle/air and liquid/air interfaces, then wetting is favored. The relative strength

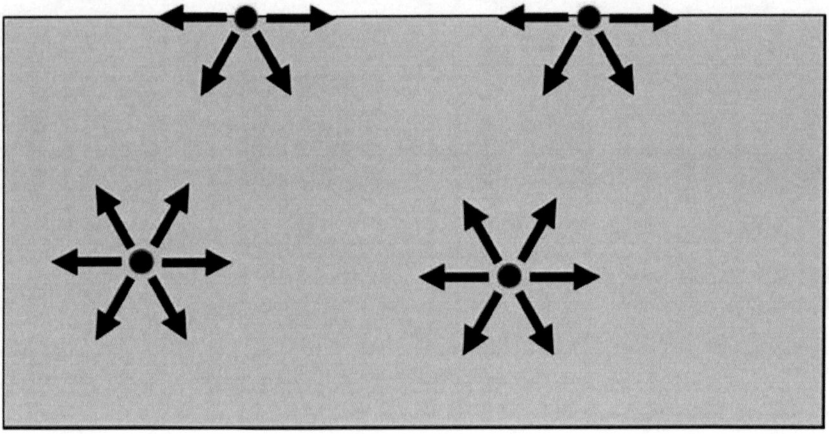

**Fig. 11.4** Balance of forces for the bulk and surface molecules in a liquid or solid

of this interaction can be quantified as the contact angle of the liquid on the particle (Fig. 11.5). This angle is determined by the surface energies of the liquid in the air ($\Upsilon_{AL}$), which is equivalent to its surface tension, the solid particle surface energy in the air ($\Upsilon_{AS}$), which is determined by the chemical composition and surface roughness of the particle, and the solid/liquid interfacial energy ($\Upsilon_{SL}$). This relationship is quantified by Young's equation (Eq. 11.1) [2].

$$\Upsilon_{AS} = \Upsilon_{SL} + \Upsilon_{AL} \cdot \cos(\theta) \tag{11.1}$$

For rough surfaces, the term cos(θ) is replaced by r• cos(θ), where r is the roughness ratio and is defined as the ratio of the true area of the solid surface to its apparent area [3]. When the liquid is water, the solid is considered hydrophilic if the contact angle is less than 90° and hydrophobic if it is greater than 90°.

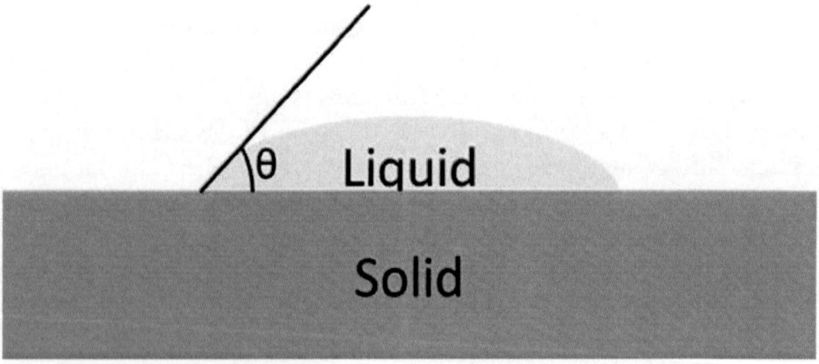


**Fig. 11.5** The wetting of a solid surface

In accordance with the laws of thermodynamics, spontaneous spreading (complete wetting) of the liquid onto a solid surface can only occur if the total free energy of the solid and liquid before spreading is greater than the total free energy after spreading. This occurs when $\Upsilon_{AS}$ is greater than the sum of $\Upsilon_{AL}$ and $\Upsilon_{SL}$. The wettability of particles by water can be enhanced with the use of wetting agents, which are a type of surfactant. Wetting agents are generally linear polymers, usually with a lower molecular weight than a dispersant (discussed below), and with a hydrophilic head and a hydrophobic tail. These molecules segregate to the air/water interface, with the hydrophilic heads immersed in the water and the tails exposed to the air. This segregation decreases the surface tension of the liquid ($\Upsilon_{AL}$), which decreases the contact angle with the particle (refer to Eq. 11.1) and so enhances wetting.

Note that dispersant molecules, which are discussed below, are similar to wetting agents in the sense that they typically have different segments that allow the molecule to simultaneously interact with two very different materials at their interface. For dispersant molecules, the two materials are the paint liquid and the solid surfaces of the particles dispersed in the paint, rather than the paint liquid and air, as is the case for a wetting agent. This difference is important as the functions of wetting agents and dispersants are quite different—wetting agents are designed to speed particle dispersion by speeding the wetting process, while dispersants are designed to attach to the surfaces of dispersed particles and enhance their stability against flocculation.

In Fig. 11.3c, we see that once the particle surface is wet, the wetting liquid must penetrate the pores of the particle agglomerates. The rate at which the liquid penetrates the loose agglomerates is described by Eq. 11.2 [4].

$$u = \frac{\gamma_{AL} cos(\theta)}{\eta} \times \frac{r}{4L} \tag{11.2}$$

Where

u is the rate of liquid penetration,

$\Upsilon_{AL}$ is the surface tension of the liquid,

$\theta$ is the contact angle,

$\eta$ is the viscosity of the liquid,

r is the average capillary radius between pigment particles, and

L is the capillary length.

From a practical standpoint, the agglomerate structure of the dry particles, which determines r, cannot be easily influenced by the paint manufacturer. Likewise, the ability to alter the viscosity of a coating is limited due to the rheology requirements of the paint at rest and during application. However, penetration rates can be increased by using previously mentioned wetting agents to lower the surface tension of the water ($\Upsilon_{AL}$), which also lowers the water contact angle and so increases $\cos(\theta)$.

Even with favorable conditions (surface tensions, surface energies, solid surface roughness, agglomerate capillary radius, and liquid viscosity), the wetting process can be slow and incomplete. As noted earlier, this process can be sped up substantially by vigorous agitation as the particles are added to the liquid.

## *Particle Separation and Dispersion*

Once fully wetted, the second step of the dispersion process is to separate the groups of particles into individual particles. This is partially accomplished spontaneously through capillary forces, [5] but for the small particles found in paints, these forces are not strong enough for full dispersion. Full dispersion is relatively easy for the large loose agglomerates that are present in the powders in the air, but requires significant energy for the agglomerates or hard aggregates holding small groups of particles together. Particle separation is also referred to as grinding or deagglomeration. This process uses mechanical force to break particle clusters into primary particles or smaller aggregates, allowing the wetting of the newly exposed particle surfaces as well as the initial separation of individual particles.

### Impact and Shear Forces

Particles can be separated with different types of mechanical force—impact (or smash) and shear (Fig. 11.6). Using impact force or smashing particles apart can be visualized as a hammering or projectile impingement process. The hammering process breaks agglomerates down as they are caught between two colliding objects. The projectile impingement process involves the agglomerates being cast forcibly against an immobile solid surface or another moving particle. These two processes

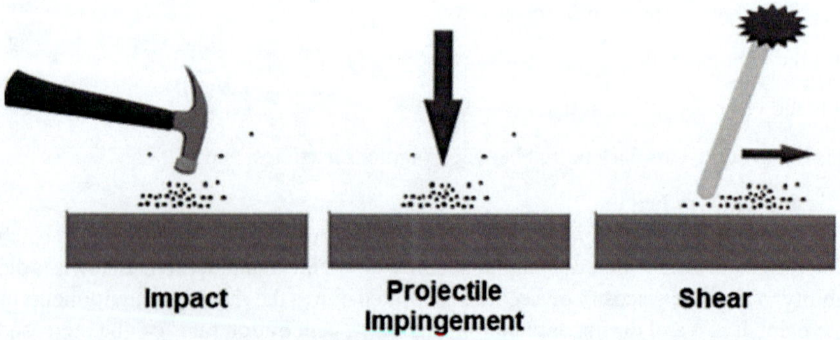


**Fig. 11.6** Examples of impact/smashing force through hammering and projectile impingement, and shear/smear force with a shearing type process

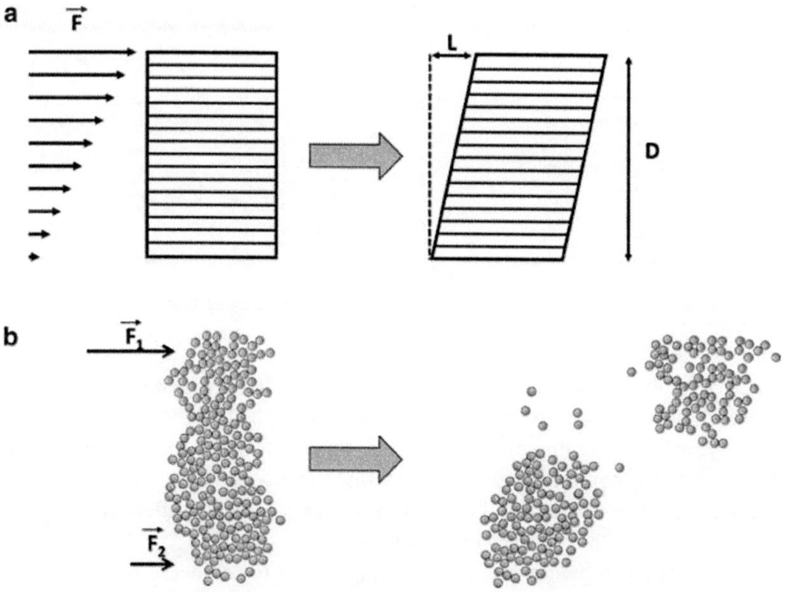


**Fig. 11.7** Shear forces. **a** Acting on a liquid. **b** Acting on an agglomerate of particles

are not fundamentally different, but instead represent different means to accomplish the same end, that is, to use impact force to smash apart clusters of particles.

Shear force is fundamentally different from impact force. In this case, a force is applied disproportionately over the extent of a body. This is typically a volume of liquid (Fig. 11.7a), or, in our case, a particle agglomerate (Fig. 11.7b). The magnitude of shear is determined by the difference in distances traveled by different elements of the material experiencing the least and greatest forces (L in Fig. 11.7a) and the distance over which the shear occurs (D in Fig. 11.7a). The shear rate is the ratio of these two distances over a unit interval of time. Shear increases when the distance D is minimized or when the velocity difference is maximized. In most milling operations, one extreme of the velocity gradient is zero (that is, the fluid is at rest against a stationary object such as a vessel wall) and so the velocity difference is simply the maximum velocity within the slurry. When shear is applied to an agglomerate, a force gradient is created across the agglomerate that separates the individual particles from one another (Fig. 11.7b).

This is similar to the spreading of a deck of playing cards on a stationary surface (such as a tabletop). When a force is applied to the top of the deck and parallel to the stationary surface, the cards will slide over one another and spread apart. Here the force applied at the top of the deck is resisted by the friction between individual cards in the deck and between the bottom card and the stationary surface. Friction is a resistive force and so is in the opposite direction as the applied force, which creates a shearing force.

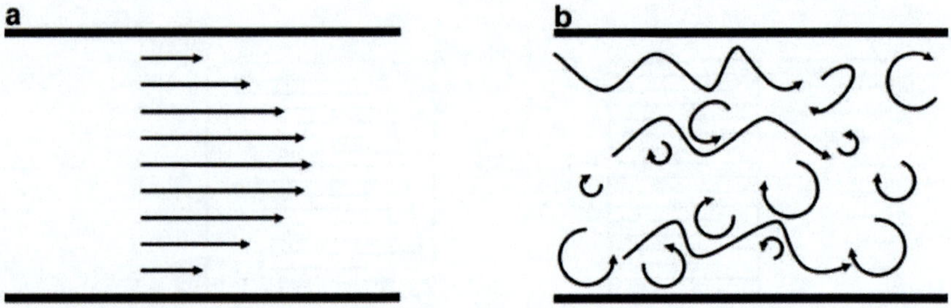


**Fig. 11.8** Flow through a pipe. Arrows indicate movement. **a** Laminar flow. **b** Turbulent flow

In this example, as well as in Fig. 11.6, shear is applied by holding one part of the body at rest on a surface while placing another part of the body in motion. However, in a slurry, very few agglomerates are close enough to a surface to be held in place by it. Instead, it is common for an agglomerate to be completely surrounded by the liquid medium. Shear is still possible in these cases because different regions of a liquid in motion can have different velocities.

There are two interconnected factors that are key to achieving the proper shear conditions for dispersing particle agglomerates in a slurry. These are the viscosity of the slurry and its flow regime. Specifically, a mostly laminar flow regime with a viscous media is required for efficient shear grinding [6].

Laminar flow is the flow regime for which different thin layers of fluid travel smoothly past one another at different velocities. Under these conditions, velocity is a smooth function of position. This can be contrasted to turbulent flow, where fluid moves chaotically or has irregular velocity fluctuations. Figure 11.8 compares the velocities in a pipe as a function of location for both laminar and turbulent flow. Dispersion in the turbulent flow regime is quite inefficient—under these conditions, the energy added to the slurry simply mixes the particle agglomerates rather than shearing them. The viscosity of the media is important because it not only provides viscous resistance, which contributes to shearing, but also because it directly impacts the flow regime of the slurry.

Flow is often described by its Reynolds number (Re), shown in Eq. 11.3 [7]. Common examples of the characteristic linear dimension (X) are the diameter of a pipe through which a fluid flows, the separation distance of two plates between which a fluid flows and, of more relevance to the coatings industry, the distance between a high-speed disperser blade and the bottom of a tank.

$$R_e = \frac{\rho \nu X}{\eta} \tag{11.3}$$

Where

Re is the Reynolds number,

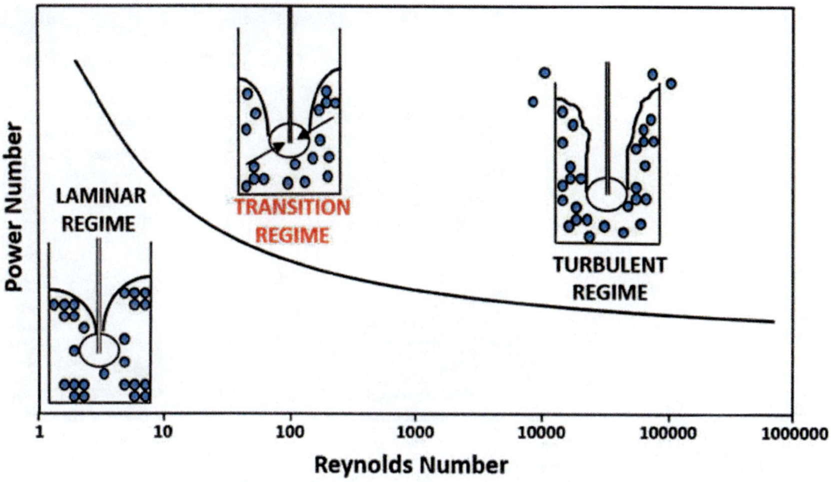


**Fig. 11.9** Change in flow regime as described by the Reynolds number from laminar to transitional to turbulent for a high-speed disk disperser

ρ is the density of the fluid,

v is the velocity,

X is a characteristic linear dimension, and

η is viscosity.

The flow regime transition from laminar to turbulent occurs at the critical Reynolds number. This number is dependent on the specific geometry of the system under investigation but is generally of the order of 2,000[1] for flow through a pipe. Under conditions close to, but above, the critical Reynolds number, the flow becomes unstable and is defined as transitional. At values significantly above the critical, flow is fully turbulent.

In the case of a commonly used high-speed disk disperser (HSD), the laminar flow regime (Re number approximately < 20) gives high viscous shear but slow particle transport. At the other extreme, turbulent flow (Re number approximately > 1000) results in fast particle transport but low viscous shear. HSD grinds are optimized in the transition regime (20 < Re < 1000) where there is a balance of viscous shear and particle transport. This example of optimizing the flow regime for a high-speed disk disperser is shown in Fig. 11.9 and the video associated with Fig. 11.10.

As will be discussed in the next section, the power input into a slurry by certain types of dispersers is maximized (and the dispersing times minimized) by increasing the velocity gradient within the slurry. This is done by increasing the velocity at the highest velocity locations within the disperser, which raises the local Reynolds number of the slurry at those locations and thereby risks the loss of laminar flow.

[1] Note that the Reynolds number is unitless.

**Fig. 11.10** Video of dispersion in the three flow regimes

The increase in Reynolds number due to increased velocity can be offset by increasing the viscosity of the slurry (refer to Eq. 11.3). This increase can be accomplished by either adding thickening agents to the slurry during the dispersion process, which creates an open network of weakly interacting thickener species, or by increasing the particle concentration (percent solids) of the slurry. The latter is usually preferred because, when thickeners are used, much of the energy added to the slurry is wasted by breaking apart the weak network of thickener species, rather than being used to overcome inter-particle attractions and separate the particles in an aggregate or agglomerate. In addition, when the viscosity increase comes from an increase in particle concentration, the increased concentration increases the rate of particle collisions, and so increases the hammering effect.

## Dispersion Equipment

As previously discussed, the separation of particles in a liquid by mechanical force is a critical step in achieving optimal dispersion, and so has a large impact on final coating performance. Many different apparatuses have been developed for the dispersion of particles in paints. These include those that operate strictly using impact or shear, as well as those that use some combination of the two. The different dispersion equipment typically used in paint manufacture and the type of force they employ are shown in Fig. 11.11.

The kinetic dispersion mill (Fig. 11.12) is an example of equipment that uses mostly impact forces. One type of kinetic mill uses a fast-moving slotted rotor to send the pigment at high velocities into a close-fitting slotted stator. The particle agglomerates impinge against the stator or each other, leading to particle separation through impact.

Another example of an impact disperser is the fluid energy mill, often referred to as a micronizer. This mill consists of a circular chamber into which a jet of particles in

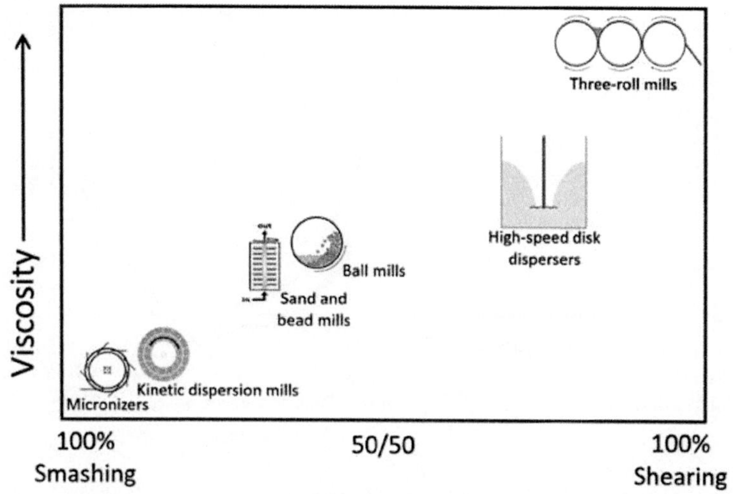


**Fig. 11.11** Classifying dispersion equipment based on the type of separation force generated and required viscosity of the mill base. Adapted from [4]

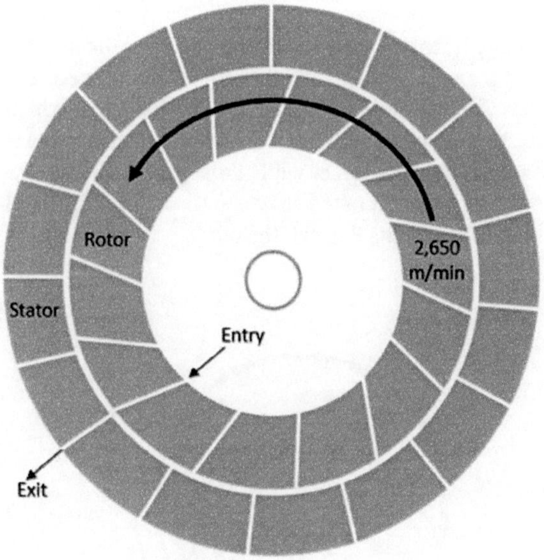


**Fig. 11.12** Overhead schematic of a kinetic dispersion or high-speed impingement mill. Adapted from [4]

a fast-moving gas—often steam—is introduced tangentially (Fig. 11.13). Additional tangential jets are distributed around the circumference of the chamber to increase the energy intensity of the grind. Particle agglomerates are attritted by impinging on one another or the chamber wall. Once below a certain size (ideally the size of the

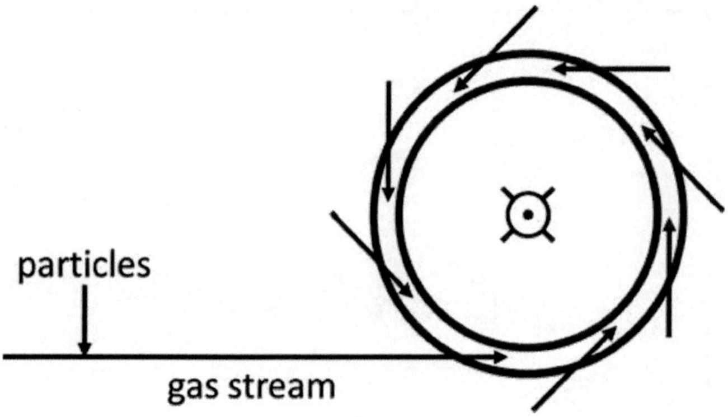


**Fig. 11.13** Overhead view of a fluid energy mill. Gas exits out of the page at the center of the chamber

primary particles), the particles are entrained in the gas as it exits the mill through an opening at the center of the chamber and are normal to it.

Ball mills use mostly impact forces, although they also impart some shear force. These mills consist of a rotating cylinder that is filled with balls (Fig. 11.14a). These balls are typically ceramic and usually have a high density, which increases the energy transfer rate between the balls and the particles. The ingredients are charged into the mill chamber and the mill is then rotated. The balls are pulled up the side of the cylinder until they reach a certain height, at which point they cascade down onto the balls lower in the chamber. The particles that are caught between the sliding balls experience shear forces while those trapped between a ball at the base of the mill and one cascading from the side experience impact forces. Flow enters from one end of the mill and exits through the other.

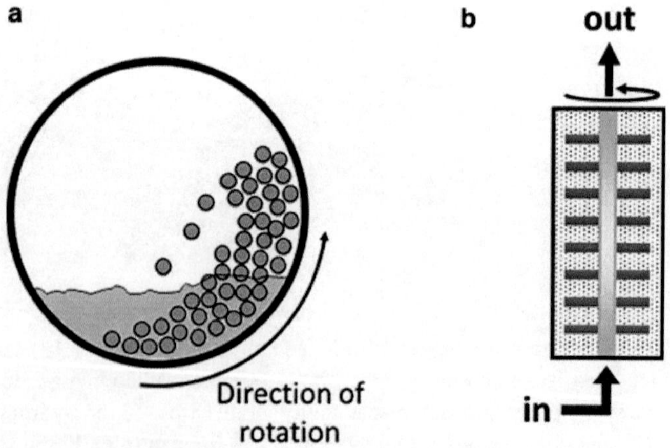


**Fig. 11.14** Schematic of the mill in operation. **a** Ball mill. **b** Sand mill

Next we consider bead mills and sand mills, which are very similar, in principle, to ball mills. The primary differences between bead and sand mills, on the one hand, and ball mills on the other, are that the media are smaller in sand and bead mills, and that rather than the cylinder rotating, as it does for a ball mill, internal disks or impellers are inserted on a shaft through the central axis of the bead and sand mills and these plates are rotated (Fig. 11.14b). The rotating plates agitate the sand or beads, resulting in grind conditions similar to those in the ball mill. Both vertical (top feed) and horizontal (side feed) configurations can be used. Energy transfer is greatest when the viscosity of the mill base (the mix of pigment, extenders, liquid, and dispersant) is low.

While ball, bead and sand mills disperse particles primarily through smashing, other types of equipment use mostly or exclusively shear forces to separate particles. The high-speed disk disperser, also called a dissolver, high-speed disk impeller, or high-speed disperser is the most prevalent of this type of equipment. A schematic of a high-speed disk disperser is shown in Fig. 11.15. The equipment consists of an upright cylindrical tank with a vertically centered disk impeller on a rotating shaft, into which the mill base is charged. The impeller disk can simply be uniformly flat,

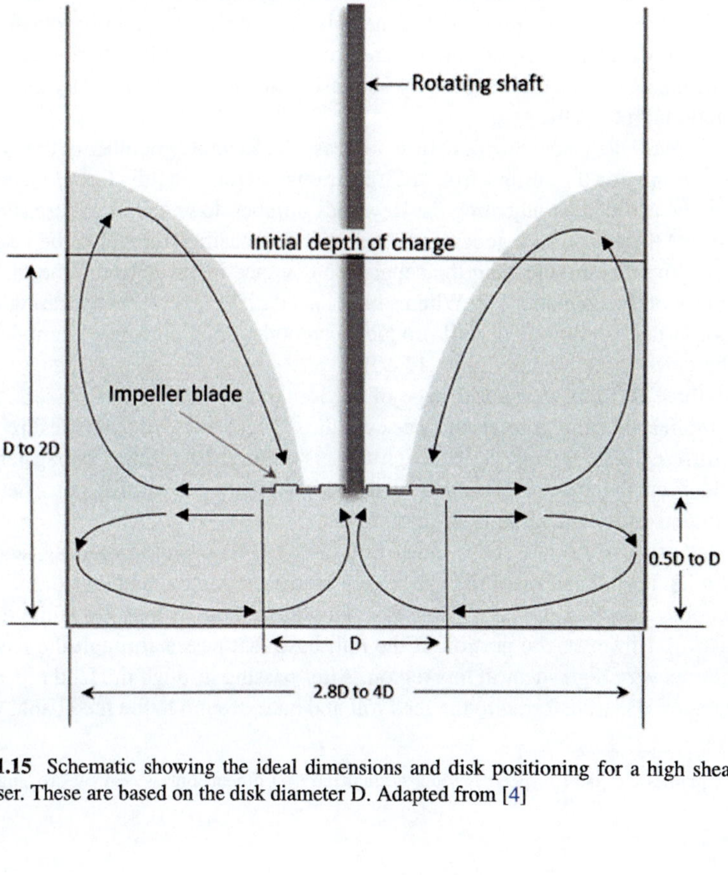


**Fig. 11.15** Schematic showing the ideal dimensions and disk positioning for a high shear disk disperser. These are based on the disk diameter D. Adapted from [4]

or, more commonly, can have serrated edges or edge sections ("teeth") that are cut to stick up, down, or out of the plane of the disk.

Only a small amount of dispersion occurs as the result of particle impact with the high-speed disk or the tank walls. Instead, most occurs by shearing in the laminar flow created by the rotating disk. This obviously requires that the milling operation remains within the laminar or transition flow regime.

To maximize dispersion efficiency at high impeller rotational speeds (20 to 25 m/s),[2] the impeller disk must be properly positioned and correctly sized for the tank. Typical configurational parameters are given in Fig. 11.15. These are approximate and must be adjusted depending on the exact dimensions of the equipment being used and the fluid properties of the slurry being dispersed. As stated previously, if the flow is turbulent, then the energy being supplied by the high-speed disk will primarily mix the slurry rather than generate the shear conditions necessary to separate particles.

As shown in Eq. 11.3, the flow regime (laminar, transitional or turbulent) is determined by fluid velocity, viscosity, and density, and by the characteristic linear dimensions of the equipment (in Fig. 11.15, it is the diameter of the impeller blade, although the distance between the impeller blade and tank bottom can also be used for this purpose). The energy transfer rate from the rotating disk to the slurry is determined in part by the velocity gradient within the tank. To increase the energy transfer rate, and shorten dispersion times, this gradient can be increased by increasing the tip speed of the disk. This can be done by increasing the disk diameter or by increasing the rotational rate of the shaft.

An increase in tip speed will, in turn, increase the Reynolds number of the system, potentially moving the system from the transitional regime to a fully turbulent regime. This can be avoided by adjusting the Reynolds number downward by increasing the viscosity of the slurry. As a general rule, to be in the transitional regime, the viscosity in poise should be higher than the separation distance in cm between the disk and the bottom of the container [4]. When a high-speed disk disperser is operating under the appropriate conditions, it will produce a smooth "doughnut-type" laminar flow (see the video associated with Fig. 11.10).

The three-roll mill is another type of predominantly shearing disperser. As the name implies, this mill consists of three cylinders. These rotate in opposite directions and at different speeds (Fig. 11.16). High shear is produced in the narrow gap between the rollers, and the shear level is controlled by adjusting the gap (nip) distance between the cylinders or by changing their rotation rates.

In this process, the mill base is fed into the area between the rotating feed and center rolls. Portions of the mill base are rejected back into the feed bank as the mill base is pulled toward the narrowing gap. This process itself leads to some degree of shearing. However, the portion of the mill base that passes through the feed nip experiences very high shear in this region. After passing through the feed nip, a part of the material is transferred to the feed roll and back around to the feed bank, while

[2] At the plant scale, the tip speed is often normalized to a dimensionless unit by dividing it by 16 m/s.

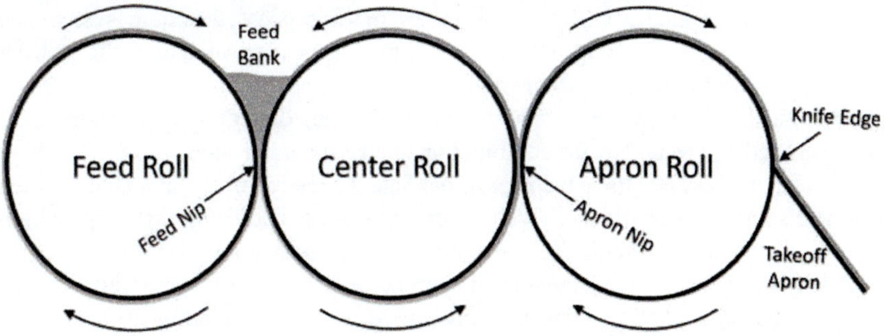


**Fig. 11.16** Schematic of a three-roll mill showing the flow of mill base (gray) through the mill. Adapted from Wikipedia Commons (license found at https://creativecommons.org/licenses/by-sa/3.0/deed.en)

the rest is transferred to the center roll. The mill base on the center roll travels around to pass the nip between the center and apron rolls. After passing through the apron nip, a part of the material remains on the center roll and returns to the feed bank, and the rest transfers to the apron roll. A knife edge pressing on the apron roll sends the material on the apron roll to the takeoff apron.

The mill base must adhere to the rollers, and therefore, must have a relatively high viscosity (generally the mill base has a paste-like consistency). Note that a portion of the material returns to the feed bank after passing through the nips, and so passes through the different nips multiple times before exiting the mill. Different roll velocities, roll diameters, and the separation distances of the rolls can be adjusted to increase dispersion efficiency.

The final dispersion process that we will consider is in-line dispersion (Fig. 11.17). While this technology is not as common as high-speed disk dispersers or media mills,

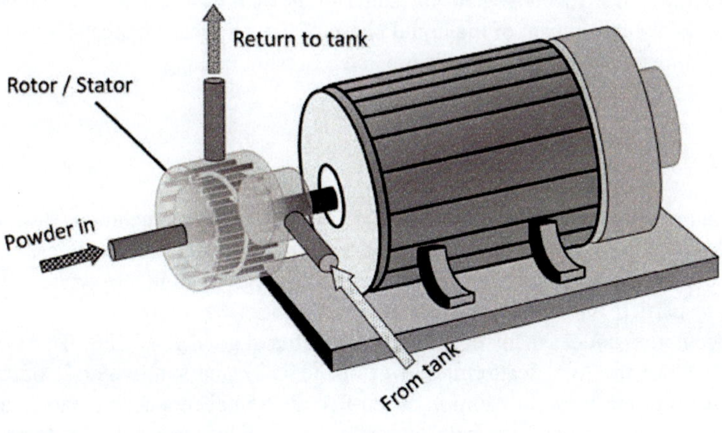


**Fig. 11.17** In-line dispersion assembly

it is growing in popularity with paint producers. In this process, the liquid is circulated in a closed system, creating a strong vacuum that is used to induct the powder ingredients into the slurry. A rotor–stator system creates this large negative pressure and a high shear energy zone where the liquid and powder combine. Agglomerates are destroyed and wetted by the combination of air expansion under vacuum and high shear. This process results in the near immediate dispersion of the inducted powder. In-line dispersers typically work well for difficult to wet and fine powders, give high yields with no dust formation, and are applicable to continuous processes.

An important aspect of all these milling processes is that there is an optimal energy input for each type of equipment. The consequences of under-grinding are obvious—incomplete dispersion, with an attendant loss in opacity (for $TiO_2$), color (for colored pigments), gloss, fineness of grind, etc. (refer to Fig. 11.1). Over-grinding, on the other hand, not only wastes energy, but can also have a negative impact on the slurry or paint, as excessively fine particles can be created, which impacts paint viscosity and dispersant requirements. In addition, this can result in a loss of color pigment tint strengh since these pigments become transparent when the particle size decreases below roughly 0.20 microns. Overgrinding can also reduce the tint strength of highly treated $TiO_2$ white pigments due to destruction of the fragile oxide coating on their surfaces.

## *Particle Stabilization*

Based on thermodynamic considerations, the particles in dispersed slurries and paints can spontaneously flocculate or re-agglomerate, but they cannot spontaneously disperse. Because of this, the greatest degree of dispersion in any paint making process will be immediately after the dispersion step, after which the state of dispersion can only degrade. The paint manufacturer must, in fact, actively prevent re-agglomeration from occurring during paint storage or as the paint dries (during which the chemical environment of the liquid phase of the paint can change significantly). These actions constitute the stabilization step in paint making.

### Flocculation

"Flocculation" is the general term used to describe the re-agglomeration of dispersed particles, and the agglomerates so formed are typically referred to as flocs. Flocculation occurs when the particles collide, either through random (Brownian) motion or due to particle attractions, and adhere to one other (Fig. 11.18).

Flocculation undoes many of the optical benefits of good dispersion. This is especially true for the light scattering power of the $TiO_2$ pigment, which is often one component of the flocs. In addition, when flocculation occurs during paint storage, the viscosity of the paint typically increases, eventually leading to gel formation. Flocculation during storage can also lead to a settling of the particles, often into a

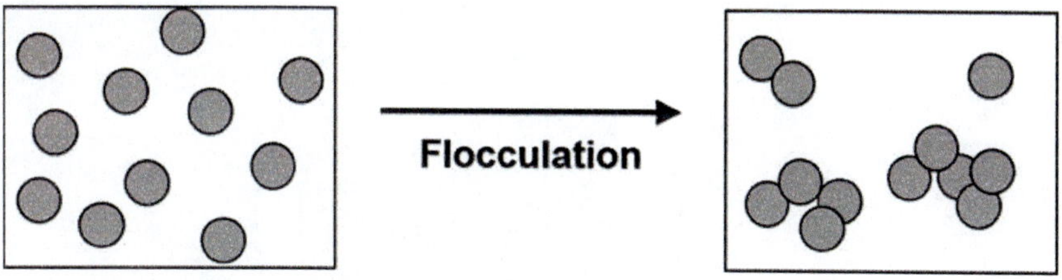


**Fig. 11.18** Flocculation of separated particles in a liquid dispersion

hard heel that cannot be reincorporated into the liquid phase of the paint by simple stirring or shaking.

In addition to flocculation during storage, flocculation can also occur as the paint dries. This type of flocculation can be detected using a paint rub-out test (Chap. 13). Here colored paint is sheared after application, usually by rubbing with a finger or stroking with a paintbrush. When the kinetics of re-flocculation are slower than the kinetics of drying, a distinct difference in color is seen between the sheared and unsheared areas of the paint film. This sort of flocculation is minimized in well formulated paints [8].

Flocculation can occur through a number of mechanisms. The most obvious mechanisms involve van der Waals or electrostatic forces. In both instances, there is an attraction between particles. Great care is taken when formulating the paint to avoid the creation of opposite charges on different particles. This is generally done by adsorbing an ionic species on the surfaces of the different particles present in the paint, which gives them a common charge sign (positive or negative). This is discussed in more detail in the next section.

The driving force for van der Waals and electrostatic flocculation is energy related (flocculation reduces the overall energy of the paint system). However, flocculation can also occur based on entropic considerations—that is, induced by an increase in entropy of the paint, rather than a decrease in enthalpy. This form of flocculation is referred to as entropic flocculation, exclusion flocculation or depletion flocculation, and its origins are in the state and concentration of the dispersant molecules that are used, paradoxically, to stabilize the paint.

Entropic flocculation occurs in a slurry that contains relatively large dissolved molecules, more specifically when these molecules are large enough that they cannot be regarded as point objects when compared to the size of the dispersed particles. We can define the locations of these large molecules based on their centers of mass, with the understanding that they occupy a finite volume around this center. These molecules can come no closer to a particle than the molecule radius. This creates an exclusion volume around the particle (Fig. 11.19) that is larger than the particle itself.

Since the exclusion volume around a particle cannot be occupied by the molecules, its presence decreases the volume available to these molecules, which raises their concentration in the remainder of the liquid phase of the slurry or paint. This increase in concentration results in a decrease in entropy. When the particles are close to one

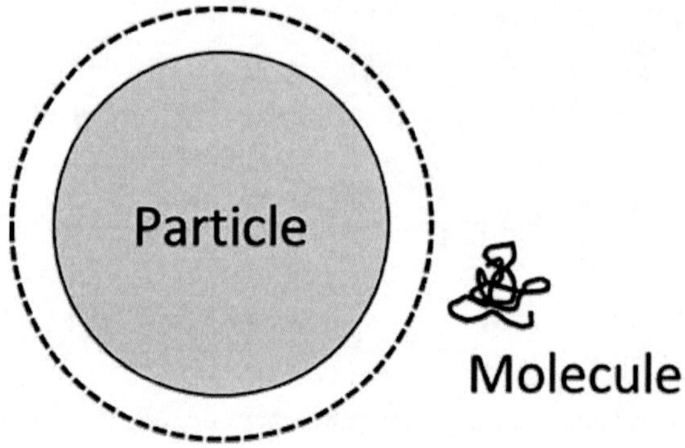


**Fig. 11.19** Exclusion volume around a particle

another (that is, flocculated), their exclusion volumes overlap, and the total exclusion volume of the system decreases—that is, the volume available to the molecules increases and so the molecules become more dilute, increasing the entropy of the system. This increase in entropy is the driving force for this type of flocculation.

An alternative explanation of entropic flocculation can be made based on osmotic pressure arguments. While the large molecules are excluded from the overlap region between two nearby particles, solvent molecules (particularly water) are typically much smaller than the large molecules and so can occupy most of the overlap region. This creates a shell of pure solvent molecules that is in contact with the remainder of the solvent portion of the slurry. When two particles are close to one another, osmotic pressure will force them together, and this pushes solvent molecules out of the region of pure solvent into the region of dissolved large molecules (said differently, it decreases the total volume of pure liquid). This lowers the concentration of the large molecules and so increases the entropy of the system. This is shown schematically in Fig. 11.20.

Finally, flocculation can occur through another form of energy minimization—the minimization of gravitational potential through sedimentation. Sedimentation is the slow downward migration of particles and their collection at the bottom of a

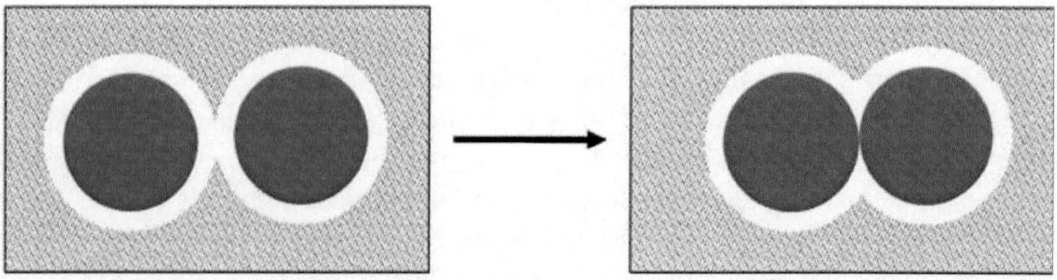

**Fig. 11.20** Entropic flocculation. The white shells around the dark particles represent regions of the pure solvent. The patterned area represents the solvent with the large dissolved molecules. The total volume of the pure solvent decreases when the particles touch one another

container or vessel. Sedimentation rates are influenced by particle mass (and therefore size) and slurry rheology. This process can become very complex but can be broadly understood by consideration of the gravity-driven settling of a single spherical particle in a viscous liquid. The rate of sedimentation can be expressed by Stokes' law Eq. 11.4 [9]. Larger and denser particles settle more quickly, and high density, highly viscous liquids reduce the settling rate of the particles.

$$v = \frac{r^2(\rho_P - \rho_L)g}{18\eta} \tag{11.4}$$

where

v is the settling rate,

r is the particle radius,

$\rho_P$ is the particle density,

$\rho_L$, is the liquid density,

g is the gravitational constant, and

$\eta$ is the liquid viscosity.

## Electrostatic Stabilization

While flocculation is complex and inevitable, there are different techniques that can be used to maintain the uniform separation of particles in a liquid for an extended period of time. The most common technique in the case of water-borne paints is electrostatic repulsions.

This method for stabilizing slurries or paints depends on the creation of a polar environment by the absorption of charge carriers onto the particle surfaces. This surface charge (zeta potential) attracts a cloud of countercharges from the surrounding solution, creating an electrical double layer that extends from the particle surfaces (Fig. 11.21).[3]

In water-borne paints, the surface charge is negative, created by the absorption of a polyacid dispersant onto the surface of the particles and the adjustment of the pH to above neutral (typically in the range of 8.5–9.0). The organic acids in the dispersant molecules are deprotonated at these pH values, resulting in a negative surface charge surrounded by a positively charged double layer. As the clouds of positive charge around nearby particles overlap, they repel one other.

---

[3] When the concentration of dissolved ionic species is very high, the double layer is quite thin, which neutralizes the charge at all but very short distances. This is referred to as a collapse of the double layer and leads to flocculation ("salting out") of electrostatically stabilized slurries.

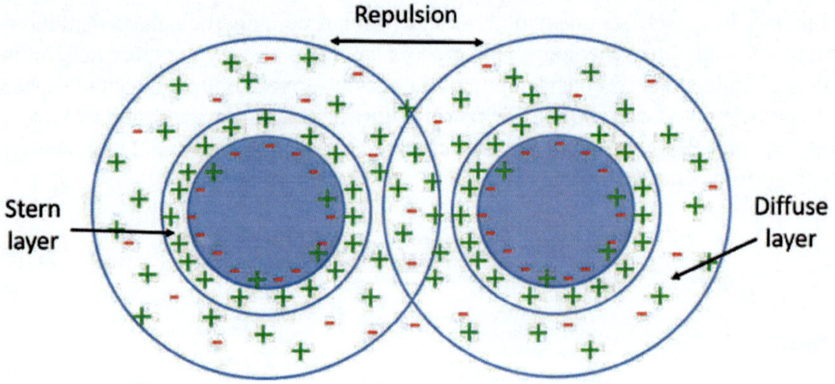


Fig. 11.21 Repulsion of particles due to electrical double layer stabilization

Electrostatic stabilization is usually described using DLVO theory, so named by the initials of the developers of this theory (Deryaguin and Landau, and, independently, Verwey and Overbeek) [10, 11]. This theory is based on the balance between the van der Waals attractive forces of the particle surfaces and the electrostatic repulsive forces of the electrical double layer. The different dependencies of these forces on particle separation distances lead to the formation of an energy maximum, or barrier, at a certain separation distance (Fig. 11.22). At shorter distances, the van der Waals attractions dominate, while at longer distances, the electrostatic forces dominate.

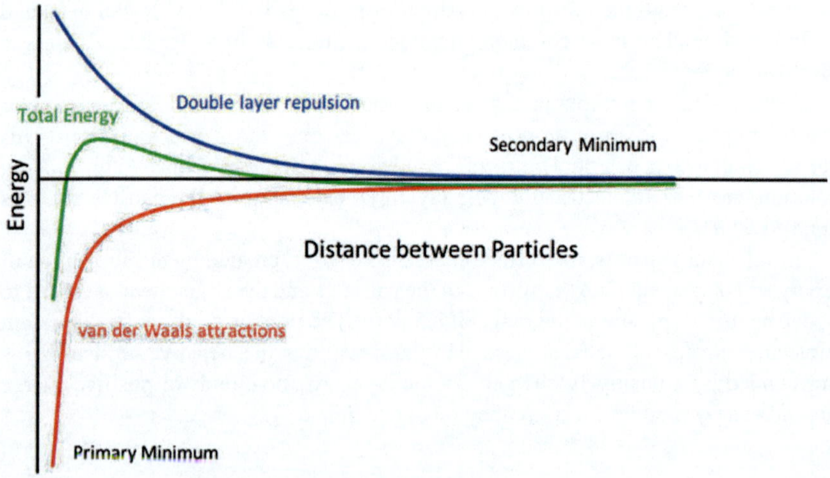


Fig. 11.22 Energy as a function of particle separation distance

The charged double layer generates an electric potential due to the unequal distribution of ions. This electric potential declines rapidly with an increase in distance away from the particle surface into the solution. Different models for an electrical potential drop-off with distance were proposed by Helmholtz, Gouy and Chapman, and Stern. Helmholtz proposed a linear fall-off. Gouy and Chapman independently proposed an exponential decay. Stern proposed a combination of the two with an initial linear drop followed by an exponential decay. In Stern's model, some of the ions adhere to the particle surface, as in the Helmholtz model, leading to the initial linear drop in potential. This internal layer of ions is referred to as the Stern layer (see Fig. 11.21). Moving beyond this layer, Stern's model has an exponential decay similar to the Gouy–Chapman model, due to the diffuse layer of ions in the region between the particle surface and the fluid bulk.

An important aspect of the energy versus separation distance curve (Fig. 11.22) is that, depending on the exact slurry or paint conditions, a weak secondary minimum can be created at relatively large particle separation distances. This results in the particles becoming weakly flocculated into a "soft cake" that can be easily redispersed by simple mixing. This condition is often desired in a slurry or paint as it prevents the particles from settling into a "hard cake" from which particles can only be reincorporated by repeating the dispersion process.

Electrostatic stabilization techniques are ineffective in solvent-borne dispersion as these systems are non-polar, and therefore, cannot support dissolved ions. Other methods for stabilizing particles dispersed in solvent must be used (discussed next).

## Steric Stabilization

Non-aqueous or solvent dispersions can be stabilized through steric (entropic) stabilization. In this type of stabilization, the dispersed particles are surrounded by polymer chains (dispersants) absorbed on the surface, Fig. 11.23. The polymer chains are fixed to the surface at one end with the remainder of the chains freely extending into the solvent. These extended chains create a steric or spatial barrier when two particles approach each other.

A close approach of particles is prevented based on entropy and osmotic pressure. As the particle surfaces approach one another, solvent molecules are squeezed out of the inter-particle region due to the increase in the concentration of the dispersant chains. This creates an osmotic pressure that forces solvent molecules into the inter-particle region, forcing the particles apart. Note that this is in some ways similar to entropic flocculation, but in this case, the large molecules are attached to the particle surfaces and so it is the solvent molecules, rather than the large dispersant molecules, that are excluded from the inter-particle region. This results in the osmotic pressure forcing the particles apart in one case (steric stabilization) and together in the other (entropic flocculation).

The proper polymeric configuration is required for optimizing particle stabilization. Coiled chains do not reach as far or move as freely as extended chains. The polymer chains also need to be well solvated, have adequate length, and cover enough

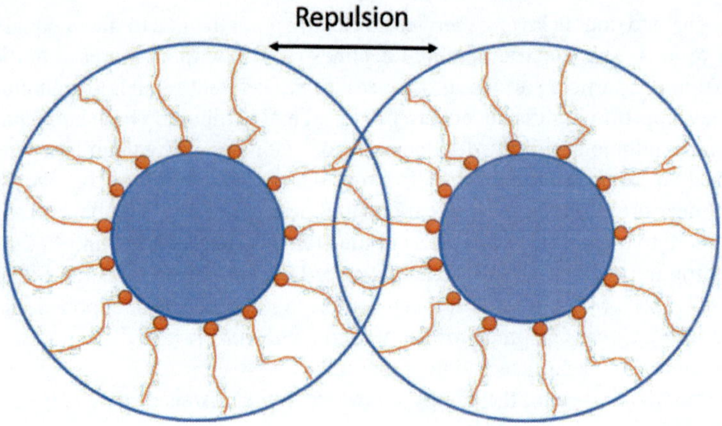


Fig. 11.23 Repulsion of particles due to steric hindrance created by polymer chains absorbed on the particle surface

of the particle surface to provide optimal stabilization. This type of stabilization is commonly used in non-aqueous dispersions, such as solvent-borne paints. It can also be used in conjunction with electrostatic stabilization in aqueous dispersions, although the steric hindrance generally plays a minor role compared to the electrostatic effects. Steric stabilization is less sensitive to the presence of ions in the system and to changes in pH. This improves the freeze–thaw stability of water-borne paint.

It is important to note that two separate stable particle dispersions may flocculate when combined. When the two dispersions are mixed the dispersant molecules can exchange between the different particles, and if the dispersant has a stronger affinity for one of the particle types, the other particle type will flocculate. This behavior is visualized in Fig. 11.24. Preventative measures such as using the same type of dispersant where possible and using enough dispersant can be effective. This destabilization of mixing can occur when a colorant dispersion is added to a dispersion

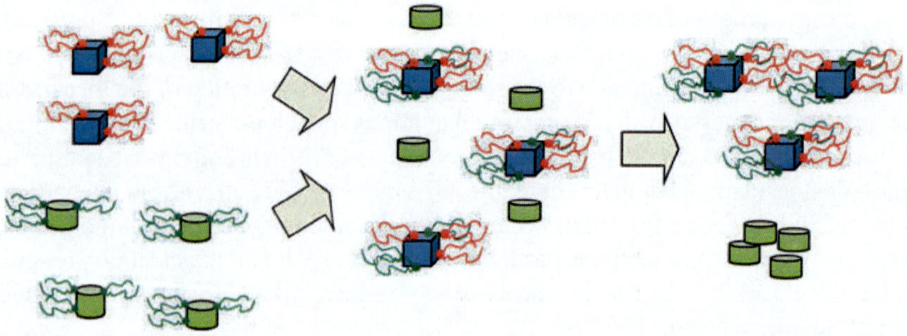

Fig. 11.24 Example of how the combination of two stable particle dispersions can lead to the flocculation of particles due to the exchange of dispersants

of white pigment and/or extender. The colorant dispersant can be absorbed by the white pigment or the extender causing flocculation of the colorant. This leads to a decrease in color strength over time.

### Particle Surface

To achieve colloidal stability through electrostatic or steric forces, the dispersant polymer or charge carriers must bind strongly to the particle surfaces. This process is referred to as anchoring, and its extent is determined by the nature of the anchoring sites on the particle surfaces and the anchoring groups on the polymer molecule. Anchoring strength can be controlled by modifying the particle surface or by modifying the chemical nature of the anchoring group. An example of the former is the deposition of hydrous alumina on pigmentary $TiO_2$ surfaces (see Chap. 7). Hydroxyl sites on the alumina chemically bond to the anchoring groups on the dispersant, which are typically carboxylic acids, significantly more strongly than do the hydroxyl sites on a bare $TiO_2$ surface.

The amount of dispersant required to stabilize a particle (that is, its dispersant demand) is determined, in part, by the strength of the anchoring bond. When this bond is weak, a high concentration of dispersant is necessary to drive an adequate amount onto the particle surfaces (that is, to shift the equilibrium toward a higher surface concentration of dispersant on the particles). An example of the importance of the strength of the anchoring bond can be seen in the incompatibility of ZnO particles in water-borne paints that contain $TiO_2$ pigment [12]. In these paints, the zinc ion is slightly soluble and quite mobile. Over time, the dissolved zinc deposits onto the $TiO_2$ pigment surface. This changes these surfaces into ZnO surfaces. Because the bonding between zinc hydroxyl groups and dispersant molecules is weaker than the bonding between the pigmentary $TiO_2$ surface, which is mostly alumina, and the dispersant molecules, the dispersant demand for the paint increases over time. Eventually, the amount of dispersant is insufficient to stabilize the particle suspension and the paint gels.

## Dispersants

Dispersants play a critical role in the successful completion of all three stages of the dispersion process. They do so by modifying the surface properties of the particles. Dispersants come in a myriad of varieties, cover a wide range of chemistries, are used optimally at different concentrations, interact in often unique ways with the other ingredients in the paint, and are tailored to different aspects of the dispersion process. Review articles and even entire books are dedicated to the different facets of these important additives [13, 14]. It is not our intention to produce here an encyclopedic assessment of all aspects of dispersant use, but instead to provide a logical framework to guide the formulator in choosing and using these materials to full advantage. That is, rather than taking a tactical approach to this issue, we will take a strategic approach. As always, our interests are particle-centered, and focus will be placed on how the dispersants affect the properties of the various particles found in the paint.

There are many sources of information regarding dispersant selection and use, including product brochures, the open literature, and prior experience. While these can be of great assistance in determining a starting point, each set of ingredients, target product performance properties and economic constraints will be unique to a given paint. It is unlikely that the optimal combination of these will be realized at the outset of a product formulation or reformulation project. Instead, a comprehensive study of each paint system must be made.

## *Dispersant Architecture*

As is true of any material, the properties of dispersant are determined by its constituent building blocks (specific atoms or chemical functional groups) and the ways that these building blocks are combined. We can use this simple principle to logically organize dispersants according to the functionality that each of these building blocks provides.

Most dispersants are organic copolymers composed of two distinct parts. The first is a segment (or segments) that attaches to the particle surfaces. These are often referred to as anchoring groups, and they are typically tailored to specific anchoring sites on the pigment or extender surface. The second is a segment (or segments) that extends into the liquid component of the wet paint and into the polymer component of the dry paint. These segments are typically tailored to the combination of the liquid carrier of the paint (water or a single or mix of organic solvents) and the chemical nature of the resin particles.

We will discuss the salient features of these two parts, beginning with the anchoring groups.

### Anchoring Groups

There are two general categories of anchoring groups: electrically neutral and electrically charged. The former group can be divided into those that attach through organic functional groups and those that are simply insoluble in the liquid paint carrier, while the latter group can be further divided into cationic and anionic groups.

Electrically neutral anchoring groups include polyols, polyamines, and polyesters. These are typically used in solvent-borne paints since the alternative—charged anchoring groups—is not supported in most organic solvents. Neutral anchoring groups can attach to particle surfaces either through hydrogen bonding or simple polar interactions. These attachments are physical and so generally weaker than ionic attachments, which can be either purely electrostatic or result from the formation of chemical bonds.

In addition to using these dispersants in solvent-borne paints, they can also be used in water-borne paints to anchor onto the non-polar surfaces of many organic pigments. In some cases, the non-ionic anchoring groups used to attach dispersant onto non-polar surfaces in water are no more than a hydrophobic chain. In this case, it is energetically more favorable for that chain to lay on a non-polar surface than to be solvated by water.

Electrically charged anchoring groups can be either cationic or anionic. Most typical of the former are various substituted amine groups or quaternary ammonium compounds. Anionic anchoring groups include carboxylic acids (typically acrylic acid or its derivatives), phosphates, and sulfates.

## Anchoring Sites

While not components of the dispersant molecules, the anchoring sites on the various particles play a crucial role in dispersant composition and so will be discussed here.

We can divide the particles to be dispersed into two groups: polar particles, which are nearly always inorganic ($TiO_2$, certain color pigments, and mineral extenders), and non-polar particles, which are nearly always organic in nature (almost exclusively color pigments).[4] The anchoring sites are different for the two-particle types, and it is often advantageous to use a mixture of anchoring groups when more than one particle type is present.

Organic particles are typically non-polar and as such are difficult to wet-in and disperse in both organic solvents and water and to stabilize once dispersed. They generally require dispersants with non-ionic anchoring groups and are often stabilized through steric repulsions. When ionic dispersants are used, they tend to be cationic rather than anionic. In all cases, it is difficult for dispersant molecules to attach to these particles due to the non-polar nature of their surfaces.

Because organic pigment particles tend to be quite small (in the tens to hundreds of nanometer range) they are difficult to separate from one another and have high surface areas (and oil absorptions). These factors result in relatively large amounts of dispersant being needed to both disperse and stabilize these particles (discussed below).

In some cases, the surfaces of organic particles can be modified to increase their polarity, and therefore, improve dispersibility and dispersion stability. As an example, some grades of carbon black—a material known to be particularly difficult to disperse—are lightly oxidized to form polar oxygen functional groups on their surfaces.

By contrast, most inorganic particles are polar. Most are oxides, hydroxides or hydrates, such as clays, or are ionic solids, such as calcium carbonate or barium sulfate. As such, a wide variety of anchoring groups can be attached to their surfaces.

[4] Resin particles in water-borne paints are also organic, but they will not be discussed here because they usually are supplied with a dispersant already. More information on resin can be found in Chap. 10.

In most cases, ionic anchoring groups are used. Often the surfaces of these particles, particularly $TiO_2$ pigments, are modified to provide even better anchoring sites.

An example of this is the surface modification of $TiO_2$ by hydrous alumina particles. The surface of untreated $TiO_2$ has relatively weak anchoring sites for common dispersants. A stable slurry of $TiO_2$, dispersant, and water can be made, but the dispersant attachments are weak and dynamic—dispersant molecules adsorb and desorb reversibly from the surface readily, resulting in a stable dynamic equilibrium. However, this equilibrium is lost when other particles are added to the slurry, such as extender or resin. These other particles often have better anchoring groups than the $TiO_2$ particles. As the dispersant molecules desorb from the $TiO_2$ surface, they adsorb onto the other particles, resulting in a destabilization of the $TiO_2$ and its flocculation.

This issue is resolved by the deposition of small islands of hydrous alumina onto the $TiO_2$ surface. This alumina has excellent anchoring sites for most dispersant molecules, and as such acts as a "glue" to permanently bind the dispersant to the pigment particle surface. This is shown schematically in Fig. 11.25. The electron micrograph shows the alumina decorations (bright areas) on the surface of a coatings

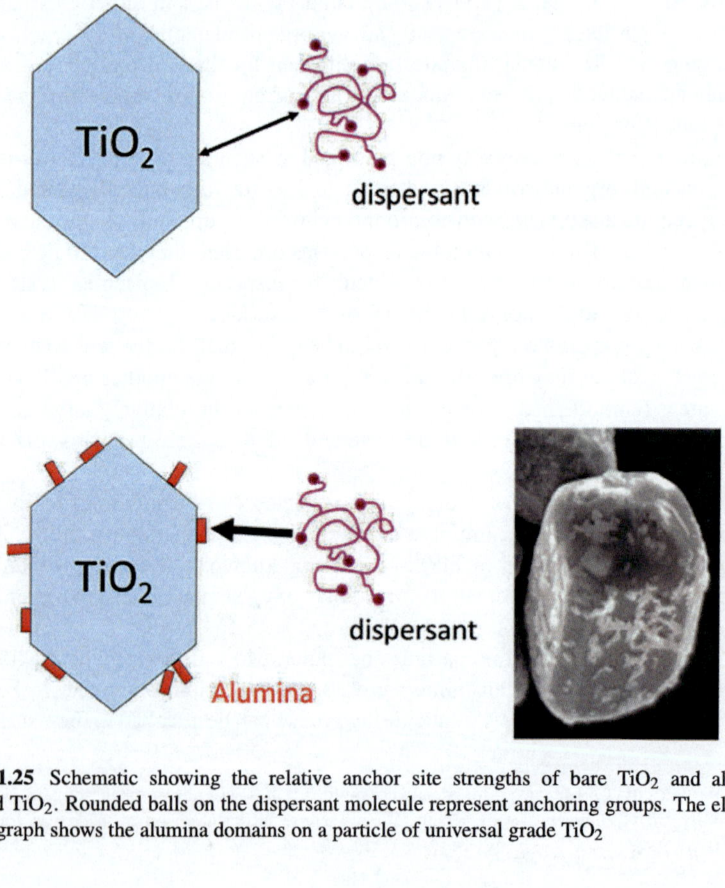


**Fig. 11.25** Schematic showing the relative anchor site strengths of bare $TiO_2$ and alumina treated $TiO_2$. Rounded balls on the dispersant molecule represent anchoring groups. The electron micrograph shows the alumina domains on a particle of universal grade $TiO_2$

grade $TiO_2$ pigment. Note that the $TiO_2$ surface does not need to be completely covered by the alumina—even at fractional coverage, there are enough anchoring sites to assure dispersion stability.

We can understand the reason alumina has better anchoring sites than bare $TiO_2$ on the basis of the acidic or basic character of the different surfaces. Anionic dispersants are typically used in waterborne paints and will, therefore, obviously, prefer positively charged anchoring sites. Pure titanium dioxide is a solid acid, and at the pH values typical of most water-borne paints (between 8.0 and 9.5), protons will leave the particle surface, resulting in a negative residual charge. This charge will repel anionic anchoring groups on the dispersant molecules. By contrast, hydrous alumina is a basic solid and has a positively charged or nearly neutral surface at typical paint pH values.

Extending this argument, we note that the surface charge characteristics of most inorganic particles in paints span a wide range, and so different types of particles may benefit more greatly from the use of different dispersant molecules. The importance of surface charge characteristics can again be demonstrated by pigmentary $TiO_2$. Hydrous alumina is not the only material deposited onto $TiO_2$. To enhance paint durability, the surfaces of $TiO_2$ pigments are sometimes coated with silica and/or zirconia (see Chaps. 7 and 14). This leads to the isoelectric points (IEPs) of coatings grade $TiO_2$ varying from roughly 5.5 to 9.0.[5]

Anchoring groups for different grades of $TiO_2$ vary accordingly. Those particles with IEP values from 7.5 to 9.0 benefit from anionic dispersants; those with IEP values between 6.5 and 8.0 benefit from neutral dispersants, and those with IEP values from 5.5 to 6.8 benefit from cationic dispersants. Note that there is some overlap in these ranges. Also, it is generally true that those $TiO_2$ pigments with silica coatings are the most sensitive to dispersant choice.

## Solvated Arms

In addition to anchoring groups, dispersant molecules also have segments that are called arms, tails or, for some arrangements, pendant groups. These are the regions of the dispersant molecule that extend into the liquid phase of the paint. These arms are not only necessary for steric stabilization, but are also crucial in most forms of electrostatic stabilization (the exception being small, highly charged ionic dispersants, discussed below). By extending the charge sites into the liquid phase, the effective size of the particle is increased accordingly. This adds to particle–particle separations in the suspension.

In order to extend into the liquid phase, the arms must be "soluble" in it. By this, we mean that the chemistry of the arm must be matched to that of the paint liquid, so that the arm can be fully solvated. Solvation occurs when three interactions are properly balanced—the interaction of the arm with itself, the interaction of the arm with the

[5] The isoelectric point is the pH at which a surface has a net zero charge. At lower pH values a particle will be positively charged and at higher pH values it will be negatively charged. This is discussed in detail in Chap. 2.

solvent, and the interaction of the solvent molecules with one another. To solvate the arms, they must first be unfurled—that is, the arms must be separated from one another. They must then extend into the solvent, creating arm-solvent interactions. An important aspect of this extension is that some solvent molecules that are touching one another must separate, in order for the arms to insert between them.

Obviously, the interactions of the arms with themselves should be minimized, and the interactions of the arms with the solvent maximized. Often the interactions between the solvent molecules are neglected in this analysis. However, these interactions are crucial contributors to the energetics of solvation. This is, in fact, the reason that hydrophobic molecules do not dissolve in water. It is not that there is a negative interaction between these molecules and water—in fact, the opposite is true. However, the interactions between water molecules, which are quite intensive due to hydrogen bonding, are greater than those between the hydrophobic molecule and water, and the energy required to disrupt the water–water interactions is greater than the energy released by the newly created molecule–solvent interactions.

In addition to the physical and chemical properties of the arms, their overall length is also important. If the arms are too short, they are not capable of extending far into the liquid phase and so their effectiveness is minimized. If they are too long, then they will hopelessly entangle themselves with other arms, also limiting their extension into the liquid phase. Finally, note that the arms will ultimately, in the dry paint film, be surrounded by resin molecules. If the arms are compatible with the resin, then they will extend into it, which enhances the adhesion of the particle with the resin. If they are incompatible, then the arms will coil onto themselves and lie tightly on the particle surface. This will decrease the interfacial strength of the particle/resin interaction. This is similar to the discussion in Chap. 9 on the organic surface treatment of extender particles to enhance their functionality in paints and plastics applications.

## Dispersant Assembly

We can consider the anchoring groups and solvated arms as the fundamental building blocks of a dispersant molecule. Like any building blocks, they can be arranged in a number of ways. This assembly of groups into dispersants is typically done in a polymerization process where different monomers are joined together.

For dispersants created in this way, we will refer to monomers that create the anchoring groups as the "A" blocks and those that created the solvated arms as the "B" blocks. We can describe these dispersants easily by the arrangements of their blocks. Common arrangements are shown in Fig. 11.26.

Note that it is important that dispersant molecules contain both building blocks—a molecule that only solvates will not attach to the particle surface, and a molecule with only attachment (anchoring) groups will not extend into the liquid phase of the wet paint or the resin phase of the dry film.

The arms that extend into the liquid phase are described here as either ionic (for electrostatic stabilization) or non-polar (for steric stabilization). In fact, most

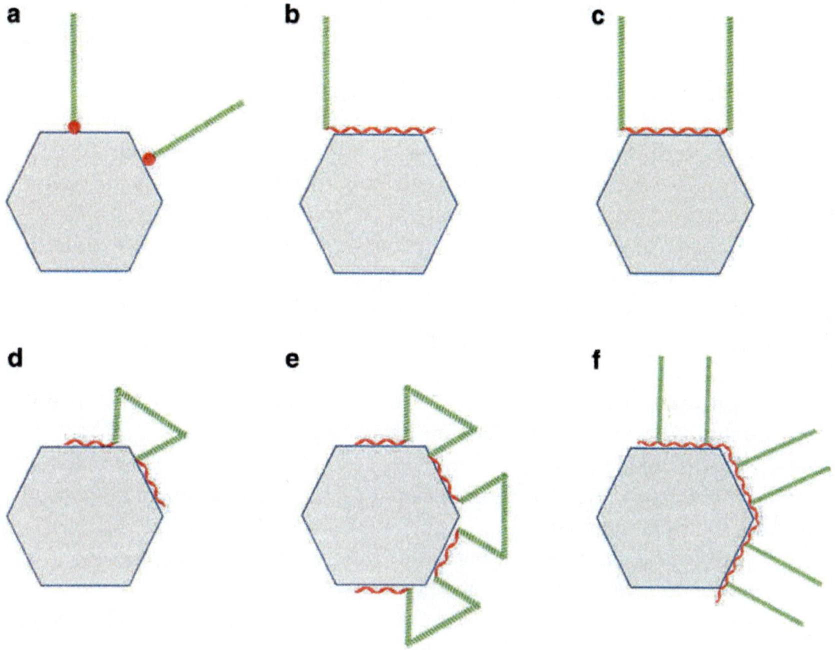


**Fig. 11.26** Common dispersant architectures adsorbed on particle surfaces. Red chains are anchor groups, green chains are solvated arms. **a** Solvated arms with a single anchor group. **b** AB dispersant. **c** BAB dispersant. **d** ABA dispersant. **e** AB random copolymer. **f** Comb copolymer

dispersants use a combination of the two stabilization mechanisms, with the balance between them being a factor that can be controlled by mixing different monomers in the desired proportions.

## *Small Ions*

In addition to polymeric dispersants, small, highly charged species, such as phosphates, silicates or citrate, are occasionally used to disperse polar particles (which are nearly always inorganic in nature). These ions can adsorb onto particles and, due to their high charge density, confer enough electrostatic repulsions at the particle surface that they overcome the need to extend from the surface into the liquid component of the paint. Specific examples are salts of phosphoric acid, dimers (tetrapotassium pyrophosphate, or TKPP), trimers (potassium tripolyphosphate, or KTPP) or salts of phosphoric acid oligomers, as well as sodium and potassium silicate. Once used extensively in coatings, these materials have, for the most part, been superseded by the more effective organic dispersants.

## *Dispersant Demand*

The amount of dispersant required to stabilize a given particle type depends on the chemical nature of the dispersant (primarily the anchoring groups), the chemical nature of the particle surface (primarily the anchoring sites), and the surface area of the particle. The exact requirements for a given combination of particle and dispersant can be determined experimentally by titrating a slurry of the particles of interest in the liquid carrier of the paint (e.g., water for latex paints) with the dispersant and following, as a function of the amount of dispersant added, a property of the wet paint or dry film that depends on the degree of particle dispersion. Slurry viscosity is typically used for this purpose, although gloss, haze, and color strength of the dry film can also be used.

A typical dispersant demand curve, in this case for a universal $TiO_2$ pigment being titrated with hydrophilic functionalized polyacrylate copolymer dispersant neutralized with ammonia, is shown in Fig. 11.27. The slight rise in viscosity after a precipitous drop is typical of these curves, and the dispersant demand is generally defined as the curve minimum. In practice, it is common to formulate with a slight excess (~10%) of dispersant as a margin of safety, since the consequences to the viscosity of under-loading dispersant are quite severe.

An insufficient amount of dispersant leads to poor paint dispersion properties while a large excess of dispersant, in addition to being uneconomical, can lead to

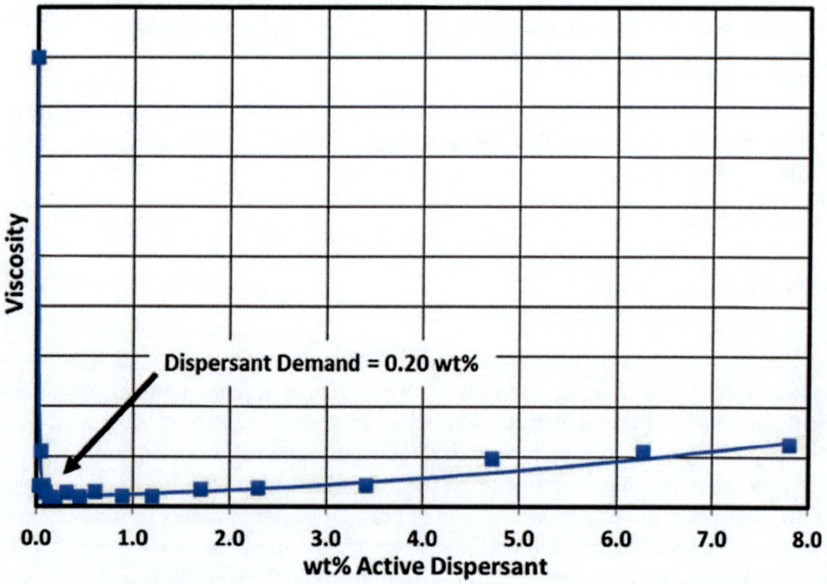


**Fig. 11.27** Dispersant demand curve of a universal grade of $TiO_2$ using a hydrophilic functionalized polyacrylate copolymer neutralized with ammonia

destabilization (as seen in Fig. 11.27 by the increase in viscosity above the demand point), poor substrate adhesion, low film hardness, and liquid water and water vapor (humidity) sensitivity. Water sensitivity causes both erosion of the paint when the soluble dispersant molecules are washed out (for example, during rain events or cleaning) and hazing of appearance.

It is useful to have a rough estimate of the dispersant requirements of a particle to use as a basis for the dispersant demand test or a starting point for paint formulation. Some dispersant manufacturers give recommended levels based on the weight of the particles. However, we would expect dispersant requirements to vary with the size of the particle being dispersed—smaller particles have more surface area and so would be expected to require more dispersant. Dispersant demand curves for a series of seven $TiO_2$ pigments with varying surface areas are shown in Fig. 11.28.

On a surface area basis, the dispersant recommendations are given by dispersant producers typically ranging from 1 to 2 mg of active[6] dispersant for every $m^2/g$ of surface area for inorganic particles to roughly 5 mg per $m^2/g$ of surface area for organic particles, reflecting their greater difficulty in dispersing and remaining stable. While these guidelines are often good starting points for the formulation, they can be exceeded significantly, depending on the exact nature of the dispersant and the particle. For example, the dispersant demand of $TiO_2$ pigments spans roughly an order of magnitude, from roughly 0.15 to 2.5 mg active dispersant per $m^2/g$.

As an example of this calculation, a dispersant manufacturer recommends 1.5 mg per square meter of the particle surface. A $TiO_2$ pigment with a surface area of 14 $m^2/g$, which is typical for universal coatings grades, will, therefore, require 21 mg of dispersant per gram of pigment, or loading of 2.1 weight percent active dispersant.

## *Dispersant Selection*

A process for choosing a dispersant follows directly from the analyses given above. The building blocks of the dispersant molecule must separately match the particle surface (for the anchor sites) and both the paint liquid and the resin chemistry (for the solvated arms). It is important that the anchoring groups are attached strongly and permanently to the anchoring sites on the particle surface, and that the arms extend deeply into the liquid phase of the paint, which requires their chemical character to match the hydrophobicity or hydrophilicity of the solvent or water.

The exact choice of these dispersant components depends on:

- The paint liquid (water or specific solvent)
- The particle (organic, inorganic, surface polarity, etc.)
- The end-use application (some dispersants are specifically formulated, for example, to give a high gloss)

[6] Dispersants are normally delivered as a solution with roughly 50% active solids by weight.

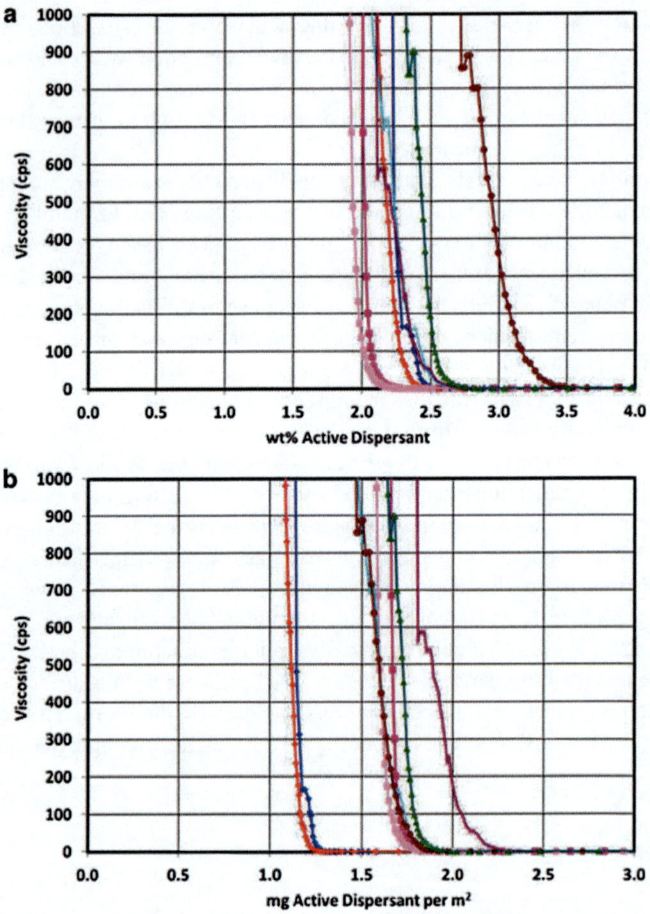


**Fig. 11.28** Dispersant demand curves for a series of seven $TiO_2$ pigments on a weight basis (**a**) and on a surface area basis (**b**)

Consideration should also be made as to which of the three dispersion steps—wet-in, grinding, and stabilizing—are particularly challenging for the particle. For example, most grades of $TiO_2$ wet-in and grind easily, so the dispersant choice should be made based on dispersion stability, even at the expense of wet-in and grind. On the other hand, carbon black, which is among the most difficult pigments to use, requires a dispersant that addresses all three steps in the dispersion process.

Finally, for certain more easily dispersed particles (such as $TiO_2$), no dispersant is needed in solvent paints. Instead, the dissolved resin fulfills the functionality of a dispersant. Although resin molecules are not designed to optimize this function, they are generally present in much greater concentrations than true dispersants.

# Summary

The quality of the dispersion of particles (pigments and extenders) in a liquid medium (aqueous or non-aqueous) directly impacts the quality of the final coating. Improper dispersion causes a number of deficiencies in the paint, including reducing the light scattering ability of pigments of $TiO_2$, decreasing the color strength of any colored pigments, and impacting the optical properties of a coating such as gloss.

The successful dispersion of particles into a liquid medium is determined by the complex properties of the particle surface, the choice of liquid carrier, and the interactions between the two. These properties affect all aspects, not only initial dispersion, but also dispersion stability.

The dispersion process can be divided into three steps: wet-in, particle separation, and particle stabilization. In the wet-in process, the air is entrained in the dry powder and any surface moisture must be displaced by the liquid carrier. The liquid must spontaneously spread over the surface of the particles and penetrate into particle agglomerates. The wettability of a particular particle by a particular liquid is dictated by the surface energy of the particle, the surface tension of the liquid, and the interaction strength between the particle surface and the liquid. Even with favorable wet-in energetics, the kinetics of wet-in can be slow and it is necessary to facilitate complete wet-in by effectively mixing the powder mass into the liquid.

In the particle separation step, mechanical force is applied to the slurry or paint to break apart agglomerates of particles into their individual constituent, as well as to expose new particle surfaces to complete the wet-in process. Two types of mechanical force can be used for particle separation: impact and shear.

Impact forces break apart particles that are caught between colliding objects or through the collision of particles themselves into an immobile object. Shear, on the other hand, is generated from uneven forces acting on a body, in our case agglomerates. A velocity gradient within the slurry or paint is used to create shear forces that pull the agglomerate apart. It is important that all velocities within this gradient are low enough to maintain transitional, rather than turbulent, flow; otherwise, there is excessive energy loss due to simple mixing rather than dispersing. The viscosity of the slurry is also important, as the onset of turbulent flow occurs at lower velocities for low viscosity fluids.

There are many different types of dispersion or grinding equipment used for particle separation. These can be classified by their use of impact and shear forces. Equipment can range from fully impact type (e.g., fluid energy mills and kinetic dispersion mills), to fully shear-type (e.g., three-roll mills). In addition, mills are available that use varying degrees of both impact and shear (e.g., ball mills, media mills, and high-speed disk dispersers).

The final step in the dispersion process is particle stabilization. Once the particles are wetted and well separated, attractions due to van der Waals forces or electrostatic charge can cause them to flocculate or re-agglomerate when they come close to one another or collide. Particles can also re-agglomerate through sedimentation, where gravitational forces slowly drive particles to the bottom of their container. Re-agglomeration is in fact thermodynamically inevitable, but the paint manufacturer can take steps to slow the delay of this for time spans measured in years.

Stabilization is typically achieved using dispersants. A wide variety of dispersants is available, giving the formulator many degrees of freedom in dispersant selection. Dispersants typically operate by one of two mechanisms (or both simultaneously). The first is through physically separating particles by attaching blocky molecules to their surfaces. This is referred to as steric stabilization and can be used in both water-borne and solvent-borne paints. The second mechanism uses surface electrostatic charges to cause the particles to electrically repel one another. This can only be accomplished in water-borne paints, since organic solvents cannot support the electrical charge. In this case, the dispersants contain functional groups that are ionic at the pH values of typical latex paints.

Most dispersant molecules are composed of segments that separately provide two important functions. The first function is to attach the dispersant molecule to the particle surface. This is accomplished by having anchoring groups on the dispersant that bind strongly to anchoring sites on the particle surfaces. The identity of the anchoring groups must be matched to the specific chemical and physical properties of the anchoring sites. The second function is to occupy as much space around the particle as possible, increasing the effective size of the particle and maximizing particle–particle separations. This is accomplished through the incorporation of polymer arms that extend into the liquid phase of the paint. These arms are electrically neutral in steric dispersants and electrically charged in electrostatic dispersants. Incorporating both anchoring groups and solubilized arms into the dispersant molecule is accomplished by copolymerization of monomers with different chemical properties.

Each of the three dispersion steps (wetting, particle separation, and particle stabilization) must be considered when formulating the paint so as to optimize paint performance. These steps build on one another—for example, particles cannot be dispersed if they are not wetted, and even a perfect initial dispersion can be spoiled by poor paint stabilization. The paint will never be as well dispersed as when it is first made, and so it is critical that the initial dispersion of the particles is done as completely as possible and that the factors driving their later flocculation to be minimized.

## References

1. Bonn, D., Eggers, J., Indekeu, J., Meunier, J., Rolley, E.: Wetting and spreading. Rev. Mod. Phys. **81**, 739 (2009)
2. Young, T.: An essay on the cohesion of fluids. Phil. Trans. R. Soc. London **95**, 65 (1805)
3. Wenzel, R.N.: Resistance of solid surfaces to wetting by water. Ind. Eng. Chem. **28**, 988 (1936)
4. Patton, T.C.: Paint Flow and Pigment Dispersion, 2nd edn. Wiley-Interscience (1979)
5. Singh, P., Joseph, D.D., Gurupatham, S.K.; Dalal, B., Nudurupati, S.: Spontaneous dispersion of particles on liquid surfaces. Proc. Natl. Acad. Sci. USA **106**(47), 19761 (2009)

6. Goldschmidt, A., Streitberger, H.-J.: BASF Handbook on Basics of Coating Technology, Vincentz Network (2003)
7. Reynolds, O.: An experimental investigation of the circumstances which determine whether the motion of water shall be direct or sinuous, and of the law of resistance in parallel channels. Phil. Trans. Royal Soc. **174**, 935 (1883)
8. Jensen, G.V., et al.: Dispersion state of $TiO_2$ pigment particles studied by ultra-small-angle X-ray scattering revealing dependence on dispersant but limited change during drying of paint coating. Prog. Org. Coat. **142**, 105590 (2002)
9. Stokes, G. G.: On the effect of internal friction of fluids on the motion of pendulums. Trans. Camb. Phil. Soc., **9**, part ii, 8 (1851)
10. Derjaguin, B., Landau, L.: Theory of the stability of strongly charged lyophobic sols and of the adhesion of strongly charged particles in solutions of electrolytes. Acta Physico Chimica URSS **14**, 633 (1941)
11. Verwey, E.J.W., Overbeek, J.T.G.: Theory of the stability of lyophobic colloids. J. Phys. Colloid. Chem. **51**(3), 631 (1948)
12. Diebold, M.P., Bettler, C.R., Mukoda, D.M.: Mechanism of $TiO_2$/ZnO instability. J. Coat. Tech **75**(942), 29 (2003)
13. Clayton, J.: Pigment/dispersant interactions in water-based coatings. Surf. Coat. Int. **9**, 414 (1997)
14. van den Haak, H.J.W., Krutzer, L.L.M.: Design of pigment dispersants for high-solids paint systems. Prog. Org. Coat. **43**(1), 56 (2001)

# Chapter 12
# Dispersion of Small Particles in Plastics

## Contents



## Introduction

The incorporation of small particles into thermoplastics is non-trivial, both in concept and in practice. Conceptually, there are certain similarities between dispersing particles in molten plastics and dispersing cocoa powder into milk to make chocolate milk, a process familiar to many [1]. In the latter process, solid fat particles (cocoa) must be suspended in water. Because the hydrophobic surfaces of the fat particles are incompatible with the polar water molecules, these particles are not wetted by the water and water does not penetrate into the gaps within particle agglomerates. This results in the initial formation of large powder pockets or capsules that, though wetted on their exterior, are completely dry in their interior.

Stirring the mixture with a spoon will eventually break the capsules apart, producing individual particles that are then dispersed throughout the liquid. Capsule disruption can be difficult when the concentration of the capsules is small—stirring simply moves the capsules through the milk without breaking them apart. However, at higher concentrations, the capsules collide with and experience shear from one



M. Diebold et al., *Pigments, Extenders, and Particles in Surface Coatings and Plastics*, https://doi.org/10.1007/978-3-030-99083-1_12

another, which disintegrates them into individual particles. Wetting and dispersion can be accelerated by smearing a capsule that is trapped between the spoon and cup wall. This efficiently generates a region of high shear within the capsule. The application of these factors to the dispersion of particles into liquid paints was discussed in Chap. 11.

The dispersion of particles in plastics is similar. In this case, however, it is the particles that are often hydrophilic and the liquid phase that is hydrophobic. However, the effect of this incompatibility on dispersion is the same. Particles tend to form relatively large capsules that resist infiltration by the liquid. At low concentrations, the addition of mechanical energy results primarily in the mixing of intact capsules, rather than in their disintegration. Regions or zones of high shear can be highly effective at breaking the capsules apart.

While wetting and separation are critical to particle dispersion in both polymer melts and liquid paints, there are some aspects of dispersion that are unique to each situation. One is that particles dispersed in liquid paints must be stabilized against flocculation, while those in a thermoplastic are frozen in place once the plastic solidifies. A second is that for thermoplastics the liquid medium must be created as part of the dispersion process through heat melting. For optimum efficiency, both in terms of energy and time, the transformation of mechanical energy must be properly balanced between the melting and dispersion processes.

## Particle–Particle Attractive Forces

The degree of particle dispersion of particles within a polymer can have a large effect on many optical, physical, and mechanical properties of the polymer. Pockets of undispersed filler particles such as calcium carbonate, can concentrate stress weak points that are exposed as the polymer matrix experiences mechanical forces (e.g., stretching, twisting, pinching, compression, etc.). Their presence leads to poor tensile and impact strengths, resulting in the polymer becoming brittle and prone to shattering. Poorly dispersed white or color pigment particles interact inefficiently with light. This leads to the need for greater particle concentrations to achieve the same appearance effect as is seen for well-dispersed pigment particles. Finally, surface appearance properties such as gloss and distinctness of image, as well as opacity and durability against ultraviolet radiation, are all degraded by poor particle dispersion.

As was the case with the particle groups encountered in dispersion into liquid paints, it is convenient to classify those encountered in dispersion into molten polymer based on their size and the strength of the forces holding these groups together. The smallest unit of these groups is the individual, or primary particles (see Fig. 12.1). Aggregates are small groups of primary particles that are held together so strongly that they are not separated during the dispersion process. This is most commonly seen with inorganic particles that are formed at high temperatures. Agglomerates, too, are relatively small groups of primary particles, but in this case, the bonds between the particles are weak enough to be broken during dispersion. Finally, clusters are very

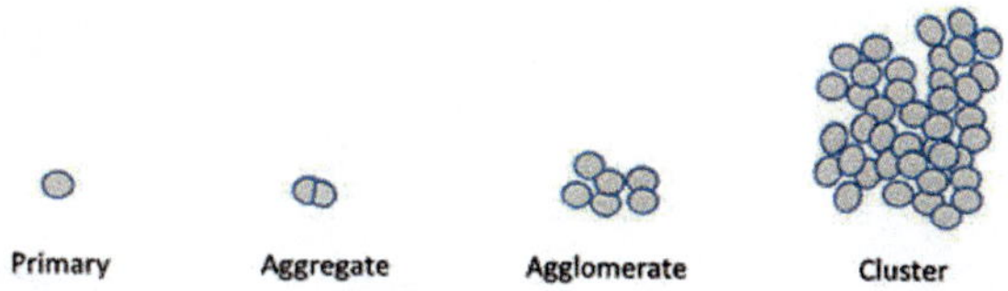


**Fig. 12.1** Particle groups relevant to dispersion in polymer melts

large groups of particles that are weakly held together. These are analogous to flocculates that form when particles in liquid paints are inadequately stabilized. Because they are so large and held together so weakly, clusters are the most susceptible of the particle groups to disintegration by shear forces.

Note that the distinction between aggregates and agglomerates is not absolute. Instead, it depends on the type of disperser and the time (and energy) spent on the dispersion process. That is, some particle groups will attrit (break apart into smaller sizes) under some dispersion conditions but not under others. Obviously, the intent of any dispersion process should be to break apart as many small groups of particles into even smaller groups. Since most particle size measurements quantify the size of the particle groups, this is equivalent to saying that the dispersion process should produce the smallest average particle size, and the narrowest size distribution, that is possible.

Note also that there is no sharp delineation between agglomerate and cluster. This is, instead, a matter of definition. However, we can generalize by saying that agglomerates are groups of no more than a few tens of primary particles, while clusters can contain many thousands of primary particles. Solid particles are to be dispersed as primaries within a thermoplastic, but they usually start in a cluster or agglomerate structure in the dry powder form. The dispersion process converts the dry agglomerates into wet primaries and aggregates.

The dispersion process involves two distinct phases. First, the particle size is reduced to the greatest extent possible, moving from right to left in Fig. 12.1. Next, the particles must be distributed as uniformly as possible throughout the polymer matrix.

The dispersion process can involve three means of cluster and agglomerate size reduction. These are erosion, rupture, and shattering (Fig. 12.2). Erosion is the gradual shearing of small fragments from the surface regions of clusters or agglomerates. These fragments are typically no larger than 10% of the original cluster [2]. Rupture is the splitting of clusters or agglomerates into fragments of comparable size to one another. Shattering is the separation of agglomerates and clusters into their constituent components in a single event or collision.

The main differences between the three processes are the number of events needed for the complete disintegration of the initial cluster and the amount of energy required for each event. A shattering event is the most energetic of the three, but it results in the most size reduction. Erosion and rupture are more gradual processes requiring many more events and greater times than shattering.

Besides the energy and time differences between these means of attrition, the intermediate particle size distributions resulting from each are different. Erosion leads

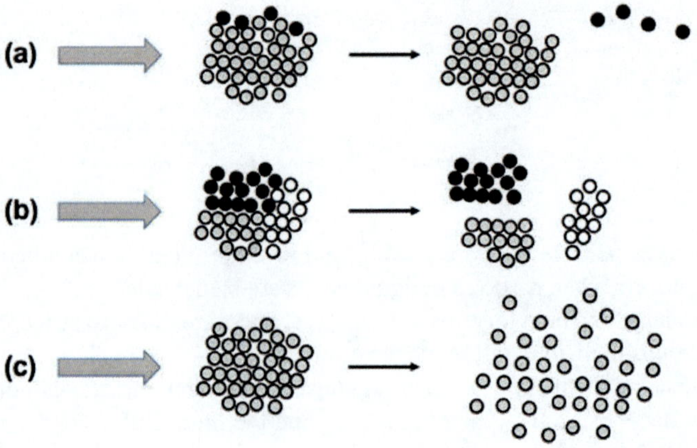


**Fig. 12.2** Disruption of agglomerates. Gray arrows denote force on the agglomerate. Particles are shaded for identification purposes only and are otherwise identical. **a** Erosion. **b** Rupture. **c** Shattering

to a bimodal distribution consisting of individual particles or small agglomerates, [2] which grow in number over time, and agglomerates or clusters, which reduce in size and number over time. Rupture leads to a monomodal distribution that shifts to lower sizes over time. This distribution can be described by a power law equation (Eq. 12.1) [3].

$$Y = kX^{\alpha} \tag{12.1}$$

where

Y is the fraction of particles with a certain diameter,

X is that diameter, and

k and α are constants.

The energy required to fragment clusters and agglomerates and to provide dispersive and distributive mixing must have a source. Compressive loads are the most common means for dispersive mixing. Particle–particle collisions and particle–wall impacts are examples of these compressive loads. Additional forces, such as those of a hydrodynamic nature, provide another source of energy for agglomerate fragmentation in thermoplastic dispersions. These forces are of a shear, rather than impact, nature. The magnitude of hydrodynamic forces is determined by the viscous shear stress.

For shear grinding, the ease with which particles will disperse is determined by the balance of the shear forces pulling them apart (the viscous shear stress) and the attractive forces holding them together (typically van der Waals forces). This balance is quantified by the fragmentation constant, $F_a$, which is simply the ratio of the two forces [4]. For agglomerate disruption to occur, the viscous shear force must be greater than the cohesive strength of the particles within the agglomerate, i.e., $F_a$ must be greater than unity.

Cohesive strength is not an absolute property of a material, such as melting point or color, but instead varies depending on the internal structure or packing of the particles, the density of the packing, and the interparticle attractions due to electrostatic charges, van der Waals forces or capillary forces arising from moisture condensed between the particles [5–8]. In most agglomerates, the cohesive force is dominated by van der Waals attractions. In this case, the particle–particle interactions can be described by the Hamaker constant, H [9]. The cohesive strength between particles can be calculated as the ratio of the Hamaker constant between two surfaces with a certain particle radius, at a defined separation distance, and with a defined solids volume fraction as shown in Eq. 12.2.

$$S_c = \frac{H}{a \bullet z^2 \bullet \rho} \tag{12.2}$$

where

$S_c$ is the cohesive strength,

H is the Hamaker constant,

a is the radius of the primary particle,

z is the equilibrium separation distance between the two-particle surfaces, and

ρ is the solids volume fraction.

Reduction of the Hamaker constant is an effective means of decreasing interparticle cohesive strength and reducing the energy needed to grind agglomerates. The Hamaker constant can be reduced by decreasing the polarity of the particle surfaces [10]. This is typically done by coating the particles with a dispersing aid (see also Chap. 11) [11, 12]. Organic surface treatments are commonly used for this purpose, [13, 14] although the inorganic surface treatments applied to $TiO_2$ particles also enhance dispersibility (Chap. 7). The organic surface treatments are not to be confused with surface treatments for adhesion promotion [15]. Although the latter type of surface modification can supply "wet-in" capabilities, their primary functions are to protect solid particles, especially those with a non-unity aspect ratio, from fracture during processing, and to strengthen the adhesive bond at the solid particle–matrix interface (see also Chap. 9).

Particle dispersion into a plastic occurs through a well-defined sequence of steps, with some steps involving a number of stages. We will discuss these steps in order, beginning with the dry blending of the solid ingredients.

## Dry Blending

The first step in particle dispersion is to intimately mix the powder particles with polymer pellets, a process known as dry blending. When done properly, dry blending reduces dispersion time and energy; the more intimate this mixing, the greater the reduction of the magnitude of these process parameters.

In addition to simply bringing the different components together, dry blending can reduce the size and number of large particle clusters. These clusters form during powder handling prior to use (e.g., pneumatic conveying, screw conveying, intentional compression to increase bulk density, subjection to a load, etc.) and are dependent on the nature of the particles themselves (e.g., particle size, surface area, geometry, surface polarity, etc.) [16–19]. This nature is often described by non-technical but very descriptive terms such as "clumpy", "lumpy", "chunky" or "dusty".

Cluster formation and growth can occur when the powder is subjected to relatively low intensity agitation. Under these conditions, agglomerates that collide have insufficient energy to attrit one another. Instead, the collision process brings agglomerate surfaces into close contact. When the attractive forces between particles on different agglomerates are strong enough, the agglomerates themselves agglomerate, forming large clusters. Similarly, if an agglomerate encounters a single particle, this particle can be incorporated into the agglomerate (Fig. 12.3). These clusters can grow further by settling under gravity (Fig. 12.4).

In addition to cluster formation during handling, clusters can also form during dry blending [20]. Since the ultimate goal of dispersion is to completely disintegrate agglomerates and clusters, care should be taken to avoid this. One means for doing this is to provide sufficient shear during dry blending, especially if the subsequent

**Fig. 12.3** Growth of an agglomerate during low intensity mixing

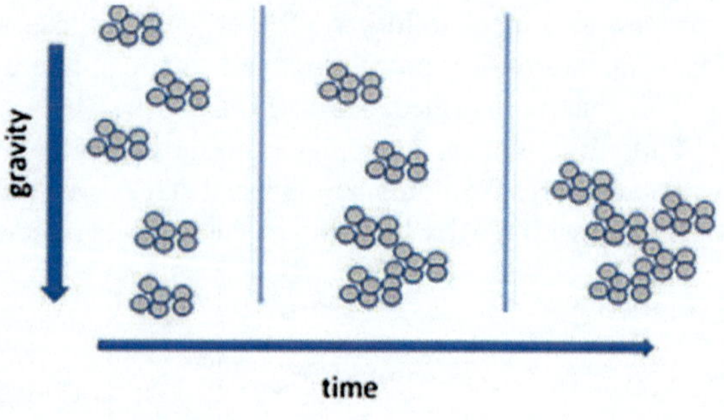


**Fig. 12.4** Cluster formation during powder settling

processing step is "flood feeding". Flood feeding is when the feed hopper sits atop the compounder's mouth and feeds the dry blend into the compounding equipment—that is, the compounding equipment regulates the feed rate. Here adequate mixing of the solids in the dry blend step is critical, particularly for downstream compounding processes that function by melting and pumping the polymer into a die with minimal mixing and devolatilization.

Minimization of agglomerate and cluster growth during dry blending can also be addressed by adding various materials during this process. The solid particles are often combined with dispersants (typically fatty acid/glycerol derivatives), other additives (typically phenolic-based anti-oxidants), and polymers during dry blending. Dispersants moderate the stickiness of solid particles and can reduce static charges that bind the agglomerates together. Other additives are mixed into the solids at this time to assure thermal stability in the subsequent processing steps, which involve melting and shearing the compound components.

Certain particles in the dry blends can stick to equipment. Friction on the wall surfaces, such as the hopper walls, varies with the solid additive as well as the roughness and frictional coefficient of both the solid and wall surfaces. In addition, some particles cannot be introduced in dry blending due to their sensitivity to processing. In particular, hollow glass microspheres, alumina flakes, and carbon or glass fibers are not recommended for dry blending as excessive shear stress can result in significant damage to the particles and a resulting loss of functionality. Finally, consideration must be given to particles with high densities, such as metallic fillers, as these can segregate in the dry blend based on their weight.

Dry blending is crucial for many plastics processes, but it is not an exact science. In general, the dry blend process and equipment must be carefully chosen to be compatible with the needs of a particular set of ingredients. Low shear, low-temperature blenders, such as tumble blenders, V-cone blenders, and ribbon blenders are often used when the formulation has only one type of solid particle and it is present at relatively high concentrations.

High-speed blenders are much more shear intensive and often require cooling to keep the pre-blend mixture below the resin melting point. These blenders are quite common in the polyvinylchloride (PVC) industry, where the dry blend mixture consists of half a dozen or more ingredients. High-speed and high shear blenders are essential for PVC compounding because the PVC resin must be softened with heat during the mixing process. During intensive blending, solid particles are often suspended in a bed of PVC using liquid ingredients (e.g., lubricants, plasticizers, liquid colorants, etc.). After cooling, the free flowing, homogenized PVC pre-blend is ready for the next step in compounding.

## Particle Wetting

As was discussed in the context of paints (Chap. 11), the first step in the dispersion process is wetting. Wetting refers to the replacement of air and other materials, such as moisture, from the particle surfaces and from the voids between them, with the melted thermoplastic resin (Fig. 12.5). This process can be likened to the burning of a candle. Wax that has been melted by the flame soaks into the porous cotton wick, where it provides the fuel needed to sustain the flame. The melted wax represents the polymer resin melt, and the candle wick represents the porous particle agglomerates.

The surface tension of the polymer melt must be lower than the surface energy of the particles in the agglomerate if the polymer melt is to wick between the particles. This condition can be achieved in two ways. First, the polymer melt surface tension can be lowered by using lubricating agents. These are analogous to the wetting agents used in the dispersion of particles into a liquid paint. Second, and more commonly, the particle surfaces can be coated with a treating agent that reduces surface energy. These agents were discussed earlier for $TiO_2$ particles in Chap. 7 and extender particles in Chap. 9.

Intuitively, we might expect the replacement of air in the voids within an agglomerate by molten resin to occur from the outside in. That is, a front of molten resin would uniformly flow through all of the outer voids, into the agglomerate interior, filling the inner voids as it progresses. However, infiltration of the molten resin into the agglomerate proceeds through a different process. This is because the pores connecting the intra-agglomerate voids have a size distribution, and liquid flow through any particular pore will be controlled by the diameter of that pore. In particular, the flow becomes hindered as the pore diameter, or cross-sectional area, decreases.

Rather than flowing uniformly into the agglomerate, the molten resin follows the path of least flow resistance, as indicated by the black line in Fig. 12.6. This is typically through the widest pores, or channels. When the particle surfaces can be

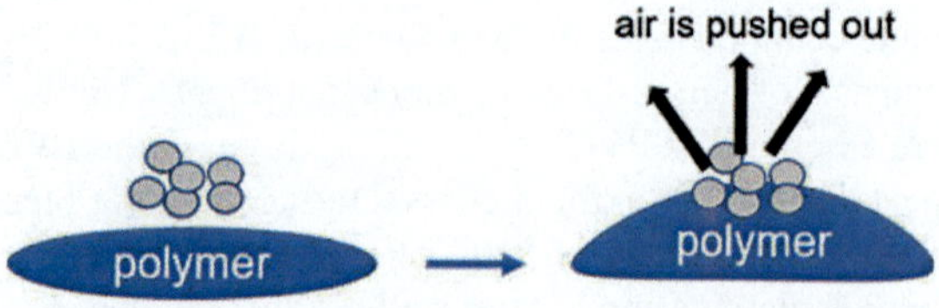


**Fig. 12.5** Wetting of a particulate agglomerate by polymer resin

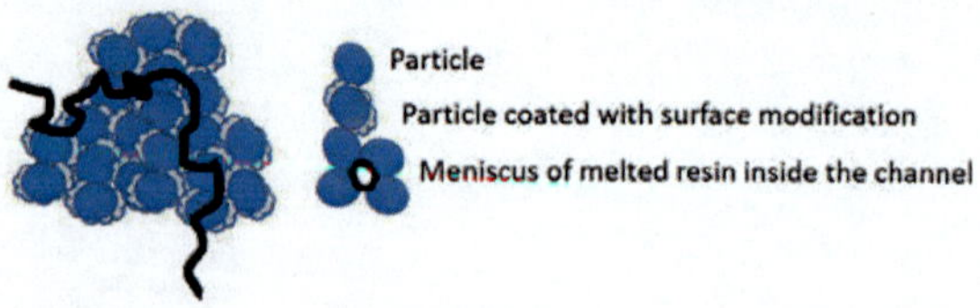


**Fig. 12.6** Flow of molten polymer into a particle agglomerate

wetted by the resin, the gaps between them do not necessarily fill initially. Instead, the resin coats the particle surfaces, and a meniscus is formed at the interface of the molten resin and air remaining within the pore [21].

Once the molten resin has infiltrated the agglomerate through the paths of least resistance, it enters into the narrower, more resistant, pores. This process proceeds primarily from the inside out—that is, resin deep within the agglomerate enters the narrower pores that connect to the larger channels [22]. This resin is replaced by additional resin from outside the agglomerate that travels along the paths of least resistance. This sets up a flow system in which the molten resin is, in effect, pumped into reservoirs within the agglomerate, and then is redistributed from these reservoirs into small pores that reside deep within the agglomerate.

The initial layers of resin on the particle surfaces create cohesive bridges between the particles, and the resulting capillary pressure actually adds to the cohesive strength of the agglomerate [23]. Figure 12.7 shows the strength of an agglomerate with the saturation of the agglomerate voids. Once saturation is reached, the structure collapses as a flood of the polymer flows from the inside to the outside. Agglomerates, therefore, do not break apart during the wetting process, but rather at its conclusion.

Many physical factors affect the facility with which the agglomerates are wetted and the strength of the agglomerates during this process. High polymer melt viscosity increases both the agglomerate strength and the time needed for complete flow into the agglomerate. Particle size and shape also affect the flow rate of the molten polymer.

The effect of particle size can be complex. In general, smaller particles have greater surface areas, and so require more resin to saturate the surface before the voids can be filled. In addition, pore size, up to a point, decreases with decreasing particle size, and this affects the velocity of the molten polymer through the pore.

However, the trend of decreasing pore size with decreasing particle size only applies when the particles pack uniformly. Once a certain size minimum is reached, this trend is reversed because, as was shown in Chap. 1, small particles pack in small groups, these groups pack together into larger groups, and so on. Large pores form between the particle groups. The sizes of these larger pores are consistent with the sizes of the particle groups. In addition, the void volume increases substantially as smaller groups of particles nestle together into large groups, and this results in the need for higher amounts of resin to saturate the agglomerate. Finally, the lower

**Fig. 12.7** Polymer saturation relationship with agglomerate/cluster strength

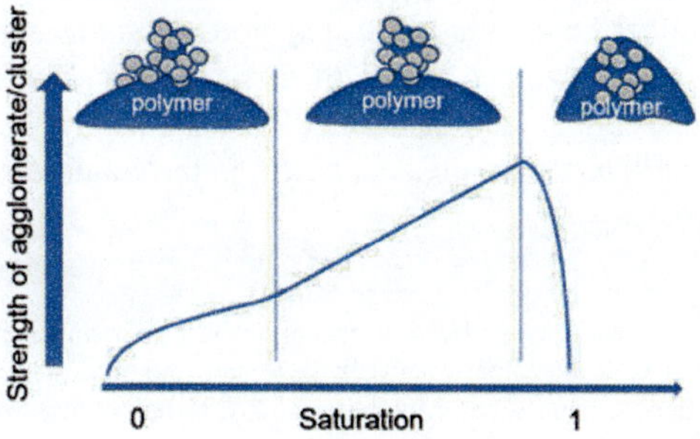

density of agglomerates of small particles decreases the cohesive strength of the agglomerates, which also affects the time necessary to completely wet the agglomerates. This situation is often encountered with carbon blacks and nanoscale fillers [24].

An additional complication is seen when the wetting process alters either the particle packing structure or the overall size of the agglomerate. This can occur when capillary forces from the molten resin pull the particles closer together. This is seen in clusters of nanoparticles, which are typically diffuse and fragile, and can collapse when wetted. In other situations, the cluster size can be reduced when the exterior of the particle becomes fully wetted and sheds from the partially dry interior. This process can repeat several times, depending on the size and structure of the agglomerate. This slows the kinetics of cluster wetting and has been reported for calcium carbonate clusters of one micron or larger [25].

In general, shear stresses induced in particle structures from the dispersion process are most efficiently transferred throughout these structures when they are completely saturated with molten resin—that is, when they are fully wetted. In the absence of this, both erosion and rupture are inefficient, and extra time and energy is required to disperse the particles into their smallest units.

## Disruption of the Agglomerate Structure Using Shear

We discussed above the necessity of energy for the dispersion process and will now consider the mechanisms by which this energy is applied to and transmitted through particle agglomerates. Recall that energy must be applied both to disperse the particles and to distribute them uniformly throughout the resin mass. These two actions can be considered independently of one another.[1]

Distributive mixing addresses the rearrangement of the mixture through an ordered or random mixing process. The goal of distributive mixing is to spread agglomerates uniformly throughout the polymer melt, but not necessarily to break them apart. In addition, distributive mixing circulates regions of the dispersion mass into, and out of, zones of high shear intensity. Dispersive mixing, on the other hand, involves the rupture and erosion of the agglomerates into their constituent units through the shear that is applied in these zones. This shear must be strong enough to overcome the forces holding the wetted agglomerates together.

The size and nature of these intense shear zones are dependent on the type of dispersion equipment, as will be shown by the example below. Regardless of how these zones are created, however, the conditions within them are similar. Dispersive

[1] The same consideration occurs in the dispersion of particles into a liquid paint (Chap. 11). In that case the balance between dispersion and mixing is controlled through the relationship between rheology and shear. More specifically, turbulent flow results in mixing while laminar flow results in dispersion.

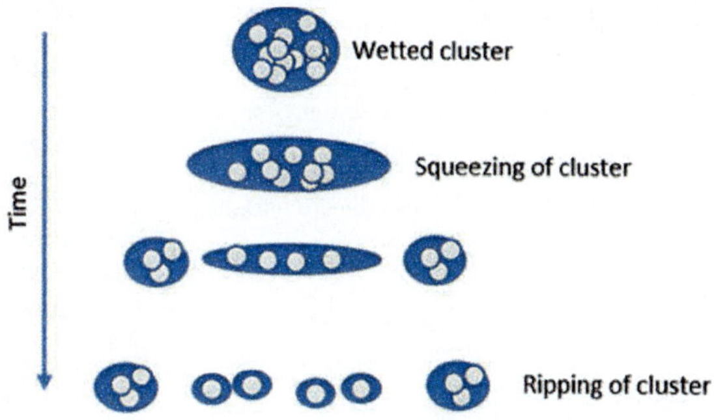


**Fig. 12.8** Squeezing and ripping of wetted clusters during dispersive mixing

mixing proceeds through repeated cycles of compression (squeezing) and rupture (ripping) of wetted clusters and agglomerates, as shown in Fig. 12.8.

There are many varieties of plastic dispersers available commercially [26–28]. The more common ones are listed in Table 12.1, along with their advantages, disadvantages, and recommended applications. For each type, there are several adjustable parameters that can be used to optimize dispersion. These include temperature, residence time, fill level, clearance between mixing elements, etc. In addition to these machine parameters, melt composition will also affect dispersion efficiency. In particular, high concentrations of solid particles will increase melt viscosity, which, in turn, raises the shear stress within the polymer (see Chap. 17).

## Examples of Particle Dispersion into Plastic Resin

Although a wide variety of equipment is used to disperse particles into the molten polymer, there are certain attributes that are common to all. Dispersive mixing and distributive mixing occur at different locations within the equipment, allowing the equipment to be configured (or reconfigured) to match an exact task. In addition, the energy required to melt the polymer is decoupled from the energy required to disperse the particles. Although frictional heat is invariably generated in the dispersion process, this amount of heat does not need to bring the temperature to exactly that which is desired. If higher temperatures are required, heat can be added through an external heater. If excess heat is created in the dispersion process, it can be removed through external coolers in much the same way.

Examples of the operations of three disperser types, a roll mill and an internal batch mixer, both of which operate in batch mode, and a screw extruder, which operates in continuous mode, are given below.

**Table 12.1** Disperser types for melt processing solid particles

| Disperser type | Advantages | Disadvantages | Typical application |
|---|---|---|---|
| Two-roll mill | • Application of shear is extremely efficient<br>• Final material is two-dimensional, i.e., a film sheet | • Successful processing is dependent on the skill of the operator<br>• Limited dispersion efficacy for particles with non-unity aspect ratios (platelets or needles)<br>• Final material is two-dimensional, i.e., a film sheet | • Flexible PVC<br>• Rubber |
| Internal mixer | • Very energy efficient (fewer kW per kg)<br>• Simplified operations<br>• Processes a wide range and combination of ingredients, from pliable to extremely rigid materials, without changes to machine parts as might be required for screw extruders | • Insufficient pressure for extrusion or pelletizing operations<br>• Batch charging and discharging which can affect productivity and product consistency<br>• Precise control of melting and mixing is difficult | • Rubber |
| Continuous mixer | • Very energy efficient (fewer kW per kg)<br>• Less shear compared to twin- screw extruders | • Insufficient pressure for extrusion or pelletizing operations<br>• Higher capital cost compared to screw extruders<br>• Sustaining a starved condition requires balance between orifice setting, rotor speed and temperature controls | • Color concentrates, masterbatches and mineral-filled concentrates |
| Single- screw extruder | • More economical based material of construction<br>• Less shear compared to twin- screw extruders<br>• Operational flexibility for a wide range of different types of materials | • Limited feed performance<br>• Limited capability for dispersive mixing<br>• Lower productivity compared to similarly powered twin-screw extruders, especially under high-pressure conditions | • Melt polymers requiring pressure (e.g., injection/blow molding, cast/blown film) |

(continued)

**Table 12.1** (continued)

| Disperser type | Advantages | Disadvantages | Typical application |
|---|---|---|---|
| Twin screw extruder | • High pumping efficiency<br>• Lower dependence on the flow properties of the material<br>• Higher mixing efficiency<br>• Heat exchange rate from barrel surface to the material is faster and more uniform | • High cost<br>• Heat generated in the barrel is low, requiring external heating of the barrel | • Rigid PVC<br>• Compounding with solid particles |

## *Dispersive/Distributive Mixing with a Two-roll Mill*

The two-roll mill is perhaps the simplest disperser. It consists of two heated (or cooled) cylinders (also called drums or rolls) that rotate at high speeds in opposite directions, that is, toward one another (Fig. 12.9). Dispersion occurs in the space between them, termed the nip gap, or simply gap, [29] in a manner similar to the three roll mill discussed in Chap. 11. The gap is the zone of high shear where dispersion takes place. The polymer melt adheres to one of the heated cylinders as a blanket and passes through the gap. In batch processing, the blanket is then rotated back into the gap several times, while in continuous processing, it is separated as a sheet from the roller.

The rotational speeds and the gap clearance are both controlled in this process. Shear rates within the gap are typically adjusted to between $10^2$ and $10^5$ $s^{-1}$. One cylinder is normally rotated at a higher speed than the other. This increases the friction that the melt experiences, and higher rotational speed differences give greater melt heating. The ratio of the peripheral speeds is referred to as the friction ratio, and a typical value would be 1:1.25.

The mix of polymer melt and solid particles is referred to as the flux. The viscosity of this flux is low enough to allow for flow when pressure is applied to it. As the flux enters the gap, the high shear rips it apart. During passage through the gap, the flux is squeezed. Mixing is accomplished through a lateral cross motion or a folding motion. This reduces the formation of streaks within the flux. Two-roll milling is a highly

**Fig. 12.9** Operation of a two-roll mill

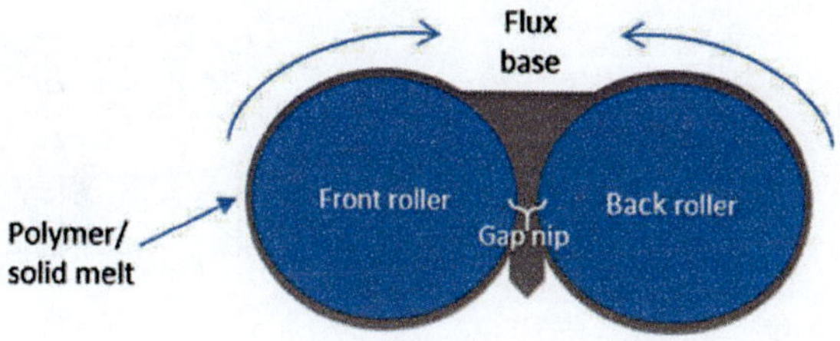

effective means of dispersing solid particles since the shear rate can be controlled over a wide range of values [30, 31].

## *Dispersive/Distributive Mixing with an Internal Batch Mixer*

The second disperser that we will discuss is the internal batch mixer, which is the oldest mixing device used for polymers and dates back to the 1820s [32]. As was true for the two-roll mill, these devices disperse particles into the resin by passing the mix through a zone of high shear. The internal batch mixer operates by the addition of a blend of the polymer and solid particles into a heated mixing chamber. Two parallel, cylindrical blades in this chamber rotate at fixed speeds (usually slightly different from one another), while the power to the blade rotors is monitored. A typical curve of power as a function of time is shown in Fig. 12.10 [33].

The consistency of the polymer changes during processing, proceeding through four phases:

(1) Freshly introduced solid pellets
(2) De-formable solid pellets
(3) Flux (i.e., a dough-like suspension of solid particles in resin)
(4) Viscoelastic fluid

The progression of these phases and their relationship to the power curve are shown in Fig. 12.11.

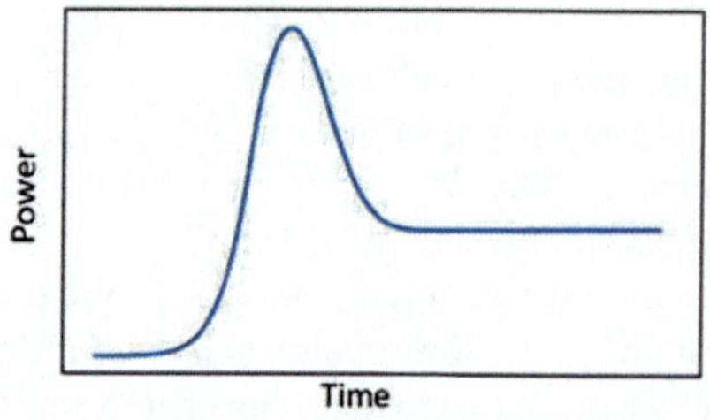


**Fig. 12.10** Evolution of power over time in an internal batch mixer

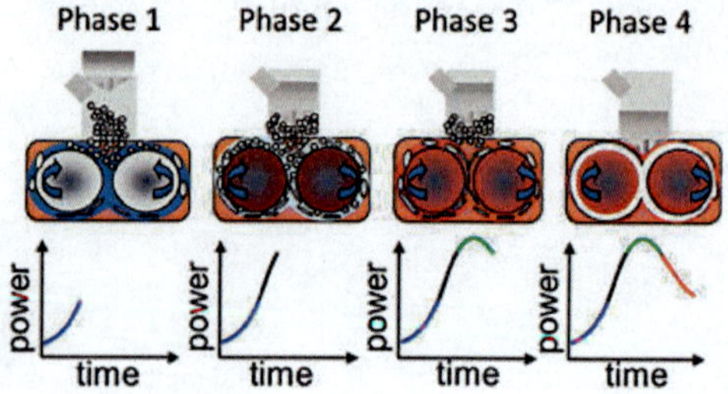


**Fig. 12.11** Progression of phases in the internal batch mixer

In the first phase, the ingredients are placed in the chamber with blades rotating. The ingredients are pushed into the chamber with a piston ram. Pressure from this piston squeezes the resin pellets, increasing the power needed to maintain a constant rotor speed. In the second phase, compression of the resin pellets is complete, and the temperature begins to increase.

Next, the polymer pellets deform, and dispersive mixing reaches its peak. In the third phase, the combination of pressure and heat causes the resin to soften. This lowers the viscosity of the mixture, which, in turn, decreases the power required to turn the rotors. At this point, the resin is completely melted and the flux of resin and solid particles is distributed throughout the chamber. Finally, in the fourth phase, the solid particles are dispersed within the flux. At the completion of this process, the melt is typically discharged into a two-roll mill or a screw extruder for further cooling and to convert the solidified material into the desired form (sheets, pellets or chunks).

Figure 12.12 shows the interplay of squeezing and ripping over the course of this process. During the initial squeezing, the solid particles embed themselves into the polymer pellets. The extent of this embedding is dependent on particle hardness. Continued application of pressure causes the softened pellets to deform. As the heat and pressure build, the lower melting point material will flow into the particle agglomerates and clusters, wetting them and forming capillary bridges with other resin pellets. The capillary bridges are then pulled apart by mechanical forces, creating smaller agglomerates that then repeat the process.

Dispersive mixing occurs in the internal batch mixer as the thermoplastic resin melts, completely or partially, and passes through the nip region of the disperser (Fig. 12.13). In this process, the nip region is the clearance between the chamber wall and rotor tip. In Fig. 12.13, as the mix of polymer and solid powder is squeezed by the clockwise rotation of the rotor, it becomes deformed. As the wetted agglomerates pass through the nip, they are pulled apart.

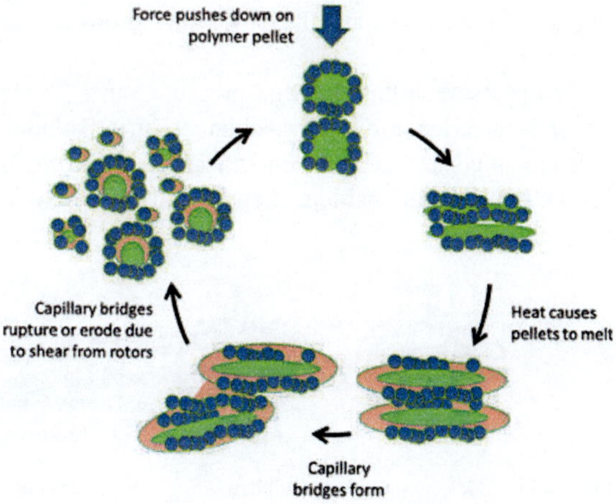


**Fig. 12.12** The cycle of squeezing resin (green) and solid particles (blue) into flux, followed by shearing the flux

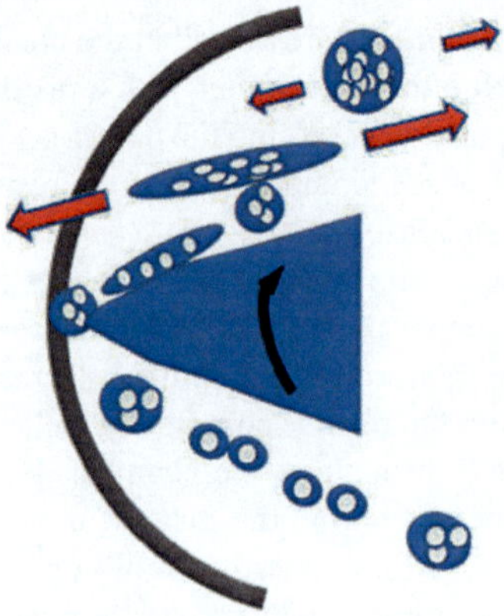

**Fig. 12.13** Nip region between the blade and wall of an internal batch mixer

After passing through the nip, the unmelted polymer is added to the flux and the process is repeated during the next rotational cycle. This sequence of squeezing and ripping provides the dispersive mixing required for compounding. Simultaneous with this dispersive mixing, distributive mixing occurs as the wetted particle agglomerate moves from one side of the chamber to the other. During this time, the consistency (viscosity) of the flux changes, with a concomitant change in the power draw of the rotors.

The shape of the rotor blades and their orientation with respect to one another ensures that all fluid particles undergo highly intensive shear forces (dispersive mixing) in the nip region [34]. The flow patterns between the rotors promote efficient laminar (distributive) mixing [35]. Figure 12.14 shows the rotors turning in the chamber as they stir the mixture of polymer and particles. The red dots show the highest-pressure zone, which is immediately in front of the face of the rotating blade. The pale blue regions in this figure represent the volume occupied by the mixture of resin and particles, and the dark blue regions represent the chamber walls and the piston ram.

The pressure within the high-pressure zones (the red dots in Fig. 12.14) depends on the blade orientation, and so changes during rotation. The greatest pressure occurs when the blades sweep past the inner wall of the chamber. This is seen in Fig. 12.14 for the left rotor in positions 3 and 4 and for the right rotor in positions 1 and 2. As

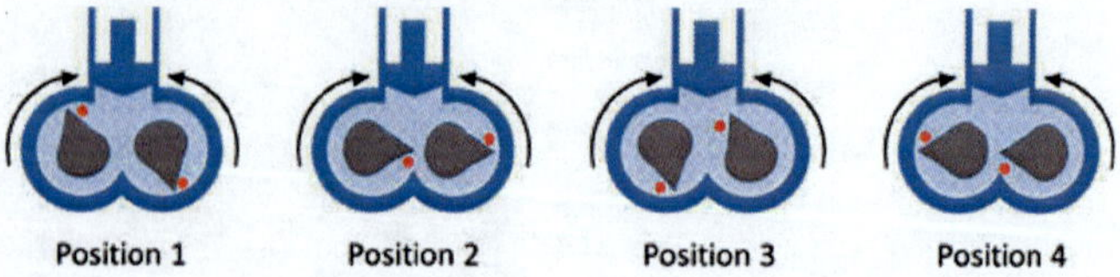


**Fig. 12.14** Side on view of internal batch mixer rotors during rotation

the red dots move into the intersection zone of the two-chamber sections, distributive mixing occurs as material is exchanged between the two rotors.

Although they are configured differently, the underlying principles of the two-roll mill and the internal batch mixer are similar. Both dispersers operate by repeatedly passing the mixture of resin and particles through regions of high shear stress, followed by intimately mixing the sheared material. The two operations differ in the location of the high shear zone (the nip). In the two-roll mill, the nip is located between the rotating drums, while in the internal batch mixer, the high shear zone is the small clearance between the rotors and the chamber wall.

Because of torque and heat transfer limitations in the internal mixer, only a small portion of the compound is subjected to high stress at any given time. However, the overall rotor configuration ensures that the entire compound mixes freely between the high stress regions, since during the rotation, one rotor blade pushes material across the intersection zone and into the path of the other rotor blade. In a two-roll mill, mixing of this kind is accomplished by manually cutting the melt off the roller and then placing the cut sections into the feed.

## *Dispersive/Distributive Mixing with a Single-Screw Extruder*

Both the two-roll mill and internal batch mixers operate in batch mode, meaning that a set quantity of material is charged into the disperser, the disperser runs to the desired degree of dispersion, and the material is discharged, leaving the disperser ready to accept a new charge. Plastics can also be dispersed in a continuous process that involves the screw conveyance of the material through a tube, or barrel, with the mixing and dispersive steps occurring in different regions of the barrel. The dispersed material, still in the molten state, is then extruded through a die for further processing.

Both single-screw and twin-screw extruders are quite common in the plastics industry. As the names imply, these differ in the number of screws used. In the twin-screw extruder, screws are placed parallel to one another, with their flights either inter-meshed or not, and rotated either concurrently or counter-currently. In the simpler single-screw extruder, a single screw with a tapered shaft is employed.

The principles of screw extrusion can be demonstrated with a single-screw extruder, with the understanding that the operations of a twin-screw extruder are more complex and have more ability to be modified for different materials. A schematic of a typical single-screw extruder is shown in Fig. 12.15. The extruder consists of

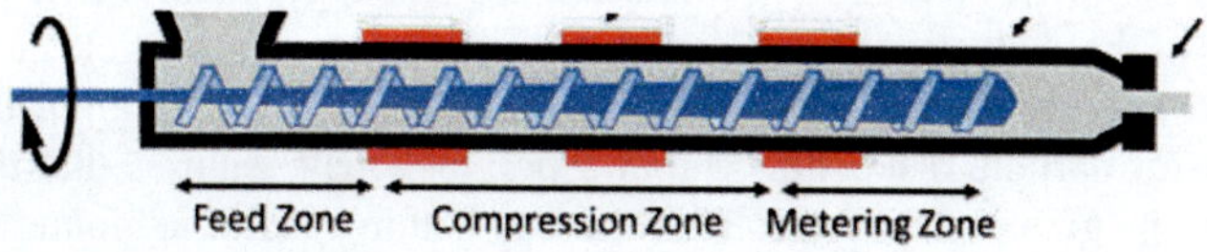


**Fig. 12.15** Single-screw extruder

a metal barrel into which the screw is centered. The barrel is typically heated at different positions along its length (the red bars in this firgure), with the ability to create a temperature gradient through it. A feed mixture is dropped into the barrel through a hopper, and the final dispersed material is extruded through a die at the end of the barrel.

The extruder can be configured in a number of ways so as to be optimized for any mixture of resin and particles. Among the most important of these is the screw itself. The shaft of the screw is tapered while the flights are not. The shaft is narrowest at the feed end of the barrel, allowing for maximum feed rates. As the material is conveyed along the barrel, the shaft gradually widens, compressing the material as it is conveyed through the barrel. At the end of the extruder the shaft thickness remains constant so as to convey the material through the die. This portion of the extruder is called the metering zone.

Material is moved through the extruder by the back pressure that develops within the barrel. The magnitude of this pressure is determined by the dimensions of the barrel and the screw, and screws are typically designed specifically for a given polymer and particle mix. Screw parameters that can be adjusted include the diameter of the shaft, the reach of the flights, and the distance between the flights. In addition to providing the force necessary to convey the material through the extruder, the back pressure also determines the degree of shear in the different regions of the barrel.

Without going into excessive detail, we can describe some ways in which twin-screw extruders are similar to and different from single-screw extruders. Twin-screw extruders are, understandably, more complex. The screws themselves are configured with multiple zones along the barrel, and by proper screw design, the material can be added into different regions along the barrel. This allows materials that are temperature or shear sensitive to be incorporated into the melt under less stressful conditions, the latter of which is particularly important because twin-screw extruders generate significantly higher shear than their single-screw counterparts (this is beneficial for mixing efficiency). In addition, the configuration along the barrel of the extruder can be more extensively modified than is the case for single-screw extruders. This is done by using modular screw and heating elements.

## Summary

Dispersion of particles and additives into polymeric resins relies on many of the same principles as dispersion of particles into liquid paint. Dispersion proceeds through the initial (and complete) wetting of particle clusters and agglomerates, followed by the disruption of these clusters and agglomerates into the smallest particle unit (primary particles or aggregates) and their intimate mixing into the dispersion medium.

Successful particle dispersion within a polymer melt requires that the viscous shear stress be greater than the cohesive forces within particle agglomerates. There are two strategies for doing this: maximizing the viscous shear stress that is applied to the agglomerates, or minimizing the cohesive forces that hold the agglomerates together. Viscous force maximization is addressed through equipment design, and many different types of dispersers are used in the plastics industry, with each type optimized for specific end-use. Alternatively, cohesive forces can be decreased by altering the polarity of the particle surfaces. This is typically accomplished by coating the particles with an organic surface treatment. These treatments also improve the compatibility of inorganic particles with the organic polymer matrix.

Particle wetting can be complex because the viscous polymer melt does not flow uniformly through a particle cluster or agglomerate. Instead, the polymer initially flows through the largest pores within the cluster, which represent the path of least resistance. Once filled, these pores, and the larger voids connected by them, act as resin reservoirs for the filling of nearby smaller pores and voids. In the early stages of wetting, the capillary forces within the agglomerates increase their cohesive strength. At the final stage of wetting, however, the cohesive strength decreases to a value lower than in the original dry agglomerate.

Dispersion is done primarily through shear. Large agglomerates can either erode from the shearing process, or can be ruptured by it. The primary difference between these is the size of the reduced agglomerates after shearing. Erosion leads to a bimodal distribution of particle sizes, with populations of primary particles or aggregates and large agglomerates, whereas rupture leads to a monomodal size distribution that shifts to lower values as the shearing process continues.

Dispersion in plastics can be done using either batch or continuous operations. Batch dispersers include the two-roll mill and internal batch mixers. These operate under the principle of having zones, or regions, of high shear, in which the agglomerates undergo dispersive mixing, and regions of low shear, in which the material undergoes distributive mixing. The zones of high shear are created by sweeping one surface closely past another.

Continuous dispersers include both single- and twin-screw extruders. In this equipment, the material to be dispersed is conveyed through a barrel using one or two screws. Zones are present along the length of the barrel where the mix can be sequentially heated, melted, dispersed, and mixed. This equipment can be easily reconfigured for different types of polymers (which may melt at different temperatures) and combinations of particles and additives.

## References

1. Mongia, G., Ziegler, G.R.: The role of particle size distribution of suspended solids in defining the flow properties of milk chocolate. Int. J. Food Prop. **3**(1), 137 (2000)
2. Cheng, Z., Redner, S.: Kinetics of fragmentation. J. Phys. A: Math. Gen. **23**(7), 1233 (1990)
3. Cheng, Z., Redner, S.: Scaling theory of fragmentation. Phys. Rev. Lett. **60**(24), 2450 (1988)
4. Iliev, P.S., Wittel, F.K., Herrmann, H.J.: Evolution of fragment size distributions from the crushing of granular materials. Phys. Rev. E. **99**(1), 1 (2019)
5. Wünsch, I., Finke, J.H., John, E., Juhnke, M., Kwade, A.: The influence of particle size on the application of compression and compaction models for tableting. Int. J. Pharm. **599**, 120424 (2021)

6. Afshar, T., Disfani, M.M., Arulrajah, A., Narsilio, G.A., Emam, S.: Impact of particle shape on breakage of recycled construction and demolition aggregates. Powder Technol. **308**, 1 (2017)
7. Rumpf, H., Schubert, H.: The behavior of agglomerates under tensile strain. J. Chem. Eng. Japan **7**(4), 294 (1974)
8. Capes, C.E.: Agglomerate bonding. In: Handbook of Powder Technology. Elsevier Science B.V. (1980)
9. Hamaker, H.C.: The London—van der waals attraction between spherical particles. Physica **4**(10), 1058 (1937)
10. Lomboy, G., Sundararajan, S., Wang, K., Subramaniam, S.: A test method for determining adhesion forces and hamaker constants of cementitious materials using atomic force microscopy. Cem. Concr. Res. **41**(11), 1157 (2011)
11. Bergstrom, L.: Hamaker constants of inorganic materials. Adv. Colloid Interf. Sci. **70**, 125 (1997)
12. Birdi, K. S.: Colloid systems and interface stability of dispersion through polymer and surfactant adsorption. In: Handbook of Surface and Colloid Chemistry. 3rd edn. CRC Press (2008)
13. Li, C.C., Chang, S.J., Tai, M.Y.: Surface chemistry and dispersion property of $TiO_2$ nanoparticles. J. Am. Ceram. Soc. **93**(12), 4008 (2010)
14. Cao, Z., Daly, M., Clémence, L., Geever, L.M., Major, I., Higginbotham, C.L., Devine, D.M.: Chemical surface modification of calcium carbonate particles with stearic acid using different treating methods. Appl. Surf. Sci. **378**, 320 (2016)
15. Zeyu, L., Letao, Z., Erlei, Y., Zhaoming, Y., Yagang, Z., Xuanchi, L., Wumanjiang, E.: Modification of glass fiber surface and glass fiber reinforced polymer composites challenges and opportunities: from organic chemistry perspective. Curr. Org. Chem. **19**(11), 991 (2015)
16. Cambow, B., Dubujet, P., Emereault, F., Sidoroff, F.: Homogenization for granular materials. Eur J. Mech. A/Solids **14**(2), 255 (1995)
17. Emereault, F., Chang, C.S.: Interparticle forces and displacements in granular materials. Comput. Geotech. **20**(3/4), 223 (1997)
18. Valverde, J.M., Ramos, A., Castellanos, A., Watson, P.K.: The tensile strength of cohesive powders and its relationship to consolidation, free volume and cohesivity. Powder Technol. **97**(3), 237 (1998)
19. Quintanilla, M.A.S., Castellanos, A., Valverde, J.M.: Correlation between bulk stresses and interparticle contact forces in fine powders. Phys. Rev. E **64**, 9 (2001)
20. Silviya, E. K., Varma, S., Unnikrishnan, G., Thomas, S.: Compounding and mixing of polymers. In: Thomas, S., Weimin, Y. (eds.) Advances in Polymer Processing. Woodhead Publishing (2009)
21. Rishi, K., Narayanan, V., Beaucage, G., McGlasson, A., Kuppa, V., Ilavsky, J., Rackaitis, M.: A thermal model to describe kinetic dispersion in rubber nanocomposites: the effect of mixing time on dispersion. Polymer **175**, 272 (2019)
22. Gopalkrishnan, P., Manas-Zloczower, I., Feke, D.L.: Effect of morphology and extent of infiltration on the cohesivity and dispersion mechanisms of particle agglomerates. Chem. Eng. Sci. **62**(14), 3740 (2007)
23. Pietsch, W., Hoffman, E., Rumpf, H.: Tensile strength of moist agglomerates. Prod. R&D **8**(1), 58 (1969)
24. Rwei, S.P., Manas-Zloczower, I., Feke, D.L.: Observation of carbon black agglomerate dispersion in simple shear flows. Polym. Eng. Sci. **30**(72), 702 (1990)
25. Levresse, P., Manas-Zloczower, I., Feke, D.L.: Dispersion studies of agglomerates in steady and dynamic flows of polymeric materials. Rubber Chem. Technol. **75**(1), 119 (2002)
26. Ess, J.W., Hornsby, P.R.: Twin-screw extrusion compounding of mineral filled thermoplastics: dispersive mixing effects. Plast. Rubb. Process Appl. **8**, 147 (1987)
27. Benkreira, H., Shales, R.W., Edwards, M.F.: Mixing on melting in single screw extrusion. Inter. Polymer Process. **2**, 126 (1992)
28. Latif, L., Saidpour, H.: Assessment of plastic mixture quality in injection moulding process. Polym. Testing **16**, 241 (1997)

29. Murakami, Y., Hirose, T., Chung, G., Tenda, T., Akimoto, Y.: "Power Consumption in a Two-Roll Mill"; *I & E Chem.* Process Design Dev. **24**(4), 1005 (1985)
30. van der Reijden-Stolk, C., Van Dam, J., Boerstoel, H.: Dispersive mixing on a two-roll mill. In: Lemstra, P.J., Kleintjens, L.A. (eds.) Integration of Fundamental Polymer Science and Technology—3. Springer (1989)
31. Yao, C.H., Manas-Zloczower, I.: Study of mixing efficiency in roll-mills. Polym. Eng. Sci. **36**(3), 305 (2004)
32. Min, K., White, J.L.: Specialty polymers and polymer processing. In: Allen, G.; Bevington, J.C. (eds.) Comprehensive Polymer Science and Supplements, Volume 7. Elsevier (1996)
33. Bousmina, M., Ait-KadiJ, A., Faisant, B.J.: Determination of shear rate and viscosity from batch mixer data. J. Rheol. **43**(2), 415 (1999)
34. Cheng, J., Manas-Zloczower, I.: Hydrodynamic analysis of a banbury mixer-2-D flow simulations for the entire mixing chamber. Polym. Eng. Sci. **29**(75), 1059 (1989)
35. Gogos, C.C., Tadmor, Z., Kim, M.H.: Melting phenomena and mechanisms in polymer processing equipment. Adv. Polym. Technol. **17**(4), 285 (1998)

# Chapter 13
# Measurement of the Optical Properties of Paints and Plastics

## Contents





M. Diebold et al., *Pigments, Extenders, and Particles in Surface Coatings and Plastics*, https://doi.org/10.1007/978-3-030-99083-1_13

## Introduction

There are three fates for light entering a paint film. It can pass through the film, allowing it to interact with the substrate, it can be redirected out of the film through light scattering, and it can be absorbed within the film. The balance between these fates in any given film is a function of three distances—the distance from the film surface to the substrate, the distance the light travels before being redirected out of the film, and the distance the light travels before being absorbed by a film component. The first of these distances is determined by the amount of the film applied to the substrate and the second two are determined mainly by the type and amounts of pigments, both white and color, within the film.

For complete opacity, the film thickness should be much greater than the distance the light travels before being redirected out of the film or being absorbed within it. Measuring these various distances provides us with a means by which we can quantify opacity. While we can measure film thickness directly, it is more difficult to measure the other two distances. However, we do not need to directly measure the exact values of these distances and can instead use indirect means to characterize them in a meaningful way.

There are several ways of measuring the average distance that light travels in a paint film, and therefore the number of light interactions within it. Four of these will be described in this chapter:

- contrast ratio, where distance is characterized based on the amount of light that travels through the film and strikes the substrate,
- spread rate, where reflectance values from films too thin to be opaque are used to calculate the area that a given volume of paint can cover at the full hide,
- tinting strength, where the distance light travels through an opaque film is determined by measuring the amount of light absorbed within the film (this can be used to characterize both the strength of $TiO_2$ light scattering and the strength of color pigment light absorption), and
- undertone, where differences in scattering distances (and so, in scattering strength) as a function of wavelength are determined by partially absorbing light as it travels through a mix of black pigment and titanium dioxide.

We will review each of these in turn, beginning with the contrast ratio.

## Contrast Ratio

Contrast ratio is perhaps the most common way of measuring opacity in the paint industry. This is for a number of reasons, including the ease of performing the test (one drawdown, two reflectance measurements), the intuitive nature of the measurement, and the ability to see differences among paints by eye. However, the test has several limitations, some that are commonly known and others that often go unrecognized.

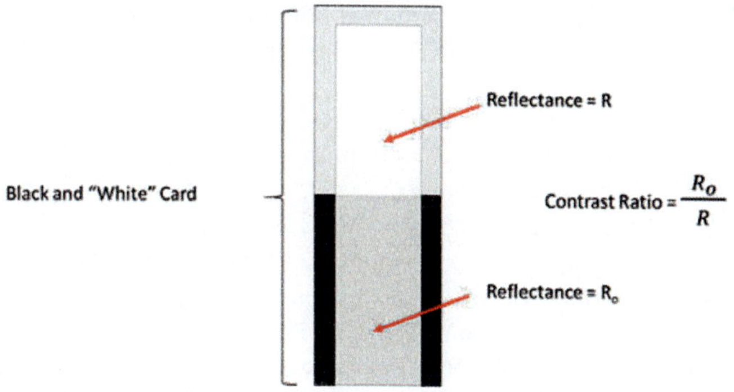


**Fig. 13.1** Definition of contrast ratio

## *Concept*

The strategy behind the contrast ratio test is straightforward. The paint of interest is drawn down on a black and white substrate (typically a paper card specially coated for this purpose).[1] After drying, the amounts of light reflected from the two areas of the card are measured (typically as tristimulus Y) and compared by taking the ratio of light reflected from the black section of the substrate ($R_o$ or "R naught") to that reflected from the white section of the substrate (R), as shown in Fig. 13.1. The contrast ratio can be reported either as a fraction (0 to 1) or a percentage (0% to 100%).

The difference in light reflectance between these areas is due to light that penetrates the entire film depth and then is either absorbed by the substrate (black area) or reflected by it (white area). The contrast ratio test measures the amount of light that penetrates at least a certain distance through the film (that distance being the thickness of the film).

For this test to be accurate, the film thickness should not be so great as to give complete opacity (defined as a contrast ratio of 0.98). In such a case, so little light travels the entire distance of the film depth that it is difficult to accurately measure that amount. Nor should the film be so thin that very little light is scattered by it. The ideal film thickness is one that gives a contrast ratio between 0.92 and 0.95.[2] Often an unknown paint is applied at a series of different thicknesses to determine the most appropriate thickness for that paint to be tested.

Contrast ratio can also be measured on plastic films by placing them over a black and white chart and measuring reflectance over each area. In this case, a thin layer of oil must be applied to the chart, the film placed onto the oil, and all bubbles removed. This is necessary to prevent light from being scattered by the air gap at the plastic / chart interface. Since oil has roughly the same refractive index as the plastic film,

[1] The reflectance of the white portion is 0.80, making it light gray in color. However, we will follow the convention of the industry in referring to this as white.

[2] Note that ASTM D2805 specifies a somewhat higher contrast ratio target of 0.97.

light passing between the two is unperturbed. This process is known as bringing the plastic film into optical contact with the opacity chart.

## *Limitations*

While this test is simple, both in concept and practice, it does make several crucial assumptions that are often violated without recognition by the tester, and it can be used in a misleading way to minimize opacity differences that may be important in actual end-use applications. The most important of these assumptions is that the film thickness is the same for all paints being tested. This is because contrast ratio is not solely a property of the paint itself but is also a property of the drawdown thickness. Obviously, a thicker film will have a higher contrast ratio than a thinner film of the same paint.

The requirement for a constant film thickness is often addressed by using the same drawdown blade for all paints. In cases where accuracy is critical, two paints will often be drawn down side-by-side on the same drawdown card. In this way, any inaccuracies due to variability in the speed at which the paint is drawn down or due to variability in drying conditions of the paint film are thought to be avoided. In addition, this facilitates direct comparison by eye between the paints.

Unfortunately, the situation is more complex than this. It is commonly believed in the coatings industry that the wet film thickness of a drawdown is determined only by the gap in the drawdown blade and that this thickness is equal to the drawdown clearance. Neither of these beliefs is correct. The actual wet film thickness of a drawdown is, in fact, roughly half the gap clearance. This would be of little consequence if that thickness was identical for all paints, but this is not the case. The rheology of the paint has a strong effect on the wet film thickness, with more viscous paints giving a thicker wet film than less viscous paints.[3]

As an indicator of how widely the wet film thickness can vary from one paint to another, we show in Fig. 13.2 a histogram of the wet film thicknesses of 246 similar paints (differing only in the relative amounts of extender and $TiO_2$ pigment) drawn down by an automated draw down apparatus using the same blade for all drawdowns. For these paints, the wet film thicknesses were between 67.0 and 86.6 microns, a range of 22%. Since total scattering (SX) is equally sensitive to scattering strength (S) and film thickness (X), a 22% variation in film thickness would be interpreted

[3] This can explain why adding nanoparticle extenders to some paints can seemingly improve opacity. The mechanism claimed for these opacity improvements is better spacing of the $TiO_2$ particles, which would increase their scattering efficiencies. However, small particles are known to increase the viscosity of paints (and are often used for this purpose), and if the paint viscosity after nanoparticle addition is not adjusted to the original viscosity, then a thicker drawdown can result. Of course, such an increase in thickness will increase the opacity of the film, but not because of better $TiO_2$ particle spacing. On a spread rate or coverage basis (square meters per liter of paint at complete opacity), there is no improvement in opacity.

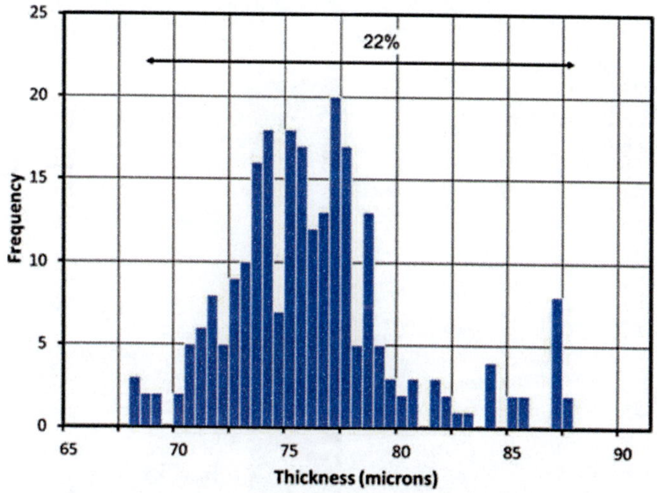


**Fig. 13.2** Histogram of the wet film thicknesses of 246 similar paints

as a 22% variation in light scattering strength if we were to assume that all film thicknesses were the same.

A second issue with contrast ratio is that while it is determined by paint opacity, it cannot distinguish between opacity due to light scattering and that due to light absorption. Two paints may have the same contrast ratio, but different brightnesses. This is illustrated in Fig. 13.3, which shows a comparison of two films with identical contrast ratios but different brightnesses. We may conclude that the performances of the two paints are identical, since the contrast ratios are the same, but such a comparison is invalid since the paints look different. Consumers don't simply desire high hiding, they desire high hiding with the specific color of their choice. As such, comparing the performance of paints with different colors (or brightnesses) is not relevant. If hiding efficiency were the only criteria for choosing a paint, then all paints would be black. This is, of course, not acceptable for most applications.

A third shortcoming of the contrast ratio test is that while one can quantify differences between the hiding abilities of different paints, these results do not give guidance as to how to change the paint film to achieve a certain level of improvement. That is, one paint film may be shown to be less opaque than another under certain application conditions, but the contrast ratio provides no quantitative information as to how much the opacity properties of the lower opacity film must be changed to equal the other film. For example, we do not learn from contrast ratio how much thicker a film must be made to achieve complete opacity, or how much stronger either the light absorption or the light scattering must be made to do so (or what combination of the two will give complete opacity).

A final problem with the contrast ratio test is that it is prone to either intentional or unintentional misuse. There is a truism that any paint or plastic film can be opaque as long as the film is thick enough. If we are comparing the opacity of a low-quality

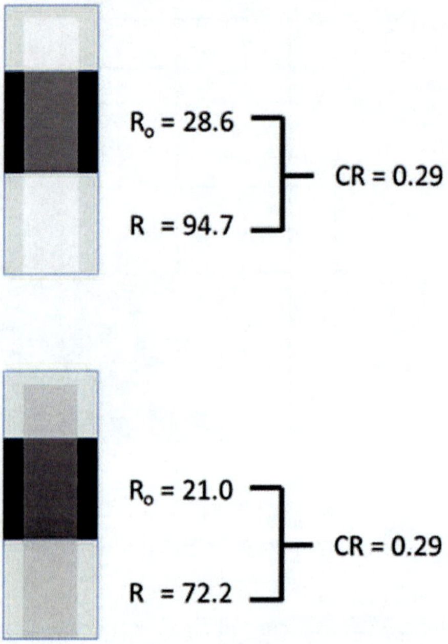


**Fig. 13.3** Two paint films with the same contrast ratio. The brightness ($R_\infty$, discussed below) of the top paint is 0.96 while that of the bottom paint is 0.67

paint to that of a high-quality paint, we might find that the two are equally opaque after six coats are applied. However, this is of no relevance to the paint consumer, who is expecting complete opacity with a single or, at most, double coating of the paint. At these lower film thicknesses, there is likely to be a significant difference in opacity between the high- and low-quality paints.

This is relevant to the contrast ratio test because the contrast ratio of two films can be close to one another even if the true opacities of the two paints, measured, for example, as square meters covered per liter of paint, are quite different. This occurs when the paints are applied too thickly. As an example, Table 13.1 shows the contrast ratios reported in an extender advertisement for a series of paints for which some of the $TiO_2$ is replaced with the advertised extender. We see that the contrast ratio of the reference paint, at 0.997, is only slightly higher than the contrast ratio of the paint made with a 10% replacement of $TiO_2$ (0.993). Such a difference in contrast ratio would be impossible to see by eye, suggesting that the opacities of the two paints are functionally equivalent.

However, it is clear from the contrast ratio values that these paints were applied at a much greater thickness than is needed for complete opacity. While the difference

**Table 13.1** Contrast ratio measurements on two paints

| % $TiO_2$ replacement | Contrast ratio |
|---|---|
| Reference | 0.997 |
| 10% | 0.993 |

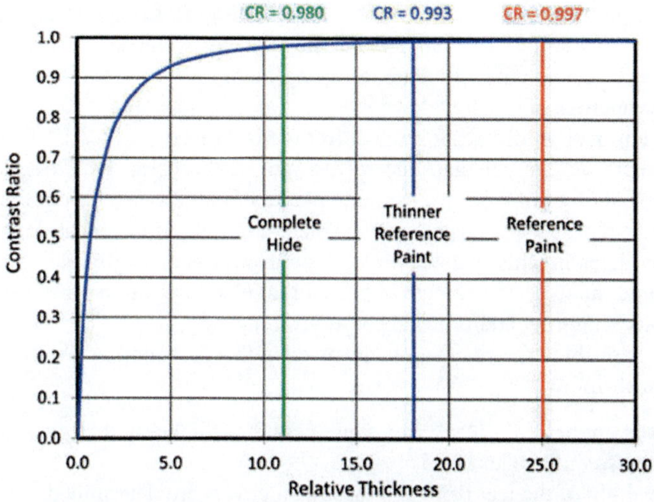


**Fig. 13.4** Contrast ratio as a function of relative film thickness for a white paint

in contrast ratios between complete hide (0.980) and measured hide for the reference paint as applied (0.997) is less than 2.0%, the applied film was more than double the thickness needed for complete hide (see below).

We can see this in Fig. 13.4. Here, we plot the calculated contrast ratio of a white paint as a function of film thickness (blue curve).[4] On this chart, we locate the original drawdown contrast ratio with a red line and see that the relative thickness is 25 units. The location of the modified paint, with a contrast ratio of 0.993, is shown with a blue line. We would achieve the same contrast ratio—0.993—by applying the reference paint at a relative thickness of 18 units. That is, we could either reduce cost by replacing 10% of the $TiO_2$ with the special extender, or by applying a film that is 28% thinner, reducing the entire cost of coverage by 28%. These two options give identical contrast ratios. To complete our analysis, we note that a relative film thickness of only 11 units (44% of the actual thickness used) is needed to provide a complete hide for the reference paint. Even this is too thick for a valid contrast ratio test—as mentioned above, the ideal thickness for this test is one that gives a contrast ratio between 0.92 and 0.95.

## Spread Rate

The various limitations of the contrast ratio test, as outlined above, can be remedied in a straightforward manner by adding two steps to the analysis of the paint drawdown. The first is to measure the thickness of the wet film, and the second is to make

[4] An $R_\infty$ value of 0.95 was used to generate this graph.

additional computations beyond simply taking the ratio of two reflectances. These additional computations are complex, but they are easily amenable to incorporation into a spreadsheet or other program, allowing them to be automatically calculated from the properties of the drawdown.

The additional information gained from this procedure, described below, is twofold: first, we can calculate the spread rate of the paint, which is defined as the area covered at full hide by a certain volume of paint, typically reported in $m^2/l$ or $ft^2/gallon$. These values were used in Chap. 4 to quantify the scattering strength of $TiO_2$ particles in white paints (where light absorption is negligible). Second, we can calculate the expected change in spread rate if we were to modify the paint by changing its scattering and/or the absorption strengths.

Before describing the calculations involved in this procedure, we will outline the steps we will follow:

1. Paint is applied to a black and white card at a thickness that gives a contrast ratio between 0.90 and 0.95.
2. The weight of the wet paint and the area it covers are determined.
3. Based on the weight of wet paint, the area it covers, and the density of the wet paint, the volume of paint applied and the wet thickness of the film are calculated.
4. Using the volume of paint applied and the area covered, the application rate of the drawdown is calculated (this is the area that a unit volume of paint, such as one liter, will cover at the same thickness as the drawdown).
5. Reflectance over the black and white portions of the card is measured and the contrast ratio is calculated.
6. These parameters are then entered into equations that are then solved to give the spread rate of the paint (the area that a unit volume of paint will cover at complete hide, defined as a contrast ratio of 0.98), as well as the absorption and scattering strengths of the paint (abbreviated as K and S).

## *Kubelka–Munk Framework*

In Chap. 3, we described the theoretical scattering power of a single, isolated particle surrounded by either air or resin (this was termed "Mie scattering"). However, in Chap. 4, we saw that scattering efficiency is significantly decreased when other particles are nearby (this was termed "dependent scattering"). To demonstrate this effect, in that chapter we showed the measured light scattering strengths of white paints (i.e., paints for which light scattering was much stronger than light absorption) as a function of a number of paint parameters. We did not, however, describe how these scattering strengths were measured. Separately, in Chap. 8, we discussed the ability of color pigment particles to absorb visible light. In that discussion, we considered the effects of the intrinsic absorption strength of the material, its concentration, and its thickness on the total amount of light absorbed.

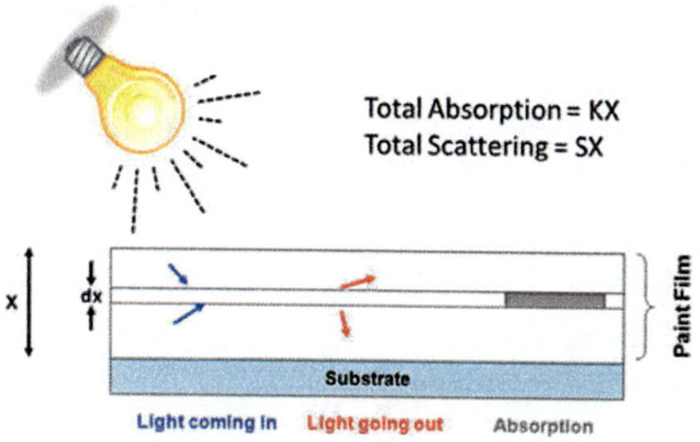


**Fig. 13.5** The Kubelka–Munk model. The total film thickness is X and the thickness of a thin layer is dX

Although we have so far considered light scattering and light absorption separately, in nearly all paints, (and plastics) both occur together, and it is their combination that determines opacity. In the 1930s, German physicists Paul Kubelka and Franz Munk developed a system for measuring light absorption and light scattering simultaneously in a paint film [1]. Their approach was to ignore the fact that both of these processes involve interactions with individual particles and instead treat them as intrinsic properties of a volume of paint film.

By making this simplifying assumption, Kubelka and Munk derived a series of powerful equations that describe the optical properties of a paint. These equations can be used to calculate the light scattering and light absorption strengths of a paint, the opacity of any thickness of that paint (or, conversely, the film thickness required for any desired opacity), and the effect of changing the light scattering or light absorption strengths on paint opacity. In addition, these equations can be used to calculate the color resulting from mixtures of individual colorants or paints (this will be discussed in detail in Chap. 15).

The Kubelka–Munk approach is based on an accounting of the amount of light entering and exiting a thin layer within a paint film over a short period of time (Fig. 13.5) [2]. During this time, light will enter this thin layer both through the lower boundary of the layer (light moving toward the substrate) and the upper boundary of the layer (light moving toward the film surface). This is shown by the blue arrows in Fig. 13.5. Similarly, some light will exit this layer through both the upper and lower boundaries (the red arrows in Fig. 13.5), and, finally, some light within the layer will be absorbed (the gray region in Fig. 13.5).

Kubelka and Munk were able to model the interactions between the paint film and light by making some simplifying assumptions. These are that the light striking and within the film is diffuse (that is, the intensity of light traveling in every direction is the same), that there are no special interactions (such as reflection) at the film/air and film/substrate interfaces, that the film is homogenous (which ignores that it is discrete particles that interact with light), that these particles are evenly distributed throughout the film, and that the film thickness is much greater than the thin layer thickness.

These assumptions are not true for every paint film, and it must be emphasized that the results of the Kubelka–Munk analysis are only as sound as the assumptions going into them. For example, the light striking a paint is rarely diffuse. In addition, very thin paint films might not be thick enough for the light to randomize. Similarly, in very dark films, all of the light could be absorbed before it is randomized. Moreover, a significant amount of light can reflect from the film surface, both as the light enters the film and as it exits the film. Despite the limitations of these assumptions, the Kubelka–Munk analysis of most paint films is valid, at least for comparison purposes.

### *Application of the Kubelka–Munk Equations to Spread Rate*

The general procedure for using the Kubelka–Munk equations to determine opacity and spread rate is given in ASTM and DIN specifications [3, 4]. Here, we will describe the underlying basis of these procedures.

The Kubelka–Munk assumptions described above lead to a set of differential equations. There are two ways of approaches for solving these equations, an exponential approach [1] and a hyperbolic approach [5]. The two approaches each result in several mathematical equations relating reflectance, film thickness, light absorption, and light scattering. For the practical application of this model to paint opacity and hiding power, we use the hyperbolic approach and many of the equations derived from it.

## Application Rate as Drawndown

The starting point for using the Kubelka–Munk equations to calculate the spread rate of a paint at complete hide is to draw a paint down at a thickness with incomplete hide on a black and white chart. This thickness is chosen to give a contrast ratio in the range of 0.90 to 0.95. In addition to measuring the reflectance values over the black and white portions of the card after the paint dries, we also must determine the coverage, or application rate, of the paint at the thickness it is drawn down. Application rate is defined as the area covered by a unit volume of paint. This is typically given in units of $m^2/l$ or $ft^2/gallon$. Note that the application rate is a function entirely of the drawdown and is not a property of the paint itself.

To calculate the application rate, we must first determine the volume of the wet paint and the area that this volume covers. We determine the volume of the film using the weight of the film and the density of the paint (which must be either measured separately or determined based on the paint composition). We measure the area covered by this volume of paint with a ruler, after the paint has dried.

From the density of the liquid paint and the weight of the wet paint applied to the panel, we can easily calculate its volume in ml:

$$\text{ml paint} = \left(\frac{\text{paint weight}(g)}{\text{paint density}(g/ml)}\right) \tag{13.1}$$

$$\text{ml paint} = 8.3454x\left(\frac{\text{paint weight}(g)}{\text{paint density}(lb/gal)}\right) \tag{13.2}$$

The thickness of the wet paint, in microns or in mils (thousandths of an inch), is

$$X(\text{micron}) = 10,000x\left(\frac{\text{volume paint }(ml)}{\text{area(square cm)}}\right) \tag{13.3}$$

$$X(mil) = 61.0273x\left(\frac{\text{volume paint}(ml)}{\text{area(square inches)}}\right)$$

$$X(mil) = 61.0273x\left(\frac{\text{volume paint}(ml)}{\text{area (square inches)}}\right) \tag{13.4}$$

Finally, we can calculate the application rate of the drawdown as follows:

$$\text{Application Rate}\left(m^2/l\right) = 0.1x\left(\frac{\text{area (square centimeters)}}{\text{paint weight}(g)/\text{paint density}(g/ml)}\right) \tag{13.5}$$

$$\text{Application Rate}\left(ft^2/gal\right) = 3.15x \left(\frac{\text{area(square inches)}}{\text{paint weight}(g)/\text{paint density}(lb/gal)}\right) \tag{13.6}$$

In these equations, a lower case "x" indicates multiplication while an upper case "X" is film thickness.

## Calculation of SX and KX

The next step in determining the spread rate of a paint is to measure the reflectance values of the dry paint over the two portions of the drawdown card. From these reflectances, we can calculate the total absorption and scattering strengths, KX and SX, by using two secondary, or "helper", parameters, *a* and *b*:

$$a = \frac{1}{2}\left[R + \frac{R_o - R + R_g}{R_o R_g}\right] \tag{13.7}$$

where

R = reflectance over the white part of the drawdown card,

$R_o$ = reflectance over the black part of the drawdown card, and

$R_g$ = reflectance of the white part of the card before application of the paint

$$b = \sqrt{a^2 - 1} \tag{13.8}$$

In Chap. 6, we measured R at each wavelength interval to characterize the absorption spectrum (and color) of a paint (this was done at complete hide; that is, we measured $R_\infty$, as described below, at each wavelength interval). In Eq. 13.7 and those that follow, the R value represents reflectance over the entire visible spectrum and is typically measured as the tristimulus Y value of the paint at full hide.

In Eq. 13.7, $R_g$ refers to the reflectance of the white portion of the drawdown card prior to paint application. It is useful that some light be absorbed by the "white" portion of this card as this will allow the amount of light that passes through the paint film to the substrate to be calculated. The value of $R_g$ for standard drawdown cards is 0.80. However, there can be variability from this value, and it is recommended that the $R_g$ for each card be measured at six locations prior to draw down and that the average of these reflectances be used in these equations.

We can calculate the value of SX as a function of *a*, *b* and the reflectance over the black portion of the card ($R_o$) as follows:

$$\text{SX} = \frac{1}{b}\text{Arctgh}\left[\frac{1 - aR_o}{bR_o}\right] \tag{13.9}$$

where "Arctgh" is the hyperbolic arc-cotangent function ($\coth^{-1}$).

We can calculate KX as

$$\text{KX} = \text{SX}(a - 1) \tag{13.10}$$

## Reflectance of a Film at "Infinite" Thickness—$R_\infty$

Before moving to the calculation of spread rate, it is helpful to introduce the concept of $R_\infty$ and the very useful diagram developed by Deane Judd, a pioneer in paint optics, in 1937 [6]. This diagram was developed prior to the ready availability of calculators and computers to graphically estimate the solutions of the Kubelka–Munk equations, greatly reducing the amount of work necessary to calculate the K, S, and spread rate properties of a paint.

To define $R_\infty$, we begin by imagining a series of drawdowns of the same paint at different film thicknesses onto black and white cards. For the thinner films, we are able to see a difference in reflectance over the two parts of the card. As the film becomes thicker, the magnitude of this difference decreases until, at some point, the two areas of the cards appear the same ($R_o$ = R). We refer to this reflectance as $R_\infty$

("R infinity") and define the thickness for this drawdown as $X_\infty$ ("X infinity"). Note that not only do the two areas of the film appear identical, but any film drawn down at a greater thickness will appear the same as the film drawn down at $X_\infty$.

The physical implication of $X_\infty$ is that all of the light entering a film of this thickness or greater is either reflected back out of the film or is absorbed by components of it before it reaches the substrate. Since no light reaches the substrate, the appearance of the film over any the substrate will look the same. The reflectance at a thickness of $X_\infty$ or above ($R_\infty$) is determined by the balance between the light absorption (K) and light scattering (S) within the film. Obviously, when the ratio K/S is high, the film is dark, while when it is low, the film is light. For this reason, the $R_\infty$ value is often referred to as the brightness or reflectivity of the film.

The relationship between K, S, and $R_\infty$ is given by Eq. 13.11:

$$R_\infty = 1 + \frac{K}{S} - \sqrt{\left(\frac{k^2}{S^2}\right) + 2\frac{K}{S}} \tag{13.11}$$

This equation can be rearranged to give perhaps the most widely known of the Kubelka–Munk equations:

$$\frac{K}{S} = \frac{(R_\infty - 1)^2}{2R_\infty} \tag{13.12}$$

It is impossible to ensure that every photon that enters a film does not pass through it and strike the substrate, regardless of how thick that film is. $R_\infty$ is therefore a value that can be approached asymptotically but never quite achieved. Because of this, no paint can have a truly complete hide. However, by long-established convention, a paint is considered to have a complete hide when its contrast ratio is 0.98 (or above) [7]. This value was chosen based on the ability of the average human eye to discern a difference in reflectance between the paint over the white section of the card and the paint over the black section. By definition, a film that has a contrast ratio of 0.98 or above is infinitely thick.

Like S and K, $R_\infty$ is a fundamental property of the paint itself and not of any particular drawdown. That is, while the R and $R_o$ reflectances of paint drawndowns will vary with film thickness, the $R_\infty$ value is the same for all drawdowns, regardless of thickness, of that paint.

## The Judd Graph

While the intention of this chapter is to show how the opacity properties of a paint film can be calculated based on that film's optical properties, it is worthwhile at this juncture to consider a very useful graphical approach to this problem, the so-called Judd Graph. This graph is shown in Fig. 13.6. Here, we plot two easily measured

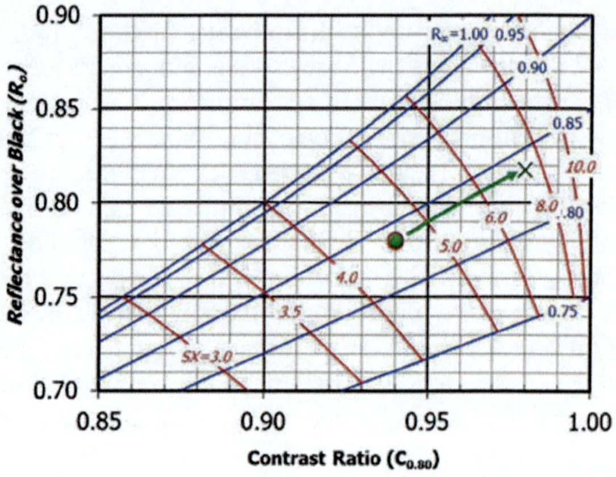


**Fig. 13.6** Judd Graph relating $C_r$ and $R_o$ to SX and $R_\infty$. The $R_g$ value is 0.80 (as indicated by the designation "$C_{0.80}$" on the X-axis)

film properties—$R_o$ on the Y-axis and contrast ratio on the X-axis—and read from the graph two properties that are difficult to calculate, at least by hand—SX and $R_\infty$.

As an example, consider a paint drawdown that has a $R_o$ value of 0.78 and an R value of 0.83 on a card with a white background reflectance value ($R_g$) of 0.80. From these reflectance values, we calculate the contrast ratio to be 0.94 (0.78/0.83). Using the black rectilinear lines in Fig. 13.6, we place this drawdown information on the graph as a green circle. We can estimate the SX value of this film using the family of red curves as a guide. This value is between 4.0 and 5.0, and closer to 5.0, and we might estimate it to be 4.6. Likewise, the $R_\infty$ value can be estimated using the family of blue lines, and here we see the data point lies between $R_\infty$ values of 0.80 and 0.85, and we might estimate it to be 0.84.

Drawdowns of the same paint at other thicknesses will all have the same $R_\infty$ value as this drawdown (recall $R_\infty$ is a property of a paint, not of a drawdown), but different SX values (because the X values are different, SX is a property of the drawdown rather than the paint). Because the SX values are different, these drawdowns will appear at different locations along the same $R_\infty$ line.

We can use this graph to estimate the spread rate of this paint at the complete hide. We do this by following the $R_\infty$ line up and to the right, until it intersects the vertical line at a contrast ratio of 0.98 (this is shown by a green arrow in Fig. 13.6). At the intersection point (green X in this figure), we use the red SX curves to estimate an SX value at the complete hide. Here, we might estimate it to be 7.5.

Before proceeding, note that the Judd Graph shown in Fig. 13.6 is calculated based on an $R_g$ value of 0.80. $R_g$ values of actual drawdown cards may vary somewhat from this value, but if the $R_g$ value is close to 0.80, this version of the Judd Graph can be used.

Since the original drawdown and the hypothetical drawdown at a contrast ratio of 0.98 (full hide) are made with the same paint, they must have the same S value (recall S is a property of a paint rather than of a specific drawdown). Thus, the difference in SX values between the drawdown (4.6) and the estimate of the paint at complete hide (7.5) is entirely due to the difference in film thickness (X). If we determined, using Eq. 13.6, that the application rate of the film as drawn down (green circle in Fig. 13.6) is 12.80 $m^2/l$, then the spread rate at complete hide is this value times the ratio of X values (4.6/7.5), or 7.85 $m^2/l$.

The historical importance of the Judd Graph is difficult to overstate. When Kubelka and Munk first published their work, it was not immediately embraced by the coatings industry [6]. This was not because the coatings industry did not think the analysis was useful, but rather because the equations involved are so complex and their solutions are far from simple to calculate, particularly with the computational tools available at the time. The effort needed to solve Eqs. 13.9–13.12 was far too great to allow them to be useful in, for example, quality control. Instead, their use was mainly limited to fundamental and applied research labs. By deriving a graphical solution to these complex equations, SX and $R_\infty$ could be determined in seconds.

Today we can quickly and accurately solve complex mathematical equations using calculators and computers and so no longer rely on the Judd Graph. That said, this graph continues to be very useful in mapping out the performance attributes of paints and in visualizing complex relationships between paints, and this diagram still has an important place in coatings science.

## Calculation of Spread Rate

In the previous section, we graphically estimated the spread rate of a paint at complete hide by using the Judd Graph to estimate the SX and $R_\infty$ value of a film drawn down at incomplete hide and then extrapolating these values to other drawdown thicknesses. In this section, we derive the equations that perform this task exactly, rather than estimate it graphically.

Our solution uses the application rate as drawndown and the R and $R_o$ values of the drawdown. We begin by solving for SX using Eq. 13.9. While it is not apparent from Eq. 13.7, *a* is a property of the paint, rather than of a drawdown. The same is true for *b*. This can be seen in these alternative equations for *a* and *b* that are based on other parameters that are properties of the paint:

$$a = \frac{1}{2}\left(\frac{1}{R_\infty} + R_\infty\right) \tag{13.13}$$

$$b = \frac{1}{2}\left(\frac{1}{R_\infty} - R_\infty\right) \tag{13.14}$$

$$a - b = R_\infty \tag{13.15}$$

$$a + b = \frac{1}{R_\infty} \tag{13.16}$$

$$a = \left(\frac{S + K}{S}\right) \tag{13.17}$$

Using the definition of contrast ratio, $C_r$ ($C_r = R_o/R$), we can substitute $R_o$ with $C_r$ times R in Eq. 13.7 and rearrange to get:

$$C_rR_gR^2 + (C_r - 2aC_rR_g - 1)R + R_g = 0 \tag{13.18}$$

Before proceeding, we must address the issue of which value of $R_g$ we should use in this (and subsequent) equations. It is important to use the measured $R_g$ value when calculating $a$ in Eq. 13.7 so as to obtain the most accurate value of SX of the film (Eq. 13.9). However, moving forward we will be *calculating* the appearances of drawdowns, rather than measuring them. When doing so, it is critical that a common $R_g$ value be used in all calculations, since the contrast ratio of a drawdown is affected by the $R_g$ value. In theory, we could choose any value for $R_g$ between 0 and 1, but for our purposes, we will use a value of 0.80, since this is the value Judd used in his graph and is typical of drawdown cards.

Returning to Eq. 13.18: The $a$ parameter is constant for a given paint, and $R_g$ is assigned a value of 0.80, for the reasons discussed above. If we specify a value for $C_r$ (typically 0.98, the condition of complete hide), then R is the only unknown. As written, Eq. 13.18 is a simple quadradic for R. We solve this using the standard procedure to give

$$R = \frac{-(C_r - 2aC_rR_g - 1) \pm \sqrt{(C_r - 2aC_rR_g - 1)^2 - 4C_rR_g{}^2}}{2C_rR_g} \tag{13.19}$$

There are two solutions for R. The equation using the positive root gives an R value greater than 1, which is physically impossible (such a condition requires more light to be reflected than actually strikes the film). We therefore use the negative root of this equation to find R at complete hide:

$$R = \frac{-(C_r - 2aC_rR_g - 1) - \sqrt{(C_r - 2aC_rR_g - 1)^2 - 4C_rR_g{}^2}}{2C_rR_g} \tag{13.20}$$

With R in hand, it is trivial to calculate $R_o$ at complete hide using the definition of contrast ratio ($R_o = RC_r$). This allows us to calculate the SX value at complete hide using Eq. 13.9, since we now know $a$, $b$, and $R_o$.

Assuming the contrast ratio of the drawdown is between 0.90 and 0.95, which is the target contrast ratio for this analysis, the SX values as drawndown and at complete hide are different from one another. Since S is constant, the X values must be different. We calculate the spread rate at complete hide by multiplying the application rate as drawndown by the ratio of SX values:

**Table 13.2** Comparison of estimated and calculated spread rates

| Quantity | Estimated from the Judd Graph | Calculated | Equation used |
|---|---|---|---|
| SX as drawn down | 4.6 | 4.562 | Equations 13.7, 13.8, 13.9 |
| $R_\infty$ | 0.84 | 0.837 | Equation 13.11 |
| Application rate as drawn down | – given as 12.80 $m^2/l$ – | | |
| SX at complete hide | 7.5 | 7.326 | Equations 13.7, 13.8, 13.9 |
| Spread rate at complete hide | 7.85 $m^2/l$ | 7.97 $m^2/l$ | Equation 13.21 |

$$\text{Spread Rate (at complete hide)} = \text{Application Rate (as drawndown)} \times \left( \frac{\text{SX (as drawndown)}}{\text{SX (at complete hide)}} \right) \quad (13.21)$$

Similarly, we can calculate the film thickness at complete hiding as

$$\text{X (at complete hide)} = \text{X (as drawndown)} \times \left( \frac{\text{SX (at complete hide)}}{\text{SX (as drawndown)}} \right) \quad (13.22)$$

Note that there is an inverse relationship between film thickness and spread rate—as film thickness increases, spread rate decreases by the same factor.

We can solve the equations in this section for the paint described in the previous section and compare the calculated values of SX, $R_\infty$, and spread rate at complete hide to those read from the Judd Graph. This comparison is made in Table 13.2.

The spread rate at complete hide estimated from the Judd Graph is within 1.5% of that calculated using the Kubelka–Munk equations. This level of agreement is similar to the experimental error expected for the spread rate procedure [3, 4], confirming the validity of the Judd Graph.

## Spread Rate at Another Value of $R_\infty$

It is often the case that we wish to evaluate the scattering abilities of a group of paints that have different $R_\infty$ values. Directly comparing these paints is problematic because their appearances will be different at complete hide (some paints will be darker than others). We could correct for differences in paint brightness by toning all paints to a common $R_\infty$, but this would be painstaking to do in practice. However, we can do this mathematically by using the Kubelka–Munk equations to determine the spread rate of the paints when toned to a common $R_\infty$ value.

This process is computationally straightforward and is based on the strategy used to determine the spread rate at complete hide (described above). We will begin by using Eq. 13.9. In the previous section, we solved this equation using *a* and *b* values from the original drawdown. We cannot do that here because our hypothetical toned paints have different $R_\infty$ values than their untoned counterparts, so the values of *a* and *b* will be different for the toned and untoned paints (Eqs. 13.3–13.16). Instead, we will calculate the *a* value of the toned paint using Eq. 13.13, and then solve Eq. 13.20 (again with $R_g = 0.80$) for the value of R of the toned paint at complete hide. We can then solve for spread rate at complete hide using exactly the same procedure as described above.

## $R_\infty$ Values Greater Than 1.0

Consider a paint drawdown with a measured R value of 0.944 and a measured $R_o$ value of 0.85, giving a $C_r$ value of 0.90. The location of this paint on the Judd graph is shown as a green point in Fig. 13.7. This data point is noteworthy as it lies above the line for $R_\infty = 1.0$. Such a result is only possible if the light absorption coefficient, K, is less than zero. This, in turn, is only possible if the film created light—that is, rather than absorbing light, the film emits it. This can occur with the incorporation of a fluorescing agent, but such a situation is extremely rare in practice.

Although it is theoretically impossible for a drawdown to have an $R_\infty$ value greater than 1.0, such a result is occasionally seen in practice for very bright white paints. This situation can arise for a number of reasons that can be classified into two groups. The first is an experimental error in measuring the reflectance values. For example, the light intensity-measuring equipment could be poorly calibrated. Even a small departure from accurate calibration could lead to an $R_\infty$ value greater than 1.0 for paints that have very little light absorption (again, for very bright white paints). Another source of experimental error is not correcting for $R_g$ values that deviate from

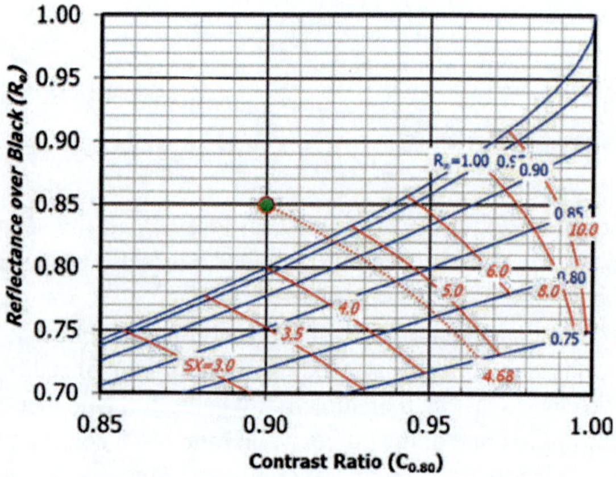


**Fig. 13.7** A data point with an $R_\infty$ value greater than 1

0.80. This source of error can be eliminated by measuring this value prior to use, as described above.

The second potential group of causes for an $R_\infty$ value being greater than 1.0 is a violation of one or more of the assumptions made in the Kubelka–Munk model. As noted above, the model does not take into account the reflection of light at the air/film and film/substrate interfaces. This can be problematic when comparing films with widely different gloss values or indices of refraction. In addition, the model assumes that the distributions of light scattering and light-absorbing particles are even throughout the film, which may not be true in flocculated systems. Finally, the model is based on the assumption that the light striking the film and within it is diffuse—that is, moving in all directions equally. Even in a perfectly dispersed paint, this assumption may be incorrect for the topmost layers of the film, where there may not have been enough scattering events to randomize the direction of the light entering the film from above. This complication would be most problematic for thin films.

Because $R_\infty$ values greater than 1.0 are seen on occasion, it is worthwhile to develop a strategy for dealing with this situation.

Before doing so, it is useful to discuss the implications of an $R_\infty$ value greater than 1.0. As noted above, the K value for the film becomes negative. In addition, we cannot calculate $R_\infty$ using Eq. 13.11, as the term within the radical (the radicand) is negative. When solving for $a$ using Eq. 13.7, we arrive at a value less than 1.0. This prevents us from solving for $b$ using Eq. 13.8, since the radicand in this equation will also be negative.

For paints that have calculated $R_\infty$ values greater than one, we can estimate SX by extrapolating the lines of constant SX in the Judd Graph to the region above $R_\infty = 1.0$. This can easily be done visually, particularly if the departure from the $R_\infty = 1.0$ curve is slight. However, calculating SX values in the $R_\infty = 1.0$ curve is more problematic, as we cannot use Eq. 13.11 due to the fact that $b$ is undefined.

This computational issue is resolved by turning to the derivation of Eq. 13.9. While this derivation is beyond the scope of this book, we note that one aspect of it involves the solution of a quadratic equation. As was the case in the section on spread rate at complete hide, there are two roots to such an equation. One root gives an $a$ value greater than one, leading to Eq. 13.9. The other root leads to an $a$ value less than one. It is this equation that will allow us to extend the lines of constant SX to the region of the Judd graph above $R_\infty = 1.0$. Solving for SX in the case of $a$ being less than one gives

$$\mathrm{SX} = \frac{1}{b}\mathrm{Arctg}\left[\frac{1 - a\mathrm{R_o}}{b\mathrm{R_o}}\right] \tag{13.23}$$

where Arctg is the arc-cotangent function ($\cot^{-1}$).

In Eq. 13.23, $a$ is calculated by Eq. 13.7, as before, and $b$ is calculated as

$$b = \sqrt{1 - a^2} \tag{13.24}$$

Equations 13.23 and 13.24 may seem familiar—as they should. We can see that they are very close to Eqs. 13.9 and 13.8, respectively, except the normal arc-cotangent function is used in Eq. 13.23 (rather than the hyperbolic arc-cotangent in Eq. 13.9), and the radicand in Eq. 13.24 is the negative counterpart of that in Eq. 13.8.

Equation 13.23 can be solved for the data point in Fig. 13.7, and its solution is depicted as the red dashed SX curve in this figure. Note that two equations are used to generate this SX curve—the region below the $R_\infty = 1.0$ curve is calculated using Eq. 13.9 while the region above the $R_\infty = 1.0$ line is calculated using Eq. 13.23. Based on these equations, we calculate the SX value of this paint to be 4.68.

The SX value of any drawdown with $R_\infty > 1.0$ can be calculated using Eqs. 13.26 and 13.27. However, solving for spread rate at complete hide ($C_r = 0.98$) when $R_\infty > 1.0$ is not always possible. For points that are far above the $R_\infty = 1.0$ curve, the radicand in Eq. 13.20 becomes negative, and R becomes undefined. We can therefore solve Eq. 13.20 only when the following condition is satisfied:

$$(C_r - 2aC_rR_g - 1)^2 \geq 4C_rR_g^2 \tag{13.25}$$

As noted above, when $R_\infty > 1.0$, the $a$ value is less than one. As we move to greater distances from the $R_\infty = 1.0$ curve, the $a$ value becomes progressively lower and eventually becomes so low that the radicand in Eq. 13.20 becomes negative (i.e., the condition described by Eq. 13.25 is no longer satisfied). This transition occurs when

$$a = \frac{2\left(\sqrt{C_rR_g^2}\right) - 1 + C_r}{2C_rR_g} \tag{13.26}$$

For an $R_g$ value of 0.80 and a contrast ratio of 0.98 (i. e., at full hide on a standard black and white card), $a$ cannot be below 0.9973974 if we are to solve for spread rate. It is the authors' experience that this situation is rarely encountered in practice. Should $a$ drop below this value, it is probably because the film deviates greatly from the Kubelka–Munk assumptions, in which case the analysis is likely to be invalid anyway.

## Examples and Commentary

By way of example, we will use the equations defined above to compare the spread rates of four hypothetical paints. Reflectivity information for these paints, and the optical properties derived from them, are given in Table 13.3. The location of these paints on the Judd Graph is given in Fig. 13.8 (note that, strictly speaking, the drawdowns of Paints C and D cannot be placed on this graph since their $R_g$ values are not 0.80. That said, the deviation of $R_g$ from 0.80 is slight enough to allow us to place these paints in Fig. 13.8).

**Table 13.3** Optical properties for four paints

| | Row | Property | Units | Paint | | | | Data source |
|---|---|---|---|---|---|---|---|---|
| | | | | A | B | C | D | |
| Paint and drawdown parameters | 1 | Paint Density | g/l | 1378 | 1378 | 1161 | 1161 | Measured |
| | 3 | Weight | grams | 3.76 | 3.76 | 3.01 | 3.01 | Measured |
| | 4 | Area | $cm^2$ | 350.0 | 350.0 | 350.0 | 350.0 | Measured |
| | 5 | $R_g$ | Unitless | 0.80 | 0.80 | 0.82 | 0.82 | Measured |
| | 6 | $R_o$ | Unitless | 0.83 | 0.85 | 0.81 | 0.82 | Measured |
| | 7 | R | Unitless | 0.87 | 0.89 | 0.90 | 0.93 | Measured |
| As drawn down | 8 | Contrast Ratio | Unitless | 0.954 | 0.950 | 0.900 | 0.882 | $= R_o/R$ |
| | 9 | SX | Unitless | 5.892 | 6.154 | 4.231 | 4.067 | Equation 13.9 |
| | 10 | KX | Unitless | 0.043 | 0.017 | −0.002 | −0.029 | Equation 13.10 |
| | 11 | *A* | Unitless | 1.0073 | 1.0028 | 0.9995 | 0.9930 | Equation 13.7 |
| | 12 | *B* | Unitless | 0.1210 | 0.0744 | 0.0305 | 0.1185 | Equation 13.8 or 13.24 |
| | 13 | $R_\infty$ | Unitless | 0.886 | 0.928 | XXX[13.1] | XXX[13.1] | Equation 13.11 |
| | 14 | X | microns | 78.0 | 78.0 | 74.2 | 74.2 | Equation 13.4 |
| | 15 | S | $microns^{-1}$ | 0.0755 | 0.0789 | 0.0570 | 0.0548 | = SX/X |
| | 16 | Application Rate | $m^2/l$ | 12.84 | 12.84 | 13.50 | 13.50 | Equation 13.5 |
| At complete hide (CR = 0.98) | 17 | Contrast Ratio | Unitless | 0.980 | 0.980 | 0.980 | 0.980 | Assumed |
| | 18 | $R_g$ | Unitless | 0.800 | 0.800 | 0.800 | 0.800 | Assumed |
| | 19 | $R_o$ | Unitless | 0.861 | 0.893 | 0.928 | XXX[13.1] | = R·CR |
| | 20 | R | Unitless | 0.878 | 0.911 | 0.947 | XXX[13.1] | Equation 13.20 |
| | 21 | SX | Unitless | 8.708 | 10.095 | 12.152 | XXX[13.1] | Equation 13.9 |
| | 22 | KX | Unitless | 0.063 | 0.028 | −0.006 | XXX[13.1] | Equation 13.10 |
| | 23 | *A* | Unitless | 1.0073 | 1.0028 | 0.9995 | 0.9930 | Equation 13.7 |
| | 24 | *b* | Unitless | 0.1210 | 0.0744 | 0.0305 | 0.1185 | Equation 13.8 or 13.24 |
| | 25 | X (thickness) | microns | 115.3 | 128.0 | 212.9 | XXX[13.1] | Equation 13.22 |
| | 26 | Spread Rate | $m^2/l$ | 8.69 | 7.83 | 4.69 | XXX[13.1] | Equation 13.21 |
| Hiding at new $R_\infty$ | 27 | Target $R_\infty$ | Unitless | 0.899 | 0.899 | 0.899 | 0.899 | Assumed |
| | 28 | Contrast Ratio | Unitless | 0.980 | 0.980 | 0.980 | 0.980 | Assumed |
| | 29 | $R_g$ | Unitless | 0.800 | 0.800 | 0.800 | 0.800 | Assumed |

(continued)

**Table 13.3** (continued)

| Row | Property | Units | Paint A | Paint B | Paint C | Paint D | Data source |
|---|---|---|---|---|---|---|---|
| 30 | $R_O$ | Unitless | 0.871 | 0.871 | 0.871 | 0.871 | = R•CR |
| 31 | R | Unitless | 0.889 | 0.889 | 0.889 | 0.889 | Equation 13.20 |
| 32 | SX | Unitless | 9.110 | 9.110 | 9.110 | 9.110 | Equation 13.9 |
| 33 | KX | Unitless | 0.052 | 0.052 | 0.052 | 0.052 | Equation 13.10 |
| 34 | *a* | Unitless | 1.0057 | 1.0057 | 1.0057 | 1.0057 | Equation 13.7 |
| 35 | *b* | Unitless | 0.1067 | 0.1067 | 0.1067 | 0.1067 | Equation 13.8 or 13.24 |
| 36 | X (thickness) | microns | 120.6 | 115.3 | 159.5 | 165.9 | Equation 13.22 |
| 37 | Spread Rate | $m^2/l$ | 8.30 | 8.66 | 6.26 | 6.04 | = Row 26 x (Row 25/Row 36) |

13.1 Incalculable

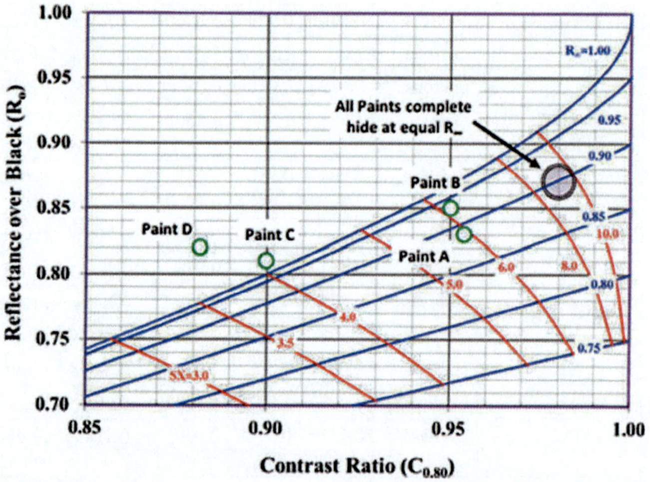


**Fig. 13.8** Judd Graph for the four paints described in the table

As can be seen in Fig. 13.8 and Row 13 in Table 13.3, and confirmed by Eq. 13.11, the $R_\infty$ values for Paints A and B are less than 1.0 while those for Paints C and D are greater than 1.0. Paint C falls close enough to the $R_\infty = 1.0$ curve that we can calculate its spread rate at complete hide, while the location of Paint D on the Judd Graph is far enough from the $R_\infty = 1.0$ curve that we cannot do this (Row 26).

The spread rates of Paints A and B at full hide (Row 26 in Table 13.3) show that, despite the fact that Paint B has greater scattering (Row 15), Paint A has a greater spread rate. This is because the $R_\infty$ values of the two paints are different (Row 13):

Paint A is darker (lower $R_\infty$) and its greater opacity is due to it having more light absorption than Paint B (Row 10).

Note that all paints, when adjusted to a common $R_\infty$ value of 0.90 and calculated at full hide (contrast ratio of 0.98), fall at the same location in the Judd Graph (large circle in Fig. 13.8). This means that the appearance (brightness) of these drawdowns would be identical to one another. As shown in Row 32, and by virtue of the fact that these drawdowns fall on the same point on the Judd Graph, under these conditions, the SX values of all the paints are also identical, as are their $R_\infty$ values (Row 27), *a* values (Row 34), and *b* values (Row 35) values. Although the SX values are identical, the S values are different. This results in the film thickness for complete hide being different (Row 36), leading to the different spread rates at equal brightness (Row 37).

## Applied Hide

In the previous section, we saw that the Kubelka–Munk equations link the reflectances of a paint film over both black and white backgrounds, the film thickness, and the absorption and scattering strengths of the paint (K and S). We used these equations to determine what we will call here the "intrinsic" hiding power of a paint, as expressed by the spread rate of the paint at the full hide. An important aspect of this analysis is that the measurements are on films that are uniformly applied by a blade or drawdown applicator. Film thickness uniformity is necessary if we are to have a single value for X with which to work.

In real-world applications, particularly those for which the paint is applied with a brush or roller, film thickness typically varies across the substrate in a series of peaks and valleys. This is seen as brush marks for paints applied by brush and stipple (also known as texture or roller marks) for paints applied by roller. Here, the meaning of film thickness and complete hide becomes uncertain—do we define spread rate based on the average film thickness needed for complete hide (i.e., measured as described in the previous section), or when the thinnest portions of the film are thick enough to obscure the substrate completely, or at some point in between? Do we report different spread rates for different types of applicators—brush versus roller, short versus long nap lengths on the roller, etc.? As we discuss this complication, we will refer to the perceived opacity of a paint film as applied by the intended applicator as being the "applied hide" of that film [8].

Intuitively we would expect that, at equal application rates, the uniformity of the film thickness would affect its perceived opacity, since a non-uniform application would result in some areas of the film being thicker than needed for complete hide and other areas being thinner than needed. The excess paint in the thick areas does not alter the appearance of those areas—as discussed when we defined $X_\infty$, all paint films above the thickness necessary for complete hide will appear the same. However, the deficit of paint in the thin areas leads to partial transparency of the film, with the substrate surface being detectible in these areas. Because of this, we expect the perceived opacity (and spread rate) from a given volume of paint to be greatest for an even film applied by a blade applicator and lower for an uneven film applied by a brush or roller.

## *Traditional Methods to Assess the Applied Hiding of an Architectural Coating*

The applied hiding power of an architectural coating can be assessed as described in ASTM D5150. Here, the paint is applied by a roller or a brush ontp a panel with six stripes going from light gray to black (see Fig. 13.9 for a white paint). The hiding power of the applied coatings is rated as the number of the darkest stripe that is completely (or almost completely) obscured, at a specified thickness or spread rate.

Although this rating appears straightforward, there are several limitations associated with the method:

(1) When one applies a paint with a roller or a brush, it is very difficult to accurately control the applied thickness, hence the hiding power might differ from one area to the other.
(2) The assessment of which strips are completely covered is determined by eye and so is very subjective.
(3) Comparing paints with different flow characteristics (different formation of peaks and valleys) is very difficult and also subjective.

In practice, paint producers have several variants of this method. One that is seen on a regular basis is that a trained applicator applies the paint on a (partially) black substrate. After drying, panels are rated versus a set of numbered reference panels

**Fig. 13.9** Application panel used to determine the applied hiding power as described in ASTM 5150. A white paint has been applied to this panel. The rating for this paint is 3

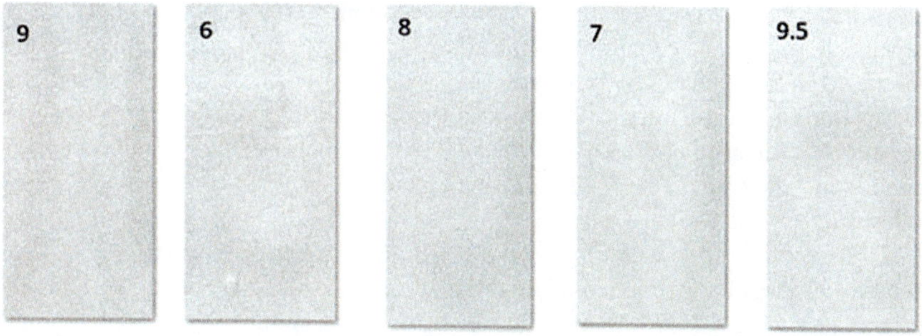


**Fig. 13.10** Applied hiding panels of the same paint, applied by five different trained painters. The numbers represent the rating given by each painter of their own panel, and the order of the panels goes from poorest hiding (left) to highest hiding (right)

that span the range from very poor hiding to very good hiding. If the result is very close to the reference panel number 8, the test panel is given a rating of 8. However, there is some ambiguity and there are inconsistencies in how to rate a test panel that lies between two reference panels—this is, these variants of the ASTM method do not overcome the limitations of the original method.

This issue is illustrated by an experiment in which five trained painters were asked to apply the same paint to a gray substrate and to independently rate their results against the same set of reference panels. The panels, as well as the ratings, are shown in Fig. 13.10. If the method was very reproducible, we would expect the same result (both in terms of visual appearance and rating) for the five painters. Clearly, this is not the case.

Two observations can be made from this small series of panels. First, although the paint, rollers, substrates, drying conditions, etc., were all identical, it is clear that there is a high degree of visual variability between the panels. The second observation is that there is not only variation in the visual appearance of the panels, but also in the perceived rating by the painters. Importantly, there is a complete lack of correlation between the visual appearance of the panel (demonstrated by the panel order from left to right in Fig. 13.10) and its assigned rating.

It is clear from even this limited assessment that the rated panel method suffers a high level of variability. To overcome this deficiency, a new test, based on the objective measurement of the panel appearance using an image scanner and algorithmic-based analysis of the measured data, was developed [8]. This test is described in the following section.

## An Alternative Method for Applied Hide

Any alternative method for applied hide analysis cannot be based on appearance as determined by the eye and must instead be read by an instrument. A typical colorimeter cannot be used to make this assessment as it measures only a limited area of the sample (the aperture of the colorimeter is typically a few $cm^2$), and our interest is in the appearance uniformity across the entire painted surface. In addition, these instruments can only give an average value of opacity over this area while we are interested instead in the variability of opacity. This variability, which is due to the peaks and valleys in the paint film, is over a much smaller scale than the colorimeter can resolve.

To overcome the low resolution and the limited investigated surface area when using a colorimeter, a commercial, high-resolution flat-bed scanner (A3 format) can be used to measure the color (reflectance) of every pixel of the painted panel. Figure 13.11 shows schematically the technique and example results.

The pixels with reflectance values equal to $R_\infty$ are fully hiding and are indicated in bright green in Fig. 13.11. As the scanner is more sensitive than the average human eye, the pixels with reflectance values close, but not equal, to $R_\infty$ will also be perceived as fully hiding. These pixels are indicated in pale green in this figure. However, as the reflectance values deviate more significantly from $R_\infty$, the pixels will be visually perceived as being different than $R_\infty$. These pixels can be grouped into three additional classes: pixels with moderate hiding power (orange in Fig. 13.11), low hiding power (red), and very low hiding power (black), and the number (or percentage) of the pixels in each class can be calculated. We now can define the applied hiding as the percentage of pixels that are at least moderately hiding (pixels in yellow, light green, and dark green). In Fig. 13.11, this represents about 80% of the covered area.

The Kubelka–Munk equations for each pixel can be solved by knowing the reflectance value of the pixel and the Kubelka–Munk parameters (K and S) from

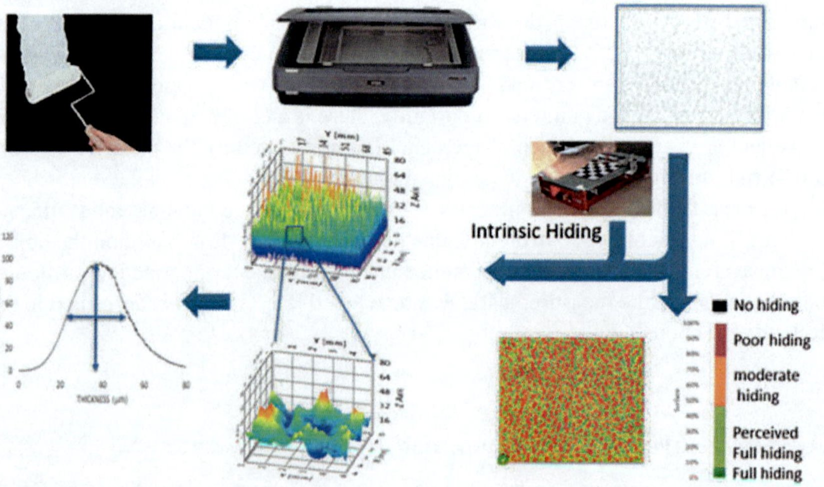


**Fig. 13.11** Schematic representation of the scanning method to quantify the applied hiding and to generate a 3D image of the scanned panel

the intrinsic hiding power test (discussed earlier in this chapter). From this, we can determine the film thickness, Xi, for every pixel. Plotting this thickness results in a three-dimensional image from which the distribution of thicknesses across the scanned surface can be calculated. Note that in this procedure, the Kubelka–Munk equations are used in two different ways: First, to determine the intrinsic hiding power, K and S, starting from a measured film thickness and reflectance values from a uniform drawdown, and second, to calculate the thickness value Xi for each pixel in the roller or brush applied film using from the measured reflectances and the calculated K and S.

As a verification of this approach, a paint was applied with a drawdown bar with stepped clearances (Fig. 13.12). Figure 13.12a shows the drawdown card used in this test, and Fig. 13.12b shows the profilometer profiles of the paint over the black and

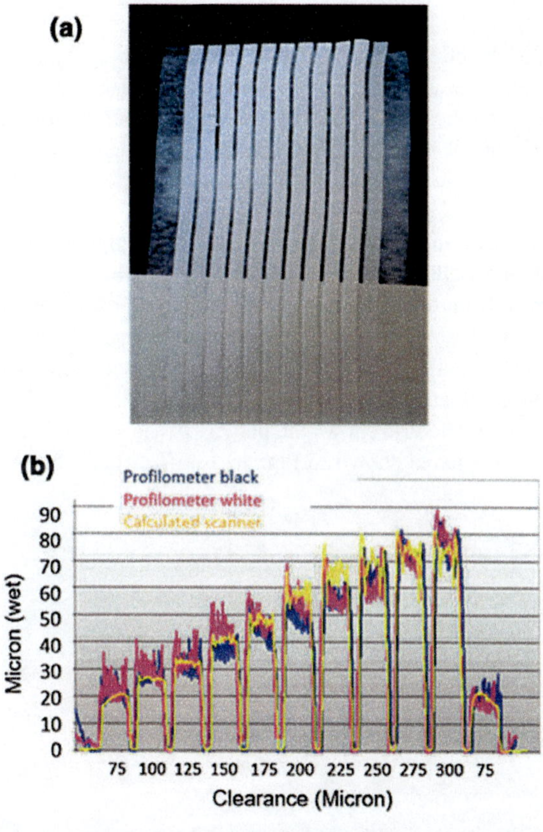


**Fig. 13.12** **a** Drawdown of an architectural coating with a drawdown bar with stepped clearances from 75 to 300 micron in steps of 25 micron. **b** Calculated wet film thickness values using the applied hide methodology and adjusted thicknesses as measured by profilometry

white portions of the card as well as the dry film thickness values calculated from the scan as described in Fig. 13.11.

As the thicknesses are all expressed as wet film thickness, a correction is applied taking into account the volume solids of the paint. As can be seen, there is excellent agreement between the thickness measured via a profilometer (after correction for the volume solids) and the thickness calculated via the scanner and the Kubelka–Munk equations. Note there is a significant discrepancy between the clearances X-axis values) and the actual wet film thicknesses (Y-axis values), as discussed in the section on contrast ratio limitations above, that is attributed to the rheology of the paint.

## *Applied Hide Example*

As an example of the effect of average film thickness on applied hide, a wall paint was loaded onto a roller and applied to a series of six black opacity charts without reloading between applications. Panels were weighed before and after application to determine the amount of applied paint. Knowing the density and the surface area of the panel, the average thickness of the paint can be calculated as indicated in Fig. 13.13.

Figure 13.13 shows that initially (left image in this figure) the paint hides the black substrate quite well. However, as the number of painted panels increases, less paint is transferred from the roller to the substrate. As a result, the thickness of the layer decreases, and the black surface becomes increasingly visible. The lower part of the figure is the translation of the visual image to the pseudo-color scale as defined in a previous paragraph and shown in Fig. 13.11.

The bar chart in Fig. 13.14a shows the percent of the surface area of each hiding classification for each panel shown in Fig. 13.13. The graph in Fig. 13.14b shows

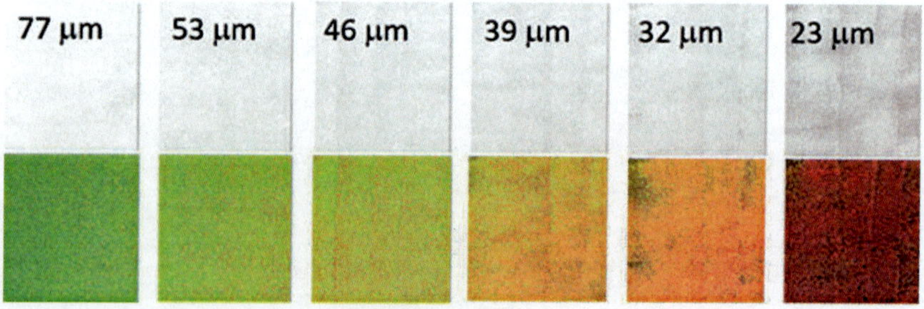


**Fig. 13.13** Application of a single roller load of a paint onto six black opacity charts. The leftmost image is the first application, and the rightmost image is the sixth. The numbers shown are the average wet film thicknesses of the coating. The upper part of each represents the visual images, and lower shows the translation of each image into pseudo-color scale (see Fig. 13.11)

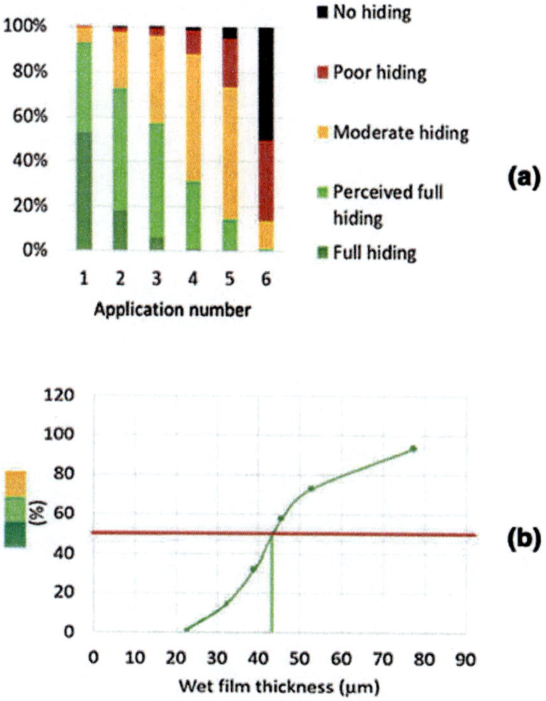


**Fig. 13.14** Image analysis results for multiple applications without reloading roller. **a** Bar chart showing the total surface of the different hiding classes. **b** Total percent of the surface area classified as moderate (yellow), perceived (light green), and full (dark green) hiding as a function of average wet film thickness for each drawdown

the total surface area covered at incomplete, perceived, and full hiding (defined in Fig. 13.11) with the average thickness of the applied paint.

Figure 14b shows that about 43 microns of wet film thickness are required to cover 50% of the surface area with at least moderate hide. We define this number as the applied hiding power. This number is characteristic of a paint and roller combination, and is independent on the amount of paint initially loaded onto the roller.

This is illustrated in a study in which two painters applied the same paint with the same type of roller. Each painter loaded the roller and made seven applications onto black opacity charts without reloading the roller. The resulting drawdowns are shown in Fig. 13.15. As can be seen visually, there are significant differences between the two drawdown series. Since the paint and roller are the same in each series, the only remaining variable is the painter. Apparently, these painters had different techniques and preferences for loading and applying paint by roller. This again illustrates the difficulty in reproducibly applying paint by roller, as was also shown in Fig. 13.10.

Although the paint series clearly differed in appearance, when the panels were scanned and the percent surface area with at least a moderate degree of hiding was plotted versus the wet film thickness (Fig. 13.16), the results for both painters fell on the same curve. In particular, the same wet film thickness—35 microns—was necessary to give 50% of the surface with at least a moderate hiding. This is not surprising since this coverage amount should be a property of the paint itself. This

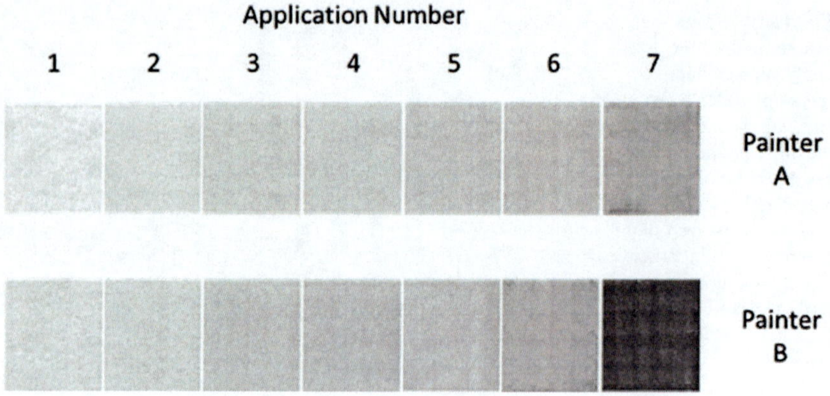


**Fig. 13.15** Multiple rollouts of the same white paint onto a black substrate using the same roller, as applied by two different professional painters

**Fig. 13.16** Percent surface area with moderate hiding power for both painters

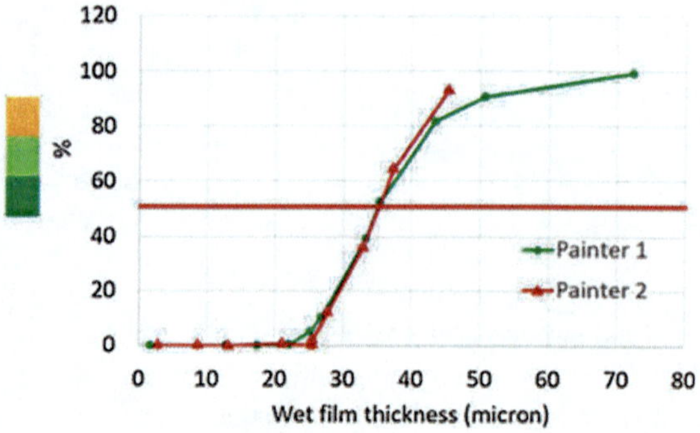


illustrates the high degree of reproducibility for this method of determining applied hide.

## *Factors Affecting Applied Hide*

We expect applied hiding power to be controlled by several paint and film parameters. Obviously, it is dependent on the intrinsic hiding power of the paint. In addition, it depends on the degree to which the applied paint thickness is uneven. We will call this property the "structure" of the dry paint film, and we note that structure depends primarily on the rheology (flow kinetics) of the paint and the exact nature of the roller that is used to apply the paint. The effect of the latter is outside the scope of this book. The drying kinetics (the so-called open times) are also important because longer drying times provide more opportunity for the initially structured paint surface,

**Table 13.4** Overview of the properties that affect applied hiding power for four commercial paints

| Property | Paint A | Paint B | Paint C | Paint D |
|---|---|---|---|---|
| Intrinsic hiding ($m^2/l$) | 14.6 | 11.6 | 15.3 | 12.3 |
| Flow characteristic | Pseudo plastic | Newtonian | Highly pseudoplastic | Pseudo Plastic |
| Wet film thickness for 50% of the surface to have moderate hiding power | 31 microns | 40 microns | 34 microns | 44 microns |

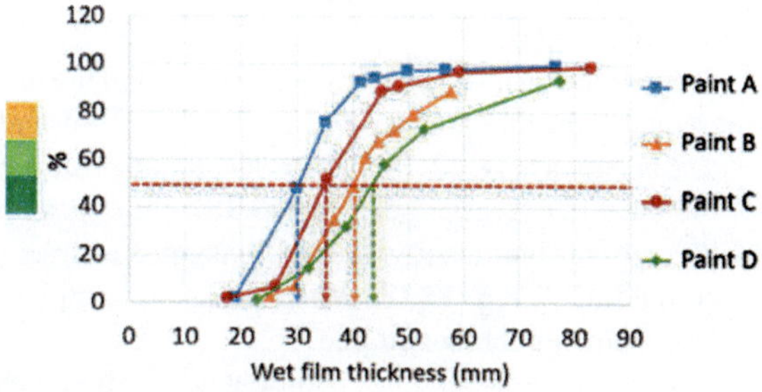


**Fig. 13.17** Percent surface area with moderate hiding power as a function of the wet film thickness for four commercial paints

as formed by the roller, to flow and become even. Although drying kinetics play an important role in the applied hide, this factor is also beyond the scope of this book and will not be discussed here.

The importance of paint flow to applied hide was investigated using four commercial paints with different flow characteristics and intrinsic hiding powers (Table 13.4). These four paints were applied and analyzed in a roll-out experiment as described in the previous paragraph. The applied hide results for these paints are shown in Fig. 13.17.

Figure 13.17 shows that the required wet film thickness to reach 50% surface area of at least moderate coverage for these four paints ranges between 31 and 44 microns.[5] Despite the high intrinsic hiding power of Paint C, more of it is required to reach moderate coverage for 50% of the surface area than of Paint A, which has a lower intrinsic hiding power. This is due to the very pseudoplastic behavior of Paint C, which results in it having a very structured surface and larger variability in film thickness. This, in turn, results in a relatively large area having poor coverage.

Paint B has the lowest intrinsic hiding but flows relatively well, resulting in an even surface and a lower required wet film thickness than expected based on the other paints. Paint A combines high intrinsic hiding with relatively good flow properties resulting in the highest applied hiding power. Paint D, by contrast, has lower intrinsic hiding and worse flow than Paint A, giving it the worst applied hiding power of the series.

[5] Note that these paints do not necessarily bracket the range generally seen for commercial paints and that these values may vary largely for other paints.

## Paint Rheology

As suggested in the paragraph above, we expect the structure to be related to the rheology of the liquid paint. If the paint flows freely under zero shear conditions (e.g., after application of the paint), then surface tension should eliminate irregularities in wet film thickness, thereby minimizing structure and improving the applied hide. However, the free flow of paint after an application is generally undesirable as it results in sagging and running of the wet paint. This severely limits the amount of paint that can be applied in a single application to vertical surfaces. The rheology required for acceptable flow and leveling creates a trade-off between these properties and applied hide.

Depending on the nature of the thickener and the interactions between the different ingredients, the rheology profile of a paint will be pseudoplastic, Newtonian, or somewhere between (Fig. 13.18). The viscosity of a true Newtonian liquid is independent of the shear rate (green line in Fig. 13.18) while for a pseudoplastic paint the viscosity will decrease significantly with increasing shear rate (shear thinning) following the solid red line in Fig. 13.18. When the shear is removed, the viscosity will build up again, following the dotted line.

The rheology profile of a paint has great significance in many of its application properties. During the painting process, the shear rate increases to between 1000 and 10,000 $s^{-1}$. A low viscosity in this domain provides easier application, but more difficultly in achieving a sufficiently thick layer of paint for complete hide. A low viscosity at low shear rates (0.1 $s^{-1}$) can cause sagging if the paint is applied too thickly while a high viscosity at low shear rates can result in poor leveling of the wet paint. In this context, the leveling can be seen as the peaks filling the valleys. High viscosity at very low shear rates (0.001 $s^{-1}$) can prevent the settling of the paint on storage. A paint formulator must invest much time and effort to find the right balance between these different rheological parameters.

To illustrate the effect of rheology on applied hide, four paints were made with exactly the same amounts of $TiO_2$, extenders, resin, etc., but differing in the rheology package. These were applied to panels, images of which are shown in Fig. 13.19.

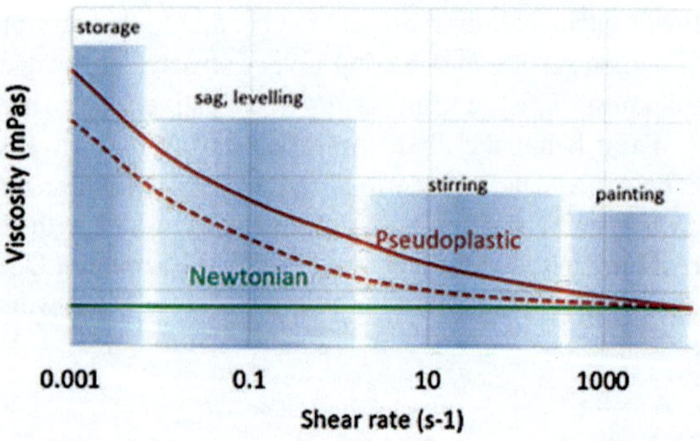


**Fig. 13.18** Schematic representation of the two archetypical rheology profiles in coatings

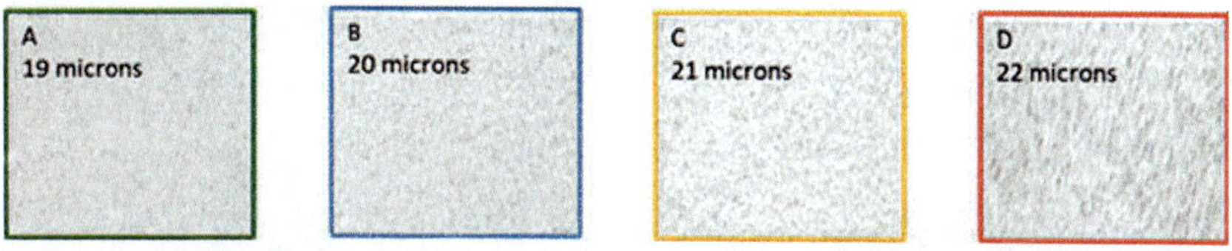


**Fig. 13.19** Application of four paints with differing in rheology only. The numbers indicate the average wet film thickness in microns

Paint A contains a Newtonian thickener, while Paint D contains a very pseudoplastic thickener. The two paints in the middle contain thickeners that give intermediate behaviors. All paints were brought to similar high shear ("ICI") viscosities and applied with the same roller type at roughly the same average wet film thickness (ranging between 19 and 22 microns).

Paint D, with pseudoplastic rheology, shows a high degree of brightness variability, with both very white areas and very dark areas. As the rheology becomes more Newtonian (moving from right to left in Fig. 13.19), the variability in brightness decreases, and the images become more uniformly gray.

Three-dimensional reconstructions of film thickness, using the same format as Fig. 13.11, are shown in Fig. 13.20. While the average wet film thicknesses of the paints are all close to 20 microns, the peaks and valleys are much more pronounced in the pseudoplastic paint (Paint D) compared to the Newtonian paint (Paint A). Paints B and C show intermediate levels of height variability. This observation is in line with the proposition that the viscosity of a pseudoplastic paint drops during the roller application (high shear) but builds after application (low shear), preventing the paint from leveling during drying. The viscosity of the Newtonian paint, on the other hand, remains relatively low throughout the application and drying process, allowing paint to flow from patches of high thickness to those of low thickness. This results in a more even thickness and so a more even opacity across the paint.

Figure 13.21 quantifies the degree of image uniformity in Fig. 13.19 using histograms of the wet film thickness values for each pixel in these images. The widths of these histograms correspond directly to image uniformity. As the rheological behaviors of the paints become more Newtonian (i.e., in going from Paint D to Paint A), the curves become sharper and the appearance of the paint more uniform.

The Newtonian paint (Paint A) has very few pixels with very thin (<10 micron) or very thick (>40 micron) calculated thicknesses. By contrast, the pseudoplastic paint (Paint D) shows a significant number of pixels at these two thickness extremes. It is worth noting that pixels at either end of the thickness extremes are undesirable. Obviously, those at the low end look dark, and so contribute directly to a non-uniform appearance and a low applied hide. This contributes to a pattern of strong contrasts in Paint D, which attracts the human eye and contributes strongly to the perception of incomplete covering. On the other hand, while excessively thick regions do not look different from other regions with acceptable hide, they represent a waste of paint.

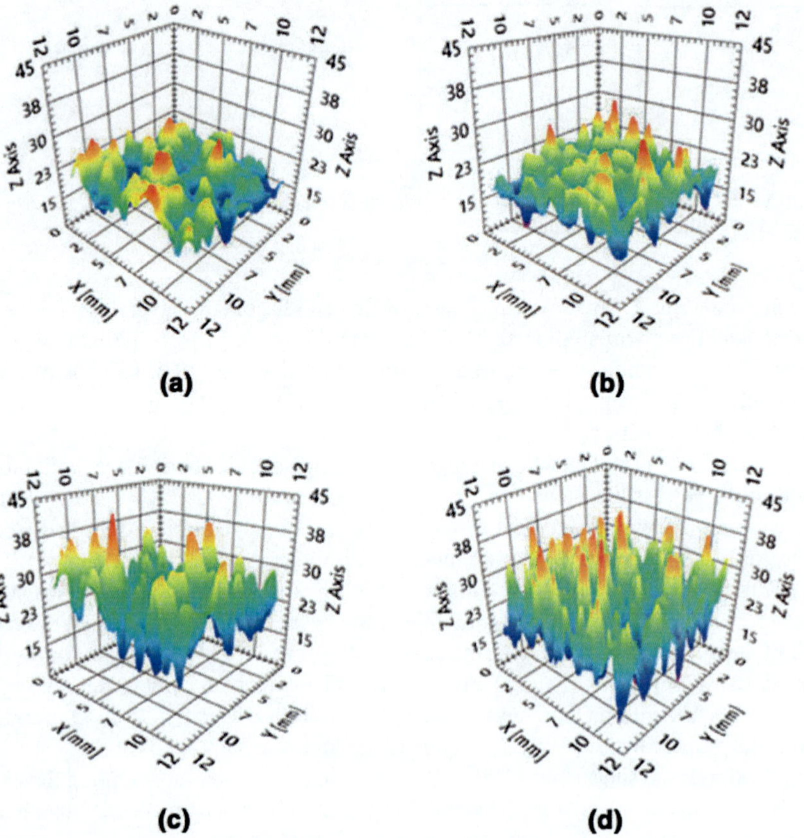


**Fig. 13.20** Three-dimensional representations of the four paints shown in Fig. 13.19. **a** Newtonian rheology. **b** and **c** intermediate rheology. **d** pseudoplastic rheology

These pixels would look the same if a portion of paint was removed from them and redistributed to areas of thin coverage.

The applied hide shown in Fig. 13.19 for Paint D is unacceptable for most paint applications. This deficiency can be overcome in one of three ways. First, we could apply a second coat of this paint. This would shift the histogram to higher thicknesses and so reduce the number of overly thin pixels.[6] While this results in acceptable applied hide, it increases both the cost of labor (to apply a second coat[7]) and the cost of coverage by the paint (to make a thicker coating).

[6] Due to the nature of counting statistics, we would expect this effect to be somewhat offset by an increase in curve width.

[7] This may not be strictly true, since paints are physically more difficult to apply at a higher build, and so this may slow the speed of the painter.

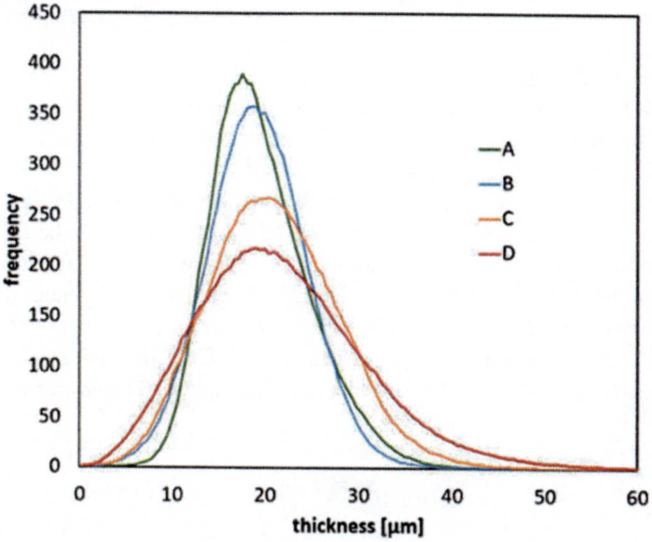


**Fig. 13.21** Histogram of the number of pixels at a certain wet film thickness for a paint with a Newtonian rheology package (A), a pseudoplastic rheology package (D), and two intermediate cases (B and C)

Alternatively, we can simply increase the film build—that is, increased the amount of wet paint applied to the substrate during a single application. While this can be done without increasing labor costs, it, too, suffers from an increased cost of raw materials, due to the increase in average film thickness.

Finally, the rheological behavior of this paint can be made more Newtonian, to sharpen the histogram curve without shifting its average. While this option maintains the original average film thickness while increasing opacity, it is less costly than the other alternatives. However, this leads to a greater sensitivity of the wet paint to sagging, which limits film build. If the film build is restricted due to sag, then it must be applied in two coats, increasing labor costs.

We see, then, that the formulator is challenged to find the best overall rheological behavior to minimize non-uniformity in film thickness while maximizing the amount of paint that can be applied in one application.

## Tinting Strength

In a completely opaque film, all light entering the film is either redirected back from it via light scattering before interacting with the substrate or is terminated within the film by light absorption. The brightness of the film is determined by, among other factors, the distance that the light travels within the film. This distance will be greater for films with weak light scattering than for films with strong light scattering, and greater for films with weak light absorption than for films with strong light absorption. Since high opacity is found when light scattering and light absorption are at their highest, we can use the distance traveled through the film as a proxy for opacity. Longer distances give lower opacity, since light penetrates more deeply into the film as the path length increases.

Opacity can therefore be measured as the distance that the light travels within a film. To directly measure this distance by tracing the paths of photons through the film is impossible. However, a quite viable alternative is available to us—light absorption. In Chap. 8, we discussed the Beer–Lambert law, which relates the amount of light absorbed by an object to three parameters—the absorption strength of a unit amount of light-absorbing material (colorant, for paints), the concentration of this light-absorbing material, and the distance the light travels. This law is summarized in Eq. 13.27:

$$A_\lambda = \varepsilon_\lambda bC \tag{13.27}$$

where $\varepsilon_\lambda$ is the absorption coefficient of the colorant, b is the path length through the object, and C is the concentration of the colorant.

The relationship given in Eq. 13.27 suggests that film reflectance (brightness) could be used as an indirect means of measuring the average distance that light travels through an opaque film. Because a tinted paint is required for this measurement, it is referred to as a tinting strength test. We could, in theory, use the reflectance measurement to calculate the actual distance that the light travels through the film. However, this is a laborious and unnecessary task. Instead, this test is almost always performed with a standard that has a defined tinting strength (typically an assigned value of 100) [9].

Both absorption strength and scattering strength determine light penetration distance, and so the tinting strength test can be used in two ways: either to compare the light scattering abilities of different $TiO_2$ samples or to compare the light absorption abilities of different color pigment samples. The critical difference between these two forms of testing is in how we attribute any differences in reflectance between a test paint and the control paint. If we are comparing the performance of $TiO_2$ pigments, say two batches of the same pigment, or samples from two suppliers, then any reflectance differences are assumed to arise from differences in scattering strength of the $TiO_2$. If we are comparing the performances of two color pigments, then we attribute any differences in reflectance to differences in the light absorption strength of the color pigment.

## *Tinting Strength of the White Pigment*

The two versions of testing are fairly straightforward to run, but with some subtle yet crucial differences in how the actual tinting strength values are calculated. To measure the tinting strength of the $TiO_2$, colored paints with the $TiO_2$ samples of interest (including a standard) are made with reflectivities around 0.40 to 0.50 (measured as Tristimulus Y). The colorant can either be one that is familiar to the experimenter or carbon black. The latter is specified for the ASTM method for this test [3]. The paints are then drawn down at a thickness great enough to ensure complete opacity, dried under controlled conditions, and the reflectance of the drawdowns measured. If a color pigment is used, we have a choice as to which reflectance values to use for this test, either the reflectance at the wavelength of maximum absorption strength of the color pigment, or the tristimulus Y value (total reflectance at all wavelengths). When a black pigment is used in the test, the tristimulus Y value is taken as the reflectance.

Since the films are drawn down at a great enough thickness to achieve complete opacity, the measured R values can be assumed to be the $R_\infty$ values for the paints. In this situation, we can apply the Kubelka–Munk theory to determine the balance between light absorption and light scattering (K/S) using Eq. 13.12.

The tinting strength of the standard paint is assigned a value of 100, and the tinting strength of an unknown is calculated as [10]

$$\mathrm{TS}_{\mathrm{Unknown}} = \frac{(K/S)_{\mathrm{Standard}}}{(K/S)_{\mathrm{Unknown}}} \times \mathrm{TS}_{\mathrm{Standard}} \tag{13.28}$$

where K/S values are calculated using Eq. 13.12 and the $\mathrm{TS}_{\mathrm{standard}}$ is normally assigned a value of 100. Assuming that the absorption strength K of the colorant is the same in all paints—that is, any differences in reflectivity are due to differences in $TiO_2$ scattering strength—Eq. 13.28 reduces to

$$\mathrm{TS}_{\mathrm{Unknown}} = \frac{1/S_{\mathrm{standard}}}{1/S_{\mathrm{Unknown}}} \times \mathrm{TS}_{\mathrm{Standard}}$$

$$\mathrm{TS}_{\mathrm{Unknown}} = 100 \times \frac{S_{\mathrm{Unknown}}}{S_{\mathrm{Standard}}} \tag{13.29}$$

A drawdown example showing under-dispersion of the $TiO_2$ pigment is shown in Fig. 13.22.

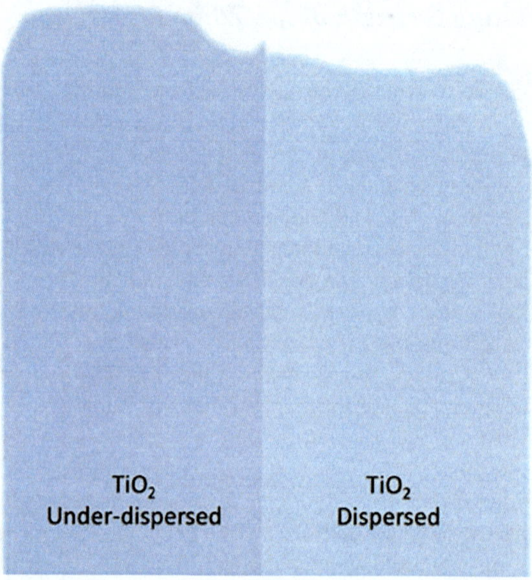


**Fig. 13.22** Tinting strength test for $TiO_2$. Difference in drawdown brightness is due to the degree of $TiO_2$ dispersion

## *Tinting Strength of the Color Pigment*

Alternatively, the tinting strength test can be used to compare the absorption strengths of two color pigment samples. Again, two paints are made. In this case, the same $TiO_2$ is used for both paints, with one paint containing the standard color pigment and the other the unknown color pigment. Dark paints are diluted with equal amounts of white paints to give a reflectance value in the range of 0.34 to 0.45. Drawdowns are again made at complete opacity, and the R, G, and B tristimulus values of the films are measured [12]. Of these three, the tristimulus value closest to the color of the paints is used to calculate the tinting strength of the unknown pigment as follows:

$$\mathrm{TS}_{\mathrm{Unknown}} = \frac{(K/S)_{\mathrm{Unknown}}}{(K/S)_{\mathrm{Standard}}} \times \mathrm{TS}_{\mathrm{Standard}} \tag{13.30}$$

Assuming that the scattering values of the paint are identical, this reduces to the counterpart of Eq. 13.29:

$$\mathrm{TS}_{\mathrm{Unknown}} = 100 \times \frac{K_{\mathrm{Standard}}}{K_{\mathrm{Unknown}}} \tag{13.31}$$

Note that Eqs. (13.28) and (13.30) are inverses of one another. In this way, a tinting strength value greater than 100 means that the optical property of interest (scattering or absorption) will exceed 100 if that property is greater in the unknown paint than in the standard. However, the meaning of a brighter test paint than control is dependent on the test type—this would indicate low tinting strength for a color pigment or high tinting strength for a $TiO_2$ pigment.

## *Color Development and Shear Strength Uniformity*

A test closely related to the tinting strength test is used to determine whether the degree of pigment dispersion in a paint is complete. Under-dispersion of either the white pigment or the color pigment will lead to the need for excessive levels of that pigment.

There can be different causes for under-dispersion. The two most common are incomplete dispersion during paint production and pigment flocculation during storage or drying. Both can apply to either the white pigment or the color pigment(s). Regardless of which type of pigment is under-dispersed, the degree of under-dispersion can be determined in the same way. The paint of interest is drawn down at complete hide and allowed to dry. Fresh paint is applied to a portion of the drawn-down film and subjected to a high level of shear. This shear should be sufficient to disperse any under-dispersed pigment. If under-dispersion is present in the paint, the appearance of the drawndown area and the highly sheared area will be different.

Historically, the additional shear was applied by rubbing the wet paint with a finger or thumb—the so-called "rub-out" test. However, the shear is not well controlled in this case, and there can be a great deal of variability in the degree to which the two portions of the drawdown film appear different, depending on the shear level. To overcome this variability, ASTM has developed a specific way of applying shear to the wet film using a brush [12]. In addition, in some procedures, the initial coat of paint is sheared after a set amount of drying, rather than allowing the first coat to dry completely and then applying a layer of fresh coat.[8]

Once the added paint has dried, the reflectance values of the two areas are compared. In some cases, appearance differences are easily seen visually. In other cases, the shift in color is too slight to be apparent. In any event, the reflectance measurements of the paint are measured and their difference, if any, is determined. This can be either as a $\Delta E$ value or as a difference in brightness (tristimulus Y).

The identity of the flocculated pigment is determined by whether the sheared area is brighter or darker than the unsheared (drawn down) area. If the $TiO_2$ is partially flocculated, the sheared area will be brighter than the drawndown area (that is, shearing improves light scattering). If the color pigment is partially flocculated, then the sheared area will be darker than the drawndown area. A shear strength

[8] Using two coats of paint is generally preferred for the following reason: The shear applied to the paint, especially if it has partly dried and is tacky, results in a non-uniform thickness. In thin areas, the background can be seen. In this case, it is better for that background to be the color of interest rather than white. In this way, if there is no dispersion after shear, the color of the sheared and unsheared portions of the drawdown will appear identical. If only one coat is applied and a portion sheared when the paint is tacky, then it will generally appear lighter than the unsheared portion for the reasons discussed in the section on applied hide.

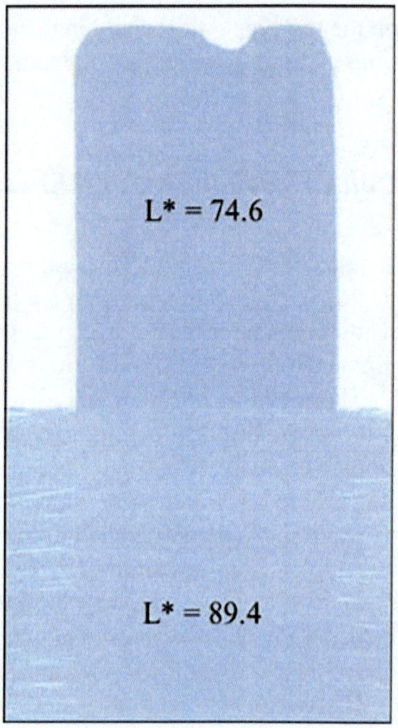


**Fig. 13.23** Shear strength uniformity drawdown for the paint shown in Fig. 13.22. The top area is as drawn down; the lower area is after shearing with a brush

uniformity drawdown card for the under-dispersed $TiO_2$ paint in Fig. 13.22 is shown in Fig. 13.23.

## Undertone

Our analyses in this chapter have so far omitted any mention of wavelength when discussing light scattering. Obviously, the wavelength is very important for light absorption by a chromatic pigment, with the absorption strength of a color pigment being a strong function of wavelength. Less obviously, light scattering is also wavelength dependent. This was touched on in Chap. 3, where we discussed the importance of matching particle size to light wavelength in determining the scattering strength of a particle. There we focused on particle size, but light wavelength can have an even stronger effect on scattering strength. This is because the match between particle size and wavelength depends equally on both and because the refractive index of a particle is wavelength dependent (for example, the refractive index of rutile $TiO_2$ in the visible region of the light spectrum drops from 2.91 at 380 nm to 2.48 at 700 nm) [13]. The combined effects of this are shown in Fig. 13.24 [2].

The importance of this has two sources. The first is that the strength of light diffraction—the primary mechanism of light scattering by $TiO_2$ particles—depends

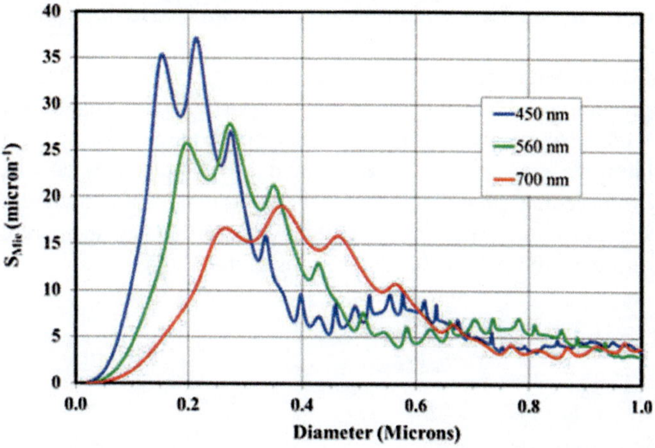


**Fig. 13.24** Calculated light scattering from a single $TiO_2$ particle as a function of light wavelength and particle size

on the ratio of wavelength and particle size, and so if we hold particle size constant, we would expect scattering strength to vary with wavelength. The second source is an indirect one: in addition to particle size, scattering strength depends on the difference in refractive indices of the scattering particle and the surrounding medium. The refractive index of $TiO_2$ varies with wavelength and is a maximum at the blue end of the visible spectrum and decreases significantly as wavelength increases to the red end. The refractive index of resin varies little over this wavelength range, so the difference in refractive indices between the particle and its surroundings decreases as the wavelength increases.

The overall effects of these two components of scattering strength are that the path length of blue light through a paint film is shorter than that of red light, at least within the range of $TiO_2$ particle sizes that maximize visible light scattering. This difference in path length is different for different $TiO_2$ particle sizes. This difference is typically characterized as the ratio of blue scattering strength to red scattering strength, and this ratio can be used to characterize the average particle size of the $TiO_2$ particles. This ratio, which is referred to as the undertone of the $TiO_2$, makes a contribution to the color of a chromatic paint, and differences in particle size between one pigment sample and another can cause a subtle, but observable, shift in paint color. In particular, smaller particles have a bluish undertone, while larger particles have a yellowish or neutral undertone.[9] For this reason, it is important that

[9] The actual diameters of what are described as "larger" and "smaller" particles are not very different from one another. In Chap. 3, we stated that the optimal size of a $TiO_2$ particle to scatter visible light is about 0.25 microns. The average diameter of a typical blue undertone $TiO_2$ sample is roughly 0.23 microns while that of a neutral undertone pigment is roughly 0.26 microns. Despite the small difference in average particle size, the effects on the appearance of the pigment are pronounced.

the $TiO_2$ pigment manufacturer tightly control the particle size for a particular grade of pigment.

In practice, $TiO_2$ undertone is typically measured by dispersing a mix of $TiO_2$ and a black pigment (typically carbon black or black iron oxide) in a suitable medium (for example, silicone or mineral oil) [14]. Once dispersed, the X and Z tristimulus values of the resulting off-gray paste are measured and the ratio of Z/X is calculated. Using standards with defined undertone values, the undertone of an unknown sample is calculated by linear interpolation. Alternatively, the yellowness index of the sample can be determined and used to characterize the pigment [15].

## Summary

Light is modified by a paint through two competing processes—light absorption and light scattering. The sum of these processes determines the opacity of the film, while their balance determines its brightness. The amounts of light absorbed and scattered depend on the details of the paint formulation and the thickness of the applied film.

The absorption and scattering character of a paint can be determined by drawing that paint down on a black and white card at a thickness below that required for complete opacity. The ratio of reflectance values is the contrast ratio of the drawdown. By measuring the wet film thickness, the Kubelka–Munk equations can be used to determine the absorption and scattering coefficients of the paint (K and S). This information can then be used to calculate the effects of changing film thickness, film absorption strength, and film scattering strength on appearance and determine the proper combination of these three parameters to give a complete hide. In addition, the spread rate of the paint—that is, the area that a unit volume of paint can cover at complete hide—can be calculated from these measurements. This is a value of direct importance to paint consumers, since it determines the cost of coverage.

The spread rate, or intrinsic opacity, of a wall paint can be measured as outlined above. In this measurement, it is important that the film be uniformly thick. However, this stipulation is not met by many end-use applications, including wall paints. These paints are typically applied by a brush or roller, each of which results in an uneven, or structured, surface. In these situations, the amount of paint required for complete hide—termed the applied hide of the paint—is greater than that indicated by the spread rate calculated on a uniformly thick paint film.

We can also determine the effectiveness of color pigments to absorb light and white pigments to scatter it by using the Beer–Lambert law. This law relates the amount of light absorbed by a material to the thickness of the material, the light absorption strength of the material, and the concentration of light-absorbing species (for paints, these species are color pigments). By measuring the fraction of light reflected from a thick film of colored paint, we can determine the relative contributions of light absorption and light scattering to opacity. This test, called the tinting strength test, is typically used to compare paints made with different $TiO_2$ pigments or different color pigments. Comparisons of the light reflectance values of a test paint to a control can indicate whether one grade of $TiO_2$ scatters light better than another, or whether one grade of color pigment absorbs light better than another.

Finally, we can use the concepts developed above to characterize the average particle size of a sample of $TiO_2$ pigment. This size characterization is possible because the scattering strengths of different wavelengths of light vary differently with particle size, giving particles of different sizes different undertones. By determining the relative path lengths of red and blue light (done by measuring their ratio, also known as the undertone of the pigment), we can determine if the average particle size of one sample of $TiO_2$ differs from another. As particle size decreases, the scattering strength of blue light is preferentially increased, giving smaller particles a bluer undertone.

## References


1. Kubelka, P., Munk, F.: Ein beitrag zur optik der farbanstriche (A Contribution on the Optics of Paint Layers). Z Tech Phys **12**, 593 (1931)
2. Diebold MP (2014) Application of light scattering to coatings. Springer
3. "Standard Test Method for Hiding Power of Paints by Reflectometry" (2018) ASTM D2805-11
4. "Analytical Colorimetry - Part 2: Saunderson Correction, Solutions of the Kubelka-Munk Equation, Tinting Strength, Hiding Power" (2018) DIN EN ISO 18314-2:2018-12
5. Kubelka P (1948) New contributions to the optics of intensely light-scattering materials. Part I" J Opt Soc Am A 38(**5**):448
6. Judd, D.: Optical specifications of light-scattering materials. J Res Natl Bur Stand **19**, 287 (1937)
7. Mitton PB (1972) "Hiding Power". In: Paint testing, 13th edn. American Society for Testing Materials
8. De Backer S (2019) New method for objective hiding assessment of wall paint. In: Proceedings of the European coatings conference
9. Mitton PB (1972) Mass color and tinting strength. In: Paint testing, 13th edn. American Society for Testing Materials
10. "Standard Test Method for Relative Tinting Strength of White Pigments by Reflectance Measurements" (2017) ASTM D2745-00
11. "Standard Test Method for Determining the Relative Tinting Strength of Chromatic Paints" (2016) ASTM D4838-88
12. "Standard Test Method for Color Development in Tinted Latex Paints" (2013) ASTM D5326-94a
13. Devore, J.R.: Refractive indices of rutile and sphalerite. J Opt Soc Am **41**, 416 (1951)
14. "Standard Test Method for Evaluating the Relative Tint Undertone of Titanium Dioxide Pigments" (2017) ASTM D6131-17
15. "Standard Practice for Calculating Yellowness and Whiteness Indices from Instrumentally Measured Color Coordinates" (2020) ASTM E313-20

# Chapter 14
# Durability of Paints, Plastics, and Paper Laminates

## Contents





M. Diebold et al., *Pigments, Extenders, and Particles in Surface Coatings and Plastics*, https://doi.org/10.1007/978-3-030-99083-1_14

## Introduction

Generally speaking, the term "durability" is used to designate some characteristic of a material that changes when exposed to environmental stresses such as heat, water, and light. In paints and plastics, these characteristics can include physical properties, such as the ability to protect a substrate or the tensile strength of the plastic article, and appearance properties, such as gloss or color. The changes in appearance properties are often determined by the particles embedded within the resin or plastic matrix, and for this reason, we will focus on appearance changes due to exposure to the elements in this chapter. We will also discuss appearance changes in paper laminates that occur on exposure to sunlight.

The primary determinant of paint and plastic durability is the chemical identity of the polymer. Some resins and polymers are quite susceptible to UV and/or moisture-mediated degradation, whereas others can endure for decades without change. It is noteworthy that there is a strong correlation between resin cost and resin durability, with high durability resins and polymers often costing many times more than their lower durability counterparts.

While resin chemistry is the key determinant of polymer durability, there are other factors that have a lesser, but still meaningful, impact on this important property. Many of these factors are within the control of the formulator, including the use of additives such as UV absorbers and hindered amine light stabilizers (HALS), and the use of color pigments that resist degradation [1]. A lesser but still important factor affecting polymer durability is the type and amount of titanium dioxide pigment used. Since nearly all exterior paints and plastics use $TiO_2$ pigments, it is worth understanding in detail the effects that $TiO_2$ can have on their durability.

The role of titanium dioxide in paint and plastics durability is complex. $TiO_2$ is both a strong absorber and a strong scatterer of UV light. The relative importance of the two types of light interaction is complicated for nanoparticles of $TiO_2$, but for pigmentary-sized particles scattering is somewhat stronger than absorption [2].

The interactions of $TiO_2$ particles with UV light result in a shielding of underlying resin and color pigments from destructive interactions with UV light. $TiO_2$ efficiently transforms most of the UV light energy into heat, which is a relatively benign form of energy. This beneficial effect is partially offset by a detrimental effect—the occasional transformation of UV light energy into chemical energy in the form of chemical radicals. These radicals are free to travel from the $TiO_2$ surface, where they are formed, into the surrounding polymer. The excess chemical energy stored in these radicals can break bonds within the polymer, degrading the film integrity. These radicals can also attack color pigments, leading to a significant shift in hue or in color intensity.

$TiO_2$ is nearly unique among the particles found in paints and plastics in its ability to absorb UV light energy and transform it into chemical radicals. In this chapter, we will discuss paint and plastic durability, focusing on the different ways that the appearance of these materials changes over time, and on how the $TiO_2$ pigment affects the rates of these changes. In addition to the involvement of $TiO_2$ in polymer degradation, we will also consider the degradation and/or color shift of color pigment particles due to UV exposure, and the effect of $TiO_2$ on the photo-graying of paper laminates.

Finally, we note that the high extinction coefficient of $TiO_2$ can useful in applications beyond paints, plastics, and paper [3]. For example, the ability to absorb wavelengths lower than 380 nm is quite useful for cosmetic applications [4].

## The Electronic Structure of Network Solids

In Chap. 8, we discussed the creation of molecular bonds through the combination of atomic orbitals on different atoms. We saw that molecular bonds can form between pairs of atoms and that, under certain circumstances, molecular bonds can be distributed over several atoms. In this chapter we consider an alternative arrangement for molecular bonds, one that is common in inorganic particles.

As before, we will begin by considering the combination of an *s* atomic orbital on each of two similar atoms. The two atomic orbitals combine to give one low energy bonding orbital and one high energy anti-bonding orbital. This is shown in Chap. 8, Fig. 8.3 and is also shown on the second column of Fig. 14.1.

Next, we add a third, identical atom to the first two, producing a triangular molecule. In this case, there is a single bonding orbital and two partially anti-bonding orbitals, with the average energy for the three orbitals being roughly equal to that of the atomic *s* orbitals on the isolated atoms.

As we continue to add atoms to our molecule, going from left to right in Fig. 14.1, we create new molecular orbitals, some bonding, some anti-bonding and some non-bonding—that is, some with energies roughly equal to that of the unperturbed atomic *s* orbital.

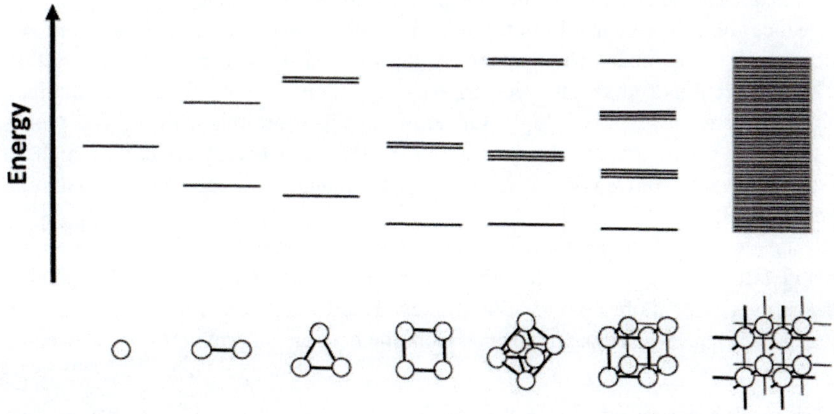


**Fig. 14.1** Combinations of *s* atomic orbitals on molecules having an increasing number of atoms

There are two trends as we increase the number of atoms in these molecules. The first is that for each added atom we add one additional *s* atomic orbital, and so one additional molecular orbital is created, consistent with the rule of orbital conservation described in Chap. 8. The second is that, as more *s* orbitals are added, the difference in energy between the lowest energy molecular orbital and the highest energy molecular orbital increases. This is because the total amount of atomic orbital overlap increases with an increasing number of atoms. However, the size of this increase becomes less and less as more atoms are added. For molecules with roughly a dozen or more atoms, the increase in energy difference effectively stops.

We can understand the reason for this by recalling, from Chap. 8, that the energy difference between the lowest and highest energy orbitals is determined, in part, by the degree to which the atomic orbitals spacialy overlap. We can see how this overlap changes from left to right in Fig. 14.1, beginning with the single central atom.

Initially, as each new atom is added, the extent of atomic orbital overlap between the central atom and those of the surrounding atoms increases, increasing the difference in energy between the lowest energy bonding molecular orbital and the highest energy anti-bonding molecular orbital derived from these atomic orbitals. However, there is a limited number of atoms that we can place in contact with the original central atom. Depending on the geometry of the molecule, a single atom can have at most 12 touching neighbors.

Once the central atom is fully surrounded by other atoms, the orbitals of any added atoms will be too far away to overlap with those on the central atom. At this point, the amount of orbital overlap that the central atom experiences is fixed, and so there is no longer an increase in the energy difference between the lowest and highest energy molecular orbitals derived from this set of atomic orbitals.

On the far right of Fig. 14.1, we see the molecular orbital diagram for a molecule composed of an infinite number of atoms. While a truly infinite molecule is an impossibility, molecules comprised of millions, or even billions, of atoms do exist. These materials are called network solids and are characterized by a network of chemical bonds that joins together all atoms in the solid particle.[1]

Since there is a nearly infinite number of atoms in these molecules, there is a nearly infinite number of molecular orbitals created from these *s* atomic orbitals. These molecular orbitals fall over a limited range of energies, for the reasons outlined above. This results in a very high number of molecular orbitals with energies within a relatively narrow range. The density of orbitals in this energy range is so high, in fact, that these orbitals can be considered a continuum. We call this continuum an energy band.

The electrons in most atoms occupy many atomic orbitals, and in the same way, the electrons in a network solid are found in many energy bands, each band being based on one or more atomic orbital types. This is shown in for sodium metal in Fig. 14.2. Here, we see bands formed from the occupied atomic orbitals, as well as

[1] This can be contrasted with "molecular solids", which are composed of discrete molecules held together by weak attractive interactions. There are no chemical bonds between atoms of separate molecules in molecular solids.

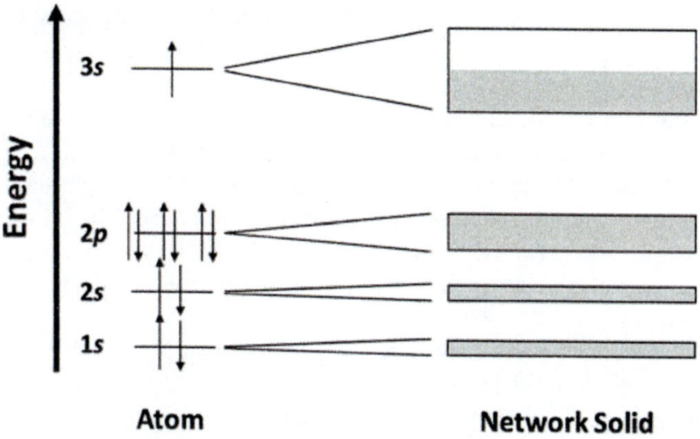


**Fig. 14.2** Electronic band structure of sodium metal

from one unoccupied orbital. Electrons are shown as shading within the bands (rather than as arrows). We see that the 3*s* atomic orbitals are only partially filled, and so the band created from these orbitals is only partially shaded. The shaded portion is at the bottom of the band, a result of the lowest energy orbitals filling first.

As was the case for molecular orbitals in small molecules (Chap. 8), the most interesting bands in network solids are the ones that are either just filled or just empty (these are sometimes referred to as frontier bands, similar to the frontier orbitals in small molecules). These are the highest energy filled bands (HOMOs), and the lowest energy empty bands (LUMOs), called, respectively, the valence band[2] and the conduction band. These are the most important orbitals because they determine many properties of the network solid.

There are two fundamental types of frontier bands. The first is a partially occupied, or incompletely filled, band, such as we saw for sodium in Fig. 14.2. Here, the energy difference between the highest energy occupied orbitals and the lowest energy unoccupied orbitals is very small, and thermal energy is energetic enough to promote electrons into the unoccupied orbitals of that band. Network solids for which the frontier band is only partially occupied are typically opaque, since any frequency of light (up to a point) will have enough energy to move an electron to a higher energy within the band. In addition, these solids are typically good conductors of electricity.

The second fundamental type of frontier band is seen when the HOMO band is filled and the LUMO band is empty. The difference in energy between these bands is called the band gap, and the band gap energy represents the least amount of energy needed to promote an electron.

---

[2] This is not to be confused with the valence orbitals on atoms or small molecules (Chap. 8), since in those cases, the term "valence" is applied to both filled and unfilled orbitals, whereas with network solids the term "valence" applies only to the HOMO bands.

We can categorize these network solids according to the magnitude of the band gap. When the gap is quite large, as it is, for example, for silica, the photon energy required to promote an electron is so high that it is outside the range that naturally occurs in sunlight or is easily generated in the lab. These network solids are non-conductive, because conduction of electrons through a network solid requires easily accessible LUMOs, and these solids are colorless since visible light does not have enough energy to promote electrons into the conduction band. The lack of conductivity makes silica ideal for use as an electrical insulator, and silica glass is used as an electrical insulator in many applications (for example, for attaching above-ground electrical wires to electric poles).

When the band gap is of intermediate energy—energy too high for thermal promotion, but low enough for visible or UV light promotion—the electrical properties of the solid are intermediate between the cases of no band gap (partially filled bands) and high band gap (electrical insulators). These materials are called semiconductors and are electrically insulating in the dark but electrically conducting in the light.

These three configurations of bands—partially filled, filled with a small band gap, and filled with a large band gap—are depicted in Fig. 14.3, along with an arrow denoting the energy of an average visible light photon.

The reason for our lengthy discussion on network solids is that $TiO_2$ is such a material, with a band gap of just slightly more energy than visible light (that is, accessible by photon energies in the near UV region of the spectrum), making it a semiconductor. We will use our knowledge of band gaps and semiconductors to understand one aspect of how $TiO_2$ pigments affect paint durability.

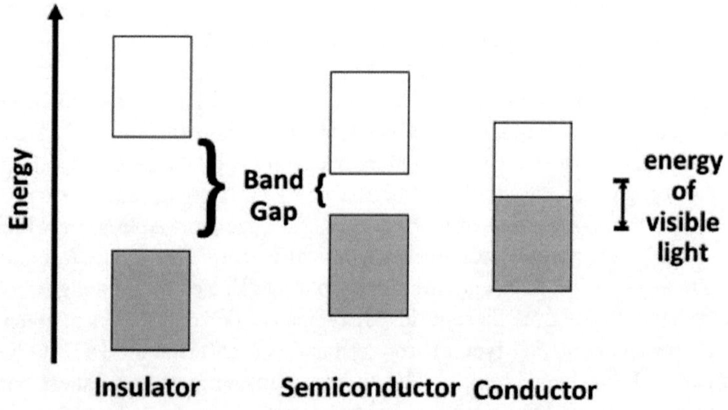


**Fig. 14.3** Characterization of band gaps

## Degradation Pathways

There is not a single chemical equation that describes polymer degradation because, for each polymer chemistry, there are numerous degradation pathways, each involving a unique set of chemicals reactions that occur in series and in parallel. However, we can make some generalizations about polymer degradation.

The first is that degradation occurs primarily due to exposure to UV light. Such photons are of sufficient energy to break carbon-hydrogen or carbon–oxygen bonds (that is, of sufficient energy to promote a bonding electron into an anti-bonding orbital). Visible light photons do not have sufficient energy to break most chemical bonds, and so have less of an effect, if any, on polymer durability. This explains why interior paints and plastics have very low durability requirements—they are generally exposed to only the small amount UV light that is present in artificial lights or in sunlight filtered through a glass window.

The next generalization regarding paint and plastics durability is that there are two fundamentally different ways for the energy of the UV photon to break a chemical bond in the polymer molecule [5]. The first occurs when the polymer itself absorbs the UV light with the concomitant breaking of a bond (this degradation route is known as direct degradation). The second is more complex. Here, the $TiO_2$ particles absorb UV light ($TiO_2$ is an excellent UV light absorber) and transform the light energy into chemical energy in the form of chemical radicals, as mentioned in the introduction to this chapter. We refer to this reaction pathway as photocatalytic degradation. These two reaction pathways are shown in Fig. 14.4 and will be discussed in more depth below.

Another generalization is that the ultimate products of these degradation reactions are primarily $CO_2$ and water. However, these reactions rarely proceed to the point where $CO_2$ and water are actually made. This is because the intermediate products tend to be water soluble. This water solubility is due to chemical differences between the initial resin (which is not water soluble) and the organic compounds that are intermediate to the complete mineralization of the resin.

Photocatalytic

$$TiO_2 + \text{UV Light} \longrightarrow e^- + \text{hole}^+$$
$$e^- + \text{hole}^+ \longrightarrow TiO_2 + \text{heat}$$
$$\text{hole}^+ + OH^- \longrightarrow OH^\bullet$$
$$e^- + O_2 \longrightarrow O_2^{-\bullet}$$
$$O_2^{-\bullet} + OH^\bullet + (\text{-}CH_2\text{-})_n \longrightarrow \text{intermediates}$$
$$O_2^{-\bullet} + OH^\bullet + \text{intermediates} \longrightarrow CO_2 + H_2O$$

Direct

$$\text{UV Light} + O_2 + (\text{-}CH_2\text{-})_n \longrightarrow \text{intermediates}$$
$$\text{UV Light} + \text{intermediates} \longrightarrow CO_2 + H_2O$$

**Fig. 14.4** Film degradation pathways

There are two factors that make the intermediates more water soluble than the initial polymer molecules. The first is that degradation reactions break carbon–carbon bonds, and so the intermediate molecules have a much lower molecular weight than the original resin molecules. Secondly, the oxygen atoms in the radicals tend to attach to the intermediates to form carboxylic acids, alcohols, esters and ketones. In fact, the formation of carbonyl functional groups during photo-oxidation can be monitored by infrared spectroscopy [6, 7]. These different oxygen functional groups increase the water solubility of the molecules.

Since the degradation intermediates are water soluble, they tend to be removed from the paint or plastic when it is intentionally washed or exposed to rain. This erodes the film, removing resin much more quickly than if the resin was required to completely degrade into $CO_2$ and water for removal.[3]

## *The $TiO_2$ Photocatalytic Cycle*

Pigmentary $TiO_2$ can have large and conflicting effects on polymer durability. Its beneficial effects will be discussed in the next section; here we will focus on its harmful effects [8–10].

In Chap. 8, we discussed the absorption of light by color pigment particles. In that chapter our interest was in the absorption process and how it altered the balance of light reflected from the paint or plastic. In this chapter we are interested in a different aspect of light absorption—the disposition of the light energy after the light is absorbed. This is because the potentially harmful effects of $TiO_2$ arise from the different ways that the energy from absorbed UV light is dissipated.

As noted above, $TiO_2$ is an excellent absorber of UV light—so much so that a single, 0.25 micron thick particle screens out essentially all of the UV light that strikes it from sunlight. The photon energy, which must be conserved, is temporarily transformed into electronic energy within the $TiO_2$ particle. In Chap. 8 we discussed the different electronic configurations that can exist in a molecule or substance. $TiO_2$ is a network solid, and so its electronic structure consists of bands of different energies [11]. UV light is absorbed by $TiO_2$ because the energies of UV photons are great enough to promote an electron across the band gap and into the conduction band. We refer to this particle as being in an electronically excited state and designate it with a superscript asterisk as $TiO_2^*$.

The energy level diagram for UV light absorption is shown on the left in Fig. 14.5, with the $TiO_2^*$ electronic structure on the right. A prominent feature of the electronic

[3] An interesting observation was made in the 1940s regarding resin erosion. A number of automotive paints were exposed to the elements in the high Andes mountains [5]. This was done to accelerate durability testing—the UV intensity of sunlight increases significantly with elevation. When the panels were returned to the laboratory, the films appeared to have changed little. However, when placed in water, the films completely disintegrated. The explanation is that there was little to no rainfall in this part of the Andes, and so the all the degradation intermediates remained in the films, holding them together, until they were submerged in water.

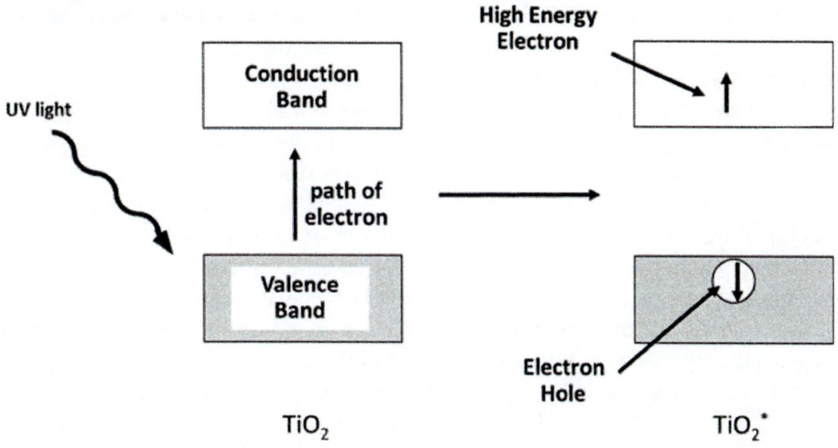


**Fig. 14.5** Promotion of an electron into a high energy band

structure of the excited molecule is the electron in the conduction band. This electron has higher energy than when it occupied the valence band, and this energy is indicated by its location in Fig. 14.5. This high-energy electron can lower its energy by returning to the valence band—something that typical happens within 10 to 20 ns of promotion and transforms the energy originally from the UV light into heat. This transformation of UV light energy into heat energy, mediated by $TiO_2$, occurs for the majority of UV absorption events.

However, approximately one time in a million, a different reaction sequence occurs. Here, the high-energy electron loses its excess energy by transferring it to another molecule. This occurs when the other molecule has an empty or partially filled orbital that is of lower energy than the conduction band in the $TiO_2$ particle. The result is the transfer of an electron from one species (the $TiO_2^*$) to another species (typically $O_2$ absorbed on the particle surface). From an energy viewpoint, this reaction converts light energy into chemical energy in the form of a radical ($O_2^-$), through the creation of an electronically excited $TiO_2$ particle intermediate.

We also see in Fig. 14.5 an open circle in the valence band in the $TiO_2^*$. This symbol represents an entity termed an electron hole. The meaning of this hole requires some explanation. Before absorbing the UV light, the valence band was completely filled. After an electron is promoted out of the valence band, it leaves behind a partially filled orbital—more specifically, it leaves behind one orbital that is only half occupied (the other orbitals in this band remain fully occupied). The half occupied orbital is unstable—it is of low energy and so is a very desirable state for an electron. So desirable, in fact, that electrons on many ions or molecules that are adsorbed onto the $TiO_2$ surface, most especially hydroxide (-$OH^-$), can transfer into the hole, losing energy (becoming more stable) in the process. This creates a hydroxyl surface radical ($OH^\bullet$).

We, therefore, see that the promotion of an electron into the high energy conduction band leads to the formation of two very reactive species—a strongly reducing electron, and a strongly oxidizing electron hole. As discussed above, for all but approximately one event in a million these two will simply recombine with one another, converting the energy that originated from the UV light into heat energy. If, however, either the electron or the electron hole reacts with a surface species, a highly reactive radical will be formed. Regardless of which species undergoes this reaction first, the partner species will eventually react with a surface species as well, since it can no longer combine with the first species that reacts.

This sequence of reactions is shown in energy level diagram form in Fig. 14.6. Note that at the end of these reactions, the $TiO_2$ electronic structure has returned to its ground state. Because of this, the $TiO_2$ particle plays the role of photocatalyst—once it has been activated by a photon, it catalyzes polymer degradation reactions and then returns to its original state. The complete reaction sequence, including the catalytic cycle, is shown in schematic form in Fig. 14.7, highlighting both direct and photocatalytic degradation reactions. These reaction sequences were also given in the conventional chemical reaction format in Fig. 14.4.

Later in this chapter, we will describe ways to treat the $TiO_2$ pigment surface to decrease its photocatalytic activity. We refer to these treated pigments as being "durable", but this designation is misleading. $TiO_2$ itself is completely durable when exposed to the elements (it is, in a sense, titanium rust). As discussed in detail above, the same cannot be said for the polymers found in most paints and plastics. When

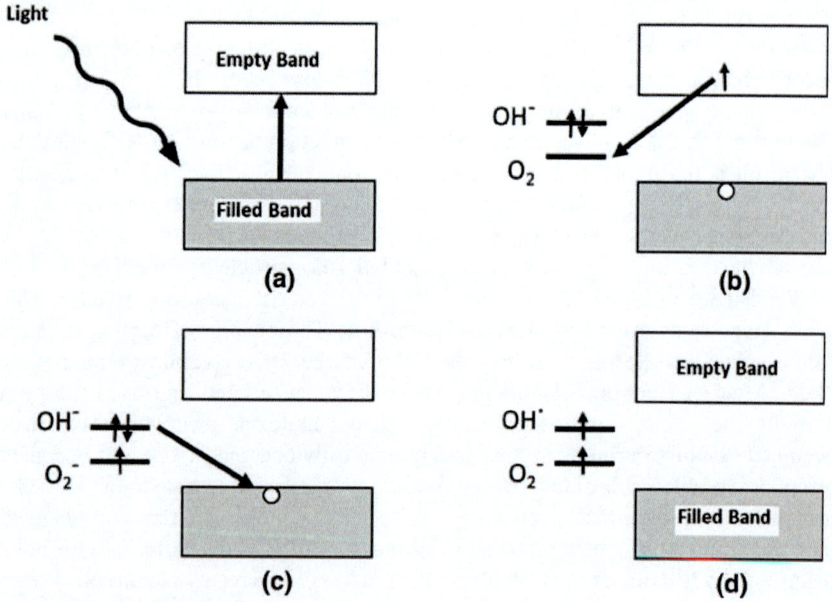


Fig. 14.6 Sequence for the photogeneration of hydroxyl and superoxide radicals

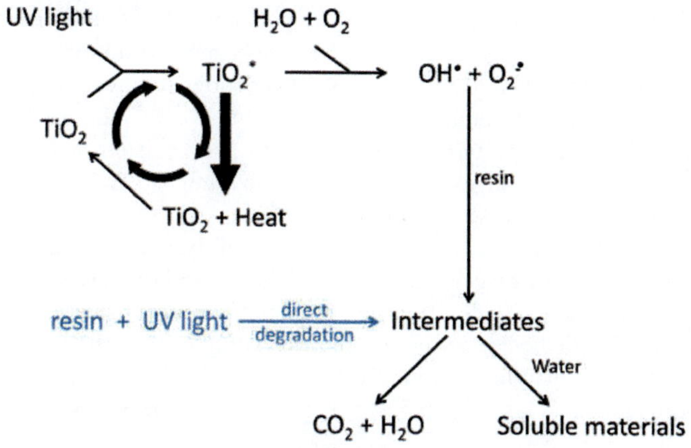


**Fig. 14.7** Photocatalytic cycle and direct degradation

we apply the term "durable" to a $TiO_2$ pigment we are referring to the effect that the pigment has on paint or plastic durability, not to the durability of the $TiO_2$ pigment itself.

## *The Effect of $TiO_2$ on Direct Degradation*

In the previous section, we focused our attention on the negative contribution that $TiO_2$ makes to the durability of a polymer due to its photo-catalytic nature. However, $TiO_2$ also makes a positive contribution to durability, although this contribution is not as well recognized in the coatings and plastics industries. This contribution is the reduction in the rates of direct degradation reactions.

Although $TiO_2$ does not participate in direct degradation, it does play a very large role in its rate. This is because of the strong absorption of UV light by the $TiO_2$ pigment. As described above, a single $TiO_2$ particle absorbs UV light so strongly that it effectively removes all the UV content from the solar spectrum. This removes one of the reactants in the direct degradation pathway (the UV photon), and so decreases significantly the rate of this undesirable reaction sequence. As we will see, the degree to which the $TiO_2$ protects the resin from direct degradation is not the same for all $TiO_2$ grades. Variability in % $TiO_2$ in the pigment, or the degree to which the $TiO_2$ particles disperse, will affect the protective ability of the pigment.

We see, therefore, that $TiO_2$ has conflicting effects on paint durability. The overall effect of $TiO_2$ durability depends on the balance between photocatalytic and direct degradation rates. Since $TiO_2$ particles absorb essentially 100% of the UV light that strikes them and very rarely transform the light energy into chemical energy in the form of reactive radicals, we would expect the presence of $TiO_2$ to be beneficial to paint or plastics durability [2]. This is the case for most polymers. However, polymers that themselves are very resistant to direct degradation benefit very little from the protective effects of $TiO_2$ (that is, since these polymers do not absorb UV light with the breaking of a chemical bond, they do not suffer from direct degradation), but are still susceptible to attack by radicals. In such cases, the $TiO_2$ has a net detrimental effect on durability, and it is important that even paints made with highly durable resins use the proper grade of $TiO_2$, so as to minimize $TiO_2$ initiated radical generation, as will be discussed below.

The consequences of the existence of two types of degradation reactions and of the $TiO_2$ playing opposing roles in them are quite far reaching. This adds significant complexity to degrees to which paints and plastics appearances change over time as a function of the ingredients used, and significant complexity to the ways that these appearance changes manifest differently under different stress conditions (specifically, under natural and artificial exposure conditions). These will be discussed in a later section.

## *Thermal Degradation*

In addition to direct and photocatalytic degradation, there is a third degradation pathway—one that does not involve UV light. This is thermal degradation. The reaction of polymers with atmospheric oxygen is thermodynamically favorable yet does not occur at appreciable rates at ambient temperatures. This is due to the energy barrier that must be overcome to initiate degradation reactions. For both direct and photocatalytic degradation, this barrier is overcome by the absorption of an energetic UV light photon. Alternatively, thermal energy can be used to overcome this barrier. This is seen in fire, where organic molecules rapidly react with oxygen at high temperatures.

Although paints and plastics in most applications are not exposed to the temperatures found in a fire, they nonetheless can thermally degrade under typical outdoor exposure conditions, especially if they are darkly colored. On a hot, bright day, the surfaces of painted metal panels can reach 90 °C or more. This is roughly 70 °C hotter than a temperate day in most localities. Since thermal reaction rates roughly double for every 10 °C increase in temperature, the thermal degradation rate on a hot, bright day will be roughly 100 times faster than on a temperate day. We will return to the issue of thermal degradation when we discuss accelerated methods for paint durability testing.

## *Ultimate Durability*

There are three ways that the direct and photocatalytic degradation rates can compare relative to one another. First, the direct degradation rate can be much higher than the photocatalytic degradation rate. As discussed above, this is the case for all paints except those with very durable resin. Alternatively, the rate of photocatalytic degradation can be much higher than the rate of direct degradation, which is true when the resin is very resistant to direct degradation. Finally, the two rates could be approximately equal. The factor that limits paint or plastic durability is different for each of these three possibilities.

## *The Concept of "Ultimate" $TiO_2$ Pigment Durability*

The fact that there are two separate polymer degradation mechanisms has many important consequences. One consequence is that, as detailed above, the choice of $TiO_2$ grade can affect paint or plastic durability by altering the rate of either or both degradation mechanisms. Another consequence is that there are certain situations wherein decreasing the rate of photocatalytic degradation, even by a relatively large amount, does not lead to a significant increase in overall paint or plastic durability. This is true when the rate of direct degradation is much greater than that of photocatalytic degradation. In this case, reducing the already minimal amount of photocatalytic degradation has an insignificant effect on paint durability [12].

This situation is shown schematically in Fig. 14.8. The chart shows the degradation rates of five hypothetical paints, all made with the same binder but with $TiO_2$ pigments possessing a wide range of photocatalytic degradation rates. For simplicity, we will assume that each of the six pigments affects the direct degradation rate equally, and that each pigment has one-fifth of the photocatalytic activity of the preceding pigment (that is, the photocatalytic degradation rate of Pigment A is five times greater than Pigment B; that of Pigment B is five times greater than Pigment C, and so on).

Figure 14.8 clearly shows that there is a point at which direct degradation overwhelms photocatalytic degradation, and further decreasing the photocatalytic activity of the $TiO_2$ no longer gives a measurable improvement in overall paint durability.

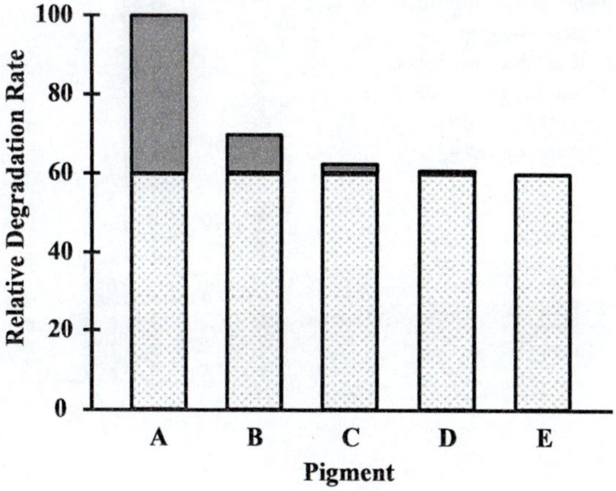


**Fig. 14.8** Total degradation rates for pigments with decreasing photocatalytic activity. Dotted areas are direct degradation rates and dark gray areas are photocatalytic degradation rates

For example, the durability of the paints made with Pigments C and D would be very difficult to distinguish despite the fact that Pigment C has a five times greater photocatalytic degradation rate than Pigment D. In fact, it is entirely possible that the paint made with Pigment D would actually be less durable than the paint made with Pigment C—despite Pigment D having only one-fifth the photochemical activity of Pigment C!

This is because less photocatalytically active pigment grades tend to have lower $TiO_2$ contents, for reasons discussed later, and also tend to be harder to disperse. Both of these tendencies decrease UV screening ability of the pigment and so increase direct degradation rates (which in Fig. 14.8 are assumed to be equal for all paints). Even a small increase in direct degradation rates in the paint made with Pigment D could entirely offset the benefit of the lower photocatalytic degradation rate of that pigment.

We will refer to the photocatalytic activity at the point where photocatalytic degradation rates are insignificant compared to direct degradation rates as the "ultimate pigment durability". This is an important level of photocatalytic activity since any improvement in pigment durability in the traditional sense (i.e., any decrease in photocatalytic activity) beyond this point will not measurably improve paint or plastics durability. The exact point at which pigment photocatalytic degradation rates are deemed insignificant compared to direct degradation rates is somewhat arbitrary, but for our purposes, we will define "ultimate pigment durability" as occurring when less than 5% of the total degradation is due to pigment photocatalytic activity. This value (5%) was chosen because it is well below the upper limit of reproducibility of most end-use durability tests [13].

One very important aspect of the ultimate pigment durability concept is that it is dependent on the reactivity of the binder involved and will therefore vary from one polymer to another. This is demonstrated in Fig. 14.9 and Table 14.1. Here, the total degradation rates of six hypothetical paints are compared. The paints are made by

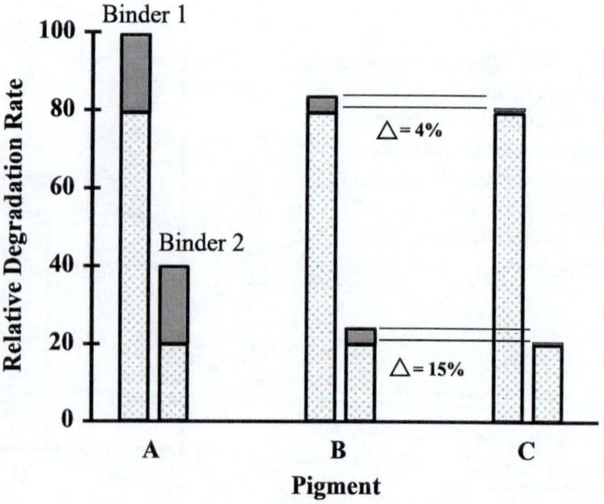


**Fig. 14.9** Comparison of paints made with different binders and pigments. Dotted areas are direct degradation rates and dark gray areas are photocatalytic degradation rates

**Table 14.1** Degradation rate values for the paints shown in Fig. 14.9

| Paint | Binder | Pigment | Degradation rates | | Percent of total degradation | |
|---|---|---|---|---|---|---|
| | | | Direct | Photocatalytic | Direct | Photocatalytic |
| 1 | 1 | A | 80.0 | 20.0 | 80.0 | 20.0 |
| 2 | 2 | A | 20.0 | 20.0 | 50.0 | 50.0 |
| 3 | 1 | B | 80.0 | 4.0 | 95.2 | 4.8 |
| 4 | 2 | B | 20.0 | 4.0 | 83.3 | 16.7 |
| 5 | 1 | C | 80.0 | 0.8 | 99.0 | 1.0 |
| 6 | 2 | C | 20.0 | 0.8 | 96.2 | 3.8 |

taking every combination of two binders (Binder 1 being reactive; Binder 2 being inert) and three $TiO_2$ pigments (Pigments A, B, and C, which differ in photocatalytic activity by factors of five, as in Fig. 14.8). As before, we will initially assume that the effects of pigment grade on direct degradation rates are equivalent (i.e., the direct degradation rates of the three Binder 1 paints are equal to one another; those of the three Binder 2 paints are also equal to one another but significantly less than those of the Binder 1 paints).

For both binders, Pigment C has ultimate pigment durability while the durability of Pigment A is clearly far from "ultimate". The characterization of Pigment B durability, however, is not so clear. In the more reactive binder (Binder 1), Pigment B has ultimate durability; in the more stable binder (Binder 2), Pigment B does not have ultimate durability. This is a simple consequence of the fact that the amount of direct degradation affects the relative contribution of photocatalytic degradation to the overall durability of the paint, and it is this relative contribution that determines whether or not a $TiO_2$ grade has ultimate durability.

The results of this example can be generalized as follows: Pigment photocatalytic activity has a greater impact on overall paint durability in those paints made with inert binders, and a lesser impact in those paints made with reactive binders.

There are many consequences of this generalization, most of which are to the detriment of the paint producer. For example, the durability of a paint made with a cheap reactive binder cannot usually be increased by increasing the durability of the $TiO_2$ pigment used in the paint. Such a paint is expected to fail rapidly, regardless of the durability of the pigment used, because direct degradation reactions will quickly degrade the binder. In such a binder, a pigment with what is normally considered medium durability will, in fact, have ultimate pigment durability. Increasing pigment durability beyond this modest level will not improve paint weatherability.

A second consequence of the above generalization is that only the most inert binders will reap the full benefits of very high durability $TiO_2$ pigments. Stated another way, only those paints formulated with high quality, inert binders will reap the full benefits of using a "super-durable" grade of $TiO_2$ pigment. A related consequence to this is that the measurement of $TiO_2$ pigments with low photocatalytic activities can only be made in highly inert binders—in more reactive binders the total degradation rates for paints made with medium and high durability $TiO_2$ pigments will be essentially the same. This increases the time required to accurately measure the durability of high durability $TiO_2$ pigments, since a paint made from an inert binder and high durability pigment will degrade very slowly on exposure.

## Manifestations of Degradation

Degradation reveals itself through a number of appearance and mechanical changes to the paint film or plastic object as it degrades. The most obvious mechanical changes are that a film can separate from its substrate through peeling, that the film can become brittle and crack, and that the film no longer protects the substrate to the same extent, if at all, as the fresh film did [14–18].

While mechanical changes can be quite relevant to consumer satisfaction with a paint or plastic film, we find that appearance changes tend to be more important. This is because appearance changes usually show themselves earlier in the degradation process than do mechanical changes, and so by the time the film tensile properties weaken, cracks or peels, it has already been deemed a failure based on appearance. In addition, understanding the progression of mechanical changes over the life of the film is quite straight-forward—they are generally a simple function of the loss of resin or time exposed. However, appearance changes can be quite complex and non-linear with exposure time. Finally, for the purposes of this book, mechanical changes in paint films tend to be less dependent on the particles in the film than in the identity and physical properties of the resin. For these reasons, we will focus attention here on the appearance changes that occur during exposure to the elements.

### *Changes at the Film Surface—Erosion and Contraction*

Before discussing the changes in appearance that take place as a paint degrades, we will consider the physical changes that occur at or near the film surface and that cause these appearance changes. Because UV light is attenuated as it travels through the film, its highest intensity is at or near the film surface, and so most degradation takes place there. This is rather unfortunate because it is the surface that determines appearance, and so resin degradation and loss at the surface is the most deleterious to appearance. The loss of surface resin is referred to as erosion [19, 20].

In addition to erosion, the film surface is roughened by degradation that occurs beneath, but not far from, the surface by the process of resin contraction. The degraded resin near the surface can escape the film, either as a gas ($CO_2$ or a volatile organic) or through washing or rain events (that is, as a water-soluble species). The departing resin leaves behind a void.

This loss in resin beneath the surface affects the top layer of the film in two ways. First, as the overlying resin creeps into the void, it leaves a depression on the paint surface (Fig. 14.10). This depression roughens the surface. Second, because

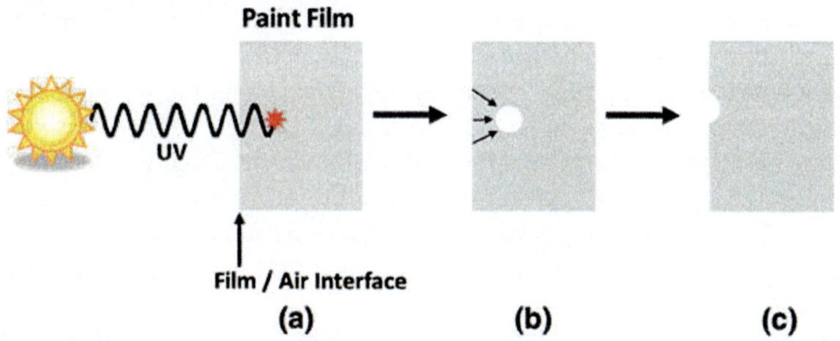


**Fig. 14.10** Film contraction and surface roughening in an unpigmented resin following an internal degradation event. **a** Destructive absorption of UV photon. **b** Void left after degradation products are released from the film. **c** Subsequent relaxation of resin into the void, leading to a depression feature on the film surface [23]. *Source* May, 2009 issue of CoatingsTech magazine, © American Coatings Association (2009). Used with permission

the particles near the surface are not lost (at least not before chalking begins, as detailed below), the particle volume concentration (PVC) near the surface increases (Fig. 14.11). Eventually, the PVC at the surface exceeds the CPVC, and air voids appear within the paint. This often coincides with the appearance of loose particles on the film surface (chalking) and with an abrupt gloss failure [21], since the air voids present in films above the CPVC cause significant surface roughening. These two processes are referred to collectively as contraction [22].

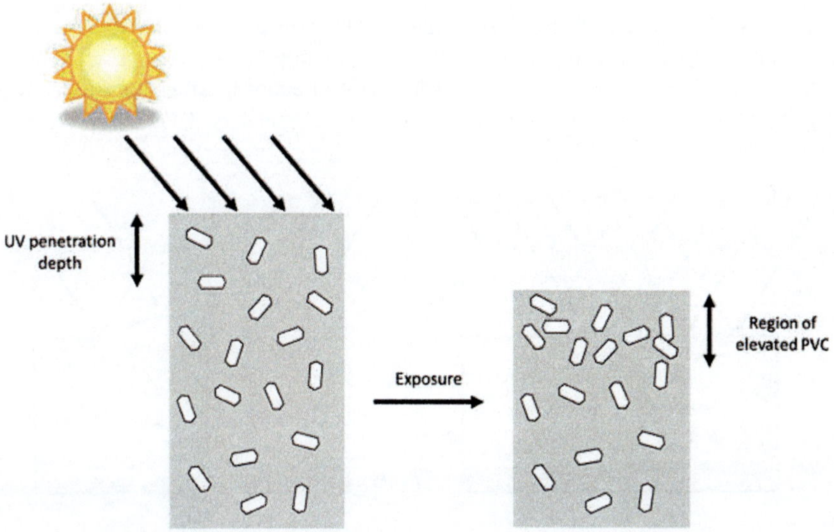


**Fig. 14.11** Formation of high surface particle concentration (PVC) as surface resin degrades and the film contracts [23]. *Source* May, 2009 issue of CoatingsTech magazine, © American Coatings Association (2009). Used with permission

There are a number of consequences of contraction. The increase in local PVC both distorts the film surface, and it increases the local refractive index at the surface. These two phenomena have opposing effects on the initial changes in gloss on exposure. Surface distortion decreases gloss, but increasing the refractive index at the surface leads to a higher proportion of the light striking the surface being reflected from it, increasing gloss. The two opposing effects occur concurrently and can lead to an initial "confused" period during which gloss can show an initial induction period of no change or, in some cases, may actually increase by a modest amount during this time.

In general, for paints with good particle dispersion, the benefit of increased local refractive index outweighs the detrimental effect of increased surface roughness, and gloss initially increases on exposure. However, for poorly dispersed paints, gloss typically decreases during initial exposure. In addition, the concentration of $TiO_2$ particles at the surface increases the UV protection that these particles provide to the paint. This lowers the rate of resin degradation as measured by weight loss rates and so can also increase gloss retention during this period of exposure.

Because resin loss occurs over a depth, rather than right at the surface, several microns of film thickness can be lost before chalking occurs—even though there is generally significantly less than one micron thickness of resin above the topmost pigment particles.

## Gloss Loss

Film gloss is a measure of surface smoothness at the optical scale—that is, a measure of the lack of surface perturbations with sizes roughly the wavelength of light or above. Such perturbations reflect light in many directions, rather than reflecting the light at the mirror image angle (Fig. 14.12). Pigmentary $TiO_2$ contributes to

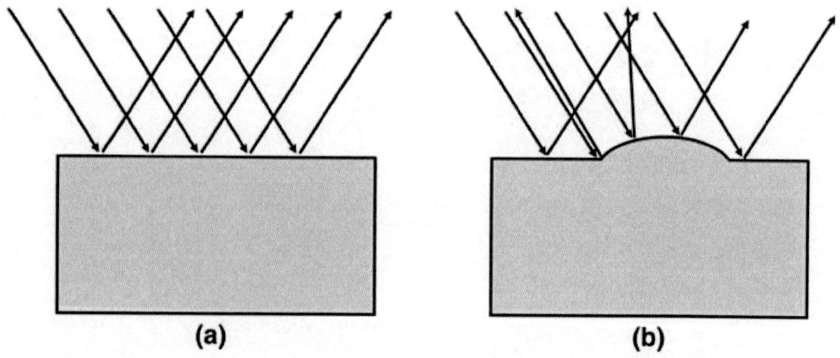


**Fig. 14.12** Surface reflection for **a** smooth and **b** rough surfaces

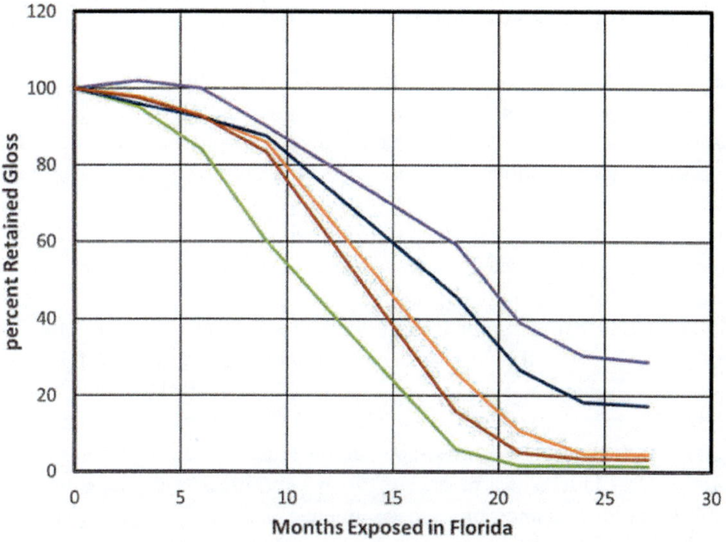


**Fig. 14.13** Typical gloss retention curves. Paints are identical except for the grade of $TiO_2$

surface perturbations of this size even though individual particles are smaller than this because incompletely dispersed agglomerates can be large enough to affect gloss.

Many exterior applications, such as automotive and architectural trim coatings, value gloss. When these coatings are exposed to the elements, their surfaces invariably roughen once past the induction period, causing a loss in gloss. Because this roughening occurs at or near the surface of the film, where degradation is the greatest, gloss loss is often the first appearance change that shows itself as the film is exposed. Even in applications where gloss is of secondary importance, gloss loss can be a leading indicator of other appearance or mechanical changes.

Gloss retention is typically measured by simply exposing painted metal panels to the elements (or in a weathering chamber) and measuring gloss periodically. In general, gloss initially falls linearly with time after the induction period (if any). At some point in the exposure the changes in gloss decrease, and eventually gloss will plateau and remain level. Gloss retention is normally reported as the time to reach a predetermined value (time to failure) or the gloss retained after a set period of time (typically from 1 to 10 years). Some typical gloss retention curves are shown in Fig. 14.13.

## *Chalking*

When a paint or plastic is exposed to the elements, the polymer in the topmost region will oxidize first. As the resin degrades and is removed from the surface, any extender or $TiO_2$ particles that had been in that portion of polymer will remain[4]—these particles are already fully oxidized, and so will not degrade on exposure. The exposed particles appear as a fine white, powdery coating. Brushing one's hand over the surface will transfer the particles, as they are only loosely held (rather than embedded) on the surface. We refer to this powdery white material as "chalk", due to its appearance, even though in most cases it is only partially, if at all, calcium carbonate.

Since any exposed extender is surrounded by air rather than polymer, it will scatter visible light and so add to the white appearance of the powder. In colored paints and plastics, in addition to giving the surface a white powdery appearance, the extender and $TiO_2$ particles that accumulate on the surface will make the film or object appear lighter. This is because these particles scatter some of the incident light before it enters the film or object and has the opportunity to be absorbed by the color pigment. This lightening effect is referred to as fade.[5]

In most cases chalking is undesirable, and the appearance of chalk is considered the point of total failure of a paint. For this reason, most exterior paints are tested for chalk or fade resistance prior to commercialization. There are two strategies commonly used to quantify chalking. The first is to color the paint in question and then follow the brightness of the paint over time [24]. As long as the color pigment is itself durable, this brightening will be due exclusively to the presence of chalk. The second strategy is typically used with white paints and involves application of sticky tape to the surface of the exposed paint using a controlled and reproducible pressure [25]. The tape is then lifted from the film and the amount of chalk or other debris adhering to it is quantified by measuring the fraction of light that is transmitted through the tape.

Film brightness over time typically follows a certain pattern (Fig. 14.14). In this figure, we quantify film brightness by the percentage of red light reflected from a dark blue paint film. Initially, brightness does not change. During this induction period, although the paint is degrading, it has not yet done so to the point that extender or pigment particles are exposed. This initial induction period is followed by a period of roughly linear increase over time, as more and more particles become exposed. Finally, the brightness will often reach an inflection point or even a plateau. This is referred to as "free chalking". During this period, additional particles become exposed, but the number of particles on the surface remains nearly constant (that is, a steady state is achieved). This is because once the particles become totally detached from the film surface, they can be removed by rain or wind.

The change in film lightness can be quantified in a variety of ways. Overall brightness can be measured as L* and followed over time. Alternatively, when a colored paint is involved, the reflectance of the paint at a wavelength of strong absorption can be used [24] For example, red reflectance can be monitored on blue panels as shown

---

[4] Depending on the identity of any color pigments in the film, these particles may also accumulate loosely on the surface.

[5] Fade is also seen when the color pigment degrades and no longer absorbs light strongly (see Chap. 8).

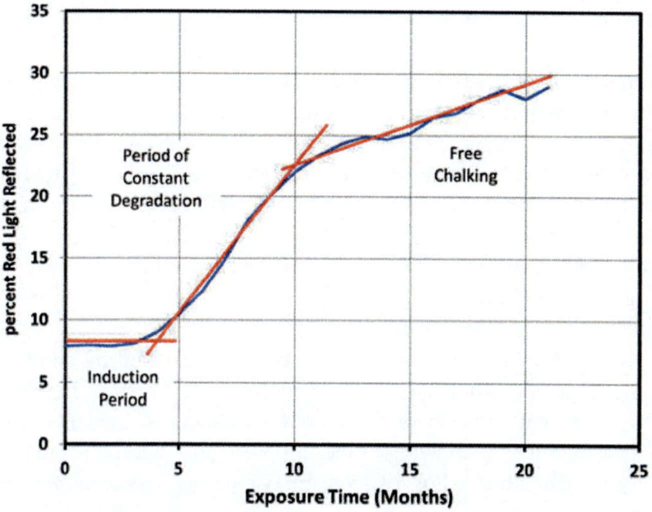


**Fig. 14.14** Three stages of color change during Florida exposure

in Fig. 14.14. Finally, color change can be measured as a ΔE value (refer back to Chap. 6). In most cases, testing is continued until the appearance change has met a certain predetermined limit—for example, an L* increase of 10 units, or a specific ΔE value—at which time the paint is considered to have completely failed.

Although steps are normally taken to minimize chalking, there is one situation for which it is desired, and this is for certain white paints that are marketed as being "self-cleaning", such as the exterior house paints so ubiquitous in Greece. One form of self-cleaning is achieved by controlled chalking. As the resin at the surface degrades, any adhered dirt will be removed when it is washed or rained on. Since the paint is white, color fade due to accumulation of chalk on the surface is not an issue. However, since the white particles travel with the wash or rainwater, they can be transported to locations where their presence is objectionable (that is, where they appear as a white stain). This strategy of self-cleaning was used in the past by the application of several layers of whitewash paint, which would flake off as the surface became dirty.

## *Color Shift*

So far we have considered the degradation of organic polymer on exposure. However, other organic materials in the paint film or plastic object are also susceptible to oxidative degradation. This is inconsequential for most non-resin organic paint components as they are present at low levels and are generally used to modify the properties of the liquid paint, rather than those of the dry film. One exception to this is organic pigments.

The strength of light absorption for organic pigments is strongly determined by the exact chemical structure of the pigment molecules (Chap. 8). Unlike resin molecules, which must undergo numerous reactions before they can cause a change in paint or plastics appearance, the creation or destruction of a single chemical bond in an organic color pigment molecule can have a dramatic effect in its absorption spectrum. This, in turn, can both shift the color of the pigment and significantly decrease its color intensity. The color shift is distinctly different from the color fade that occurs on chalking, and the durability of color pigments in this context is referred to as "lightfastness", to differentiate it from film durability.

Color shift is especially an issue when a color is achieved by combining more than one color pigment—for example, if a paint has both yellow and green pigments, and the green pigment degrades over time while the yellow pigment does not, then there will be a significant yellowing of the sample. This situation is seen in Chap. 8, Fig. 8.25, reproduced here as Fig. 14.15. Here two paints are shown before and after Florida exposure. The color pigment in the purple paint is stable, and there is little difference in hue between the before and after images. However, the color pigment in the green paint is not stable, and this panel shows an obvious shift in hue.

Because of the decrease in color intensity on exposure, color pigment degradation can often be misattributed to chalk-induced fade, and vice-versa. While the appearance effects for the two types of degradation can be similar, it is important to properly ascribe color loss on exposure according to its source, as this will dictate how it can be mitigated.

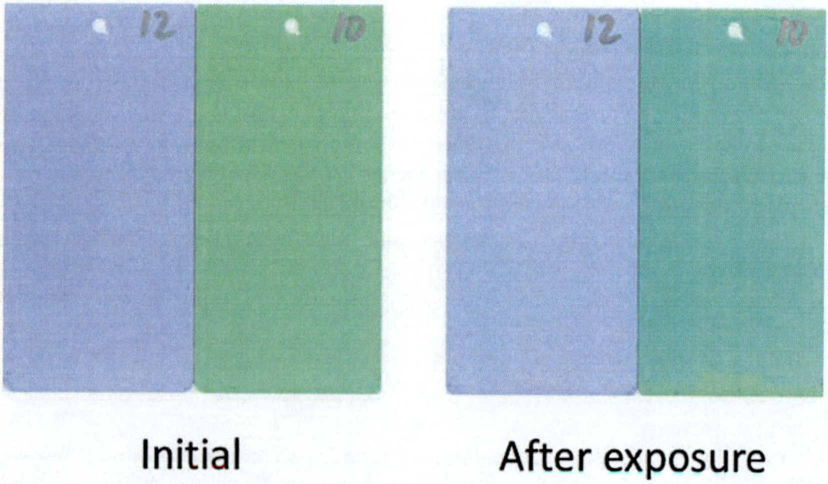


**Fig. 14.15** Color shift on exposure. Color pigments in the blue paint are stable while the yellow pigment in the green paint is not

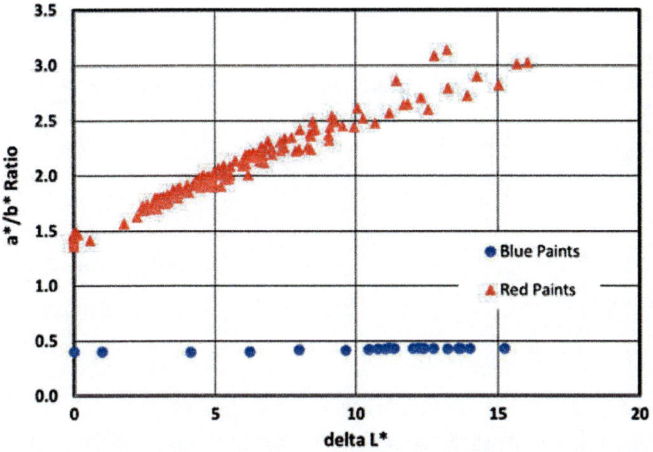


**Fig. 14.16** Color shift on exposure. The pigment in the red paint is unstable while that in the blue paint is stable

Perhaps the most straightforward means of accomplishing this is to monitor a* and b* in addition to L* (or ΔE). The shift in absorption spectra that accompanies color pigment degradation will typically affect the balance between a* and b*. For example, a shift towards red will increase a* without affecting b*. This type of analysis is shown in Fig. 14.16. In this figure, we show the shift in the ratio of a* to b* as a function of L* for blue and red paints during exposure in Florida. This ratio is stable for the blue paint but shifts significantly for the red one.[6]

In addition to color shifts caused by degradation of pigments, paints will also yellow over time. This is generally due to the resin degradation intermediates having a slight light absorption in the blue and violet region of the visible spectrum. While this shift can occur in all paints, it is most obvious in whites, since the shift away from a neutral tone to a yellow tone can be quite apparent. As is the case for the degradation of color pigments, the amount of discoloration depends on the organic identity of the material (resin, in this case).

## *Plastics Yellowing*

Changes in color due to photo-induced oxidation can also involve formation of chromophores within the polymer matrix. A good example of this is in the color shifts that occur in plastic matrices using phenolic-based chemicals—yellowing due to additive

[6] The time element of these tests is not shown directly in Fig. 14.16. However, the delta L* shifts increase with time, and so when moving from left to right in this graph we are moving from short exposure times to longer ones.

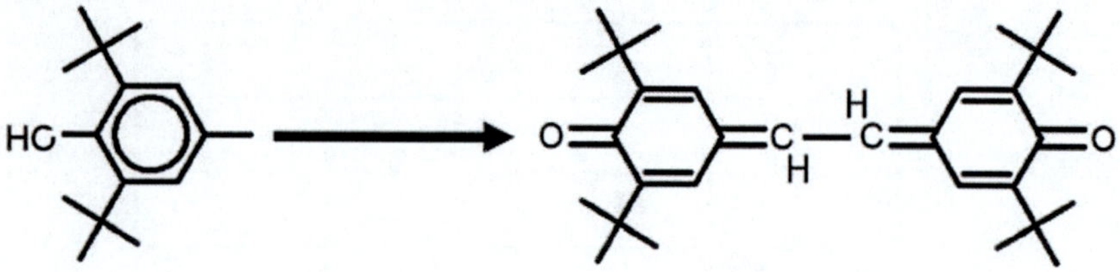


**Fig. 14.17** Color formation in plastics

degradation [26]. Certain plastics can contain organic molecules that serve as antioxidants. An important one of these is BHT (butylated hydroxytoluene). While colorless, when BHT is attacked by chemical radicals it can dimerize into stilbenequinone, a canary yellow compound (Fig. 14.17). Other antioxidants that are based on similar chemistry can also form colored quinones analogous to stilbenequinone [27].

The mechanism for phenolic-based yellowing requires two components, an oxidizing reagent and a base to deprotonate the phenol functionality. The role of $TiO_2$ in this mechanism is as a catalyst to provide a potent photo-generated oxidizing reagent, peroxyl and hydroxyl radicals. $TiO_2$ can also be a source of buffer, modifying the $H^+/OH^-$ ratio to match the $pK_a$ value of the phenolic additive. The role of $TiO_2$ as a photocatalysis is the more important, however. For example, $TiO_2$ catalyzes color formation under low level UV source in the presence of gaseous $NO_x$ compounds, which are by-products of natural gas combustion. When the UV source is removed, formation of yellow chromophores occurs due to reaction with $NO_x$ at reduced rates.

However, $TiO_2$ pigment, which is found in many of these articles, can also contribute to yellowing. The magnitude of this contribution is very dependent on the particular grade of $TiO_2$ used, as is true of the contribution of $TiO_2$ to the durability of paints. The $TiO_2$ pigments that are used in most of these applications are treated to minimize appearance change. These treatments, and other determinants of the effect of $TiO_2$ on appearance changes during weathering, are considered in a later section of this chapter.

## Photodegradation of Polyvinyl Chloride

The sensitivity of polyvinyl chloride (PVC)[7] towards photo-degradation is primarily due to the presence of defects within the polymer backbone. Bulk PVC rarely consists

---

[7] Care must be taken in this section and the next to not confuse the acronym PVC for being "pigment volume concentration", which is how this acronym is used most commonly in this book.

of a polymer with only $-(CH_2\text{-}CHCl)_n-$ repeating monomers [28, 29]. The extrusion and molding operations subject the defects within the molecular structure to breakdown from mechanical stresses. The term "mechano-degradation" is often used to describe the cascade of events that ultimately result in scission of the PVC polymer and produce macro-radicals. These macro-radicals are seeded with polymer empirical defects, fertilized by mechanical stresses, and grow as initiators for thermal oxidation.

A popular mechanism for mechano-degradation has been proposed [30] that highlights the role of the polymer defects serving as moieties with high absorption coefficients for UV wavelengths under 270 nm [31, 32]. The photo-oxidation mechanism most commonly accepted was proposed by Gardette and Lemiare [33, 34]. The photo-oxidation of PVC defects is initiated by Cl atoms and catalyzed by HCl. The first step involves absorption of light by the chromophore defect in the PVC polymer backbone. The alpha-chlorinated diene defect, if not oxidized immediately (i.e., in the absence of oxygen), propagates down the PVC polymer backbone to form HCl and a conjugated backbone. With the formation of additional unsaturation, the adsorption shifts to longer wavelengths and moves into the visible range [35]. As a result, the PVC plastic changes color, starting from yellow and moving to dark brown as the conjugation length grows with continued UV exposure.

In the presence of oxygen, photo-bleaching reactions occur to form alpha chloroketones, acid chlorides and beta chlorinated ketones. The products of the photo-bleaching reactions are signals for deterioration of the PVC mechanical properties such as impact strength or tensile strength. Additionally, gloss loss begins as the surface of the PVC article roughens during this photo-oxidation degradation (similar to gloss loss in paints). This gloss loss is typically termed as "chalking" within the PVC industry, a term often used to describe degradation of the paint matrix in coatings applications (discussed above).

Depth profile studies have shown photo-oxidation degradation products at depths as high as 100 microns [36, 37], depending on the presence of other reagents, such as plasticizer, in the PVC matrix. Oxygen solubility in the polymer is believed to restrict the depth of photo-bleaching reactions to the top 100 microns of the sample. By contrast, the formation of a hyper-conjugated backbone, which occurs in the absence of oxygen, is seen at even greater depths (up to 400 microns).

Certain solid particles in the PVC matrix serve as catalysts for the initiation or regulation of hyperconjugation and photobleaching. The solid particles that are the most likely to do this are those that are photoactive and that produce an oxidative environment on UV exposure. In most plastics applications, these would be titanium dioxide particles. Several reviews have described $TiO_2$ as playing a dual role in PVC photo-degradation. First, these particles, in the presence of light of wavelength less than 380 nm and in the absence of oxygen, accelerate hyperconjugation because they accelerate radical formation. In the presence of oxygen, the titanium dioxide particles accelerate the formation of radicals that oxidize hyperconjugation, which deteriorates the mechanical strength of the PVC. Hence, $TiO_2$ both promotes polymer backbone hyperconjugation and accelerates its destruction in the presence of oxygen.

As with coatings application, proper surface treatment on the titanium dioxide particle hinders its role as a photo-catalysis and decreases the rate at which conjugation formation/bleaching occurs.

As described earlier, $TiO_2$ has two opposing effects on organic degradation. It promotes degradation due to the photo-initiated formation of chemical radicals, but it also lessens degradation due to its high UV light attenuation. Care must be taken when balancing these effects in PVC applications. We can better understand the durability benefit of $TiO_2$ pigments by examining the depths at which the PVC radical chemistry occurs in the presence and absence of $TiO_2$. Typically, hyperconjugation and photo-bleaching reactions are limited to a much shallower region near the surface of plastics containing $TiO_2$ than of those without $TiO_2$ due to the screening effect of the $TiO_2$ particles [38, 39]. The depth of the polyene maxima concentration is from 2 to 5 times greater in the absence of $TiO_2$ that in its presence, depending on the concentration of the $TiO_2$ pigment (the lower the concentration, the greater the depth). Hence, the particle is very beneficial to durability even if it is photoactive because it severely limits penetration of UV light into the polymer.

Limiting the UV light penetration distance is especially beneficial to the retention of mechanical properties. On the other hand, optical properties such as gloss and color changes are more dependent on degradation at the PVC plastic surface. Therefore, gloss measurements for chalking can be used to establish a time scale for surface chemistry, but this may or may not align with the time scale for changes in mechanical strength.

Other solid particles can be used in PVC formulations to provide other durability functions. Certain particles in PVC applications can serve as an ion exchange agent for HCl during hyperconjugation formation. Alkaline earth carbonates are the obvious solid particles for the neutralization of HCl. These particles are present as "extenders" and play a minor role in retarding the formation and growth of unsaturation in the polymer [40]. Hydrotalcite particles are more efficient for the removal of HCl due to their higher surface area, which gives them a greater capacity for HCl neutralizing. These hydrated magnesium–aluminum hydroxycarbonate materials [41] exchange two chlorides for one carbonate. As such, these hydrated clay structures remove the Cl ions that "fertilize" the formation of PVC backbone hyper-conjugation.

## *Photodegradation in Polyolefin*

The influence of solid particles on non-halogenated polymers is somewhat less involved. Based on the discussion of PVC degradation, we may be concerned with radical formation due to "mechano-stresses", "polymer defects", and "macro radical formation". However, the degree to which unsaturated bonds are formed is less and, as such, degradation chemistry is less complicated. That said, photo-initiated degradation does occur, and this results in a decay of mechanical and optical properties. As was seen for PVC, certain solids, such as titanium dioxide and zinc oxide [42], act as

UV screens, preventing deep penetration of UV light and the attendant formation of radicals. In contrast to PVC, however, the photodegradation of polyolefins does not involve polymer hyperconjugation. Hence, solid particles such as ion exchangers do not affect durability.

The effect of photodegradation on the behavior of semi-crystalline polyolefins such as polypropylene has been the subject of active investigation by many scientists [43–45]. Their studies show that photo-oxidation is the principal cause of the aging effects due to direct exposure to solar radiation. Photodegradation governed by diffusion of $O_2$ in the polypropylene irreversibly modifies the structure and the behavior of this material as well as the polymer surface property that is termed "gloss" [46].

Oxidation in the macromolecular chain of polypropylene is initiated by decomposition with hydroperoxides. This reaction self-accelerates by following a closed loop. Photo-oxidation, through the formation of free radicals, causes molecular scissions that lead to changes in the physical and mechanical properties of the polypropylene. One such property is the gloss of the surface. Due to the oxygen diffusion-controlled nature of polypropylene photodegradation, a thick sample, after exposure to ultraviolet radiation, can be described as a ductile substrate (the undecomposed regions deep within the substrate) with a fragile surface layer (the decomposed regions near the surface).

As the photodegradation of the surface proceeds, the amorphous regions of the polymer oxidize, in addition to undergoing molecular scission. As scission proceeds, the surface undergoes a chemi-crystallization and the surface layer contracts (chemi-crystallization refers to crystal growth via molecule segments that are released by the scission of molecules). The most important practical consequence of this is the formation of cracks at the polymer surface [47–49]. The presence of surface cracks is indeed one of the main reasons for the embrittlement of ductile semi-crystalline polymers such as polypropylene. This embrittlement causes significant deterioration of the mechanical properties of the polymer after short-term exposure.

Ageing of thick polypropylene samples by photo-oxidation can be divided into three well-defined periods:

1. *Incubation/induction period*: During this period, the effects of photodegradation of polypropylene are still invisible. Cracks are absent from the surface and the gloss is unchanged.
2. *Period of surface pit formation*: This is characterized by the formation of a net of spontaneous pits on the sample surfaces. These pits cause surface gloss loss (similar to gloss loss in coatings).
3. *Period of chemi-crystallization on the surfaces*: At this stage of ageing, the effect of molecular chain scissions in the amorphous phase at the sample surface (chemical degradation) becomes very important, activating the process of rearrangement of the cut molecular chains into a crystalline phase. During this period, cracks at the sample surface create the optimal conditions for the development of chemi-crystallization [50].

During chalking, the polypropylene molecules nearest the titanium dioxide particle surfaces undergo oxidation, leaving behind a pit around the particle. The pit causes a roughening of the rigid polypropylene surface that is manifest as gloss loss, similar to what is seen in coatings.

Photodegradation has two opposing effects on chemi-crystallization. The shorter chains produced by scission events can crystallize readily, but crosslinks and molecular defects (carbonyls, etc.) that form during oxidation are not able to crystallize and so are rejected from the newly formed crystals. The chain scissions and surface shrinkage that occur due to polypropylene photodegradation lead to the development of tensile residual stresses near the surface [51]. The formation of a damaged surface layer (or "skin") is accompanied by "chalking".

## *Paper Laminate Photo-Graying*

Although paper laminates do not generally experience resin degradation in the same way that coatings and plastics do, there is still an appearance change that can occur on exposure to UV light. This appearance change is called photo-graying, and the rate of graying is dependent on the grade of $TiO_2$ used. Although this change is caused by UV light striking the $TiO_2$ particles, its underlying mechanism is different from that of coatings or plastics.

Photo-graying occurs when the photocatalytic cycle depicted in Fig. 14.7 is interrupted. In our discussion of this figure, we noted that the order of radical formation (superoxide radical first or hydroxyl radical first) is unimportant because, once one radical forms, the other radical must form, since high energy electrons and electron holes are produced in the $TiO_2$ lattice in pairs. In the case of paints or plastics, the immediate environment around the $TiO_2$ particles has enough oxygen and water vapor that the particle surfaces are always saturated, and so the radical formation reactions are nearly instantaneous.

The environment of the particles in a paper laminate is much different. Here, the surfaces of the $TiO_2$ particles are still saturated with water, since the surface oxides are hydrous. However, the thick melamine layer on top of the paper is impermeable to oxygen. This results in the pigment surface being oxygen starved. In this situation, the hydroxyl radical can readily form, injecting an electron into the $TiO_2$ lattice, but not the superoxide radical. As a result, the $TiO_2$ particles will accumulate an excess of electrons (that is, they become partially reduced).

The excess electrons occupy the conduction band, which is composed primarily of titanium 3d atomic orbitals (refer back to Chap. 8 for a description of atomic orbitals).[8] Being in a partially filled band, these electrons can readily absorb visible light. Visible light absorption is so strong that it causes these particles to turn black—even when the excess is only a few electrons per particle. As a result, the paper laminates gray upon exposure to UV light. An example of this is shown in Fig. 14.18.

[8] This reaction is often described as the formation of a reduced titanium atom in the + 3 oxidation state ($Ti^{3+}$). However, these excess electrons are actually delocalized over the entire $TiO_2$ lattice.

**Fig. 14.18** Table with a paper laminate surface after 1 day exposure to sunlight. Here the tabletop was partially covered, resulting in light regions (original color) surrounded by dark regions (photo-grayed)

This picture shows the photo-graying of a table that was exposed outdoors on a bright day. The table was partially covered, as can be seen by the lighter areas.

Unlike resin degradation in coatings and plastics, this effect is not permanent. Given enough time, oxygen can diffuse through the melamine layer and oxidize ("bleach") the affected pigment particles. In the case of the table shown in Fig. 14.18, the table was returned to its original appearance after approximately one month.

Because the mechanism involved in photo-graying is different than that involved in resin degradation, $TiO_2$ grades that have been developed for high durability coatings and plastics applications do not necessarily perform well in paper laminates. This can be seen in Fig. 14.19. Here, laminate coupons that were made with a high durability coatings grade $TiO_2$ and a grade developed specifically for laminate applications (this grade facilitates recombination of surface radicals) were exposed to high-intensity light. The coatings grade pigment showed unacceptable graying while the laminate grade pigment remained unchanged.

## *Paper Laminate Durability in Compact Board Applications*

Paper laminates used in outdoor applications experience the effects of weather as well as ultraviolet light exposure, and this can degrade appearance over time. Obviously, $TiO_2$ light stability and the durability of any color pigments used in the inks applied to the décor paper component of the laminate play a dominate role in color stability.

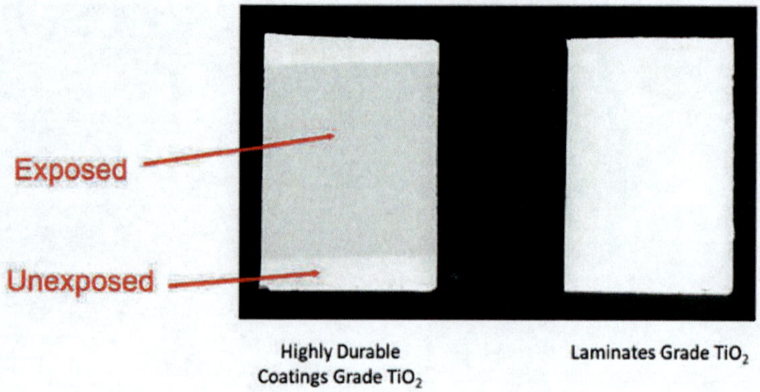


**Fig. 14.19** Laminate coupons made with different grades of $TiO_2$ after exposure to high-intensity UV light

However, gloss changes can also occur in these materials. These gloss changes are manifest as a haze appearance that grows over time.

The appearance of haze has been attributed to $TiO_2$ photocatalytic degradation. However, the fact that the $TiO_2$ grades used in these applications do not discolor over time, and that radicals must travel a significant distance from the $TiO_2$ particle surface to the laminate surface, both indicate that this cannot be the mechanism responsible for haze. Instead haze is the result of micro-crack formation in the cured melamine resin. The melamine resin used in this application is thermosetting and so is inflexible. When stress is applied, as it is when the surface resin reacts with ambient water, the surface cracks. This issue can be avoided by applying a topcoat of resin cured with an electron beam.

## Factors that Determine the Effect of $TiO_2$ Pigment on Polymer Durability

Many $TiO_2$ grades are available to the coatings or plastics manufacturer, with each grade reflecting a certain balance in performance properties. Included in these grade-dependent properties is the effect of $TiO_2$ on paint or plastic durability. As discussed earlier, the $TiO_2$ pigment can both enhance paint or plastic durability (by absorbing UV light) or can harm it (by producing destructive chemical radicals).

The importance of the effect of $TiO_2$ pigment on polymer durability is such that different $TiO_2$ grades are specifically treated for exterior applications to reduce the harmful effects while maintaining the beneficial effects.

The effect of $TiO_2$ grade on appearance change can be seen in Fig. 14.20. Here, we show five paints that have been exposed for 1 year in Florida. The only difference between the paints is the grade of $TiO_2$ used. The photocatalytic activities of the

**Fig. 14.20** Color fade of panels exposed for 12 months in Florida. Paints made with different grades of $TiO_2$ pigment. The topmost portions of each panel were protected from UV light and show the initial color of the panels

paints decrease from left to right in this figure. The $TiO_2$ used in the paint at the far left in this figure is not recommended for exterior applications, while the pigment used in the far right paint is considered a "superdurable" grade.

## *Encapsulating Surface Coating*

$TiO_2$ affects the degradation of paints and plastics through the photocatalytic oxidation of organic molecules near the particle surface. As discussed above, the photocatalytic degradation of polymer by $TiO_2$ proceeds through a sequence of steps (Fig. 14.4). If we slow or prevent any one of the steps in this sequence, we will slow or prevent the entire degradation process. One such step is the exchange of electrons between the $TiO_2$ lattice with the oxygen and water molecules that are absorbed on the particle surface.

One strategy that the $TiO_2$ manufacturer can take to slow this electron exchange step is to completely encapsulate the $TiO_2$ particle with a layer of an electrically insulating material (refer to Chap. 7). Recall that an electrical insulator is characterized as having a large band gap—larger than that of $TiO_2$, which is a semiconductor. For most electrical insulators, the energy of the conduction band is greater than that of $TiO_2$ whereas the energy of the valence band is lower (Fig. 14.21).

As depicted in this figure, an electron that has been promoted by UV light absorption into the conduction band will generally not have enough energy to transfer into the insulator conduction band. In the same way, the electron hole in the $TiO_2$ valence band cannot enter the insulator valence band and travel to the particle surface because to do so an electron would need to move from the insulator valence band into the $TiO_2$ valence band. This requires additional energy. The encapsulating insulator coating, therefore, prevents the transfer of an electron from a surface water molecule into the

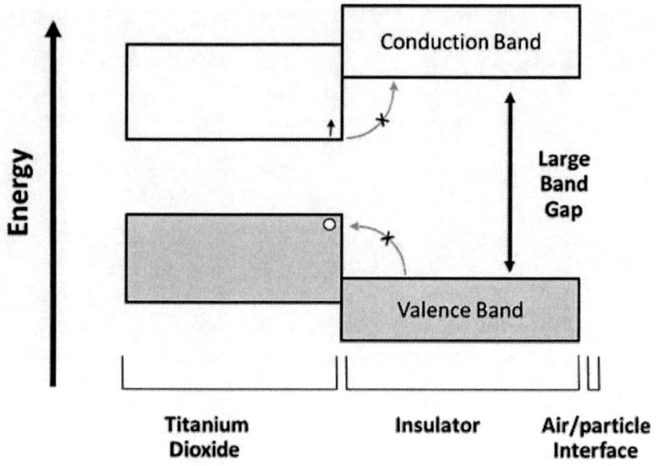


**Fig. 14.21** Electronic configuration of an insulator covering a $TiO_2$ particle

$TiO_2$ lattice and the transfer of an electron from the $TiO_2$ lattice to a surface oxygen molecule.

In theory, any electrical insulator can be used for this purpose, but there are many practical considerations that limit those that are used in practice. Specifically, the insulator must be insoluble in water and solvent, must be non-toxic, colorless, inexpensive and relatively easy to apply. Most commercial durable grades of $TiO_2$ use one or both of two materials: hydrous silica and hydrous alumina (Chap. 7). Both of these coatings are added to the pigment surface through a precipitation reaction. For silica, sodium silicate is added to a water slurry of the untreated $TiO_2$ at high pH conditions (where the silicate is soluble). The pH of the slurry is then decreased with an acid, leading to a precipitation of the silicate as hydrous silica (Eq. 14.1) [52].

$$Na_2SiO_3 + 2HCl + nH_2O \rightarrow SiO_2 \cdot (n + 1)H_2O + 2NaCl \tag{14.1}$$

Similarly, to deposit hydrous alumina, sodium aluminate is added to a water slurry of the $TiO_2$ particles and acid added to precipitate the alumina (Eq. 14.2). Alumina can also be deposited by adding an acidic source of alumina (such as aluminum sulfate) to the slurry and neutralizing with a base (Eq. 14.3) [53] Although both oxides can give a high durability $TiO_2$ pigment, only silica is able to give the highest level of durability ("super durability") [54].

$$2NaAlO_2 + 2HCl + nH_2O \rightarrow Al_2O_3 \cdot (n + 1)H_2O + 2NaCl \tag{14.2}$$

$$2Al_2(SO_4)_3 + 12NaOH + nH_2O \rightarrow 2Al_2O_3 \cdot (3 + \tfrac{1}{2}n)H_2O + 6Na_2SO_4 \tag{14.3}$$

Note that in all reactions the resulting oxide (silica or alumina) is hydrated by an undefined number of water molecules that varies with the specific deposition conditions. Because of this, the silica and alumina content of most $TiO_2$ pigments are typically reported as %$SiO_2$ and %$Al_2O_3$, based on the weight of the $TiO_2$—that is, as if present as the anhydrous oxide rather than a hydrous oxide.

For this strategy to work, the $TiO_2$ surface must be completely encapsulated by the insulating oxide. Even a small gap in coverage can lead to a significant amount of electron exchange and radical formation. A crystalline material—even if an insulating solid—cannot be used for this purpose because there will inevitably be gaps between the discrete crystals, no matter how well they are packed on the $TiO_2$ surface. These surface coatings must instead be amorphous. By definition, an amorphous solid has no preferred shape and so is free to follow the contours of the $TiO_2$ surface, ensuring no gaps in coverage.

Hydrous silica, such as that precipitated in Eq. 14.1, is naturally amorphous. However, the hydrous alumina deposited in Eqs. 14.2 and 14.3 is prone to crystallization. To prevent this, a hetero-species, such as $Zr^{4+}$, $Sn^{4+}$, $F^-$, $SO_4^{2-}$ or $PO_4^{3-}$, can be added in small amounts into the amorphous alumina structure [53]. These hetero-species prevent crystallization because they must be ejected from the lattice before alignment of the alumina octahedra into a regular arrangement can occur. Energy is required to break the chemical bonds that hold the hetero species in the lattice, and this energy represents a barrier to crystallization.

Although a single monolayer of encapsulating oxide may seem enough to completely prevent $TiO_2$ photocatalytic degradation (approximately 0.4% for $SiO_2$ on $TiO_2$), significantly higher levels are used in practice. This reflects the fact that the silica does not deposit evenly, but rather randomly on the pigment surface. In addition, due to quantum tunneling, a single layer of insulating oxide will not completely prevent all electrons from passing between the $TiO_2$ and the water or oxygen molecules adsorbed onto the pigment surface.

We may, therefore, expect that the durability of a paint will increase with increasing silica amount. This is true, but only up to a point. This is shown in Fig. 14.22, where a series of $TiO_2$ pigments were prepared with a wide range of silica contents and tested for chalk resistance (higher numbers signify greater resistance to chalk formation). In this paint, there is a maximum chalk resistance around 8% $SiO_2$, beyond which durability actually decreases. This is because, in this particular paint system, $TiO_2$ with 8% silica has ultimate durability. Going above this value decreases durability because there is a lower $TiO_2$ content in the pigment and so less beneficial shielding of the resin from UV light. For this reason, the highest level of silica typically found on $TiO_2$ pigments that are used for durable paints or plastics is typically between 7 and 8%.

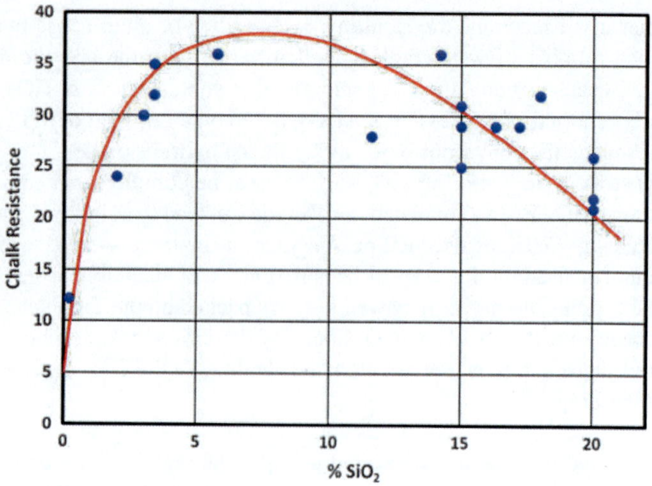


**Fig. 14.22** Effect of silica level on chalk resistance

## *The Effect of $TiO_2$ Dispersion on Chalking*

While perhaps not intuitive, the degree of particle dispersion can affect paint durability in some cases [55]. Gloss retention and chalking are, in fact, both hurt by poor pigment dispersion, but for different reasons.

We will first consider the effects on chalking (these effects carry over to other durability properties that are controlled by the total rate of resin loss). We will begin by considering two paints made with the same $TiO_2$ pigment, but with different degrees of dispersion. In the first paint (Fig. 14.23a), the $TiO_2$ is optimally dispersed. In the second paint (Fig. 14.23b) the $TiO_2$ is under-dispersed and is found in agglomerates.

Consider the dark gray particle in both films. When UV light strikes this particle, it is absorbed, shielding the region beneath. When the $TiO_2$ is well dispersed (Fig. 14.23a), the particle shields the resin, and so decreases the rate of direct degradation. When the $TiO_2$ is poorly dispersed (Fig. 14.23b), the particle shields other $TiO_2$ particles, and so decreases the rate of photocatalytic degradation. Therefore, the degree of $TiO_2$ dispersion affects the balance between direct and photocatalytic degradation. In most paints, direct degradation is more damaging than photocatalytic degradation, and so complete $TiO_2$ dispersion is desired.[9]

The effect of $TiO_2$ dispersion on the chalk resistance of 25 blue automotive topcoat paints that differ in degree of $TiO_2$ dispersion is shown in Fig. 14.24. Here, we use the initial film brightness of the paints as a proxy for the degree of $TiO_2$ dispersion (better dispersion gives greater light scattering, so the better dispersed paints have higher

[9] Of course, complete dispersion is desired for other reasons as well.

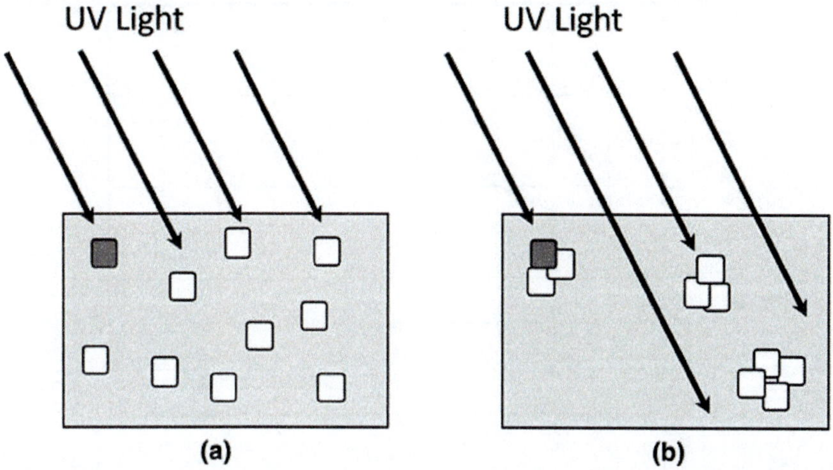


**Fig. 14.23** UV screening by a $TiO_2$ particle in a film. **a** Well dispersed pigment. **b** Poorly dispersed pigment

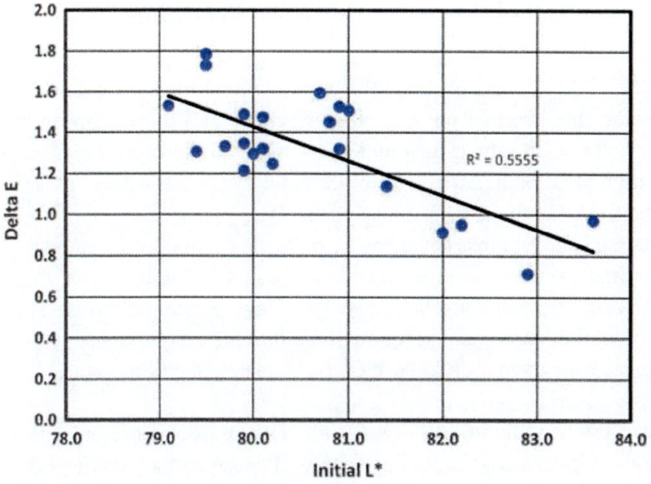


**Fig. 14.24** Effect of $TiO_2$ dispersion (measured as initial brightness) on color fade from exposure

brightness).[10] While there is some scatter in the chalking data, there is a clear trend towards higher chalk formation (measured as $\Delta E$) as degree of dispersion decreases.

[10] This is analogous to the tinting strength described in Chap. 13.

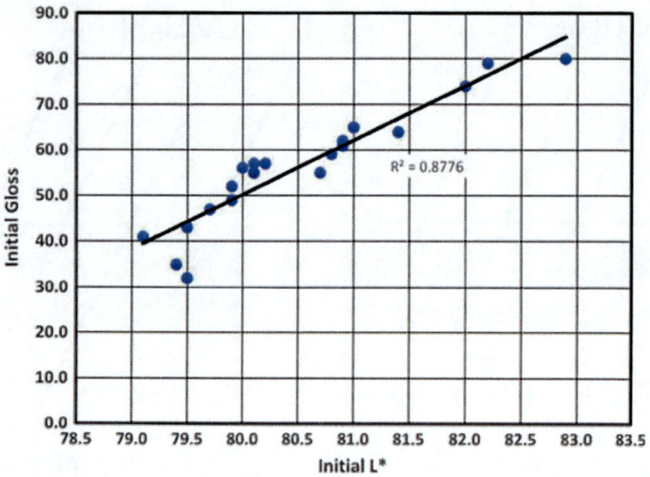


**Fig. 14.25** Correspondence between initial brightness and initial gloss for the paints shown in Fig. 14.24. Both are determined by the degree of $TiO_2$ dispersion

## *The Effect of $TiO_2$ Dispersion on Gloss Retention*

The degree of $TiO_2$ dispersion also affects gloss retention, but for a different reason. To understand this, we first consider the effect of $TiO_2$ dispersion on initial gloss. Different grades of $TiO_2$ have different gloss potentials—some grades can be formulated into high gloss paints, while for other grades only a mid-level gloss is possible. This can be traced to the dispersibility of the $TiO_2$ pigment.

As discussed above, gloss is a measure of surface smoothness at the optical scale. For visible light, surface features, such as dips, pits or protrusions, that are larger than roughly one micron cause a decrease in gloss. Since individual $TiO_2$ particles are slightly smaller than the surface features that affect gloss, any perturbations to the film surface due to an individual $TiO_2$ particle near the film surface will not affect gloss.

However, this situation is different if the $TiO_2$ particles are present as agglomerates. Agglomerates near the film surface will cause surface features that are large enough to decrease gloss, and this is the reason that different grades of $TiO_2$ give different gloss properties to paints (that is, different pigment grades will disperse to a different degree). This suggests that for the paints shown in Fig. 14.24, there should be a correlation between initial brightness (L*) and initial gloss, since both are affected by $TiO_2$ dispersion. This is, in fact, the case (Fig. 14.25).

In a fresh film, the effect of agglomerates on surface roughness is somewhat mitigated by the surface tension of the film, which pulls the agglomerates near the surface down, in a similar way that a wrinkled sheet can be partially leveled by pulling on the edges. Agglomerates in the unexposed film, therefore, cause a relatively minor surface feature compared to their overall size.

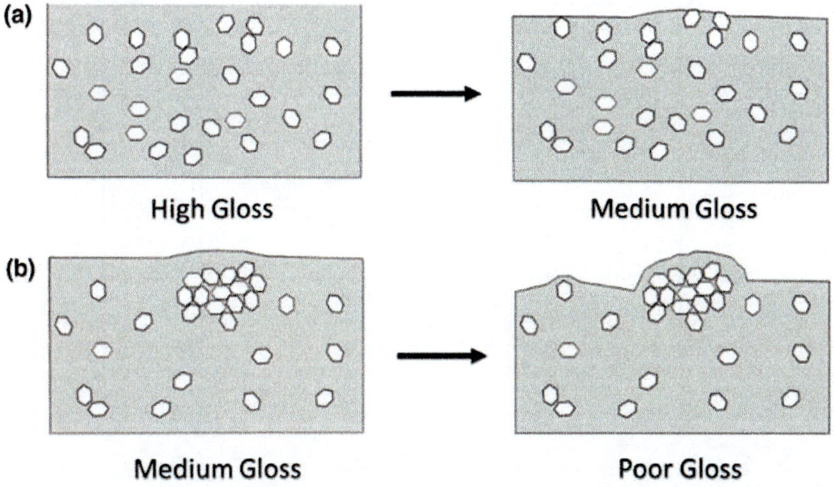


**Fig. 14.26** Effect of surface film loss on surface smoothness. **a** Film with an initially high gloss. **b** Film with an initially medium gloss. After exposure, the films have medium gloss and low gloss, respectively

This changes when the film is exposed and the surface resin erodes (Fig. 14.26). Because surface erosion is random, it creates a slightly uneven surface for the initially smooth film (Fig. 14.26a), leading to a minor to moderate change in gloss. However, in a film with agglomerates, the initially slight roughness caused by the subsurface agglomerates becomes quite pronounced after exposure (Fig. 14.26b) leading to a major change in gloss.

This suggests that the initial gloss of a paint film would be an indicator of gloss retention, all else being equal. This has been found to be true, to a greater or lesser extent, in many paint systems [23]. An excellent example of this is shown in Fig. 14.27, where the initial gloss and gloss retention values for a series of alkyd paints are shown to correlate highly with one another.

## *The Two-Component Approach to Gloss Retention*

The impact of initial gloss on gloss retention adds complexity to the role of the titanium dioxide pigment in gloss changes that occur on exposure. One way of interpreting this is that there are two components to gloss retention [23]. The first is that which is recognized widely, which is the rate of resin degradation and loss. The second can be described as the sensitivity of gloss to this resin loss. In one paint system, it could be that the loss of even a small amount of resin leads to a large decrease in gloss, while the gloss of another paint system could be insensitive to

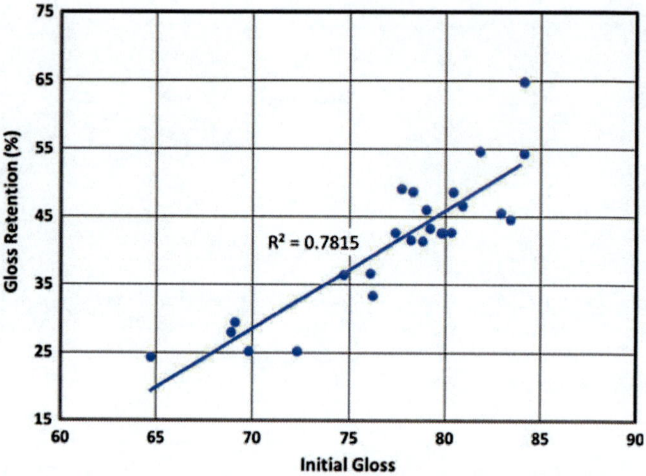


**Fig. 14.27** Correlation between initial gloss and 18-month gloss retention for a series of paints made with different grades of $TiO_2$ pigment [23]. *Source* May, 2009 issue of CoatingsTech magazine, © American Coatings Association (2009). Used with permission

resin loss, leading to high gloss retention despite resin loss. We will refer to this as the sensitivity of gloss to resin loss.

The contribution of the $TiO_2$ pigment to the first of these components—the rate of resin loss—is straightforward: high photocatalytic degradation rates lead to rapid resin loss. This component can be quantified by a chalking measurement, since chalking depends only on the total rate of resin degradation. A series of pigments can be tested for chalking in one paint system and the relative rates of photocatalytic degradation—which should be the same for all other paint systems—can be determined.

The sensitivity of gloss to resin loss will vary from one paint system to another. Based on the above discussion on the effect of $TiO_2$ dispersion on both initial gloss and gloss retention, we see that initial gloss can be a good proxy for gloss sensitivity—paints made with those pigments that give a high initial gloss will be the least sensitive to resin loss, since these paints have the least number of near-surface agglomerates.

To test the validity of this concept, paints for an aliphatic coil formula were made using 20 $TiO_2$ pigment grades that had a wide range of initial gloss and of chalking values, were exposed in Florida and the 18 month gloss retention values were determined. The predictive power of the initial gloss, of the $TiO_2$ chalking value, and of a two-variable regression analysis using both of these properties are demonstrated in Fig. 14.28. For this formula, the chalking value appears to be of moderate predictive power ($R^2 = 0.51$), but this correlation is strongly affected by a single data point that lies far from the region of the others (the $R^2$ value drops to 0.23 when this point is excluded). The initial gloss value is of only moderate predictive power ($R^2 = 0.42$). However, using both of these properties to predict

**Fig. 14.28** Correlation for a series of 20 $TiO_2$ pigments in a coil formulation between 12-month gloss retention and **a** initial gloss, **b** chalking value of the $TiO_2$ pigment, and **c** a predictor based on both the initial gloss and the chalking value [23]. *Source* May, 2009 issue of CoatingsTech magazine, © American Coatings Association (2009). Used with permission

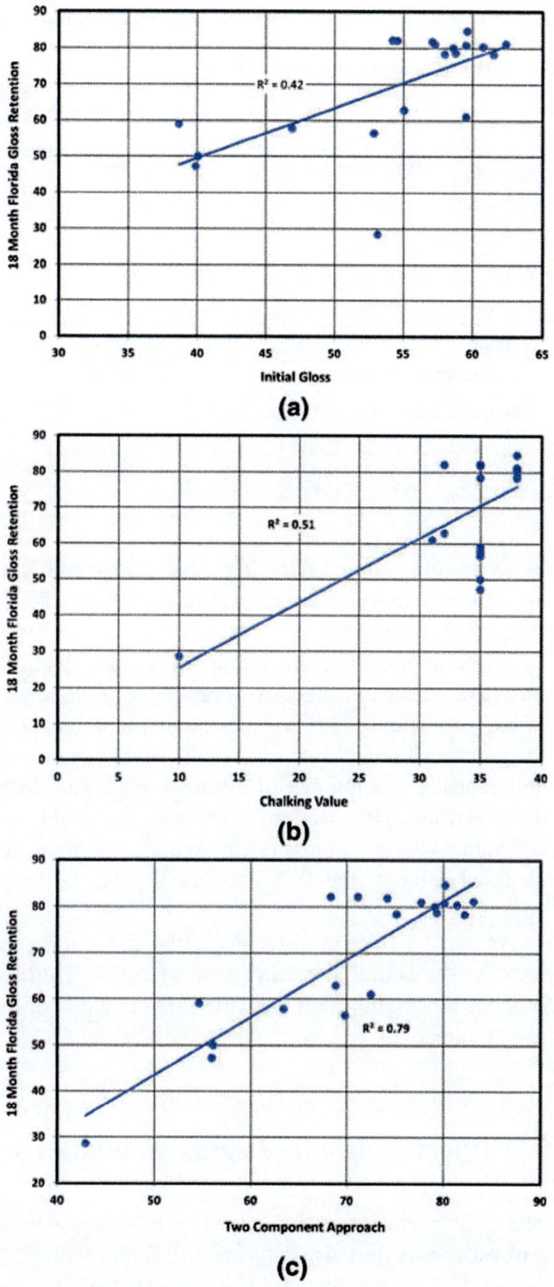

**Table 14.2** Correlation coefficients ($R^2$) between the gloss retention values, chalking values, initial gloss and a two-component analysis for 25 $TiO_2$ pigments in five paint systems Values in bold indicate a negative or counter correlation

| | Chalking value | Initial gloss | Two component |
|---|---|---|---|
| Medium oil alkyd—Xenon | 0.02 | 0.75 | 0.79 |
| Colored Coil—Florida | 0.05 | 0.02 | 0.06 |
| Colored Coil—Xenon | 0.05 | **0.52** | 0.90 |
| White Coil—Xenon | 0.40 | 0.90 | 0.91 |
| Aliphatic Coil—Florida | 0.51 | 0.42 | 0.79 |
| Aliphatic Coil—Xenon | 0.00 | 0.32 | 0.47 |
| Aliphatic Coil—UV chamber | **0.44** | 0.41 | 0.70 |
| Aliphatic Coil—Natural Accelerated | 0.00 | 0.87 | 0.88 |
| Automotive Topcoat—Florida | 0.23 | 0.36 | 0.62 |
| Automotive Topcoat—Natural Accelerated | 0.11 | 0.13 | 0.47 |

gloss retention increases the $R^2$ value to 0.79, which is probably close to the highest agreement achievable when the relative errors of the chalking and gloss retention values are considered.

While the two-component approach works well in this particular paint system, there are others for which it works no better than either the chalking value or the initial gloss alone. This is demonstrated in Table 14.2, which shows an analysis of the gloss retention results for 25 $TiO_2$ pigments incorporated into five paint systems and exposed in a number of different ways (the different exposure conditions are described in a later section). Entries in this table show the correlation coefficient ($R^2$ values) between gloss retention under the indicated exposure condition and the chalking value of the $TiO_2$, the initial gloss of the paint and the two-component approach.

As can be seen in Table 14.2, the two-component analysis can significantly increase the ability to predict gloss retention. However, this table also shows that it is not successful in all paint systems. In some of these systems, there is another as-yet unrecognized contributor to gloss retention.

## *The Effect of Relative Degradation Rates on Gloss Retention*

The chalk performance of a series of $TiO_2$ pigments will generally fall in the same rank order when tested in paints with different resins [56] This is not unexpected—the rate of chalk formation should be related to the total rate of resin degradation, and the relative $TiO_2$ photocatalytic contribution of the different pigments to this degradation should be the same in every resin system. However, gloss retention values for different $TiO_2$ pigments often do not rank the same in different resin systems. This is because

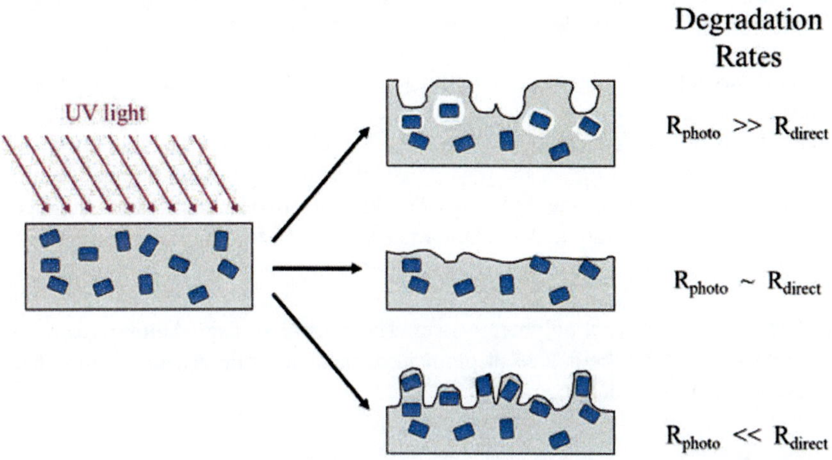


**Fig. 14.29** Consequences of different balances of direct and photocatalytic degradation rates

gloss retention does not depend on the total rate of degradation (that is, the sum of direct and photocatalytic degradation rates), but instead on the balance of the rates of these two types of polymer degradation [57].

To understand this, we will consider three limiting cases. In the first, the photocatalytic degradation rate is much greater than the direct degradation rate. In the second, the two rates are equal, and in the third, the photocatalytic degradation rate is much less than the direct degradation rate. These three scenarios are shown schematically in Fig. 14.29.

For this analysis, we will assume that after some period of exposure the three paints have lost the same amount of resin—that is, the total degradation rate for the three are the same; only the balance between the photocatalytic and direct degradations rates are different. Because the total degradation rates are the same, we would expect similar chalk performance for these three paints. However, the gloss retention performances of the paints will be different. When photocatalytic degradation is much greater than direct degradation, holes form around the reactive $TiO_2$ particles. Eventually, these holes reach the film surface, where they become pits. These pits significantly roughen the surface, leading to poor gloss retention.

A similar outcome is seen when the direct degradation rate is much greater than the photocatalytic degradation rate. In this case, the $TiO_2$ particles protect the underlying resin from degradation. Surface regions with no protective $TiO_2$ particles degrade more rapidly, with the result being that pedestals capped with $TiO_2$ particles form [58]. Again gloss retention is poor because these pedestals roughen the surface significantly.

The situation is different when the rates of the two types of degradation are similar. Here, the amount of degradation around and beneath the $TiO_2$ particles is roughly matched by the degradation of the resin that is not near the $TiO_2$ particles, resulting in a much smoother overall surface and a higher gloss retention than seen for either of the two more extreme cases. For high gloss retention, we therefore strive for a balance in degradation rates, rather than a lower total rate.

This partially explains why the rankings of gloss retention values for a series of $TiO_2$ pigments that cover a wide range of photocatalytic reactivities can be different in one resin system than another. When changing resin systems, we typically change the direct degradation rate of the resin,[11] and this changes the balance between the two degradation rates for the different $TiO_2$ pigments. In one resin system, pigments with a certain photocatalytic degradation rate will match the direct degradation rates and give the highest gloss retentions. In another resin system, the $TiO_2$ pigments with photocatalytic degradation rates that best match the direct degradation rates will be different than in the first polymer system. The balance of degradation rates is also partially affected by the degree of pigment dispersion, as described in more detail above in relation to chalking rates.

## Measuring Degradation

Premature failure of paints due to degradation can be a great cause of dissatisfaction with an exterior paint. The cost of such failures goes beyond the cost of the paint itself, as paint application is often a much greater expense for structures such as bridges, tall buildings and water towers. Because of the importance of paint longevity in most exterior applications, it is critical that the durability of exterior paints is tested in some way before application.

There are two main factors that affect the rate of paint degradation—the innate resistance of the paint to environmental stresses, and the severity of the exposure conditions. For many reasons, the overall durability of a given paint or set of paints is often difficult to predict. The literature has many examples of "reversals"—a change in the durability ranking of paints exposed under different exposure conditions. In addition, the effect of ingredient choice on durability (resin supplier, $TiO_2$ grade, extender particle size, etc.) can often affect different paints differently. For these reasons, it is critical that exterior paints be tested for durability before they are qualified for use. Similar considerations apply to plastics, since these factors apply equally to them.

There is a wide and rich literature on paint and plastics durability, a reflection of its great importance in many end-use applications. In addition to individual research reports, there are many reviews of this subject [56, 59, 60]. We will not attempt to summarize this work, but instead will focus on one aspect of durability—the contribution of the solid particles within the organic matrices. Within this narrow segment of durability science, we will focus, although not exclusively, on the contribution of the titanium dioxide pigment, since this particle type has a greater effect on durability than others.

[11] This was discussed this earlier in the section on ultimate durability.

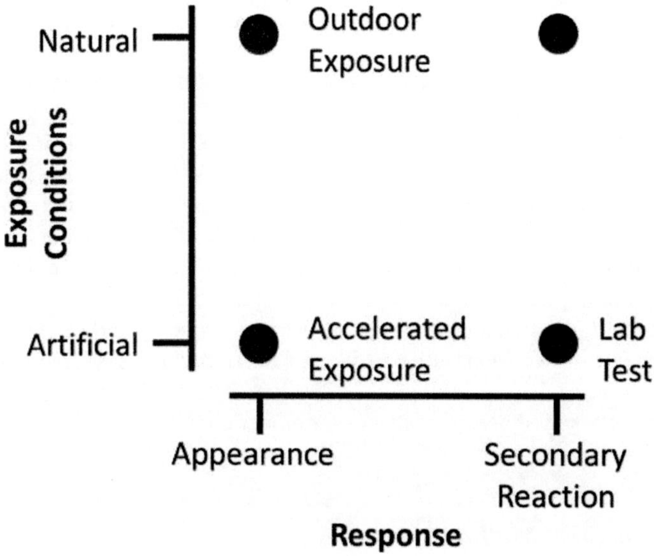


**Fig. 14.30** Categorization of exposure test strategies

We can group most durability tests into one of four categories, depending on exposure conditions and on the type of exposure-induced changes that are being monitored. There are two exposure conditions commonly used—natural exposure and accelerated (high light intensity) exposure. There are also two types of changes—appearance changes (which are physical) and chemical changes—that can be monitored as the polymer degrades. The distinguishing features of these four categories are shown in Fig. 14.30.

## *Natural Test Methods*

The most obvious way of determining the durability of a paint or plastic, or the effect that a certain ingredient or formulation factor has on that durability, is to expose that paint or plastic to natural conditions and monitor for appearance changes that would be important to the ultimate end user (i.e., the consumer). The merits of natural tests are that they exactly duplicate the conditions under which the paint will be exposed under normal use and they measure the exact appearance changes of concern to the paint consumer. However, these merits come at a significant cost in terms of test duration—coatings formulated to last for decades will, by their nature, require decades to test. In addition, weather is uncontrollable and never reproduced, and the consistency of results from one exposure series to another can be poor.

Because of this variability, it is essential that standards and/or controls be included in every outdoor exposure study, and that the durability of test samples be based on their relative performance versus these standards or controls, rather than on an absolute basis. Despite this, many international paint durability standards are, in fact, based on absolute performance over time. For example, standards may specify that gloss loss after two years can be no more than 20%, even though the exposure intensities of no two-year time periods are ever the same. This creates a moving target that is all but impossible to hit on a reliable basis [60–62].

One compromise to the desire for shorter test durations under natural conditions is to expose the paints at locations with harsh environments. Even the environment for Florida testing—the benchmark of durability—is harsher than that of the locations of most outdoor applications. Exposures in more severe natural environments, such as in Arizona or Australia, can shorten exposure times compared to Florida, with a minimal compromise of exposure conditions. That said, even these shorter times can be longer than practical for new product development, and on occasion, the results correlate poorly with those generated under milder climates.

## *Accelerated Test Methods*

In response to the desire for shorter test turn-around times, coatings formulators, plastics manufacturers and the exposure testing industry have long sought an accelerated testing procedure that will, after a short exposure time, accurately predict long-term performance. Much effort has been spent developing these tests [56], and the voluminous literature on accelerated weathering is a testament to the difficulty of this pursuit and to the lack of any general solution to it. Simply stated, there are no accelerated exposure techniques that reliably predict outdoor exposure results on a universal basis [63–65]. That said, when comparing similar paints, certain accelerated technologies can give an indication of the relative durability of the different paints.

There are several factors that complicate the development of a universal accelerated test. The first is that long-term appearance changes are not always linear with exposure time. Often, we will see an induction period during which the paint is degrading, but not yet changing in appearance. A good example of this is chalking—chalk does not appear for several months, during which time the top layer of resin is degrading, but this degradation has not yet come to the point of exposing particles. This was shown in Fig. 14.14. Similar induction periods can be seen for gloss changes in certain paints (Fig. 14.13).

A second complication is that, as mentioned above, the environmental stresses of natural exposures are uncontrollable and are highly variable on any time scale—from day to night, from month to month, from season to season, from year to year and even from decade to decade. Durability results for the same paint or plastic exposed at different dates can vary quite significantly [61]. Even the position on a test rack can affect the degradation rate of a paint, particularly for racks with black backgrounds [66].

A third complication is that under accelerated conditions, resin degradation can follow new pathways that are not seen under natural exposure. For example, the UV-B and UV-C portions of sunlight are removed in the upper atmosphere and so the reactions they initiate are not important under natural exposure conditions. Inclusion of this highly destructive radiation in certain accelerated tests leads to faster times-to-failure, but at the expense of being meaningless to the process that occurs under natural exposure conditions.

Even if new reaction pathways are not created, the relative rates of the different reactions that occur during degradation can change, with reactions that are of little importance under natural exposure conditions accelerating at a greater extent than reactions that are of high importance under natural exposure conditions. This shift in relative importance can skew the results of the accelerated test.

An often-made comparison of this latter issue to the incubation of a chicken egg. An egg held at 38 °C for 22 days will produce a chick. This occurs as the result of hundreds of inter-connected reactions that transform the protein in the egg into the different body components of the chick, similar in concept to the large number of inter-connected reactions that occur as a paint or plastic degrades. Using the theoretical rule that reaction rates double with every increase in temperature of 10 °C, we calculate that the reactions requiring 22 days to go to completion would require only 3 minutes at 172 °C.

However, reality is much different from theory, and treating an egg to such temperatures would not result in a chick, but instead in a fried egg. Here, as with paints, we find that the added stress does not accelerate all reactions equally—in this case, reactions that denature the protein completely dominate at the higher temperatures while they are negligible at lower temperatures. We will discuss the difficulties of using high stress accelerating conditions to rapidly predict durability results under natural conditions below.

## *Methods Based on Early Results from Natural Weathering Exposures*

The simplest means of shortening test durations is to use early changes in appearance to predict long-term performance. This has the benefit that the exposure conditions are exactly those of the intended end-use, as discussed above, which can be a significant advantage over methods based on an increase in exposure stress of intensity.

When degradation occurs at a constant pace, it is possible to extrapolate early appearance changes to longer exposure times and predict the time to failure. To do this, paints or plastics must have completed their induction period, if any (refer to Fig. 14.14). Because gloss loss tends to follow a curved decrease over time and often experiences abrupt failures, it is often difficult to extrapolate early gloss retention data to estimate later values [61]. However, once the induction period has passed for chalking or color fade tests, lightness or color tends to change linearly.

When the film is in the early times of the second stage of color shift, it is possible to make a relatively accurate prediction of the rate of change and thus the time needed to reach a certain color shift (assuming this is attained prior to free chalking).

## *Methods Based on Increased Exposure Intensity*

Perhaps the most obvious way to decrease durability test times is to artificially increase the stresses that degrade polymers. These include UV light intensity, heat, water (both as a vapor and as a liquid), and thermal cycling. Two strategies can be taken for this approach. The first is to increase the intensity of natural sunlight through the use of mirrors, and the second is to use weathering chambers for which exposure conditions can be accurately controlled.

Before discussing these chambers, it is worthwhile to consider some general principles that are recommended regardless of the exposure specifics. First, the stresses should not exceed the maximum stress seen in natural exposure—they should instead subject the samples to this level of stress for a greater proportion of time. For example, the UV light intensity should not exceed that of a typical testing location, such as South Florida. In a similar way, the maximum panel temperature should not exceed that of an outdoor panel at the hottest part of day.

The spectral composition of the UV light used in these chambers is also important. The atmosphere screens out the highest energy UV wavelengths (UV-B, from 280 to 315 nm, and UV-C, from 100 to 280 nm), but not the lower energy UV-A (315 to 400 nm).[12] Including the lower wavelength (higher energy) UV-B and UV-C radiation in an exposure test has the danger of skewing the test results for the reasons discussed above. The ability of a UVA-340 lamp to match the UV component in sunlight is shown in Fig. 14.31.

Overall, acceleration is best achieved instead by extending the fraction of time that the samples spend at these extreme conditions, rather than increasing the intensities above these levels. This avoids the occurrence of degradation reactions that only proceed under higher light intensities or more elevated temperatures relative to these maxima since these would be unimportant in actual end-use applications.

Two types of weathering chambers are most commonly used for accelerated exposures. In the first, samples are arranged in a circle around the light source (normally a filtered xenon arc tube) and rotated during exposure. This type of chamber has the ability to cycle conditions, to turn the light on and off, and to spray the samples with water (both to cool them during the dark cycle and to wash away soluble degradation intermediates), as well as the ability to control temperature at elevated levels. Standard methods have been developed using this type of chamber, both by ASTM [68] and ISO [69].

[12] For reference, the minimum UV wavelength required to break a typical C–C single bond ranges from 315 to 345 nm, a C-F bond requires approximately 250 nm, a C-H bond requires approximately 290 nm, and an ether bond requires approximately 355 nm.

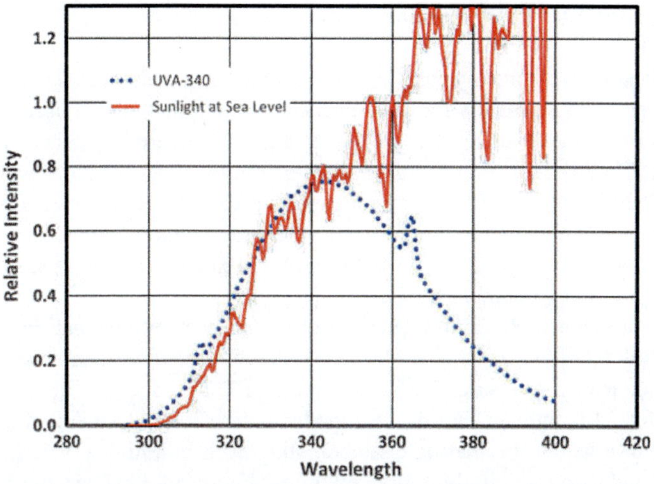


**Fig. 14.31** Comparison of the spectral power distributions of natural sunlight [67]. (red solid line) and a UVA-340 lamp (blue dotted line) (Color figure online)

In the second chamber type, samples are arranged linearly along a fluorescent source of unfiltered, high-intensity UV light [67]. Samples are held in closer proximity to this light than in the first chamber type, resulting in a significantly higher light intensity per unit area of sample. The light is typically on continuously, the temperatures are usually quite high, and the samples are often not sprayed with water. Because of the higher stresses in this type of chamber, especially the inclusion of highly destructive UV-A and UV-B radiation, test durations are typically much shorter than the xenon-based weathering chamber. While shorter test times are desired, they often come at the expense of accuracy, as will be demonstrated below.

There is an alternative strategy for increasing the intensity of the light falling on a sample. In this method, the intensity of natural sunlight falling on a sample is increased using mirrors. A row of samples is arranged on a special rack with their backs to the sun and facing an array of mirrors arranged in a semi-circle. The rack and mirrors are mounted on a motor assembly that is programmed to follow the sun during the day so that the mirrors constantly reflect sunlight onto the samples [70, 71]. While this equipment can be sited at any location, it is typically found in regions with high UV light intensity and minimal periods of cloud cover.

This method typically reduces exposure times by a factor of between 3 and 5, depending on the time of year during which it is run, compared to natural exposure. Exposure is quantified by the amount of UV light striking the specimen (typically measured as $mJ/m^2$) or as equivalents of time for a typical Florida exposure (defined as 280 $mJ/m^2$ annually). Samples are naturally heated due to the increase in thermal (infrared) radiation directed at the samples, which also accelerates degradation.

While this method has the benefit of increasing light intensity while maintaining an exact match to the terrestrial sunlight, it suffers from sample overheating and excess levels of thermal degradation, which is particularly problematic when the intention of the test is to determine the contribution to durability due to UV light absorption. This can be partially overcome by spraying the backs of the panels with water, but excessive temperatures ultimately limit the intensity of light that can be used. This issue has been overcome by using mirrors based on light interference, rather than simple reflection [72–75]. Interference mirrors preferentially reflect certain wavelengths of light while allowing others to pass through. In this case, UV light is reflected back to the sample while visible and infrared light (thermal radiation) is not. This can allow UV light intensities approaching 100 times that of natural light without the sample being heated to unacceptable levels.

The degrees to which results from these different acceleration conditions correlate with natural exterior exposure results can be demonstrated with an example of coil paints made from a series of 23 different grades of $TiO_2$. Panels of these paints were exposed for 24 months in Florida, 10 months under natural accelerated conditions, 125 days in a xenon weathering chamber, and 750 h in a high UV light intensity chamber, and then measured for gloss retention. Figure 14.32 shows the correlation between the Florida gloss retention results (which are the target values for the other exposure tests to match) and results for the three forms of accelerated testing, and Table 14.3 gives the correlation coefficients between these results. As can be clearly seen in the table and figure, the more severe the exposure conditions, the faster the results but the poorer the correlation and the predictability of the test. For this particular paint series, results from the high-intensity UV chamber are completely incapable of providing guidance as to which paints will perform well outdoors.

## *"Reversals" and the Difficulty of Accelerated Weathering Tests*

Perhaps the most important conclusion of the above discussion is that accelerated weathering tests will, under certain conditions, give entirely erroneous results. This is true for both accelerated chalking and accelerated gloss retention tests. In our experiences, accelerated tests can normally distinguish a non-durable grade of $TiO_2$ pigment from a durable one, but often rank very durable pigments as having equal or even lower durability than medium durability pigments. There are sound theoretical reasons for this behavior.[13]

Although high UV light intensities increase the rates of both direct degradation and photocatalytic degradation, they do so by different amounts. Egerton and King showed that the direct degradation rates increase linearly with light intensity whereas the photocatalytic degradation rates increase as the square root of light intensity [5]. These conclusions were confirmed by Cutrone, Patel and Martin [76–78].

[13] There are no doubts that many reasons exist for poor correlation between accelerated and outdoor exposure test results. The following discussion details just one of these reasons and will not apply to every case of poor correlation.

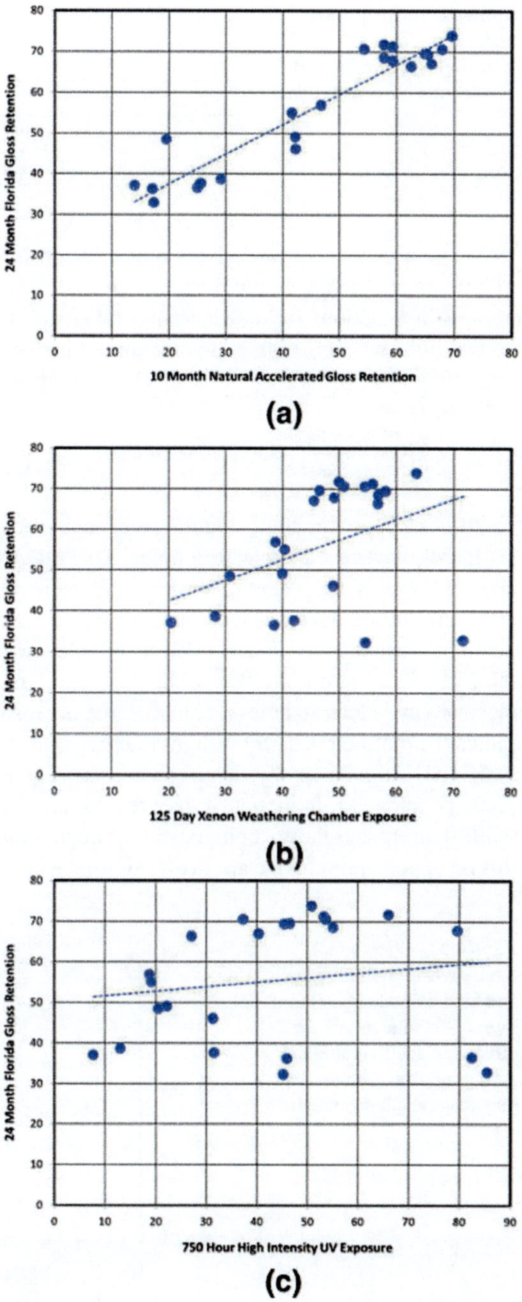


**Fig. 14.32** Correlation between Florida gloss retention values to values measured under acceleration. **a** Natural accelerated conditions. **b** Weathering chamber with filtered xenon radiation. **c** Weathering chamber with high-intensity unfiltered UV radiation

**Table 14.3** Correlation coefficient between Florida 2-year gloss retention and accelerated gloss retentions

| | Exposure Time | $R^2$ |
|---|---|---|
| Florida | 2 years | – |
| Accelerated natural | 10 months | 77.1% |
| Xenon weathering chamber | 125 days | 20.6% |
| High-intensity UV chamber | 750 h | 2.5% |

While a mathematical derivation of this non-intuitive result is beyond the scope of this paper, it can be qualitatively understood on the basis of there being two competing reactions for the electrons and electron holes which form on absorption of UV light by $TiO_2$ pigments: reaction with surface species to form radicals, and recombination with release of heat. These reactions are summarized in Eq. 14.4.

$$TiO_2 \underset{\text{heat}}{\overset{+\text{UV light}}{\longleftrightarrow}} \text{electron} + \text{electron hole} \underset{H_2O}{\overset{O_2}{\longrightarrow}} \text{surface radicals.} \tag{14.4}$$

To a first approximation, doubling the intensity of the UV light doubles the concentration of electrons and electron holes. This *quadruples* the rate of recombination, since this rate is dependent on the concentration of both species (that is, the reaction rate will double once because of the increase in electron concentration, and then again because of the increase in electron hole concentration), while it only doubles the rate of formation of either surface radical. Under conditions where most of the electrons and electron holes recombine, the net result is a square root dependence of radical formation rates on UV light intensity.

High UV light intensity, therefore, upsets the balance between direct and photocatalytic degradation. In particular, it over-emphasizes the effects of direct degradation while it minimizes those of photocatalytic degradation (see Fig. 14.33). The effects of this on appearance change are two-fold. First, since gloss retention is determined by

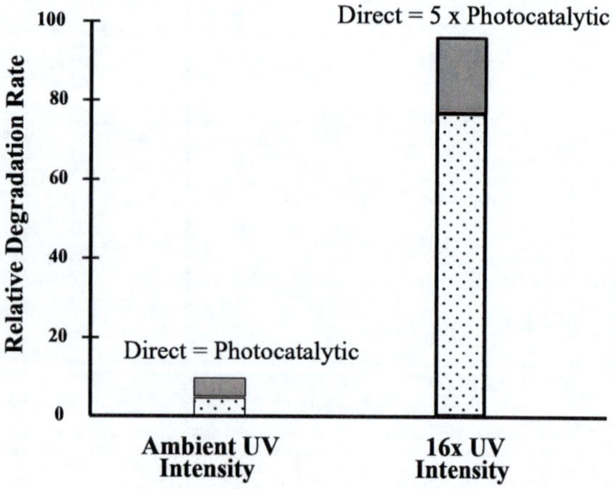


**Fig. 14.33** Effect of UV light intensity on individual and total paint degradation rates. Dotted areas are the direct degradation rates and dark gray areas are the photocatalytic degradation rates

the balance between photocatalytic and direct degradation rates (discussed above), by changing this balance, the ordering of gloss retention performance will change with UV light intensity. This explains the results in Fig. 14.32.

The second effect, as detailed in the discussion on ultimate $TiO_2$ durability, is that the increase in the relative rate of direct degradation makes the effect of photocatalytic degradation less important. That is, it becomes difficult or impossible to distinguish chalking performance between and among medium and high durability pigments. This may lead to the conclusion that two or more $TiO_2$ pigments are equal in durability when, under natural conditions, they may be quite different. This is similar to the reason why the durability of paints made with mid and high durability $TiO_2$ grades are the same as one another in low durability resins, as discussed in the section on ultimate $TiO_2$ pigment durability.

## $TiO_2$ Lab Tests

When the interest in durability is focused on the contribution of $TiO_2$ to degradation rate, which in turn is mainly controlled by $TiO_2$ photocatalytic activity, an alternative strategy for estimating paint durability—or, at least, the contribution of $TiO_2$ photocatalytic activity—can be used. This situation often arises when a formulator wishes to determine which grades of titanium dioxide are acceptable for certain exterior applications.

### *Measurement of Reaction Products*

One form of laboratory testing is the measurement of the concentration of a reaction product over time (refer back to Fig. 14.4). To be of practical use, this reaction product must be detectable more quickly than the appearance changes in the paint film. The logic behind this test is that it can detect the rate of chemical changes occurring during the induction period, before appearance changes have yet occurred [79].

An example of this is the formation of carbon–carbon double bonds during the degradation of an aliphatic resin [80], or the formation of carbonyl groups ($C = O$) in low-density polyethylene [2]. The concentration of new functional species can be conveniently monitored using infrared spectroscopy. While this strategy is appealing since it shortens the time needed to measure paint degradation under natural exposure conditions, it, too, has complications that can cause misleading results.

A schematic is given in Fig. 14.34 showing the degradation reactions in a film. The "new moiety" for our example would be the alkene species, and the rate coefficient for creation of this moiety from the resin is given as $k_1$. This is the rate of interest since it is governed by the photocatalytic reactivity of the pigment. However, in most cases, the new moiety will further react to give other intermediates (Fig. 14.4) and, in limited cases, can back-react to create more reactant.

$$[\text{New Moiety}]_{steady\ state} \propto \frac{k_1}{k_{-1} + k_2}$$

**Fig. 14.34** Formation of a new chemical moiety as resin degrades. The various $k$ values represent the rate coefficients of the reactions

The amount of the new moiety present in the film at any given time depends not only on the rate of the initial reaction, the determination of which is the goal of the test, but also on the rates of reaction of the new moiety to either create additional degradation product or, in some cases, reactant. If we compare the rates of moiety formation between two paints, we have no way of discerning whether any differences are due to differences the in rate of formation of the moiety or to differences in the rate of destruction of the moiety.

This dilemma can be resolved by measuring the rate of creation of the final products in the series of degradation reactions—water and carbon dioxide [81]. Because water can also be a reactant in the degradation reactions (for example, as the source of hydroxide for the formation of hydroxyl radical on the pigment surface), it is more practical to measure the rate of carbon dioxide formation. Since carbon dioxide is the thermodynamic endpoint of the carbon in the resin, its rate of formation can be unambiguously assigned to the rate of paint degradation.

## *$TiO_2$ Photocatalytic Rate of a Simple Redox Reaction*

An alternative simple strategy for estimating the effect of pigment grade on paint durability in the laboratory is to measure the ability of the $TiO_2$ to photocatalyze a simple redox reaction. The benefits of this approach are that it can be done under carefully controlled conditions, that it does not require paints to be made with the pigments of interest, that it avoids the complexity of paint degradation, including the inter-relatedness of a multitude of degradation pathways, and that it can be completed in a few hours at most.

A number of redox reactions have been used in these laboratory tests. These include the reduction of a lead salt ($Pb(CO_3)$) to lead metal [82], the reduction of a silver salt ($Ag(NO_3)$) to silver metal, [83, 84], the discoloration of a blue wool sample (used for paper laminate applications) [85] and the oxidation of iso-propanol to acetone [86]. In the first two of these reactions, a paste is made of the appropriate metal salt, the $TiO_2$ pigment of interest, and a liquid in which to suspend the two. The reactions are quantified based on the rate of graying when exposed to UV light

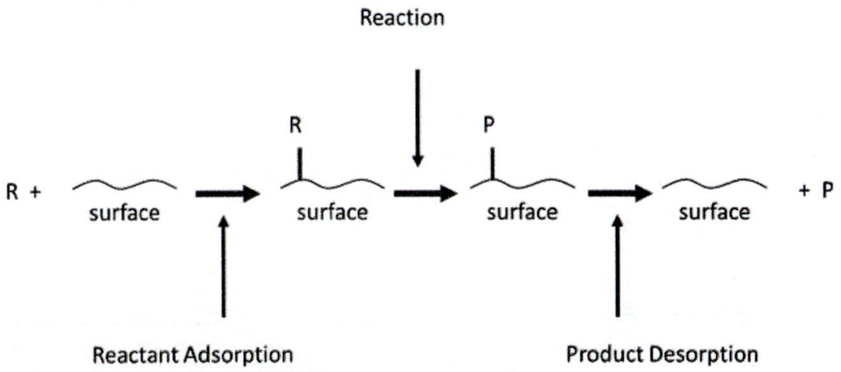


**Fig. 14.35** Schematic of a representative laboratory test. "R" represents reactant (for example, silver ion) and "P" represents product (for example, silver metal)

under oxygen-free conditions (the salts are white while the metals are black).[14] For the iso-propanol test, acetone concentrations are followed by gas chromatography.

While many aspects of these tests are appealing, they are based on several assumptions that are often not correct, and in many cases, their results can be misleading. These assumptions include that the complex, inter-related reactions that occur as a paint decomposes can be accurately modeled by a single, simple reaction, and that the factors that are important to paint degradation (and that differentiate the durability performance of different $TiO_2$ grades) are also important to the laboratory reactions.

Violations of the latter assumption can be understood as follows: Even though the laboratory reactions are simple when compared to those occurring in an exposed paint film, they still involve a number of steps, some of which are not photocatalytic but nonetheless affect the overall rate of photocatalysis. A schematic of these steps in shown in Fig. 14.35.

This reaction sequence begins the absorption of the reactant onto the $TiO_2$ pigment surface, which is necessary for the transfer of electrons into and out of the $TiO_2$ crystal. Next, the $TiO_2$ is irradiated with UV light and the adsorbed reactant is transformed into an absorbed product. Finally, the adsorbed product is detached from the pigment surface, regenerating the empty surface site onto which the reactant will adsorb.

The reaction rate of interest is the rate of surface reactant transforming into surface product. However, any step in this reaction sequence can affect the rate of product appearance. As an example, a series of $TiO_2$ pigments was made with surfaces being either acidic or basic. These pigments were incorporated into a paint and exposed for 18 months in Florida to determine their true durability. At the same time, these pigments were made into a paste with silver nitrate. The pastes were exposed to high-intensity UV light for two hours and the brightness values of the pastes were measured. From these values a relative rate of silver reduction was determined.

[14] This is similar to the darkening seen when décor paper is exposed to UV light—see Fig. 14.18.

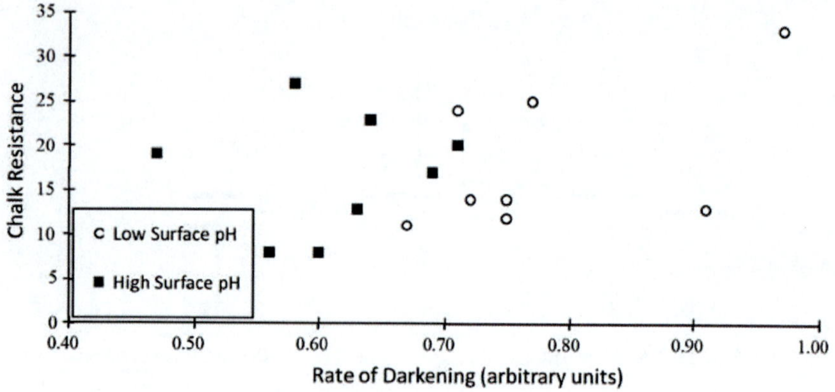


**Fig. 14.36** Comparison between rate of silver reduction (measured as paste darkness; higher values represent higher reactivities) for a series of pigments and the chalk resistance paints made with these pigments (higher numbers indicating better durability) after 18 months of Florida exposure

In Fig. 14.36, we compare the rate of silver photoreduction rates of these pigments with the chalk resistance of the paints made from them. We see that there is little, if any, correlation between the two, and also that, as a group, the pigments with low surface pH values darkened to a greater extent than those with high surface pH values. That is, surface pH affected the lab test but not the outdoor paint exposure test. This implies that adsorption of the silver ion onto the pigment surface is controlled by surface pH, and that by varying surface pH we can vary the amount of silver available for the reaction without varying the actual pigment photocatalytic activity.

In addition to surface pH, surface composition can also affect the adsorption strength of the reactant to the pigment surface. Most grades of $TiO_2$ have a combination of alumina, silica and/or zirconia to decrease pigment photocatalytic activity, and it can be the case that laboratory tests can differentiate pigments with a similar surface composition while not correlating between pigments with different surface compositions. An example of this is shown in Fig. 14.37, where we see a good correlation between the iso-propanol photo-oxidation rate for pigments with 5% surface silica and the exterior chalk value for paints made from them, but no correlation for the pigments made at the 1% silica level.

## *Indirect Tests for $TiO_2$ Durability*

For those $TiO_2$ grades that have enhanced photo-durability achieved through a coating of hydrous silica, rather than from a form of stabilized amorphous alumina, an indirect test can be used to assess the quality of the silica encapsulation layer. This test, called the acid solubility test, makes use of the fact that titanium dioxide is soluble in hot sulfuric acid whereas silica is not. In this test, the pigment of interest is slurried in

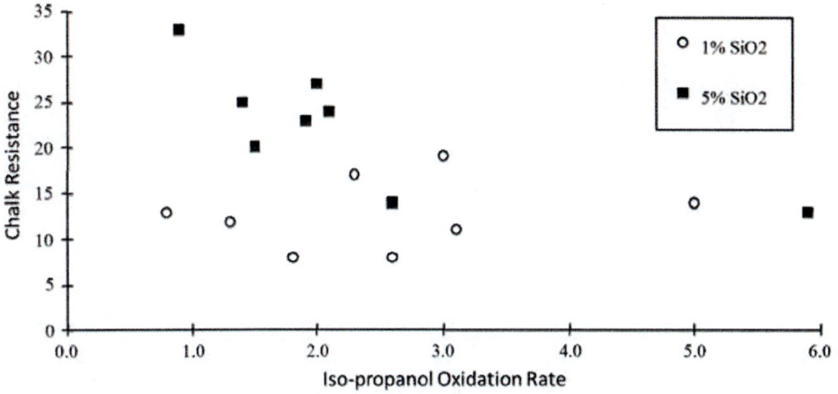


**Fig. 14.37** Correlation between the lab iso-propanol oxidation rates for two series of $TiO_2$ pigments and 18 month Florida chalk values for paints made from them

concentrated sulfuric acid and heated to near boiling for one hour [87]. At the end of this time, the slurry is quenched with ice, the solids separated from the liquor, and the amount of titanium dissolved in the liquor is measured.

Since silica is insoluble in this slurry, any $TiO_2$ particles for which the silica encapsulation is complete will not dissolve. However, if there are any gaps or cracks in the silica layer on a given particle, even if quite small, then the entire $TiO_2$ content of that particle will dissolve. The concentration of titanium in the sulfuric acid solution is therefore an indicator of the quality of the silica encapsulation, which in turn determines the photocatalytic activity of the pigment in question.

This test is very sensitive because even if only a few percent of a $TiO_2$ particle surface is not covered, 100% of the $TiO_2$ in that particle will dissolve, thus amplifying any systemic gaps in coverage. This in fact reflects the importance of complete coverage for preventing film degradation—even small gaps in silica coverage can allow all of the photo-generated high energy electrons and electron holes to react with surface water and oxygen.

A comparison of acid solubility results with outdoor chalk results is shown in Fig. 14.38 [88]. Here, the average values for several grades of $TiO_2$ with different silica levels are shown. The error bars represent the 95% confidence intervals of a single measurement. Note that the error bars for the acid solubility test are much narrower than for the chalking test, reflecting the inherent variability in chalk results due to the uncontrolled nature of natural exposure conditions. Because of its speed and its high precision, the acid solubility test can be used as a release test for highly durable or superdurable grades of $TiO_2$.

While this test is an excellent predictor of the photocatalytic activities of silica-coated $TiO_2$ grades, it cannot be used on $TiO_2$ grades that derive their durability from stabilized amorphous alumina. This is because alumina is quite soluble in hot sulfuric acid, and so offers no protection to dissolution of the encapsulated $TiO_2$ particles.

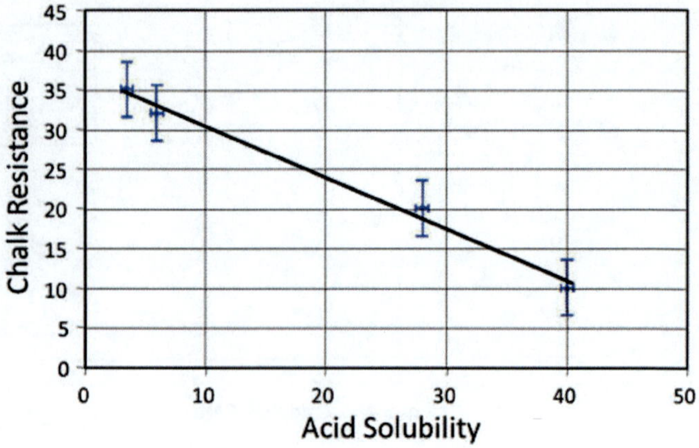


**Fig. 14.38** Correlation between chalk resistance (2-year Florida exposure) and acid solubility. Bars represent the 95% confidence intervals for a single measurement

Also, the presence of crystalline alumina on the $TiO_2$ surface might complicate the interpretation of the results.

## Summary

Paint or plastic degradation is a significant issue for materials that are exposed to the elements. The issue is quite complex—degradation occurs as the result of a complex web of inter-related chemical reactions that occur both in series and in parallel with one another. Each reaction has its own sensitivity to specific exposure conditions, such as heat, temperature cycles and exposure to sunlight. The ultimate origin of these degradation reactions is the instability of organic molecules to atmospheric oxygen. Under ambient conditions there is a significant energy barrier to these reactions, slowing their rates and extending the materials useful lifetimes. However, heat and, to a greater extent, ultraviolet light can provide the energy needed to overcome these reaction barriers, causing significant polymer degradation on exposure to the elements.

While durability is determined primarily by the chemical composition of the polymer, it is also affected by the presence and type of titanium dioxide used. $TiO_2$ has a two-fold impact on polymer durability, both resulting directly from its strong UV light absorptivity. This strong absorption provides the benefit of protecting the underlying polymer from the UV portion of sunlight. Acting against this benefit, a small portion of the UV light energy absorbed by $TiO_2$ particles is converted into reactive chemical energy in the form of chemical radicals. These radicals form on the $TiO_2$ surface and are highly destructive to polymer molecules when they migrate from the particle surface and into the matrix.

The photocatalytic activity of titanium dioxide stems from its electronic structure. This material is a network solid, resulting in the bonding and anti-bonding electrons being confined to discrete energy bands. Promotion of an electron from one band to another can lead to the exchange of electrons between the $TiO_2$ particle and the oxygen and water molecules adsorbed to its surface. Pigment manufacturers can slow this reaction by encapsulating the $TiO_2$ particles with an enveloping layer of an electrical insulator—either hydrous alumina or hydrous silica.

Polymer degradation is manifested in several appearance and physical changes on exposure. Because appearance changes tend to show earlier in the exposure process, it is important that they be monitored and quantified. The most significant appearance changes are a color shift, typically an increase in brightness as the white particles within the polymer become exposed, and gloss loss, due to roughening of the paint film or plastic surface. Each of these is affected differently by the grade of $TiO_2$ used as well as the conditions under which the samples are tested. Although gloss loss occurs in compact board paper laminate applications, it occurs by a different mechanism (surface stress) that is specific to thermosetting resins such as the melamine used in this application.

There are many test conditions and protocols available for quantifying durability, with the prime differences being two related and opposing parameters—the test duration, and the predictive ability of the test results to actual end-use applications. Ideally, we would test paints and plastics under conditions that represent completely the actual conditions that the sample would be exposed to in its intended application. However, test durations are very long under these conditions—for example, a paint or plastic warrantied for a 20-year lifespan would require 20 years to test.

To overcome this issue, various technologies have been developed to increase the stress on the samples and so deliver the same overall destructive force over a shorter period of time. This can be done by increasing the sunlight striking a panel using mirrors or by creating weathering chambers with the ability to increase temperature and, in particular, UV light intensity and quality, to levels many times greater than the sample would see under normal in-use conditions. There is a general trend with these different acceleration strategies, which is that the shorter the test duration, the less reliable the results are in terms of predicting in-use performance. We can understand these difficulties in terms of the factors that affect the way that degradation occurs and is manifest as a function of stress intensity. By knowing this, we can develop better acceleration methods or better interpret the results we get from existing ones.

## References


1. Allen, N.S., Bullen, D.J., McKellar, J.F.: Photo-oxidation of commercial polyethylene containing titanium dioxide (rutile)/antioxidant systems. J Mater Sci **12**, 1320 (1977)
2. Egerton TA (2014) UV-absorption — the primary process in photocatalysis and some practical consequences. Molecules 19:18192
3. Cabrera, M.I., Alfano, O.M., Cassano, A.E.: Absorption and scattering coefficients of titanium dioxide particulate suspensions in water. J Phys Chem **100**(51), 20043 (1996)
4. Egerton, T.A., Tooley, I.R.: UV absorption and scattering properties of inorganic-based sunscreens. Int J Cosm Sci **34**(2), 117 (2011)
5. Egerton, T.A., King, C.J.: The influence of light intensity on photoactivity in $TiO_2$pigmented systems. JOCCA **62**, 386 (1979)
6. "New Test Method for Determining the Carbonyl Index of a Polyolefin Plastic Material using Infra-Red Spectrometry (FT-IR)", ASTM WK65360
7. Almond J, Sugumaar P, Wenzel MN, Hill G, Wallis C (2020) Determination of the carbonyl index of polyethylene and polypropylene using specified area under band methodology with ATR-FTIR spectroscopy. e-Polymers 20(1):369
8. Diebold, M.P.: The causes and prevention of titanium dioxide induced photodegradation of paints, part 1. Surf Coat Int **1995**(6), 250 (1995)
9. Diebold, M.P.: The causes and prevention of titanium dioxide induced photodegradation of paints, part 2. Surf Coat Int **1995**(7), 294 (1995)
10. Balfour, J.G.: Back to basics: durability and titanium dioxide pigments. JOCCA **78**, 478 (1990)
11. Qu, Z.-W., Kroes, G.-J.: Theoretical study of the electronic structure and stability of Titanium Dioxide Clusters $(TiO_2)^n$ with n= 1–9. J Phys Chem B **110**(18), 8998 (2006)
12. Diebold MP (1999) Unconventional effects of $TiO_2$ on paint durability. In: Proceedings of 5th nurnberg congress (1999), pp 371–389
13. Grossman, G.W.: Correlation of laboratory to natural weathering. J Coat Technol **49**, 45 (1977)
14. Johnson, R.: An overview of degradable plastics. J Plast Film Sheet **4**(2), 155 (1988)
15. Xingzhou, R.: Wavelength sensitivity of photo-oxidation of polyethylene. Polym Degrad Stab **55**(2), 131 (1997)
16. Chew, C.H., Gan, L.M., Scott, G.: Mechanism of the photo-oxidation of polyethylene. Eur Poly J **13**(5), 361 (1977)
17. Yang, X., Yu, J., Liu, Y., Wang, K.: Effects of inorganic fillers on the natural photo-oxidation of high-density polyethylene. Polym Degrad Stab **88**(2), 333 (2005)
18. Davidson, R.S., Meek, R.R.: The photodegradation of polyethylene and polypropylene in the presence and absence of added titanium dioxide. Eur Poly J **17**(2), 163 (1981)
19. Colling, J.H., Wilkinson, T.W.: Implication of the paint film contraction theory. J Oil Col Chem Assoc **58**, 377 (1975)
20. Colling, J.H., Dunderdale, J.: The durability of paint films containing titanium dioxide – contraction, erosion and clear layer theories. Prog. Org. Coat. **9**, 47 (1981)
21. Faucheu, J., Wood, K.A.: Relating gloss loss to topographical features of a PVDF coating. J Coat Technol Res **3**(1), 29 (2006)
22. Braun JH, Cobranchi DP (1995) Durability and gloss. J Coat Technol 67(851):55
23. Diebold MP (2009) Effect of $TiO_2$ pigment on gloss retention: a two-component approach. JCT Coatings Tech 5(239):32
24. Daiger, W.H., Madson, W.H.: Chalk-fade evaluations of pigmented finishes by use of instrumentation and computer analysis. J Paint Technol **39**(510), 399 (1967)
25. "Standard Test Methods for Evaluating the Degree of Chalking of Exterior Paint Film" (2015) ASTM D4214-07
26. Allen, N.S., Bullen, D.J., McKellar, J.F.: Photo-yellowing of a phenolic anti-oxidant in the presence of various stabilizer/titanium dioxide pigment combinations in polyethylene. J Mater Sci **13**, 2692 (1978)
27. Pospíšil J, Habicher WD, Cayambe J, Nešpůrek S, Kuthan J, Piringer G-O, Zweifel H (2002) Discoloration of polymers by phenolic antioxidants. Polym Degrad Stab 77(3):531

28. Starnes, W.H., Jr.: Structural defects in poly-vinyl-chloride and the mechanism of vinyl chloride polymerization: comments on recent studies. Procedia Chem **4**, 1 (2012)
29. d'Antuono, P., Botek, E., Champagne, B., Wieme, J., Reyniers, M., Marin, G.B., Adriaensens, P.J., Gelan, J.M.: A joined theoretical-experimental investigation on the $^1$H and $^{13}$C NMR signatures of defects in poly(vinyl chloride). J Phys Chem B **112**, 14804 (2008)
30. Scott, G.: Developments in the photo-oxidation and photo-stabilisation of polymers. Polym Degrad Stab **10**(2), 97 (1985)
31. Al-Taa'y W, Nabi M, Yusop RM, Yousif E, Abdullah B, Salimon J, Salih N, Zubairi SI (2014) Effect of Nano ZnO on the optical properties of poly(vinyl chloride) films. Int J Poly Sci 1
32. Kayyarapua, B., Kumar, M., Mohommada, H.B., Neeruganti, G., Chekuria, R.: Structural, thermal and optical properties of pure and $Mn^{2+}$ doped poly(vinyl chloride) films. Mat Res **19**(5), 1167 (2016)
33. Gardette, J., Lemaire, J.: Photo-oxidation of poly(vinyl chloride): part 3—influence of photocatalytic pigments. Polym Degrad Stab **33**(1), 77 (1991)
34. Anton-Prinet C, Mur G, Gay M, Audouin' L, Verdu CJ (1998) Photoageing of rigid PVC- part IV. Effects of titanium dioxide. Polym Degrad Stab 61(2):211
35. Summers, J.W., Rabinovitch, E.B.: The chemical mechanisms of outdoor weathering in polyvinyl chloride. J Vinyl Tech **5**(3), 91 (1983)
36. Gardette, J.-L., Gaumet, S., Phillipart, J.: Photoageing of rigid PVC—II. influence of exposure conditions on the thickness distribution of photoproducts. J Appl Polym Sci **48**, 1885 (1993)
37. Hollande, S., Laurent, J.L.: Study of discolouring change in PVC, plasticizer and plasticized PVC films. Polym Degrad Stab **55**(2), 141 (1997)
38. Anton-Prinet C, Mur G, Gay M, Audouin L, Verdu J (1998) Photoageing of rigid PVC—III. Influence of exposure conditions on the thickness distribution of photoproducts. Polym Degrad Stab 60(2–3):283 (1998)
39. Anton-Prinet C, Mur G, Gay M, Audouin L, Verdu J (1998) Photoageing of rigid PVC—IV. Effects of titanium dioxide. Polym Degrad Stab 60(2–3):265
40. Karayildirim, T., Yanik, J., Yuksel, M., Saglam, M., Vasile, C., Bockhorn, H.: The effect of some fillers on PVC degradation. J Anal Appl Pyrol **75**(2), 112 (2006)
41. Sauerwein, R.: New acid scavenger enhances PVC stabilization. Plast Addit Compd **11**(4), 16 (2019)
42. Ammala, A., Hill, A.J., Meakin, P., Pas, S.J., Turney, T.W.: Degradation studies of polyolefins incorporating transparent nanoparticulate zinc oxide UV stabilizers. J Nanoparticle Res **4**(1), 167 (2002)
43. Grabmayer, K., Beißmann, S., Wallner, G.M., Nitsche, D., Schnetzinge, R.K., Buchberger, W., Schobermayr, H., Lang, R.W.: Characterization of the influence of specimen thickness on the aging behavior of a polypropylene based model compound. Polym Degrad Stab **111**(21), 185 (2015)
44. Yakimets, R., Lai, D., Guigon, M.: Effect of photo-oxidation cracks on behavior of thick polypropylene samples. Polym Degrad Stab **86**(1), 59 (2004)
45. Allen, N.S., Fatinikun, K.O., Henman, T.J.: Some important factors which influence the photo-oxidation of polypropylene. Polym Degrad Stab **4**(1), 59 (1982)
46. Girois, S., Delprat, P., Audouin, L., Verdu, J.: Oxidation thickness profiles during photooxidation of non-photostabilized polypropylene. Polym Degrad Stab **56**(2), 169 (1997)
47. Langlois, V., Audouin, L., Courtois, P., Verdu, J.: Change of mechanical properties of crosslinked polyethylene during its thermo-oxidative aging. Ang Makro Chem **208**(1), 47 (2003)
48. Blais D, Carlsson DJ, Wiles DM (1972) Surface changes during polypropylene photo-oxidation: a study by infrared spectroscopy and electron microscopy. J Poly Sci A Poly Chem 10(4):1077
49. Schoolenberg GE, Meijer HDF (1991) Ultra-violet degradation of polypropylene 2. Residual strength and failure mode in relation to the degraded surface layer. Polyhedron 32(3):438
50. Rabello MS, White JR (1997) Crystallization and melting behavior of photodegraded polypropylene—I. Chemi-crystallization. Polyhedron 38(26):6379

51. Shyichuk, A.V., Turton, T.J., White, J.R., Syrotynska, I.D.: Different degradability of two similar polypropylenes as revealed by macromolecule scission and crosslinking rates. Polym. Degrad. Stab. **86**(2), 377 (2004)
52. Iler RK (1969) Product comprising a skin of dense, hydrated amorphous silica bound upon a core of another solid material and process of making same, US 2,885,366
53. Howard PB (1977) Treatment of Pigment, US 4,052,224
54. Kinniard SP, Campeotto A (2002) Improved method for manufacturing high opacity, durable pigment, EP 1 373 413
55. Balfour, J.G.: The cost of flocculation. JOCCA **60**, 365 (1977)
56. Nichols ME (2018) Paint weathering tests. In: Kutz M (ed) Handbook of environmental degradation of materials, 3rd edn. Elsevier
57. Diebold, M.P.: A comprehensive understanding of '$TiO_2$ pigment durability.' Paint Coat Ind **21**(7), 90 (2005)
58. Kampf, G., Paneroth, W., Holm, R.: Degradation processes in $TiO_2$–pigmented paint films on exposure to weathering. J Paint Technol **46**(598), 56 (1974)
59. In: Kutz M (ed) Handbook of environmental degradation of materials, 3rd edn. Elsevier (2018)
60. Hunt, F.Y., Galler, M.A., Martin, J.W.: Microstucture of weathered paint and its relation to gloss loss: computer simulation and modeling. J Coat Technol **70**(880), 45 (1998)
61. Rommens J, Michiels G, Diebold M, Verhaege A (2017) $TiO_2$ impact on paint weather resistance. Coatings World
62. Rommens, J., De Backer, S., Gijsman, P., Molhoek, L.: The lasting impact of titanium dioxide. Eur Coat J **4**, 30 (2020)
63. Wood, K.: Evaluation of the ASTM D7869–13 test method to predict the gloss and color retention of premium architectural finishes – I. J Coat Technol Res **15**(5), 933 (2018)
64. Association of Automobile Industries: Comparison of outdoor and accelerated exposure methods – results of a round-robin Test. J Coat Technol **58**(598), 56 (1974)
65. Crewdson, M.: Outdoor weathering must verify accelerated testing. Surf Coat Int **91**(5), 260 (2008)
66. Fischer, R.M., Ketola, W.D., Murray, W.P.: Thermal variability in outdoor exposure tests. Prog Org Coat **19**, 151 (1991)
67. "Standard Practice for Operating Fluorescent Ultraviolet (UV) Lamp Apparatus for Exposure of Nonmetallic Materials" (2016) ASTM G154-16
68. "Standard Practice for Operating Xenon Arc Light Apparatus for Exposure of Non-Metallic Materials" (2013) ASTM G155-13
69. "Plastics — Methods of Exposure to Laboratory Light Sources — Part 2: Xenon-Arc Lamps" (2013) ISO 4892-2:2013
70. Qin, J., et al.: Sunlight tracking and concentrating accelerated weathering test applied in weatherability evaluation and service life prediction of polymeric materials: a review. Polym Test **93**, 106940 (2020)
71. "Standard Practice for Performing Outdoor Accelerated Weathering Tests of Plastics Using Concentrated Sunlight" (2013) ASTM D4364-13
72. Nichols, M., et al.: An improved accelerated weathering protocol to anticipate florida exposure behavior in coatings. J Coat Technol Res **10**(2), 153 (2006)
73. Wu H (2017) Highly accelerated UV weathering: when and how to use it. In: White CC, White KM, Picket JE (eds) Service life prediction of polymers exposed to outdoor weathering. William Andrew
74. Hardcastle, H.K., Jorgensen, G.J., Binham, C.E.: Ultra-accelerated weathering system I: design and functional considerations. J Coat Technol Res **7**(8), 28 (2010)
75. "Standard Practice for Performing Accelerated Outdoor Weathering of Materials Using Concentrated Natural Sunlight" (2017) ASTM G90-17
76. Cutrone L et al (1989) Studies on the effect of titanium dioxide on durability - part 2. Pig Res Tech 16
77. Patel, J.N.: Artificial weathering of paints. JOCCA **79**, 104 (1991)

78. Chin, J., Nguyen, T., Byrd, E., Martin, J.: Validation of the reciprocity law for coating photodegradation. J Coat Technol Res **2**(3), 499 (2005)
79. Zhang, W.R., Hinder, S.J., Smith, R., Lowe, C., Watts, J.F.: An investigation of the effect of pigment on the degradation of a naturally weathered polyester coating. J Coat Technol Res **8**(2), 329 (2011)
80. Bauer DR, Gerlock JL, Mielewski DF (1990) Predicting long-term durability and quality of automotive coatings. In: XX'th FATIPEC congress, p 225
81. Jin, C., Christensen, P.A., Egerton, T.A., Lawson, E.J., White, J.R.: Rapid measurement of polymer photo-degradation by FTIR spectrometry of evolved carbon dioxide. Polym Degrad Stab **91**(5), 1086 (2006)
82. Braun JH (1990) Titanium dioxide's contribution to the durability and degradation of paint film II. Prediction of catalytic activity. J Coat Technol 62(785):37
83. Sbrolli, W., Sbrolli, E.B.: Measurement of the photochemical activity of titanium dioxide. Ann Chim Rome **49**, 143 (1959)
84. Ohtani, B., Kakimoto, H., Miyadzu, H., Nishimoto, S., Kagiya, T.: The photocatalytic effect of surface-adsorbed 2-propanol on the photocatalytic reduction of silver and/or nitrate ions in acidic titania suspension. J Phys Chem **92**, 5773 (1988)
85. "Textiles — Tests for Colour Fastness — Part B02: Colour Fastness to Artificial Light: Xenon Arc Fading Lamp Test" (2020) ISO 105 B02
86. Cundall, R.B., Hulme, B., Rudhan, R., Salim, M.S.: The photocatalytic oxidation of liquid phase propan-2-ol by pure rutile and titanium dioxide pigments. JOOCA **61**, 3510 (1978)
87. Jacobson HW (1988) Titanium dioxide pigment coated with boria-modified silica, US Patent 4,781,761
88. Diebold, M.P., Kwoka, R.A., Mehr, S.R., Vargas, R.W.: Rapid assessment of $TiO_2$ pigment durability via the acid solubility test. JCT Coat Tech **1**(3), 239 (2004)

# Part V
# Formulation

# Chapter 15
# Formulating with Color

## Contents



## Introduction

Color is of high importance in almost any paint application, and paint producers have developed technologies to satisfy strong consumer demands for an expansive color palette and for the ability to match the color of an existing paint or object. The importance of color to the consumer can be seen in the architectural (décor) paint marketplace. A customer in a paint store expects to see a display with hundreds of colors, and to find, behind the counter, a color reader and tint station that allow the store clerk, who likely has no experience whatsoever in color formulation, to match




M. Diebold et al., *Pigments, Extenders, and Particles in Surface Coatings and Plastics*, https://doi.org/10.1007/978-3-030-99083-1_15

virtually any color. Such an operation would not be possible without a comprehensive understanding of the science behind color mixing in systems that simultaneously scatter and absorb visible light.

Color matching is one of the most difficult issues encountered in paint formulation. To color match experimentally typically requires several cycles of mixing colorants at various concentrations into a white paint and measuring the tristimulus values of the resulting paint. The selection of starting point can be aided by the intuition of an experienced color formulator. While this empirical approach can ultimately be successful, the time and material costs are prohibitive for most applications.

As an alternative, we can measure the spectral properties of all the colorants and of the white base paint and, based on theoretical calculations and with the spectrum of the color we wish to match, quickly determine the amounts of different pigments needed to match, or nearly match, the color in question. In most cases, the initial formula is good enough to allow the formulator to skip further rounds of color optimization [1].

As we have seen in prior chapters, additive color mixing is relatively straight forward, and the *xy* chromaticity diagram can be treated as a road map for color matching, However, although color mixing rules for subtractive color mixing, which is how we create colors in paints, exist, they are more complex, and rely on more simplifying assumptions, which often break down in practice. We will discuss these rules, and their shortcomings, after a brief introduction to light absorption.

## Comparing Additive and Subtractive Mixing

The development of XYZ tristimulus values from color matching experiments relied on mixing lights of different colors to create new colors or match existing ones. Based on the fundamental properties of light mixtures, and the color sensations that we experience from them, we were, in Chap. 6, able to develop a means of measuring and quantifying color sensations—the Standard Observer and the accompanying *xy* chromaticity diagram.

However, while we perceive and measure color based on additive mixing, this is not how we create it in a paint. When we additively mix light sources of different colors, we begin with a black (unlit) background and add color to it with selective wavelengths of light. When we subtractively make color by mixing paints and colorants, on the other hand, we begin with a white background (the white base paint) and selectively remove certain wavelengths using colored pigment (Fig. 5.2 from Chap. 5, reproduced here as Fig. 15.1). The laws governing the appearance of color mixtures are very different for subtractive mixing than for additive mixing.

An example of this difference is the mixing of red and green. When red and green lights are mixed (such as on an electronic screen), yellow results. When red and green paints are mixed, however, the result is a dark, muddy brown color. This can be seen in Fig. 15.2. This figure shows a screen capture of these two colors being additively combined on an LED monitor, while Fig. 15.2b was created by printing a

**Fig. 15.1** Color mixing. **a** Additive; **b** subtractive

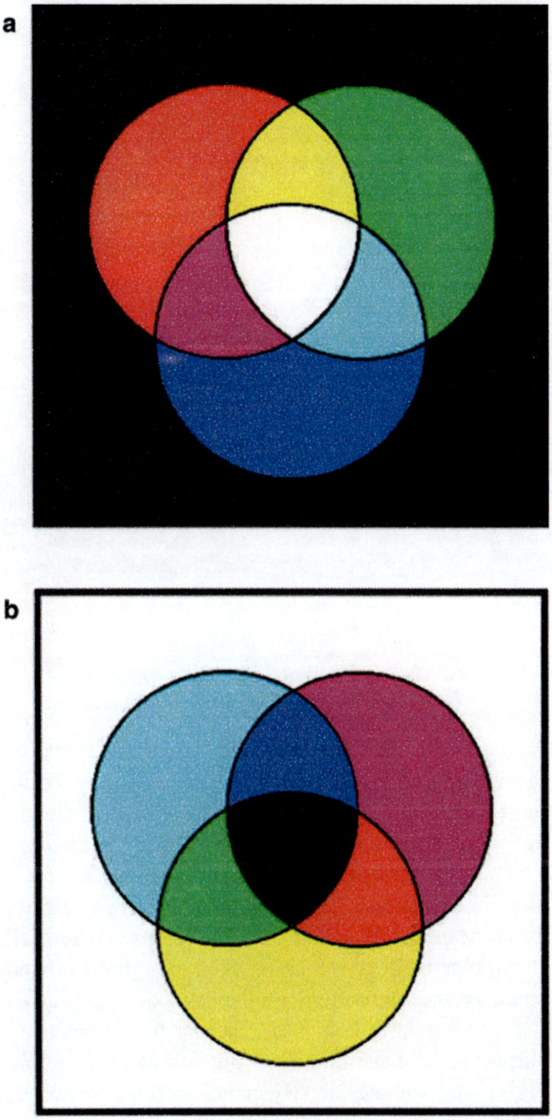


red rectangle on a piece of paper, then returning the paper to the printer and printing the partially overlapping green rectangle. The overlap region is, therefore, covered by a mixture of the inks used to make the red and green rectangles.

One of the most important aspects of additive mixing is that such mixing follows Grassmann's laws. These laws allow us to predict the color sensations of mixtures of different colors by adding their reflectances as a function of wavelength. This, in turn, led to the development of the *xy* chromaticity diagram (Fig. 6.11 in Chap. 6,

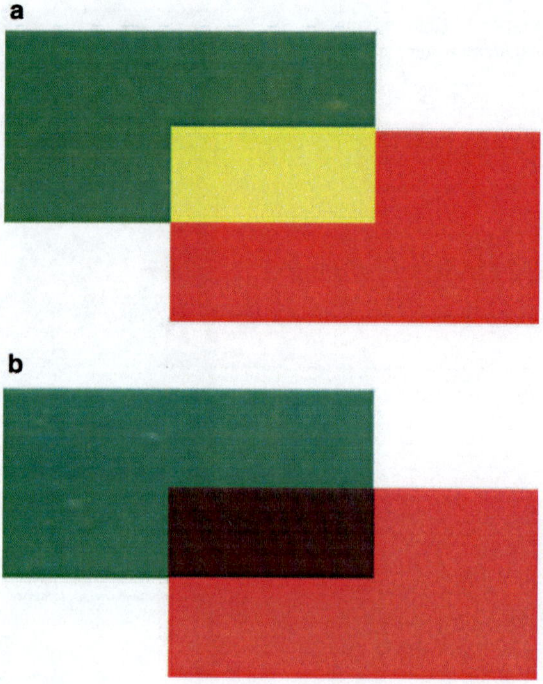


**Fig. 15.2** Different ways to mix colors. **a** Additive color mixing. **b** Subtractive color mixing

reproduced here as Fig. 15.3). The great usefulness of this diagram is that colors on the line joining any two colors can be created by mixing these two colors. In practice, this result is normally extended to three colors, allowing us to create any color within the gamut defined by the triangle that has the three primary colors as its vertices.

The aspect of Grassmann's laws that allows for this is the additivity principle of color sensations. Of special interest here, a pair of metameric colors, which have different spectra but create the same color sensation, will continue to match when each is combined with a new color, even though the reflectance spectra of the metameric colors (and of their mix with the new color) are different.

Such a mixing law does not exist for subtractive color mixing. For illustration purposes, we have contrived the following example to demonstrate this fact. In Fig. 15.4a, we see the reflectance spectra of two hypothetical metameric paints. Their color is a greenish yellow, as indicated by the inset in this Figure. Although the colors appear the same, they are derived in different ways. Paint A reflects a mix of green and red, while Paint B reflects a single band of wavelengths centered at the yellow/green interface.

Next, we add an identical cyan pigment to both paints. The new pigment gets its color by reflecting all wavelengths except those in the red portion of the visible spectrum (Fig. 15.4b). This is equivalent to saying that this pigment blocks red light and reflects all other colors. The result of adding this pigment to the reflectance spectra of the two paints is shown in Fig. 15.4c.

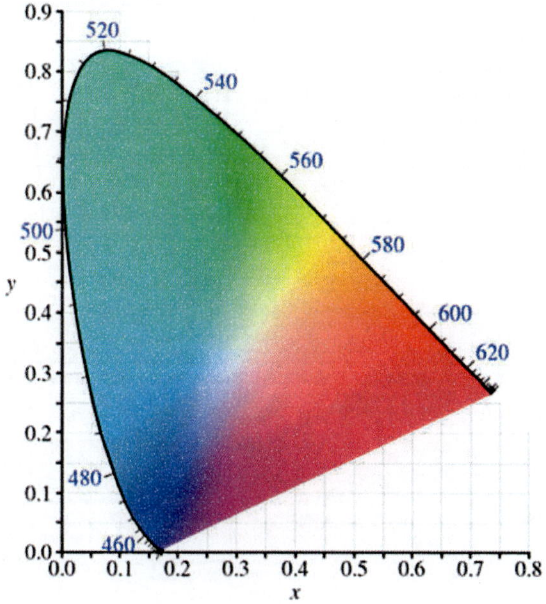


**Fig. 15.3** *xy* chromaticity diagram

For Paint A, the red portion of the reflectance curve is nearly completely removed, while the green portion remains mostly unchanged. This results in a green appearance for Paint A (see inset in Fig. 15.4c). However, for Paint B, the reflectance spectrum has not shifted greatly, and as a result remains greenish yellow (see inset in Fig. 15.4c). From this example, we can clearly see that Grassmann's laws do not apply to subtractive color mixing.

## The Principles of Simple Light Absorption

Light is both absorbed and scattered by most paints. The ramifications of the simultaneous occurrence of these processes can be quite intricate, and the analysis of paint interactions with light can be very complicated. Understanding these interactions in colored paints begins with understanding light scattering and light absorption separately from one another. In Chap. 3, we discussed light scattering in the absence of absorption. In this section, we will review the principles of light absorption in the absence of light scattering (what we term "simple light absorption"), and then later in this chapter, we will discuss simultaneous scattering and absorption.

The fundamental unit of light is the photon, and all photons have well-defined energies that are determined by their wavelength. When the energy of a photon matches the difference in energy between two energy states of an atom or molecule, that photon can be absorbed by that atom or molecule, a process that moves the

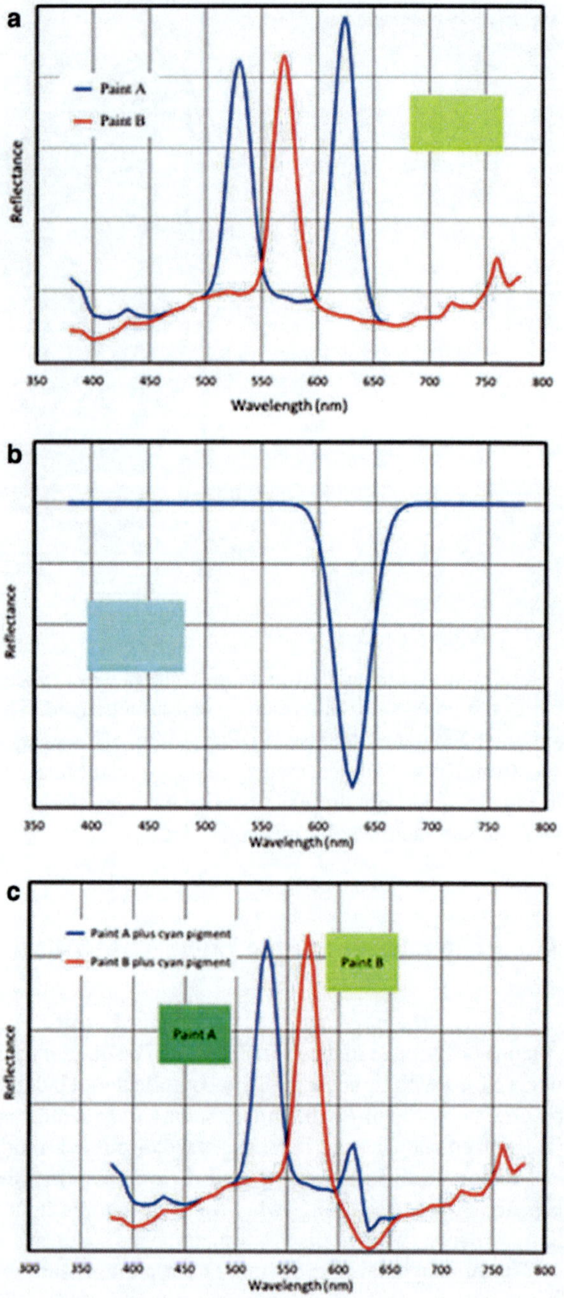


**Fig. 15.4** Combination of metameric colors **a** with a second light absorber, **b** to create reflectance spectra of different colors, **c**. Approximate colors are shown in insets

absorbing species to the higher energy state, with the transformation of light energy into another form of energy, as discussed in Chap. 8.

The amount of energy required to change the energy state of a molecule or atom depends on the form of energy into which the photon is transformed. One of the lowest energy state changes transforms light energy into rotational energy in a molecule. The differences in energies between the rotational ground state and rotational excited states match light in the microwave region of the spectrum. Energy levels associated with vibrations between atoms within a molecule are generally matched to infrared light. Ultraviolet light has enough energy to break chemical bonds (the subject of Chap. 14), while X-rays are energetic enough to completely remove electrons from an atom or molecule.

For some molecules, visible light is energetic enough to move an electron from one orbital to another (higher energy) orbital, creating an electronically excited molecule without irreversibly breaking chemical bonds or removing electrons altogether (Chap. 8). This is typically a temporary transition and nearly always the high energy electron will quickly relax into its initial (ground) state, shedding the electronic energy as thermal energy—i.e., the molecule heats up.

The relaxation of electronically excited molecules back to their initial state is of critical importance to color science since it means that the light absorbing molecules are not consumed or in some way permanently altered in this process. As such, this process can be repeated indefinitely. Such is not the case for higher energy absorptions, which irreversibly change the absorbing species.

## *Quantifying Light Absorption*

Even when a photon has the proper energy to change some energy state of an atom or molecule, there is no guarantee that it will in fact be absorbed when it strikes that atom or molecule. Instead, there is a probability that any given interaction will result in absorption. This probability is not the same for all absorbers; those species for which it is high are strong absorbers, while those species for which it is low are weak absorbers.

As discussed in Chap. 8, the total amount of light that is absorbed by a material depends on three factors: the absorption coefficient(s) of the light absorbing material(s) in it, the concentration(s) of these materials, and the distance over which the light travels. The dependence of absorption on these three properties individually is linear, and the total absorption of light passing through a material is summarized by the well-known Beer-Lambert law:

$$A_\lambda = \varepsilon_\lambda bc$$

where:

$A_\lambda$ is the total amount of light absorbed,

$\varepsilon_\lambda$ is the absorption coefficient,

b is the pathlength of the light through the materials, and

c is the concentration of the absorbing specials.

Note that absorption (A) and absorptivity ($\varepsilon$) are wavelength dependent, which is reflected in this equation and others by the subscript $\lambda$.

The amount of light absorbed by a material can also be quantified by the amount of light transmitted through it or reflected by it. The relationship between the amount of light absorbed and transmitted is given by the following equations:

$$\% T_\lambda = 10^{(2-A)}$$
$$A_\lambda = 2 - \log(\% T_\lambda).$$

If transmittance is expressed as a fraction rather than a percent, the number "2" is removed from these equations. When our interest is in the light reflected, rather than absorbed, we can replace $\%T_\lambda$ with $\%R_\lambda$ in these equations.

## *Light Absorption by Non-Scattering Mixtures*

Paints are often tinted with more than one light absorbing species (i.e., more than one colorant), and so we must consider the light absorption of mixtures to fully understand color formation and color control in paints. In Fig. 15.5, we shine white light (a combination of red, green, and blue lights) through combinations of two differently colored glass filters, one cyan and the other magenta. As we will see in the next section, these are two of the three subtractive primaries, and these filters selectively absorb red and blue light, respectively. In Fig. 15.5a and b, we see white light passing through these filters sequentially, in each case resulting in the transmittance of green light. In Fig. 15.5c, we have fused the filters together and allowed the colored pigments to diffuse randomly between them.

Intuitively, we might expect the amount of light ultimately transmitted through the filters in Fig. 15.5a and b to be the same, and they are. This is because the light interacts independently with the pigments in the two filters, which is expected since the two pigments are separated from one another. Perhaps less intuitively, the situation in Fig. 15.5c results in the transmittance of the exact same green light as in the other two configurations.[1] We can infer from this that the two pigments interact independently of one another even when they are in close proximity and intimately mixed.

---

[1] Note that the concentrations of the colorants in the fused filter are half that in each unfused filter, but the thickness of the volume containing each colorant has doubled, resulting in the same total absorption. This is consistent with the facts that the number of absorbing species is the same in the fused filter as in the individual filters, and that the absorption strength of an individual particle does not change with particle concentration or the presence of other types of particles.

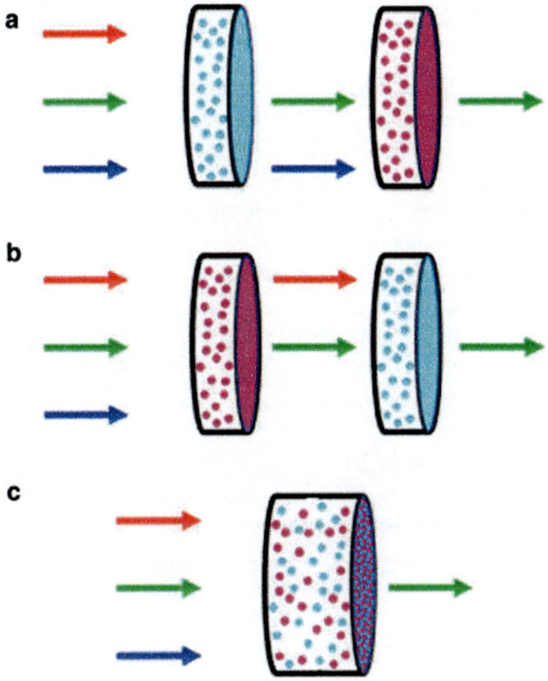


**Fig. 15.5** White light transmitted through color filters

In Fig. 15.5, each filter absorbed a different region of the visible spectrum and so there was no overlap between them. However, in the vast majority of cases where two (or more) colorants are combined, there is absorption overlap. Of concern to us is how we can calculate the spectra of such combinations. In this case, we cannot simply add or average percent transmittance curves. Instead, the percent transmittance at a particular wavelength is the product of the percent transmittance of the individual colorants at that wavelength. That is, if one colorant transmits 50% of incident light and another transmits 80%, then the total transmission is 80% of 50%, or 40%.

Because of the power relationship between percent transmittance and absorbance, multiplying transmittance values together is the equivalent of adding absorption values together. In the example above, the absorbance values for the individual components of the mix are 0.301 (for 50% T), 0.097 (for 80% T), which sum to an overall absorption value of 0.398 (equivalent to 40% T). We will see throughout this chapter the use and usefulness of the additivity of absorption values.

Finally, by definition, no light is transmitted through an opaque object such as a fully hiding paint film. Our interest here, however, is on the light that is reflected from the film, not light that is transmitted through it. That said, the same rules and equations apply to light reflected from an opaque sample as to light transmitted through a translucent one.

## Subtractive Primaries

We can divide the visible spectrum into three broad regions: a long wavelength red region, a mid-wavelength green region and a short wavelength blue region. Roughly speaking, we can say that a representative wavelength from each of these regions can be used as a primary color for additive color mixing (as is the case for RGB color space). Choosing one primary from each region gives us a wide color gamut.

For the subtractive primaries, we begin with a background illuminated by white light, and subtract light from each of the three regions. We can think of these primaries as "white light without red" (that is, blue mixed with green, giving cyan), "white light without green" (that is, blue mixed with red, giving magenta) and "white light without blue" (that is, green mixed with red, giving yellow). We create a wide color gamut when we remove the appropriate amount of color represented by each of these primaries from white. These primaries, and the colors made from them, were shown earlier in Fig. 15.1b.

A significant difference between additive and subtractive color mixing is that for additive mixing we can use a single wavelength for each primary. However, for subtractive mixing, we must use primaries that absorb a range of wavelengths. This is because removing only one wavelength of light has a very small effect on color. Colors span a range of wavelengths, and large regions of this range must be removed to have the desired effect on color.

## Metamerism Revisited

In Chap. 5, we discussed metamerism, which is the phenomenon by which two different stimulus spectra give us the same color sensation. Our discussion was in the context of additive color mixing, where three well-chosen primaries can combine to reproduce most color sensations, even those for which the stimulus spectrum is quite complex. In this context, metamerism is very useful—it is, in fact, the basis by which all color screens and monitors operate, and this form of metamerism makes possible a gamut of millions of colors using only three primary lights.

There are two other forms of metamerism, and these are far less desirable. The first is observer metamerism. In this situation, two different stimulus spectra appear as identical colors to one observer, but different colors to another, even under the same viewing conditions. This comes about because visual perception differs from person to person. Of particular importance are the exact absorption characteristics of the color-sensing chemicals (opsins) in the cone cells, the distribution of these cells on the retina, and age-related yellowing of the lens. All of these vary slightly from one person to another, resulting in color matching functions that vary slightly from one person to another. As an extreme example, a person with color blindness may judge green and red panels to look the same (that is, to be metameric), while someone with more normal vision would see a sharp difference between the two.

The final form of metamerism that we will consider is termed "illuminant metamerism". This form of metamerism can be a serious issue for the paint maker. Here, we find that two colors match when illuminated by one light source but appear different when illuminated by another. This occurs when the reflectance spectra of the two objects are different, but in such a way that when we multiply them by a certain source spectrum, to give a stimulus spectrum, the resulting color sensations are the same (that is, the stimulus spectra have the same tristimulus values). When illuminated with a different source, the two objects no longer have the same tristimulus values and so their color sensations will be different.

A visual example of this can be seen in Chap. 5, Fig. 5.11, reproduced here as Fig. 15.6. In this figure, a dark blue square was printed on a sheet of white paper, and

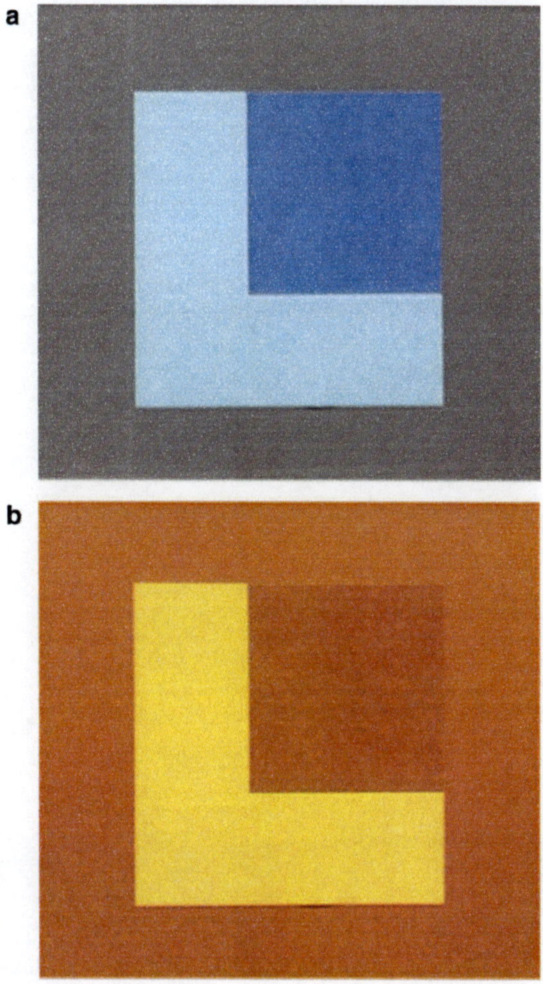


**Fig. 15.6** Photographic example of metamerism. Both images are of the same composition. **a** Illuminated by white light; **b** Illuminated by yellow light. Note that the color of the light itself is seen in the "L" shaped region

this paper was laid over a mauve background. Figure 15.6a shows the appearance of the paper, the printing, and the background under normal fluorescent lights. Here the difference in color between the blue square and the mauve background is readily apparent. However, in Fig. 15.6b, the same scene is illuminated by a yellow light, the color of which can be seen in the white portion of the paper. Under this light, the blue square and mauve background give nearly identical color sensations.

We can demonstrate this form of metamerism using the reflectance curves from earlier (Fig. 15.4). In that figure, we showed the reflectance curves for two different paints as viewed under Illuminant E. As noted earlier, the tristimulus values for these curves are equal, and so the color sensations of these paints were the same. However, if we change illuminants, we shift these color sensations differently. Figure 15.7 shows the approximate colors of these paints under CIE Illuminants E, $D_{65}$ and FL1. Under Illuminant E the color sensations are equal, as shown in Fig. 15.4a. Under Illuminant $D_{65}$ the color sensations are close, but not quite the same. However, under Illuminant FL1, the color sensations of the two paints are quite noticeably different.

Illuminant metamerism in the coatings industry typically occurs for one of two reasons. The first is when a paint color is matched to a non-paint color. For example, portions of a car exterior are painted while other portions are made of colored plastic

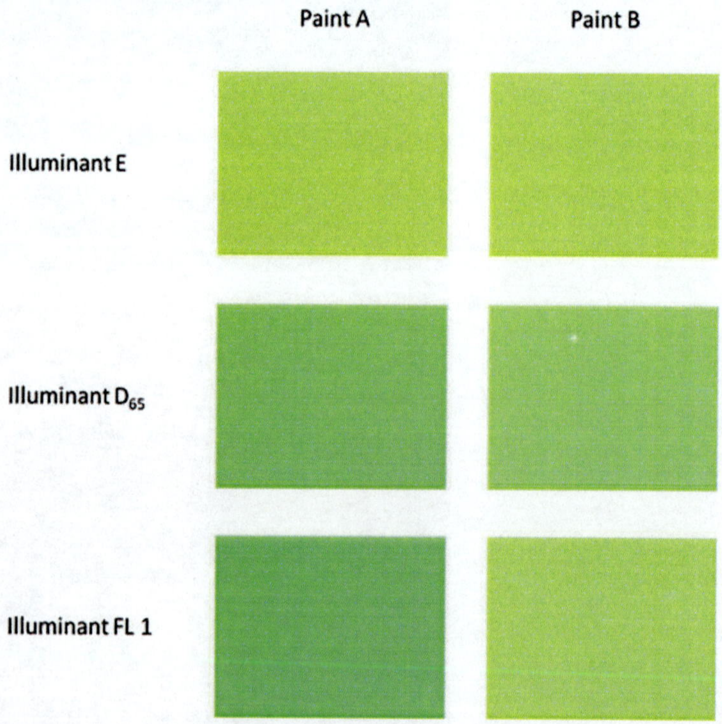


**Fig. 15.7** Approximate colors of paints in Fig. 15.4a illuminated by three different light sources

(e.g., the bumper). The colors of the paint and plastic must match one another under any viewing condition. Simply having the same pigments in the same proportions does not guarantee a color match. In addition, different pigments are often used by the plastics and coatings industries. Often the automobile industry will simply paint exterior plastic parts rather than try to attain a so-called "in-mold" color match, in which the polymer is pigmented such that an accurate and non-metameric match can be made.

The second common occurrence of illuminant metamerism occurs when the color of a paint made with one set of pigments is matched to a paint made with a different set of pigments. For example, a paint formulator might wish to duplicate the green color of a certain automotive paint for spot or collision repair but does not have the same green pigment as was used in the original paint. In this case, a match can be successfully made under one type of illumination, only to look different under another illumination. Because different green pigments have different reflectance spectra, it is generally not possible to make a match using two different colored pigments that is not metameric under some lighting condition.

## Application to Paints

We discussed light absorption in the absence of scattering above, and light scattering in the absence of absorption in Chap. 3. While some paints interact with light in one of these ways but not the other, most paints—all those that are colored and opaque—scatter and absorb light simultaneously. As might be anticipated, combining these types of light interactions complicates the analysis of color and of color mixing.

The basis for understanding color mixing in paints is a theoretical framework reported by Kubelka and Munk in the early 1930s [2]. They approached the issue of light interactions in paint films by treating a paint film as a homogenous material with certain light scattering and light absorption properties. This ignores the fact that paints are composed of discrete particles that interact with light and that are embedded in a continuous phase of nearly transparent resin. In the Kubelka–Munk model, the mechanisms of light scattering and light absorption are unimportant—what is important is that these processes occur and that they follow certain rules [3].

The Kubelka–Munk approach to light interactions in films containing only scattering particles was discussed in detail in Chap. 13. Here, we will focus on a different aspect this approach—how we can predict color resulting from the mixing of differently colored paints, or when more than one colorant is added to a white base paint. Our goals are to develop a procedure for matching a specific color with a set of colorants and to predict the color of a mixture of colorants or colored paints.

A recurring quantity in the Kubelka–Munk analysis is the ratio K/S, that is, the balance between the amount of light scattered out of an opaque film (S) and the amount of light absorbed by that film (K). In the general case of light interactions with a film, there is a third possible fate of a photon—exiting the bottom surface of the paint film (where it will interact with the substrate). By definition, films at complete hide have no light crossing their lower surface (the film/substrate interface).

In the analyses of one-constant Kubelka–Munk theory below, we consider only opaque films, and all photons striking such films are either absorbed or reflected back through the film surface. Under these conditions, it is possible to measure the ratio of K/S, but not K and S values separately. The "one constant" in the one-constant Kubelka–Munk analysis is the ratio of K/S. In two-constant Kubelka–Munk theory, also discussed below, we use the individual values of K and S (these are the "two constants"), which are calculated from the optical properties of films that are not thick enough to give complete hide.

S and K are already familiar to us from Chap. 13. The total amount of scattering and absorption in an opaque paint film is simply S times X and K times X, respectively, where "X" is the film thickness (analogous to "b" in the absorption equation). In colored paints both K and S vary over the wavelength range of visible light, and we typically measure them at wavelengths from 400 to 700 nm, in intervals of 5 nm or 10 nm.

The balance between the total amount of light absorbed by a paint film and the total amount of light scattered by it is KX/SX, which simplifies to K/S. Paints with large K/S ratios are dark and have low reflectance values, whereas those with small K/S ratios are light and have high reflectance values. Kubelka and Munk showed that the equations governing the wavelength-dependent relationship between K/S and reflectance, measured on an opaque film, are[2]:

$$(K/S)_\lambda = \frac{(1 - R_\lambda)^2}{2R_\lambda} \tag{15.1}$$

$$R_\lambda = 1 + (K/S)_\lambda - \sqrt{(K/S)_\lambda{}^2 + 2(K/S)_\lambda} \tag{15.2}$$

In these equations and those that follow, reflectance values must be expressed as fractions, from 0 to 1, rather than as percentages, from 0 to 100. That said, reflectance spectra will be shown in various figures in this chapter as percentages, as this conforms to the typical way that absorption is measured and reported.

Equation (15.1) provides us with a simple way to determine the ratio K/S from measured reflectance values at complete hide. This information is sufficient for calculating the color mixing behavior of paints and colorants that contain only one scattering component (typically $TiO_2$). However, the color mixing behavior of paints

[2] Note that what we are calling "$R_\lambda$" here is the reflectance, at each wavelength, of the paint when it is applied thickly enough to give complete opacity. In Chapter 13, we saw similar equations that referenced "$R_\infty$", which is the total reflectance over the entire visible spectrum (that is, the tristimulus Y value), as determined on a film thick enough for complete hide, rather than the reflectance at specific wavelengths of light. As such, $R_\infty$ is not a function of a particular wavelength of light but is instead a measure of the overall lightness of the paint.

and colorants with more than one light scattering material is controlled by the individual values of K and S. Determining these values separately is more difficult than determining their ratio K/S—the calculations involved are detailed in Chap. 13.

## *Colored Paints*

In our discussion above we showed that light absorption values are additive—that is, we can determine the combined light absorption of a mixture of materials (e.g., different colorants) by adding together their separate absorbances. The contribution to absorbance by a given species is simply its unit absorption coefficient k multiplied by its concentration C.[3] We can therefore express mathematically the absorption value of a mixture as:

$$K_{mixture,\lambda} = C_1k_{1,\lambda} + C_2k_{2,\lambda} + C_3k_{3,\lambda} + \cdots = \sum C_nk_{n,\lambda} \quad (15.3)$$

As in previous equations, and in those that follow, the symbol "λ" indicates values that are wavelength dependent and so must be measured or calculated across the visible light spectrum. The $C_n$ values are the concentrations of each absorbing species, and the $k_{n,\lambda}$ values are the unit absorptivities of each species at each measured wavelength.

Similarly, the scattering value (S) of a mixture is determined by simply adding the individual scattering values for the constituent materials, also adjusted for concentration:

$$S_{mixture,\lambda} = C_1s_{1,\lambda} + C_2s_{2,\lambda} + C_3s_{3,\lambda} + \cdots = \sum C_ns_{n,\lambda} \quad (15.4)$$

The K/S ratio for the mixture is simply the sum of K values for the individual components (adjusted for concentration) divided by the sum of S values (also adjusted for concentration) [4]:

$$(K/S)_{mixture,\lambda} = \frac{C_1k_{1,\lambda} + C_2k_{2,\lambda} + C_3k_{3,\lambda} + \ldots}{C_1s_{1,\lambda} + C_2s_{2,\lambda} + C_3s_{3,\lambda} + \ldots} = \frac{\sum C_nk_{n,\lambda}}{\sum C_ns_{n,\lambda}} \quad (15.5)$$

---

[3] We will follow the established convention of notating the total absorption and scattering of a paint using capital (upper-case) letters K and S while using the lower-case letters k and s to designate unit absorption and scattering coefficients. The latter parameters must be multiplied by concentration to give the total amounts of absorption and scattering. That is,

$K_{i,\lambda} = Ck_{\lambda}$

$S_{i,\lambda} = Cs_{i,\lambda}$

for each component of the paint system.

## *Color Matching*

Matching a paint to a specific color is a multi-step process [5]. It begins with the generation of a library of the spectrum of each available constituent that interacts with light, including all colorants and the white base paint that is to be used for the color match. In addition, the spectrum of the color to be matched is measured.

Next, spectra are calculated for combinations of selected colorants, and these spectra are compared to that of the target color. Colorant selection is chosen based on the target color, either by prior experience or through an automated selection process. A least-squares analysis is used to quantify the degree to which the spectrum of particular combinations matches the target spectra [6]. This program can optimize color in one of two ways: either to get the least deviation between the target and calculated spectra (for example, minimizing the area of a graph of the difference between the two spectra), or to get the closest color match based on the CIE X, Y and Z tristimulus values calculated for the spectra [1, 5].

Unless the exact same colorants are used in the test paint and target paint, it is likely that there will be some degree of metamerism between the two (if the same colorants are used, then an exact match can be made, resulting in no metamerism). Optimizing for the least deviation between the two spectra typically results in less severe metamerism, while optimizing on tristimulus values typically results in a better match under a single lighting condition but is more prone to the generation of metameric colors. In addition, minimizing the number of colorants used to match the spectra decreases the likelihood and severity of metamerism.

## *One-Constant Kubelka–Munk Theory and Practice*

Equation (15.5) simplifies considerably when only one of the mixture components contributes to scattering. This situation describes the coating formulations for which $TiO_2$ is the only light scattering material (that is, scattering from extenders and colorants is insignificant due to their much lower indices of refraction). This leads us to the so-called one-constant equation Kubelka–Munk equation [7]:

$$\begin{aligned}(K/S)_{mixture,\lambda} &= \frac{C_w k_{w,\lambda} + C_1 k_{1,\lambda} + C_2 k_{2,\lambda} + \ldots}{C_w S_w} \\ &= (k_w/S_w)_\lambda + \frac{C_1}{C_w}(k_1/S_w)_\lambda + \frac{C_2}{C_w}(k_2/S_w)_\lambda + \ldots \end{aligned} \quad (15.6)$$

where the subscript “w” refers to the white pigment. The term “one constant” refers to the fact that in this procedure we use the ratio K/S for each component and do not calculate K and S separately. We can measure this ratio for a given paint by measuring the reflectance of that paint at complete hide (Eq. 15.1). To measure K and S separately, as is done in the two-constant approach, we must measure the reflectances of paint films that are not at complete hide.

In many paints, the white pigment is incorporated into a base paint to which colorant is added. Assuming that the volume of colorant added is small compared to the volume of white base, the concentration of the white pigment will not change when tinted. We will therefore define the unit concentration of the white pigment as its concentration in the base ($C_w = 1$). This simplifies the above equation to:

$$(K/S)_{mixture,\lambda} = (K/S)_{w,\lambda} + C_1(k_1/S_w)_\lambda + C_2(k_2/S_w)_\lambda + \ldots \quad (15.7)$$

## *Calculating the Colors of Paint Mixtures*

When mixing colorants into a white base paint, we can use Eq. (15.7) above to model and predict the color resulting from a colorant mixture, if we know the reflectance spectrum of each colorant when incorporated individually into the base paint. To demonstrate this, we will consider three color mixing examples that are representative of this practice.

For these examples, we will use the same white base paint. We will make the assumptions that all scattering comes from the white base, that the white base may absorb some light (that is, it may not be a true white), that the volume of colorant added to the base paint is insignificant compared to the volume of the base paint itself, that the drawdowns we analyze are made at sufficient thickness to give us complete hide at all visible light wavelengths, and that there are no interactions between pigments that would alter either the s or the k of any species. Finally, we will assume that the unit absorptivity k for each colorant is constant over all concentrations considered. Of these assumptions, this final one is the most likely to be violated in practice, resulting in computational inaccuracies for dark paints. In those cases, an equation should be developed relating k to concentration, and this equation used in place of a constant k value in any calculations.

## *The Saunderson Correction*

In addition to these assumptions, there is an implicit assumption in most reflectance measurements that there is no loss of light intensity due to reflection at the surface of the film when light passes between air and the film, or when light travels from the film into the substrate. Since our concern is with the reflectance events that occur within the paint film ("internal reflectance"), we may wish to correct the measured R values so as to remove the effects of any reflectances that occur at the air/film or film/substrate interfaces and avoid this implicit assumption. While this is an important consideration in some situations and studies, for others the measured reflectances are sufficient for analysis [8].

The factor controlling whether or not corrected reflectances are needed is the type of analyses under consideration. When analyzing the reflectances of paints of different colors, but otherwise similar compositions and properties (i.e., same gloss, same $TiO_2$ content, same resin, etc.), the effects of surface reflectance nearly cancel, and we can safely use the measured reflectance values rather than corrected ones. We will test this assumption in one of our worked examples by comparing $R_\lambda$ values calculated on both a corrected and uncorrected basis. On the other hand, if the motivation of the work is to match measured reflectance data to theoretical data, to measure reflectance with high accuracy, or to derive new insights in light interactions in a paint film, then the measured data must be corrected for surface reflectance.

Reflection at surfaces was first studied by Fresnel. Surface reflection occurs whenever there is a discontinuity in refractive index—that is, whenever light passes from a material with one refractive index into a material with a different refractive index. The magnitude of Fresnel reflectance is determined by the degree of difference between the refractive indices (see also Chap. 3, Eq. 3.2)[4]:

$$\text{Reflected intensity} = \left[\frac{n_1 - n_2}{n_1 + n_2}\right]^2. \tag{15.8}$$

For our purposes, $n_1$ is the refractive index of air (1.0) and $n_2$ is the refractive index of the film, which is typically assumed to be 1.5 for most paints.

There are some situations where the effects due to reflection when comparing two films do not adequately cancel. In these situations, a correction can be made to the measured reflectance values that remove from them the contribution of surface reflectance. This correction was developed by Saunderson [9], and takes into account the surface reflection that occurs both when the light enters the film and when internally reflected light leaves through the top surface of the film (since we are only considering films with complete hide, none of the light will leave through the bottom surface, or back, of the film). This correction is:

$$R_{\lambda,i} = \frac{R_{\lambda,m} - K_1}{1 - K_1 - K_2 + K_2 \cdot R_{\lambda,m}} \tag{15.9}$$

where $R_{\lambda,i}$ is the internal reflectance within the film (which is the reflectance value used in the Kubelka–Munk equations), $R_{\lambda,m}$ is the measured reflectance, $K_1$ is the Fresnel reflectance (Eq. 15.8), which corrects for light reflecting when traveling from air to the film, and $K_2$ is a correction for the reflection of light that occurs at the film/air interface when the light exits through the top surface of the film. Note that

[4] This equation applies only to light that is incident normal to the surface. The relationship between the fraction of light reflected and the angle of incidence is complex.

the symbol "K" here is used much differently than in the Kubelka–Munk equations, where "K" is the absorption coefficient.

Assuming the light within the film is completely diffuse (that is, its direction is completely randomized), $K_2$ can be calculated by integrating the Fresnel equations from 0 to 90 degrees. The value for $K_2$ generally falls between 0.4 and 0.6. A $K_2$ value of 0.40 has been reported to give good results [10], although some workers prefer to use $K_2$ as an adjustable parameter to optimize the fit between a calculated reflectance spectrum and its measured counterpart [7, 11]. In any event, the actual value used for $K_2$ is less important than is using the same value in all calculations [12].

The Saunderson correction is applied as follows [5]: The measured reflectance values are transformed into internal reflectance values ($R_{\lambda,i}$) using Eq. (15.9). These reflectances are used to determine the $(K/S)_\lambda$ values for the film, using Eq. (15.1). The $(K/S)_\lambda$ values are then processed (by the procedure shown in the worked examples) to give the internal reflectance spectra of mixed color paints. These calculated internal reflectance values are then be back-transformed into corrected measured reflectance values ($R_{\lambda,\mathrm{corrected}}$), using this equation [9]:

$$R_{\lambda,corrected} = K_1 + \frac{(1 - K_1) \cdot (1 - K_2) \cdot R_{\lambda,i}}{1 - K_2 \cdot R_{\lambda,i}} \tag{15.10}$$

## *Worked Examples*

There are several different situations where we would like to accurately predict the reflectance spectrum of a colored paint. The two most common are matching a specific color using a limited number of colorants with known reflectances, and predicting the color formed when two or more colorants or paints with known reflectance spectra are mixed.

This process begins by measuring the optical properties of the colorants or colored paints containing the colorants. These measurements are then combined mathematically to determine the expected spectra of paint mixtures.

The steps for this process are:

1. Measure the reflectance spectrum of the paints that will be the basis of our mixtures.
2. If desired, apply the Saunderson correction to the measured reflectance values using Eq. (15.9).
3. Convert the reflectance values into K/S values using Eq. (15.1).
4. Combine the K/S values measured for the basis paints for the mixture of interest using Eq. (15.6).
5. Convert the K/S values for the mixture to reflectance values using Eq. (15.2).

6. If the Saunderson correction was used (step 2), back-transform the reflectance values using Eq. (15.10).
7. The calculated reflectance spectrum for the mixture can be compared to the measured reflectance spectrum if desired. The color sensations of each can be converted to L*a*b* values by applying the procedure described in Chap. 6. The difference in calculated and measured colors can be quantified as ΔE and used to determine the validity of the calculated spectrum.

In the examples below, we will apply Saunderson correction to the measured reflectance values using 0.04 for the $K_1$ value and 0.40 for the $K_2$ value. In these examples, we will evaluate four situations that are commonly encountered in color mixing—mixing two paints of different color, diluting a colored paint with its white base paint, adding more colorant to a paint that already has that same colorant (i.e., increasing the concentration of that colorant), and adding a colorant to a paint that already has a different colorant (that is, to a paint of a different color). Because the same base paint is used in all of these examples, the $TiO_2$ content will be the same for all mixtures of paints, making the scattering ($S_w$) and absorption ($K_w$) contributions of the white base paint the same in all colored paints. Note that this would not be true if we were mixing paints with different $TiO_2$ concentrations.

Our examples will make use of two colorants—one purple, the other yellow—and four paints—a white base paint, and a purple paint, a yellow paint and a gold paint (the latter made by doubling the yellow colorant amount in the yellow paint). The visible light reflectance curves for the four paints, measured in 10 nm intervals, are shown in Fig. 15.8. We see that the white paint reflects only ca. 90% of the incident light, rather than 100%, which technically makes this paint a very light gray color,

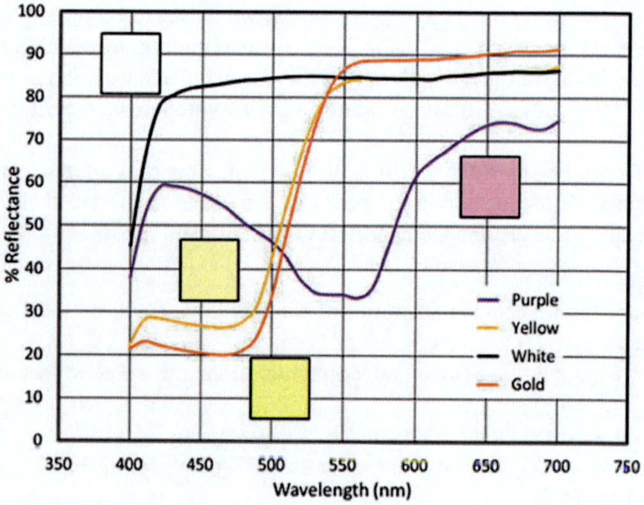


**Fig. 15.8** Reflectance spectra of the four paints used in the color mixing studies. Insets show paint color

rather than true white. This is not uncommon for "white" paints, and most people would consider this level of brightness to signify white, rather than gray.

In our first example, we will mix the yellow paint with the purple paint. We will not initially choose specific amounts of the two paints for this mixture, but instead solve the problem more generally. We begin by defining the K/S values for the initial paints, starting with the yellow paint. This paint is a mixture of yellow colorant (yc) and white paint (w). Since we are deriving a general solution to such mixing, we will leave these concentrations as parameters rather than numeric values.

Substituting into Eqs. (15.3) and (15.4), we get:

$$K_{yellow\ paint,\lambda} = C_{yc}k_{yc,\lambda} + C_w k_{w,\lambda} \tag{15.11}$$

$$S_{yellow\ paint,\lambda} = C_{yc}s_{yc,\lambda} + C_w s_{w,\lambda} = C_w s_{w,\lambda} \tag{15.12}$$

Note that we have removed the term $C_y s_{y,\lambda}$ from the scattering component (Eq. 15.12) because we are assuming that the colorant does not scatter light (if the colorant scatters light, then the two-constant Kubelka–Munk approach must be used, as outlined below). Although the ideal white paint would not absorb any light, as mentioned above and clearly seen in Fig. 15.8, the white base paint in our case does absorb some light and so we must include the contribution $C_w k_{w,\lambda}$ in the absorbance component of this paint.

From Eqs. (15.5) and (15.11) we define the K/S value for the yellow paint as:

$$(K/S)_{Yellow,\lambda} = \left(K_{yellow\ paint,\lambda}/S_{yellow\ paint,\lambda}\right) = \left(K_{yellow\ paint,\lambda}/C_w s_{w,\lambda}\right) \tag{15.13}$$

Similarly, for the purple paint:

$$(\mathrm{K}/S)_{purple,\lambda} = \left(K_{purple\ paint,\lambda}/C_w s_{w,\lambda}\right) \tag{15.14}$$

We next consider the mixture of A parts yellow paint and B parts purple paint. Substituting Eqs. (15.11) and (15.12), and their purple counter-parts, into Eq. (15.5) gives, for the mixture,

$$\begin{aligned}(K/S)_{mixture,\lambda} &= \frac{\sum C_n k_{n,\lambda}}{\sum C_n s_{n,\lambda}} = \frac{A \cdot K_{yellow\ paint,\lambda} + B \cdot K_{purple\ paint,\lambda}}{A{\cdot}C_w s_{w,\lambda} + B{\cdot}C_w s_{w,\lambda}} \\ &= \frac{A \cdot K_{yellow\ paint,\lambda} + B \cdot K_{purple\ paint,\lambda}}{(A+B){\cdot}C_w s_{w,\lambda}} \\ &= \left(\frac{A}{(A+B)}\right) \cdot \frac{K_{yellow\ paint,\lambda}}{C_w s_{w,\lambda}} + \left(\frac{B}{(A+B)}\right) \cdot \frac{K_{purple\ paint,\lambda}}{C_w s_{w,\lambda}}\end{aligned} \tag{15.15}$$

Substituting Eqs. (15.13) and (15.14) into Eq. (15.15) gives:

$$(K/S)_{mixture,\lambda} = \left(\frac{A}{(A+B)}\right) \cdot (K/S)_{Yellow,\lambda} + \left(\frac{B}{(A+B)}\right) \cdot (K/S)_{purple,\lambda}$$

or

$$(K/S)_{mixture,\lambda} = C'_{y} \cdot (K/S)_{Yellow,\lambda} + C'_{p} \cdot (K/S)_{purple,\lambda} \quad (15.16)$$

where $C'_y$ is the fractional concentration of the yellow paint and $C'_p$ is the fractional concentration of the purple paint. Note that, since $C'_y$ and $C'_p$ add up to 1.0, the scattering component of all mixtures, regardless of the values of A and B, are the same and identical to the scattering component of the white base paint.

Equation (15.16) can be worded as "the K/S for the mixture is the weighted sums of the (K/S) values of the components". This is something we may have expected intuitively based on the additive natures of the absorption and of scattering.

A comparison of the calculated reflectance values of a 3:1 mix of the purple and yellow paints to the measured values for a physical mix of these paints is shown in Fig. 15.9. As can be seen, there is excellent agreement between the calculated and measured reflectance values. Such agreement confirms the validity of this type of analysis as well as the complete compatibility between the different color pigments.

We next look at the dilution of a colored paint with its white base paint. Here, we will consider a series of three dilutions of the gold paint with the white base paint, as well as the gold and white base paints themselves. Such a series of paints is called a "tint ladder" and is very useful in determining whether the k and s values

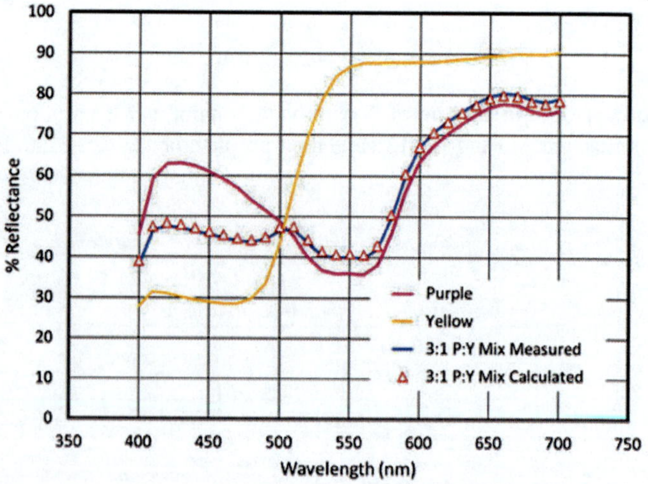


**Fig. 15.9** Reflectance spectra for purple and yellow paint mixture (3:1 by volume)

(that is, the unit absorptivity and unit scattering constants) for the colored components of the paint are constant over the range of concentrations that they bracket. This is a basic assumption in paint color mixing but is often violated, especially at high colorant concentrations. If this assumption is found to be violated, then a mathematical expression of the k or s values as a function of concentration must be determined, and this expression is used in place of the constants k and s in the color mixing equations. Obviously, this adds much complexity to these calculations.

The following five paints are used in this example:

- undiluted gold paint
- 75% gold paint, 25% white base
- 50% gold paint, 50% white base
- 25% gold paint, 75% white base
- white base.

Note that the third paint is identical to the yellow paint used in the previous example (i.e., it is the paint with the 2 × concentration of yellow pigment that is then diluted by one half). In this example, we will calculate the predicted spectra both with and without the Saunderson correction. We can then compare the calculated results to the measured results to determine the magnitude of this correction for these paints.

To calculate the predicted reflectance spectra of the different dilutions, we will use Eq. (15.16), but with the yellow and purple paints being replaced by the gold and white paints (Eq. 15.17). Importantly, to calculate the optics of, for example, the gold paint diluted to 75% with the white base paint, we do not simply take three quarters of the K/S value of the gold paint. Doing so would ignore the contribution that the added white base makes to the absorption of the mixture.

$$(K/S)_{mixture} = C'_y \cdot (K/S)_{gold,\lambda} + C'_p \cdot (K/S)_{white,\lambda} \tag{15.17}$$

The measured reflectance values of the five paints are shown as solid lines in Fig. 15.10. In this figure, the reflectance values predicted for the three mixtures are shown as symbols and the measured reflectance values are given as lines. Figure 15.10a shows the results when the Saunderson correction is omitted and Fig. 15.10b shows the results with the Saunderson correction included. In this case, the fit of predicted and measured reflectances improves with inclusion of this correction, although even the uncorrected values are close to those measured.

We next consider the addition of colorant into a paint already colored by this same colorant. In this case, we will calculate the reflectance spectrum of the yellow paint to which a second, identical dose of yellow colorant was added (that is, we will calculate a predicted spectrum for the gold paint, based on the addition of more yellow colorant to the yellow paint).

Our first inclination may be to simply double the K/S value for the initial yellow paint (that is, we double K and leave S unchanged). Doing so, however, will also double the contribution of the white base paint to the total absorption, which is incorrect since the concentration of the white pigment in the double dosed paint is the same as in the single dosed paint.

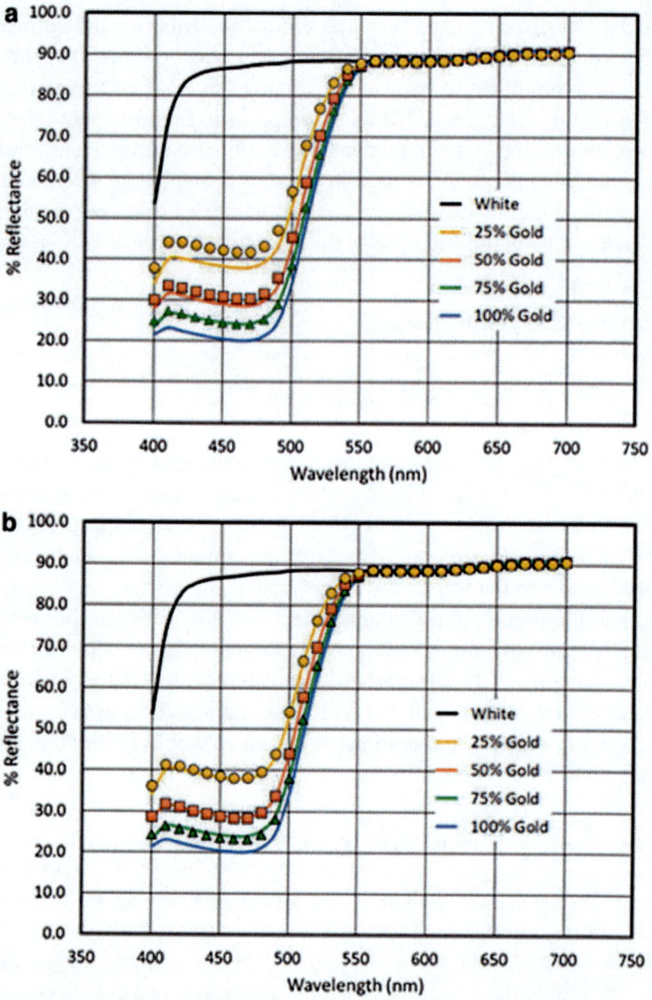


**Fig. 15.10** Reflectance spectra of diluted gold paints. **a** Uncorrected. **b** With Saunderson correction

Instead we begin with Eqs. (15.3)–(15.5) and define the individual absorption and scattering contributions of the components of the double dosed paint, which gives us:

$$K_{double\,dose,\lambda} = \left(C_{yc}k_{yc,\lambda} + C_w k_{w,\lambda}\right) + C_{yc}k_{yc,\lambda} \tag{15.18}$$

$$S_{double\,dose,\lambda} = C_w s_{w,\lambda} \tag{15.19}$$

$$(K/S)_{double\,dose,\lambda} = \frac{\left(C_{yc}k_{yc,\lambda} + C_w k_{w,\lambda}\right) + C_{yc}k_{yc,\lambda}}{C_w s_{w,\lambda}}$$

$$= \frac{\left(C_{yc}k_{yc,\lambda} + C_w k_{w,\lambda}\right)}{C_w s_{w,\lambda}} + \frac{C_{yc}k_{yc,\lambda}}{C_w s_{w,\lambda}} \tag{15.20}$$

The terms in parentheses in Eqs. (15.18) and (15.20) are the K contribution of the original paint, and the $C_{yc}k_{yc,\lambda}$ term is the contribution from the added colorant.

Equation (15.20) is partially solvable with the information we have on hand because we know, from

Equations (15.11) and (15.12), that $(K/S)_{yellow\ paint}$ is:

$$(K/S)_{Yellow,\lambda} = \frac{C_{yc}k_{yc,\lambda} + C_w k_{w,\lambda}}{C_w s_{w,\lambda}} \tag{15.21}$$

We can combine Eqs. (15.20) and (15.21) to give:

$$(K/S)_{double\,dose,\lambda} = (K/S)_{Yellow,\lambda} + \frac{C_{yc}k_{yc,\lambda}}{C_w s_{w,\lambda}} \tag{15.22}$$

We have the measured reflectance curve of $(K/S)_{yellow\ paint}$ (Fig. 15.8), but we do not have the absorption values for the yellow colorant. We can measure this separately on the colorant alone—this is referred to as a masstone—but in the present case, we can calculate the absorption spectrum of the colorant based on the absorption spectra of the white paint and yellow paint. To do this, we modify the right side of Eq. (15.20) by adding and subtracting $C_w k_{w,\lambda}/C_w s_{w,\lambda}$ to it:

$$(K/S)_{double\,dose,\lambda} = \frac{\left(C_{yc}k_{yc,\lambda} + C_w k_{w,\lambda}\right) + C_{yc}k_{yc,\lambda} + C_w k_{w,\lambda}}{C_w s_{w,\lambda}} - \frac{C_w k_{w,\lambda}}{C_w s_{w,\lambda}} \tag{15.23}$$

This simplifies to:

$$(K/S)_{double\,dose,\lambda} = \frac{2 \cdot \left(C_{yc}k_{yc,\lambda} + C_w k_{w,\lambda}\right)}{C_w s_{w,\lambda}} - \frac{C_w k_{w,\lambda}}{C_w s_{w,\lambda}} \tag{15.24}$$

The first term on the right-hand side of Eq. (15.24) is simply double the K/S value of the yellow paint (refer to Eq. 15.21). The other term in Eq. (15.24), $C_w k_{w,\lambda}/C_w s_{w,\lambda}$, is simply the (K/S) value of the white base paint. Based on these observations, we can restate Eq. (15.24) as:

$$(K/S)_{double\,dose,\lambda} = 2 \cdot (K/S)_{Yellow,\lambda} - (K/S)_{w,\lambda} \tag{15.25}$$

In Fig. 15.11, we see a comparison of the spectrum calculated from Eq. (15.25)

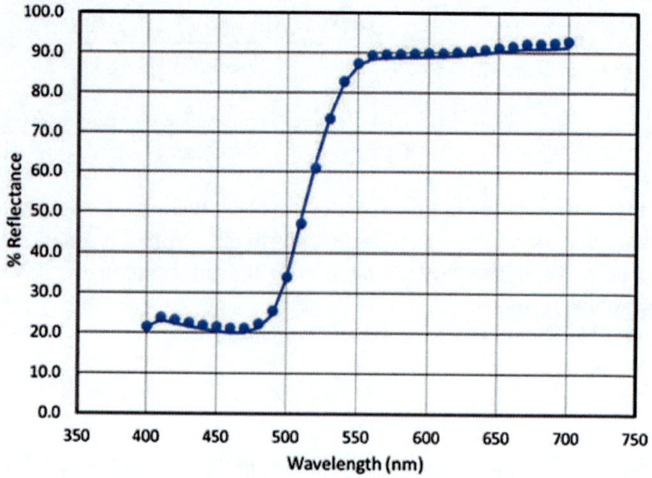


**Fig. 15.11** Comparison between measured (line) and calculated (circles) reflectance values for a yellow paint with additional yellow colorant (i.e., the measured results are for the gold paint)

and the spectrum measured for the gold sample. In this case, the two spectra match nearly perfectly.

Our final example is for the addition of the yellow colorant into the purple paint. Here we will add the same amount of yellow colorant as used to make the yellow paint. This example is similar to that above, where yellow colorant was added to a yellow paint, but here with the complication that there are two colorants involved, rather than one.

Again we begin with Eqs. (15.3)–(15.5):

$$K_{yellow\ dose\ into\ purple,\ \lambda} = \left(C_{pc}k_{pc,\lambda} + C_w k_{w,\lambda}\right) + C_{yc}k_{yc,\lambda}$$

$$S_{yellow\ dose\ into\ purple,\ \lambda} = C_w s_{w,\lambda}$$

$$(K/S)_{yellow\ dose\ into\ purple,\ \lambda} = \frac{\left(C_{pc}k_{pc,\lambda} + C_w k_{w,\lambda}\right) + C_{yc}k_{yc,\lambda}}{C_w s_{w,\lambda}} \tag{15.26}$$

We can rewrite Eq. (15.26) as:

$$(K/S)_{yellow\ dose\ into\ purple,\ \lambda} = (K/S)_{purple,\lambda} + \frac{C_{yc}k_{yc,\lambda}}{C_w s_{w,\lambda}} \tag{15.27}$$

As was done with in the double dose of yellow colorant, we add and subtract $C_w k_{w,\lambda}/C_w s_{w,\lambda}$ to the right-hand side of Eq. (15.27):

$$(K/S)_{yellow\ dose\ into\ purple,\ \lambda} = (K/S)_{purple,\lambda} + \frac{C_{yc}k_{yc,\lambda} + C_w k_{w,\lambda}}{C_w s_{w,\lambda}} - \frac{C_w k_{w,\lambda}}{C_w s_{w,\lambda}} \tag{15.28}$$

We can substitute terms on the right side of Eq. (15.28) with Eq. (15.21) and with the definition of the K/S for the white paint ($C_w k_{w,\lambda}/C_w s_{w,\lambda}$):

$$(K/S)_{yellow\ dose\ into\ purple,\ \lambda} = (K/S)_{purple,\lambda} + (K/S)_{yellow,\lambda} - (K/S)_{white,\lambda} \tag{15.29}$$

Notice that we arrive at the exact same Equation if we were to dose the purple colorant into the yellow paint, as expected since these paints would have identical compositions.

The results of this calculation are shown in Fig. 15.12, along with the measured values for this paint and the measured spectra for a mix of the yellow and purple paints together in equal proportions, for reference. As can be seen, mixing the yellow and purple paints together gives a different reflectance spectrum than adding yellow colorant to the purple paint. This is because we are diluting the purple colorant when we mix the yellow and purple paints but not when we add the yellow colorant into the purple paint.

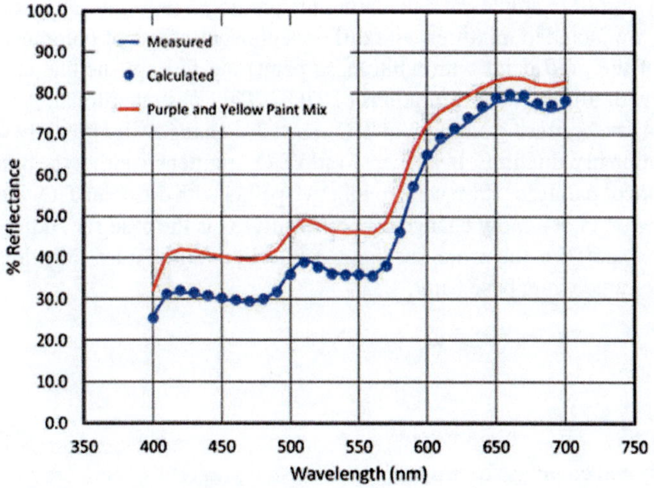


**Fig. 15.12** Comparison between measured (blue line) reflectance values for a yellow paint with additional colorant, the calculated values for this paint (blue circles) and the measured values for a mix of the purple and yellow paints (red line)

## *Two-Constant Kubelka–Munk Theory*

In the analyses outlined above, we assumed that the only significant light scattering came from the $TiO_2$ pigment. However, some colorants provide a non-insignificant amount of scattering in addition to their light absorption properties. When this is the case, we can no longer use Eq. (15.6) and those equations derived from it. Instead, we must use Eq. (15.5). This situation is referred to as the two-constant Kubelka–Munk analysis. A detailed study of this approach, as well as approaches that are hybrids of the one-constant and two-constant approaches, is beyond the scope of this book, but such studies are detailed elsewhere [13–18].

## *Limitations of the Kubelka–Munk Analysis of Color Matching*

The equations and procedures that follow from the Kubelka–Munk analysis of colored paints are based on a number of assumptions, many of which were mentioned in the text. As such the values computed using these equations are only as good as the assumptions supporting them. It is important that, when using these equations to predict the colors of paint mixtures, the calculated values are confirmed by experimental values at the extremes of colorant concentrations.

Finally, in our examples of calculating the predicted reflectance spectra of paint mixtures, we included mixtures with different concentrations of colorants (e.g., the dilution of the gold paint with white base paint) but did not include examples of mixtures with different concentrations of $TiO_{15.}$ This was intentional. As discussed in Chap. 4, the scattering strength of $TiO_2$ particles changes significantly with $TiO_2$ concentration in paint films. If we were to add $TiO_2$ pigment to either the base paint or to a mixture of paints, or if we were to mix two paints with different $TiO_2$ content, the $S_\lambda$ value would not simply change proportionally, as is the case for adding colorant to the base paint. We, therefore, see that we must limit all mixtures of paints to those having the same white base paint.

## Inks

While inks and coatings share a number of related properties, there are some important differences between them. Inks are typically applied at a much lower thickness than paints, are typically applied only to white substrates (uncolored paper) and are not applied at complete hiding (in fact, many inks have minimal S values, making complete hide impossible except for black). Because inks are not applied at complete hide, some of the optical properties of the substrate (paper) show through the ink and so contribute to appearance. This is intentional as it can lead to brighter colors than if the ink was opaque.

Printer inks normally consist of four different colorants—the three subtractive primaries (cyan, magenta, and yellow) and black. These are collectively referred to as the CMYK colorants.[5] Black ink is included for two reasons—it provides a truer black than a mix of the three colorants gives, and the colored pigments are generally more expensive than the black pigment, resulting in a higher cost for creating black by combining the three subtractive primaries.

Bright shades of colors are often needed in printing. In paints we create brightness by increasing the white pigment concentration, which increases the scattering of the film and so decreases the amount of light absorbed by it. In inks a different approach is used. Rather than mixing white pigment into the inks, the inks are applied thinly enough that the brightness of the underlying paper shows through [5]. As a general rule, the laws of optics are such that a higher brightness can be made by applying a thin coating of a non-light scattering colorant over a white background than can be made by combining the colorant with white. This rule is commonly used to increase brightness in ink applications but is seldom used for paints. The distinguishing factor being the color of the substrate—in inks the substrate is almost always white, whereas in paints it can be of any color. Paints are, of course, specifically designed to completely hide the substrate, whatever its color may be, and so cannot benefit from the substrate brightness.

Inks also differ from coatings in another fundamental way, and that is in how colors are mixed. In some printing processes, inks are applied as dots that are too small to be resolved by the human eye. When dots of different colors are placed one on top of the other, the colors subtractively mix, but when dots are placed separately but so close to one another that they cannot be resolved, the colors mix additively. This latter situation is quite analogous to the creation of color on an electronic screen by the activation of different colored dots that are too close together to be resolved.

There is a style of artistic painting that, like inks, exhibits both subtractive and additive color mixing. This style is call "pointillism" and entails placing small drops of color—too small to resolve by eye—next to one another. As in inks, this gives a different color sensation than if the colored paints were mixed prior to application.

The recognition that closely juxtaposed dots could create colors additively was first made by Chevreul in the mid-nineteenth century and was later popularized by the artist Georges Seurat at the end of that century. Seurat's most famous painting, "A Sunday Afternoon on the Island of La Grande Jatte", was created in this style (Fig. 15.13). This painting is quite large and meant viewed from a distance, which causes the colors of neighboring dots to blend with one another rather than to be seen discretely. On close inspection, the dots of paint are quite distinctive Fig. 15.13 inset. Seurat did not simply happen to paint in this manner—he was instead quite deliberate in its style, which at the time represented an innovation in the way that colors could be created with paints.

[5] Here, "K" refers to the black colorant; "B" is not used as there is an opportunity to misinterpret this abbreviation as blue.

**Fig. 15.13** "A Sunday Afternoon on the Island of La Grande Jatte", G. Seurat (Art Institute of Chicago)

## Summary

An object can interact with light by absorbing it, scattering it, or both. The laws of light absorption follow from the conjecture that light encountering a given absorption center has a specific chance of being absorbed by it, and that this chance does not change with concentration of the absorption centers. Light absorption can be characterized on either a percentage basis (percent reflectance) or a logarithmic basis (absorbance). For the color properties of paints, absorption is typically used because the absorbance of a mix of materials is simply the concentration weighted sum of the individual absorbances.

Most paints interact with light through both absorption and scattering. Predicting the color of a paint or colorant mixture, or matching a specific color, can be done using the Kubelka–Munk framework. In this framework, a unit volume of paint film is characterized as having certain scattering (S) and absorption (K) strengths without regard to the physical processes that give rise to these interactions. Many optical properties of the film are controlled by the relative strengths of absorption and scattering (S and K), and so their ratio (K/S) is of high importance in the various Kubelka–Munk equations.

Because both S and K are additive, the S and K values for mixtures of colors are simply the weighted average values for the component colors. This provides a means of predicting the reflectance spectrum of a color mix based on the reflectance spectra of the component colors. The reflectance values of the component spectra, at each wavelength of interest, are converted into K/S values, and then the K/S values of the components are weighted according to concentration, then added together. This gives the K/S of the mixture at that wavelength. The K/S values for each wavelength are then converted to reflectance values, resulting in the predicted reflectance spectrum for the color mixture.

The Kubelka–Munk equations are based on reflectances within the paint film, and these differ from the measured reflectance values due to light reflecting at the air/film interface. When comparing paints that differ in color but are otherwise similar (i.e., the same $TiO_2$ content, gloss, etc.), surface reflectance can be ignored without great loss of accuracy. However, in cases where the paints differ significantly, or when the goal is to compare measured reflectance values to a theory, then corrections to the reflectance values should be made.

## References


1. Johnston, R.M.: Color Theory. In: Patton, T.C. (Ed.) The Pigment Handbook, Vol III. Wiley-Interscience (1973)
2. Kubelka, P., Munk, F.: Ein Beitrag Zur Optik der Farbanstriche (A Contribution on the Optics of Paint Layers). Z. Tech. Phys. **12**, 593 (1931)
3. Kubelka, P.: New contributions to the optics of intensely light-scattering materials. Part I J. Opt. Soc. Am. A **38**(5) 448 (1948)
4. Duncan, D.R.: The colour of pigment mixtures. Proc. Phys. Soc. **52**(3), 390 (1940)
5. Berns, R.S., Mahammadi, M.: Single-constant simplification of Kubelka-Munk turbid-media theory for paint systems—a review. Color. Res. Appl. **32**(3), 201 (2007)
6. Mitton, P.B.: Opacity, hiding power, and tint strength. In: Patton, T.C. (Ed.) The Pigment Handbook, Vol III. Wiley-Interscience (1973)
7. Saunderson, J.L.: Calculation of the color of pigmented plastics. J. Opt. Soc. Am. A **32**(12), 727 (1942)
8. Mudgett, P.S., Richards, L.W.: Kubelka-Munk scattering and absorption coefficients for use with glossy, opaque objects. J. Paint Technol. **45**, 43 (1973)
9. Okumura, Y.: Developing a spectral and colorimetric database of artists paint materials. MS Thesis, Rochester Institute of Technology (2005)
10. Billmeyer, F.W., Jr; Abrams, R. L.: Predicting reflectance and color of paint films by kubelka-munk analysis I. Turbid-medium theory. J. Paint Technol. **45**, 23 (1973)
11. Berns, R.S.: Billmeyer and Saltzman's Principles of Color Technology. Wiley (2000)
12. Judd, D.B., Wyszecki, G.: Color in Science, Business and Industry, 3rd edn. Wiley Series in Pure and Applied Optics (1975)
13. Walowit, E.: Spectrophotometric color formulation based on two-constant Kubelka-Munk theory. MS Thesis, Rochester Institute of Technology (1985)
14. Allen, E.: Basic equations used in computer color matching, II. Tristimulus match, two-constant theory. J. Opt. Soc. Am. **64**, 991 (1974)

15. Abed, F.M., Berns, R.S.: Linear modeling of modern artist paints using a modification of the opaque form of Kubelka-Munk turbid media theory. Color. Res. Appl. **42**(3), 308 (2017)
16. Nickols, D.G., Orchard, S.E.: Precision of determination of Kubelka and Munk coefficients from opaque colorant mixtures. J. Opt. Soc. Am. **55**(2), 162 (1965)
17. Berns, R.S., Mahammadi, M.: Evaluating single- and two-constant Kubelka-Munk Turbid media theory for instrumental-based inpainting. Stud. Conserv. **52**, 299 (2007)
18. Yang, H., Zhu, S., Pan, P.: On the Kubelka-Munk single-constant/two-constant theories. Text. Res. J. **80**(3), 263 (2010)

# Chapter 16
# Formulating Paints with Small Particles

## Contents





M. Diebold et al., *Pigments, Extenders, and Particles in Surface Coatings and Plastics*, https://doi.org/10.1007/978-3-030-99083-1_16

## Introduction

Paints are a complex mixture of many ingredients, each having a specific purpose and contributing to different aspects of film performance. It is the challenge of the formulator to find the combination of ingredients that give the desired set of paint characteristics at a minimal cost. Very often this is a balancing act, as an ingredient might have a positive effect on some properties but a detrimental effect on others. Therefore, the development of a new paint can be a trial-and-error process that may involve several iterative steps. A comprehensive knowledge of the raw materials and their interactions with each other are key elements in the formulator's skill set. Here, we detail many of these elements as well as a mixture design method that is a completely different approach to the development of paints.

While some paint reformulation efforts are made to produce a new line of paints, it is more common to reformulate an existing paint rather than formulating a new one. Such reformulation is mainly done with the goal of matching the properties of an existing paint, rather than changing or improving these properties. In this case, the goal is to change one or more ingredients, typically to a lower cost or more readily available alternative, without changing the nature of the paint. While it may seem that reformulation would be easier than the original formulation of a paint, it can often be more difficult because the attributes of the reformulation often must exactly match those of the original formulation, rather than being allowed to vary within the confines of overall acceptability.

## Composition of Paints

The exact compositions of paints vary largely depending on their intended use, but generally speaking, they all contain a resin (binder), solvent (or water), pigments, extenders, and additives. Exceptions are powder coatings or radiation-cured coatings that do not contain water or solvent and clear coats and varnishes that do not contain pigments.

The properties of pigments and extenders were described in earlier chapters of this book. In this chapter, we will limit ourselves to a high-level introduction of the main characteristics and functions of the other ingredients, as a detailed description is beyond the scope of this book but available elsewhere [1, 2].

Regardless of the application of the paint, one of the first and most important component choices is the liquid vehicle, water, or solvent, as this will determine the type of binder that is required. For a traditional solventborne system, a solvent solution of a resin is used, while for a waterborne coating, one needs a dispersion of the resin. For both waterborne and solventborne coatings, the essential function of the resin is to bind all the ingredients together in the dried paint and to provide gloss for glossy paints and varnishes.

The correct chemistry and amount of binder are determined by balancing the required properties against cost. Binder amounts vary from over 50% in a varnish to a few per cent in a lower quality architectural coating. The properties and cost of the resin are determined to a large extent by the exact composition of the monomer, and these are matched to the end-use requirements of the coating. For example, automotive topcoats are often based on a polyurethane system that is known to have excellent mechanical properties and UV durability, while a primer might use an epoxy resin that is known for excellent adhesion properties.

Acrylic resins will often be found in exterior architectural coatings, while for interior coatings, styrene-acrylic resins are preferred for cost reasons. Ethylene-Vinyl Acetate (EVA) resins are often used in Europe for interior architectural coatings due to their balance of mechanical properties and hiding power. Alkyd systems are often used for wood coatings due to their high gloss and excellent application properties. Most resins systems are organic in nature, but silicate resins are often used in Central Europe for exterior walls, showing that this is not universally the case.

Additives have an extremely wide variety of functions and chemistries but are always added in very small amounts (typically below a few percent). Despite these small dosages, their effect on the paint properties can be very large. Typical examples are dispersing agents that make the solid particles more compatible with the other ingredients of the paints (see Chap. 11). Other additives reduce the formation of foam, increase the wetting of the pigments and of the substrate, provide storage stability and stability when subjected to freeze/thaw cycles, and aid in coalescing of waterborne resins.

Most waterborne paints provide an excellent medium for bacterial growth since they are typically formulated around a pH of 8 and contain high concentrations of organic materials [3]. Biocides are therefore added to most paints. In addition to in-can protection, biocides can be added to avoid the growth of mildew after the application of an exterior architectural coating. Alternatively, paints made with silicate binders are formulated at high pH and so provide in-can preservation without added biocide. In addition, these binders are known for their extreme durability.

The final additive family that we will consider are the co-solvents or coalescing agents used in waterborne coatings. These are water-soluble organic solvents that are added to the system to aid in the formation of a strong and coherent film from the individual resin particles, as discussed in Chap. 10. In addition to aiding in film formation, these materials have an impact on drying behavior and therefore open time (this is the length of time a coat of paint can be spread after application). In some cases, these organic compounds are meant to evaporate at the final stage of drying. They therefore contribute to the volatile organic content (VOC) of the paint. Other compounds (plasticizers) are meant to remain in the paint film after drying. They often give a softness or pliability to the film that tends to cause the films to be tacky and so to retain dirt in outdoor applications.

## Important Paint Properties

One of the main reasons to paint a substrate, besides giving it protection, is to make it more aesthetically pleasing. A few of the more important properties that describe the aesthetics of a coating, such as hiding power and color, were already discussed in previous chapters. Before addressing the practice of formulating paints, we will discuss some additional paint properties and methods to test them. Given the wide application area of coatings, this list cannot be seen as complete. It is instead intended to give a flavor to the various appearance aspects that might be considered when developing a paint.

### *Opacity, Color, and Brightness*

Opacity, color, and brightness are three fundamental properties of great importance in any paint. These three properties derive from the interactions of particles in a paint with light. Brightness relies on light scattering, color on light absorption, and opacity on both. The importance of these interactions with light is reflected in this book by the fact that a total of four chapters are devoted to them (Chaps. 3 and 4 for light scattering and brightness, and Chaps. 6 and 15 for light absorption and color). Formulating for color is relatively straightforward and was discussed in depth in Chap. 15 and so will not be repeated in this chapter. Formulating for brightness and white opacity is more complex and will be discussed in greater detail later in this chapter.

### *Gloss*

Gloss is the measure of surface smoothness at the optical scale and is described by the fraction of light reflected from a surface at the mirror image angle [4]. For visible light, gloss is controlled by surface features, such as dips or protrusions, that are larger than roughly one micron. Light reflected from these features is not reflected at the same angle as light reflected from a flat surface and instead is scattered in all directions (Fig. 16.1).

In practice, a formulator will often use particles with a larger particle size (above several microns) to create a rough surface when that is desired or will avoid these larger particles when a smooth glossy appearance is desired. The materials used to decrease gloss are referred to as flattening or matting agents.[1] A flattening agent is

[1] There is an opportunity for confusion regarding the word "flat" because this word has very different meanings when applied to the gloss of a paint and the smoothness of a surface. Flat surfaces do not give flat glosses, instead a rough surface has a flat appearance, while a flat surface has a glossy appearance.

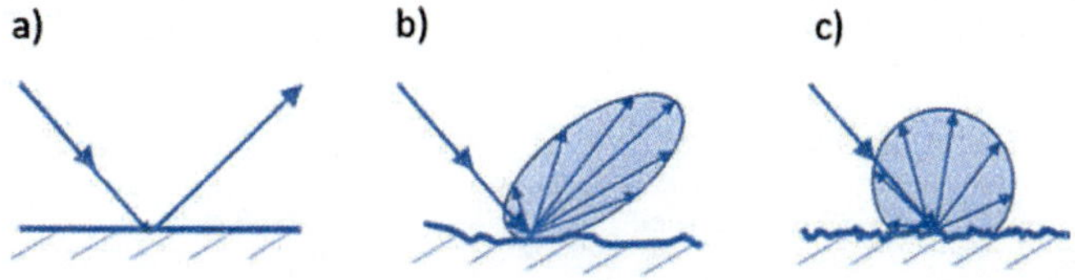


**Fig. 16.1** Gloss as a function of surface roughness. **a** Smooth high gloss surface. b Rough semi-gloss surface. **c** Very rough low gloss (flat or matte sheen) surface

typically a large-size extender or other particles, including diatomaceous earth, silica gel, and fumed silicas, and used in levels no higher than a few percent. Fumed silica flattens paints because these small particles remain mobile in the drying film for more time than larger particles. Convection during the latter stages of drying brings these particles to the film surface, where they roughen the surface [2]. In addition, certain waxes can also be used as flattening agents.

Gloss is typically measured at 20°, 60°, or 85° from vertical, depending on the overall gloss of the sample (Fig. 16.2). High gloss films are measured at lower angles and low gloss samples at higher angles as these are the best angles to differentiate gloss measurements in those ranges. Values measured at 85° are often referred to as "sheen" rather than "gloss".

Considerable shrinkage occurs when traditional waterborne and solventborne paints dry, causing significant surface features to form. For this reason, it is often difficult to achieve a high gloss with these paints, but relatively simple to achieve a mid to low gloss level. Universal grades of $TiO_2$ are typically designed to give high gloss because of the ease with which gloss can be lowered. Increasing the gloss of a low gloss paint made from a low gloss potential $TiO_2$ pigments, which are typically those pigments that do not disperse well, is much more difficult. Solvent-free (100% solids) and radiation-cured paints, on the other hand, shrink less on film formation and so here the challenge is, when desired, to formulate a paint with very low gloss.

As discussed above, the gloss level can be controlled by the addition of a matting agent and by the choice of $TiO_2$ pigment grade. In addition, because particles roughen the paint surface, film gloss can often be controlled through the paint PVC. As PVC

**Fig. 16.2** Schematic representation of the different angles that are typically used to measure gloss

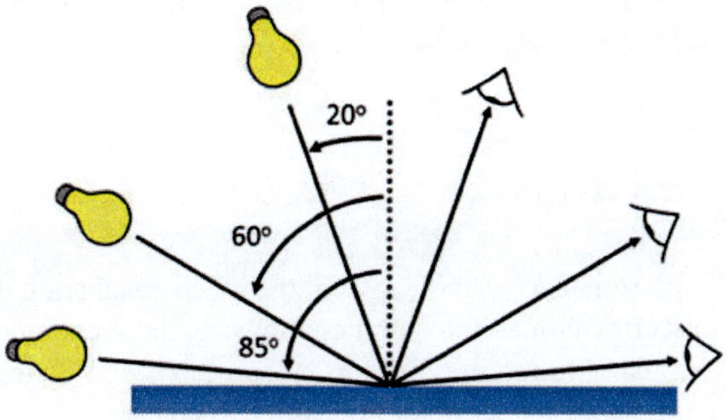

**Table 16.1** Paint gloss types and their typical PVC values

| Sheen level | Typical PVC range |
|---|---|
| Gloss | 15 to 25 |
| Semi-gloss | 30 to 45 |
| Eggshell | 45 to 60 |
| Flat | > 60 |

increases, the number of particles that perturb the film surface also increases, and so gloss goes down. This effect is greatest for large particles, such as extenders. A rough correspondence of paint PVC and gloss is given in Table 16.1.

As a general rule, the highest gloss paints contain no extender because most extenders, even those with small particle sizes, roughen the film surface. In addition, the lowest sheen (i.e., flattest appearance) often requires the paint to be formulated above its CPVC value.

## *Mudcracking*

During the drying process, the volume of a wet film reduces significantly. A paint with a solids volume of 50%, which is not unusual, will experience a volume reduction of 50% when we exclude the volume of pores, if present, in the dry film. As the paint shrinks, tension develops within the film. If the tension is larger than the cohesive forces in the coating, the film will crack. Apart from the loss of aesthetics, these cracks provide a path for air and water to reach the substrate, resulting in a strong reduction in the protective function of the coating.

This phenomenon becomes more important at higher film thicknesses, where moisture gradients and the stresses they cause are the greatest, although the drying conditions (temperature and humidity) and the nature of the substrate also play very important roles. To verify the sensitivity of a paint to cracking, a wedge of the paint is applied and allowed to fully dry (note that it can take a few days for the cracks to appear). The maximum film thickness that is free of cracks is reported. If a paint is sensitive to mudcracking, platelet fillers like talc can be added to reinforce the film and to increase the ability to dissipate the tension generated by volume shrinkage during the drying process.

## *Scrub Resistance*

Scrub resistance is indicative of the mechanical strength of a paint film. It is usually only critical in architectural coatings, as these coatings can be quite fragile and are also likely to require cleaning at some point in their service life. Scrub resistance is measured by scrubbing the fully dried paint with an abrasive pad and washing liquid in a specially designed device. Note that film forming must be complete, a process that varies from paint to paint but can take as long as 28 days. There are two variants of the test methods that quantify scrub resistance. In the ISO 11998 and ASTM D4213 methods, the number of microns that are removed after 40 or 200 wash cycles, depending on the required performance, is measured. By contrast, in the ASTM D2486 test, the paint was applied to a black substrate and scrubbed until the substrate becomes visible due to the removal of the paint. The number of wash cycles to reach this failure point is reported.

Although both methods are used widely, recent improvements in the scrub resistance of latex paints have resulted in many paints falling into the same ISO category (that is, they show little loss after the limited number of scrub cycles in this test). These paints can be differentiated by the ASTM D2486 method since this method runs to failure.

There are several ways to improve poor scrub resistance. Typically, this property is improved by decreasing the PVC (i.e., increasing the amount of resin and decreasing the amount of air in the film, if it is above the CPVC) or adjusting the extender package. Hard and lamellar extenders like mica increase the mechanical strength of the film while extenders with a high oil absorption increase the extent of pore formation and so weaken the paint.

## *Degree of Dispersion*

There are two length scales for which degree of dispersion is important. The first is on the scale of several tens of microns, corresponding to large agglomerates of particles that are visible to the naked eye. Quantification of dispersion at this scale is typically done using a metal block into which a tapered groove has been machined. These blocks are often referred to as "Hegman" gauges, although, strictly speaking, this designation applies to only a single type of gauge with a specific geometry. Details of this form of dispersion characterization are given in Chap. 2.

The second length scale of importance is centered at the wavelength of visible light (roughly from 0.3 to 1.0 microns). The interactions of particles with light are very sensitive to particle sizes in this range, and so the degree to which particles of these sizes can be dispersed is critical to the optics of the paint. Dispersion in this length scale is, in fact, best measured using the appearance attributes that are affected by particle size, such as a hiding power or tint strength test. These tests are discussed in detail in Chap. 13.

## *Durability*

Paint durability refers to the resistance of paint to appearance or mechanical changes over time [5]. This resistance is primarily determined by the resin chemistry, but

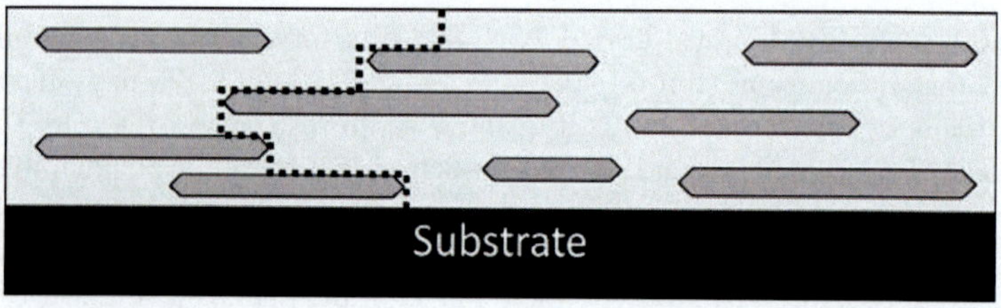


**Fig. 16.3** Tortuous path (dotted line) of water or oxygen through a film with plate-like particles

appearance changes are also dependent, in part, on the pigment and extender particles present (see Chap. 14). Colored pigment particles, particularly organic pigments, can be oxidized by the combination of atmospheric oxygen and solar irradiation, resulting in color fade and/or color shift (see Chap. 8). Paints containing $TiO_2$ pigment are subject to UV-generated radical formation at the $TiO_2$ surface. These radicals can then attach to the organic components of the paint (resin and any colored pigments). As a result, $TiO_2$ grades intended for exterior applications are surface treated to minimize radical formation (see Chap. 7). Finally, high PVC paints have less resin, and thus, for an equal resin loss, experience a greater change in properties.

## *Substrate Protection*

Many paints are applied to protect the substrate from chemical attacks due to the environment. This can be from rain, humidity, or oxygen during ambient exposure or from chemicals specific to a particular application.

The main determinant of substrate protection is the porosity of the paint film. Obviously, a porous film will have little to offer in the way of long-term substrate protection. Since porosity is generally determined by the PVC of the paint, there is a correlation between protection and PVC [6]. In addition, incorporating plate-like extenders, such as mica, improves substrate protection because during the drying process they tend to orient parallel to the surface of the film. Since the oxygen and water molecules cannot pass through these particles, but instead must travel around them, oriented plate-like particles increase the distance over which these molecules must travel to reach the substrate surface (see Fig. 16.3). This property is sometimes termed the "tortuosity" of the film. This slows the rate of substrate degradation.

## *Rheology*

Rheology is a critical factor in paint formulation, and a variety of different, mostly organic materials are available to control it. These include Hydrophobically modified

Ethoxylated Urethane Resins (HEURs), acrylics, Alkali Swellable Acrylics (ASAs), Hydrophobic Alkali Swellable Acrylics (HASEs), and cellulosic thickeners. But, as described in Chap. 9, some inorganic materials have interesting and useful effects on the rheology of the paint (e.g., certain clays or fumed silicas).

A variety of rheological behaviors are possible, depending on the exact chemistry of the modifier. For example, some modifiers increase paint viscosity at low shear conditions (in the can) but decrease it when the paint is sheared during application (shear thinning), while others increase the viscosity regardless of the shear conditions (Newtonian thickeners). It is important to achieve a balanced viscosity across the different domains of shear.

Low shear conditions are seen in paints during storage, such as in the can. A relatively high viscosity is typically preferred under these conditions, as gravitational settling of the pigments and extenders can occur if the viscosity is too low. By contrast, during application, the paint experiences high shear, and here, a lower viscosity is typically desired as this improves the flow and leveling of the paint immediately after the application. Viscosity should build up again after application so as to avoid the formation of runners, drips, or tears, particularly on vertical surfaces such as walls. For a manual application with a roller or a brush, the viscosity at high shear will also determine the "feel" of the application and the ease with which the paint can be applied.

## *Environmental Footprint*

Although not a physical property of the paint, the environmental cost of coverage is an important consideration when formulating or reformulating a paint [7]. The environmental costs associated with a given paint can be divided into three categories: the environmental impact of making and delivering the raw materials, the environmental impact of the paint making and application processes, and the service life of the paint (i.e., how often a coating must be applied).

Raw material suppliers are becoming more transparent about the environmental impact of their products. As this trend continues, it will be easier to include this as a formulation factor when developing a new paint or modifying an existing one. The impact of paint making is typically approached from an energy efficiency standpoint, and of paint application through a minimization of volatile organic compounds (VOCs).

The effect of service life on the environmental impact of a coating is often neglected in these analyses, although it can be as impactful as the other factors listed here. Obviously, if the time between repainting is doubled, the environmental costs of the paint will be halved.

This environmental impact of a paint is often assessed through a Life Cycle Analysis (LCA), as discussed in Chap. 19.

## Strategies for Optimizing Hiding Power

Paint formulation is complex due to the many often conflicting performance properties that must be simultaneously satisfied. Balancing the performance of many of these properties is most effectively addressed using a statistical Design of Experimentation (DOE) approach, as will be illustrated in the final section of this chapter.

In this section, we will discuss in detail one of these performance properties, opacity, because it is of critical importance in virtually all paint types and applications. We will focus particularly on maximizing light scattering because $TiO_2$ is a relatively expensive ingredient in comparison to other paint components. It is important that it be used at its highest efficiency and that partial replacement options are taken advantage of when practical. Before discussing these details, however, we will discuss some basic principles by which paints should be formulated or reformulated and certain complicating factors that must be kept in mind when interpreting the results of these efforts.

### *Principles of Reformulation*

As mentioned previously, many formulation efforts are directed at reformulating an existing paint rather than formulating a new paint ab initio. Many times, these reformulations are carried with the intent of replacing one ingredient with another similar (in-kind) ingredient (e.g., to replace one extender with another), either without changing paint performance or to increase some aspect of it. Alternatively, these reformulations are commonly made with the intention of partially reducing the $TiO_2$ content without losing opacity.

#### General Guidelines

There are numerous opportunities for misinterpreting paint test results when formulating or reformulating a paint, the most important of which are discussed throughout this section. Such misinterpretations can lead to significant costs in terms of both lost time and wasted resources. These costs can often be avoided by applying some general guidelines, both individually and in combination, to the interpretation of reformulation results.

The first of these is quite simple: paints should be compared at equal volume solids. Obviously, increasing the solids level of any paint will increase its cost per unit volume and spread rate by the same amount, just as decreasing the solids level of a paint will decrease its cost per unit volume and spread rate by the same amount.

The second guideline is that the mechanism by which a formulation change causes a change in performance should be understood. Because paints are so complex, changing one aspect of the formula (for example, replacing one ingredient with another) can often impact a performance attribute in more than one way. In such a case, it is important to understand the mechanism by which this change improves paint properties as this can give insights into other, more desirable, ways to bring about this improvement.

For example, the incorporation of a special, small-size extender that is claimed to improve $TiO_2$ particle spacing may result in an opacity improvement by transitioning the paint from below the CPVC to above it, rather than by better spacing the $TiO_2$ particles. If this change is acceptable, then by knowing the mechanism of opacity improvement, the formulator could investigate alternative, less expensive, or more effective ways to transition the paint above the CPVC. Misinterpreting the cause of the improved hiding would lead to an incorrect focus on alternative ways to improve particle spacing.

Finally, it is important that the comparison of paints is done using the appropriate test and the appropriate test conditions. For example, drawdowns made with the same drawdown bar do not always have the same wet film thickness, and so a proper test involving drawdowns must account for thickness differences. A common example of this is that a new ingredient can increase the contrast ratio of a paint but do so by increasing the drawdown film thickness rather than by increasing the intrinsic hiding power of the paint. The paint formulator can avoid wasted effort by understanding the mechanism of the contrast ratio change from the outset. Similarly, using a test outside of its intended limits may give incorrect results that lead to erroneous conclusions as to the effectiveness of particular reformulation on paint performance.

Example applications of each of these guidelines are discussed in greater detail below.

## Percent Solids Changes

Simply replacing one material with another often results in a change in the percent solids of the paint. As mentioned above, any change in the solids content of a paint will lead to an equivalent change in the spread rate without actually changing the intrinsic hiding ability of the paint.

In general, such formulation changes are trivial and will not be discussed as a method of reformulating. That said, there are situations where the solids level of a paint is inadvertently changed without recognition that this has occurred. In such situations, care must be taken to avoid attributing an increase in spread rate to an increase in the intrinsic hiding power of the paint film rather than an increase in solids content. This most often occurs when paints are reformulated on a weight basis rather than a volume basis.

## Weight Versus Volume Replacements

Replacing one type of particle with another on an equal weight basis can be a convenient way to reformulate, since ingredients are purchased on a weight basis and, in production, added on a weight basis. However, if there is a difference in density between the materials involved in this replacement, then changing a formulation on a weight basis will also change the total volume of the formulation.

For example, if a paint contains 15 weight percent of $TiO_2$ (density 4.2 g/ml), and 3 PVC of that is replaced on a weight basis with a calcium carbonate extender (density 2.71), then the volume of paint will increase by 4.7%. With this increase comes a change in percent solids of the paint. If we are to compare the spread rate of the reformulated paint to that of the original paint, then we need to correct for the opacity increase that is due simply to this increase in volume—without this correction, we would be comparing the number of square meters covered by 1.0 L of the original paint to the number of square meters covered by 1.047 L of the modified paint.

## Comparing Paints with Different Levels of Brightness

Opacity can always be increased by adding a light absorber to a paint. In paint reformulation, this can occur either intentionally or inadvertently when a new ingredient added to a paint is even slightly dark. For example, in some cases, an off-color extender may allow for a decrease in $TiO_2$ content while maintaining opacity by replacing light scattering with light absorption. In a white paint, the decrease in perceived brightness can be surprisingly small, and the opacity boost from the new extender may be misattributed to, for example, better $TiO_2$ spacing. If the slightly lower brightness is acceptable, then alternative ways of darkening the original paint should also be investigated, as bringing in absorption with the new ingredient may not be the most cost-effective way of making this change.

This is illustrated with an example in which a high-quality $TiO_2$ is replaced with a lower quality counterpart. Often such a replacement can occur with little or no change in opacity. This does not mean that the lower quality pigment performs identically to the original pigment, however. Instead, we find that low-quality pigments typically scatter light more poorly than their higher quality counterparts, but also are slightly darker (off white) due to impurities. Exchanging the high-quality pigment for the lower quality counterpart in effect replaces some amount of light scattering with an offsetting amount of light absorption.

Although the reformulated paint is as opaque as the original paint, it is also less bright. If this lower brightness is acceptable, then the formulator should consider what would happen if the original paint was toned to the same brightness. This can result in a significant increase in opacity that could be leveraged to a greater value than the replacement of the higher cost high-quality pigment with the lower cost, lower quality counterpart.

**Table 16.2** Optical properties of paints with different scattering and absorption strengths

| Description | Paint A incumbent | Paint B alternate $TiO_2$ | Paint C At equal brightness |
|---|---|---|---|
| R | 0.855 | 0.846 | 0.845 |
| $R_o$ | 0.775 | 0.763 | 0.769 |
| Contrast ratio | 0.906 | 0.902 | 0.910 |
| Brightness ($R_\infty$) | 0.880 | 0.866 | 0.866 |
| L* | 95.2 | 94.5 | 94.5 |
| SX | 3.82 | 3.63 | 3.82 |
| KX | 0.0307 | 0.0376 | 0.0418 |
| Spread rate ($m^2/l$) | 9.21 | 9.20 | 9.79 |

As an example, we will consider the optical characteristics—brightness and opacity—of three related paints. Paint A is the incumbent paint made with a high-quality $TiO_2$. Paint B is identical to Paint A except the $TiO_2$ has been replaced on an equal basis with a lower quality pigment that costs 5% less than the incumbent pigment. Drawdowns are made of both paints, and the optical parameters measured on these are shown in Table 16.2.

As can be seen from the $R_\infty$ and spread rate values, Paint B is less bright than Paint A but has the same opacity. The SX and KX values calculated from the reflectance measurements confirm that this reformulation reduces light scattering but increases light absorption.[2] We can therefore decrease the $TiO_2$ pigment cost of the paint by 5% while maintaining opacity. Since $TiO_2$ costs can be as high as 40% of the total raw materials cost of a paint, this exchange could decrease the total ingredient costs of the paint by as much as 2%.

If the slightly lower brightness of Paint B is acceptable, then the formulator has the option to tone paint A with, for example, carbon black, to a lower (just acceptable) brightness. Paint C is simply Paint A that has been toned with carbon black to the same brightness as Paint B. This modification does not change the scattering strength of the paint (SX) but increases the absorption strength (KX). Paint C differs from Paint B, then, in two important ways—stronger scattering and stronger absorption—both of which increase opacity. As can be seen in Table 16.2, this increases the spread rate by 6.4%.

We can compare the cost advantages of Paints B and C to one another. As calculated above, Paint B provides a total ingredient cost savings of 2% compared to Paint A. However, because Paint C is 6.4% more opaque than Paint A (and Paint B), it decreases the cost of coverage ($\$/m^2$) by 6.4%. The greater savings, then, is to maintain the high-quality pigment but to tone the paint made with it to match the lower brightness of paint made with the low-quality pigment.

---

[2] These calculations are described in Chapter 13.

## Using Contrast Ratio to Quantify Opacity

Contrast ratio is commonly used to quantify the opacity of a paint. Unfortunately, this test is not appropriate for this purpose, and in Chap. 13, we detail its shortcomings and outline reasons why spread rate, rather than contrast ratio, should be used for this purpose.

Briefly, the issue with the contrast ratio is that it assumes that paints drawn down with the same drawdown blade will have the same wet film thicknesses. This assumption seems reasonable, but it is in fact often false. Instead, the physics that applies to the formation of a thin liquid film drawn down from a relatively viscous liquid dictates that the wet film thickness will be roughly half the clearance of the drawdown applicator. The exact film thickness depends on a number of factors. For paints, the most impactful of these factors is viscosity. Simply stated, thicker (more viscous) paints give thicker (deeper) films.

The degree to which film thickness can vary between paints with similar formulas was shown in Chap. 13, Fig. 13.2, which is reproduced here as Fig. 16.4. These drawdowns were all made using an automated applicator and the same drawdown blade. Were one to assume that the wet film thicknesses were identical, then the apparent hiding power of the thickest paint would appear to be 22% greater than that of the thinnest paint if, in fact, the paints had equal opacity.

In this regard, it is worthwhile to mention that small (nano) particles added to a liquid paint, even at low levels, can cause a large increase in paint viscosity. These materials are, in fact, often used as thickening agents, as discussed earlier in this chapter. Quite often the addition of nanoparticles to a paint is claimed to improve the paint opacity by better spacing the $TiO_2$ particles (see the section below on nano-spacer extenders for more details). While these particles can increase the contrast

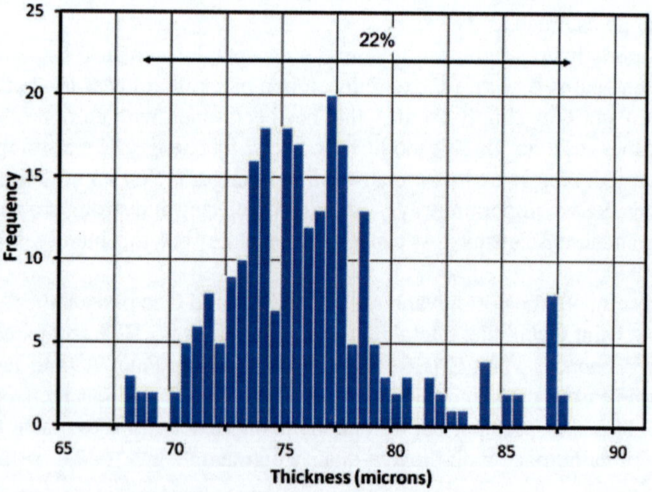


Fig. 16.4 Histogram of the wet film thicknesses of 246 similar paints

**Table 16.3** Results for three paint formulations

| Paint | wt % $TiO_2$ | wt % $CaCO_3$ | Contrast ratio | SX ($m^2/l$)[a] | Relative hide |
|---|---|---|---|---|---|
| A | 23 | 16 | 0.989 | 9.4 | 100 |
| B | 20 | 29 | 0.986 | 8.7 | 92.6 |
| C | 17 | 22 | 0.985 | 8.5 | 90.4 |

[a] From Fig. 16.5

ratio of a paint, it is often because they increase the viscosity of the paint and so the thickness of the drawdown, rather than the intrinsic hiding ability of the paint.

## Measuring Opacity on Excessively Thick Films

There is a general rule in coatings science that any paint—even a clear coat—will be opaque as long as it is applied at a great enough thickness. As an application of this rule, consider the following scenario: a paint customer finds that a can of premium paint costs 50% more than a similar can of economy paint. The customer asks the store clerk whether the paints give the same degree of hiding. The clerk's response is that, after applying seven coats of paint, the two look the same. The customer, obviously, would not receive much reassurance from this response nor have any confidence in the economy paint itself! It is the appearance after one or two coats of paint that is important to the customer, not the appearance after seven coats.

While this last observation seems obvious, it is sometimes the case that the opacities of paints are compared at such great thicknesses that all paints being compared have full hiding, even if the spread rate of the paints at a thickness merely sufficient to give complete hide is quite different.

This mistake is often made when using contrast ratio as a measure of opacity. As an example, a recent article in a respected paint journal compared the opacities of three paints—an original paint and two formulations for which a portion of the $TiO_2$ was replaced with a calcium carbonate extender. In all cases, the brightness of the paint was approximately 0.85. The composition and contrast ratio results for these paints are given in Table 16.3 (note that these paints were reformulated on a weight basis, rather than a volume basis, so there are more differences between them than a simple replacement of $TiO_2$ by $CaCO_3$). Although the authors followed an ISO procedure for measuring hiding power, in this case, the procedure produced contrast ratio values in excess of that for complete hide.

The implicit assumption made in the article was that the three paints have essentially equal opacity—after all, the contrast ratios differed by only 0.4% and, in any event, all paints had full opacity. However, as detailed in Chap. 13, contrast ratio values are most meaningful when measured between 0.92 and 0.95.[3] Because contrast ratio values cannot exceed 1.0, when the values are close to 1.0, as they are here,

[3] Note that ASTM D2805 specifies a somewhat higher contrast ratio target of 0.97.

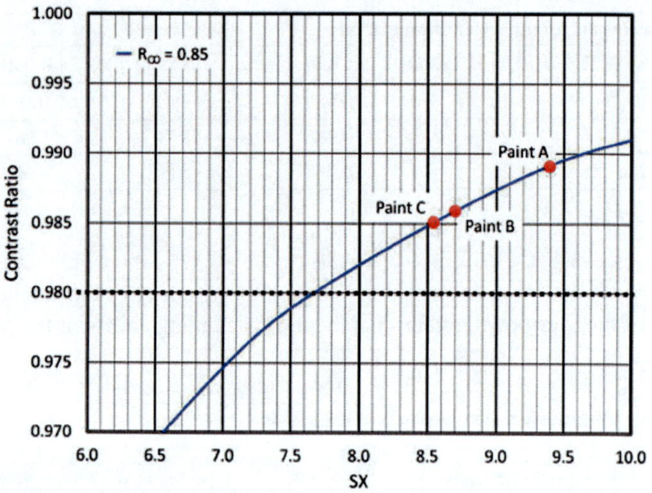


**Fig. 16.5** Relationship between contrast ratio and SX for paints with a reflectance value of 0.85. Dotted line indicates full hide

a large change in paint scattering power (SX) results in a small change in contrast ratio. Looking at this from the opposite perspective, a small change in contrast ratio equates to a large change in scattering power.

The relationship between contrast ratio and SX value for paints with brightness values of 0.85 and 1.00 will be shown later in Fig. 16.12. We will defer a detailed discussion of this figure until a later section, but will use it to estimate the SX values of the three paints in Table 16.3. An enlargement of the relevant region of Fig. 16.12 is shown in Fig. 16.5. Here, we see that these small differences in contrast ratio equate to very significant differences (up to 9.6%) in opacity when the paint is applied at full hide (CR = 0.98). The assumption that the replacement of $TiO_2$ by extender did not result in a loss of hiding power is therefore refuted.

### Large Differences in Surface Roughness (Gloss)

As discussed in Chap. 13, paint opacities are calculated using reflectance measurements, in particular, the fraction of light reflected from the film surface and captured by a light detector. As discussed earlier, gloss paints reflect a high proportion of light at the mirror angle to the incident light, whereas flat paints scatter the light diffusely (at all angles). By simply increasing the surface roughness of a glossy paint, the amount of reflected light that is measured by the light detector can decrease, depending on the geometry of the detector. This decrease in reflectance value can be incorrectly interpreted as a change in opacity of the paint, rather than a change in the distribution of the reflected light. Because of this, directly comparing the opacities of paints with very different gloss values can be problematic [8–10].

As an example of this, Legrix measured the opacities of three paints that differed in the presence of an extender [9]. Compared to a paint made with $TiO_2$ as the only particle type, similar paints containing precipitated calcium carbonate or ultrafine clay showed contrast ratio increases of 0.3 and 1.4.[4] However, the three paints had very different 60° gloss values (81, 32, and 41, respectively). She then added a clear coat to the drawdowns. This brought the 60° gloss values in closer alignment (56, 41, and 52, respectively) without altering actual paint opacity (no additional light scattering or absorption). When the opacities of the clear-coated drawdowns were measured, the paints made with precipitated calcium carbonate and ultrafine clay showed slight reductions in opacities (contrast ratios were reduced by 1.7 and 1.2, respectively).

## Transitioning Above the CPVC

As a final example of an artifact that can cause a change in opacity, small particle size (nano) extenders are sometimes formulated into paints that are close to, but below, their CPVC value. Because of the high resin demand of small particles, their inclusion in the paint, even when simply replacing another particle on an equal volume basis (i.e., when keeping the paint PVC the same), can cause the paint to transition from below the CPVC to above (again changing more than one paint property at once).

As discussed below, this increases the opacity of the paint film. While it is correct to ascribe this increase to the small particle size extender, it may not be due to the improved spacing of the $TiO_2$ particles, but rather to the addition of air to the film. In this case, the correct control paint is a modification of the original paint for which the CPVC has shifted by a similar amount, but using conventional, high resin demand extenders such as diatomaceous earth.

This consideration is not entirely limited to paints formulated below the CPVC. Using fine particle extenders will, for the reasons outlined above, result in an increase in the air content (porosity) of a paint that is already above the CPVC [11]. This will increase hiding power, as discussed in detail below, but at the cost of reduced film strength properties and increased stain pick-up. A paint that has already been formulated at the limits of these properties may not be able to tolere the increase in the air that accompanies the incorporation of small extender particles into the film. As is the case for any reformulation effort, all properties of the paint must be considered when assessing the acceptability of any given change.

[4] All contrast ratio measurements were between 89.7 and 93.8, a range in which meaningful differences in opacity are apparent.

## *Conventional Extenders*

As discussed in detail in Chap. 4, the light scattering strength of $TiO_2$ pigment particles decreases as the concentration of these particles (their PVC) increases. This is due to the overlap of scattering volumes when particles are crowded together. It is therefore essential to provide situations that favor good particle spacing. Large extender particles limit the available space for the $TiO_2$ particles, and so do not allow good $TiO_2$ particle spacing. This is illustrated in Fig. 16.6 where in the upper left figure we see the wet film with $TiO_2$ particles and coarse extenders (for example $CaCO_3$). In the wet film, the entropy of the system is very high allowing a good spacing of the $TiO_2$ particles.

During the drying process, the volume of the film reduces due to the evaporation of the water or solvent. As the large extenders occupy a large fraction of the available film volume, the options for placing the $TiO_2$ particles are limited and it becomes necessary for the $TiO_2$ particles to be arranged in very close proximity to one another in the regions between the extender particles. This results in a lower efficiency of the $TiO_2$, and hence lower hiding power (upper right in Fig. 16.6).

If we replace the large extenders with an equal volume of small extenders and keep all other parameters constant, the $TiO_2$ particles in the wet state are again well spaced (lower left of Fig. 16.6). However, in the dry state, there are more open regions into which the $TiO_2$ particles can be placed compared to the situation with the coarse extenders. This results in the better spacing of the $TiO_2$ particles and higher opacity. In addition, the surface is now much more even, resulting in a higher gloss. Depending on the desired gloss level, this effect may not be desired.

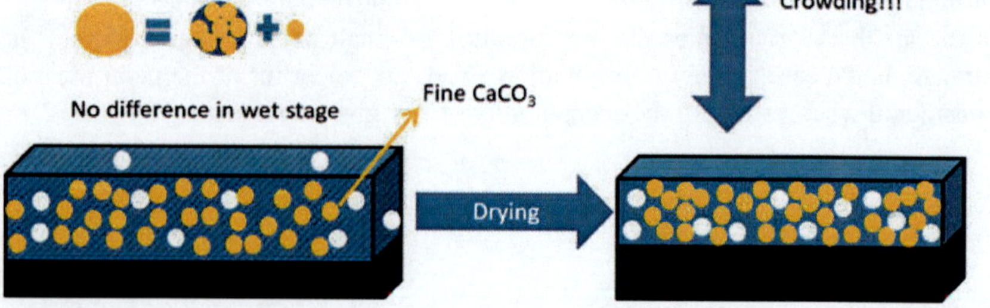


**Fig. 16.6** Schematic representation of a coating with a large extender ($CaCO_3$) in the wet and the dry state (above), and a coating with a fine extender ($CaCO_3$) (below)

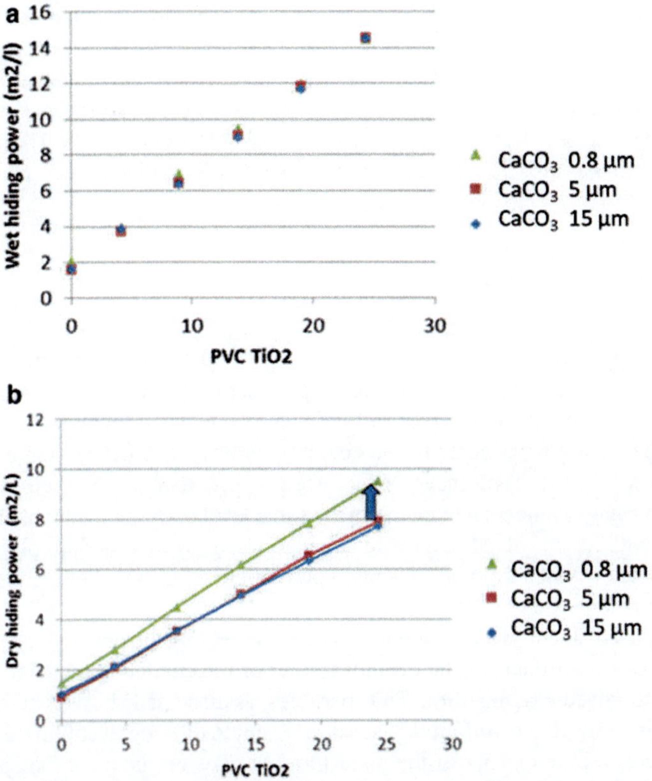


**Fig. 16.7** Wet (a) and dry (b) hiding power of paints with $CaCO_3$ with a median particle size of 0.8, 5, and 15 microns at different $TiO_2$ levels

This analysis is illustrated experimentally with a series of paints made below their CPVC at increasing levels of $TiO_2$ and with three types of $CaCO_3$ extenders that differ in median particle diameter (0.8, 5, and 15 microns). Figure 16.7 shows that hiding power increases with increasing $TiO_2$ concentrations in both the wet and the dry states, as expected. In the wet state, the increase in opacity does not depend on the median particle size of the $CaCO_3$, showing that under all circumstances the $TiO_2$ is well spaced. In the dry state, we find that the 5 micron and 15 micron $CaCO_3$ particles have the same degree of negative effect on dry opacity. This is because, in the context of $TiO_2$ particle crowding, both particle sizes can be considered "large" and so they affect $TiO_2$ crowding similarly.[5]

[5] Typically, the crowding effect of "large" particles applies to any extender particle larger than about 3 microns. "Mid-sized" particles, which partially affect $TiO_2$ spacing, range from roughly 0.8 to 3.0 microns, and extender particles below 0.8 microns are considered "small" and typically have no effect on $TiO_2$ particle spacing.

In the dry state, the hiding power of the films is always lower than that seen for the wet films. This is because, as the paint volume shrinks during the drying process, the average distance between the $TiO_2$ particles reduces and so the scattering efficiency of the $TiO_2$ is reduced. The effect is most obvious for the coarser types of $CaCO_3$, illustrating that these large particles reduce the efficient spacing of the $TiO_2$ particles. This concept was also discussed in Chap. 4, but in a simplified model paint system.

## *Nano-Spacer Extenders*

There is a prevailing theory within the coatings industry that very small extender particles—as small or smaller than the pigmentary $TiO_2$ particles—will position themselves between $TiO_2$ particles, enforcing a spacing between them. The "spacer" concept has many proponents in the coatings industry, but not all agree with this mechanism [12–15]. There have in fact been reports that, at best, replacement of resin with small extenders will maintain the same level of opacity, but only at modest replacement levels [16]. Monte Carlo calculations show that random placement of nanoparticles within a field of $TiO_2$ pigment particles does not alter the distribution of the pigment particles [17].

Examples of small (nano) extender particles increasing the opacity of a paint film are often due to artifacts of the opacity testing or reformulation techniques, rather than the improved spacing of the $TiO_2$ particles, as discussed above. These artifacts include viscosity effects (different viscosities result in different wet film thicknesses, which increases the opacity of the paint film but does not increase the spread rate of the paint), the addition of a small amount of light absorption, and changes in surface roughness (gloss), all of which affect apparent opacity [9]. In some cases, reformulation has been made on a weight basis rather than a volume basis, which makes a direct comparison of results nearly impossible, or the concentration of solids is increased when nanoparticles are present.

In other cases, an unreasonably thick film is used (given enough thickness, all paints will eventually provide complete opacity, regardless of their scattering and absorption strengths). Finally, because of their small size, nanoparticles have a higher oil absorption value than the $TiO_2$ particles they are replacing, and if the original paint is formulated near the CPVC, the reformulated paint can be above the CPVC and so enjoy an opacity enhancement due to air voids, rather than to better spacing of the pigment particles.

The flaw with the nanoparticle spacing concept is that randomly distributed nanoparticles are seldom positioned between $TiO_2$ particles. Instead, they are most commonly found in regions that do not have nearby $TiO_2$ particles and so cannot providing spacing benefits (Fig. 16.8a) [17]. In fact, finding small particles exactly in between two larger particles is quite rare while finding small particles in regions that are mostly resin is common. Quite simply, the random positioning of the smaller particles does not preferentially place them in the region of the film that would benefit most from their presence.

**Fig. 16.8** Arrangement of spacer particles in paint films. **a** Random placement. **b** Targeted placement

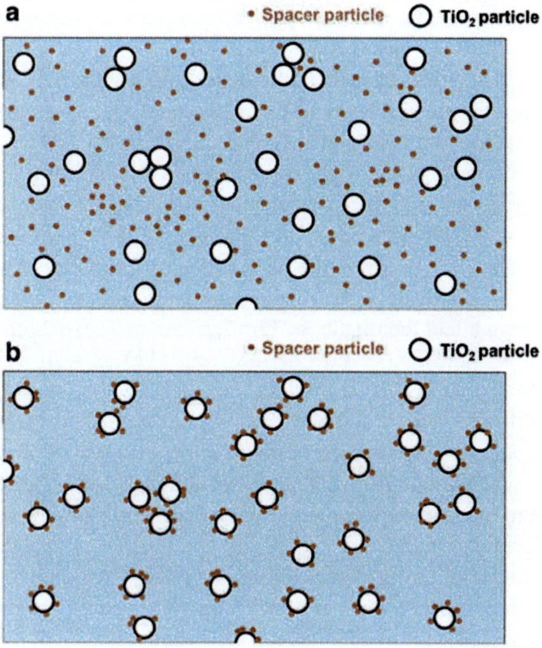


Although the random distribution of nanoparticles does not improve the spacing of pigmentary $TiO_2$ particles, there is a scenario in which nanoparticles are effective spacers. This is when the nanoparticles are not randomly distributed but are instead attached to the surface of the pigment particles, a situation known as targeted placement (Fig. 16.8b). This creates a composite particle (discussed below). In this case, we are assured that all the nanoparticles are positioned correctly to space the pigment particles. This has been achieved using nanoparticles of zinc oxide [18], and a similar strategy is followed with small resin particles that attach to the $TiO_2$ pigment surfaces, creating $TiO_2$-polymer composite particles [19].

## *Resin Particles*

Resin (latex) particles are an important determinant of paint properties for a number of reasons. These reasons are described in detail in Chap. 10, but here we will review the most salient of them.

Resins are chiefly characterized by the chemical nature of the polymeric material from which they are made. While this strongly impacts many important paint end-use properties, such as gloss and durability, our focus is on the particulate nature of the latex in the liquid paint and how these particles affect film formation and paint properties. Particle size is of importance in this regard, and, as an example, smaller resin particles allow for better $TiO_2$ particle spacing, which improves the opacity of the film (see also Chap. 4, Fig 4.10).

The effect of resin particle size on $TiO_2$ particle spacing can be quantified by the binding power index (BPI). This index describes the fraction of the resin particle volume that is available to push between the $TiO_2$ particles and separate them. The remainder of the resin volume is unavailable for this task and so does not improve the spacing of the pigment particles. BPI can vary from zero (a completely non-coalescible polymer particle) to one (a completely coalescible particle, or one with a diameter of zero—that is, a solvent paint polymer molecule).

In addition to the size of the resin and its BPI, the thermal properties of the resin affect the quality of film formation. These thermal properties include the glass transition temperature of the resin (Tg) as well as the minimum film formation temperature (MFFT). These are controlled by the resin chemistry (typically through different combinations of resin monomers) and can be modified by the amount of coalescing agent present in the film.

The relationship between Tg and the end-use application temperature is an important consideration when choosing a resin. Paints with Tg values below the end-use application temperature are soft and flexible, but are difficult to clean and suffer from blocking.[6] Both of these issues are related to the soft, tacky nature of the film surface. On the other hand, paints with Tg values above the end-use application temperature are hard and durable but can crack and flake.

## *Composite Particles*

To better space $TiO_2$ particles in paints, other particles, typically the same size or smaller, are often attached to their surfaces. This results in the formation of a composite particle—a particle made by combining particles of different types. In the section on nano-spacer particles, we discussed two examples of composite particles—$TiO_2$ to which nano resin particles are attached, and $TiO_2$ to which nano zinc oxide particles are attached. We can also consider highly treated $TiO_2$ pigment grades as composites of $TiO_2$ particles and porous aluminosilicate particles.

Composite particles have a rich history in the early days of modern coatings. At that time, the quality of pigmentary $TiO_2$ was low, and instead of offering pure $TiO_2$ pigments, pigment manufacturers produced composite particles of $TiO_2$ and an extender. This was viewed as a cost-effective way of providing opacity. However, once the production of optimized $TiO_2$ particles became common, most of these composite particles were discontinued [20]. In addition to the excessive costs of these materials on a coverage basis [12], at roughly the same time the industry migrated

[6] Blocking refers to the propensity of painted surfaces to adhere to one another when pressed together. An example is the paint on a window sash and frame. In the case of strong blocking, the two would be held in place and it would not be possible to open the window.

from solventborne paints to waterborne paints; many of the particles used in these composites were slightly water soluble, leading to instability in latex paints [21].

Most composites disappeared from the paint industry many decades ago. However, there is one class of composite that is still available—composites made of pigmentary $TiO_2$ and calcium carbonate particles. Many such combinations are described in the literature [22, 23], but of these only a few are commercially available [24]. The $TiO_2$ and calcium carbonate particles in these composites are roughly the same size and shape—so much so that they are difficult to distinguish in electron micrographs (Fig. 16.9).

These materials are similar in principle to the highly treated $TiO_2$ grades discussed in Chaps. 4 and 7. However, in this case, the percent $TiO_2$ by weight is significantly less, since the highly treated grades of $TiO_2$ are coated with a porous (low density) material whereas the calcium carbonate particles in these composites are solid. Based on their similarities, we find these composite particles to also be used primarily in highly crowded (high PVC) paints.

## *Encapsulated Air*

Entraining air into a paint film improves light scattering by two mechanisms. First, regions of air of an appropriate size will scatter light in the same way that $TiO_2$ particles do. In this case, however, these regions scatter light because their refractive index (1.0) is below that of resin (~1.5), rather than above, as is the case with $TiO_2$ (2.7). Second, the presence of air decreases the average refractive index of the matrix, increasing the difference in refractive indices between the $TiO_2$ particles and their surroundings [25]. A commonly used way to bring air into the paint is to formulate above the CPVC, as will be discussed in detail later. However, an alternative is to introduce opaque hollow sphere polymer particles (HSOPs) into the paint [26–28]. This has the optical advantages of formulating above the CPVC without the disadvantages of creating extensive pore networks.

Figure 16.10 shows schematically the use of a fine $CaCO_3$ (0.8 micron) combined with HSOP (compare to Fig. 16.6), with experimental results for such combinations shown in Fig. 16.11. It is clear that in the wet state the HSOP particles do not contribute to hiding power (the pores are still filled with water), and this experiment and the one reported in Fig. 16.7 show that a white pigment such as $TiO_2$ is necessary to get acceptable hiding power in the wet state. Wet hide is important in many applications, and loss of wet hide often limits the extent to which $TiO_2$ can be removed or replaced in a paint system, either by air voids or extender particles of any size [29, 30]. However, in the dry state, hiding power is significantly enhanced by the presence of the discrete air voids.

Encapsulated air can also be added in the form of flash calcined clay [31]. Many clays consist of sheets of hydrated oxides held together by waters of hydration. When these particles are subjected to flash calcination, the water between the sheets volatilizes violently, pushing the sheets apart (a process known as exfoliation). The

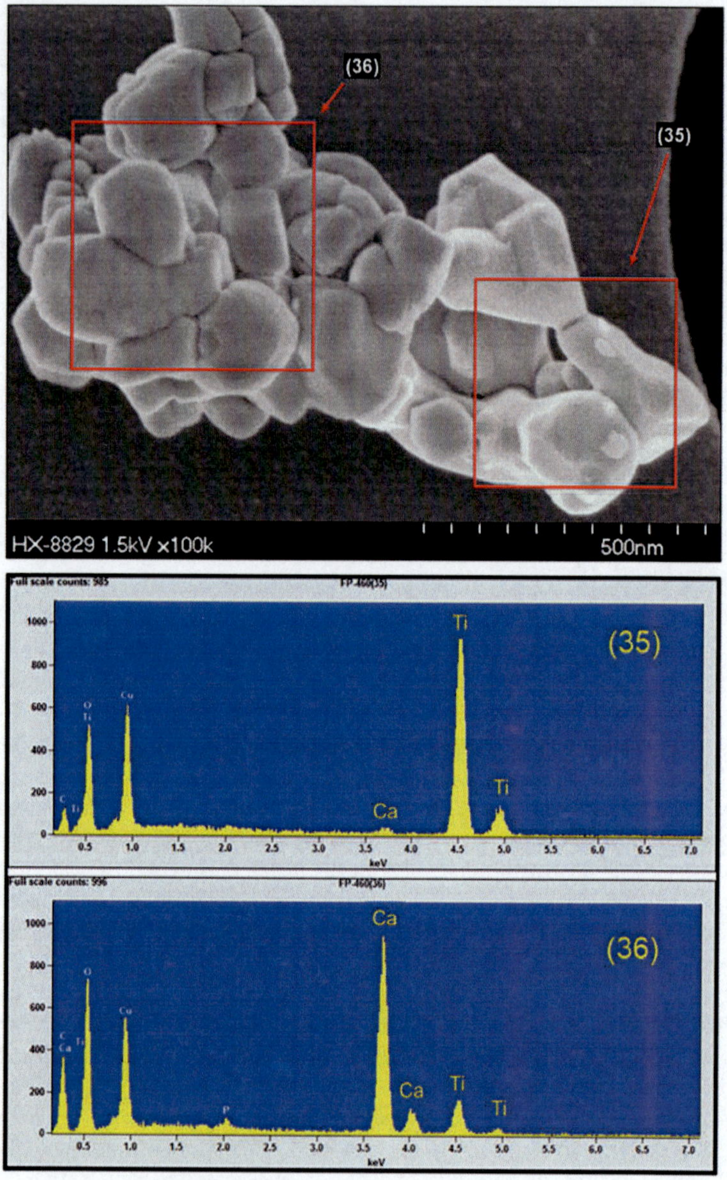


**Fig. 16.9** Calcium carbonate/$TiO_2$ composite particles. The top shows an image of a composite particle with two regions identified. The bottom shows the elemental analysis (EDAX) for each region, allowing for the distinction between calcium carbonate and $TiO_2$

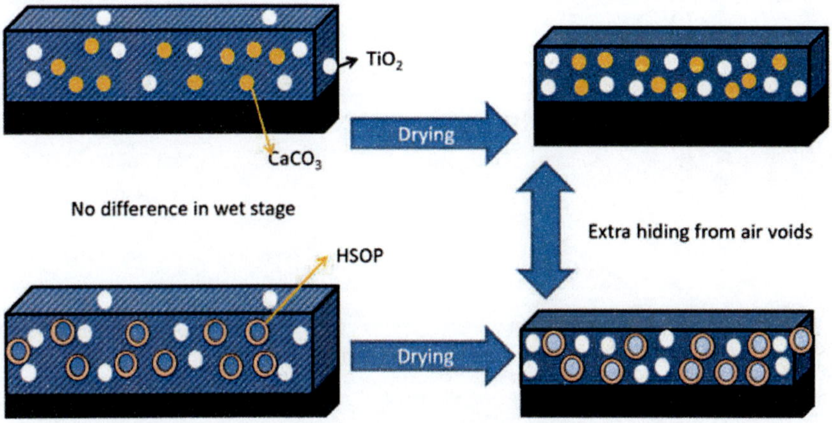


**Fig. 16.10** Schematic representation of a coating containing a fine $CaCO_3$ in the wet (left) and dry (right) states, and a coating where the $CaCO_3$ is replaced by HSOP

result is the presence of thin layers of air within the particles. These air layers are typically sealed by a vitreous surface layer formed during calcination. In any event, the thickness of these air layers is generally too low to allow the resin to infiltrate, and as a result, these particles bring air voids into the paint.[7]

## *Light Absorption*

Earlier we noted that in some instances a slightly discolored $TiO_2$ pigment can enhance opacity by light absorption. There we saw that even a small amount of discoloration can improve hiding by a meaningful amount. This effect can be used beneficially, when carefully controlled, to increase opacity even in bright paints. This may seem a contradiction, but unless a paint must be the brightest possible, it can absorb a small portion of light while still being considered bright (or white). In the language of Kubelka and Munk, it is quite rare that a paint film requires an $R_\infty$ value of exactly 1.0, or, in the language of CIE color space, it is quite rare that a paint film requires a tristimulus Y value of 100. In fact, for architectural (décor) applications,[8] paints with a tristimulus Y value greater than approximately 80 ($L^*$ = 91.7) are considered white, while those with tristimulus Y values between 75 and 80 ($L^*$ from 89.4 to 91.7) are considered "off white" (particularly if they are

[7] Note that this is different than the voids in diatomaceous earth particles (Chapter 9), which are large enough for resin infiltration, resulting in the high resin demands of these materials.

[8] In architectural (décor) applications, brightness is often quantified as the "light reflectance value" (LRV). This is the tristimulus Y value of the paint at full hide and is often referred to simply as "brightness".

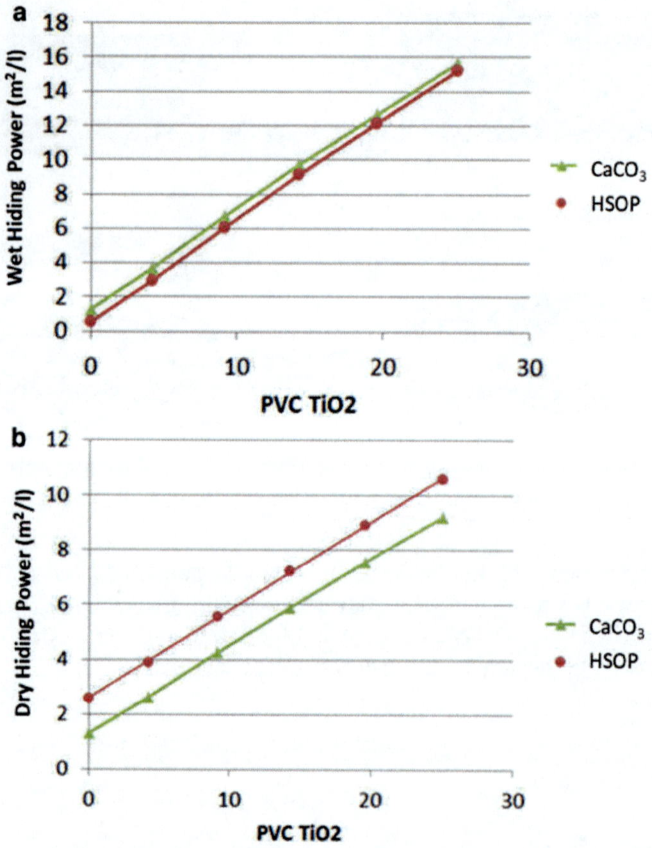


**Fig. 16.11** Wet and dry hiding power of paints with fine $CaCO_3$ (0.8 micron) and HSOP

achromatic). Architectural paints with tristimulus Y values above 90 ($L^* = 96$) are rare, as such a brightness is generally considered "harsh" or "sterile".

A decrease in brightness within these ranges can be very advantageous to the opacity of a paint film, and even small amounts of carbon black can add significant opacity to an otherwise "white" paint [32, 33]. This is due to the shape of the opacity versus scattering (SX) curve (Fig. 16.12). Because the contrast ratio is bound at a value of 1.0, it takes increasingly higher increments in SX to create even a modest change in contrast ratio once the contrast ratio exceeds roughly 0.96. By decreasing the reflectance of the film from 1.00 to 0.85 (i.e., to a tristimulus Y value of 85, or an $L^*$ value of 93.9), the thickness of the film can be reduced by over a third (from an SX value of 11.8 to one of 7.7).[9]

[9] An argument is often made that a poor-quality $TiO_2$ pigment, with low light scattering and low brightness, can replace a high-quality pigment in these applications since high brightness is not required. However, even though a paint with the same brightness can be made with either pigment,

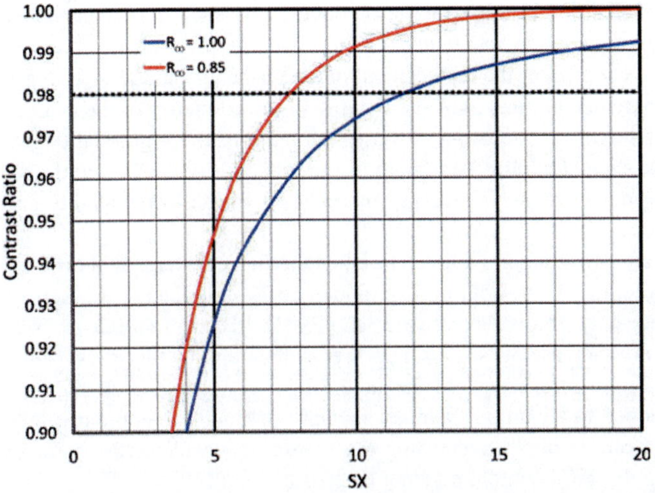


**Fig. 16.12** Contrast ratio as a function of total scattering (SX) for paints with two different $R_\infty$ values. A contrast ratio of 0.98 is considered complete hide

## *Cost Optimization*

The importance of low manufacturing costs to the economic success of a coating is obvious, and much formulation and reformulation in the coatings industry are directed at cost reduction. Cost reduction is most significant when reducing the costs of the highest cost ingredients. These are typically the $TiO_2$, the resin, and the color pigment.

### Basis for Cost Optimization

When discussing the cost of a paint, we must clearly define the cost basis—either as the cost of a certain volume of paint, or as the cost to cover a certain area. Simply replacing one paint component with a less expensive alternative will decrease the cost on a volume basis, but care must be taken that cost on an area of coverage basis does not increase. For example, we could replace half of the solids in a latex paint with water. This will reduce the cost of the paint by roughly one half. However, it will also decrease the coverage rate by one half. The same dry film thickness is required for full opacity, but now it is obtained by doubling the number of coats of paint, thereby doubling application costs.

---

the paint made with the low-quality pigment will have a reduced coverage rate because it both scatters and absorbs less light than the paint made with the high-quality pigment that has been toned down to the same brightness.

The conclusions for this example are obvious, but the full implications of focusing on cost per unit volume, rather than per unit coverage, are typically much more subtle. For example, consider the replacement of a high-quality grade of $TiO_2$ with a lower quality grade that scatters light at only 90% of the efficiency of the incumbent grade. If the low-quality grade is only 85% of the price of the high-quality grade, it may seem obvious that adding 10% more of the low-quality grade would result in a 5% reduction in $TiO_2$ costs. However, the real cost of the lower quality grade can be much more.

There are two strategies that can be taken for this replacement of high-quality $TiO_2$ with low-quality $TiO_2$, both having hidden costs associated with them. In the first strategy, we may simply increase the $TiO_2$ PVC by 10% on a relative basis. However, this does not increase the scattering power of the paint by that same percentage (see Chap. 4, Fig. 4.8). Because of dependent light scattering, each additional portion of $TiO_2$ added to a paint has a lower scattering strength than the previous portions. Indeed, in our example, increasing the scattering power of the paint by 10% by increasing the $TiO_2$ content may not be possible at all.

In the second strategy, we may keep the paint formulation the same, but increase film thickness in order to achieve full opacity. In this case, film thickness would need to be increased by 10%. This not only increases the $TiO_2$ coverage by 10%, but it also increases the coverage of all other paint ingredients by 10%. While we reduce the $TiO_2$ costs by 5%, we increase the other costs, including resin, by 10%. In most cases, this more than offsets the $TiO_2$ cost savings.

Of course, a hybrid strategy could be followed in which the $TiO_2$ content is increased by a certain percent and the film thickness by a certain percent, in order to return to full opacity. The costs of this hybrid strategy will be somewhere between the costs of each strategy on its own. Since each strategy on its own can increase the cost of coverage, we would expect any hybrid approach to also increase the cost of coverage.

## Reducing $TiO_2$ Costs

There are two approaches for reducing the $TiO_2$ content in a paint while maintaining other properties (in particular, spread rate). These are the partial replacement of $TiO_2$ by another light scattering particle, and the optimization of the $TiO_2$ dispersion and dispersion stability. We will consider each in turn.

Perhaps the most common approach for cost reduction is a reduction of the $TiO_2$ content of the paint. This approach benefits from the fact that $TiO_2$ effectiveness decreases as $TiO_2$ content increases. In effect, we can consider the last few PVC of $TiO_2$ being added to the paint as being the least effective for light scattering (this was touched on in our analysis above on the cost-effectiveness of non-optimal $TiO_2$

grades). This is because as the $TiO_2$ concentration increases, so too does the loss of opacity due to particle crowding.[10]

Viewed in the other direction, the first few PVC of $TiO_2$ taken out of a paint is also the least effective and so the most easily replaced by a light scattering alternative such as flash calcined clay or HSOP.[11] For this reason, it is possible to reduce the $TiO_2$ content by using these replacements. As a general rule, the replacement should be done on a 1:3 or 1:4 volume basis (i.e., 1 PVC of $TiO_2$ is replaced by 3 PVC of HSOP). The exact replacement levels, and recommended reformulation to accompany such replacements, are specific to specific products, and the manufacturer of the replacement particles will typically provide guidance to reformulation. Note, however, that removing or replacing even small quantities of $TiO_2$ from a paint can result in a significant loss of wet hide, which is of importance in many paint applications [29, 30].

The alternative means of reducing the $TiO_2$ content of a paint, while retaining opacity, is to improve the degree of $TiO_2$ dispersion in the final paint film. This is only possible when the $TiO_2$ is under-dispersed in the original paint. This can be determined in the laboratory through a dispersion ladder where the level of dispersion energy or dispersion time is sequentially increased, and paint opacity is measured at each level. This opacity level can be compared to that obtained in the paint plant to determine if better dispersion is required. When interpreting the grind ladder data, it is important to keep in mind that it is possible to over-disperse highly treated (flat grade) $TiO_2$ pigments. This causes the oxide surface treatments—which are present to improve particle spacing—to grind from the pigment surface. This is only an issue with highly treated grades of $TiO_2$, since the treatment on these particles is porous and so is relatively fragile. Certain color pigments also lose effectiveness when over-ground.

Finally, as noted in the earlier section on light absorption, most "white" commercial paints have brightness values (tristimulus Y or $R_\infty$ values) of the order of 85. We can contrast this with the laboratory paints used to measure light scattering (Chap. 13). To simplify this measurement, the paints are formulated at very high brightness values (thus decreasing the significance of any light absorption). Such paints are brighter than most applications require, and paint opacity is significantly enhanced by toning the paint to a brightness of 85.

Related to this is the importance of comparing paint opacity at equal brightness. Often a new raw material will be darker than the material it replaces and so will increase opacity due to higher light absorption. As an extreme example, we can replace all of the $TiO_2$ in a typical paint with a much lower concentration of carbon black and still achieve complete hide. Obviously comparing the costs of a white paint to a black paint is not relevant because in most paint applications the end-user would not be willing to use black and white paints interchangeably. Consumers typically specify the desired color or brightness of paint and it is up to the paint producer to

[10] Although this reduces the scattering strength of all $TiO_2$ particles, it is convenient to instead treat the loss as if it is all due to the added $TiO_2$ having lower scattering power.

[11] We will consider replacement by air voids in the section on formulating above the CPVC.

meet this demand, rather than to persuade the customer to change their preference. For this reason, cost comparisons of paints must be made for paints of the same color or brightness if they are to be relevant.

The implications of toning paints to equal brightness or opacity can be understood as follows: at the same brightness, the ratio of light absorption to light scattering (K/S) in the paints is the same. However, each of these factors can separately be higher in one paint than in the other. By toning a brighter paint down, we increase K while holding S constant. This results in the sum of K and S increasing. Since opacity is a direct function of this sum, the opacity of the toned paint (i.e., the spread rate of the paint) will be higher after toning than before.

## Considerations When Formulating Above the CPVC

Although all coatings end-use applications are unique in one sense or another, flat house paints (also known as architectural or décor paints) are fundamentally different from other paint types because these paints are generally formulated above the CPVC and so, by design, contain air voids. These air voids offer two means of lowering paint costs.

First, because air voids within a polymer matrix scatter visible light, we can reduce paint costs by replacing some of the titanium dioxide pigment with air. This opacity benefit is somewhat offset by the high degree of titanium dioxide particle crowding that is an unavoidable consequence of formulating above the CPVC. That said, in most cases, the cost savings of reducing $TiO_2$ content is greater than the loss of scattering power due to crowding.

The second cost advantage of formulating above the CPVC is due to resin replacement by lower cost materials. To exceed the CPVC, we must replace a large amount of resin with particles. In theory, any particle can do this, but, for economic reasons, extender particles exclusively are used for this function, resulting in cost savings. In addition to replacing resin with an extender, above the CPVC we are also replacing resin with air (this is, in fact, how PVC is increased in paints above their CPVC value). This, too, reduces the cost of the paint.

### *The Consequences of Air Pores*

While we do not typically consider them as such, the pores or voids in paint films can be thought of as "particles" of air. The presence of pores in the dry paint film can be determined immediately by comparing the paint PVC with the CPVC of the particle mix: paints formulated below their CPVC do not contain pores, while those formulated above their CPVC do. This is shown schematically in Fig. 16.13.

Paint gloss is negatively affected by pores. As is shown in Fig. 16.13, the surface of a low PVC paint can be very smooth, resulting in a high gloss, while for coatings

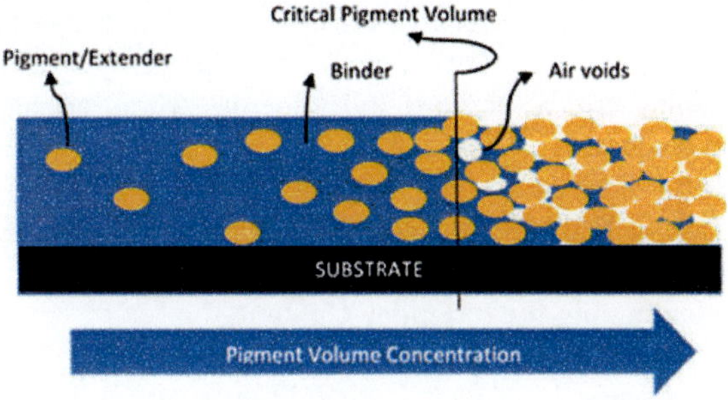


**Fig. 16.13** Schematic representation of a cross section of a film with varying PVC. Left low PVC, right high PVC

formulated at high PVC values, due to the excess of solid particles, the surface is very irregular. This necessarily results in a matt or flat finish. In addition, this form of opacity comes at a cost of compromised mechanical strength, dirt pick-up resistance, cleanability, and burnish resistance. Due to these limitations, industrial coatings such as automotive, coil, can, and powder coatings, as well as most premium architectural coatings, are formulated below their CPVC value.

Counterbalancing these considerable negative properties, pores can contribute significantly to the hiding power of a coating. This follows from the very low refractive index of air. Pores with a diameter of roughly 0.3 to 0.4 microns scatter visible light by a very similar mechanism as light scattering by $TiO_2$ particles, but less intensely [8]. However, since air is a "raw material" that is cost free and readily available, economy lines of paints are formulated above the CPVC whenever possible.

## *Cost Optimization Above the CPVC*

Raw material cost optimization is less straightforward for paints formulated above the CPVC than for paints formulated below the CPVC. Below the CPVC, there is generally a balance point between the cost of the raw ingredients and the opacity of the film. However, such a balance point does not exist for paints formulated above the CPVC. This is because increasing the concentration of air within the film increases the opacity of the film without any additional cost, since air is free. This leads to the most cost-effective film being one that consists of only of extender and the absolute minimum amount of resin required to hold the extender together.

While this paint composition provides the lowest monetary cost of coverage, it comes with an extreme cost in terms of film integrity and durability. Such a paint would not be able to withstand any sort of contact without burnishing or disintegrating, would stain easily, and would detach from the substrate when washed. $TiO_2$-free wall paints, which get their opacity from high concentrations of air voids, are occasionally seen in the marketplace. However, their poor performance under normal use has prevented them from achieving widespread acceptance.

The challenge for cost-effective formulation above the CPVC, therefore, is determining the composition that maximizes extender and air content while still maintaining critical non-opacity properties within acceptable ranges. Maximizing the extender and air contents within these constraints are considered separately in the sections below.

### Maximizing the Extender Content

As discussed in Chap. 4, CPVC values vary from one paint system to another and depend on the distribution of sizes of the various particles present in the paint. With the freedom to control CPVC, we may ask, is there an advantage to formulating paints at a high CPVC value or a low one? We can answer this question by imagining a paint that is below the CPVC, then progressively adding extender particles to it. Until the CPVC is reached, the extender replaces resin. Once the CPVC is reached, adding more extenders results in the resin being replaced by air. The factor that limits how much extender can be added is the level of air that the film can tolerate. It is therefore advantageous to replace as much resin as possible with the extender before reaching the CPVC, since we are limited on how much can be replaced by air above the CPVC. For this reason, a high CPVC value is desired. This occurs when the proportions of extender and pigment particles are adjusted to give maximum packing efficiency.

Particle packing efficiency is maximized when particles of more than one size are used. Ideally, the largest particles will pack at their highest efficiency throughout the film, and the smaller particles will pack at their maximum efficiency within the interstices between the larger particles. For this reason, nearly every well-formulated paint above the CPVC contains two or more extender particles that differ significantly in particle sizes. These are typically referred to as coarse and fine particles.

The relative proportions of the different particle types that maximize the CPVC can be determined by measuring the oil absorption values for mixtures of coarse extender, fine extender, and $TiO_2$ (see Chap. 2 for a description of the oil absorption test). Low oil absorption values indicate high CPVC, and so the paint should be formulated with the particle mixture that gives the minimum oil absorption. However, in some cases, the maximum CPVC occurs for mixtures with higher $TiO_2$ content than desired. In this case, the particle mix should be adjusted such that the proportion of coarse extender is kept constant, the volume of the $TiO_2$ lowered to the desired level, and the fine extender volume raised to compensate for the decrease in $TiO_2$ volume (that is, fine extender should replace $TiO_2$ on a volume basis until the desired $TiO_2$ PVC is reached).

### Optimizing the Air Content

The porosity index should be used as a guide to how much air can be formulated into a paint (Eqs. 16.1 and 16.2). As described in Chap. 4, paints with equal porosity indices generally have similar physical properties, such as stain resistance and scrub performance, particularly when the same resin is used.[12] When reformulating an existing paint, the first step should be to match the porosity index of the original paint and then make adjustments for specific properties as necessary. When formulating a new paint, or replacing the resin in an existing one, a porosity index ladder should be made and the paints tested for film strength (e.g., by scrub resistance). The porosity range that gives acceptable performance can then be used to guide the paint formulation.

$$PI = \frac{\text{air volume}}{\text{air volume } + \text{resin volume}} \quad (16.1)$$

$$= 1 - \frac{\text{CPVC}(1 - \text{PVC})}{\text{PVC}(1 - \text{CPVC})} \quad (16.2)$$

where the PVC and CPVC values in Eq. 16.2 are expressed as fractions (from zero to one).

## Highly Treated TiO2 Grades

The extreme particle crowding in paints formulated above the CPVC causes a significant loss in $TiO_2$ scattering efficiency when compared to paints formulated below the CPVC, even though the $TiO_2$ content of paints formulated above the CPVC is typically less than that of paints formulated below their CPVC. This loss in scattering efficiency can be reduced by using a highly treated grade of $TiO_2$ pigment. Formulation strategies based on this type of pigment were discussed in detail in Chap. 4 and so will only be summarized here.

At and above the CPVC, the $TiO_2$ particles are crowded into the small regions between larger extender particles. The packing in these regions is similar to the packing of $TiO_2$ particles at the CPVC of a paint containing only $TiO_2$ and resin. The scattering power of the $TiO_2$ pigment at this concentration will vary depending on the pigment grade. This scattering power is of little to no importance for commercial paints formulated below the CPVC, since the $TiO_2$ content of such paints will be well below the CPVC value. However, since the crowding conditions of $TiO_2$ particles in paints formulated above the CPVC are similar to the conditions of the $TiO_2$ at the CPVC of the paint made with only $TiO_2$ and resin, $TiO_2$ grades with high scattering at their CPVC will scatter light more efficiently in paints formulated above their CPVC.

Highly treated $TiO_2$ grades are specifically designed to maximize scattering at their CPVC condition. The thick, porous coatings on these particles prevent the $TiO_2$ cores from coming close to one another and so prevent excessive scattering volume overlap at the CPVC. This can be seen in Fig. 16.14 (reproduced from Chap. 4,

[12] This is true only to a first approximation for scrub resistance. Pore size can also affect scrub resistance, as will be shown in a later section of this chapter.

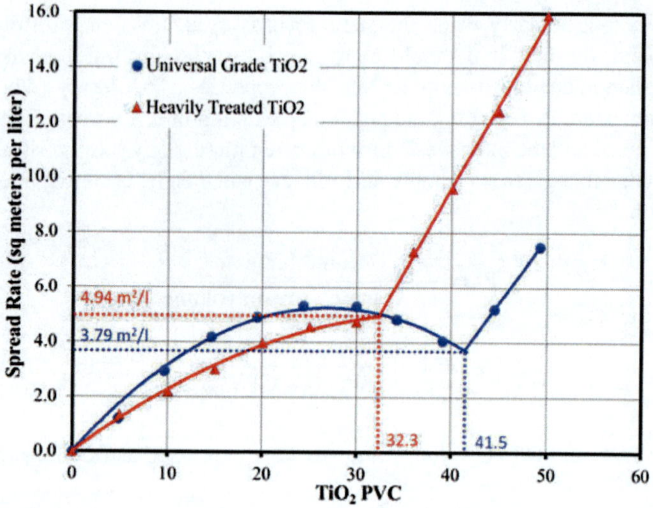


Fig. 16.14 Comparison of the opacity versus PVC curves for a universal grade of $TiO_2$ and a highly treated grade of $TiO_2$

Fig. 4.23), which shows that although there are fewer particles present at the CPVC condition for the highly treated grade compared to the universal grade, the scattering advantage of being more distantly spaced is large enough to more than offset the scattering loss due to fewer particles being present. Said differently, the heavily treated grade has higher opacity at its CPVC than the universal grade does at its CPVC.

The paints used to make Fig. 16.14 were laboratory paints developed to highlight certain principles of light scattering in crowded systems. In particular, these paints did not contain any extender particles. Opacity results for more realistic paints that are formulated with high levels of extender also show the hiding power advantage of the highly treated pigment (Fig. 16.15) [1, 2]. These paints were made with three different extender particles (clay, fine calcium carbonate, and medium calcium carbonate), and in each case, the highly treated $TiO_2$ gave a higher opacity than the universal $TiO_2$.

In addition to improved hide, the highly treated grade of $TiO_2$ gave improved scrub insensitivity compared to the universal grade of $TiO_2$, measured by the ISO 11998 method (microns removed after 200 scrub cycles), as shown in Fig. 16.16. This was attributed to a more optimal air void size in the paints made with the highly treated grade of $TiO_2$.

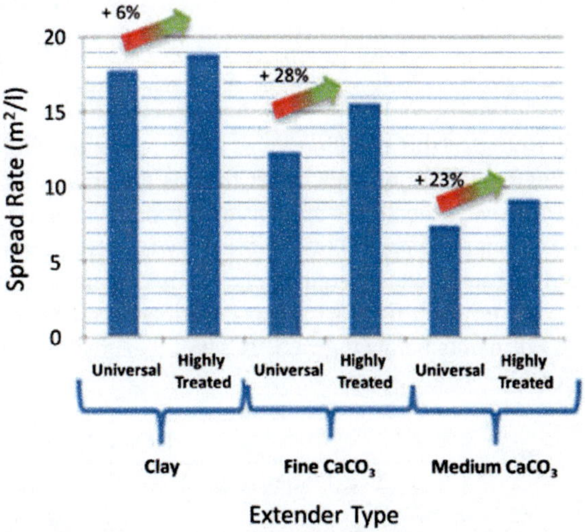


**Fig. 16.15** Comparison of the opacities of paints formulated above the CPVC as a function of $TiO_2$ type and extender particle size

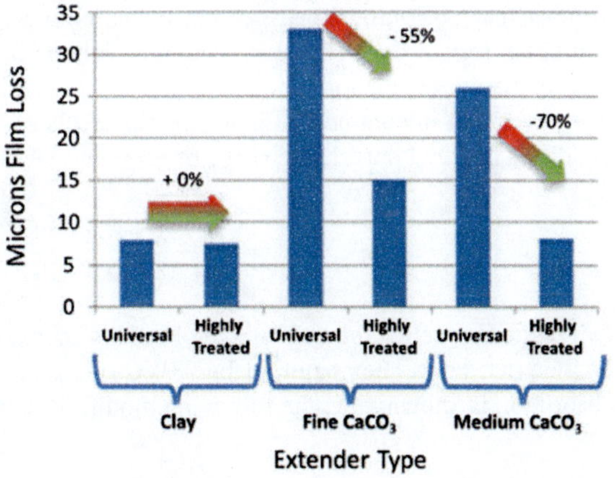


**Fig. 16.16** Scrub sensitivity of the paints shown in Fig. 16.15

# Semi-Empirical Approach to Optimization

Stieg has proposed a semi-empirical equation to predict paint opacity based on the physical characteristics of the ingredients used [25]. These characteristics are the binding power index of the resin, (BPI—see Chap. 10), the porosity index of the paint (if formulated above the CPVC), and the CPVC of the paint (for a solvent-based paint, this can be calculated directly from the oil absorption; for a latex paint, it must be measured or calculated from the oil absorption of the particles and the BPI of the resin).

The Stieg equation is shown as Eq. 16.3:

$$\text{Spread Rate}\left(\frac{m^2}{\text{kg TiO}_2}\right) = 907.4 \cdot f \cdot \left(1 - \left(\frac{PVC_{eff}}{0.74}\right)^{1/3}\right) \quad (16.3)$$

Note that this equation gives the spread rate in terms of the amount of $TiO_2$ (in kilograms) per unit area rather than as the amount of paint (in liters) per unit area. In this way, the solids content of the paint is automatically taken into account.[13] Expressing the spread rate in this way is useful to the paint formulator as the $TiO_2$ cost per unit area covered can be easily calculated from this spread rate value.

The terms in Eq. 16.3 are as described in the following paragraphs:

For paints below their CPVC, $PVC_{eff}$ is the effective $TiO_2$ PVC. From Chap. 4, Eq. 4.9:

$$PVC_{eff} = \frac{\text{TiO}_2\text{ volume}}{(\text{small extender volume}) + (\text{TiO}_2\text{ volume}) + (\text{resin volume})} \quad (16.4)$$

For paints with a $TiO_2$ $PVC_{eff}$ value that is above the CPVC of the $TiO_2$ by itself, the "$PVC_{eff}$" term should be replaced with the $TiO_2$ CPVC value [34]. This reflects the fact that $TiO_2$ crowding does not increase above its CPVC and so there is no additional loss of opacity due to scattering volume overlap. While this condition is seen in many of the laboratory test paints considered in this chapter and in Chap. 4, it is never met in commercial paints as the levels of $TiO_2$ necessary to satisfy this condition are well outside the range of economic viability.

$$f = \left(\frac{1.22 + 0.5 \cdot PI}{4.22 - 0.5 \cdot PI}\right)^2 \quad (16.5)$$

f is the Fresnel factor and PI is the Porosity Index, as expressed in Eqs. 16.1 and 16.2 earlier. For a latex paint, if the CPVC of the paint is not measured but the oil absorption is known, then the following modification to the Eq. 16.2 is necessary:

$$PI = 1 - \frac{CPVC_{oil}(1 - PVC)}{PVC(1 - CPVC_{oil})} \cdot BPI \quad (16.6)$$

The BPI concept and measurement are discussed in detail in Chap. 10 and reviewed in the section of this chapter on resin particles. Here, we will note that it represents the fraction of the resin particle that is available to coat the pigment and extender particles and fill the voids between them and that, as such, it is one factor that determines the CPVC of a latex paint.

Equation 16.3 can be tested using results for the laboratory paint series discussed in other sections of this book. This paint series contains $TiO_2$ and resin only and is used to make the spread rate versus PVC curve shown in, among other locations in this book, Chap. 4, Fig. 4.8. Using the paint formulation, we can calculate the

[13] We can always change the spread rate of a paint by changing its solids content. Such trivial changes are not of interest to us here.

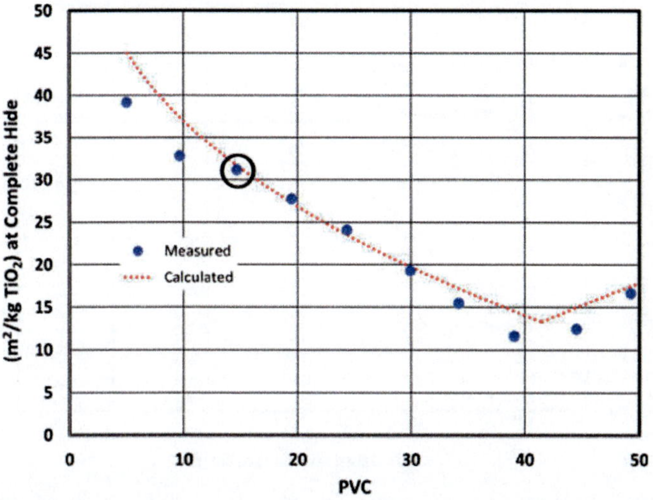


**Fig. 16.17** Experimental and calculated spread rate values (in square meters covered at full hide per kilogram of $TiO_2$) for a laboratory paint. Calculated values are based on Eq. 16.3. For example, one kilogram of $TiO_2$ in a paint made at 14.6 PVC (circled point) can cover 31.1 square meters of substrate at complete hide

spread rate at full opacity, in square meters per kilogram of $TiO_2$, for these paints. The results of this calculation, and of Eq. 16.3, are shown in Fig. 16.17. As can be seen, there is excellent agreement over the PVC range typical of commercial paints.

Stieg pointed out that these equations may be more useful for paint reformulation rather than initial formulation [35]. In this case, the equations would be used in a relative, rather than absolute, fashion. More specifically, he noted that Eq. 16.3 is dependent on only a few factors—spread rate, the Fresnel factor (which is the same for all paints below the CPVC), the effective $TiO_2$ PVC, the porosity index, and the CPVC. When reformulating by adjusting one of these factors, then the paint can be returned to its initial opacity performance by adjusting a different factor by the same proportion.

Note that Eq. 16.3 was developed for multi-purpose or universal grades of $TiO_2$ and tends to deviate from practice when highly treated $TiO_2$ grades are used. That said, even for the highly treated grades, the equations can give useful guidance as to a starting point formulation or the effect of reformulation on opacity. In this case, it is best to treat the $TiO_2$ as consisting of two components—$TiO_2$ itself, which is typically present at about 85% by weight, and small-size extender particles (density 2.7 g/ml) for the remaining 15%.

This situation is shown in Fig. 16.18. Here, the measured spread rates of six paints, all formulated above the CPVC and using a highly treated $TiO_2$ grade, are compared to the values calculated from Eq. 16.3. As can be seen, the fit of the data is best when the actual $TiO_2$ content of the pigment is used in these calculations.

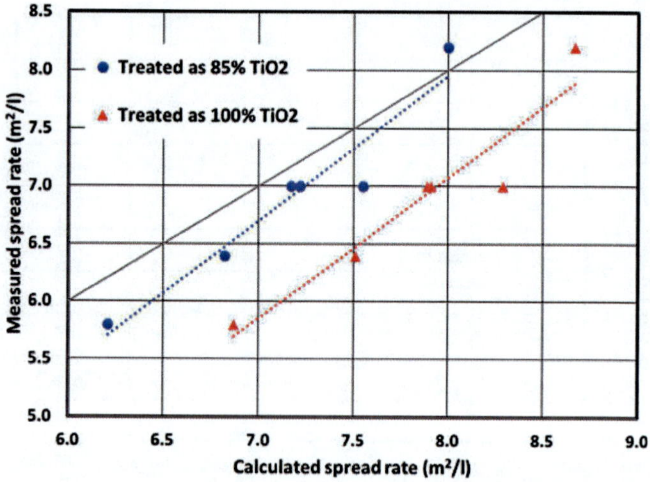


**Fig. 16.18** Comparison of calculated (Eq. 16.3) and measured spread rates for six paints formulated with a highly treated $TiO_2$ grade above the CPVC. The calculations were made based on the particles being treated as 85% $TiO_2$ (circles) or 100% $TiO_2$ (triangles). Gray line indicates perfect agreement between calculated and measured values

Although Eq. 16.3 accurately predicts opacity, it has several restrictions that limit its utility to the paint formulator. These are mostly tied to the necessity of knowing the CPVC value for a given particle mix. This value is necessary for correctly calculating the porosity index. The CPVC value for a particle mixture cannot be calculated as the weighted average CPVC values of the individual components, as explained in Chap. 4. Instead, it must be measured either through a PVC ladder, or, if the BPI of the resin is known, using oil absorption (Eq. 16.6).[14]

Because of this, we can only compare and predict the performance of paints with a specific composition of particles—that is, we cannot freely change the proportions of the different particles without the need to redo the entire experiment. This means that we need to know the composition of the particle mix before formulating the paint. This is the exact opposite of the normal formulation process, where a number of different compositions are investigated, and from this, an optimal composition is determined.

Finally, this equation addresses only one paint property, and formulation must take all properties of importance to the end-use into consideration. As a result of these limitations, it is in most cases simpler and more effective to take a DOE approach to paint formulation. This is discussed in the following section.

[14] The CPVC values of the paints shown in Fig. 16.18 were individually determined.

## Development and Optimization of Paint Formulas

A recurring theme of this chapter is that commercial paints are not required to satisfy a single attribute but instead must satisfy a number of different, often conflicting, performance requirements. Thus, in many ways, the science of formulation is an exercise in compromise. This requires a balanced approach that addresses each of these performance requirements. Of course, not all requirements are of equal importance, but even the lower priority ones often have a minimum requirement below which the paint is unacceptable, regardless of how well it satisfies the more important requirements.

The analyses both in the preceding sections of this chapter and in Chap. 4 are of a more theoretical nature. They are useful as they indicate the various factors that can affect paint properties, particularly opacity, and give a general understanding of how these factors interplay with one another. However, from a more practical viewpoint, it is often more efficient to design a set of experiments for which the factors that are presumed to be important to performance (as identified from the theory in this chapter and Chap. 4) are varied in a systematic way [36].

The traditional approach for evaluating the effect of a new ingredient in a paint formula is to add this ingredient to the formula, eventually compensating by (partially) leaving out another ingredient and testing the new paint for a set of properties. The formulator builds, with only a few experiments, a feel for the potential of this new ingredient. Although this trial-and-error approach can be very useful for initial screening, it can also be very time-consuming for fully optimizing paints as many parameters must be balanced via a back-and-forth process. In addition, very often compromises must be made in either cost or performance.

As an alternative to this trial-and-error approach, an experimental mixture design approach, typically referred to as a DOE (Design Of Experimentation), can be followed [37–39]. In this approach, the boundaries for the compositions of the paints are determined initially (i.e., minimum and maximum values of all the ingredients). The levels of ingredients are changed in a systematic way, making sure that the entire formulation space is covered and with an additional requirement that the amounts of all the ingredients sum to 100%. Typically, a computer program is used to generate the customized "optimal" design, so that with a minimal number of experimental points the desired amount of information can be extracted from the data [40].

The minimum number of experiments (n) needed to find the link between k ingredients and the properties of the paint is given by the following:

$$n = k*\frac{k+1}{2} + l + r \tag{16.7}$$

where $l$ stands for the number of extra combinations to check for the lack of fit (typically 5) and r stands for the number of replicas to assess the repeatability of the experiment (typically 5 as well). The number of required experiments increases very quickly with an increasing number of varied ingredients (k). A study that varies, for example, 10 ingredients, will need at least 65 experiments to generate enough data to feed the model.

Once the experimental runs are determined, paints are prepared and tested for the properties of interest. This leads to an extensive experimentally generated dataset of paint compositions and properties. Through statistical regression analysis, mathematical relationships between properties and composition are determined. Paint properties can be predicted for an arbitrary composition within the formulation space studied using these mathematical relationships.

Despite this large number of experiments, it will normally take less time to come with a final formulation using a DOE approach than a random one. The main limitation of the DOE approach is that it is not possible to add a new ingredient to the model without doing a large number of additional experiments.

## *An Example of a DOE*

As an example of this process, we will discuss the development of a model that allows us to predict the effect of changing the PVC and the type and amount of some typical solid particles in a paint on different paint properties. The approach that we will take is to first design a balanced set of experiments, then execute the experiments, and finally build a mathematical model to predict the different performance properties. We can then use these models to guide us to a formulation that satisfies all the requirements of the paint and is of minimal cost.

We used a formula based on an ethylene-vinyl acetate copolymer dispersion (EVA), as a starting point, to which we will add several families of extender particles (see Chap. 9), each representing a different technology and mechanism for improving hiding power: diatomaceous earth (DE), zeolite, calcium carbonate, calcined clay (CC), and flash calcined clay (FCC). The maximum amounts of the functional fillers were calculated based on the recommendations provided by the respective suppliers.

The starting formulation and all experimental paints are formulated above the CPVC. This is because there are typically many more degrees of freedom above the CPVC than below. This DOE, therefore, shows the highest level of complexity that a formulator will typically encounter.

To ensure that the boundaries for the amounts of resin, calcium carbonate, clays, silicates, and $TiO_2$ were consistent with the values that are commonly used in the industry, more than 150 commercial paints above the CPVC from Turkey, Germany, France, and Russia, spread over different quality segments, were analyzed. Based on this study and on discussions with co-suppliers, the type, minimum, and maximum level of the different raw materials were determined. An overview of the materials used in the formulations can be found in Table 16.4.

Let $X_1$ through $X_7$ denote the weight proportion of each of the seven components shown in Table 16.4 present in a given mixture, and let $X_8$ denote the balance (water plus additives) needed to bring the sum of ingredients to 100%. If the cost per unit weight of each component is $c_1$ through $c_8$, then the cost of any mixture of these

**Table 16.4** Overview of the relevant characteristics of materials used in the study. Minimum and maximum values for the experimental design are also given

| | $TiO_2$ | $CaCO_3$ | Resin | Zeolite | DE | CC | FCC |
|---|---|---|---|---|---|---|---|
| Commercial name | Ti-Pure® R-902 + | Calcilit® M4 | Mowilith® LDM1871 | Zeocross® CG 180 | CelTiX™ | Polestar 200P | Opacilite™ |
| Oil absorption[a] | 24 | 18 | N/A | 60 | 100/220[b] | 50 | 74 |
| Particle size (d-50, μ) | 0.3 | 4 | N/A | 1.2 | 1.7–11.0[c] | 2 | 1.6 |
| Refractive index | 2.7 | 1.65 | 1.5 | 1.5 | 1.43 | 1.56 | 1.56 |
| Min (mass %) | 0 | 15 | 4.8 | 0 | 0 | 0 | 0 |
| Max (mass %) | 25 | 47.5 | 19.2 | 6 | 5 | 8 | 12 |
| Proposed mechanism of opacity differences | High refractive index | NA | NA | Improved spacing of $TiO_2$ | Decrease CPVC Added Air voids | Improved spacing | Sealed air voids |

[a] Expressed as g oil/100 g solids, measured according to ASTM D281-95
[b] Value depends on the source
[c] Depending on the measurement technique: according to the supplier, 1.7 is measured by Sedigraph and 11.0 by laser diffraction

eight components is directly calculated by

$$cost = c_1X_1 + c_2X_2 + \ldots + c_8X_8$$

Our assumption is that the physical properties of interest are affected by the relative amounts of the different ingredients (i.e., by $X_1$ through $X_8$). The relative importance of a given variable to any given property is expected to be different for each material and each property. A model for each property can be envisioned using a coefficient for each ingredient ($b_1$ through $b_8$). In addition, non-linear blending terms (in the form of cross-products) should be included in these models, so each property Y is modeled by

$$Y = b_1X_1 + \ldots + b_8X_8 + b_{12}X_1X_2 + b_{13}X_1X_3 + \ldots + b_{78}X_7X_8,$$

This model therefore has 36 unknown coefficients (8 linear blending terms plus 28 non-linear blending terms).

## Building the Model

To assess the goodness of fit of the model, a customized "optimal" design of 40 different formulations to test was generated, a (different) control sample was added, and finally 6 of these 41 different formulations in the 47-run experimental design were replicated. Partial tables of the experimental design (*X*s) and measured properties (*Y*s) are shown in Table 16.5 and Table 16.6, respectively.

The dispersant (a commonly used acrylic dispersant) was kept constant for the 47 formulations. However, the level was adjusted based on the amount of dispersed pigment and extender and on their respective oil absorption values. The amount of thickener also was optimized to guarantee a good dispersion and an acceptable application of the paint by an automated applicator.

The applied paints were tested according to the test methods listed in Table 16.7. As all the paints are formulated above the CPVC, pores are present in the dry film. The opacity due to these pores can be quantified by oiling the dried paint film. This fills the pores with an oil that has a refractive index very close to that of the resin, removing the effect of the pores on hiding power. Consequently, the oiled hiding power removes the hiding power contribution of the pores in the film (i.e., the dry flat hiding) and therefore is a good measure for the $TiO_2$ contribution in the total hiding power. A large difference between dry and oiled hiding power indicates a large opacity contribution of pores. As discussed above, a high pore content has a negative impact on the overall mechanical strength of the film. This will translate into a greater scrub sensitivity (microns of paint removed after a certain number of scrub cycles). In this study, the logarithm of the scrub sensitivity was used to characterize performance, as this increased the accuracy of the predictions in the lower value range.

In addition to the experimental data, properties such as formulation cost, PVC, and CPVC of the modifications were calculated and modeled. As these parameters are purely calculated values, establishing the fits is trivial (similar to the fit for cost, as previously discussed) and so is not further discussed here.

A forward selection mixture regression approach revealed the mathematical relationships between the amounts of the different ingredients and the corresponding

**Table 16.5** Partial overview of the experimental runs (paint modifications) that were tested in the laboratory and used to feed the model. All amounts are in weight %

| Mod | $TiO_2$ | $CaCO_3$ | Resin | Zeolite | DE | CC | FCC | Balance | PVC | CVPC | Rel. cost |
|---|---|---|---|---|---|---|---|---|---|---|---|
| 1 | 22.75 | 15.00 | 7.92 | 0.00 | 1.00 | 0.00 | 12.00 | 41.33 | 83 | 47 | 1.00 |
| 2 | 0.00 | 41.87 | 4.81 | 0.00 | 0.00 | 2.00 | 8.00 | 43.32 | 90 | 56 | 0.24 |
| 3 | 0.00 | 31.00 | 19.23 | 6.00 | 0.00 | 8.00 | 0.00 | 35.77 | 66 | 55 | 0.47 |
| 4 | 25.00 | 15.00 | 19.23 | 0.00 | 0.00 | 0.00 | 3.00 | 37.77 | 60 | 53 | 1.11 |
| 46 | 25.00 | 15.00 | 4.81 | 3.00 | 5.00 | 0.00 | 4.92 | 42.27 | 89 | 45 | 1.07 |
| 47 | 12.00 | 15.00 | 19.23 | 6.00 | 0.00 | 0.00 | 12.00 | 35.77 | 66 | 47 | 0.85 |

**Table 16.6** Partial overview of the measured properties of the paint modifications shown in Table 16.5

| Mod | Dry hiding power ($m^2/l$) | Oiled hiding power ($m^2/l$) | Scrub sensitivity (log µ) | Tinting strength (L*) | Dirt pick-up (a.u.) | Oil pick-up (a.u.) | Gloss | b* | L* | BI[a] |
|---|---|---|---|---|---|---|---|---|---|---|
| 1 | 18.61 | 11.00 | 2.03 | 57.01 | 5.37 | 16.93 | 12.3 | 1.92 | 97.68 | 1.0 |
| 2 | 6.44 | 1.11 | 3.08 | 39.46 | 7.04 | 22.37 | 8.1 | 2.04 | 96.46 | 1.7 |
| 3 | 2.19 | 0.87 | 1.24 | 27.08 | 7.15 | 18.02 | 4.0 | 2.67 | 95.99 | 1.5 |
| 4 | 8.28 | 8.49 | 0.92 | 50.91 | 7.27 | 16.17 | 7.8 | 1.68 | 97.40 | 0.6 |
| 46 | 20.92 | 11.74 | 2.31 | 59.00 | 6.10 | 19.46 | 7.9 | 2.20 | 97.90 | 1.3 |
| 47 | 10.55 | 7.55 | 0.91 | 48.00 | 8.60 | 16.00 | 10.2 | 2.20 | 97.40 | 1.0 |

[a] Burnish index

**Table 16.7** Overview of the test methods and the statistical parameters evaluating the quality of the statistical model for the different tests

| Tested parameter | Test method | $R^2_{adj}$ (%) | $R^2_{pred}$ (%) |
|---|---|---|---|
| Dry hiding power | Reference [41] | 99.59 | 99.06 |
| Oiled hiding power | See Chap. 4 | 99.90 | 99.57 |
| Scrub sensitivity | EN ISO11998 | 97.31 | 95.55 |
| Tinting strength | Internal method[a] | 99.78 | 95.66 |
| Dirt pick-up | Internal method[b] | 92.94 | 85.88 |
| Oil pick-up | Internal method[c] | 79.86 | 72.98 |
| Gloss | EN ISO2813 | 94.11 | 70.86 |
| b* | ASTM E1347 | 96.44 | 89.74 |
| L* | ASTM E1347 | 97.40 | 93.94 |
| Burnishing index (BI) | ASTM D6736 | 96.42 | 91.57 |

[a] Tinting strength was evaluated by mixing 13 g of carbon black (PBk 7) in 70 ml of the white paint, followed by the measurement of the L* value of the applied paint

[b] Dirt pick-up is measured by scattering a mixture of silica gel, aluminum oxide powder, iron oxide, and lampblack powder on the applied paint. After the removal of the excess mixture, the L* was measured. The lower the value, the more the paint is sensitive to picking up dust particles

[c] Oil pick-up: similar to dirt pick-up but here carbon black is mixed in petroleum jelly and applied on a painted panel. The lower the L* value, the more the paint is sensitive to oil/fat stains or fingerprints

properties. These equations include all linear terms and the most influential non-linear blending terms of the ingredients. The statistical summary parameters, shown in Table 16.7, demonstrate that the data is accurately described (high $R^2_{adj}$) and that the model makes accurate predictions of performance (high $R^2_{pred}$). For some properties, such as oil pick-up, dirt pick-up, and burnishing index, lower $R^2$ values are observed. This can be explained by the slightly larger experimental errors that are associated with these tests. The lower $R^2$ value for gloss (particularly the $R^2_{pred}$ value) is somewhat unexpected, but it can at least in part be attributed to the narrow range of gloss values seen in these flat paints. Since flattening agents can be used to adjust gloss rather easily and without a significant impact on other properties, the low $R^2$ values for gloss are of minor concern.

## Using the Model to Achieve Balance

One of the most important aspects of formulating (or reformulating) a paint is the need to balance performance properties against one another. It is quite rare that more than one property is optimized at the same location in formulation space. Instead, each property must be modeled in formulation space, and, because any paint can

**Table 16.8** Overview of different paint modifications and some calculated parameters. All amounts are in weight %

| Mod | $TiO_2$ | $CaCO_3$ | Resin | Zeolite | DE | CC | FCC | Balance | PVC | CVPC | Rel cost |
|---|---|---|---|---|---|---|---|---|---|---|---|
| 1 | 15 | 15 | 16 | 0 | 0 | 0 | 11 | 43.0 | 66 | 49 | 1.00 |
| 2 | 0 | 22 | 6 | 6 | 5 | 8 | 12 | 41.0 | 88 | 43 | 0.54 |
| 3 | 0 | 15 | 19.2 | 5.35 | 5 | 8 | 12 | 35.5 | 68 | 40 | 0.72 |
| 4 | 20 | 30 | 14 | 0 | 0 | 0 | 0 | 36.0 | 71 | 58 | 1.00 |
| 5 | 16 | 30 | 14 | 0 | 0 | 0 | 0 | 40.0 | 69 | 60 | 0.83 |
| 6 | 16 | 30 | 14 | 3 | 0 | 0 | 0 | 37.0 | 71 | 57 | 0.89 |
| 7 | 16 | 30 | 14 | 0 | 1.7 | 0 | 0 | 38.3 | 70 | 56 | 0.86 |
| 8 | 16 | 30 | 14 | 0 | 0 | 3.0 | 0 | 37.0 | 71 | 57 | 0.87 |
| 9 | 16 | 30 | 14 | 0 | 0 | 0 | 2.6 | 37.4 | 71 | 57 | 0.87 |

exist at only one point in formulation space, a compromise must be made as to the balance of properties of that paint. Less important properties must be allowed to decrease in order to maximize the more important properties.

Two scenarios were created to illustrate the potential uses of this model developed in the last section and the need to balance one property against another. In the first scenario, the objective was to formulate a paint with a hiding power of 10 $m^2/l$ and less than 20 micron loss of paint after 200 scrub cycles (scrub class 2 per ISO 11998). Three hypothetical modifications to the paint were conducted to obtain the desired result (Modifications 1 to 3 in Table 16.8).

As shown in Table 16.9, the predicted values from the model indicate that, with the help of $TiO_2$, several combinations are possible (e.g., Modification 1; more combinations are possible, but only one is shown). Modification 2 demonstrates that without $TiO_2$, it is possible to get the required hiding power, but scrub sensitivity is far from acceptable (338 micron > > 20 micron). Increasing the amount of resin (modification 3) in order to decrease the scrub sensitivity results in a significant loss of hiding power. This demonstrates the impossibility of formulating a paint with a good scrub resistance in combination with a good hiding power in the absence of $TiO_2$.

In a second, less extreme scenario, the effect of reducing the $TiO_2$ content from 20% was investigated. Without compensation with extenders, this has a negative impact on the hiding power (Modification 4 versus Modification 5 in Table 16.8). However, by adding the correct amount of extender, the model predicts that the dry hiding power can be increased to the same level as in the original paint formula (Modifications 6 through 9), but at the cost of a decrease in performance for the other parameters, as shown in Table 16.9.

To independently verify the values predicted by the model, Modifications 1, 2, 4, 5, and 7 were prepared in the laboratory and the performance values were experimentally determined. Figure 16.19 and Table 16.9 indicate that the predicted values correlate very well for all the verified paint properties.

**Table 16.9** Overview of the predicted and experimentally determined results for each paint modification

| Mod | Dry hiding power ($m^2/l$) | | Oiled hiding power ($m^2/l$) | | TS[a] (a.u.) | | Scrub (micron removed) | | Oil pick-up (a.u.) | | Burnishing index (a.u.) | | b* | |
|---|---|---|---|---|---|---|---|---|---|---|---|---|---|---|
| | Pred | Exp | Pred | Exp | Pred | Exp | Pred | Exp | Pred | Exp | Pred | Exp | Pred | Exp |
| 1 | 10.2 | 10.6 | 7.9 | 8.4 | 48.6 | 50.4 | 22 | 9 | 16 | 21 | 1.0 | 1.2 | 2.2 | 2.1 |
| 2 | 9.5 | 9.1 | 1.1 | 1.2 | 44.8 | 46.4 | 392 | 396 | 26 | 25 | 1.4 | 1.5 | 2.4 | 2.5 |
| 3 | 6.2 | – | 1.3 | – | 37.4 | – | 28 | – | 22 | – | 1.2 | – | 3.2 | – |
| 4 | 8.3 | 7.6 | 7.2 | 7.1 | 49.8 | 51.44 | 25 | 11 | 16 | 19 | 0.7 | 0.6 | 1.9 | 1.8 |
| 5 | 6.7 | 6.5 | 6.2 | 6.5 | 46.7 | 47.6 | 31 | 10 | 17 | 21 | 0.8 | 0.8 | 1.8 | 1.9 |
| 6 | 8.3 | – | 6.7 | – | 48.0 | – | 28 | – | 17 | – | 0.9 | – | 1.8 | – |
| 7 | 8.3 | 8.1 | 6.7 | 6.5 | 47.8 | 48.7 | 34 | 11 | 18 | 23 | 0.9 | 0.9 | 2.1 | 2.2 |
| 8 | 8.3 | – | 6.7 | – | 48.1 | – | 30 | – | 18 | – | 0.8 | – | 1.9 | – |
| 9 | 8.3 | – | 6.9 | – | 48.0 | – | 29 | – | 17 | – | 0.8 | – | 2.0 | – |

[a] Tinting strength

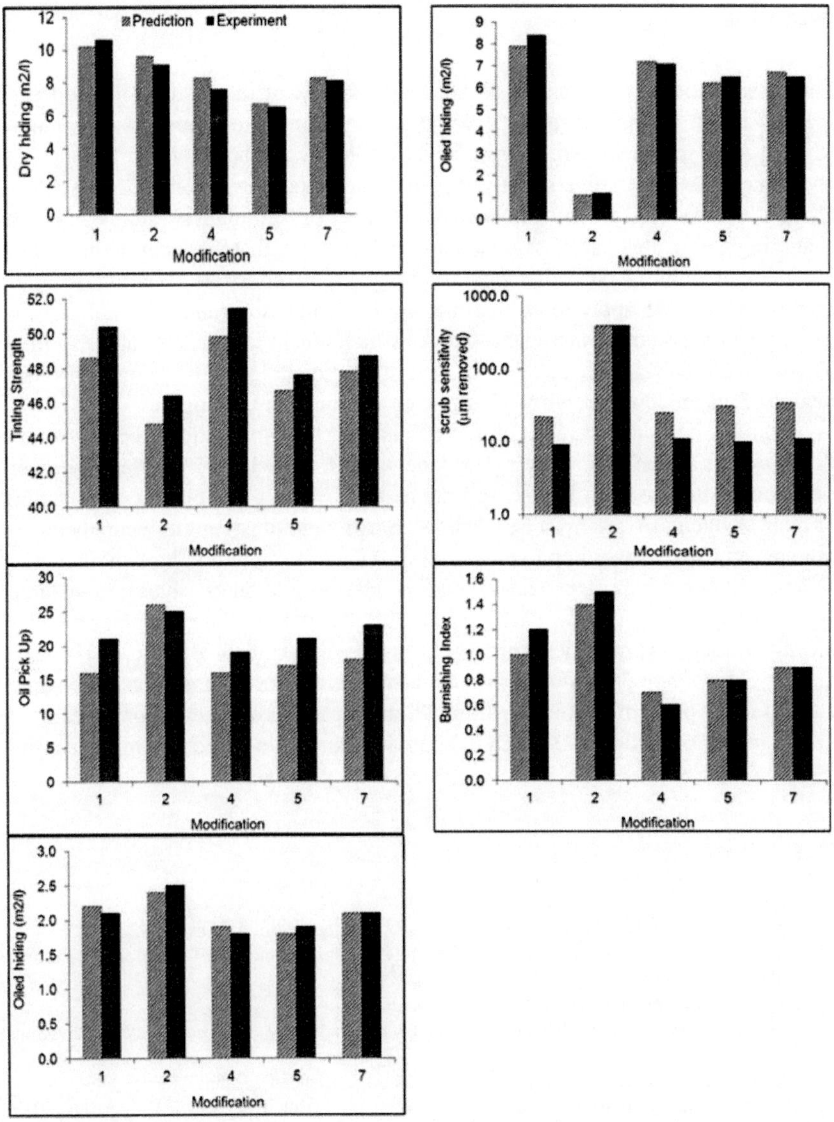


**Fig. 16.19** Correlation between predicted and experimentally determined values of dry and oiled hiding power, tinting strength, scrub sensitivity (logarithmic scale), oil pick-up, burnishing index, and b*

## Summary

Paint formulation is a complex task. There are often many important paint properties that must be met simultaneously. A large number of paint components can be used to meet these properties when formulated in the correct proportions.

Perhaps the most universal of required paint properties is complete opacity. In Chap. 4, we investigated the influence of a number of formulation variables on light scattering and white opacity using laboratory test paints that were not optimized for commercial viability, but rather were formulated to demonstrate specific principles. In this chapter, we apply these findings to the formulation (and reformulation) of paints that are commercially viable and for which other requirements are specified. These other requirements include gloss, resistance to mudcracking, scrub resistance, degree of dispersion, durability, substrate protection, and rheology.

A number of paint parameters can be adjusted in the formulation process. Among these are the use of nano-spacing extenders, resin particles with different properties, composite particles, light-absorbing materials, and gloss modifiers (flattening or matting agents). Care must be taken when using or adjusting these parameters as often an adjustment that improves one performance attribute will hurt another.

The number of important paint properties and formulation variables is quite large, and the trial-and-error study to delineate the effects of all of these variables on all of the important paint properties is not practical. Instead, a statistical approach (DOE) is the most effective means of determining the regions of formulation space for which all important properties are met. In addition to paint properties, economics is an important consideration and the DOE approach can also be used to minimize cost.

## References

1. Mueller, B., Poth, U.: Coatings Formulations: 3rd Revised Edition, Vincentz (2017)
2. Jones, F.N., Nichols, M.E., Pappas, S.P.: Organic Coatings: Science and Technology 4th Edition, Wiley and Sons (2017)
3. Sauer, F.: Microbicides in Coatings, Vincentz (2017)
4. Hunter, R.S., Hunter, R.W.: The Measurement of Appearance, Second Edition. Wiley and Sons (1987)
5. Weldon, D.G.: Failure Analysis of Paints and Coatings, Revised Edition. Wiley and Sons (2021)
6. Uemoto, K.L., Agopyan, V., Vittorino, F.: Concrete protection using acrylic latex paints: effect of the pigment volume content on water permeability. Mater. Struct. **34**, 172 (2001)
7. De Backer, S., Diebold, M.P.: Influence of paint quality on the environmental footprint of architectural paints. Euro. Coat. J. **30** (Jan 2022)
8. Crowther, J.A., Johnson, R.W.: Shedding Light on White - Eight Fundamental Factors that Govern Efficiency of $TiO_2$. Euro. Coat. J. (June 2008)
9. Legrix, A., Moitrier, S., Pamplin, J., Riflet, R., Wood, N.: Titanium Dioxide Extension in Low PVC Decorative Paints – A Novel Approach. ECS congress, Nuremberg (April 2013)
10. Cuppo, F.L.S., Garcia-Valenzuela, A., Olivares, J.A.: Influence of surface roughness on the diffuse to near-normal viewing reflectance factor of coatings and its consequences on color measurements. Color. Res. Appl. **38**(3), 177 (2013)
11. Stieg, F.B.: Dilution efficiency of extenders. J. Coat. Tech. **53**(680), 75 (1981)

12. Stieg, F.B.: Author's criticism of extender 'Spacer Theory' called into question. J. Coat. Tech. **58**(740), 81 (1986)
13. Hook, J., Cutrone, L.: Conclusions on 'Crowding/Spacing Theory' clarified. J. Coat. Tech. **58**(742), 81 (1986)
14. Cutrone, L.: The influence of Fine-Particle Size Extenders on the optical propertied of latex paints. J. Coat. Tech. **58**(736), 83 (1986)
15. Stieg, F.B.: Ending the 'Crowding/Spacing Theory' debate. J. Coat. Tech. **59**(748), 96 (1987)
16. Cutrone, L.: Influence of Fine-Particle Size Extenders on the optical properties of latex paints. J. Coat. Technol. **58**(736), 83 (1986)
17. Diebold, M.P.: A Monte Carlo determination of the effectiveness of nanoparticles as spacers for optimizing $TiO_2$ opacity. JCT Res. **8**(5), 541 (2011)
18. Tarng, M.-R., Chu, K., Steffenhagen, M., Friedman, S.: Pigment spacing. US 7,727,323 (2010)
19. Trapani, A., Hook, J., Fitzwater, S., Reffner, J., Bors, D.: Optimising waterborne paints vs $TiO_2$-polymer composites. Polym. Paint. Colour J. **20** (Jan 2012)
20. Stieg, F.B.: Calcium-based titanium pigments: death of a giant. J. Paint Technol. **44**(565), 63 (1972)
21. Braun, J.H., Baidins, A., Marganski, R.E.: $TiO_2$ pigment technology: a review. Prog. Organic Coat. **20**, 105 (1992)
22. Yan, Q., Lei, Y., Yuan, J.: Preparation of titanium dioxide compound pigments based on kaolin substrates. J. Coat. Technol. Res **7**, 229 (2010)
23. Sun, S., Ding, H., Hou, X.: Preparation of $CaCO_3$-$TiO_2$ composite particles and their pigment properties. Materials **11**(7), 1131 (2018)
24. Wilkenhoener, U., Mersch, F.: Composite pigments comprising titanium dioxide and carbonate and method for producing. US 8,858,701 (2014)
25. Stieg, F.B.: Air as a dispersion medium. Ind. Eng., Prod. Res. Develop. **13**(1), 41 (1974)
26. McDonald, C.J., Devon, M.J.: Hollow latex particles: synthesis and applications. Adv. Colloid. Inter. Science **99**, 181 (2002)
27. Hungenberg, K-D, Jahns, E.: Trends in emulsion polymerization processes from an industrial perspective. In: Polymer Reaction Engineering of Dispersed Systems. Springer (2017)
28. Jiang, S., Van Dyk, A., Maurice, A., Bohling, J., Fasano, D., Brownelf, S.: Design colloidal particle morphology and self-assembly for coatings applications. Chem. Soc. Rev. **46**(12), 3792 (2017)
29. Dietz, P.F.: The effect of Fine-Particle-Size Extenders and entrapped air on $TiO_2$ in emulsion paints. Paint Coat. Ind. **19**(9), 28 (2003)
30. Dollani, H., Elton-Legrix, A.: "Coverage Uncovered." Euro. Coat. J. **10**, 40 (2021)
31. Narayan, R., Kothapalli, R.V.: The use of calcined clay as part replacement of titanium dioxide in latex paint formulations. J. Appl. Polym. Sci. **77**, 1029 (2000)
32. Li, T., Sheng, D., Xiong, Z., Gao, X., Ji, F., Yang, Y.: Effect of titanium dioxide ($TiO_2$) distribution and minute amounts of carbon black on the opacity of PVDF based white composite films. J. Appl. Polym. Sci. **133**, 43064 (2015)
33. Simpson, L.A.: An attempt to develop a one coat emulsion paint. J. Coat. Technol. **56**(715), 57 (1984)
34. Stieg, F.B: Volume relationships. In: Woodbridge, R. (ed.) Principles of Paint Formulation. Blackie and Son, Ltd. (1991)
35. Stieg, F.B.: The ABC's of white hiding power. Pigm. Resin Technol. **7**(1), 5 (1978)
36. De Backer, S., Diebold, M.P., Bailey, S.P.: Predicting properties of architectural coatings by modeling data generated from a designed mixture experiment. JCT Coatings Tech. **11**(4), 36 (2014)
37. Piepel, G.: Models and Designs for Generalizations of Mixture Experiments Where the Response Depends on the Total Amount. Dissertation Discovery Company (2019)
38. Cornell, J.: Experiments with Mixtures: Designs, Models, and the Analysis of Mixture Data 3rd Edition. Wiley and Sons (2002)
39. Anderson, M.J., Whitcomb, P.J., Bezener, M.A.: Formulation Simplified: Finding the Sweet Spot through Design and Analysis of Experiments with Mixtures. Productivity Press (2018)

40. Snee, R.D., Hoerl, R.W.: Strategies for Formulations Development: A Step-by-Step Guide Using JMP Revised Edition. SAS Institute (2016)
41. Standard Test Method for Hiding Power of Paints by Reflectometry. ASTM D2805–11 (2018)

# Chapter 17
# Formulating Plastics with Small Particles

## Contents



## Introduction

Pure plastics are typically transparent to semi-transparent materials with varying levels of mechanical properties and permeability. The underlying structure of most plastics is a hydrocarbon-based polymer. This is a macromolecule constructed from a defined repeating unit (the monomer). As the number of repeating units increases, the polymer structure begins to evolve in two or three dimensions versus its approximately one-dimensional starting block. These macromolecules have either a natural origin (e.g., silk, proteins, cellulose, etc.) or are produced synthetically from hydrocarbon feedstocks (e.g., polyethylene, polyvinyl chloride, polystyrene, etc.). Polymer chemistries incorporate a wide spectrum of architectures to provide a broad range of thermoplastics and thermosets. Thermoplastics are those polymers that can be softened with heat and pressure, and, once softened, molded into the desired final shape by cooling. The chemical identity of the polymer is not altered during this physical transformation. Thermoset polymers, on the other hand, undergo a chemical reaction during the molding or post-molding step and cannot be reshaped once formed.




M. Diebold et al., *Pigments, Extenders, and Particles in Surface Coatings and Plastics*, https://doi.org/10.1007/978-3-030-99083-1_17

For most petroleum-based thermoplastics (e.g., polystyrene, polyvinylchloride, polyolefins, etc.), formulations of the polymer typically contain a mixture of the dry polymer with some or all of the following: additives, particulate solids, colorants, and liquids containing plasticizers, lubricants or other processing aids. Additives in formulations typically provide enhanced properties derived from the inherent polymer structure of composition (e.g., rheology control, surface tension, thermal stability, UV stability, etc.). These added features are described using terms such as melt flow rate (described in more detail below), weathering stability (the ability to remain unchanged upon exposure to the elements, including solar ultraviolet radiation, heat, various pollutants, and other aspects of outdoor environments), tensile strength, oxygen transmittance rate (an aspect of permeability that is of particular importance to films and other packaging), fire retardancy, and impact strength, among others. Hence, plastic materials can be utilized in a wide variety of end-use applications based on the ingredients used in their formulation. These modifications provide the versatility to allow these materials to be used in many different applications in the everyday world, including such diverse applications as shopping bags, food packaging, window profiles, detergent bottles, intravenous fluid bags, mobile phone casings, toothbrush handles, and many more.

Thermoplastic formulations take advantage of highly desirable plastic properties like corrosion resistance, low density, and ease of processibility as melts. Solids are added to thermoplastics in order to mitigate two drawbacks—low stiffness and poor overall mechanical strength. Solids are added to plastic resins to improve stiffness in applications such as highway structures (guardrails and signboards), drainage systems, bridge decks, utility poles and pipelines for gas, water and sewage. Magnetic solids can be added to plastics for magnetic decals and credit card strips.

By contrast, thermosets become "set" once they are cured to the solid state. Instead of softening when heated, these materials simply burn or char. Examples of thermosets include melamine, polyurethane, epoxy, and urea formaldehyde. Particles are added to thermosets to modify mechanical and physical properties in much the same way as for thermoplastics. However, there are several important differences between these two types of polymers:

- Thermoplastics are more easily recycled
- Thermoplastics are more easily customized for application by the use of solid additives
- Thermoplastics are tolerant of solids incorporation
- Thermosets are more resistant to heat and provide higher inherent structural integrity

## *Description of Solids in Plastics*

The term particle requires more definition than simply its chemical composition. For example, when combined with plastic resins, large pieces of marble behave quite differently than marble dust. Both forms of marble are chemically identical, but the physical properties are nonetheless quite different. When added to plastics, inorganic solids are often termed extenders or fillers. The terms are often used interchangeably. To be more precise, extenders are solid particles that provide a desired function to the matrix while fillers are solid particles used solely to replace the polymer in the resulting matrix. Other particles commonly used in plastics are colorants such as dyes and pigments. These provide the ability to control the color characteristics of the finished items manufactured from the polymer formulation [1]. To avoid confusion, the terms solids or particles will be used in this chapter.

The dominant parameter for a formulation involving a thermoplastic and a solid particulate is the solids volume. The focus of this chapter is inorganic particles that have very limited water solubility and are suspended in polymer systems that are traditionally petroleum-based materials. Hence, this chapter will focus on polymers such as polyethylene, polyvinyl chloride, and polystyrene. Within this arena, the solid characteristics of the greatest interest are particle geometry and size distribution.

## Particle Size for Solids

The first criterion by which we will classify the nature of particle size in polymers is whether the particles exhibit Brownian or non-Brownian motion [2–5]. We make this distinction based on the time scale for diffusion of particles based caused by thermal fluctuations within the polymeric matrix, either in the molten or solid state. Diffusion is most meaningful in the molten state. In typical plastics processing, the formulation is processed as a highly viscous fluid that is pumped and molded.

Many particle scientists use the Peclet number to define Brownian versus non-Brownian motion in a suspension [6, 7]. The Peclet number is the ratio of the shear rate experienced by the particle to the particle diffusivity [8]. The particle diffusivity is inversely proportional to the particle diameter. The rate of diffusion decreases as the solid particles become larger in one or more dimensions. When the Peclet number is greater than one, the diffusivity value is low and the suspension is typically considered non-Brownian. When Peclet values are less than one, the particles are considered more Brownian, and the diffusivity factor becomes larger. This makes the particles more sensitive to thermal fluctuations.

Suspensions of nanoparticles (i.e., particles with at least one dimension less than 0.1 micron) in molten polymer are considered Brownian [9, 10]. Suspensions with particles from one to one hundred microns are less Brownian. Particles of intermediate range, less than one micron but greater than 0.1 micron, are of the most interest to us. Note that particles in the nanometer range (e.g., carbonaceous materials, fumed silica, organoclays, etc.) are not within the scope of interest since their typical particle size is below the intermediate range.

This narrow range of particle size generates much interest because these intermediate particles, when suspended in polymer matrices, are controlled mostly by Brownian motion derived from hydrodynamic and interparticle forces. Brownian motion will randomize particle motion when the solid content in the suspension is low; as the solid content increases, hydrodynamic forces begin to influence the behavior of the particles. At high solids content, interparticle forces begin to influence particle motion.

## Aspect Ratio for Solid Particles

Particle aspect ratio is another parameter that can be used to guide formulation. The aspect ratio is defined as the length of the longest axis divided by the axis or axes perpendicular to the long axis. The aspect ratio determines whether the shape of the particle is a needle (ratio greater than 1), a platelet (ratio less than 1), or a sphere (ratio equal to 1). The aspect ratio affects the ability of particles to orient when suspended in fluids as well as the maximum packing fraction for the particles. In general, as the aspect ratio deviates from unity in either direction, the maximum packing fraction decreases. As a result, increasing the concentration of particles with non-unity aspect ratios increases polymer viscosity to a much greater extent than particles with aspect ratios close to unity (Fig. 17.1). Platelets and needles often show the same viscosity at one-third of the volume concentration as do spheres with similar diameters [11–13].

Figure 17.2 is a schematic of the flow around two particles with different aspect ratios. Spherical solids rotate, spin, and flip in all directions with minimal resistance. However, when the aspect ratio is not unity, the force required to spin, rotate, or flip the particles in one or more of the three directions is different. This results in aspect-ratio-related differences in relative viscosities and in the maximum volume packing.

The aspect ratio of a particle affects many fundamental properties of the material in which it is dispersed. For example, in a paper matrix, thin platelets are valued

**Fig. 17.1** Relationship of aspect ratio and viscosity at constant solids content

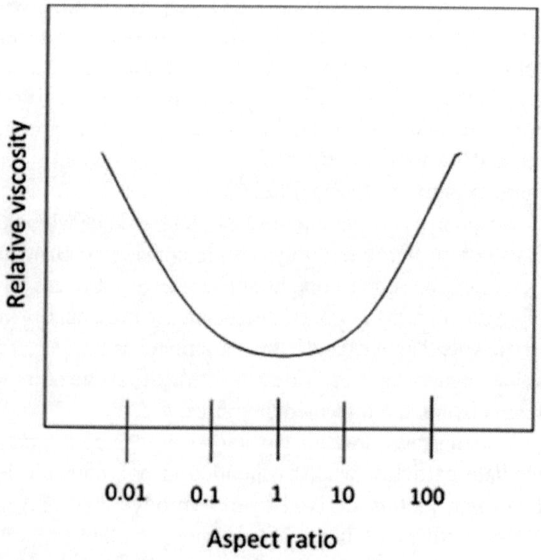

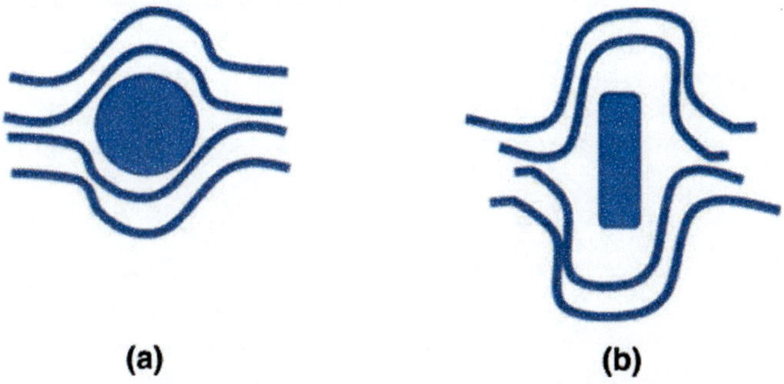


**Fig. 17.2** Flow around solid of different aspect ratio. **a** Aspect ratio is unity. **b** Aspect ratio is notunity

because thinner particles can overlap more easily and provide calendared structures that give improved opacity, while overlapping platelets in paints can improve gas barrier properties because they create a more tortuous path through the paint film. Overlapping platelets can also improve the gas barrier properties of thermoplastics for the same reason.

The aspect ratio can also influence the flow of a solid suspension in a thermoplastic. As the aspect ratio deviates from unity, the effective volume fraction of the solids in the suspension increases compared to that possible for a spherical particle. This results in an increase in melt viscosity due to increased friction between the particles. Higher effective volumes can also improve properties such as bulk magnetism, thermal conductivity, and electrical conductivity by reducing the maximum packing fraction and changing the percolation limit in comparison to spherical particles at the same concentration. Some particles with high aspect ratios, such as certain organic pigments, are often difficult to disperse. This can prevent these pigments from reaching their full color saturation potentials unless sufficient energy is applied to allow full dispersion to be achieved.

A vivid demonstration of particle aspect ratio in plastics can be seen in the injection molding process for thermoplastics. The primary reasons for introducing particles into the plastic are to reduce structural limitations or to introduce a colorant. The particles are first combined with the plastic and blended as a feedstock for the molding process (this is often called a concentrate). At this stage, spherical particles are advantageous because they provide less increase in viscosity at high concentration than particles that have aspect ratios that differs from unity. The resultant lower viscosity enables the suspension to flow through the injection mold more easily and to better conform to the shape of the mold. However, aspect ratios that differ from unity can be advantageous for the release of the plastic article from the mold and to achieve improved mechanical properties. Acicular (needle-like) particles are of particular use in the latter regard.

**Fig. 17.3** Relationship of solids volume as a function of the solid blend

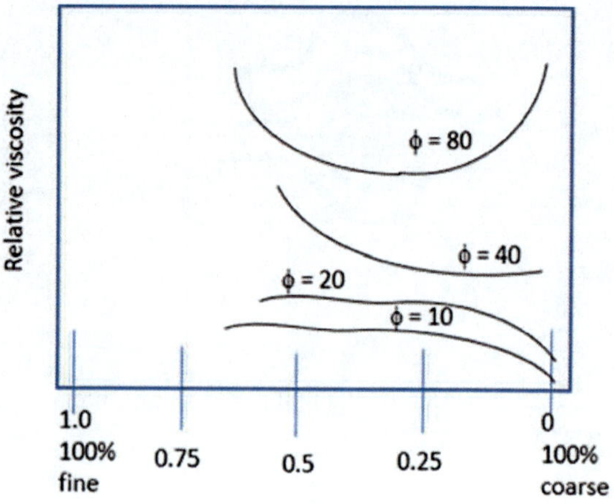


## Particle Size Distribution for Solids

In addition to size and aspect ratio, particle size distribution impacts viscosity. As is the case in paints (Chap. 4), the maximum packing volume for identically sized particles (a monodisperse distribution) is less than that for a blend of large and small size particles (a polydisperse distribution), for which the small particles can fill the voids between the larger ones [14–16]. The intent of blending particles of different sizes is to maximize the solids content while minimally influencing flow properties. This is analogous to the situation of small ball bearings allowing larger particles to roll more easily over one another. In practice, suspension mixtures with mainly intermediate particle size, as defined earlier, and only a few large particles, do not behave significantly differently than suspensions with only intermediate particles. The effect of particle size distribution is greater when a majority of the particles are large (non-Brownian sizes). The effect of the balance between large and small particles on viscocity, as a function of the solids concentration ($\varphi$, expressed as a percent), is shown in Fig. 17.3.

## Formulation with Solid Particles

Knowledge of the volume concentration at which interactive forces begin to increase (and hydrodynamic forces decrease) is critical to an understanding of solids formulation in plastics. Literally hundreds of empirical, semi-empirical, and theoretical relationships have been developed to describe the flow characteristics of the particle suspension as a function of solids volume, size, shape and, to some degree, shear rate. All of these relationships are demonstrated by conducting a titration of volume occupied by solids as a function of a property.

Properties such as opacity, rheology, color, and mechanical strength are often manipulated by adjusting solids volume concentration. Typically, this titration is only linear for a tightly defined volume range, and the slope of the response changes as the particle volume concentration (PVC) increases. Mathematical manipulations are required to arrive at linear equations to describe the entire particle/volume titration. For thermoplastics, the viscosity changes as a function of solids content serves as a useful starting point to address the behavior of the plastic when the concentration of particles of intermediate size increases.

An example of the relationship of intermediate particle volume content and viscosity is seen in pigmentary titanium dioxide particles dispersed in low density polyethylene. In this example, melt flow rate (MFR) is used as a measure of viscosity. MFR is a simplified version of more sophisticated rheometric techniques, as this technique measures viscosity at only one shear rate per sample. The test uses a specific orifice size in a heated barrel, and a piston that matches the barrel dimensions. The polymer formulation is packed into a clean barrel with the piston. After a specified heating period to melt the resin, a weight is placed on the piston, and the molten material is forced through the die. The melt flow index (MFI) is the amount, measured in grams, that exits the die in 10 min. This measurement can obviously be made both on neat thermoplastics as well as systems containing solid components.

Each resin system has a particular orifice length and diameter, barrel temperature, and piston load, as specified by the ASTM D1238–86 and ISO R1133 test methods. MFR measurements were not originally intended to be viscosity indicators, but workers in the plastics industry have developed correlations of MFR with properties that are associated with the system rheology. MFR and MFI can reveal the general impact on viscosity of increasing the particle volume concentration in a plastic [17]. Note that lower MFR and MFI values indicate higher viscosity. A common practice is to normalize MFR by calculating the ratio of the MFR value of the filled system to the MFR value of the neat polymer. This ratio is referred to as the MFR index.

MFR evaluations can be used to determine the effect of particle volume concentration on viscosity when designing formulations with thermoplastic resins such as polyethylene. As the particle volume concentration increases, the MFR of the filled system drops (i.e., the system becomes more viscous), and the MFR index increases. Figure 17.4 shows the increase in the melt flow index value as the titanium dioxide pigment volume increases in low density polyethylene (LDPE). The relative MFR value, $MFR_{natural}/MFR_{filled}$, increases with concentration in a linear fashion until a certain point is reached at which the polymer melt flow behavior changes. Eventually, as the particle volume is increased further, the flow ceases and the system has an MFR value of zero or an MFR index value of infinity.

More sophisticated rheological techniques have been used to demonstrate similar trends, [18] especially for titanium dioxide pigments, which are an ideal example of intermediate particles. Titration experiments such as that shown in Fig. 17.4 are necessary to establish predictable models for intermediate particles in plastics.

A refinement of the titration experiment is to use capillary rheometry to characterize the impact of solids volume concentration. Figure 17.5 shows a typical

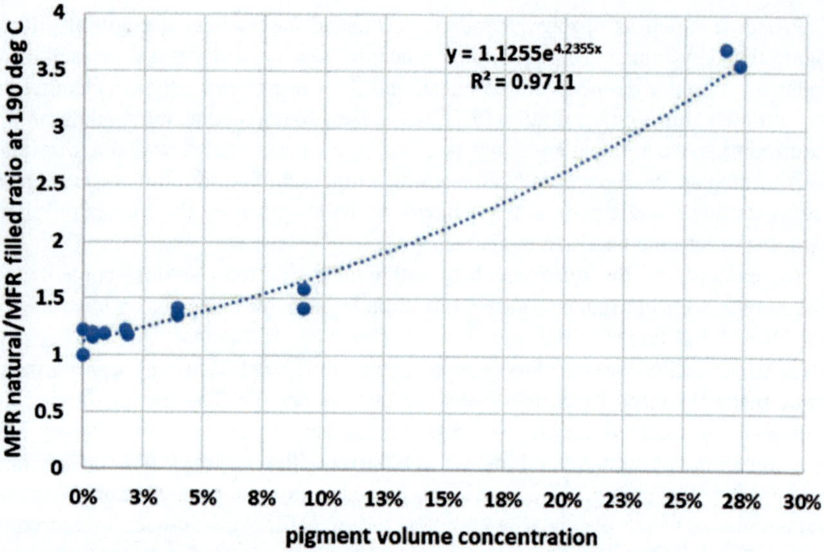


**Fig. 17.4** MFR values for various volume% of titanium dioxide in LDPE (MFR = 13.5)

**Fig. 17.5** Generic viscosity relationship for a plastic containing solid particles

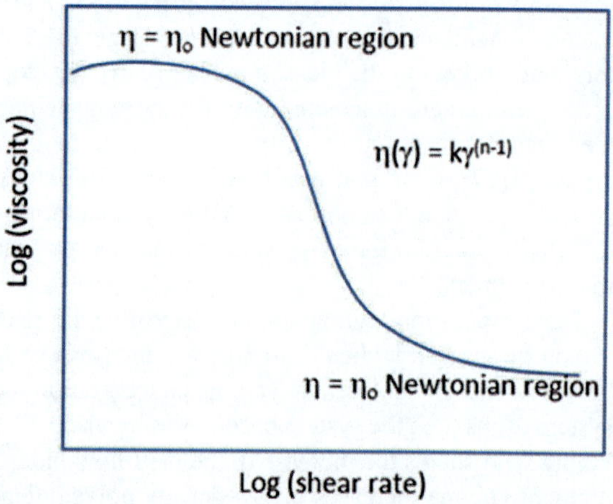


relationship of viscosity, η, with shear rate, γ. At very low shear rates, the thermoplastic behaves as a typical Newtonian fluid, but it becomes shear-thinning at higher shear rates. At these higher shear rates, the viscosity relationship is best described with a power law curve fit.

Many text and literature references present the theory of rheology changes as solids content increases in a polymer melt [19–24]. The most recognized equation (Eq. 17.1) relating viscosity to the volume of solids is Einstein's equation from 1906 [25].

$$\eta_{\text{filled}} = \eta_{\text{unfilled}}(1 + 2.5\ V_s) \tag{17.1}$$

where $V_s$ is the solids volume fraction.

Einstein made several assumptions in order to obtain a simple equation. The equation assumes an infinitely-dilute suspension of rigid, neutrally buoyant, identically spherical, smooth particles in an incompressible Newtonian fluid. Einstein also assumed there was no particle migration or rotation, no slip between the particles and the fluid, and no influence of the walls. Although such conditions are never encountered in real-world systems, this equation is the basis of viscosity predictions in many suspensions at low concentrations. Additional modifications can be made to the general Einstein equation at higher solids concentrations. Mooney developed a more generally-used equation that relates viscosity to particle volume concentration, Eq. 17.2 [26].

$$\eta = \eta_o \cdot e^{\left[2.5/1-(\Phi/\Phi^*)\right]} \tag{17.2}$$

where φ* is the so-called critical concentration. The physical meaning of φ* is that it is the limiting value for the degree of solid particles filling. This is similar to the CPVC concept in paints (Chap. 4). This value depends on both the shape of the particles and their mode of packing, i.e., the coefficient of packing. For instance, φ* is typically 0.74 for hexagonally packed solid spherical particles.

Percolation theory is commonly used to explain the change in viscosity with solids content. This theory is often used to link the power law index, n,[1] to the maximum volume that can be occupied by particles. When the solids/polymer melt exhibits Newtonian behavior, the composite can relieve any applied stress via fluid dissipation. In other words, the particles tumble freely as the molten polymer moves around them. Here the power law index n is one. As the solids content increases, the particles begin to interact with each other and form cluster structures to relieve the stress. In this situation, the power law index begins to approach zero. The cluster formation manifests itself as a shear thinning.

In theory, as the particles cluster together, regions of pure resin are created, allowing the polymer melt to flow. One hypothesis proposes that the power law region of the viscosity curve in Fig. 17.5 reflects a change in volumetric energy dissipation that is related to the solids concentration and the shear rate [27]. In other words, when the clusters form, there is a loss in surface area for shearing from the polymer, which thins the solids/polymer melt. Eventually, as the solids content increases, the particle clusters form a network. This is referred to as the percolation threshold, $\phi_c$.

Figure 17.6 shows how the solid particles start to form clusters that permit fluid to flow. Eventually, the particles align at the percolation threshold, $\phi_c$. At $\phi_c$, the

[1] n can take on values between zero and one. It represents the fractional loss of volume that is capable of viscosity energy dissipation.

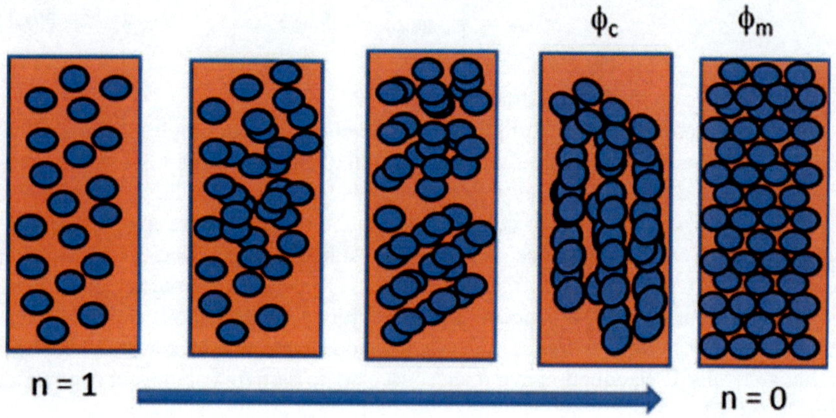


**Fig. 17.6** Diagram showing the change in power index, n, with increasing solids volume

particles may or may not be at the maximum volume, $\phi_m$ (this was called $\varphi^*$ above). Theoretically, $\phi_c$ and $\phi_m$ should be the same. However, in real particle/polymer systems with high solids levels, they are often not identical. For repulsive particle–particle interactions the value of $\phi_m$ should be close to the maximum random packing density ($\approx 0.638$) and for attractive interactions, $\phi_m$ is the value at which the suspension reaches its percolation threshold. The exact value of $\phi_m$ can be determined from a plot of the square root of relative viscosity versus the filler volume fraction. For example, in a study involving pigmentary titanium dioxide, [27] the percolation model provided a reasonable fit to actual capillary rheometer data for %volume ranges of 0 to 49%. The calculated $\phi_m$ was 0.98 and $\phi_c$ was 0.26.

The hypothesis of volumetric energy dissipation has also been evaluated using calcium carbonate particles in an organic solvent [27].

This theoretical discussion shows that one formulation component, the solid volume concentration, is a dominant factor, particularly when the flow characteristics of the particle/polymer melt are of importance. The effect of this is so great that additional ingredients must be added to the formulation in order to modify particle/polymer melt flow and other end-use properties. This modification is of particular interest in many plastics processing steps that are pressure-driven and that would ideally exhibit plug flow. Plug flow is a simple model of the velocity profile of a fluid flowing in a pipe. An advantage of the plug flow model is that no part of the solution can be perpetuated upstream. Plug flow is contrasted to laminar flow, which has a parabolic velocity profile (Fig. 17.7), with zero flow at the pipe walls and maximum flow through the center.

Particle/polymer melt separation, or lack of plug flow, causes particle concentration gradients to form in pressure-driven systems [28, 29]. These concentration gradients can manifest themselves as extrudate distortion. They can also impact performance parameters such as tensile strength, sharkskin surface appearance, die

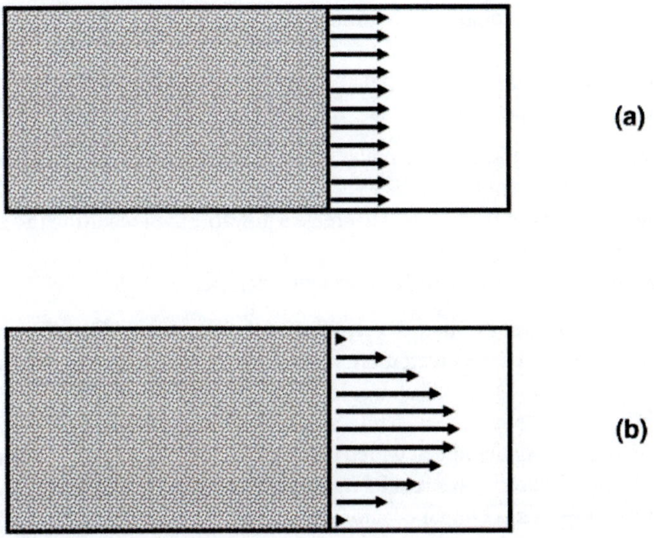


**Fig. 17.7** Velocity profiles of flow through a pipe. **a** Plug flow. **b** Laminar flow

lines in film extrusion, inhomogeneity of electrical conduction, and other aspects of performance [30].

This example showing viscosity changes with solids volume concentration demonstrates the transition from hydrodynamic forces being dominant to interparticle forces controlling the system. Particles of intermediate size in a polymer melt have the ability to behave in a Brownian fashion in one formulation but non-Brownian in another, depending on the solids volume concentration. The general guideline for the formulation of intermediate particle size is that once the solids volume concentration exceeds 20%, interactive forces become largely dominant. The interparticle forces that regulate these interactions are van der Waals interactions, steric hindrance, and electrostatic interactions between the particle surfaces. Studies of pigmentary and nanosized titanium dioxide particles in polypropylene melts indicate that the distance at which particles interact with one another is of the same magnitude as the particle diameter [31]. This particle–particle distance results in $\phi_c$ being near 52% volume.

## *Additives that Influence Viscosity at High Solids Volume*

Additives, such as lubricating aids, are commonly formulated into plastic systems to compensate for this type of behavior, especially in high solid materials [32]. Measurement of the impacts of lubricants in polymer processing has traditionally required the use of an internal batch mixing device. Batch mixers have been used extensively to study dispersive mixing of solid fillers in polymer melts [33, 34].

The time required to achieve polymer melt flow or flux is an important parameter in internal batch mixing. The sequence of events in a batch mixer is

- incorporating the solid into the melt,
- dispersing/distributing the solids,
- final latent melting of the polymer to create a flowable melt resulting in a solid/melt composite, and
- discharging the polymer melt from mixing chamber.

The total time for all events is referred to as the flux time. There is an optimum level of lubricant for a given formulation. Adding too much lubricant can increase the flux time by slowing the discharge of the material from the equipment due to resin adhering to the equipment, while adding too little lubricant can increase the flux time by decreasing the rate at which mechanical/blending energy is transformed into heat through friction. This results in longer melt times. The volume of material within the extruder is a secondary factor that affects flux time. It is essential that the ideal %volume occupied in the batch mixer is known if the proper shear stress in the polymer melt at a fixed shear rate is to be achieved. The ideal %volume occupied in the mixer is commonly referred to as the degree of fill [35, 36]. Hence, to achieve the minimum flux time, optimization of the lubricant dose and the degree of fill is required.

As an example of this, the design of experiment techniques described in Chap. 16 was used to systematically determine the effects of fill and zinc stearate (lubricant) concentration on flux time for the production of a low-density polyethylene (MFR 12) containing 70% by weight titanium dioxide pigment (33.5% pigment volume concentration) in a laboratory scale internal mixer. A regression analysis quantified the relationship between these variables and flux time and revealed the conditions for minimum flux time. The regression fit of the data is given in Eq. 17.3, where the experimentally derived constant was 63.2, $\alpha$ was 0.17, $\beta$ was 0.34, and $\tau$ was 0.002.

$$\text{Flux in sec} = \text{constant} + \alpha - \beta - \tau, \tag{17.3}$$

where $\alpha$ is the concentration of zinc stearate in ppm, $\beta$ is the degree of fill in per cent, and $\tau$ is the product of the concentration of zinc stearate and the degree of fill.

The main output from the regression is the extent to which zinc stearate increases flux time. This additional time can be offset with proper fill selection. Therefore, in this example, formulation design for lubrication involves selecting process conditions for optimum melt behavior in the mixer. The overall optimization reduced the flux time by 15%. The optimization for degree of fill and viscosity modification with the lubrication additive is critical; while the ingredients matter, the manner in which the ingredients are combined is equally important.

## *Mixing Particles of Different Chemical Composition (Calcium Carbonate with Titanium Dioxide)*

Calcium carbonate, $CaCO_3$, is a common ingredient in many plastic applications. This material is commonly referred to as a filler, but this does not reflect some of the useful properties of this material. $CaCO_3$ is quite useful in increasing plastic processing productivity due to its good thermal conductivity. The cooling behavior in the mold of an injection molding machine is influenced by the thermal properties of the filler, and a filler with high thermal diffusivity can thus increase the efficiency of the molding process. Thermal diffusivity is highest in materials with high thermal conductivity, low density, and low heat capacity. In radiant heat transfer processes, a polymer/mineral composite blend cools more slowly than the neat polymer. This is particularly important in extrusion processes [37–40]. While talc can also be used to improve thermal insulation to facilitate the molding operation, calcium carbonate is the most widely used filler in thermoplastics.

In addition to improving the thermal properties of a plastic, calcium carbonate can provide opacity in cavitated applications [41], and it impacts surface gloss to improve surface finish [42–45]. In more typical applications its utility as an opacifier is quite limited due to its low refractive index of only 1.63.

## *Example of Formulation Using Particle Size for Optical Properties*

Pigmentary titanium dioxide is typically formulated in thermoplastics to achieve desired optical properties such as color and opacity. Titanium dioxide is an efficient provider of brightness and a relatively subtle means of adjusting hue, thus allowing for maximum color saturation. In Chap. 3, we discussed the fundamental factors that distinguish titanium dioxide versus other materials from an optical perspective. One of the key factors for optical performance is particle size. Most of the titanium dioxide grades used in plastics have a relatively small particle size to ensure maximum light scattering. Usually, the median particle size of titanium dioxide for coatings applications ranges from 0.2 to 0.4 microns, while the particle sizes used in plastics applications ranges from 0.15 to 0.3 microns (see Chap. 7 for a discussion on the effect of pigment crowding on the optimal size for $TiO_2$ light scattering). At these sizes the $TiO_2$ particles preferentially enhance the scattering of shorter wavelengths, giving the polymer a blue undertone. This blue undertone can offset the yellow appearance of some resins that are prone to forming chromophores with high extinction coefficients.

Even within titanium dioxide grades, there are degrees of attainable colors that are strongly associated with the particle diameter. To demonstrate the different ways that titanium dioxide particles can influence the CIE L*, a*, b* color coordinates in the same black plasticized PVC formulation, an experiment was conducted in which the titanium dioxide particle size distribution was varied. The median particle size was used to quantify the differences in the particle size of three titanium dioxide pigments. The typical median diameters of the three pigment types, as measured by X-ray centrifuge measurement techniques, were between 0.28 and 0.3 microns for Type A pigment, between 0.25 and 0.28 microns for Type B pigment and between 0.22 and 0.25 microns for Type C pigment [46]. Type A pigment attenuates the longest wavelengths of light and Type C pigment the shortest wavelengths. The common nomenclature for the three types of pigments refers to the wavelength that is attenuated to the greatest extent. Type A pigment is often referred to as the "yellow undertone titanium dioxide", Type B pigment is "neutral undertone titanium dioxide", and Type C pigment is "blue undertone titanium dioxide".

The impact of the different titanium dioxide pigments on the final color of the polymer can be plotted using the CIE L*, a*, and b* color axes [47]. Figure 17.8 highlights the color progression of the black PVC formulation as the $TiO_2$ content is increased for the three $TiO_2$ types. Each increment of titanium dioxide added to the PVC formulation results in an increase in L*, a move outward or inward from zero b*, and forward or backward toward zero a*. The differences among the b* values highlight the nomenclature assigned to each titanium dioxide type. For example, the yellow titanium dioxide has the most positive b* value at a given concentration of titanium dioxide. There is also an a* difference among the three types of pigments, but at smaller increments than seen for b*.

To increase L*, the titanium dioxide concentration must be increased. By adding more titanium dioxide, L* increases and the a* and b* values move accordingly. A key aspect of Fig. 17.8 is that each pigment has a unique set of color characteristics that cannot be matched by the others at the same concentration. At no point will the yellow Type A pigment be able to provide the same b* value as the blue Type C pigment.

The "Whiteness Index" (WI) is a measure of the degree to which a surface resembles a perfect white. The ASTM E313 whiteness index used for measuring near-white,

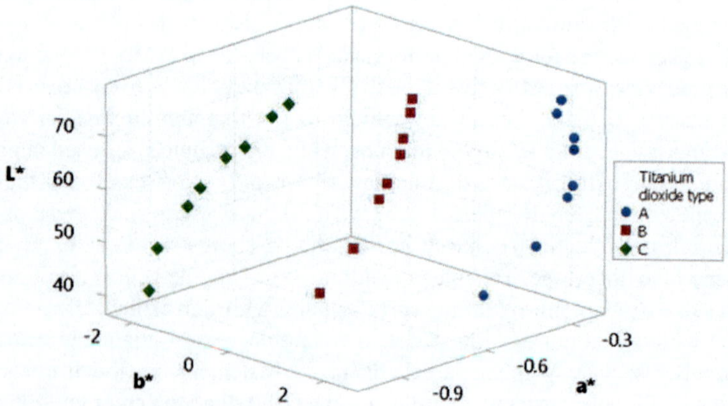


**Fig. 17.8** Scatterplot of L* a* b* for different titanium dioxide materials at different concentrations

opaque materials is calculated as shown in Eq. 17.4.

$$WI = Y + (WI, x)(x_n - x) + (WI, y)(y_n - y) \tag{17.4}$$

where, *x, y, z* are the chromaticity coordinates of the specimen, Y is the tristimulus Y value, and $x_n$ and $y_n$ are the chromaticity coordinates for the CIE standard illuminant and source used (see Chap. 6), and (WI,*x*) and (WI,*y*) are numerical coefficients that are equal to 800 and 1700 for most illuminants. The *x, y, z* chromaticity coordinates are calculated as shown in Eqs. 17.5 through 17.7 [48].

$$x = \frac{X}{X + Y + Z} \tag{17.5}$$

$$y = \frac{Y}{X + Y + Z} \tag{17.6}$$

$$z = 1 - x - y \tag{17.7}$$

Tristimulus X, Y, and Z values can be calculated from CIE L* a* b* values by the following equations when L* > 8.0 [49].

$$\mathrm{X} = \mathrm{X}_n\left[\left(\frac{\mathrm{L}* + 16}{116}\right)^{1/3} + \left(\frac{\mathrm{a}*}{500}\right)\right]^3 \tag{17.8}$$

$$Y = Y_n\left(\frac{\mathrm{L}* + 16}{116}\right)^3 \tag{17.9}$$

$$\mathrm{Z} = \mathrm{Z}_n\left[\left(\frac{\mathrm{L}* + 16}{116}\right)^{1/3} - \left(\frac{\mathrm{b}*}{200}\right)\right]^3 \tag{17.10}$$

In these experiments, the CIE L* a* b* measurements were made using a $D_{65}$ illuminant/10° Observer (CIE 1964), for which, the chromaticity coordinates are

| | |
|---|---|
| $X_n$ | 94.8 |
| $Y_n$ | 100.0 |
| $Z_n$ | 107.3 |

Figure 17.9 shows the WI for the different titanium dioxide grades at different loading levels. As expected, the WI increases with increasing titanium dioxide content. Significant differences between WI are seen for the different titanium dioxide grades. This example demonstrates that formulation for color can be finely adjusted by using similar titanium dioxide materials that have different particle sizes.

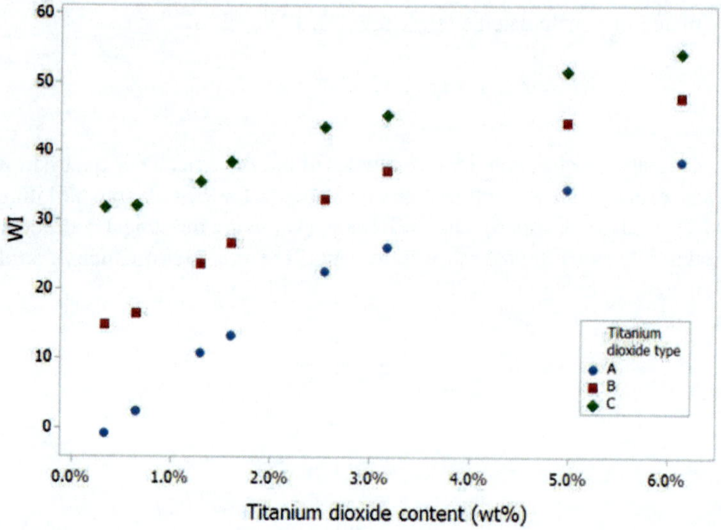


**Fig. 17.9** Whiteness index of compounded black PVC as titanium dioxide content increases

## Miscellaneous Properties Affected by Particles

Polymeric materials exhibit rather low thermal conductivity (TC), in the range from 0.01 to 0.5 W $(mK)^{-1}$, due to their lack of free-moving electrons (as well as low electrical conductivity—these materials are insulators). However, TC can be enhanced by the incorporation of particles. Polymers with enhanced TC are typically used for LED devices, electronic packaging, batteries, and solar applications to improve their thermal management. The enhancement of TC can be achieved via the incorporation of metal powders (e.g., aluminum, copper, or nickel), carbon additives (e.g., graphite, graphene, carbon nanotubes, or carbon black), and synthetic or mineral fillers (e.g., boron nitride, aluminum nitride, or talc).

Solid particles are also used in plastics to improve flame retardancy. The mineral fillers most commonly used for this purpose are metal hydroxides and metal carbonates that decompose between 180 and 400 °C. A common example of this is magnesium hydroxide (MNH), which is one of the most-used flame retardant additives for polyethylene (PE), polypropylene (PP), ethylene–vinyl acetate copolymer (EVA), and other polymers. Aluminum hydroxide ($Al_2O_3{\cdot}3H_2O$ or AlO(OH)), often called alumina trihydrate (ATH)), is commonly incorporated into the plastic fibers used in carpet backing, furniture upholstery, and polyethylene-coated cables for flame retardancy. The most commonly used carbonate for fire retardancy is magnesium carbonate, although calcium carbonate is also used in combination with other fillers (e.g., silicones) and activating materials for wire and cable compounds.

## Summary

The melt flow behavior of solids formulations in plastics is dominated by the solids volume concentration factor. The particle size, aspect ratio, and size distribution of the incorporated solids are means of modifying the polymer melt flow without affecting the molecular structure of the polymer. Knowledge of the flow characteristics of a formulation and its dependence on the behavior of the particulate ingredients is critical to the understanding of the various reasons solid particles are widely used in plastics. The most common reason is to improve mechanical properties, including dimensional stability. Attenuation of gas and liquid permeation, improvements in heat deflection/fire proofing, enhancement of electrical properties, control of color and opacity as well as magnetic properties are typical reasons for solids incorporation into a plastic formulation. The judicious selection of these particles can add significant value to the overall plastic composite.

In thermoplastic applications, the technology of solids formulation hinges on the ability to manage the rheology of the particle/polymer suspension such that the necessary physical properties are achieved and that the performance of the material within processing equipment can be readily managed. Knowledge of the sensitivity of system rheology to solid particle formulation effects may be considered as the combination of practical knowledge with theory. To the theoretical formulator, the creation of the equation and calculations of fundamental constants are of great importance while to the practical formulator, the overriding concern is the actual behavior observed.

## References

1. Berns, R. S.: Billmeyer and Saltzman's Principles of Color Technology, 4th edn. John Wiley & Sons Inc. (2019).
2. Pusey, P.N.: The dynamics of interacting brownian particles. J. Phys. A: Math. Gen. **8**, 1433 (1975)
3. Hess, W., Klein, R.: Generalized hydrodynamics of systems of brownian particles. Adv. Phys. **32**(2), 173 (1983)
4. Foister, R., Ven, T.: Diffusion of brownian particles in shear flows. J. Fluid Mech. **96**(1), 105 (1980)
5. Zahmatkesh, I.: On the importance of thermophoresis and brownian diffusion for the deposition of micro- and nanoparticles. Int. Commun. Heat Mass Transf. **35**(3), 369 (2008)
6. Bossis, G.: The rheology of brownian suspensions. J. Chem. Phys. **91**, 1866 (1989)
7. Yang, H.-G., Li, C.-Z, Gu, H.-C., Fang, T.-N.: Rheological behavior of titanium dioxide suspensions. J. Colloid Interf. Sci. **236**(1), 96 (2001)
8. Maldonado-Camargo, L., Yang, C., Rinaldi, C.: Scale-dependent rotational diffusion of nanoparticles in polymer solutions. Nanoscale **9**(33), 12039 (2017)
9. Batchelor, G.: Brownian diffusion of particles with hydrodynamic interaction. J. Fluid Mech. **74**(1), 1 (1976)
10. Ounis, H., Ahmadi, G., McLaughlin, J.B.: Brownian diffusion of submicrometer particles in the viscous sublayer. J. Colloid Interf. Sci. **143**(1), 266 (1991)

11. Barnes, H.A.A.: Handbook of Elementary Rheology. University of Wales, Institute of Non-Newtonian Fluid mechanics, Aberystwyth (2000)
12. Kitano, T., Shirota, T.: An empirical equation of the relative viscosity of polymer melts filled with various inorganic fillers. Rheol. Acta. I **20**, 207 (1981)
13. Genovese, D.B.: Shear rheology of hard sphere, dispersed and aggregated suspension, and filler matrix composites. Adv. Colloid Interf. Sci. **171–172**, 1 (2012)
14. Hanemann, T.: Influence of particle properties on the viscosity of polymer–alumina composites. Ceramics Int. **34**, 2099 (2008)
15. Shewan, H. M., Stokes, J. R.: Analytically predicting the viscosity of hard sphere suspensions from the particle size distribution. J. Non-Newtonian Fluid Mech. **222**, 72 (2015)
16. D'Haene, P., Mewis, J.: Rheological characterization of bimodal colloidal dispersions. Rheola Acta **33**, 165 (1994)
17. Shenoy, V., Saini, D.R.: Thermoplastic Melt Rheology and Processing. Marcel Dekker Inc., New York (1996)
18. Minagawa, N., White, J.L.: The influence of titanium dioxide on the rheological and extrusion properties of polymer melts. J. Appl. Poly. Sci. **20**, 501 (1976)
19. Pal, R.: New generalized viscosity model for non-colloidal suspensions and emulsions. Fluids **5**, 150 (2020)
20. Rueda, M.M., Auscher, M.C., Fulchiron, R., Périé, T., Martin, G., Sonntag, P., Cassagnau, P.: Rheology and applications of highly filled polymers: a review of current understanding. Prog. Polym. Sci. **66**, 22 (2017)
21. Andrew, D., Jones, R., Leary, B., Boger, D.V.: The rheology of a concentrated colloidal suspension of hard spheres. J. Colloid Int. Sci. **147**(2), 471 (1991)
22. Dealy, J.M., Wissbrun, K.F.: Melt Rheology and Its Role in Plastics Processing. Chapman & Hall, New York (1995)
23. Gupta, R.K.: Polymer and Composite Rheology, 2nd edn. Marcel Dekker, New York (2000)
24. Macosko, C.W.: Rheology Principles Measurements and Applications. VCH Publishers, New York (1994)
25. Einstein, A.: Eine Neue Bestimmung der Molekuldimensionen. Ann. Phys. **19**, 289 (1906)
26. Mooney, M.: The viscosity of a concentrated suspension of spherical particles. J. Colloid Sci. **6**, 162 (1951)
27. Campbell, G.A., Wetzel, M.D.: Characterizing the flow of slurries using percolation theory-based functions. Polym. Eng. Sci. **54**(4), 403 (2017)
28. Newman, S., Trementozzi, Q.A.: Barus effect in filled polymer melts. J. Appl. Polym. Sci. **9**, 3071 (1965)
29. Utracki, L.A.: The shear and elongational flow of polymer melts containing anisometric filler particles; Part I. Rubber Chem. Technol. **57**(3), 507 (1984)
30. Shore, J.D., Ronis, D., Piché, L., Sagui, C., Grant, M.: Flow of polymer melts. Phys. Can. **53**, 166 (1997)
31. Acierno, D., Filippone, G., Romero, G., Russo, P.: Rheological aspects of PP-$TiO_2$ micro and nanocomposites: a preliminary investigation. Macromol. Symp. **247**, 59 (2007)
32. Ram, A.: Compounding and Processing of Plastics: Fundamentals of Polymer Engineering. Springer, Boston, MA (1997)
33. Richardson, J., Matchett, A.J., Coulthard, J.M., Gibbon, S., Wilson, C., Watson, C.: The characterization of pigment powders for titanium dioxide/polymer dispersions by the 'Masterbatch' process. Chem. Eng. Res. Des. **78**(1), 39 (2000)
34. Ozkoc, G., Bayram, G., Quaedflieg, M.: Effects of microcompounding process parameters on the properties of ABS/Polyamide-6 blends based nanocomposites. J. Appl. Polym. Sci. **107**(5), 3058 (2008)
35. Erol, M., Kalyon, D. M.: Assessment of the degree of mixedness of filled polymers. Int. Polym. Process. **20**(3), 228 (2005)
36. Hyvärinen, M., Jabeen, R., Kärki, T.: The modelling of extrusion processes for polymers-a review. Polymers **12**, 1306 (2020)

37. Bartzcak, Z., Argon, A.S., Cohen, R.E., Weinberg, M.: Toughness mechanism in semi-crystalline polymer blends: II. HDPE toughened with calcium carbonate filler particles. Polymer **40**, 2347 (1999)
38. Weindenfeller, B., Höfer, M., Schilling, M.: Thermal conductivity, thermal diffusivity, and specific heat capacity of particle filled polypropylene. *Composite: Part A* **35**, 423 (2004)
39. Zaman, H.U., Beg, M.D.H.: Effect of $CaCO_3$ contents on the properties of polyethylene nanocomposites sheets. Fibers Polym. **15**, 839 (2014)
40. Zhang, X., Zhang, W., Su, J., Wang, M., Lu, C.: Preparation, characterization, and properties of polyethylene composites highly filled with calcium carbonate through co-rotating conical twin-screw extrusion. J. Vinyl Add. Tech. **20**(2), 108 (2014)
41. Brunner, M., Tinkl, M., Brookes, D., Schulz, K.: Calcium carbonate as cavitation and voiding agents in biaxially oriented films. POLYMERS Commun. 87 (Oct–Nov 2017)
42. Rothon, R. (Ed.): Fillers for Polymer Applications, Polymers and Polymeric Composites: A Reference Series. Springer International Publishing Switzerland (2017)
43. Jimoh, O.A., Ariffin, K.S., Hussin, H.B.: Synthesis of precipitated calcium carbonate: a review. Carbonates Evaporites **33**, 331 (2018)
44. Sun, S., Ding, H., Hou, X.: Preparation of $CaCO_3$-$TiO_2$ composite particles and their pigment properties. Materials **11**(7), 1131 (2018)
45. Passaretti, J.D., Young, T.D., Herman, M.J., Duane, K.S., Evans, D.B.: Application of High-Opacity Precipitated Calcium Carbonate. Tappi J. **76**(12), 135 (1993)
46. Stanley-Wood, N., Lines, R. W.: Particle Size Analysis. The Royal Society of Chemistry, Cambridge, U.K. (1992)
47. Berns, R.: Billmeyer and Saltzman's Principles of Color Technology, 3rd Edn. J. Wiley and Sons (2000)
48. "Standard Practice for Calculating Yellowness and Whiteness Indices from Instrumentally Measured Color Coordinates", ASTM E313-10 ASTM International, West Conshohocken, PA.
49. Pascale, D.: A Review of RGB Color Spaces, The Babel Color Company Application notes (2003)

# Chapter 18
# Application to Décor Paper for Use in Laminates

## Contents



## Introduction

Laminate flooring and furniture products are familiar to all. These materials provide an attractive appearance while being less costly, easier to work with, and, often, more durable than natural wood. The natural wood appearance of a laminate is created by printing a colorful wood pattern onto a sheet of high-quality décor paper.

 
M. Diebold et al., *Pigments, Extenders, and Particles in Surface Coatings and Plastics*, https://doi.org/10.1007/978-3-030-99083-1_18

**Fig. 18.1** Components of a laminate floorboard

This paper is then impregnated with an uncured thermosetting resin, either melamine formaldehyde (MF), urea formaldehyde (UF) or a combination of the two. Finally, the impregnated resin is pressed against the substrate (usually particle board) and heated. This simultaneously adheres the printed paper onto the substrate and cures the resin, giving the laminate strength and stability. The resulting product is shown in Fig. 18.1.

The unique environment of the décor paper within the finished laminate creates challenges for the paper maker. Décor paper, like graphics and writing paper, must be opaque. However, the opacifying materials are different in the two paper types. In graphics and writing paper, the voids between the particles and fibers are filled with air, which scatters light. In addition, extender particles, as well as the fibers themselves, scatter light when surrounded by air and so contribute to opacity. In practice, extender particles and paper fibers can scatter light so strongly that $TiO_2$ pigment is not needed in many graphics and writing paper applications (e.g., newsprint, magazines, office paper, etc.).

By contrast, most extender particles do not scatter light in paper laminates. This is because the pores within the paper are filled with resin during the lamination process, resulting in the loss of air void scattering and in the extender particles being surrounded by resin, rather than air.[1] Since resin and fibers have roughly the same refractive index as the extender particles, these particles do not scatter light in the laminate. This situation is similar to that encountered in paints. As was seen in Chap. 3, Fig. 3.7, reproduced here as Fig. 18.2, extender particles are transparent when embedded in resin.

That said, there are certain forms of extenders that are sometimes seen in décor paper. These are engineered extenders that bring with them entrained air. This same strategy is used in coatings (Chap. 7). However, not all extenders used in coatings can be used in paper laminates. This is because not all such extenders can survive the high temperature (180 °C) and pressure (90 bar) experienced during the lamination process (see below). For example, flash calcined kaolin can be used in laminates while some forms of hollow sphere opacifying pigments cannot be used.

Because of this, light scattering in décor papers is almost exclusively provided by pigmentary $TiO_2$. When extenders are found in décor paper, it is often for non-opacity functions such as printability and abrasion reduction. We might expect that the $TiO_2$

[1] There is speculation that some component of air voids remains after lamination process, but, if so, certainly not as much air in a graphics and writing paper.

**Fig. 18.2** Opacity contribution of different white powders to paint opacity

particles would be dispersed as completely as possible in the paper, so as to maximize scattering efficiency. However, the state of $TiO_2$ dispersion also affects the retention of the $TiO_2$ particles in the paper. If the $TiO_2$ pigment was dispersed as individual particles, it could easily pass through the large voids and channels between the cellulose fiber particles and be removed from the paper during the dewatering step. To prevent this, the $TiO_2$ particles and other particles are flocculated into agglomerates that are large enough to be captured by the fibers prior to dewatering.[2] As a result, the crowded pigment particles scatter light inefficiently and so must be present at loadings as high as 40% by weight for a basis weight of 80 g/m$^2$ (gsm) in order to ensure complete opacity.

Figure 18.3 shows a scanning electron micrograph of a cross section of white décor paper. Pulp fiber, air (within cut sections of pulp fiber), and $TiO_2$ are clearly distinguished in this image. The image shows that décor paper is not homogenous in the same way that paints and plastics are. Instead the $TiO_2$ particles are present in large clusters trapped within the pulp fibers.

[2] Other measures are also taken to maximize $TiO_2$ retention in the paper. These will be discussed later in this chapter.

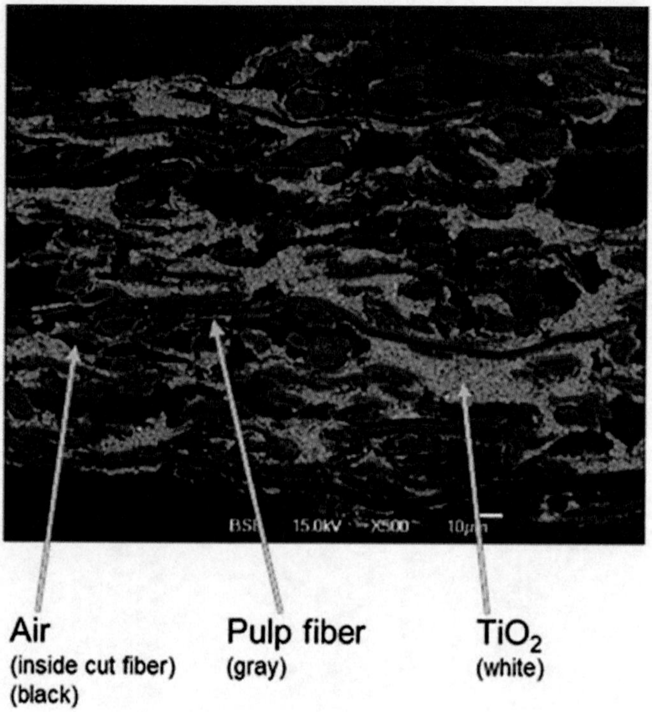


**Fig. 18.3** Electron micrograph cross section of a white décor paper cross section

The demanding requirements placed on décor paper, and the unique environment in which it is found in the laminate, have led to the development of new analytical techniques to characterize the distribution of both the cellulose fiber and the pigment particles in the paper. As we will see, many of these techniques rely on technologies that have only recently become available and that can provide unique insights into fundamental aspects of this application.

## Décor Paper Properties and the Impact of $TiO_2$

Décor paper affects many important properties of the laminate into which it is incorporated. These include appearance properties (e.g., opacity and print quality), physical properties (e.g., paper strength and integrity), and chemical properties (e.g., reactivity with the resin and resistance to discoloration under ultraviolet light). Obviously, this places many requirements on the décor paper itself. If we are to control the factors that affect these requirements, we must first understand the role that the décor paper plays both in the laminate production process and in determining the finished laminate properties.

## *The Role of Décor Paper in the Lamination Process*

Décor paper performs several important functions in a laminate. First, it is opaque and so hides the substrate. Second, it provides the white 'canvas' onto which a pattern or picture is printed. Finally, the paper carries the thermosetting resin precursor that, when heated and put under pressure, cures into a hard and protective surface layer.

Because of the various demands placed on décor paper, it is subject to tight specifications for several parameters. Porosity and pore structure are the most important of these because the pores are responsible for carrying the uncured resin into the system. During impregnation the pores are filled with all of the resin that will be present in the final laminate panel, and inadequate pore volume or structure will result in unacceptable performance of that panel.

The second most important parameter is hiding power or opacity. Paper laminates replace wood veneer, and one important role of veneer is to hide the substrate (typically particle board). High opacity is therefore necessary for proper appearance and aesthetics. The lack of air voids and crowded conditions for the $TiO_2$ make it challenging to meet the opacity demands of this application. This issue will be discussed in detail in the next section.

The paper strength is also an important property, as is color stability of the paper when exposed to ultraviolet light. As discussed in Chap. 14, $TiO_2$ particles darken when exposed to ultraviolet light in the absence of oxygen. The laminate resin is intentionally designed to be impervious to liquids and gases. As such there is little oxygen present within the laminate and so the $TiO_2$ particles must be treated to resist darkening. This property is referred to as light stability.

Finally, the paper should facilitate the curing of the resin. During curing, the MF (melamine–formaldehyde) and UF (urea–formaldehyde) resins polymerize in a condensation reaction. This reaction is conducted at high temperatures, and the water generated by it is released as steam. If not constrained, the escaping steam would exit through the laminate surface, creating a rough texture. A counter pressure is needed to prevent this from happening and so to keep the surface flat.

The curing step is done in a laminate press under controlled temperature and pressure. Curing is typically done with the temperature between 150 and 190 °C and the pressure from 20 to 90 bars. At this pressure, the steam is forced to exit through the open sides of the press rather than its surface. The curing step polymerizes the resin into a single, highly cross-linked molecule. This creates a very hard and durable surface.

The role of $TiO_2$ in laminates, as in coatings and plastics, is to provide opacity. Figure 18.4 demonstrates the opacity demand of décor paper compared to conventional office printing paper. A sheet of each of these papers was impregnated with resin and laminated onto a chipboard. The conventional paper contains 40% $CaCO_3$ in a 100 g per square meter (gsm) sheet while the décor paper contains 40% $TiO_2$ in a similar 100 gsm sheet. The opacity difference between the two papers is quite clear. The calcium carbonate in the conventional paper provides adequate opacity in

**Fig. 18.4** Effect of $TiO_2$ on décor paper opacity. On the left is a laminate panel consisting of impregnated décor paper covering a chipboard. On the right is a similar laminate panel made with impregnated conventional office printing paper

air, but in the absence of air the sheet is virtually transparent. Adequate hiding of the substrate is only achievable by using high levels of $TiO_2$.

## *Comparing $TiO_2$ Crowding in Laminates to Crowding in Coatings and Plastics*

Light scattering in crowded systems was discussed in detail in Chap. 4. In Fig. 4.5, recreated here also as Fig. 18.5, we saw that the scattering strengths of $TiO_2$ particles decrease linearly with their concentration in paints. We also saw that it is not the average concentration of the particles that determines their scattering strength, but instead their local concentration. Light scattering decreases when particles approach one another due to scattering volume overlap. This overlap only occurs for neighboring particles, and so is local in nature.

The effect of $TiO_2$ is more complex in décor paper. In paints (and plastics), the goal is to disperse individual particles to the greatest degree possible. However, in décor paper, the particles must be agglomerated in order to be retained in the fiber network. $TiO_2$ particles show a very high scattering efficiency at low concentrations, but at

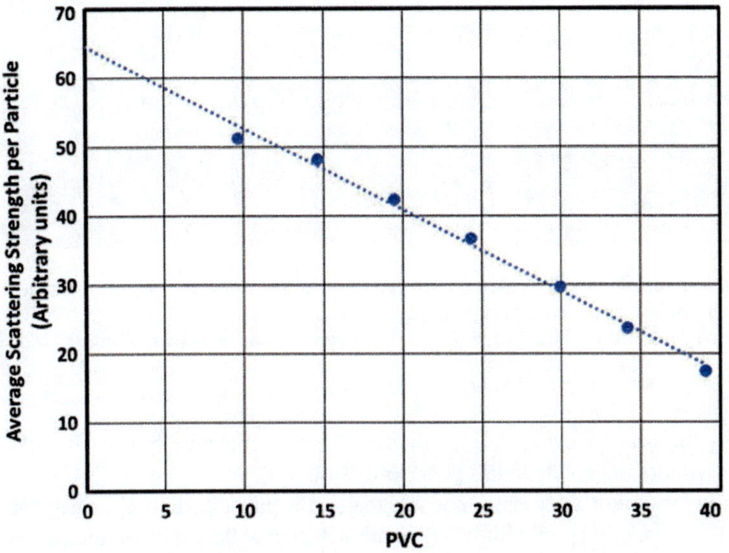


**Fig. 18.5** Impact of $TiO_2$ pigment volume concentration (PVC) on light scattering strength in paints

concentrations between 20 and 40%, $TiO_2$ scattering efficiency decreases linearly with concentration. As a consequence of pigment flocculation that occurs in the paper making process, the measured scattering efficiency of $TiO_2$ particles at 40% loadings is roughly ten times lower than at very low $TiO_2$ concentrations.

## *Appearance Uniformity and Formation*

The term "formation" is used in the paper industry to describe the uniformity of the distribution of various components in a paper, particularly as this uniformity relates to the appearance of the paper or to its basis weight (gsm). Paper is formed by filtering a slurry of fibers, $TiO_2$ pigment, and any other components through a screen. In some cases, these components are not evenly spread on the screen, resulting in non-uniformity. For paper laminates, appearance uniformity is determined both by the distribution of the fibers and the distribution of the $TiO_2$ particles. Poor formation is seen when either is flocculated, as this leads to regions of high and low density within the sheet.

Light microscope images of poor and good formation, at the cm scale, along with an electron micrograph of a laminate paper cross section, at the micron scale, are shown in Fig. 18.6. In Fig. 18.6a, we see the cross section of décor paper that shows non-uniformity (poor formation) at the micro-scale. Figure 18.6b shows that the non-uniformity carries over to the appearance of the sheet at the macro-level—the scale

**Fig. 18.6** Formation of two difference décor papers. **a** SEM micrograph cross section in décor paper. **b** Optical microscope image of poor formation. **c** Optical microscope image of good formation. The width of the image in **a** is 300 microns and the widths of images **b** and **c** are 8 cm

seen by the unaided eye. The appearance of this sheet can be compared to the sheet shown in Fig. 18.6c, which has good formation.

Good formation is necessary not only to give the paper itself a uniform appearance, but also to give good print uniformity in the next step of the paper laminate process. In addition, good formation is necessary to optimize $TiO_2$ scattering efficiency. As we will discuss later, and can be seen in the optical microscope images in Fig. 18.6a and

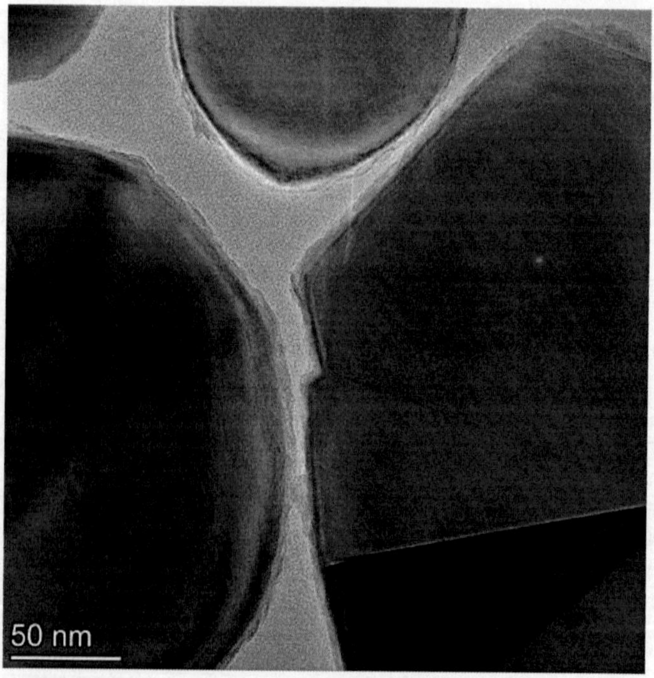


**Fig. 18.7** TEM micrograph of surface treated pigmentary $TiO_2$ used in laminate applications

b, uniformity at the macro-level is determined by uniformity at the micro-level. This latter size domain is that at which $TiO_2$ crowding and loss of scattering efficiency occur.

When formation is poor, appearance uniformity can be improved by increasing the $TiO_2$ content of the paper. This ensures that even those regions with the lowest local concentrations of $TiO_2$ will have sufficient $TiO_2$ to give full opacity. However, this comes at the obvious penalty of a higher cost per unit area.

## *Laminate Grade $TiO_2$ Pigments*

Special grades of $TiO_2$ pigment have been developed specifically for paper laminate applications. These grades are designed for optimal opacity and maximum light stability (resistance to darkening) when the laminate is exposed to ultraviolet light. These properties are created through the use of certain inorganic and/or organic surface coatings on the $TiO_2$ particles (see also Chap. 7). Opacity is controlled by modifying the pigment dispersibility and surface charge, while light stability is improved by altering the kinetics of electron transfer into and out of the pigment particles (see Chap. 14). A transmission electron microscope (TEM) image of a typical $TiO_2$ particle used in laminate applications is shown in Fig. 18.7.

## Surface Chemistry

The décor papermaking process is outlined in Fig. 18.8. Pulp, wet strength resin, $TiO_2$, and water are blended together into what is known as furnish, and then are mechanically processed in the refiner in order to cut open and decluster the pulp fibers. This makes the initially stiff fibers more pliable and increases the fiber surface

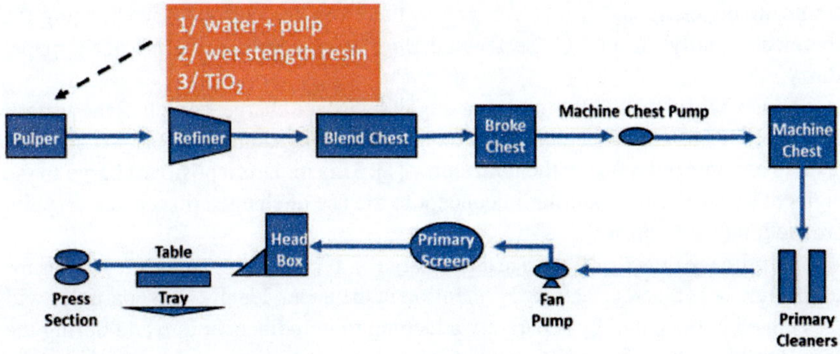


**Fig. 18.8** Schematic of papermaking process

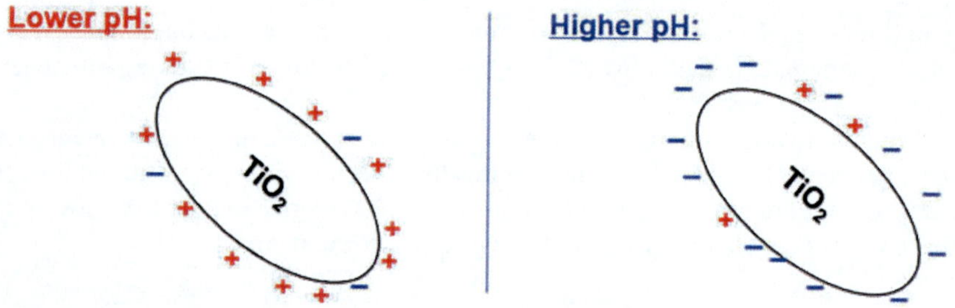


**Fig. 18.9** Formation of $TiO_2$ surface charges by pH adjustment

area, which improves fiber bonding later in the process. Additional chemicals can be added in the blend chest, and wastepaper from the process is recycled into the broke chest. Material is inventoried in the machine chest, allowing for an averaging of the pulp slurry made over a period of time, which evens out inconsistencies that may enter earlier in the process. Finally, the slurry is cleaned, then screened as it is pumped into the headbox. The headbox is under pressure and pushes the slurry evenly onto a screen. The material on the screen is dewatered and dried to form the paper.

Particle surface charges are especially important in décor paper production as they can affect several aspects of the papermaking process and impact laminate panel appearance. As discussed in Chap. 2, the surfaces of particles suspended in water can be altered by adding certain ions to the slurry suspension. For the surfaces typically found on $TiO_2$ particles, these ions are protons (for imparting positive surface charge) or hydroxide (for imparting negative surface charges). The pigment surfaces can therefore be charged by adjusting the pH of the slurry. This is shown schematically in Fig. 18.9.

Not all grades of $TiO_2$ have the same surface composition, and so not all grades show the same surface charge response to changes in pH. This response is shown in Fig. 18.10 for four grades of $TiO_2$ pigment slurried in water. At neutral pH materials with acidic surfaces are negatively charged while materials with basic surfaces are positively charged. Surface charge can therefore be controlled by changing the chemical identity of the $TiO_2$ surface coating, or by adjusting the pH of the paper slurry.

There is a third means of modifying the $TiO_2$ surface charge, as well as the surface charge on other particles and fibers, which results from adding cationic wet strength resin (a cationic polymer) to the pulp slurry [1]. This imparts a positive charge to the pigment surface. This is similar in concept to the use of charged dispersants in paint production (see Chap. 11).

The primary effect of $TiO_2$ surface charge in papermaking, and our reason for controlling it, is that it affects $TiO_2$ retention in the paper. Ideally, $TiO_2$ particles will be retained in the paper by separately attaching to pulp fibers before and during the dewatering process. This is accomplished by placing a positive surface charge on the $TiO_2$ particles so that they will be attracted to the negatively charged fiber surfaces.

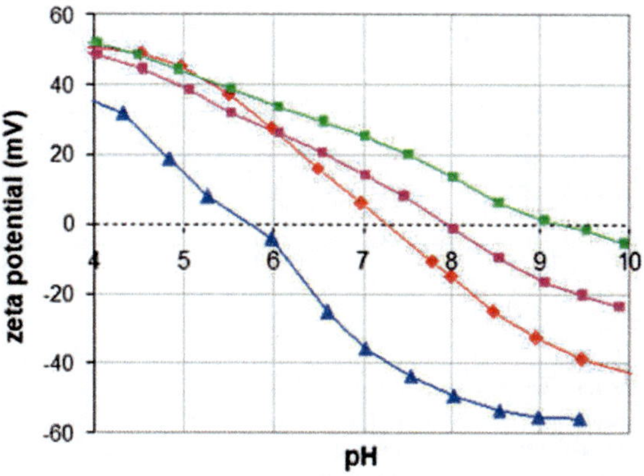


**Fig. 18.10** Surface charge as a function of pH for four $TiO_2$ pigments

By adding a wet strength resin to the paper, which adsorbs onto the $TiO_2$ surfaces, the pigment attains the desired positive charge.

Ideally, all of the $TiO_2$ particles would attach to the fibers in this way. However, in reality, most do not, and a second mechanism for retaining the $TiO_2$ particles in the paper is employed. Here the $TiO_2$ particles that are coated with wet strength resin, and so are positively charged, flocculate with the uncoated $TiO_2$ particles, which, at the papermaking pH, can be negatively charged, depending on the grade of $TiO_2$ (Fig. 18.10). This flocculates the $TiO_2$ particles into clusters that can be caught in the void regions between the fibers and so retained in the paper.

These two mechanisms for $TiO_2$ retention—adhesion of individual particles to the fibers and entrapping particle clusters within the pores—occur simultaneously in the furnish. Slight modifications to pH can be made in the headbox, immediately prior to the screening step, as a final opportunity to optimize retention (Fig. 18.8).

In addition to pigment retention, the surface treatment on the $TiO_2$ particles can also affect other décor paper and laminate properties, such as pigment dispersion, ink retention, resin retention, resin cure and, as mentioned earlier, light stability. This is because the $TiO_2$ particles interact physically and chemically with their surrounding environment through the particle surface. Altering the surface can optimize these interactions without harming the intrinsic scattering ability of the particles, since these are determined by refractive index, which is a bulk property rather than a surface property.

## $TiO_2$ Dispersion

The degree of $TiO_2$ particle dispersion in paper applications can be quantified using an optical density test. In this test, the amount of light that is scattered when traveling through a highly dilute slurry of the $TiO_2$ (20 ppm) is determined using a transmission measurement. If the particles are well dispersed, the light is scattered strongly and

**Table 18.1** Effect of slurry pH on optical density and paper opacity

| $TiO_2$ slurry (50% dry solids) | | | Handsheet | | High Pressure Laminate (LPL) Panel | | |
|---|---|---|---|---|---|---|---|
| pH | Optical density (%)[a] | Slurry viscosity (mPa sec) | Basis weight (dry g/m$^2$) | $TiO_2$ content (dry g/m$^2$) | S($TiO_2$) (m$^2$/g)[b] | Laminate opacity (contrast ratio)[c] | L* (over black) (%) |
| 7.0 | 81 | 1521 | 99.2 | 38.5 | 0.105 | 91.3 | 90.0 |
| 9.0 | 100 | 29 | 100.9 | 40.1 | 0.116 | 93.0 | 91.0 |

[a] Optical density is a measure of $TiO_2$ scattering potential at very low dilutions and is considered the entitlement opacity

[b] S($TiO_2$) is the scattering power of the $TiO_2$ in the end-use application (here the HPL panel). The $TiO_2$ light scattering efficiency is normalized to the $TiO_2$ content. This value is similar to the discussion in Chap. 4 on the $TiO_2$ scattering efficiency as a function of $TiO_2$ PVC

[c] Laminate opacity is a contrast ratio, calculated as follows: Laminate opacity (%) = 100 × Y(over black backing)/Y(over white backing)

so transmission is relatively low. On the other hand, if the particles are flocculated, less light is scattered and so more is transmitted through the sample.

Because the particles in the $TiO_2$ slurry are stabilized by surface charges, we would expect their dispersion quality to vary with pH. This is in fact the case, as shown in Table 18.1. Here we compare slurries with pH values of 7.0 and 9.0. The isoelectric point (point of zero surface charge) for this grade of $TiO_2$ is 7.35, and so we would expect that the pH 7.0 slurry would be flocculated while the pH 9.0 slurry would not be. This is confirmed by both the lower optical density and the higher viscosity of the pH 7.0 slurry (flocculation results in high viscosity). More specifically, the optical density of the slurry at pH 9.0 has a 19% advantage over that of the pH 7.0 slurry.

To improve $TiO_2$ retention in the paper, a flocculating agent is added to the $TiO_2$ slurry. Because the particles are flocculated, we may expect that the degree of dispersion in the initial $TiO_2$ slurry is not relevant to the opacity of the final laminate panel. However, this is not the case. As can be seen in Table 18.1, roughly half of the opacity advantage of the better dispersed $TiO_2$ slurry is retained in the final laminate panel (here measured as the scattering strength S of the $TiO_2$ particles).

This is an important finding as it indicates the necessity of getting good $TiO_2$ dispersion prior to the paper formation step. As is the case in coatings after grinding, the degree of $TiO_2$ dispersion will never improve over its initial level, and all subsequent process steps can only degrade the quality of the $TiO_2$ dispersion.

## *$TiO_2$ Retention*

The paper maker is confronted with a dilemma with regard to the state of $TiO_2$ dispersion. On the one hand, to achieve the highest retention, the $TiO_2$ particles

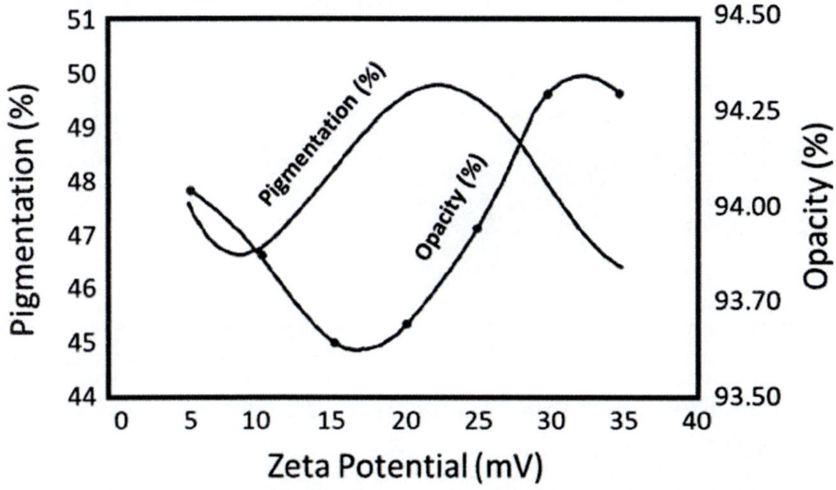


**Fig. 18.11** Effect of $TiO_2$ zeta potential on retention and opacity ( Adapted from PWA-Décor, now Ahlstrom-Munksjö)

must be flocculated into clusters large enough to be trapped by the paper fibers during dewatering. On the other hand, as was discussed in Chap. 11, $TiO_2$ scattering efficiency is decreased significantly by flocculation due to the negative crowding effects on the total scattering volume of the particles. These conflicting aspects of $TiO_2$ usage are shown in Fig. 18.11, where we see the highest $TiO_2$ efficiency at the same zeta potential values as the lowest retention, and vice versa.

In décor paper at high wet strength resin concentration, the $TiO_2$ clusters become positively charged and break apart into smaller groups. This is because of the strong electrostatic repulsion resulting from all of the particles being positively charged within the cluster: we observe de-flocculation at large positive zeta potential (25–30 mV). This is the 'sweet spot' where the paper machine operates at its highest efficiency. This is very much desired, as it improves the $TiO_2$ efficiency, due to smaller $TiO_2$ cluster size—and it results in an opacity increase (without adding extra $TiO_2$). This is well-known in the industry and was first reported in 1995.

## Tools for Analyzing Décor Paper and Laminate Panels

Optimizing laminate appearance properties requires the use of many analytical techniques and tools. These include those that quantify the appearance properties of the décor paper and the laminate into which it is incorporated (i.e., appearance uniformity and overall opacity), as well as those that reveal the distributions of the different components of the laminate (i.e., fiber, $TiO_2$ particles, air voids, and resin pockets),

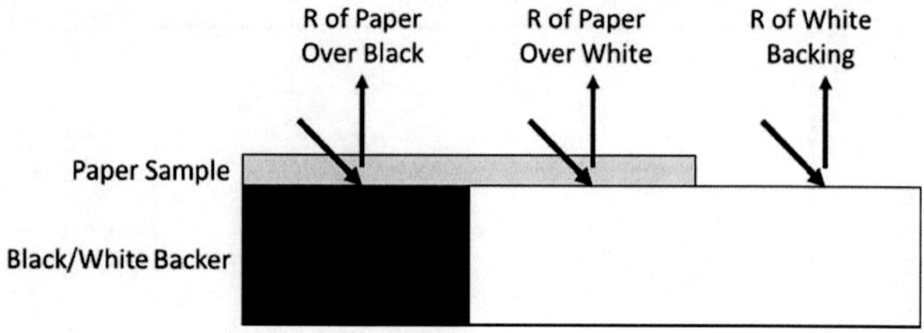


**Fig. 18.12** Kubelka–Munk method for determining the optical properties of a paper

since the uniformity of these distributions directly affect the appearance properties of the laminate.

## *Opacity and $TiO_2$ Efficiency*

In Chap. 13, we discussed the measurement of paint and plastics opacity. There we used the Kubelka–Munk framework to quantify light scattering (S) and absorption (K) per thickness of the sample (X). We can use a similar technique to quantify the optical properties of a paper.

In this test, a sheet of paper (or stack of sheets) is laid over a black and white chart and the reflectances over the black area, over the white area, and of the uncovered white portion of the chart are measured (Fig. 18.12). The paper must be oiled for this measurement. This removes all air, replacing it with an oil that has a similar refractive index as the laminate resin. In this way, the scattering environment of the paper is matched to its end-use application.

$TiO_2$ efficiency can be described as the scattering strength of the paper as a function of $TiO_2$ coverage on an area basis. $TiO_2$ coverage is measured by incinerating a sample of the paper with known area and weighing the remaining ash, which is entirely $TiO_2$ in décor paper.[3]

An important aspect of the effect of $TiO_2$ dispersion on opacity can be seen by comparing the opacity of a décor sheet that has been oiled to the opacity of a sheet that has been formed into the laminate. Oiling will replace the air voids within the décor sheet with a material that has a very similar refractive index to the resin (this is similar to oiled hide for a paint—see Chap. 4) and so removes the contribution of air void scattering to the opacity of the décor sheet (there is no such opacity contribution in the laminated paper). Subsequently, we would expect the opacities to be similar. In fact, the opacity of the laminate is significantly lower than that of the oiled sheet.

[3] This can be contrasted with graphics and writing paper, which have high levels of extender particles and little or no $TiO_2$.

This demonstrates that the pressing process not only removes air from the sample, but it also increases the crowding of the $TiO_2$ particles.

## *Formation*

The term "formation" refers to the effect that the distribution of components within the paper has on appearance uniformity across the paper sheet. In graphics and writing paper, this is determined primarily by the distribution of the fibers. If the fibers are flocculated, then there will be regions within the paper of high fiber concentrations (and low air content) and other regions of low fiber concentrations (and high air content). These will scatter light differently and so create a non-uniformity of appearance.

Formation of impregnated and laminated décor paper is more strongly determined by the distributions of fibers and $TiO_2$ particles. There is a counteraction between these distributions because a bad distribution of fibers prevents a good distribution of $TiO_2$ particles. This is due to the fact that in décor paper any voids between the fibers that escape compression during pressing are filled with resin, rather than air, and so do not scatter light (fiber and resin refractive indices are nearly the same whereas fiber and air refractive indices are not). That is not to say that the distribution of fibers is unimportant in décor paper appearance uniformity—the $TiO_2$ particles are mostly found as clusters trapped within the voids between the fibers, and so, in a sense, their distribution is like the photographic negative of the fiber distribution.

Formation is most often measured by one of two procedures. In the first, hard radiation is directed at the paper and the proportion of transmitted radiation is measured by a radiation detector on the opposite side of the paper. The amount of absorbed radiation depends on the identity of the absorbing species, with greater absorption for heavier elements such as titanium, and on its concentration. The radiation is scanned over an area of the paper and the variability of the distribution of $TiO_2$ particle concentrations is determined.

### Appearance Based Formation Analyses

While the above procedure can quantify formation, it is not ideal because it does not directly measure the property of importance to the paper manufacturer—the appearance of the paper. Instead some mathematical transformations must be done to link the breadth of $TiO_2$ particle distributions to appearance. The second procedure for measuring formation does not suffer from this issue because it is based on appearance directly. In this procedure, an image of the paper is analyzed for uniformity at different length scales. These length scales correspond to observation distances—the scale of uniformity for a laminate that is intended to be viewed from, for example, a distance of three meters is larger than for a laminate that is intended to be viewed from a distance of a few tens of centimeters.

**Fig. 18.13** Appearance analyzer employing a CCD camera to image 7 cm × 8 cm laminate sample

An approach for measuring image uniformity has been developed using a CCD camera to image a 7 cm × 8 cm sample of laminate (Fig. 18.13). Two histograms of grayscale values are calculated from this image: one for 1 mm × 1 mm areas and the other for 8 mm × 8 mm areas. These spatial distances correspond to what the eye can see at a distance of 30 cm—typical reading distance, which is representative of the distance at which graphics and writing paper is usually viewed. The width, or standard deviation, of each histogram is then taken as a measure of uniformity.

## *Nano Computer Tomographic Analysis*

Computer tomography is a long-established method of generating a three-dimensional image of a sample. Perhaps the most familiar forms of three-dimensional imaging are medical applications, where CT scanning shows many advantages over traditional X-Ray imaging. Not only do CT scans give three-dimensional images, but they also allow for a differentiation in the chemical composition of the material being scanned.

Through various technical advances, the resolution of CT imaging has become fine enough to resolve regions of roughly one cubic micron (one micron in each of the three dimensions). We will refer to imaging at this resolution as nano CT. This resolution matches the micro-structure dimensions of décor papers and paper laminates and so is ideal for visualizing the internal structure of these materials.

In this process, an X-Ray beam interacts with the sample and then is subsequently detected. The measured intensities are processed using advanced imaging software, and the data is transformed into a three-dimensional representation of the sample. The equipment used is shown in Fig. 18.14 and an example two-dimensional section

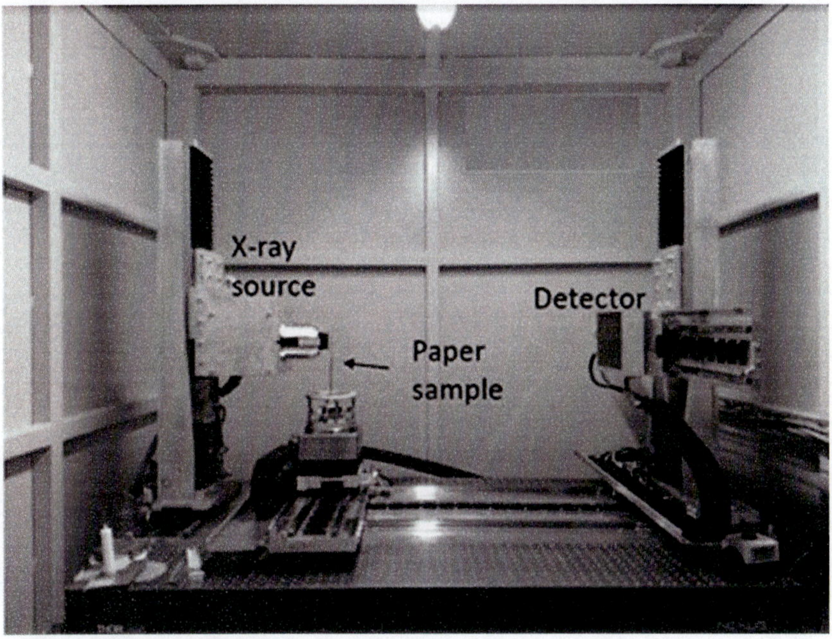


**Fig. 18.14** Nano CT instrument

of an unimpregnated décor paper is shown in Fig. 18.15.

The image shown in Fig. 18.15 is very similar in appearance to the electron microscopic images shown earlier in this chapter, and we may question the benefit of

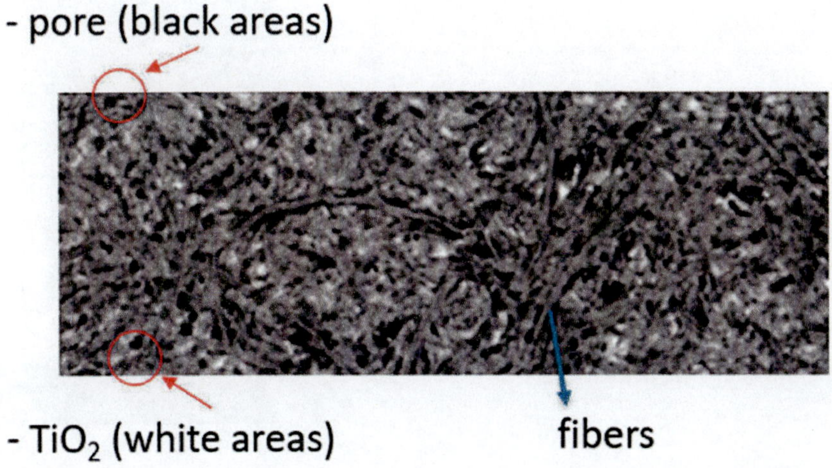


**Fig. 18.15** Image from a nano CT scan showing resolution and appearance of various components

using nano CT to analyze décor paper and laminate samples. The advantage of nano CT is that the entire sample is imaged, rather than just the surface presented in the electron microscope. This allows for three-dimensional representations of the sample to be generated and manipulated, with the ability to rotate the image freely on the computer monitor. In addition, this technique is ideal for analyzing non-homogenous materials, such as paper laminates, while electron microscopy is adequate for more homogenous samples such as plastics.

Since the X-ray absorption strength of a material is determined by the atomic weight of its constituents, the nano CT analysis can readily differentiate fibers, melamine resin, air, and $TiO_2$. This is represented in the images as an increasing brightness from air (black in the images), through resin and fibers (gray), and to $TiO_2$ (white). Individual $TiO_2$ particles cannot be resolved, but, as described above, the $TiO_2$ is mostly present as large clusters situated between the fibers, and these clusters are clearly distinguished in the images. An example of a three-dimensional representation of the $TiO_2$ distribution in a décor paper, color coded by cluster size, is shown in Fig. 18.16.

**Fig. 18.16** Distribution of $TiO_2$ clusters in a décor paper. Color indicates cluster size (size increases from blue to green to orange to red)

Images such as these are made by setting a threshold for voxel[4] brightness and only displaying those voxels that exceed this threshold. Images of the air pore distribution can be similarly made, but in this case the threshold is set so that only those voxels with brightness values beneath the threshold are displayed.

In addition to imaging the sample, nano CT probes a large enough volume to statistically determine the size distribution of the various décor paper and laminate components. This is particularly useful for quantifying the distribution of $TiO_2$ cluster sizes, typically expressed as a histogram. The variation in cluster sizes in different parts of the décor paper (i.e., surfaces or internal) can be studied in this way.

## Application of Nano CT

Imaging is invaluable for the analysis of complex systems such as papers. Traditionally, paper samples have been imaged extensively by optical and scanning electron microscopies (SEM). While the images from these techniques are useful, they only show a two-dimensional view of the paper. For other applications of particles embedded in polymers—paints and plastics—this is adequate because the particles in these materials are generally distributed evenly throughout. As such, any two-dimensional image is representative of the entire material.

This situation is quite different in paper applications, particularly décor paper. As discussed earlier, the particles found in décor paper are intentionally flocculated to ensure retention in the porous fiber structure. Because of this, each three-dimensional section of the paper is unique, and to properly understand the distribution of particles within the paper—and how this distribution affects properties such as printability—a three-dimensional rendering is required.

Nano CT could potentially satisfy this need. However, this is not a certainty. Since three-dimensional renderings of paper have not been made before, there is no road map as to how they can be useful to the paper maker, if at all.

A number of step-wise studies have been made to investigate the utility of nano CT technology to décor paper. These studies, which will be reviewed below, confirm that this technology is useful if applied appropriately. The strength of this technology is "thresholding"—that is, the ability to isolate the location of large particle groups using the intensity of the image. This is done by creating a simple black and white representation of the grayscale images given by the CT software. By adjusting the threshold intensity for defining which voxels are black and which are white, a clear understanding of the $TiO_2$ groups found within the paper, and the relationship between groups, can be achieved.

[4] A voxel is a three-dimensional analog of the two-dimensional pixel.

### *Study 1—Effect of Microstructure on Formation*

In the first step of determining the utility of the Nano CT technology, two HPL (High-Pressure Laminate) samples were imaged by both traditional SEM microscopy and the nano CT method [2]. The samples differ in appearance due to differences in formation. As will be shown, the nano CT method detects important differences in the $TiO_2$ clusters that are not evident in SEM images. This demonstrates that differences in the micro-level ($TiO_2$ distribution) can be linked directly back to differences at the macro-level (optical properties).

This study demonstrated that, through proper software analysis, certain visual differences in HPL quality can be related to the microstructure of the décor paper. Such a relationship is not evident in SEM image analysis.

The ability of nano CT to distinguish differences in handsheets[5] and the high-pressure laminates made from them was investigated for two décor papers. These papers had similar $TiO_2$ levels but differed in the amount of wet strength resin used in their production. The difference in wet strength resin resulted in the sample having less wet strength resin showing clearly better formation than the other (both as observed visually and as measured by an appearance analyzer). The properties of these two sheets and the laminates made from them are shown in Table 18.2. Images of the laminate panels are shown in Fig. 18.17.

SEM analysis of the décor sheets did not reveal any differences, even when using advanced image analysis statistics. Small samples (3 mm × 6 mm) were cut from the laminates and analyzed by nano CT. Three-dimensional renderings of the samples, highlighting the $TiO_2$ clusters, are shown in Fig. 18.18. Although these renderings appear similar to the eye, a statistical analysis of them reveals two important differences.

Histograms showing the size distribution of the $TiO_2$ clusters, quantified as the size of a sphere of equivalent volume to the cluster, are shown for each laminate in Fig. 18.19. The X-axes show the spherical equivalent size of the clusters both in terms of sphere diameter and sphere volume. Although there is much similarity between the histograms, there is an important difference in number of large clusters (equivalent spherical diameters larger than 50 microns).

In addition, the volumes occupied by the clusters that were detectable in the nano CT analysis were found to be significantly different; the $TiO_2$ clusters occupied 3.48 volume % of the laminate with bad formation but only 3.01 volume % of the laminate with good formation. This is a relative difference of 13.5%. Referring back to Table 18.2, we see that the actual volume relative difference was only 2.4% (4.50% and 4.39%, respectively). This is significant because the nano CT method cannot resolve individual $TiO_2$ particles or particle clusters smaller than one micron. When the volume contribution from individual particles or small clusters is accounted for, then the difference between the measured volumes (Table 18.2) and the volumes detected in the nano CT analysis can be reconciled.

[5] A handsheet is a single sheet of paper made in the laboratory.

**Table 18.2** Properties of handsheets and laminates for décor papers with different formations

| | Handsheets | | | | | HPL | | | |
|---|---|---|---|---|---|---|---|---|---|
| | $TiO_2$ in sheet | gsm $TiO_2$ | $TiO_2$ in sheet | Impregnated sheet weight | $TiO_2$ in impregnated sheet | $S(TiO_2)$ | Laminate Opacity | L* over black | Appearance over black |
| Formation | weight (%) | gsm | volume (%) | gsm | volume (%) | ($m^2/g$) | (%) | | DAV‡ |
| Bad | 40.1 | 40.6 | 10.15 | 223 | 4.50 | 0.113 | 93.2 | 90.8 | 60 |
| Good | 39.8 | 41.0 | 10.25 | 227 | 4.39 | 0.129 | 95.5 | 91.0 | 45 |
| Difference | | | 1.0% | | 2.4% | 12.4% | 2.4% | | |

‡ Lower values indicate better formation (lower appearance variability across the sheet)

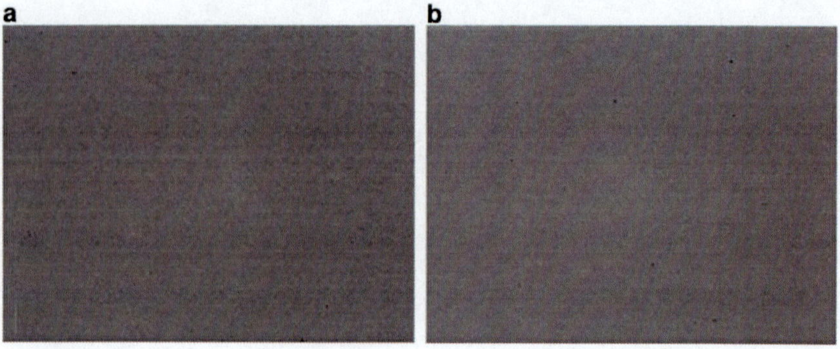


**Fig. 18.17** Appearance of laminate panels. **a** Bad formation décor handsheet. **b** Good formation décor handsheet

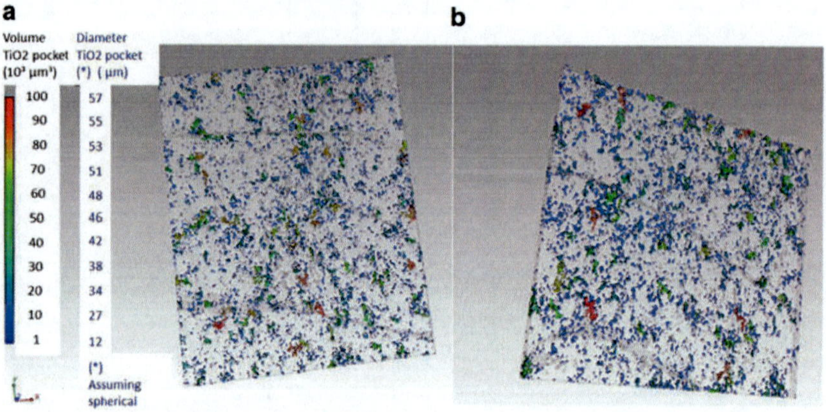


**Fig. 18.18** Three-dimensional rendering of the $TiO_2$ clusters (pockets) in the two laminates. **a** Bad formation décor handsheet. **b** Good formation décor handsheet

The differences seen in the nano CT analysis are consistent: there are more very large $TiO_2$ clusters and fewer very small $TiO_2$ clusters in the sample with bad formation. One consequence of this is seen in poorer sheet appearance (formation quality). Another is seen in the improved opacity of the laminate made from the handsheet with good formation—the $S(TiO_2)$ value is 12.4% higher and the total opacity 2.4% better in this sheet, despite the slightly lower $TiO_2$ content (Table 18.2).

We can conclude that there is a link between the microscopic properties of a handsheet (in particular, $TiO_2$ cluster size distribution) and the macroscopic appearance of the laminate (formation). This link is not apparent in electron micrographs of the sheets because it is dependent on a three-dimensional property of the sheet—the volume of the $TiO_2$ clusters—and electron micrographs are two-dimensional in nature.

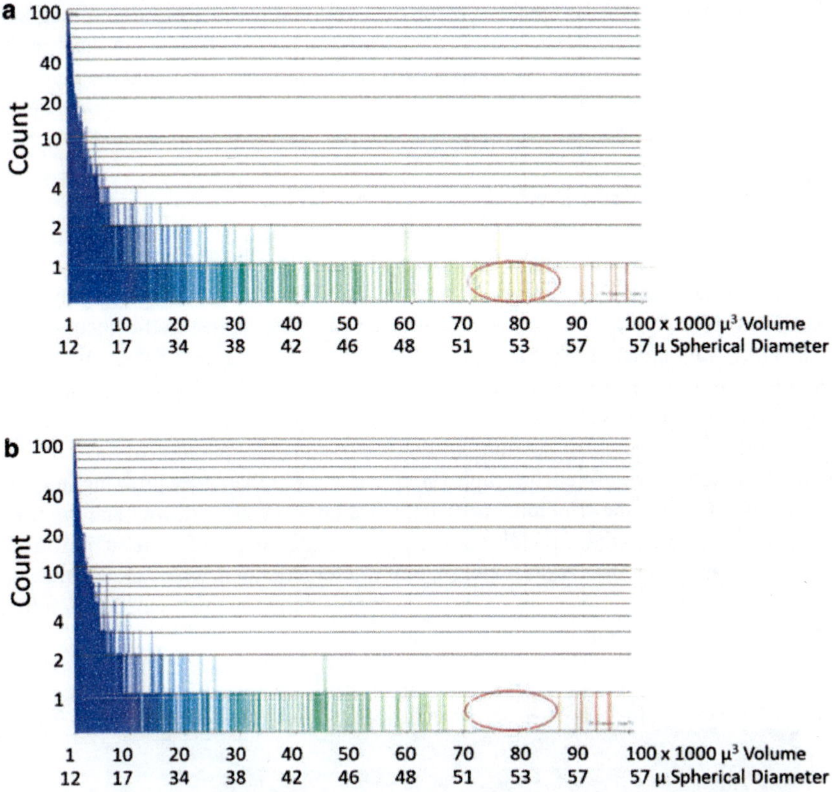


**Fig. 18.19** Histogram of $TiO_2$ cluster size in the two laminates. **a** Bad formation décor handsheet. **b** Good formation décor handsheet

## *Study 2—Development of Structure Through the Production Process*

In a follow-up of the work described above, this analysis was extended to production-made décor paper [3]. An important finding was that meaningful differences in the $TiO_2$ cluster structures were evident by the proper use of image thresholding—that is, by defining which voxels are occupied by $TiO_2$ and which are fiber or resin—by properly adjusting the minimum intensity of those voxels defined to be $TiO_2$. Importantly, it was found that the threshold value can vary for different stages in laminate construction (décor paper formation, impregnation, and pressing). A comparison of the histograms of cluster sizes showed consistent differences between the samples, and a clear correspondence was seen between the distribution within the décor paper itself and in the final laminate made from it.

In this analysis, small samples (1 mm × 1 mm) were cut from a graphics and writing décor paper, both before and after impregnation, and from a laminate made from the impregnated paper. Each sample was then characterized using the nano CT technique.

### Two-Dimensional Cross Section and Surface Images

Z-direction cross-section images for the three samples are shown in Fig. 18.20. Here the voids (dark), the $TiO_2$ (white spots), the fibers (gray), and the resin (gray) can be clearly differentiated. As anticipated, the air voids are largest in the décor paper, next in the impregnated dry film, and substantially reduced in the laminate panel. We also observe that the resin in the impregnated paper is dominant at the surfaces (both top and bottom/web-side).

In Fig. 18.21, we show sections in the XY plane (i.e., parallel to the surface) taken from the interior of each sample. Here we observe that, after pressing into the final laminate state, the structure is denser, since air has been removed and the fiber network has compressed. In addition, the air voids are large and more numerous in the décor paper, but after resin impregnation the remaining air voids are smaller. In the final laminate panel they are further reduced, as expected due to the temperature and pressure applied during the laminate formation step.

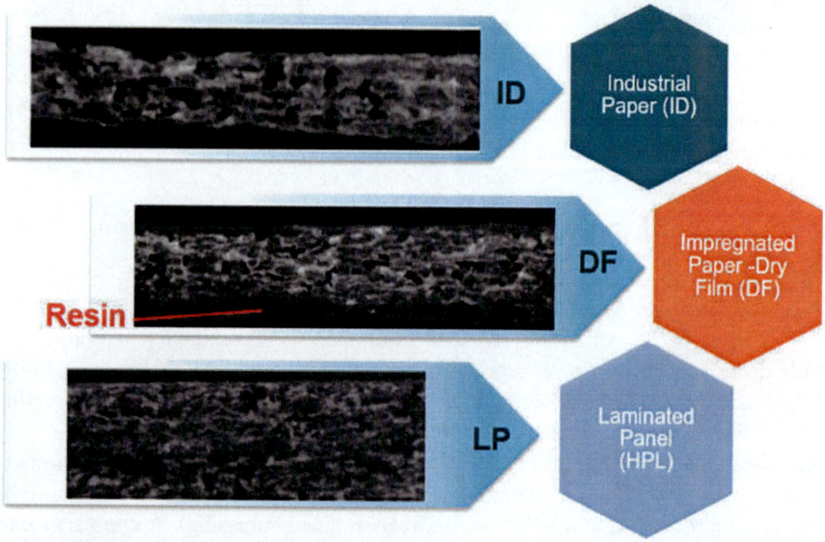


**Fig. 18.20** Z-direction cross sections of the décor paper (top), the impregnated paper (middle), and laminate panel (bottom)

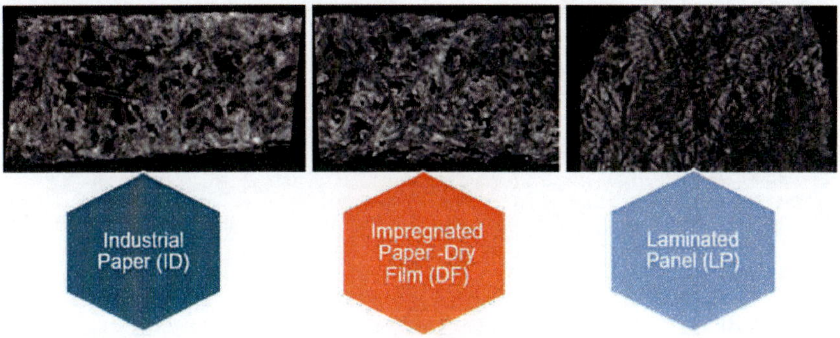


**Fig. 18.21** Images in the XY plane of internal regions of the samples

## Quantitative Analysis of Structural Features

The raw data used to construct the images shown Figs. 18.20 and 18.21 was subjected to statistical analysis to better characterize the nature of these materials. This is done by setting a threshold for selecting the voxel intensities to be analyzed. This threshold is not necessarily the same for each sample but instead is determined such that the total volume of voxels exceeding the threshold matches the known fill factor for that sample, defined as the volume concentration of the $TiO_2$. In our case, the fill factor was constant across the samples since they are of the same décor paper at different stages of processing.

When comparing similar types of samples (e.g., a series of laminates), the thresholds are typically similar to one another; however, while comparing different sample types, as we are doing here for the unimpregnated and impregnated décor paper and the laminate, the threshold values vary over a wider range. We can differentiate not only the $TiO_2$ from the other components, but also the other components from one another. Due to the similarity in chemical composition of these components, they are more difficult to differentiate. Figure 18.22 shows the grayscale histograms for the three samples, along with the thresholds used in the image analysis.

## $TiO_2$ Distribution

In Fig. 18.23, we compare the visual appearance and numerical data for the three samples. The images on the left are taken from a plane within the sample and parallel to the sample surface (these are taken from Fig. 18.21). On the right hand are curves showing $TiO_2$ content as a function of location in the sample. The red curves show $TiO_2$ concentrations in the X-direction (i.e., over the length of the sample) while the blue curves show the concentrations in the Z-direction (i.e., perpendicular to the image, and from the top to the bottom surface).

Overall, there is a fair but varying degree of non-homogeneity in the $TiO_2$ distributions through the samples. The blue curve shows that the two sides of the paper are

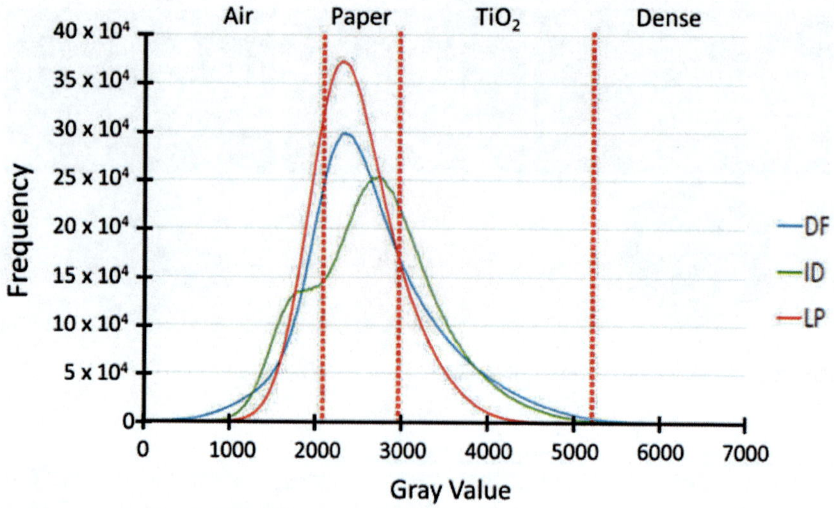


**Fig. 18.22** Grayscale distributions for the three samples, with thresholds for the different components. "Paper" refers to resin and fiber

different—the top surface contains substantially more $TiO_2$ than the bottom surface (against the Fourdrinier machine web). This is due to the $TiO_2$ particles being leached out as white water from the regions of the paper nearest the web. We do not anticipate that this will affect end-use opacity since light travels in this direction, essentially superimposing the particles from a top-down perspective. The distribution of particles in the Z-direction is therefore not as important as the uniformity in particle coverage per unit area in the XY plane.

In addition to this overall gradient in $TiO_2$ in the Z-direction, we see there is relatively little $TiO_2$ in the thin, outermost layers of the impregnated paper. Instead, resin dominates these regions. This reflects the manner in which resin is impregnated into the décor paper. During this process, the paper is submersed into a liquid resin bath. Resin penetrates the paper, starting from the exterior surfaces (top and bottom) towards the inside of the paper. This process does not go to completion during the limited time spent in the resin bath, and hence air voids remain in the dry film and laminate panel.

The red curves in Fig. 18.23 indicate an uneven distribution of the $TiO_2$ across the paper (i.e., in the X-direction). This is due to these particles being present in clusters rather than being discretely distributed throughout the paper. The $TiO_2$ variability in this dimension is the least in the laminate panel, suggesting some redistribution of these clusters, or of the $TiO_2$ particles within them, when they are squeezed by the lamination press.

The distribution of $TiO_2$ particles and clusters is even more evident in three-dimensional renderings of the samples (Fig. 18.24). Here, the location of the clusters is determined by the voxel intensity. The size of the clusters is determined using

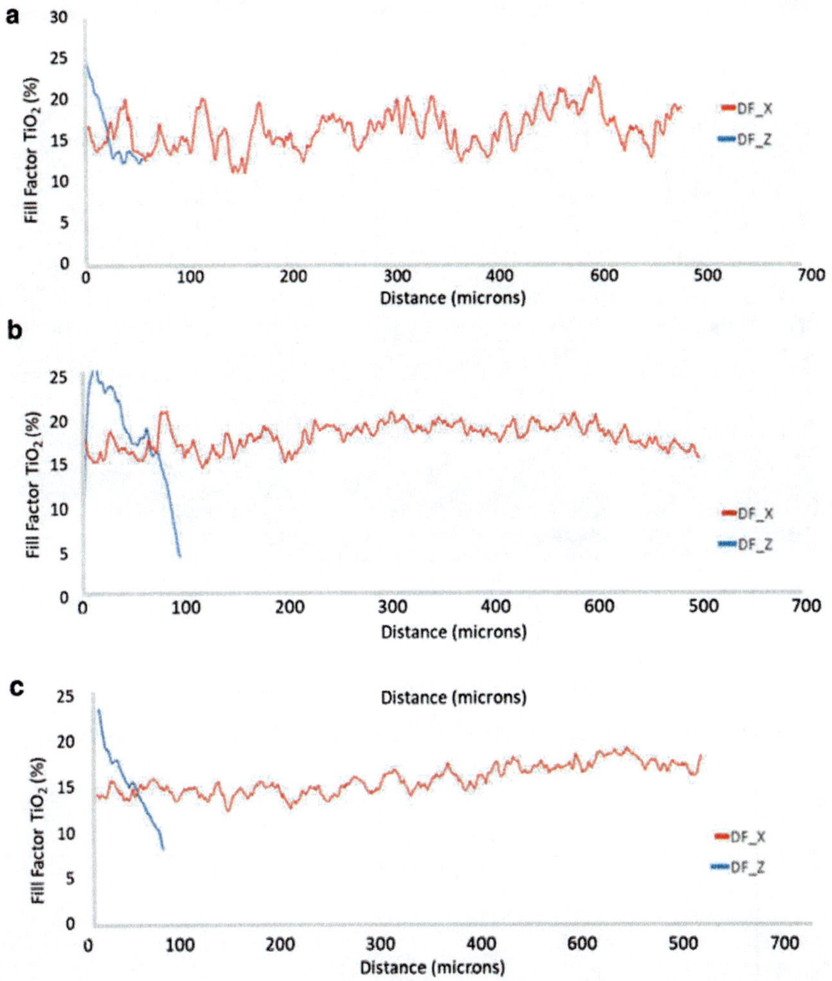


**Fig. 18.23** Comparison of visual and numeric information from the nano CT scans. The red curves show the concentration distribution parallel to the surface (X-direction) while the blue curves show the distribution from top to bottom (Z-direction). **a** Industrial décor paper. **b** Dry film. **c** Laminate panel

a computer algorithm, and in this figure is indicated by color, with blue being the smallest clusters and red the largest.

Once the clusters are identified, their size distributions can be analyzed and compared between the three samples (Fig. 18.25). The mean sizes for the three samples are quite similar, varying from 20 to 24 microns. However, there is a significant level of variability for the curve maxima. The maximum for the décor paper is at 37 microns, while that of the impregnated paper is 32 microns and that of the

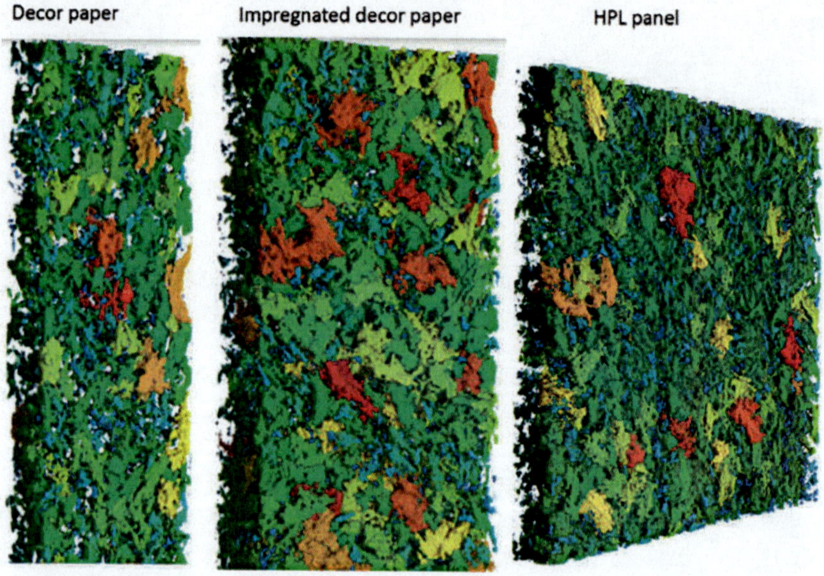


Fig. 18.24 Three-dimensional distribution of $TiO_2$ clusters in the three samples. Cluster size is denoted by color; the smallest clusters are blue, followed by the green, yellow and red clusters

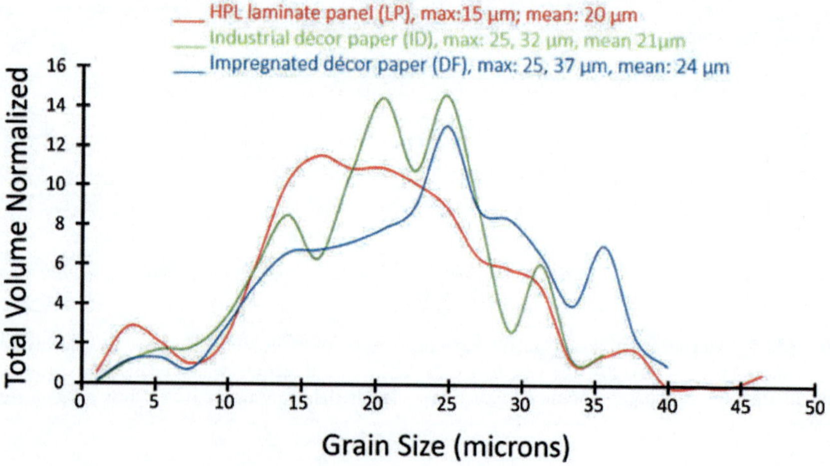


Fig. 18.25 Size distribution of the $TiO_2$ clusters in the three samples

laminate itself is only 15 microns. The modest drop in maximum seen when the paper is impregnated is most likely due to mobility of the $TiO_2$ clusters under the forces that develop in the liquid absorption step, during which fibers expand, and drying step, during which the structure compresses. The large drop seen after the lamination process is due to flow under pressure of the hot resin. This distributes the $TiO_2$ more evenly, resulting in improved formation, which results in more efficient $TiO_2$ light scattering and higher sample opacity. This demonstrates that formation is scale invariant—that is, there is the same level of variance at the micro and macro-scales, and so poor distribution of particles at the micrometer scale results in poor distribution at the scale at which formation is judged.

## *Study 3—Effect of $TiO_2$ Level on Formation and Pore Content*

In the first two studies, the distribution of $TiO_2$ clusters throughout the décor paper was determined by judicious use of thresholding. These studies were further extended by quantifying the distribution of the $TiO_2$ clusters in one direction—from the bottom of the sheet to the top of the sheet (i.e., in the Z-direction). This type of analysis revealed different information than the three-dimensional analyses described in the first two studies. In particular, it clearly showed the non-uniformity of $TiO_2$ content and pores through the sheet, with more $TiO_2$ and fewer pores at the top 20 micron surface of the paper than at the bottom. This is entirely consistent with differences in printability seen for the different sides of the paper (measured separately). In addition, by quantifying the amount of $TiO_2$ observable in the nano CT, we can determine the proportion of $TiO_2$ particles that present clusters too small to resolve (<1 micron) and those that are larger.

In Study 3, the effect of $TiO_2$ loading on pore structure and volume was investigated. Décor paper sheets were made at three different $TiO_2$ contents: 10%, 25%, and 40%, all at 100 gsm. Two samples were made at the 40% $TiO_2$ level that differed in the $TiO_2$ slurry pH. In one sample the slurry pH (7.0) was near the iso-electric point of the $TiO_2$, resulting in particle flocculation, while in the other sample the slurry pH was higher (9.0), which gives better $TiO_2$ dispersion in the slurry. The slurry pH values for the 10% and 20% samples were also 9.0, to ensure optimal formation and higher opacity. A summary of the physical and optical properties of the sheets and the laminates made from them is given in Table 18.3 and the $TiO_2$ scattering efficiencies of the sheets are given in Fig. 18.26.

From this table and figure, we see that $TiO_2$ scattering efficiency decreases significantly as the $TiO_2$ content increases (e.g., by 28% for the sheets made with pH 9.0 $TiO_2$ slurry). This shows that the crowding of the $TiO_2$ particles increases significantly with $TiO_2$ content. The $TiO_2$ particles are assumed to be in clusters in all samples, in which case the number of clusters would increase with increasing $TiO_2$ content, but not the crowding within the cluster (it is this crowding that determines the $TiO_2$ scattering efficiency). We will return to this question below. Note the loss

**Table 18.3** Physical and optical properties of the décor paper sheets

| | | | Dry basis weight | Ash | gsm $TiO_2$ | $TiO_2$ retention | S $TiO_2$ 550 nm | Relative $TiO_2$ light scattering efficiency | Laminate opacity | L* over black | Appearance variation | |
|---|---|---|---|---|---|---|---|---|---|---|---|---|
| | Target % $TiO_2$ | ID | ($g/m^2$) | (%) | ($g/m^2$) | (%) | ($m^2/g$) | (%) | (%) | | 2 mm | 8 mm |
| $TiO_2$ slurry pH 9.0 | 10 | Sample A | 100 | 10.2 | 10.3 | 74 | 0.142 | 128 | 72.2 | 79.6 | 108 | 53 |
| $TiO_2$ Slurry pH 9.0 | 25 | Sample B | 101 | 25.4 | 25.9 | 73 | 0.118 | 107 | 88.3 | 87.3 | 78 | 37 |
| $TiO_2$ Slurry pH 9.0 | 40 | Sample C | 98 | 39.0 | 38.9 | 66 | 0.111 | 100 | 92.8 | 90.3 | 57 | 32 |
| $TiO_2$ Slurry pH 7.0 | 40 | Sample D | 99 | 39.0 | 39.1 | 66 | 0.108 | 98 | 91.9 | 90.1 | 59 | 35 |

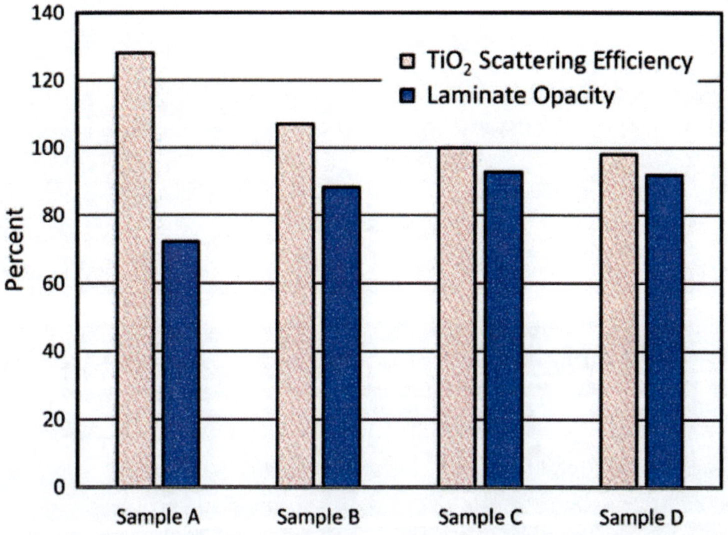


**Fig. 18.26** Relative $TiO_2$ scattering efficiencies and laminate opacities for the four samples

in $TiO_2$ scattering efficiency is more than offset by the increased number of $TiO_2$ particles, and the opacity of the laminates themselves increases with $TiO_2$ content.

We also see the effect of incomplete $TiO_2$ dispersion in the $TiO_2$ slurry on formation. Comparing the final two samples we see that the flocculated slurry (pH = 7.0) has slightly lower opacity and higher appearance non-uniformity than the panel made from the well-dispersed $TiO_2$ slurry (pH = 9.0). Based on the lower opacity of the panel made from the flocculated slurry, we would expect to see more agglomerates in Sample D than in Sample C.

### Visual Analysis

Figure 18.27 shows nano CT cross sections of the XY plane on the top sides of the four décor paper sheets. These images are from an interior section of the paper, which is inaccessible by other imaging techniques. The increase in $TiO_2$ content is clearly evident in Samples A through C. In addition to the increase in $TiO_2$ clustering with increasing $TiO_2$ content, the clustering is more pronounced in the paper made from the flocculated $TiO_2$ slurry than its well-dispersed counterpart. This is consistent with our earlier observation that poor initial $TiO_2$ dispersion leads to lower opacity, even though the $TiO_2$ is flocculated before the décor sheet is made (Table 18.1).

Figure 18.28A shows three-dimensional renderings of the $TiO_2$ distributions in the décor sheets, showing cluster location only. The same images are shown in Fig. 18.28B, but color coded to indicate cluster size (the smallest clusters are blue; the largest red). The only large (red coded) clusters were seen in Sample D, indicating

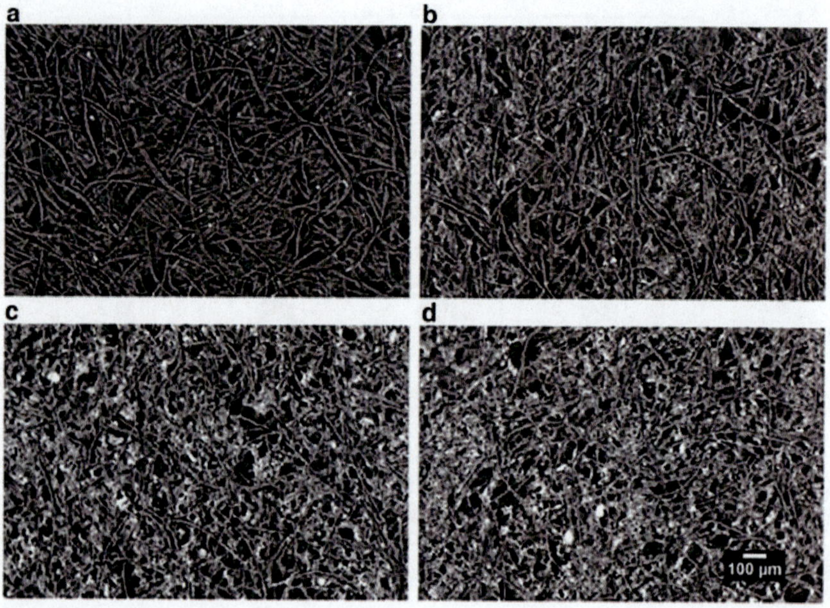


**Fig. 18.27** Nano CT images of the four décor papers

the carry-over of the degree of $TiO_2$ dispersion in the $TiO_2$ slurry to the distribution of the $TiO_2$ particles in the paper sheet.

### Quantitative Analysis

The equivalent spherical diameters[6] of the $TiO_2$ clusters in the four décor sheets are shown in histogram form in Fig. 18.29. The size of the clusters at the peaks of the respective sample curves increases substantially with increasing $TiO_2$ content (from 7 microns for Sample A to 21 microns for Sample D). This increase in average size is quite significant and shows that adding more $TiO_2$ does not simply increase the number of $TiO_2$ clusters; it also increases the size of the individual clusters. The mechanism by which this occurs will be proposed below.

The distribution of pores and $TiO_2$ clusters from the top of the sheets to their backs is shown in Fig. 18.30. We again see that the $TiO_2$ is not distributed uniformly in the Z-direction but is instead more concentrated at the top of the sheets than at the bottom. In addition, we see that regions near both sheet surfaces have high pore content, and that the pore content within the sheets decreases with increasing $TiO_2$ loading. This is understandable since the $TiO_2$ clusters occupy these pore spaces, and as the volume of clusters increases, the volume of unoccupied pores decreases.

---

[6] These are the diameters of spheres that would have the same volume as the cluster.

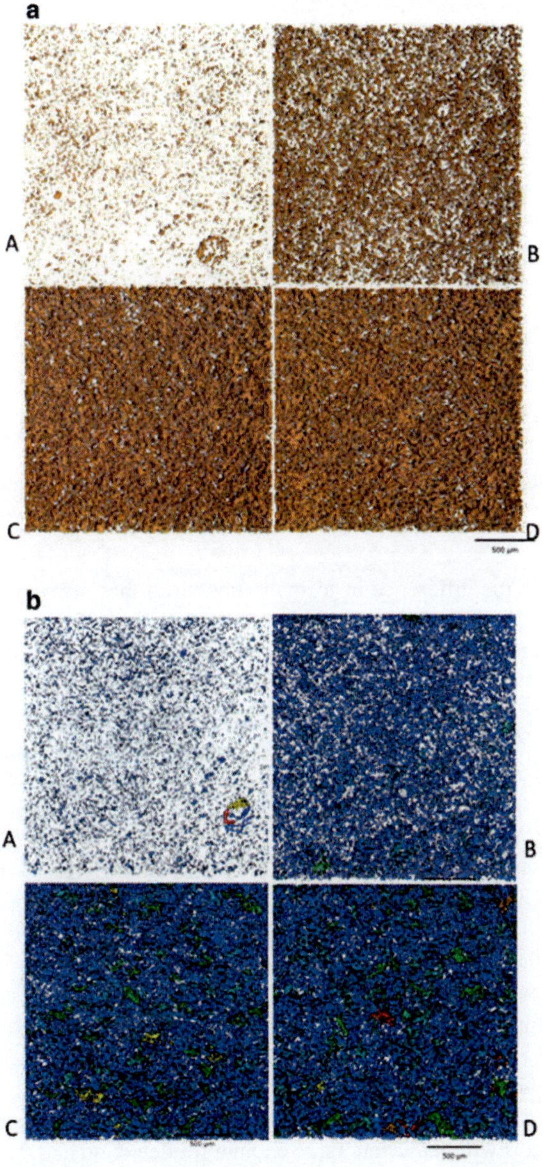


**Fig. 18.28** Three-dimensional renderings of the $TiO_2$ cluster distribution in the four décor paper samples. **a** Cluster structure. **b** Color coded according to cluster size

The differences seen in the pore distribution in the Z-direction are quite important. There is a correlation between high $TiO_2$ volume percent and lower pore volume percent.

We also see in Fig. 18.30 a significant increase in $TiO_2$ agglomeration between the two samples made at the highest $TiO_2$ loadings. This difference corresponds

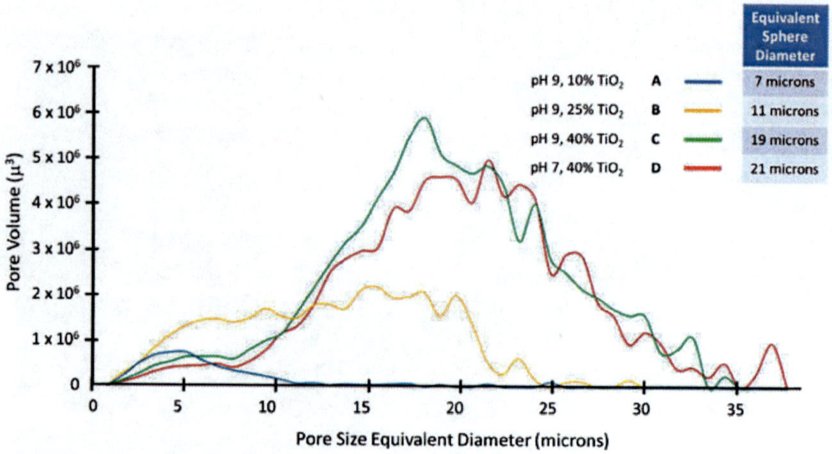


**Fig. 18.29** Histogram of cluster content as a function of cluster size, as measured by the equivalent spherical diameter. Inset shows the total volume and average cluster sizes for the four samples

to the difference in formation between these two samples (refer to the appearance variations shown in Table 18.2). The results from an analysis of the nano CT data are therefore quite sensitive to differences in formation.

## *An Alternative Method to Quantify $TiO_2$ Clustering*

We noted two important findings in our analysis. The first was that the average scattering strength of the $TiO_2$ particles decreases as the $TiO_2$ content increases (Fig. 18.26). This is important because we might expect the number of $TiO_2$ clusters to increase with increasing $TiO_2$ content, but not the degree of crowding within the clusters, since the $TiO_2$ particles are presumed to be at or near their maximum packing density within the clusters. In addition, we saw that the average size of the clusters increased with increasing $TiO_2$ content.

These findings can be understood if we take into consideration the inability of the nano CT technique to image individual particles or clusters that are 1 micron in dimension or less. This gives us a means to determine the proportion of particles within clusters large enough for detection and the proportion of particles either individually dispersed onto the fibers or present in clusters too small for the nano CT technique to detect. To do this we determine the total volume of the $TiO_2$ by ashing and compare this value to the volume of $TiO_2$ detected in the nano CT scan. The difference between these will be the volume of individual particles or small clusters. The proportion of $TiO_2$ found in large clusters, rather than in small clusters or as individual particles, provides a powerful way of characterizing the severity of flocculation in the décor paper.

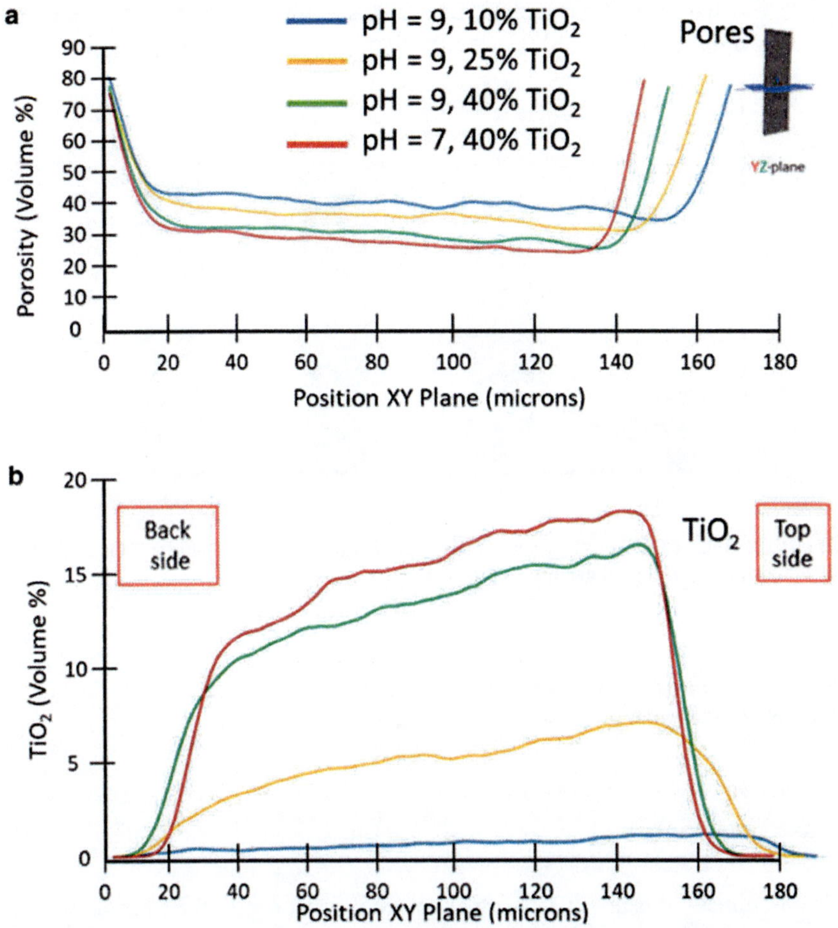


**Fig. 18.30** Pore volumes **a** and $TiO_2$ content **b** of the four décor sheets in YZ cross sections

This comparison is made for the four décor sheets in Fig. 18.31. As can be seen by comparing the green and red values in this Figure, the fraction of dispersed $TiO_2$ is greatest in the sample with the lowest $TiO_2$ content (Sample A). The majority of $TiO_2$ particles in this sample are found, in fact, in the dispersed, rather than clustered, state. However, as the $TiO_2$ content of the paper increases (i.e., going from Samples A to C), the proportion of dispersed particles decreases significantly (i.e., the proportional difference between gsm $TiO_2$ and the amount of $TiO_2$ seen in the nano CT decreases as the $TiO_2$ content increases). The proportion of particles found in clusters is the highest in the final sample, made from the flocculated $TiO_2$ slurry and at highest $TiO_2$ content.

This at least partially explains the finding that the cluster particle sizes increase with increasing $TiO_2$ content, as the dispersed particles would be expected to add

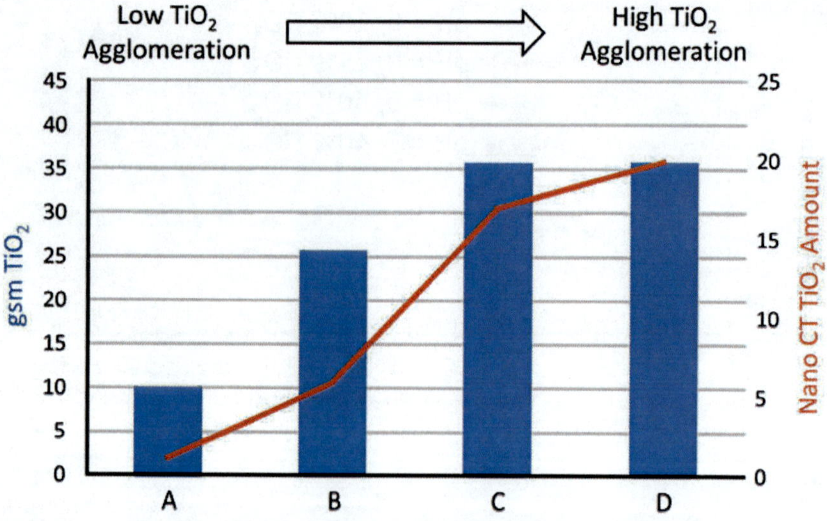


**Fig. 18.31** Total $TiO_2$ content, as determined by ashing (blue bars), and cluster $TiO_2$ content, as determined by nano CT (orange line) for the four décor sheets

to existing clusters at the higher $TiO_2$ levels. Finally, in comparing the results for Samples C and D, we again see the high sensitivity of the nano CT technique to differences in formation.

## Case Histories

The usefulness of nano CT technology to décor paper applications can be seen in two case histories in which this technology brought unique insights to commercial production issues. In particular, the findings that the pore structure in the top 20 microns of the paper, and the overall distribution of pores and $TiO_2$ clusters in the Z-direction, are applied to product consistency issues that can occur in standard production.

In the first case history, paper manufactured to the same specification by two different paper producers had different impregnation behaviors, despite the similarity in overall porosities. Here, it was found that the porosity at the top region of the papers were different and that this had an important effect on impregnation.

In the second case, a paper manufactured at different sites by the same producer had the same appearance but different print qualities. Once again there was a correlation between the microstructure of the paper as seen by nano CT and the macrostructure as evident from the visual appearance of the printed sheet.

## *Impact of $TiO_2$ and Pore Content on Impregnation Behavior*

A laminator found that décor paper manufactured to the same specifications ($TiO_2$ content, basis weight, and sheet porosity) by two paper producers showed different impregnation behaviors. This resulted in a decreased rate of impregnation for one of these papers. Conventional analysis of the two papers did not reveal differences that would explain this difference in impregnation rates.

Nano CT was used to determine if there were differences in pore distribution and $TiO_2$ clustering between the two sheets that would account for this behavior. In addition, the link between the quality of formation and the pore structure was investigated.

Images of the samples in the XY plane showed little differentiation between the two (Fig. 18.32). A cross section for the fast-impregnating sample, taken in the Z-direction (from the top surface to the bottom), shows that the distributions of $TiO_2$ clusters, paper fibers, and pores are not uniform throughout the paper (see Fig. 18.33 for the distribution profile of Sample 1). Here, we see that the top of the sheet has more $TiO_2$ than the bottom (as was seen in other examples discussed above) and that the regions of the paper near the surface are dominated by pores.

A comparison in the distributions of the pores and $TiO_2$ clusters in the two décor sheets revealed important differences in the way that the components of the paper are situated within the sheet. To illustrate this, in Fig. 18.34 we plot the volume percentages of the pores and clusters in the portion of Fig. 18.33 within the red box, and show the same information for the sheet with poor impregnation speed (Sample 2). Although the pore distributions are similar, the distributions of the $TiO_2$ clusters are not. In particular, the top surface of the poorly performing sheet has a higher concentration of $TiO_2$ clusters. These clusters impede the penetration of the resin

**Fig. 18.32** Nano CT image in the XY plane (parallel to the paper surface) for the two décor paper samples. **a** Sample 1, décor sheet with rapid impregnation. **b** Sample 2, décor sheet with slow impregnation

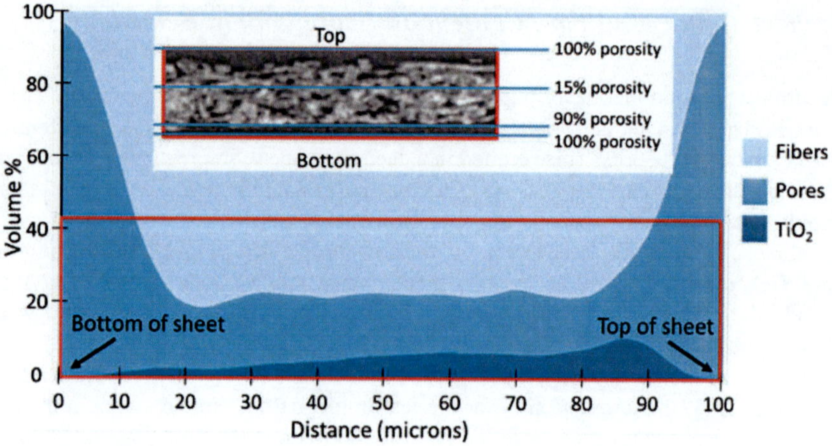


**Fig. 18.33** Distribution of paper components in the Z-direction for Sample 1

into the sheet, slowing the impregnation process. In addition, differences in pore diameter and the identities of the fibers could also play a role.

While the overall porosities of décor paper sheets have long been recognized as being important to sheet impregnation, we see here that the exact distribution of the pores at the microscopic scale is also an important factor.

## *Impact of $TiO_2$ and Pores on Print Quality and Resin Impregnation Speed*

In our second example of the application of nano CT to paper laminates, two décor papers, made by the same producer, with the same formulation, but at different plant sites, were tested for site-to-site consistency [4]. The papers looked identical before printing and had similar properties as determined by traditional paper tests (Fig. 18.35). However, significant differences in appearance developed after printing and impregnation (Fig. 18.36).

### Traditional Testing

Optical microscopy on the samples after printing showed clear differences in appearance (Fig. 18.37), with the colors on the second sample appearing more intense than on the first. Differences were also seen in the topography of the base papers, with Sample 2 showing more surface features than Sample 1 (Fig. 18.38). While these differences were evident, they did not themselves suggest a cause for this behavior.

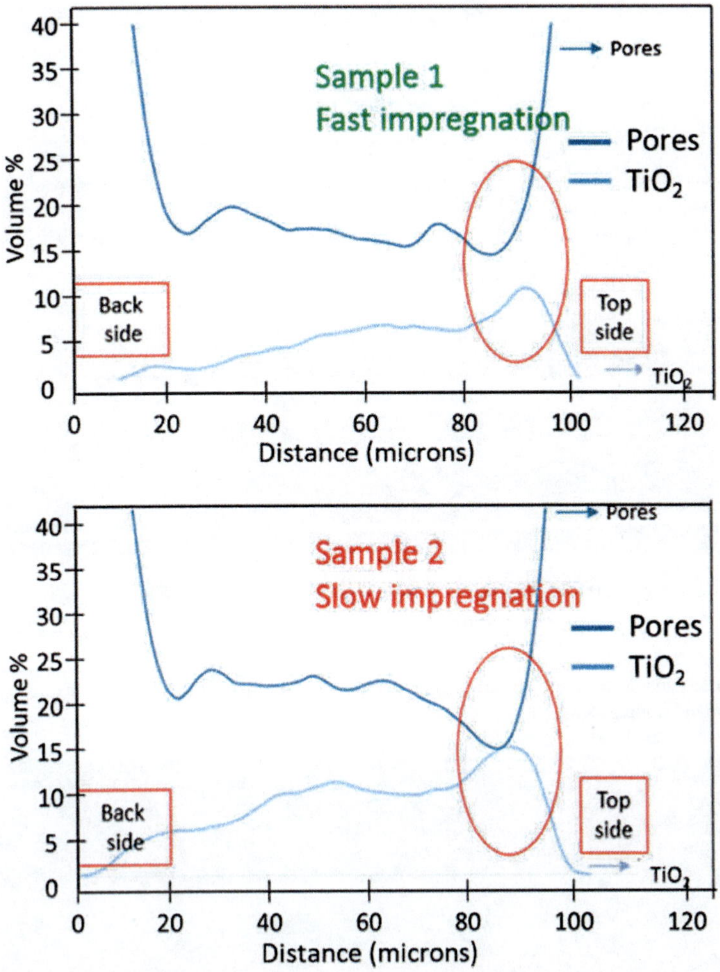


**Fig. 18.34** Comparison of pores and $TiO_2$ cluster distributions in the two décor paper sheets

## Nano CT Analysis

The samples for the two papers were subjected to nano CT analysis. A third paper was included for reference in this study. This is a resin saving paper (RSP), which shows different impregnation behavior than traditional décor paper, due to many of the pores in this paper being closed and so not available for resin. This paper had a similar basis weight as the other two (~75 gsm), but significantly more $TiO_2$ (the ash content was 42.8%, compared to 24.9% and 26.8% for Samples 1 and 2, respectively). The RSP sheet was included as a means to validate the nano CT method.

Sample 1
76.2 gsm / 24.9% ash

Sample 2
76.3 gsm / 26.8% ash

**Fig. 18.35** Visual appearance and physical properties of the two décor paper samples before printing and impregnation

Sample 1
- Poor print quality (different from master)
- Fast impregnation

Sample 2
- Good print quality (similar to master)
- Slow impregnation

**Fig. 18.36** Visual appearance after printing and impregnation

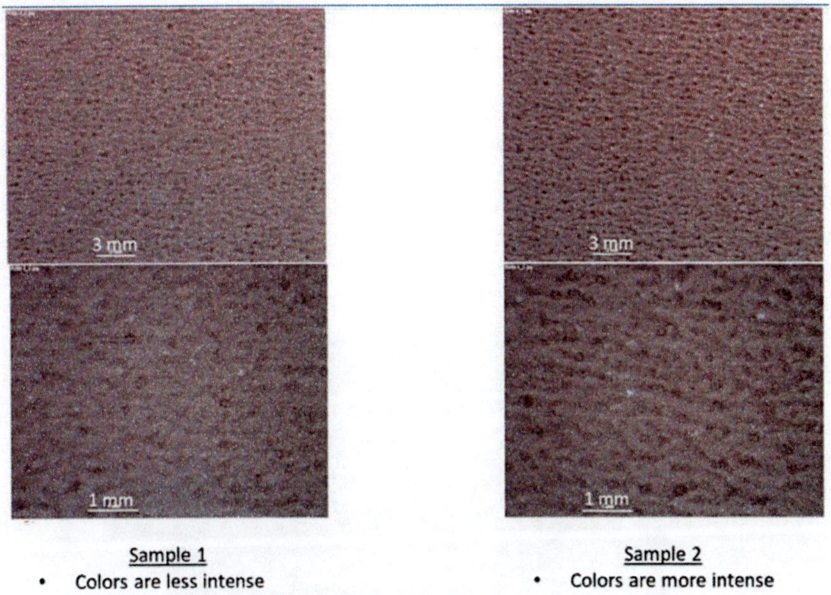


**Fig. 18.37** Optical microscope images of the printed papers

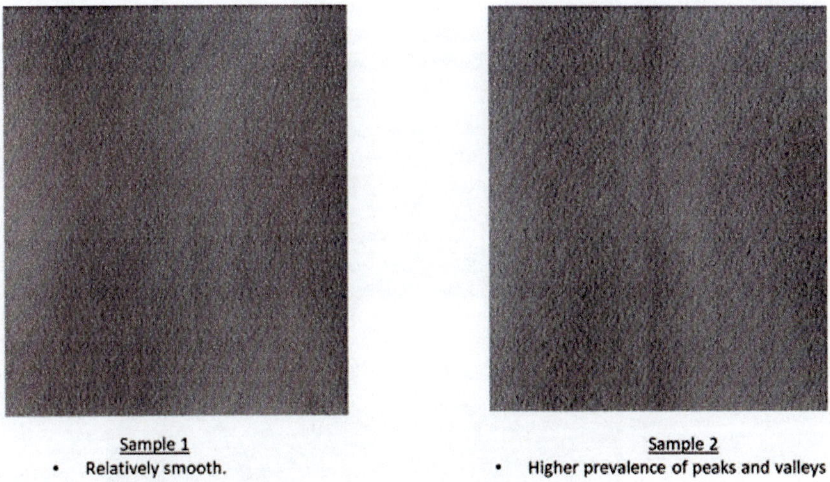


**Fig. 18.38** Topographic comparison of the two base papers

Images from the XY planes of all three samples are shown in Fig. 18.39. Here we see that the distributions of fibers, particles and voids are similar in Samples 1 and 2 while the RSP sample shows smaller air voids. Profiles of the pore and $TiO_2$ distributions in the X- and Y-directions for Samples 1 and 2 were generally unremarkable and similar to the one another.

The situation is quite different when we examine the samples in the Z-direction (Fig. 18.40). The $TiO_2$ distribution for Sample 1 is uneven, with a higher concentration near the top surface than at the bottom. This is consistent with the observations made previously. A gradient in the $TiO_2$ distribution was also seen in Sample 2

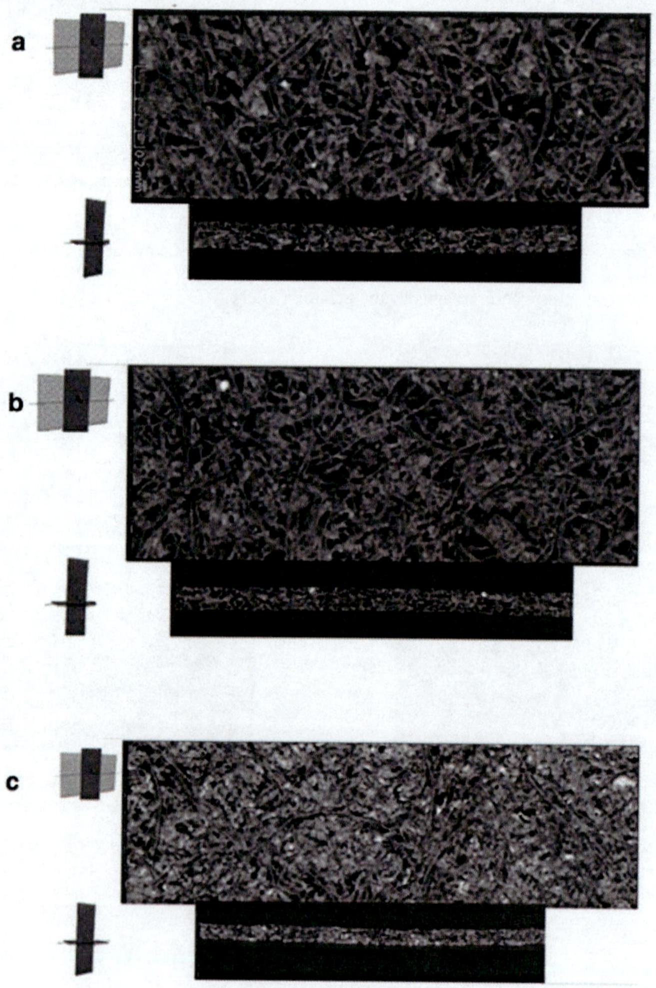


**Fig. 18.39** XY plane images of the three décor paper samples. **a** Sample 1. **b** Sample 2. **c** RSP sample

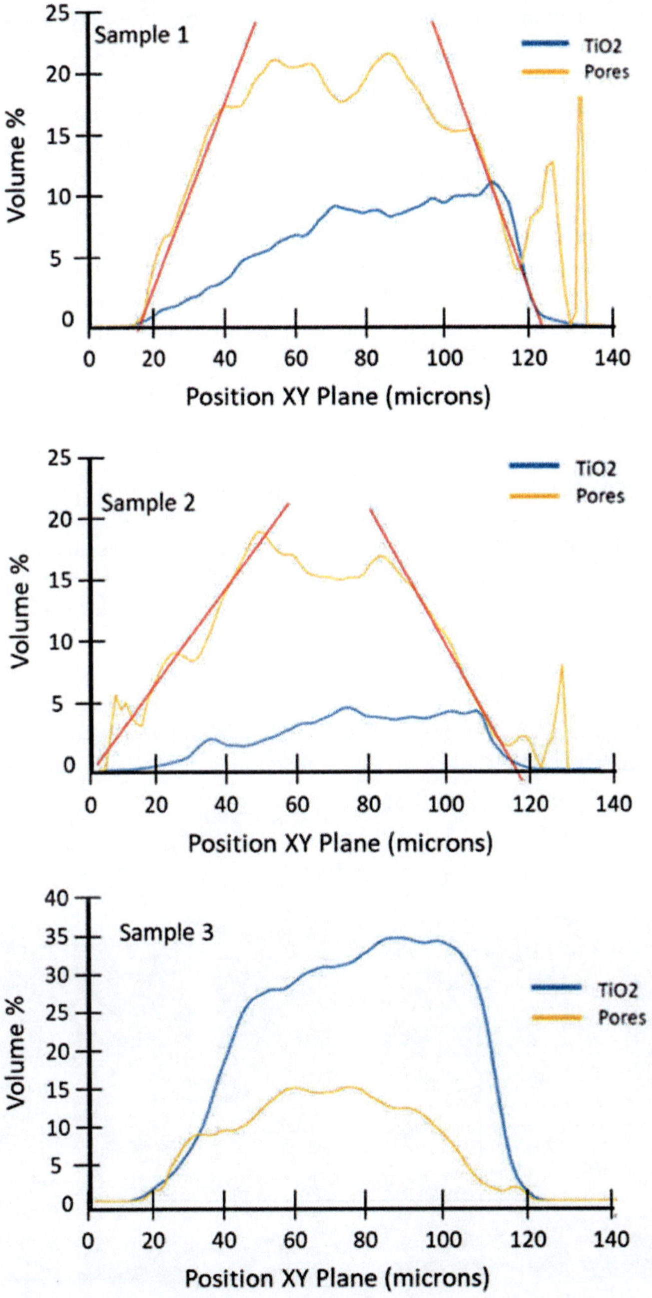


**Fig. 18.40** Nano CT profiles for the three samples in the Z-direction. In all cases, the bottom of the sheet is on the left and the top on the right. The jagged features in the pore distributions seen at the top of the sheet in Samples 1 and 2 are due to fibers protruding from the sample surface

but was less pronounced. The pore volume distributions of the two samples were qualitatively similar, with the highest pore concentrations in the center of the sheets. However, the concentration gradients (i.e., the slopes of the red lines in Fig. 18.40) of the two sheets are different. Sample 2 has a gentler slope than Sample 1. We believe that this decreases the impregnation rates while providing for better ink hold-out (the ability to keep ink at the paper surface), explaining the more intense color in Sample 2.

Although Samples 1 and 2 were intended to be of identical composition, the amounts of $TiO_2$ seen in the nano CT profiles are different. This could be because the samples truly have different $TiO_2$ levels, despite their equal ash values (ashing does not discriminate between $TiO_2$ and other non-combustible paper components), or because the $TiO_2$ particles in Sample 2 are better dispersed. Better dispersion would increase the number of individual particles or very small flocculates that cannot be detected by the nano CT technique.

The $TiO_2$ content in the RSP sheet (Sample 3) is significantly greater than in Samples 1 and 2, consistent with its higher ash value. By contrast, the pore content of this sheet is significantly lower than in the other two, reaching a maximum value of 15% in the center of the sheet, compared to roughly 20% for the other two. The observations are easily explained on the basis of the RSP sheet having sealed pores. Resin Saving Paper is designed to seal the pores through the use of sizing. This results in reduced pore volume and no fibers sticking out. The top and bottom of this sheet are designed for printing and impregnation, respectively.

Figure 18.41 shows three-dimensional renderings of the three papers. We can clearly see, on the left, that the RSP sheet (Sample 3) has a higher $TiO_2$ content while, on the right of this figure, we see it has significantly less pore volume. We also

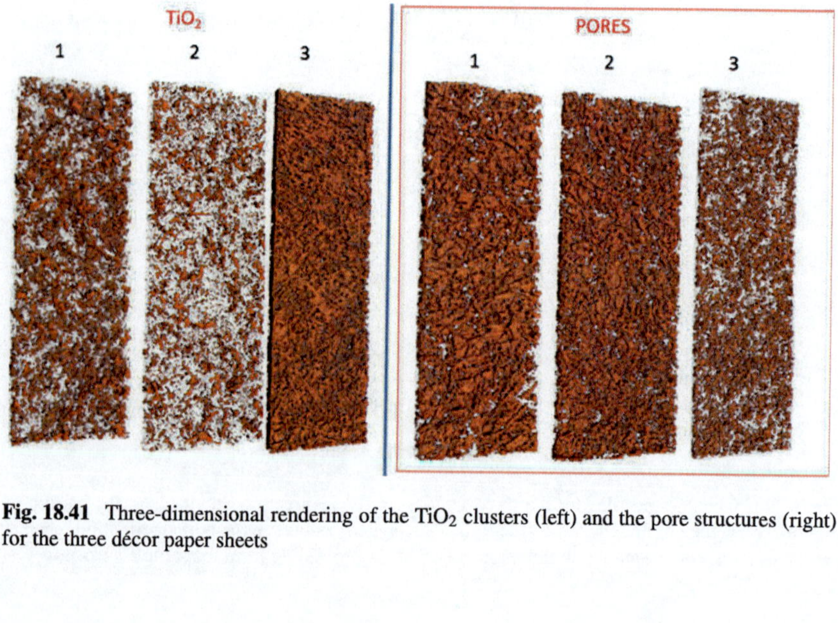


**Fig. 18.41** Three-dimensional rendering of the $TiO_2$ clusters (left) and the pore structures (right) for the three décor paper sheets

see that while the pore structures of Samples 1 and 2 are similar, the distributions and amounts of $TiO_2$ clusters are different.

To further illustrate the pore distributions of these samples, as well as their size, a color-coded rendering of the pores for the three samples is shown in Fig. 18.42. The pores with the smallest volume are in blue, next green, and the largest are red. In addition, Fig. 18.43 shows the pore size as the diameter of the equivalent sphere volume for the three samples. Here we see that the average pore diameters, over the entire volume of the samples, are similar for Samples 1 and 2, but shifted to a lower

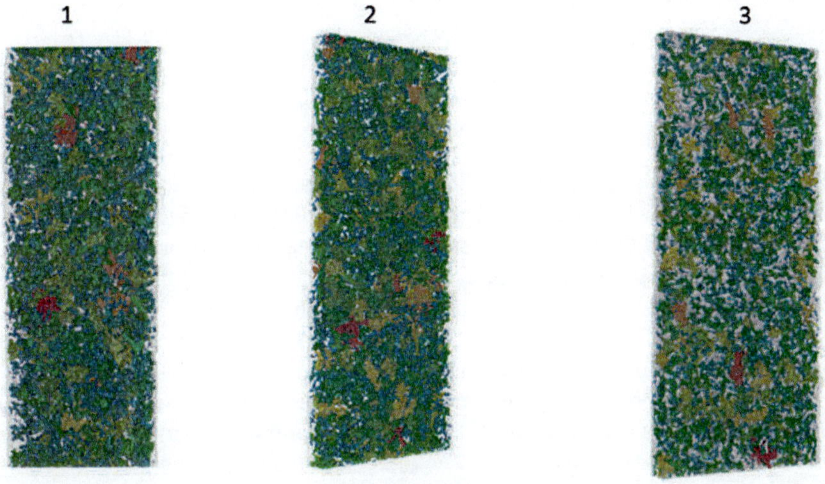


Fig. 18.42 Color coded renderings of the pores in the three samples

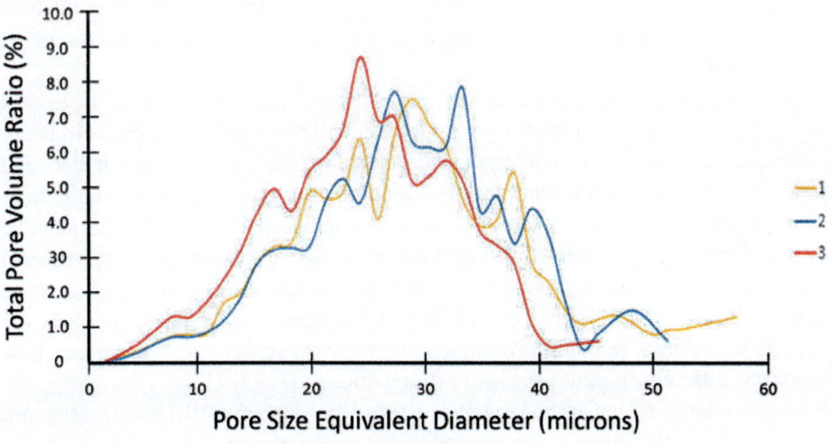


Fig. 18.43 Distribution of pore sizes for the three samples

value for Sample 3. This is consistent with the pore volumes in resin saving paper being reduced by the starch film (sizing).

This shows that the overall porosity is not as important as the porosity in the top 20 microns of the paper surface. This implies that maintaining constant total paper porosity is not as important for product consistency as controlling the porosity at the print surface: surface porosity is key. While this is well understood in the industry, measuring surface porosity has been elusive. This technique therefore fills a significant gap in the tools available to the paper décor formulator. Not only is the surface porosity imaged by this technique, but simultaneously the $TiO_2$ distribution, in the same volume of paper, is also imaged.

## Summary

Décor paper is one of the most challenging paper types to produce, in view of the many requirements it must meet, all of which are within narrow specifications. These parameters impact the use and performance of the décor paper in the downstream value chain: at the printer, the impregnator, the laminator, the furniture manufacturer or the laminate floor manufacturer, the retailer, and the final user.

We have seen that tight quality control starts with knowing which parameters are relevant and then putting a process in place to control these. Controlling variation is at the heart of producing good quality décor paper. This starts with minimizing the variation of the raw materials: the pulp used (a natural product) and the $TiO_2$ (a chemically purified product).

In order to maintain some flexibility in pulp selection, the $TiO_2$ selected must have low variability in its various performance parameters that affect appearance. These include color consistency, pigment dispersion, and surface chemistry. Surface chemistry affects not only the dispersion behavior and particle crowding in the final laminate, but also its light stability when exposed to ultraviolet radiation (as discussed in Chap. 14).

Conventional methods of paper analysis—basis weight, overall porosity, optical microscopy, and SEM microscopy—sometimes fail to explain differences in décor paper quality. Nano CT technology can be used in these cases to better understand the source of these differences. In this technique, the composition of one micron cubes within the paper are determined by X-ray analysis. The resulting data can be used to show cross sections as well as three-dimensional renderings of the paper.

Application of this technique to a series of real-life issues has shown it to be an effective measure of paper consistency at the micron scale. In many cases the surfaces of different samples look similar, but a virtual cross section will often show important differences in the sizes and distributions of $TiO_2$ clusters and air pores, despite similarities seen at the larger scale (e.g., the same overall $TiO_2$ or total pore content). This technique can also be used to distinguish the amounts of $TiO_2$ found in $TiO_2$ clusters from the amounts found in the dispersed state on the fibers. Increasing the $TiO_2$ content of the paper shifts these amounts to favor clusters over dispersed particles. This decreases the scattering efficiency of the $TiO_2$.

Other findings were that both the $TiO_2$ clusters and the air voids are not distributed evenly in the Z-direction of the paper. The $TiO_2$ concentration is highest near the top of the paper and lowest at the bottom. The details of these distributions affect the ability of resin to infiltrate the paper, causing, in some cases, excessively slow impregnation. They can also affect the uniformity of printing on the paper prior to impregnation and can explain how some sheets appear similar before printing, but after printing can appear quite different.

## References

1. Linhart, F.: The practical application of wet-strength resins. In: Au, C. O.; Thorn, I. (eds.) Applications of Wet-End Paper Chemistry. Springer (1995)
2. Nicolai, N.; Vanhecke, F.: Laminate color uniformity: impact of $TiO_2$ throughout the laminate value chain. In: 2014 TCM European Laminate Conference, Vienna (2014)
3. Borrelli, C.: Evolution of the laminates market and of the understanding of the formation: laminates quality parameter. In: 2017 TCM Decorative Surfaces Conference, Hamburg (2017)
4. Vanhecke, F.: Characterization of décor paper using nano computer tomography. In: 2018 TCM Decorative Surfaces Conference, Dresden (2018)

# Part VI
# Future Looking

# Chapter 19
# Issues and Emerging Trends

## Contents



## Introduction

Quiet, and not so quiet, revolutions in the way materials are designed and manufactured are underway in the general global economy, and these revolutions touch on the ways that the coatings, plastics, and paper laminates industries are evolving in response to new requirements, both in terms of product performance, and in long-term environmental and societal impacts. In the following sections, we review a subset of these changes and how they apply to the use of small particles embedded in polymers. This review is not intended to be comprehensive, but rather to be a broad survey of the changing landscape of the industries that use small particles.

 
M. Diebold et al., *Pigments, Extenders, and Particles in Surface Coatings and Plastics*, https://doi.org/10.1007/978-3-030-99083-1_19

## Nanoparticles

The common definition of a nanoparticle is one that has at least one dimension in the range of 1–100 nm [1–3]. The number of small dimensions and the morphologies of nanoparticles can vary, leading to nanofibers, nanotubes, and nanoplates. The properties of these particles are often determined by quantum effects, which do not operate at the scale of microparticles (see Chap. 1). Nanoparticle science is a relatively young area of research and application. While the properties of natural nanoparticles were unknowingly used in construction materials for more than 4500 years, and as pigments in ceramics and glass dating back to the ninth century, nanoparticle science has been recognized as its own field since only the second half of the last century [4–6].

Nanoparticles generally have different properties than larger particles of the same material. These differences include enhanced reactivity, greater strength, higher surface area, and, in some cases, improved stability. The size, surface area, and quantum effects of nanoparticles primarily affect the optical, electrical, and magnetic properties of these materials. These behavioral differences have been applied to such areas as catalysis, medicine, energy production and storage, and advanced materials, among others [7]. While nanoparticles show much promise in these many technologies, their use is currently limited in many applications by cost, as well as by both safety and sustainability concerns.

A commonly given example of the potential utility of nanoparticle is in quantum dots, which are nanosized semiconductors. Quantum dots are characterized by certain optical quantum confinement properties. This allows these materials to be tuned to emit different colors of light based on their size and composition [8]. This property can enhance applications such as solar cells, bioimaging, QLEDs, single-electron transistors, photocatalysis, and lasers [9]. Quantum dots can be elemental or compound systems, often metal chalcogenides. Some examples are CdS, graphene, CdSe, ZnS, Ge, carbon, Si, PbSe, and CdTe.

While quantum dots are not typically used in organic matrices, such as coatings, plastics, and paper laminates, other nanoparticles are. This is particularly true for higher value applications such as pigments. Nanosized carbon blacks and other color pigments have existed for some time, and new materials are being explored. These include the use of ceramic nanopigments based on metal oxides in applications such as ink-jet printing and color changing coatings (see Chap. 8) [10, 11].

Carbon nanotubes can be used as specialty black pigments in coatings. Vertically aligned carbon nanotube array (VANTA) pigments are currently recognized as the blackest material known. These pigments provide exceptionally broad, uniform, and low reflectance [12]. Nanosized carbon blacks are also widely used in plastics as pigments as well as fillers [13], and carbon nanotubes are poised to replace conventional carbon black in the rubber tire industry, although there is some concern as to the release of nanoparticles into the environment as these tires wear [14].

Nanosized $TiO_2$ particles have been studied extensively for applications in paints and plastics [15, 16]. Nanoparticles of $TiO_2$ are too small to efficiently scatter visible light and so cannot be used as a white pigment. This can be contrasted to color pigment particles, for which decreasing particle size almost always increases the light absorption strength and so improves pigmentary efficiency (see Chap. 8). However, nanosized $TiO_2$ particles have potential application for non-pigmentary applications such as "smart coatings".[1] In addition, these particles are strong ultraviolet light attenuators, and their presence can generally improve the durabilities of paints and plastics as a result. They are also potentially useful for self-cleaning or antimicrobial coatings [13, 17]. Their proposed use to detoxify air pollutants is discussed separately in the next section.

There are also potential uses for nanoparticles that do not interact with light (that is, nanoextender particles) in paints and plastics. As discussed in Chap. 9, some fumed and precipitated silicas, as well as some precipitated calcium carbonates, fall into the nanosize range [13]. A chief benefit of these materials is that, in crowded systems such as high PVC paints, they allow for better spacing of pigment particles than is possible with their conventional microparticle counterparts. This increases the effectiveness of both white and color pigments. Balanced against this better pigment particle spacing, nanoextenders are typically more expensive than larger extenders, and can be more difficult to initially disperse and to stabilize against flocculation.

## Photocatalytic Destruction of Pollutants

In Chap. 14, we discussed the negative impact that titanium dioxide photocatalytic activity can have on resin or polymer matrices. Titanium dioxide particles absorb ultraviolet light, which is beneficial to the durability of the underlying polymer, but a small fraction of that light is converted into chemical energy in the form of chemical radicals. These radicals attack the polymer and so harm the durability of the paint or plastic. Note that these reactions are not seen for particles that absorb visible light, because the energy of visible light photons is not great enough to break the bonds in resin molecules, and so even if they did cause radical formation, the radicals would have too little energy to degrade the polymer [18].

While measures are taken in most durable paint and plastics applications to decrease $TiO_2$ photoactivity (see Chap. 7), proposals have been made to instead maximize photogenerated radicals for the purpose of reacting with—and destroying—harmful molecules in the environment [19]. This has been promoted particularly for the catalytic destruction of $NO_x$, the pollutant responsible for smog. The oxidation of $NO_2$ molecules into nitrates (Eq. 19.1) is one example of atmospheric detoxification.

$$NO_2 + OH^\bullet \rightarrow HNO_3 \tag{19.1}$$

[1] Smart coatings are those that change one or more properties in response to environmental changes such as a change in temperature or applied lighting.

This concept is especially attractive because it harnesses sunlight, which is both free and ubiquitous, as the power source for these reactions. Paints and plastics with photocatalytic particles would clean the atmosphere passively, with no external resource requirements beyond their production and application. In addition, the particles proposed for this task, titanium dioxide (especially in the anatase form) and zinc oxide, are non-hazardous, relatively inexpensive and readily available.

Much attention has been given to this concept, particularly in academia. Initial laboratory studies using paints made with photocatalytic $TiO_2$ were promising, with reports of $NO_x$ reductions of several tens of percent. However, these values were seen after extended exposure times (hours) of small quantities of gas. Although similar reductions have been reported in the field, the rapid turnover rate of air above a painted surface would imply that these reductions would need to occur on a time scale of a fraction of a second, which is improbable, if a measurable decrease in pollutant level is to be made.

An additional concern is that the radicals generated by the photoreactive pigments—superoxide ($O_2^-$) and hydroxyl (OH)—are non-specific with regard to the materials that they attack.[2] In particular, they readily react with the organic resin in a paint film (see Chap. 14). This material is present at mass concentrations that are approximately ten orders of magnitude greater than that of $NO_x$ in highly polluted air.[3] We would therefore expect that for every reaction of one molecule of $NO_x$ with the photogenerated radicals, there would be roughly ten billion reactions with the resin in the paint film. This would decompose the film long before a significant number of $NO_x$ molecules had been destroyed.[4] This concern can be overcome by using inorganic matrices, such as cement, rather than organic matrices [22, 23].

One often used method of quantifying the destruction of $NO_x$ is measurement of the formation of nitrate ($NO_3^-$). However, recent research suggests that this nitrate can be formed by an alternative reaction—the reaction of photogenerated radicals with atmospheric $N_2$ [24]. Although this reaction has a very high reaction barrier, the concentration of $N_2$ in air (78%) is approximately ten million times greater than that of $NO_x$.

Many researchers have found that, while nitrate is the eventual product of the destruction of $NO_x$, toxic intermediates, such as hydrogen peroxide, ozone, and formaldehyde and other VOCs are formed [25, 26]. This calls into question whether these reactions, if they are occurring at meaningful rates, cause more harm than good [27].

---

[2] Direct reaction with surface adsorbed $NO_x$ is also possible. However, competition with water, which is highly abundant in the atmosphere, for surface sites on the $TiO_2$ particle makes this unlikely.

[3] The concentration of $NO_2$ in highly polluted air is approximately 200 parts per billion (ppb) [20]. The density of air is 1.2 kg/m$^3$, giving an $NO_2$ density of $2.5 \times 10^{-7}$ kg/m$^3$. By contrast, the density of paint resin is approximately 1,000 kg/m$^3$.

[4] Along these lines, sunscreens containing nano $TiO_2$ or ZnO particles that were inadvertently transferred from the hands of workers to painted metal roofs were found to accelerate the photodegradation of these highly durable paints by a factor of 100 [21].

Finally, field testing results have been mixed. Many show no efficacy of photoactive paints in reducing $NO_x$ air pollutants. For example, in one test, a panel of eleven air quality experts found levels of $NO_2$ on a roadway to be 7.32 ± 0.3 ppb in the presence of a photocatalytic paint compared to 7.35 ± 0.3 ppb in its absence [28]. Studies in the UK found that $NO_x$ conversion in the field could occur under optimal conditions, but that these conditions were so infrequent that this technology produced no overall positive effect on $NO_x$ concentrations [29, 30]. Similar findings of lack of efficacy were reported in Holland [31, 32].

With that said, there is still considerable interest in overcoming the challenges posed by these considerations. Research in air pollution reduction is currently focused on increasing the quantum yield of the particles (that is, increasing the likelihood that a light absorption event would initiate a desired reaction). The solid particles used in this process traditionally have low quantum yields (roughly 5%) [33–39]. In general, small (nano) particles are more photoactive than large (micro) particles, and the photoactivity of $TiO_2$ can be further increased by doping the lattice with a metal cation [40] or by the use of dyes [41]. Many of these materials are also under investigation for their potential to split water or to reduce $CO_2$ [42].

The use of this technology appears more promising for the destruction of organic pollutants in water than in the atmosphere [43]. For example, photoreactors using slurries of particles have been demonstrated to purify wastewater streams from pharmaceutical manufacturing facilities [44]. Large-scale testing of photoreactors using solid particles for more general wastewater treatment has shown commercial potential [45, 46]. These studies showed an 80% reduction of pesticide concentration in wastewater after four hours of exposure to natural sunlight. However, caution about efficacy for pollution control is advised. For example, the rate of photodegradation of pesticides in one waste stream was three times slower during the winter versus the summer months [47].

## Digitization, Machine Learning, Artificial Intelligence, and Industry 4.0

The paint and plastics industries are traditionally viewed as being relatively conservative in some ways, with production methods and research and development approaches changing only slowly over time. An example of this from the paint industry is its acceptance of titanium dioxide as a white pigment. Despite the great performance advantage that titanium dioxide has for opacity, the coatings industry did not use titanium dioxide in significant amounts until it was forced to abandon the incumbent lead oxide pigments in the United States (and elsewhere) by legislation.

However, this perception of these industries no longer applies, and most manufacturers are rapidly automating their production procedures and are processing great amounts of data to further streamline their operations. These activities are indicators of the so-called third industrial revolution, the digital revolution [48]. Some manufacturers are evolving to the fourth industrial revolution, commonly known as Industry 4.0. This involves the development and adoption of smart technologies, machine-to-machine communications, and other electronic connectivities allow for the analysis, diagnosis, and correction of production issues without the need of human intervention. While these smart machines show much promise, it is still too early to predict their full impact on the paint, plastics, and paper laminates industries.

The application of artificial intelligence to research is another still early development that has a significant potential for disrupting current practices. Here, historical data is mined to develop relationships between, for example, manufacturing parameters, material compositions, and product performance, that are otherwise hidden. This process, like any learning process, proceeds through stages. In the first stage, a limited number of data points are used to create an imperfect model from which predictions can be made regarding product improvements. As these predictions are acted on, and additional information is generated, the model is refined. These refinements lead to more accurate predictions, and the cycle between model generation and data collection is repeated, creating a self-learning model that is smarter and more powerful than those generated from statistically designed experiments.

A strong benefit of artificial intelligence is that new raw materials and finished products can be added in a very natural way. This is in contrast to the statistical DOE approach introduced in Chap. 16, where changing input parameters, such as adding a new raw material, requires the execution of many experimental trials. This makes the DOE approach static and unable to nimbly adjust to changing conditions. That said, the eventual goal of both approaches is to reduce laboratory work to a simple verification process that, in some cases, can be eliminated.

Another developing trend in the coatings, plastics, and paper laminates industries is that, as these industries become more mature, it becomes more difficult for producers to innovate and differentiate their products in isolation from other parts of the value chain. Collaborations between raw material suppliers, their customers and the eventual end-use producer are now common, reflecting the benefit of a systems approach in which materials from different parts of the value chain are customized to work well together [49–51]. This also applies to ideas for new products that can overcome the status quo. In addition, open collaborations with universities and government agencies are an area of high activity and great interest to all industries [52].

## Sustainability Challenges

Research into improved product sustainability has accelerated over the past two decades [53]. The goal of sustainability—reducing the impact of commercial and societal activities on the environment—is being addressed in the paint and plastics industries in a number of ways. Many petroleum-based raw materials have transitioned to renewable, bio sources. Attention has also focused on the concepts of "reduce, re-use and recycle", with the intent of minimizing not only the material requirements of a product but also the amount of material destined for disposal, usually in a landfill. Finally, concepts such as cradle-to-grave life cycle assessment (discussed below) and carbon footprint continue to gain relevancy in many diverse manufacturing technologies [54, 55], and ambitious environmental goals such as carbon neutrality by 2050 are gaining societal and governmental acceptance.

The particles used in paints and plastics are critical elements of many sustainability activities. Not only are the sustainability issues associated with the particles themselves of interest, but also their effects on the sustainability of other components. Most commonly these other components are the matrix surrounding these particles, although in some cases they include additives to the paints or plastics.

## *Legislative and Regulatory Considerations*

Government agencies are responding to society demands for safer chemicals and greater environmental protections by legislating more stringent regulations and requirements on the production and properties of many of the materials made by the coatings and plastics industries. The European Union (EU) is, in many ways, at the forefront of sustainability, environmental goal setting, and chemicals legislative regulations. For example, the EU Registration, Evaluation and Authorization of Chemicals (REACH) legislation, implemented in 2010, has inspired other jurisdictions around the world to also enact REACH-like legislation (e.g., Korea REACH and Turkey REACH). More recently, in response to the growing concern for the environment and for climate change, the European Commission launched the ambitious European Green Deal Strategy in 2019 [56]. Because the EU is leading many sustainability initiatives, it is instructive to detail the current status of regulation in the EU as this will likely be indicative of future activity in other regions of the world.

The European Green Deal sets ambitious goals for no net greenhouse gas emissions by the year 2050, with an interim goal of reducing emissions of these gasses by 55% (based on 1990 levels) by the year 2030, all while transforming the EU into a fair and prosperous society with a modern, resource-efficient competitive economy. A key aspect of this is the Chemicals Strategy for Sustainability, which is designed to protect both citizens and the environment by boosting the innovation of chemicals that are designed to be safe and sustainable throughout their life cycle [57].

Small particles play both direct and indirect roles in the enactment of these programs. This can be demonstrated by considering one type of small particle—titanium dioxide. In 2020 the EU classified titanium dioxide as a suspected carcinogen (category 2) by inhalation under the EU's Classification and Labelling (CLP) Regulation [58]. The suspected hazard described for $TiO_2$ is not specific to this substance but is instead common to all dusts and powders that fall under the category of "poorly soluble low-toxicity substances" (PSLTs). As such, this classification for $TiO_2$ may one day extend to other PSLT's. The role such future classifications may play on the use of these materials going forward is uncertain.

Balanced against this classification are the benefits that pigmentary $TiO_2$ bring to the sustainability of the materials in which it is found. For example, the strong UV light absorption properties of $TiO_2$ protect most polymers into which it is incorporated from photodegradation by sunlight, as discussed in Chap. 14. This extends the life of these materials and therefore decreases the overall life cycle impact of them. In addition, the amounts of $TiO_2$ needed to achieve paint, plastics, and paper laminate opacity, and the thickness of paint required for full hide, are significantly less than those needed for alternative white opacifiers. This enables producers to attain target performance with lower amounts of raw materials. Finally, titanium dioxide pigments can be used to reduce electrical consumption of both air conditioners and interior lighting, due to its light reflective properties, and $TiO_2$ is considered a key material in the EU's Renovation Wave Initiative to deliver higher energy efficiency [59].

There is also growing concern specifically regarding the effects of nanomaterials on the environment and human health. The EU REACH legislation has been amended with specific nanomaterial requirements that were recognized in December, 2018 and placed in force in January, 2020 [60]. These new nanoparticle annexes require registrants to provide additional physical and chemical characterization information, additional human health data, and detailed environmental endpoints.

## *Life Cycle Assessment*

One way of monitoring or accounting for the environmental burden of a product or service is through a Life Cycle Assessment (LCA). The basic principle of an LCA is the determination of the total environmental footprint of a product from the extraction of its raw materials to its ultimate disposition after reaching the end of its useful life. This is described as a "cradle-to-grave" analysis and can be further divided into a "cradle-to-gate" component, that focuses on the activities needed to produce a product (the "gate" in this case being the factory gate), and a component determined by the environmental burdens encountered during of the use and disposal of that product.

Paints can be used as an example of the interplay of the two components of an LCA [55]. A high quality paint will likely have a greater cradle-to-gate footprint than a low quality paint because the energy required to make and process high quality ingredients is greater than that required for lower quality ingredients. However, this must be balanced against the greater paint durability and longer service life of high quality paints. A proper LCA will include the environmental benefits of less frequent paint application when determining the overall effect of the paint on the environment.

Another way of categorizing emissions is by scope [61]. Scope 1 emissions are those generated at the production site of the reporting organization (e.g., the paint producer). Scope 2 emissions are those necessary to produce the heat and electricity that are generated outside the reporting organization and used by it (e.g., emissions generated by the power company). Scope 3 emissions are those that are a consequence of the activities of the reporting company (e.g., the paint producer) but that are not controlled by the company. These latter emissions are typically linked to extraction and production of purchased materials, transportation, business travel, use of the produced goods, etc.

## Reduce

Reduce, re-use and recycle are three tactics for implementing sustainability strategies. As an example of the first: in certain situations, nanoparticles are more efficient than traditional microparticles in providing product performance attributes, and as such allow for a decrease in the particle content of the material while maintaining the same performance. This leads to a reduction in the sustainability footprint of the particles.

Improved particle efficiency can also allow for a reduction in the amount of resin or polymer used in the product, since one function of resin and polymer is to hold particles in place. This further improves the environmental footprint of the material. This concept is being actively pursued in the auto industry, where "light weighting" is seen as a means of improving distance traveled per unit amount of energy.

We see, then, that nanoparticles address the "reduce" aspect of sustainability for both the particle and the matrix components of a material. Acting against these improvements, however, are limitations that arise due to the different physical properties of nanoparticles in comparison to microparticles (these were discussed in detail in Chap. 1). In particular, factors such as particle packing efficiency and slurry (or paint) rheology change significantly when transitioning from the micro-regime to the nanoregime. These factors limit, for example, the application of nanoparticles in high solids paints.

## Re-Use

The sustainability potential for particle "re-use" has many practical challenges. Re-use of the particles themselves requires several steps. Waste containing the particles must be collected, sorted and undergo physical processing, such as shredding. This produces a consistent feed for the next step, particle liberation from the matrix. This is most commonly done by pyrolysis of the matrix and collection of the solid residue for further purification [62].

The economic viability of particle re-use is a significant limitation in the broad adoption of this technology. The infrastructure for collection and sorting of matrices with particles remains limited and often requires governmental support for execution. As an example of the economic challenges of this technology for paints and plastics, a comparison can be made to the recovery of precious metals from discarded electronic equipment. Despite the high value of the recovered metals, only 10% of the discarded cell phones in the world are processed for this purpose. Even then, if the desired metals are embedded in a matrix of other metals or of polymer, the economic gain in recovering these metals is often minimal.

## Recycle

Both re-use and recycle involve the creation of a "second life" to a material. In our context we distinguish the two as follows: recycle is the use of a final product a second time (e.g., converting an unwanted paint into a useful one) while re-use is the restoration of a material to its initial state as a raw ingredient (e.g., the extraction of the particles in an unwanted paint to produce a new paint). The important practical difference between these is that in recycle the particles and the matrix are repurposed in their combined state while in re-use the particles are separated from the matrix prior to repurposing.

In plastics, recycle is most popular for "one matrix" systems, i.e., those consisting of one polymer with a single type of dispersed particle. For example, polyester recycling with solid particles is common for drinking bottles [63]. Examples of recycled mixed suspensions have been demonstrated, but process limitations present challenges [64, 65]. Typically, these challenges are overcome with compatibilizing chemicals that can emulsify different polymers into a new matrix. The recycling of plastics that contain particles is rarely a true "recycle", in the sense that the recycled material has an altered chemistry and a different end-use application. For example, opaque PET drinking bottles are currently recycled into PET carpet fibers, and HDPE milk jugs are currently recycled into plastic park benches.

Recycle of unused architectural or décor paints represents an even greater challenge due to the great diversity in physical and performance attributes of these paints. Current estimates are that as much as 10% of architectural paints are not used [66]. Recycling this unused paint is difficult since paints differ in many aspects. For example, they can differ in the chemistries used to stabilize them and to achieve good film formation. These differences can lead to gross incompatibilities when these paints are combined. In addition, paints differ widely in appearance properties such as color and sheen. As such, simply mixing unused paints together does not give a desirable or consistent product. Finding solutions to these barriers is an area of active research within the coatings industry.

## Lower VOC Paints

The VOC, or volatile organic content, of a paint is of great importance with regards to its impact on the environment, or, more specifically, on air quality. The transition of architectural paints from solventborne technology to waterborne technology that has occurred over the last five decades has greatly decreased the amount of organic material released into to the atmosphere by paints. This transition was revolutionary, rather than evolutionary, since waterborne paints use a completely different technology for creating a paint film from the suspended components in a liquid paint than is used in solventborne paints (see Chap. 10).

While the transition to waterborne technologies significantly reduced air emissions from architectural paints, they have not eliminated them completely. Most waterborne paints require an organic co-solvent, or coalescing agent, to soften and partially solubilize the outer shell of polymer in the latex particles. This is necessary if the individual latex particles are to coalesce into a coherent film (see Chap. 10).

Currently, the reduction of coalescing agent is most commonly accomplished by lowering the Tg (glass transition temperature) of the resin particles. This increases the effectiveness of the coalescing agent because these particles are inherently "softer" than their higher Tg counterparts. However, this shift in Tg is accompanied by an increase in the stickiness of the film surface, which increases the dirt pick-up of these films and makes them more difficult to clean completely.

While the reduction in coalescing agent levels is important to lowering paint VOC, there is a second, often overlooked source of emissions in many of these paints. This is the organic solvent used to suspend color pigment particles in colorant concentrates. Depending on the paint color and line of paint, organic emissions from the colorant itself can be greater than that of the uncolored paint. This issue is being addressed by the development of more easily dispersible color pigments [67]. Other efforts to decrease organic emissions in the paint industry include the drive to higher solids paints and to UV curable paints, for which all organic components become a part of the film on curing. Both of these technologies have requirements that go beyond those of traditional paints (e.g., minimal UV light absorption).

## *Biocide-Free Coatings*

The use of alternatives for in-can preservatives in waterborne architectural coatings is another trend driven by environmental considerations. The biocides that are traditionally used to prevent "in can" growth of bacteria in waterborne liquid paints may, in some cases, cause skin sensitization resulting in allergic reactions. Paint producers are therefore investigating alternatives to these biocides. A promising possibility is the formulation of these paints at very high pH values (above 11.0). These conditions are severe enough that bacteria will not grow, yet mild enough to avoid skin irritation. However, many types of particles are reactive under such high pH conditions, and specialized particles may be necessary to assure long-term storage stability.

## *Heat Reflective Coatings*

Much energy is expended globally to keep the interiors of our buildings at a comfortable temperature. For example, approximately 55% of the residential energy consumed in the United States is used for this purpose [68]. A sizable fraction of this energy is used to cool structures such as homes and other buildings during the summer. During these periods, it is very desirable to limit the degree to which the sun heats a structure. This desire is not unique to our time—these needs have applied to regions with hot climates since antiquity, and have generally been addressed by painting structures white, as evidenced by the traditional white exteriors of homes along the Mediterranean and the white adobe buildings in the southwestern United States (Fig. 19.1).

**Fig. 19.1** **a** White houses on the Mediterranean coast of Greece (*Source* Michelle Raponi, pixabay.com). **b** Adobe house in the American Southwest

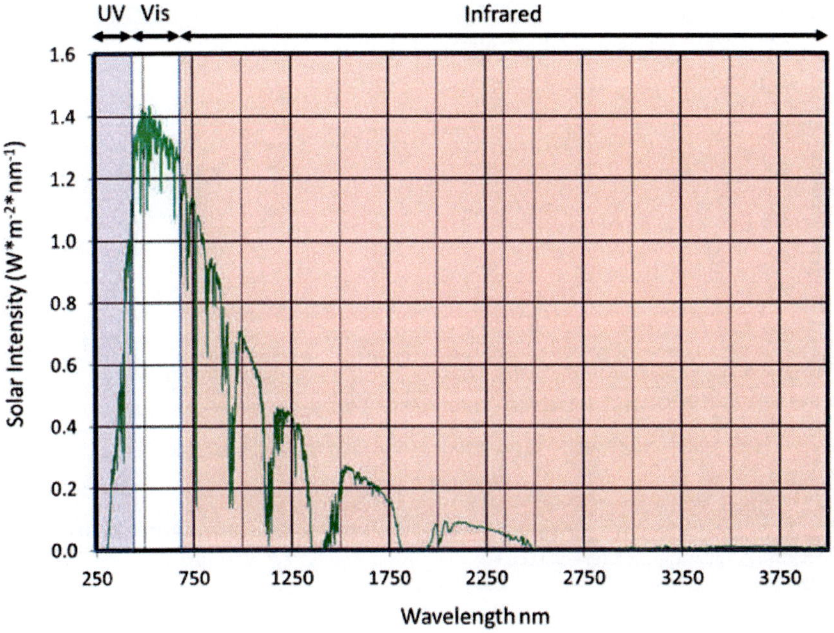


**Fig. 19.2** Terrestrial solar spectrum [69]

To put this empirical practice onto a scientific basis, it is useful to divide the terrestrial solar energy spectrum into three regions (Fig. 19.2). In typical paint and plastics applications, we are concerned with the reflectivity of the visible light only, since this provides opacity and creates color. However, our interest here is in the total energy of sunlight, and only about 45% of the energy of terrestrial sunlight is in the visible region, while slightly more than this (about 50%) is in the infrared (IR) region. For this reason IR reflectance is of great importance for keeping structures cool.

The brightness of a paint or plastic in the visible region is characterized by its tristimulus Y value (see Chap. 6). This value is simply the sum of the reflectance of light at a given wavelength weighted by the sensitivity of the eye to that wavelength, over the visible light range, and standardized to give a value of 100 for a perfect light reflector.

Solar reflectance is characterized in a related way. Here the light of all wavelengths represented in the solar spectrum is used (from 250 to 2500 nm), and the weighting function is the intensity of terrestrial solar radiation at that wavelength rather than the sensitivity of the eye. The resulting value, given as a fraction between 0 and 1, is referred to as the total solar reflectance (TSR) value for that surface.

As an example of the impact of the shape of the light reflectance curve on the TSR value of a material, the reflectance spectra of four paints, and their TSR values, are shown in Fig. 19.3. The paints in this figure come in two pairs, one pair

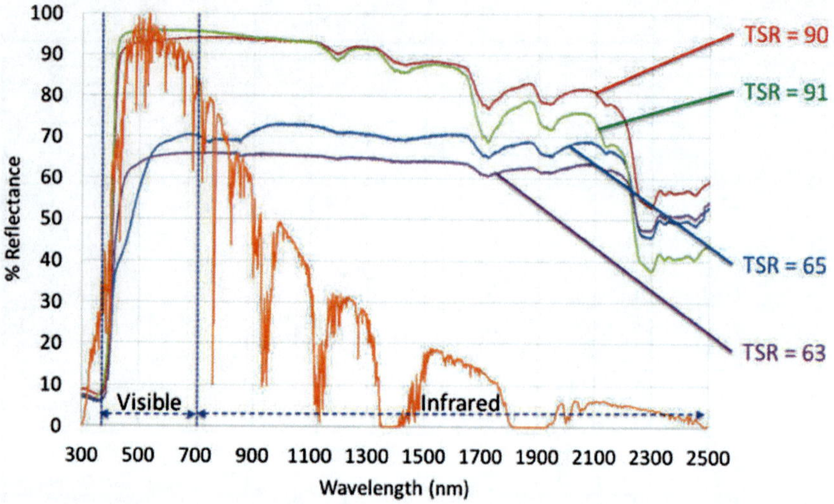


**Fig. 19.3** Solar reflectance curves for four paints, with TSR values indicated. The terrestrial solar spectrum is superimposed for reference

with a maximum reflectance of approximately 95% and the other with a maximum reflectance of approximately 70%. Within each pair, one paint shows a slightly higher reflectance in the visible portion of the spectrum and a significantly lower reflectance in the high energy IR portion of the spectrum. In each case the slightly higher reflectance in the visible region completely offsets the significantly lower reflection in the IR region. This is because of the significantly higher light intensity in the visible region of the solar spectrum compared to the IR region (Fig. 19.2).

## White Roof Initiatives

The historic means of addressing excessive solar heating have been to paint the roofs of buildings white. This technology has recently received renewed public and scientific attention as an inexpensive means of decreasing the electricity loads of air conditioners, particularly during periods of peak consumption in the summer months [70].

The white roof concept is attractive for a number of reasons. First, it is very easily understood by the layman. In addition, the cost of a white roof is essentially the same as any other color, and so there is no added expense in using this technology. Finally, it is passive, meaning that it does not require power and has no mechanical maintenance needs.[5]

[5] That said, to achieve the full effectiveness of this strategy, the roof must be thoroughly cleaned regularly.

Acting against these attractive features, cooling effectiveness is decreased substantially when the roof soils. Both dirt accumulation and algae growth rapidly darken any exterior, light colored surface (compare, for example, the brightnesses of the building in Fig. 19.1b to those in Fig. 19.1a). Periodic cleaning is expensive and loses effectiveness as the roof ages. By one estimate, roofs would require repainting every three to five years to maintain the benefits of this strategy.

In addition, it is very difficult to estimate the electricity savings for these roofs. These savings are highly dependent on environmental factors such as the ambient temperature, the intensity of the sunlight striking the roof, wind conditions, and the proximity of nearby structures, as well as on the architectural details of the building (e.g., the number of stories). Exaggerated claims, such as the complete elimination of air conditioning or the measurable cooling of the planet by the radiation of heat into space, are often made and are difficult for the layman to judge. This has led in some instances to optimistic estimates that did not materialize in the field, leading to consumer frustration and skepticism about the concept.

Perhaps the most quantitative report of energy savings was from a field study performed in Orange County, situated in Southern California [71]. Here researchers found that changing the solar reflectivity (albedo) of a roof from approximately 20% to approximately 80% resulted in a decrease in annual electricity usage of between 6 and 15 kWh/m$^2$ of roof. At the energy costs at the time of the study (US$ 0.10/kWh), it was estimated that the annual savings for a typical one-family home could be as much as approximately US$ 224.[6]

## Solar Reflectivity and IR Reflective Pigments

A cursory examination of roofs in most cities and towns shows that the white roof concept has yet to be broadly embraced by society. Part of this is no doubt due to the high cost of changing or modifying roof materials. A residential roof may be expected to last for twenty or thirty years, and the annual energy savings from replacing a dark roof before the end of its useful life are typically much less than the cost associated with replacing the fully functional roof early.

Color preference is also a factor. Residential roofing in many regions has been traditionally dark for this reason. This may be the result of an individual homeowner's preference or the requirements of a homeowners' association that has the power to regulate the appearance of homes within their jurisdiction. Regardless of the source, color preferences generally disfavor white roofs.

However, even dark roofs can benefit from increased solar reflectivity.[7] Pigments that reflect IR light strongly can decrease heat absorption by the roofing material. This, in turn, leads to a lower temperature during hot days. These pigments are invariably metal oxide based (i.e., are complex inorganic color pigments—see Chap. 8)

[6] This assumes the average residential roof to have an area of 160 m$^2$ [72].

[7] Note that a white roof will always have a greater TSR than a colored roof, even if that colored roof employs IR reflective pigments.

[73], since organic pigment particles embedded in an organic matrix do not significantly scatter IR light. These IR reflective pigments are claimed to significantly decrease the temperature of a dark roof during a hot, sunny day.

### Brighter Interiors for Reduced Electricity Usage

Room lighting is another significant consumer of electricity. Efforts to decrease power consumption for lighting have driven the conversion of incandescent lighting to more energy efficient technologies such as LED and fluorescent. However, even these technologies require relatively high levels of electricity. It is estimated that, globally, lighting currently accounts for roughly 15% of total electricity consumption, an amount that is expected to rise by 50% in the next decade [74]. Further electricity reductions can be realized by increasing the reflectivities of the interiors of buildings. This is because the light intensity in a room is controlled, in part, by the amount of light reflected from the walls, floors and ceilings [75]. That is, a black room will require more electricity to light to the same level as a white room.

In one case study, the energy required to light an interior room in an office building that was sequentially painted white, two shades of gray, and black was measured to determine the benefits of bright interior walls [76]. The room was of standard size and illuminated with fluorescent light bulbs. Even though this form of illumination is relatively energy efficient, the electricity savings seen when repainting the darkest paint with the brightest paint were found to be of similar magnitude to the energy savings of painting a dark roof white. Significant savings were also seen when comparing the white room to rooms painted with "colors of the year", which are traditionally dark.[8] In a separate study, the electricity savings seen when increasing the wall brightness in a warehouse was of similar magnitude [77].

## Summary

Numerous issues and trends are emerging regarding the use of particles in coatings, plastics, and paper laminates. These can be divided between technology and sustainability trends, although these are intertwined since technological innovations are the most likely means of resolving sustainability issues. Technological advances are being made in the general use of nanoparticles in conventional particle applications, as well as the development of new applications such as heat reflection and pollution remediation. In addition, new ways of designing, developing and manufacturing materials are being actively pursued. These include innovations such as machine

[8] A "color of the year" is a color chosen annually by an architectural (décor) paint producer as their estimation of the most stylish color for the upcoming year. These colors vary greatly from producer to producer but are generally bold.

learning, artificial intelligence, and the adoption of Industry 4.0 practices at all levels of the value chain.

## References


1. Nanotechnologies—Vocabulary—Part 2: Nano-objects. ISO/TS 80004-2:2015 (2015)
2. Standard Terminology Relating to Nanotechnology. ASTM E2456-06 (2020)
3. Recommendation on the Definition of a Nanomaterial. 2011/696/EU (2011)
4. Feynman, R.P.: Plenty of room at the bottom. Am. Phys. Soc. Meet. (1959)
5. Heiligtag, F.J., Niederberger, M.: The fascinating world of nanoparticle research. Mat. Today **16**(7–8), 262 (2013)
6. Talapin, D.V., Shevchenko, E.V.: Introduction: nanoparticle chemistry. Chem. Rev. **116**(18), 10343 (2016)
7. Ramsden, J.: Nanotechnology: An Introduction, 2nd edn. Elsevier (2016)
8. Pu, Y., Fuhong, C., Wang, D., Wang, J.-X., Chen, J.-F.: Colloidal synthesis of semiconductor quantum dots toward large-scale production: a review. Ind. Eng. Chem. Res. **57**(6), 1790 (2018)
9. Bera, D., Qian, L., Tseng, T.-K., Holloway, P.H.: Quantum dots and their multimodal applications: a review. Materials **3**, 2260 (2010)
10. Cavalcante, P.M.T., Dondi, M., Guarini, G., Raimondo, M., Baldi, G.: Colour performance of ceramic nano-pigments. Dyes Pigm. **80**, 226 (2009)
11. Kim, J.B., Lee, S.Y., Lee, J.M., Kim, S.-H.: Designing structural-color patterns composed of colloidal arrays. ACS Appl. Mater. Interfaces. **11**(16), 14485 (2019)
12. Lehman, J., Yung, C., Tomlin, N., Conklin, D., Stephens, M.: Carbon nanotube-based black coatings. Appl. Phys. Rev. **5**, 011103 (2018)
13. Rothon, R.: Fillers for Polymer Applications, Polymers and Polymeric Composites: A Reference Series, Chapter 23. In: Nanofillers, vol. 463. Springer International Publishing (2017)
14. Kim, J., Yang, S.I., Moon, H., Hong, J., Hong, J., Choi, W., Son, H., Lee, B.-C., Kim, G.-B., Kim, Y.: Potential release of nano-carbon black from tire-wear particles through the weathering effect. J. Ind. Eng. Chem. **96**, 322 (2021)
15. Wang, X.-Q., Zhang, J.-P., Zhu, W.-D., Shi, L.-Y.: preparation and properties of nano-$TiO_2$ modified interior wall paint. J. Shanghai Univ. **11**, 432 (2007)
16. Morsy, F.A., El-Sherbiny, S., Samir, M., Fouad, O.A.: Application of nanostructured titanium dioxide pigments in paper coating: a comparison between prepared and commercially available ones. J. Coat. Technol. Res. **13**(2), 307 (2016)
17. Dubbert, W., Schwirn, K., Volker, D., Apel, P.: Use of Nanomaterials in Coatings. Umwelt Bundesamt. Federal Environment Agency (2014)
18. Sterhov, A.I., Loshkarev, I.Y.: Determination of the proportion of natural light in solar radiation using the method of conversion of lighting units into energy. J. Phys. Conf. Ser. **1353**, 012002 (2019)
19. Chen, H., Nanayakkara, C.E., Grassian, V.H.: Titanium dioxide photocatalysis in atmospheric chemistry. Chem. Rev. **112**(11), 5919 (2012)
20. Nitrogen Oxides: https://scied.ucar.edu/learning-zone/air-quality/nitrogen-oxides
21. Barker, P., Branch, A.: The interaction of modern sunscreen formulations with surface coatings. Prog. Org. Coat. **62**, 313 (2008)
22. Students Develop Titanium Dioxide Roof Tile Coating to Remove Smog-Causing Nitrogen Oxide. Coatings World. (2014)
23. Yoon, Y.-S., Kim, H.-J., Park, J.-H.: $NO_x$ reduction performance in cement mortar with $TiO_2$ treatment and mineral admixture. J. Kor. Recycled Constr. Res. Inst. **8**(4), 506 (2020)
24. Shi-Jie, Y., Chen, J.-J.: Nitrate formation from atmospheric nitrogen and oxygen photocatalysed by nano-sized titanium dioxide. Nat. Commun. **4**, 2249 (2013)

25. Gandolfo, A., Marque, S., Temime-Roussel, B., Gemayel, R., Wortham, H., Truffier-Boutry, D., Bartolomei, V., Gligorovski, S.: Unexpectedly high levels of organic compounds released by indoor photocatalytic paints. Environ. Sci. Technol. **52**(19), 11328 (2018)
26. Truffier-Boutry, D., Fiorentino, B., Bartolomei, B., Soulas, R., Sicardy, O., Benayad, A., Damlencourt, J.-F., Pepin-Donat, B., Lomard, C., Gandolfo, A., Wortham, H., Brochard, G., Audemard, A., Porcar, L., Gebel, G., Gligorovski, S.: Characterization of photocatalytic paints: a relationship between the photocatalytic properties—release of nanoparticles and volatile organic compounds. Environ. Sci.: Nano **4**(10), 1998 (2017)
27. Fletcher, H.: Smog eating paint does more harm than good. Chem. World (2017)
28. Monk, P.: Paints and surfaces for the removal of nitrogen oxides. Air Quality Expert Group, Department for Environment, Food and Rural Affairs; Scottish Government. https://uk-air.defra.gov.uk/assets/documents/reports/cat11/1604130958_PB14425_Paints_and_Surfaces_for_the_Removal_of_Nitrogen_Oxides.pdf (2016)
29. 2011 Air Quality Progress Report for Salford City Council: https://www.yumpu.com/en/document/view/25963615/2011-progress-and-action-plan-report-adobe-pdf-format-29mb (2011)
30. Moorcroft, S., McCrae, I., Boulter, P., Laxen, D.: Local Measures for $NO_2$ Hotspots in London—Final Report. https://www.aqconsultants.co.uk/CMSPages/GetFile.aspx?guid=2b5163c4-a64a-482a-930a-d0d99ee52b84 (2010)
31. Dutch Air Quality Innovation Programme Concluded. https://infrastructure.planninginspectorate.gov.uk/wp-content/ipc/uploads/projects/TR010019/TR010019-002594-Dutch%20Air%20Quality%20Innovation%20Programme%20concluded.pdf (2010)
32. Invloed $TiO_2$ Catings op de Luchtkwaliteit. www.verkeerenwaterstaat.nl (actueel/publicaties) (2009)
33. Gaya, U.I., Abdullah, A.H.: Heterogeneous photocatalytic degradation of organic contaminants over titanium dioxide: a review of fundamentals, progress and problems. J. Photochem. Photobiol. C: Photochem. Rev. **9**(1), 1 (2008)
34. Bedjanian, Y., El Zein, A.: Interaction of $NO_2$ with $TiO_2$ surface under UV irradiation: products study. J. Phys. Chem. A **116**(7), 1758 (2012)
35. Monge, M.E., et al.: Ozone formation from illuminated titanium dioxide surfaces. J. Am. Chem. Soc. **132**(24), 8234 (2010)
36. Ndour, M., d'Anna, B., George, C., Ka, O., Balkanski, Y., Kleffmann, J., Stemmler, K., Ammann, M.: Photoenhanced uptake of $NO_2$ on mineral dust: laboratory experiments and model simulations. Geo. Res. Lett. **35**, L0512 (2008)
37. Bedjanian, Y., El Zein, A.: Interaction of $NO_2$ with $TiO_2$ surface under UV irradiation: measurements of the uptake coefficient. Atmos. Chem. Phys. Discuss. **11**, 27861 (2011)
38. Gustafsson, R.J., Orlov, A., Griffiths, P.T., Cox, R.A., Lambert, R.M.: Reduction of $NO_2$ to nitrous acid on illuminated titanium dioxide aerosol surfaces: implications for photocatalysis and atmospheric chemistry. Chem. Comm. (37), 3936 (2006)
39. Gligorovski, S., Strekowski, R., Barbati, S., Vione, D.: Environmental implications of hydroxyl radicals (·OH). Chem. Rev. **115**(24), 13051 (2015)
40. Gopinath, K.P., Madhav, N.V., Krishnan, A., Malolan, R., Rangarajan, G.: Present applications of titanium dioxide for the photocatalytic removal of pollutants from water: a review. J. Environ. Manage. **270**(110906), 1 (2020)
41. Pelaez, M., Nolan, N.T., Pillai, S.C., Seery, M.K., Falaras, P., Kontos, A.G., Dunlop, P.S.M., Hamilton, J.W.J., Byrne, J.A., O'Shea, K., Entezari, M.H., Dionysiou, D.D.: A review on the visible light active titanium dioxide photocatalysts for environmental applications. Appl. Catal. B: Environ. **125**, 331 (2012)
42. Gusain, R., Gupta, K., Joshi, P., Khatri, O.P.: Adsorptive removal and photocatalytic degradation of organic pollutants using metal oxides and their composites: a comprehensive review. Adv. Colloid Interface Sci. **272**(102009), 1 (2019)
43. Moles, S., Berges, J., Ormad, M.P., Nieto-Monge, M.J., Gomez, J., Mosteo, R.: Photoactivation and photregeneration of $TiO_2$/PAC mixture applied in suspension in water treatments: approach to a real application. Environ. Sci. Pollut. Res. Int. **28**(19), 24167 (2021)

44. Talwar, S., Sangal, V.K., Verma, A.: Feasibility of using combined $TiO_2$ photocatalysis and RBC process for the treatment of real pharmaceutical wastewater. J. Photochem. Photobiol. A Chem. **353**, 263 (2018)
45. Zapata, A., Oller, I., Sirtori, C., Rodriguez, A., Sanchez-Perez, J.A., Lopez, A., Mezcua, M., Malato, S.: Decontamination of industrial wastewater containing pesticides by combining large-scale homogeneous solar photocatalysis and biological treatment. Chem. Eng. J. **160**, 447 (2010)
46. Vela, N., Calín, M., Yáñez-Gascón, M.J., Garrido, I., Perez-Lucas, G., Fenoll, J., Navarro, S.: Photocatalytic oxidation of six pesticides listed as endocrine disruptor chemicals from wastewater using two different $TiO_2$ samples at pilot plant scale under sunlight irradiation. J. Photochem. Photobiol. A Chem **353**, 271 (2018)
47. Kushniarou, A., Garrido, I., Fenoll, J., Vela, N., Flores, P., Navarro, G., Hellin, P., Navarro, S.: Solar photocatalytic reclamation of agro-waste water polluted with twelve pesticides for agricultural reuse. Chemosphere **214**, 839 (2019)
48. Schwab, K.: The fourth industrial revolution: what it means and how to respond. Foreign Aff. **95**(6) (2015)
49. Belderbos, R., Gilsin, V., Lokshin, B., Carree, M., Sastre, J.F.: The antecedents of new R&D collaborations with different partner types: on the dynamics of past R&D collaboration and innovative performance. Long Range Plan. **51**(2), 285 (2018)
50. Zacharia, Z.G., Nix, N.W., Lusch, R.F.: An analysis of supply chain collaborations and their effect on performance outcomes. J. Bus. Logist. **30**(2), 101 (2009)
51. Moreira, A.C., Silva, L.F.: Involving suppliers in collaborative new product development: comparing large and small firms. Int. J. Val. Chain. Manag. **12**(1), 1 (2021)
52. Zhou, C., Etzkowitz, H.: Triple helix twins: a framework for achieving innovation and UN sustainable development goals. Sustainability **13**, 6535 (2021)
53. Pilcher, G.R.: Sustainability in the paints and coatings industry: far more than just a 'good idea.' JCT CoatingsTech **18**(7), 25 (2021)
54. Anon: Life-cycle assessment of architectural coatings: a world without preservatives. JCT CoatingsTech **18**(5), 44 (2021)
55. De Backer, S., Diebold, M.: How paint quality impacts carbon footprint. Europ. Coat. J. **30** (2022)
56. Communication from the Commission to The European Parliament, The European Council, The Council, The European Economic and Social Committee, and The Committee of the Regions: The European Green Deal. https://ec.europa.eu/info/sites/default/files/european-green-deal-communication_en.pdf (2019)
57. Communication from the Commission to The European Parliament, The European Council, The Council, The European Economic and Social Committee, and The Committee of the Regions: Chemicals Strategy for Sustainability Towards a Toxic-Free Environment. https://ec.europa.eu/environment/pdf/chemicals/2020/10/Strategy.pdf (2020)
58. Commission Delegated Regulation (EU) 2020/217 of 4 October 2019 Amending, for the Purposes of its Adaptation to Technical and Scientific Progress, Regulation (EC) No 1272/2008 of the European Parliament and of the Council on Classification, Labelling and Packaging of Substances and Mixtures and Correcting that Regulation. Off. J. Europ. Union, **L44**(63), 1 (2020)
59. Titanium Dioxide in the Sustainable Built Environment. https://tdma.info/titanium-dioxide-in-the-sustainable-built-environment/ (2020)
60. Commission Regulation (EU) 2018/1881 of 3 December 2018, Amending Regulation (EC) No 1907/2006 of the European Parliament and of the Council on the Registration, Evaluation, Authorisation and Restriction of Chemicals (REACH) as regards Annexes I, III,VI, VII, VIII, IX, X, XI, and XII to Address Nanoforms of Substances. Off. J. Europ. Union **L308**(61), 1 (2018)
61. The Greenhouse Gas Protocol: A Corporate Accounting and Reporting Standard. World Resource Institute and World Business Council for Sustainable Development. https://ghgprotocol.org/sites/default/files/standards/ghg-protocol-revised.pdf (2015)

62. Karlsson, M.C.F., Abbas, Z., Bordes, R., Cao, Y., Larsson, A., Rolland, A., Taylor, P., Steenari, B.M.: Surface properties of recycled titanium oxide recovered from paint waste. Prog. Org. Coat. **125**, 279 (2018)
63. Loaeza, D., Cailloux, J., Pérez, O.S., Sánchez-Soto, M., Maspoch, M.L.: Impact of titanium dioxide in the mechanical recycling of post-consumer polyethylene terephthalate bottle waste: tensile and fracture behavior. Polymers **13**(2), 310 (2021)
64. Matxinandiarena, E., Múgica, A., Zubitur, M., Yus, C., Sebastián, V., Irusta, S., Loaeza, A.D., Santana, O., Maspoch, M.L., Puig, C., Müller, A.J.: The effect of titanium dioxide surface modification on the dispersion, morphology, and mechanical properties of recycled PP/PET/$TiO_2$ PBNANOs. Polymers **11**(10), 1692 (2019)
65. Tramis, O., Garnier, C., Yus, C., Irusta, S., Chabert, F.: Enhancement of the fatigue life of recycled PP by incorporation of recycled opaque PET collected from household milk bottle wastes. Waste Manag. **125**, 49 (2021)
66. The Product Stewardship Institute. https://www.productstewardship.us/page/Paint
67. Barrineau, J.: Perspectives on the continued pursuit of sustainable solutions. JCT CoatingsTech **18**(7), 22 (2021)
68. https://rpsc.energy.gov/energy-data-facts (2021)
69. Standard Tables for Reference Solar Spectral Irradiances: Direct Normal and Hemispherical on 37° Tilted Surface. ASTM G173-03 (2020)
70. Pearce, F.: Urban heat: can white roofs help cool World's warming cities? Yale Environ. 360 (2018)
71. Akbari, H., Levinson, R., Rainer, L.: Monitoring the energy-use effects of cool roofs on california commercial buildings. Energy Build. **37**, 1007 (2005)
72. https://www.roofingcalc.com/roof-replacement-cost (2021)
73. Bendiganavale, A.K., Malshe, V.C.: Infrared reflective inorganic pigments. Recent. Pat.S Chem. Eng. (1), 67 (2008)
74. Rise and Shine—Lighting the World with 10 Billion LED Bulbs. https://www.energy.gov/articles/rise-and-shine-lighting-world-10-billion-led-bulbs) (2015)
75. De Vires, A., Souman, J.L., de Ruyter, B., Heynderickx, I., de Kort, Y.A.W.: Lighting up the office: the effect of wall luminance on room appraisal, office workers' performance, and subjective alertness. Build. Environ. **142**, 534 (2018)
76. De Backer, S., Diebold, M., Milone, M.: Effect of Wall brightness on room illumination. JCT CoatingsTech **16**(9), 32 (2019)
77. Challener, C.: Coatings industry innovations enable energy conservation. JCT CoatingsTech **17**(5), 46 (2020)